Extreme Natural Hazards, Disaster Risk and Societal Implications

Extreme natural events, such as the 2012 Hurricane Sandy and the 2011 Japanese earthquake and tsunami, increasingly dominate our global headlines, and understanding their causes, risks, and impacts, as well as the costs of mitigation, is essential to managing hazard risk and saving lives. Now, this book presents a unique, interdisciplinary approach to disaster risk research, combining cutting-edge natural science and social science methodologies.

Bringing together leading scientists, policy makers and practitioners from around the world, the book presents the risks of global hazards such as volcanoes, earthquakes, landslides, hurricanes, precipitation floods and space weather, and provides real-world hazard case studies from Latin America, the Caribbean, Africa, the Middle East, Asia and the Pacific region. Avoiding complex mathematics, the authors provide insight into topics such as the vulnerability of society, disaster risk reduction policy, relations between disaster policy and climate change, adaptation to hazards, and (re)insurance approaches to extreme events.

This is a key resource for academic researchers and graduate students in a wide range of disciplines linked to hazard and risk studies, including geophysics, volcanology, hydrology, atmospheric science, geomorphology, oceanography, geography and remote sensing. It is also an important reference for professionals and policy makers working in disaster prevention and mitigation.

Alik Ismail-Zadeh is Senior Scientist at the Karlsruhe Institute of Technology, Germany and Chief Scientist at the Institute of Earthquake Prediction Theory of the Russian Academy of Sciences. His scientific interests cover studies of dynamics of the lithosphere and upper mantle and their surface manifestations including seismicity, seismic hazard, and risk. He is Secretary-General of the International Union of Geodesy and Geophysics (IUGG), and has also served as President of the Natural Hazards Focus Group of the American Geophysical Union (AGU), President of the IUGG Union Commission of Geophysical Risk and Sustainability, and the leader of the ENHANS project. He is a recipient of the Academia Europaea Young Scientist Award, the AGU International Award, and the AOGS Ian Axford Lecture Award on Natural Hazards.

Jaime Urrutia Fucugauchi is Professor of Geophysics at Instituto de Geofisica, Universidad Nacional Autonoma de Mexico (UNAM), and is a leading expert in geomagnetism and volcanism. He is President of the Mexican Physics Society and a member of the Governing Board of the American Institute of Physics, the National University of Mexico, and the Science Advisory Council of Mexico. Urrutia-Fucugauchi has been awarded the Mexican Government's National Science Award, the Organization of American States' 'Manuel Noriega Morales' Prize, and the AGU International Award.

Andrzej Kijko is the Director of the University of Pretoria Natural Hazards Centre, Pretoria, South Africa, and also a Professor of the University of Pretoria. As an internationally acclaimed researcher, he has been active in engineering geophysics and seismology for about 40 years. Kijko's interests and responsibilities have taken him around the world, and he has a wide range of experience from the various research posts and consulting positions he has held.

Kuniyoshi Takeuchi is the founding Director of the International Centre for Water Hazard and Risk Management (ICHARM) in Tsukuba, and is also Professor Emeritus of the University of Yamanashi in Kofu, Japan. He is a distinguished hydrologist whose research interests cover hydrological modelling and forecasting, water resource systems analysis, flood and disaster management, and societal sustainability issues. Takeuchi is Chair of the IUGG GeoRisk Commission and Vice Chair of the Scientific Committee of the Program 'Integrated Research on Disaster Risk' (IRDR). He has received numerous awards including the IAHS-UNESCO-WMO International Hydrological Prize.

Ilya Zaliapin is Associate Professor of the Department of Mathematics and Statistics, University of Nevada, Reno, USA. His broad research interests cover problems of self-similarity, network transport, aggregation processes, delay equations, multiscale methods of time series analysis, random sums of heavy-tailed variables, geo-statistics, extreme value theory with applications to atmospheric, hydrological and seismological research. He has been Secretary of the AGU Natural Hazards Focus Group, and Secretary of the IUGG Commission on Mathematical Geophysics.

Special Publications of the International Union of Geodesy and Geophysics Series

The International Union of Geodesy and Geophysics (IUGG) promotes and co-ordinates international scientific studies of Earth (physical, chemical, and mathematical) and its environment in space, aiming to apply this knowledge to societal needs such as mineral resources, mitigation of natural hazards, and environmental preservation. The series is co-published by the IUGG and Cambridge University Press, providing researchers and graduate students with authoritative insights into major scientific developments and state-of-the-art research.

TITLES IN PRINT IN THIS SERIES

A. Ismail-Zadeh, J. Urrutia Fucugauchi, A. Kijko, K. Takeuchi, and I. Zaliapin, *Extreme Natural Hazards, Disaster Risks and Societal Implications*

Extreme Natural Hazards, Disaster Risks and Societal Implications

EDITED BY

ALIK ISMAIL-ZADEH
Karlsruhe Institute of Technology, Karlsruhe, Germany
Russian Academy of Sciences, Moscow, Russia

JAIME URRUTIA FUCUGAUCHI
Universidad Nacional Autónoma de México, Mexico City, Mexico

ANDRZEJ KIJKO
University of Pretoria Natural Hazard Centre, Pretoria, South Africa

KUNIYOSHI TAKEUCHI
International Centre for Water Hazard and Risk Management under the auspices of UNESCO, Public Works Research Institute, Tsukuba, Japan

ILYA ZALIAPIN
University of Nevada, Reno, NV, USA

CAMBRIDGE
UNIVERSITY PRESS

CAMBRIDGE
UNIVERSITY PRESS

University Printing House, Cambridge CB2 8BS, United Kingdom

Cambridge University Press is part of the University of Cambridge.

It furthers the University's mission by disseminating knowledge in the pursuit of education, learning and research at the highest international levels of excellence.

www.cambridge.org
Information on this title: www.cambridge.org/9781107033863

First published 2014

Printed in the United Kingdom by TJ International Ltd. Padstow Cornwall

A catalogue record for this publication is available from the British Library

Library of Congress Cataloging in Publication Data
Extreme natural hazards, disaster risks and societal implications / edited by Alik Ismail-Zadeh and others.
pages cm – (Special publications of the international union of geodesy and geophysics; 1)
ISBN 978-1-107-03386-3 (Hardback)
1. Natural disasters. 2. Environmental disasters. 3. Environmental risk assessment. I. Ismail-Zadeh, Alik.
GB5014.E984 2014
363.34–dc23 2013038646

ISBN 978-1-107-03386-3 Hardback

Contents

Preface

Great advances in understanding planet Earth and its environment allow its geophysical processes and phenomena including their extreme manifestations, which lead to floods, hurricanes, earthquakes, tsunamis, volcanoes, landslides, severe space weather, wildfires and other natural hazard events, to be studied. There is a deep belief in the community of natural scientists that with progress in science and the scientific ability to predict extreme events the problem of disaster risk reduction will be resolved. That is perhaps true to some extent. Extreme events (e.g. the 2004 Indian Ocean and 2011 Great East Japan earthquakes and tsunamis, the 2005 Hurricane Katrina or the 2010 Haiti earthquake) showed that despite the significant scientific knowledge and predictive capacity that has been accumulated, we could not prevent and even mitigate the related disasters. Social scientists explain that both the physical and the social vulnerability of society to extreme events and significant exposure in the regions prone to natural hazards play a crucial role in the disaster occurrences. Disaster risk can be estimated as a convolution of natural hazards with vulnerability and exposed values, where vulnerability depends on the resilience of society. Meanwhile the theory of disaster risk developed by natural and social scientists becomes useful, if and only if it is properly implemented. Important players, who could help scientists to implement and to communicate to the public the scientific knowledge on disaster risk, are engineers, lawyers, disaster management authorities, policy makers, representatives of the (re)insurance industry, international and intergovernmental organizations, media, and other stakeholders dealing with natural hazards and disaster risk reduction. An advanced effective link between all stakeholders should be established and this book promotes this idea.

Extreme Natural Hazards, Disaster Risks and Societal Implications is intended to bridge two major themes, natural hazards and disaster risks, and presents the ways in which scientific knowledge could help to reduce risks of disasters for the benefit of society. The book presents the topics discussed at the scientific meetings held in the framework of the trans-disciplinary, international project 'Extreme Natural Events and Societal Implications – ENHANS' co-sponsored by several international, transnational, and intergovernmental organizations (see Chapter 1 for more detail on the ENHANS project).

The principal goal of the book is to highlight the importance of the links between natural and social scientists, engineers, and experts dealing with political, legal, economic and insurance issues related to extreme natural events and disaster risks at local, national, and international levels in order to make the world safer (without disasters) and to contribute to the sustainable development of society.

The book is organised into the following parts: (i) fundamental knowledge of natural hazards and disaster risks; (ii) regional and local hazards and risks; and (iii) risk and societal implications. It consists of 30 peer-reviewed chapters. Chapters 1 and 2 introduce the ENHANS project and the grand challenges of integrated research on disaster risk. The first part of the book dedicated to extreme volcanism (Chapter 3), seismicity (Chapter 4), landslides (Chapter 5), hydro-meteorological events (Chapters 6–8), space weather (Chapter 9) and related disaster risks. Chapter 10 discusses the theoretical background for predictability of extreme events.

The next several chapters are organised 'geographically' to highlight the specific problems of a particular region: earthquake, volcanic and tsunami hazards and disasters risks in Latin America and the Caribbean (Chapters 11 and 12) and in Africa (Chapters 13–17); earthquake, tsunami, and flood hazards and risks in the Middle East (Chapters 18–20), and hydro-meteorological and geological hazards and risks in Asia and the Pacific region (Chapters 21–25). The remaining chapters deal with disaster risk reduction policy (Chapter 26), disaster policy and climate change (Chapter 27), vulnerability and adaptation to sea level related hazards (Chapter 28), and (re)insurance approach to extreme natural events (Chapters 29 and 30).

The book is intended for academic researchers and graduate students from a broad spectrum of Earth, environmental, and social science disciplines interested in modern aspects of natural hazard and disaster risk research. We believe that this book will complement other publications on the topic in terms of the linkage of disaster risk research with the policy of disaster risk reduction and mitigation. We hope that the book will form an important reference for researchers dealing with hazards and risk analysis, disaster prevention and mitigation as well as for policy makers, insurance companies, and the media.

We hope you enjoy reading this book!

Acknowledgments

The book could not have been written without the generous support of the ENHANS project by several international, transnational, and intergovernmental organizations (Fig. P1): the International Council for Science (ICSU) with its regional offices for Africa (ROA), Asia and the Pacific (ROAP), and Latin America and the Caribbean (ROLAC); the International Union of Geodesy and Geophysics (IUGG); the American Geophysical Union (AGU); the International Geographical Union (IGU); the International Society for Photogrammetry and Remote Sensing (ISPRS); the International Union of Geological Sciences (IUGS); the International Union of Theoretical and Applied Mechanics (IUTAM); the Scientific Program 'Integrated Research on Disaster Risk' (IRDR) co-sponsored by ICSU, the International Social Sciences Council (ISSC), and the United Nations International Strategy for Disaster Reduction (UNISDR); the Global Ocean Observing System (GOOS) of the Intergovernmental Oceanographic Commission (IOC) of the United Nations Educational, Scientific and Cultural Organization (UNESCO); the United Nations Platform for Space-based Information for Disaster Management and Emergency Response (UN-SPIDER), and Munich Reinsurance Company.

We thank very much all authors of the book chapters for their excellent contributions. We are very grateful to our colleagues (listed below) for their in-depth and constructive reviews of the book's chapters, which improved the original manuscripts and resulted in the volume that you hold in your hands. The reviewers of the book were: Thorkild Aarup, Amir AghaKouchak, Jerome Ballet, Ian Barton, Tom Beer, Raul Castro-Escamilla, Leonid

Figure P1. Graphical representation of the Supporting Partners of the ENHANS project. For colour version, see Plates section.

Chubarov, Christophe Cudennec, Susan Cutter, Sunil Kumar De, Damien Delvaux, Alberto Flandes, Karen Fontijn, Jeff Freymueller, Eulàlia Gràcia, Harsh Gupta, Chris Hartnady, AW Jayawardena, David Johnston, Vladimir Kossobokov, Allan Lavell, Cinna Lomnitz, Forrest Masters, Steve McNutt, Joan Martí Molist, Juan Carlos Mora, David Novelo, Efim Pelinovsky, David Pyle, Rado Randriamboarison, Kenji Satake, Takahiro Sayama, Soroosh Sorooshian, Peter Suhadolc, Dario Tedesco, Georges Mavanga Tuluka, Nasim Uddin, Jacques Varet, Victor Velasco, Zhang Yongxian, and Jacques Zlotnicki. Special thanks go to Susan Francis for her efforts to set up the IUGG Special Publications series and for great assistance at the stage of book preparation. Also we are grateful to Kirsten Bot, Laura Clark, Vania Cunha, Caroline Mowatt, Rosina Piovani, Fiona Saunders, Suresh Thiruvengadam and Martin Tynan for their excellent technical assistance and to Maureen Storey for her thorough copy-editing of the book.

We would be very grateful for comments, inquiries, and complaints on the content of the book, which can be sent to us by e-mail (aiz@ipgp.fr).

Contributors

Amir AghaKouchak
University of California, Irvine, CA, USA

Abdulaziz M. Al-Bassam
King Saud University, Riyadh, Saudi Arabia

Irasema Alcántara-Ayala
Universidad Nacional Autónoma de México, Mexico City, Mexico

José L. Alonso
Simón Bolívar University, Caracas, Venezuela

Keith Alverson
UNEP Division of Environmental Policy Implementation, Nairobi, Kenya

Daniel N. Baker
University of Colorado, Boulder, CO, USA

Talfan D. Barnie
University of Cambridge, Cambridge, UK

Tom Beer
CSIRO, Aspendale, VIC, Australia

Dan Braithwaite
University of California, Irvine, CA, USA

Sálvano Briceño
Retired from the United Nations Office for Disaster Risks Reduction (UNISDR)

Amy Donovan
University of Cambridge, Cambridge, UK

Stephen Dovers
Australian National University, Canberra, Australia

Kerry Emanuel
Massachusetts Institute of Technology, Cambridge, MA, USA

David J. Ferguson
Columbia University, New York, NY, USA

Jaime Urrutia Fucugauchi
Universidad Nacional Autónoma de México, Mexico City, Mexico

Andrei Gabrielov
Purdue University, West Lafayette, IN, USA

Mohsen Ghafory-Ashtiany
International Institute of Earthquake Engineering and Seismology, Tehran, Islamic Republic of Iran

Harsh K. Gupta
National Disaster Management Authority, New Delhi, India

Viacheslav K. Gusiakov
Russian Academy of Sciences, Novosibirsk, Russia

John Handmer
RMIT University, Melbourne, Australia

Ailsa Holloway
Stellenbosch University, Stellenbosch, South Africa

Kuolin Hsu
University of California, Irvine, CA, USA

Mohammad T. Hussein
King Saud University, Riyadh, Saudi Arabia

Alik Ismail-Zadeh
Karlsruhe Institute of Technology, Karlsruhe, Germany; Russian Academy of Sciences, Moscow, Russia

Jamie M. Jackson
Patton Boggs LLP, Denver, CO, USA

Vladimir Keilis-Borok
University of California, Los Angeles, CA, USA

Akio Kitoh
Meteorological Research Institute, Tsukuba, Japan

Sucharit Koontanakulvong
Chulalongkorn University, Bangkok, Thailand

T. Srinivasa Kumar
Indian National Centre for Ocean Information Services, Hyderabad, India

Nils Lenhardt
University of Pretoria, Pretoria, South Africa

Ning Lin
Princeton University, Princeton, NJ, USA

Petra Löw
Munich Reinsurance Company, Munich, Germany

Tengfei Ma
China Earthquake Administration, Beijing, China

Thomas Mahl
Munich Reinsurance Company, Munich, Germany

Brassnavy Manzunzu
Council for Geoscience, Pretoria, South Africa

Gordon McBean
Western University, London, Ontario, Canada

Vunganai Midzi
Council for Geoscience, Pretoria, South Africa

Shailesh Nayak
Earth System Science Organisation, New Delhi, India

Phu Nguyen
University of California, Irvine, CA, USA

Sayaka Olsen
University of Nevada, Reno, NV, USA

Clive Oppenheimer
University of Cambridge, Cambridge, UK

Gerassimos A. Papadopoulos
National Observatory of Athens, Athens, Greece

Antonia Papageorgiou
National Observatory of Athens, Athens, Greece

Omar J. Pérez
Simón Bolívar University, Caracas, Venezuela

Mahefasoa T. Randrianalijaona
University of Antananarivo, Antananarivo, Madagascar

Carlos Rodríguez
Simón Bolívar University, Caracas, Venezuela

Kenji Satake
University of Tokyo, Tokyo, Japan

Scott Sellars
University of California, Irvine, CA, USA

Anselm Smolka
Munich Reinsurance Company, Munich, Germany
(now: GEM Foundation)

Soroosh Sorooshian
University of California, Irvine, CA, USA

Lauren K. Thompson
University of Colorado Law School, Boulder, CO, USA

Erik Vanmarcke
Princeton University, Princeton, NJ, USA

Angelika Wirtz
Munich Reinsurance Company, Munich, Germany

Zhongliang Wu
China Earthquake Administration, Beijing, China

Sibel Yildirim
Ludwig-Maximilians-University, Munich, Germany

Gezahegn Yirgu
Addis Ababa University, Addis Ababa, Ethiopia

Faisal K. Zaidi
King Saud University, Riyadh, Saudi Arabia

Ilya Zaliapin
University of Nevada, Reno, NV, USA

PART I

Introduction

1

Extreme natural hazards and societal implications – the ENHANS project

Alik Ismail-Zadeh

1.1 Introduction

The beginning of the twenty-first century has been marked by a significant number of extreme natural events and disasters associated with floods, hurricanes, severe storms, earthquakes, tsunamis, wildfires, landslides, volcanoes, extreme space weather, and extra-terrestrial hazard. Extreme events may cause devastation resulting in a loss of human lives, great environmental damage, and a partial or total loss of infrastructure. The principal signature of such events is that their probability decreases rapidly with magnitude, while the inflicted damage rapidly increases, and so does the cost of the respective protection measures. Extreme events and disasters remind us once again that there is a strong coupling between complex natural and social processes. A holistic approach is required to understand the natural phenomena and related social disasters associated with extreme events.

Humankind will never be able to prevent the occurrence of natural phenomena entirely. However, scientists are able to gain better understanding of the complex mechanisms of extreme natural events that cause the disasters, to study the physical and the social vulnerability of society to the extreme events, to share their knowledge with disaster management agencies who need the scientific information to be prepared to cope with such extreme events. Scientists need also a deeper understanding, based on work across disciplines, of all of the processes that are involved. They must be mindful of public concerns and the underlying risk perceptions. Communication between the various stakeholders that deal with natural hazards and disaster risk reduction needs to be strengthened.

Keeping this in mind, the International Union of Geodesy and Geophysics (IUGG) initiated the 'Extreme Natural Hazards and Societal Implications – ENHANS' project, which was enthusiastically supported by the International Council for Science (ICSU) and several international scientific unions, professional societies, and interdisciplinary and intergovernmental bodies. Among the Partners were the American Geophysical Union (AGU), the International Geographical Union (IGU), the International Society for Photogrammetry and Remote Sensing (ISPRS), the International Union of Geological Sciences (IUGS), the International Union of Theoretical and Applied Mechanics (IUTAM), the Integrated Research on Disaster Risk (IRDR) Programme of ICSU, the International Social Sciences Council (ISSC), and the United Nations International Strategy for Disaster Reduction (UNISDR) as well as the Global Ocean Observing System (GOOS) of the Intergovernmental Oceanographic Commission (IOC) of UNESCO. Representatives of the United Nations Platform for Space-based Information for Disaster Management and Emergency Response (UN-SPIDER), and the Munich Reinsurance Company also took part in the project. The ICSU Regional Offices for Africa (ROA), Asia and the Pacific (ROAP), and Latin America and the Caribbean (ROLAC) have been involved in the relevant parts of the project providing links to the regions. The project has been managed by IUGG and an international scientific committee: R. Abler (IGU, USA), A. Abreu (ICSU ROLAC, Brazil), O. Altan (ISPRS, Turkey), K. Alverson (GOOS, France), P. Bobrowsky (IUGS, Canada), J. U. Fucugauchi (AGU, Mexico), N. Hasan (ICSU ROAP, Malaysia), A Ismail-Zadeh (Project Leader, IUGG/AGU, Germany), E. Madela-Mntla, (ICSU ROA, South Africa), G. McBean (IRDR, Canada), and K. Moffatt (IUTAM, UK).

The principal goals of the ENHANS project have been to

(i) improve understanding of critical phenomena associated with extreme natural events and to

Extreme Natural Hazards, Disaster Risks and Societal Implications, eds. Alik Ismail-Zadeh, Jaime Urrutia Fucugauchi, Andrzej Kijko, Kuniyoshi Takeuchi and Ilya Zaliapin. Published by Cambridge University Press.

analyse the impacts of the natural hazards on the sustainable development of society;
(ii) promote studies on the prediction of extreme events and on natural hazards mitigation;
(iii) disseminate scientific knowledge and data on natural hazards and disaster risks for the advancement of research and education in general and especially in developing countries; and
(iv) establish links and networks with the organizations involved in research on extreme natural hazards and their societal implications.

The goals of the ENHANS project have been achieved via scientific meetings and open forums that brought together research experts, decision makers, disaster managers, agents of the insurance industry, and mass media representatives. The project placed a special emphasis on the importance of research on extreme natural hazards and disaster risk mitigation in the most vulnerable regions of the world, particularly in Latin America and the Caribbean, Africa, the Middle East, Asia and the Pacific region.

The project contributed to and promoted research activities in natural hazard and disaster risk reduction, in particular via implementation of the IRDR Programme. The project revealed the existing knowledge, potentials, and scientific expertise in the relevant topics of extreme natural hazards and disaster risk research in many regions of the world. The project events were held around the world, each focusing on extreme natural hazards and disaster risks in a particular region (Fig. 1.1):

- Latin America and the Caribbean (Iguassu, Brazil, August 2010);
- North Africa (Cairo, Egypt, November 2010);
- North America and Europe (San Francisco, USA, December 2010);
- Sub-Saharan Africa (Pretoria, South Africa, January 2011);
- Middle East (Antalya, Turkey, May 2011); and
- Asia and the Pacific region (Melbourne, Australia, July 2011).

Moreover, the advanced school (Trieste, Italy, October 2011) and scientific session (San Francisco, USA, December 2011) on predictability of extreme events have been parts of the ENHANS project. The following sections describe the ENHANS events and results of the project.

1.2 Focus on Latin America and the Caribbean Region

The first ENHANS events took place at the Meeting of the Americas in Foz do Iguassu, Brazil, on 9–11 August 2010. The event consisted of a symposium and a town hall meeting. The symposium was convened by J. U. Fucugauchi (National Autonomous University of Mexico (UNAM), Mexico) and A. Ismail-Zadeh (Karlsruhe Institute of Technology, Germany/Russian Academy of Sciences, Moscow, Russia/Institut de Physique du Globe de Paris, France). The role of the ICSU Regional Office for Latin America and the Caribbean in developing research projects on natural hazards and disaster risks in the region was discussed by A. Abreu (ICSU Regional Office for Latin America and the Caribbean, Brazil) who introduced a new multidisciplinary research programme on Integrated Research on Disaster

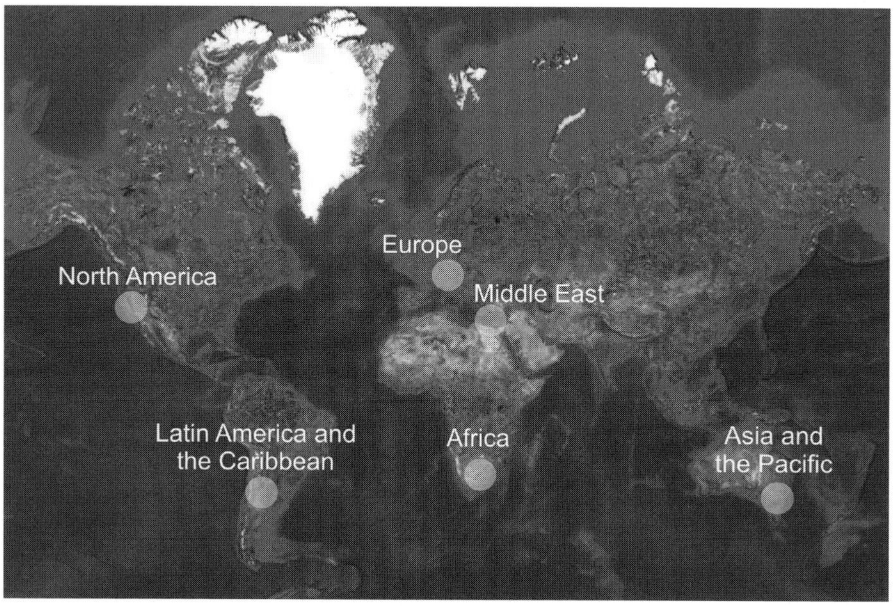

Figure 1.1. Geographical illustration of the ENHANS project meetings. For colour version, see Plates section.

Risk of ICSU and its Partners including the content and significance of the programme for Latin America and the Caribbean region. Meteorological hazards and associated risks in the Caribbean region were the subject of the talk by T. Gibbs (Consulting Engineers Partnership Ltd, Barbados), and volcanic hazards and risks in the region were presented by S. De La Cruz-Reyna (UNAM, Mexico). O. Perez (Universidad Simón Bolivar, Caracas, Venezuela) spoke on earthquake activity and associated hazards in South America and the Caribbean and on the socio-economic impact of great earthquakes in the region. The anatomy of landslide disasters was analysed by I. Alcantara-Ayala (UNAM, Mexico) with case studies from Mexico and other South American countries. O. Cardona (National University of Colombia, Manizales, Colombia) spoke on indicators of earthquake disaster risk and risk management in the Americas, while F. Romanelli (University of Trieste, Italy) presented a novel seismic hazard assessment methodology and its application to hazard evaluation in Valparaiso (Chile). Two oral sessions of the symposium were continued by a poster session with 25 papers.

The presentations by invited speakers and subsequent discussion at the town hall meeting addressed three questions: (i) How can science (both natural and social) and society form partnerships for disaster reduction? (ii) How can the science and society partnership convert natural disaster risk to opportunity? (iii) What are the urgent issues for disaster risk in Latin American and the Caribbean cities and regions under intensifying natural and social pressure? A. Lavell (Latin American Social Sciences Faculty – FLACSO and LA RED, Costa Rica) spoke on increasing the salience of disaster risk management on the political agenda. He said that 'Disaster Risk Management is still not of sufficient social and political salience given current and future predicted disaster trends and costs, and which climate change probably adds to'. The talk highlighted why this may be so and what the principal ways to place the theme more firmly on the political and social agenda are. K. Alverson (Director, Global Ocean Observing System – GOOS, Intergovernmental Oceanographic Commission) spoke on GOOS implementation in Latin America and the Caribbean. The talk focused on the importance of sustained ocean observing systems being 'in place both to prevent and mitigate disasters, where possible, as well as to rapidly bring observing assets to bear in post disaster relief efforts'. Both natural coastal inundation hazards, such as storm surges and tsunamis, as well as anthropogenic hazards, such as oil spills, were discussed, focusing where possible on Latin America and the Caribbean, including the Gulf of Mexico. P. Boccardo (Politechnico di Torino, Italy, and an ISPRS representative) discussed disaster management in the case of the 2010 Haiti earthquake. His presentation highlighted issues and challenges associated with emergencies related to disasters, particularly, data acquisition, processing, and information extraction when a humanitarian response is involved. Two panellists of the discussion, I. Alcantara-Ayala (IGU Vice-President) and M. McPhaden (AGU President), mentioned how international unions and professional societies could assist in research on natural hazards and in the mitigation of disasters. The speakers and panellists answered questions from the attendees. The discussion was moderated by J. U. Fucugauchi and A. Ismail-Zadeh.

1.3 Focus on Africa

The African meetings of the ENHANS project took place in Cairo, Egypt, on 1 November 2010 and in Pretoria, South Africa, from 17 to 20 January 2011. The symposium on Extreme Natural Hazards and Climate Changes in North Africa was organized by the Egyptian Academy of Scientific Research and Technology. A. Ismail-Zadeh introduced the ENHANS project and emphasized the importance of integrated research on natural and human-induced hazards and disasters. T. Beer (CSIRO, Melbourne, Australia) discussed the problems of environmental risk management that are related to climate change and geophysical hazards. H. Gupta (National Geophysical Research Institute, Hyderabad, India) discussed the problem of induced seismicity. D. Jackson (University of California, Los Angeles, USA) reported on global earthquake forecasts and how the forecasts help to understand seismic hazard. S. Riad (Cairo University, Egypt) discussed climate-related hazards and their impacts on Egypt. A round table discussion took place after the presentations.

The ENHANS International Workshop 'Extreme Natural Hazards and Disaster Risk in Africa' (Pretoria, South Africa, 17–20 January 2011) provided an opportunity for the research community of the African countries and international experts to discuss and analyse major topics related to extreme events and disaster risks. The workshop served as a platform to establish links and networks between African experts and relevant international organizations. The workshop was organized by A. Kijko (University of Pretoria, South Africa) and hosted by the Aon Benfield Natural Hazard Centre, University of Pretoria. More than 40 delegates, representing various international, intergovernmental, African, and other organizations, attended the workshop. The workshop presentations can be downloaded from http://www.enhans.org/showcasing/workshop_jan2011.php (accessed on 23 August 2013). The workshop's participants adopted the recommendation (see Appendix 1.1) to governments and funding institutions in cooperation with the relevant

Figure 1.2. Lake in the centre of Tswaing meteorite crater. For colour version, see Plates section.

ICSU bodies, United Nations agencies and other international entities. The Pretoria workshop was highlighted in the *UN SPIDER Newsletter*,[1] the *Insurance Journal*,[2] and in the *Newsletters of the University of Pretoria*.[3]

The workshop participants visited the Tswaing crater (1.13 km in diameter and 100 m deep), an extreme event site associated with a meteorite that hit the Earth some 200 000 years ago (Fig. 1.2). K. Moffatt (University of Cambridge, UK), a workshop participant, wrote a poem about the tour (see below)

Black Swan On First Looking into the Tswaing Meteorite Crater

In single file we climbed the narrow trail,
Through bramble thicket to the crater's rim.
Where warming sun cast shadows on the brim,
And bathed the bushveld scrub and scattered shale.

Concealed beneath the gently waving grass,
By saline lake, the haunt of duck and plover,
Lay snakes and lizards in the rain-soaked clover,
On diaplectic quartz and feldspar glass.

In prehistoric stone-age time of yore,
At hypersonic speed a chondrite fell
On thunderstruck impala and gazelle,
And vaporized upon the forest floor,

Black swans are not so rare, I heard you say!
Beware! One may befall this very day!

Keith Moffatt

1.4 Focus on North America and Europe

The symposium 'Natural Extreme Events: Modeling, Prediction and Mitigation' and related events were held at the Fall Meeting of the American Geophysical Union on 13–15 December 2010 in San Francisco, USA. The symposium consisted of three scientific sessions. The Union-wide session attracted more than 300 experts in various fields of extreme natural hazards. The session was focused mainly on North America and Europe. I. Zaliapin (University of Nevada, USA) and A. Ismail-Zadeh, organizers of the symposium, welcomed the participants and presented the ENHANS project. D. Baker (Colorado University at Boulder, USA) spoke on predictability and mitigating impacts of extreme space weather events. A storm surge, as a globally distributed risk, was the topic of the talk by H. von Storch (University of Hamburg, Germany). U. Lall (Columbia University, USA) explained why flooding was severe in 2010. He considered several case studies and discussed whether this is a coincidence or a predictable climate phenomenon and how to respond to these extremes. T. Jordan (University of Southern California, USA) spoke on new large-scale numerical simulations to forecast extreme earthquake ground motions, and S. Sparks (University of Bristol, UK) gave a talk on extreme volcanic eruptions and discussed their return period, impact, and implications. R. Douglas (Willis Research Network, UK) spoke on how (re)insurance and public science interact to manage the risk of extreme events for societal benefit. According to R. Douglas 'natural hazard modelling, married to risk, financial and economic modelling, created a new scientific and social organism – a rich and diverse community representing how nature, property and populations perform at the extremes'. N. Lin (Princeton University, USA) spoke on hurricane risk assessment related to wind damage and storm surge, while K. Klima (Carnegie Mellon University) spoke on tropical cyclones and presented her approach to a decision-analytic assessment of cyclone hazards. Extreme precipitation events in the European Alpine region were discussed by N. Awan (University of Graz, Austria). J. Rundle (University of California, Davis, USA) spoke on precursory activation and quiescence prior to major earthquakes. F. Guzzetti (University of Perugia, Italy) discussed landslide hazard, vulnerability, and risk assessment and emphasized the importance of methodology for risk assessment, its limits and challenges. K. Satake (University of Tokyo, Japan) spoke on tsunami modelling, forecasts, and early warning. A. Smith (National Oceanographic and Atmospheric Administration, USA) examined insurance loss return periods with extreme event intensity thresholds across the USA. In the poster session a variety of topics related to natural hazards, extreme events, theory, hazard and risk modelling, prediction and mitigation were presented. The culminating event of the symposium was a keynote lecture on 'Society's Growing Vulnerability to Natural Hazards and Implications for Geophysics Research' by J. Slingo (Met Office, UK).

1.5 Focus on the Middle East

The ENHANS Session 'Natural Hazards and Disaster Risks in the Middle East Region' was held on 7 May 2011 in Antalya, Turkey. The session was organized by A. Ismail-Zadeh and J. Rovins (USA/China, IRDR Executive Director), who introduced the IRDR programme and highlighted recent developments and preparation for the IRDR Conference in Beijing in 2011.

M. Ghafory-Ashtiany (International Institute of Earthquake Engineering and Seismology, Tehran, Iran) presented a talk on the topic of natural hazard and risks in Iran and the risk reduction capacity building in the country. Ghafory-Ashtiany said that 'to reduce and mitigate the unavoidable risk of natural hazard, scientists, engineers, government officials and the general public must all be involved in the process of finding realistic, achievable and appropriate ways of applying scientific knowledge to everyday life. Only by capacity building and working together, a nation can mitigate the impact of natural hazards on human life and society, and solve the risk puzzle'. He considered that Iran's experience in reducing earthquake disaster risk could be applied to many developing countries. Ghafory-Ashtiany argued that the main reasons for high human and property losses in the past were the significant vulnerability of the built environment, the rapid growth of population, and incompatible urban development.

C. Tuzun and M. Erdik (Bogazici University, Istanbul, Turkey) presented a new project, Earthquake Model of Middle East region (EMME), aimed at the assessment of seismic hazard, the associated risk in terms of structural damages, casualties, and economic losses and also at the evaluation of the effects of relevant mitigation measures in the Middle East region. The region is located at the junction of major tectonic plates, namely the African, Arabian, and Eurasian plates, resulting in very high tectonic activity. The speaker mentioned that 'some of the major earthquake disasters in human history occurred in the Middle East, affecting most countries in the region. Being one of the most seismically active regions of the world, the Middle East, extending from Turkey to India, is also a key region in terms of urbanization, energy reserves and industrialization trend.'

A. Al-Bassam (Saudi Arabia) presented natural hazards affecting the Kingdom of Saudi Arabia. The northwestern region of the country is prone to earthquakes and volcanic hazards, whereas the central and the western regions of the country are subject to flooding, especially during events of heavy rainfall. Landslides are a common phenomenon in the inhabited mountainous regions in the southwest. Dust storms are quite common in the central and the eastern parts of the Kingdom. Different government agencies and various universities having been working on these issues to mitigate these hazards and also to educate the people.

T. Merabtene (University of Sharjah, Dubai, United Arab Emirates) spoke on flood risk management in the Middle East and North African region (MENA region). He considered that 'the magnitude and impact of climate change increasingly take new catastrophic dimensions under the current paradigm shift of the MENA cities if disaster risk management actions and appropriate disaster policies are further delayed'. Flash floods in many countries in the MENA region have caused loss of life and have disrupted business activity resulting in great public and private economic damage. He argued that the increasing frequency of floods in the region is an indicator that urban flood management has potential implications for the sustainable development of current and future urban infrastructure in the MENA region. Merabtene emphasized the institutional deficiencies behind the current situation and discussed various strategies for keeping flood risk management high on the political agenda in MENA countries.

1.6 Focus on Asia and the Pacific region

Three major ENHANS events took place between 29 June and 2 July 2011 at the XXVth General Assembly of the International Union of Geodesy and Geophysics in Melbourne, Australia: (i) the symposium 'Grand Challenges in Natural Hazards Research and Risk Analysis', (ii) the symposium 'Earth on the Edge – Recent Pacific Rim Disasters', and (iii) the open forum 'Natural Hazards: From Risk to Opportunity by Partnership of Science and Society'. The events were organised by K. Takeuchi (ICHARM, Japan) and A. Ismail-Zadeh.

1.6.1 The symposium 'Grand Challenges in Natural Hazards Research and Risk Analysis'

Each year thousands of people across Asia and the Pacific region are killed and many more affected by floods, storms, earthquakes, drought, volcanoes and other such hazards. These hazards are only potentially damaging and the disasters occur when they impact on vulnerable communities, which are highly concentrated in poorer countries with weaker governance. G. McBean (University of Western Ontario, Canada; Chair of the IRDR Scientific Committee and ICSU President-elect) introduced the IRDR programme, which integrates research across hazards, natural, socio-economic, engineering, and health sciences disciplines and geographical regions. The research focuses on: the characterization of hazards, vulnerability, and risk; effective decision making; and knowledge-based actions leading to major reductions in future impacts and loss of lives. McBean considered that 'forensic investigations that will delve in greater depth into the root causes of disasters is one initial thrust. The second, risk interpretation and action, will examine how human actions, conditions, decisions and culture, and how people choose, or feel they have no choice but, to live and work in areas at higher risk, change vulnerabilities and contribute to disasters.'

J. Schneider (Geoscience, Australia) stated that 'effective disaster risk reduction is founded on knowledge of the underlying risk'. While methods and tools for assessing risk from specific hazards or to individual assets are generally well developed, our ability to holistically assess risk to a community across a range of hazards and elements at risk remains limited. Developing a holistic view of risk requires interdisciplinary collaboration amongst a wide range of hazard scientists, engineers, and social scientists, as well as the engagement of a range of stakeholders. Schneider presented some of the challenges sampled from a range of applications addressing earthquake, tsunami, volcano, severe wind, flood, and sea-level rise from projects in Australia, Indonesia, and the Philippines.

S. Sorooshian (University of California, Irvine, USA) considered that capturing the spatial and temporal distribution of precipitation in fine scales is critical to hydrologic, climatic, and ecological applications. The development of satellite remote sensing techniques has provided a unique opportunity for better observation of precipitation from space and overcomes some of the limitations of ground measurement. Sorooshian reviewed some of the development in satellite-based precipitation observation methodologies.

A. Kitoh (Meteorological Research Institute at Tsukuba, and Kyoto University, Japan) said that 'high-resolution models are vital to project reliable and possible future changes in weather extremes such as tropical cyclone and heavy rainfall'. He reported that unprecedented super high-resolution atmospheric models are being used for global warming projections. Kitoh described how projections on tropical cyclones reveal marked future increases in precipitation and surface wind velocity fields at the inner-core region within 150 km from the tropical cyclone centre, implying an increase in disaster risk induced by tropical cyclones in the future. He said that information on the uncertainty of future projection is significant for any decision-making processes and for various application studies on disaster prevention.

P. Linden (University of Cambridge, UK, and an IUTAM representative) presented an ongoing scientific and engineering project to prevent flooding in Venice. The construction of the flood protection gates at the entrance to the Venice Lagoon which are due to be commissioned in 2014 provides an unprecedented opportunity to manage the lagoon and its interaction with the Adriatic Sea. It also raises important questions about the scientific monitoring and decision-making process to enable the sustainable development of the lagoon and the historic city of Venice. Linden discussed the past and future monitoring and assessment of the physical and ecological states of the Venice Lagoon and described proposed structures for its management.

In the *Handbook of Disaster and Emergency Policies and Institutions* by J. Handmer and S. Dovers (Earthscan, 2007), a lack of previous focus on the policy and institutional aspects of disasters and emergencies is identified. S. Dovers (Australian National University, Canberra and an IGU representative) discussed the increasing emergence of climate change as a major influence on thinking about disasters, and the lessons from events such as the floods and cyclones in Australia. Dovers emphasized particularly the degree to which the existing agendas of policy and institutional reform, and the existing institutional capacities provide a basis for coping with what is likely to be a future punctuated by an increasing number of serious disasters and emergencies.

J. Eichelberger (US Geological Survey, and an AGU representative) discussed how several observatories in the Pacific region monitor volcanoes and warn of impending or ongoing eruptions. Many of the real-time monitoring data are available to the public through observatory websites, and citizen reports on volcanic activity and ash falls are solicited. A close link between Russian and the Alaska observatories tracks ash threatening North Pacific air routes. A challenge is to maximize the societal benefit from this expanded hazard community. Indeed, accessibility of real-time data makes the concept of cloistered observatories outmoded. Observatories will still be dedicated to monitoring a limited number of volcanoes for their neighbouring populations and will be the sole authoritative voice for hazard warnings, but must also become nodes for data and knowledge exchange for the larger scientific community. The Eyjafjallajokull eruption showed that such networking should be international in scope.

The Indian Ocean tsunami of 26 December 2004 was the most devastating in the world for the past 40 years. H. Gupta (President of the Asian Oceanic Geoscience Society, and a representative of the ICSU Regional Office for Asia and the Pacific) discussed how India planned to develop an integrated mitigation system for tsunami and storm surges in the northern part of Indian Ocean region with the ultimate goal of saving lives and property.

S. Zlatanova (Delft University of Technology, The Netherlands, and an ISPRS representative) described how 'A proper management increases the situational awareness after a disaster happened, gives a better overview of available data (especially dynamic data), facilitates the access to a desired piece of information, and contributes to automatic data processing. Consequently the information can be used more efficient in the decision-making process'. Well-structured data can support cost-benefit analysis in the post-disaster period, provide a strong foundation for effective mitigation measures and improve preparedness. The variety of approaches reveals the complexity of the problem. Emergency response is very much nationally and even locally (district) organized. Furthermore each country is prone to a specific set of hazards and organizes its management procedures according to the recognized vulnerability. Another complicating factor is the disaster type, which may require different data sets and procedures for management. Zlatanova analysed the challenges in data management and presented some dynamic data models.

1.6.2 The Symposium: Earth on the Edge – Recent Pacific Rim Disasters

The symposium was dedicated to the recent disasters in the Pacific Rim region. K. Takeuchi (International Centre for Water Hazard and Risk Management, Tsukuba, Japan) introduced the total picture of the Great East Japan Earthquake and Tsunami with the latest figures for damage. He said that a concatenation of events occurred from the earthquake to the tsunami and from the tsunami to nuclear meltdown. The impact spread through the industrial supply chain first locally, then nationally, and to nation and finally internationally. He also described the newly enacted Reconstruction Law and its basic policy. He emphasized that no disaster should be 'beyond expectation' in disaster risk management however unexpected the hazards are.

The Tohoku earthquake that occurred on 11 March 2011 off the Pacific coast of northern Honshu caused the worst tsunami disaster in Japan's history. K. Satake (Earthquake Research Institute, University of Tokyo, Japan) reported that this giant earthquake was a plate boundary rupture along the Japan trench, where interplate earthquakes ($M < 8$) occurred at an interval of a few decades and that the probability of one occurring in the next 30 years was forecast as >80%. The March earthquake was, however, much larger than forecast. The GPS network recorded large coseismic displacements (>5 m horizontal and >1 m subsidence), and the tsunami inundation extended several kilometres inland. The fault rupture was about 400 km long, while the aftershocks extended nearly 500 km, including three $M > 7$ events within an hour of the main shock. A similarly large earthquake and tsunami in AD 869 (called the Jogan earthquake) was recorded in a Japanese historical document, which mentioned a thousand casualties and the tsunami inundation extending several kilometres inland. Studies have clarified that the Jogan and older tsunami deposits were distributed several kilometres from the present coast with recurrence intervals of several hundreds to thousand years.

F. Imamura (Tohoku University, Japan) said that the Great East Japan Earthquake took more than 27 000 lives. Several types of tsunami impact were reported, among them inundation over a large area, destruction of buildings and infrastructure and a change of topography due to erosion and deposition. Although the observation system for tsunamis was heavily damaged along the coast, some data were recorded by the tidal gauges, GPS buoys, and deep sea pressure sensors. The extent of the affected area was quite large and limitations and difficulties for ground survey persist. Imamura reported the results of field surveys as well as satellite image analysis, which allows the extent of tsunami inundation and land use change to be identified. The distribution of tsunami run-up heights measured along the coast ranged from 7 to 15 m in Sendai and Fukushima with a simple beach geometry and from 10 to 30 m in Sanriku with the complex geometry of the Rias coast.

S. Kiefer (University of Illinois, USA) discussed how the dynamic features of the tsunami waves changed. As the tsunami approached and then flowed onto the shore of northern Honshu, the character of the waves changed several times. Deep-water waves changed to multiple shallow water waves as the tsunami approached the shore. When these waves encountered coastal cities, they broke into individual hydraulic streams channelled by the infrastructure. From videos posted on the Internet, Keifer and her group constructed hydrographs for sites in Kesennuma, Oirase, Sendai, and Kamaishi.

K. Irikura (Aichi Institute of Technology, Japan) discussed how the huge tsunami generated by the Great East Japan Earthquake led to the accidents at the nuclear reactors. When the earthquake happened, all of the reactor units at four nearby plants were automatically shut down and began to be cooled by cooling systems until they were attacked by big tsunami waves. All the units at the Onagawa and the Tokai No. 2 nuclear power plants escaped problems because the heights of the tsunami waves were lower than the altitudes of the plant sites. However, the Fukushima No. 1 and the Fukushima No. 2 plants were damaged by big tsunami waves, because the tsunami heights were much higher the altitudes of the plant sites. At the Fukushima No. 1 plant, external electric power was cut, water-tanks were broken, and all of the independent power generation systems were damaged. At the Fukushima No. 2 plant, some of the independent power generation systems were not broken because they were on higher ground, so that the cooling systems at the Fukushima No. 2 plant soon recovered. The severe accidents at the Fukushima No. 1 plant were caused by a deficiency of multifaceted protective mechanisms, not only by the tsunami.

The deadly and damaging Christchurch Earthquake of 22 February 2011 is part of the Canterbury earthquake sequence which started with the M_w 7.1 Darfield Earthquake in September 2010. K. Furlong (Penn State University, USA) showed that this sequence occurred on previously unrecognized fault(s) and was significantly distant from the main components of the plate boundary system through South Island, New Zealand. The initial rupture in the September event and subsequent aftershocks have delineated a linear (nearly east–west in orientation) trend extending over 80 km from the foothills of the Southern Alps to the Pacific coast, east of the city of Christchurch. Understanding the relationships between fault segments, regional geologic structure, and crustal stresses associated with regional plate interactions is key to placing these events into a context that allows us to apply the lessons learned to future events.

N. Nicholls (Monash University, Melbourne, Australia) discussed the role of the climate and weather in the January 2011 Brisbane floods, and the potential role of climate/weather forecasts in avoiding or reducing flood damage. Since the last major floods in the Brisbane River (1974) there have been major advances in the detail, timeliness, delivery, and quality of weather and climate predictions on all these timescales. On some timescales weather and climate forecasts were not even available for previous floods. The skill of forecasts on the synoptic timescale has improved dramatically over the past few decades. The models used to make projections on climate change timescales have become more complex than those available in the early 1970s, and the likely consequences of global warming are now better appreciated.

1.6.3 The open forum: Natural Hazards: From Risk to Opportunity by Partnership of Science and Society

The open forum attracted the attention of not only the participants of the IUGG General Assembly but also representatives of media. S. Briceño (UNISDR Director, Geneva, Switzerland) welcomed the participants in his video message and presented key issues related to disaster risk reduction and mitigation. T. Beer (IUGG President, Australia) presented several challenges in natural hazards research and risk analysis, especially those related to climate change. R. Kuroda (ICSU Vice President, Japan) talked about the partnership of science and society. She mentioned that since 2000, science itself had changed: it had become inter- and trans-disciplinary and more competitive. Meanwhile international collaboration through mega-projects had become an inevitable part of research. The most profound change for science, however, has been its stronger interdependence with society. The outcome of science and technology very quickly spreads into our everyday life and can and has changed socio-economic structures and our way of thinking dramatically. Kuroda proposed that scientists should be educated in social literacy and the public in science literacy. A. Ismail-Zadeh and K. Takeuchi moderated the forum and discussions on topics related to natural hazards, extreme events, disaster risk and the link between science and society.

The open forum was followed by an online briefing on predicting disasters, which was organized by the Australian Science Media Centre. Four experts spoke at the news briefing: J. Goff (Australian Tsunami Research Centre and Natural Hazards Research Laboratory, University of New South Wales); A. Ismail-Zadeh (International Institute of Earthquake Prediction Theory and Mathematical Geophysics, Russian Academy of Sciences, Moscow); T. Jordan (Southern California Earthquake Center, University of Southern California); and Brian Kennett (the Australian Academy of Science Committee on Earth Science and Research School of Earth Sciences, Australian National University in Canberra). The briefing addressed the following questions: Can anyone really predict earthquakes, tsunamis, and volcanic eruptions? If not, why not and will we ever be able to do so? Knowing the epicentre of a shake seems to be only part of the problem, can we say how much damage will be caused? Australia has been affected before, how likely is it that it will experience another earthquake or tsunami? Can we be forewarned enough to organize mass evacuations days in advance? Or to accurately predict that an event will happen in a few-year span? The briefing can be heard at the web-page of the Australian Science Media Centre.[4]

1.7 Focus on education

The ENHANS project promoted the activity of the Abdus Salam International Centre for Theoretical Physics (ICTP) in Trieste, Italy, and coorganised an advanced school on 'Understanding and Prediction of Earthquakes and Other Extreme Events in Complex Systems'. The need for a scientifically literate populace is increasingly recognized as critical in many countries, as they face the consequences of increasing population pressures, limited resources, and environmental degradation. Basic science literacy, coupled with scientific 'ways of knowing – namely drawing conclusions based on observation, experiment and analysis – provides citizens with the tools needed for rational debate and sound decision-making based on scientific knowledge' (from the 2011 report of the ICSU Ad-Hoc Review Panel on Science Education). Geoscience education, and particularly education on extreme natural events and disaster risks, is profoundly important for the basic science literacy of the population living at risk. Moreover, special attention should be given to economically less-developed countries.

The advanced school held between 26 September and 8 October 2011 in ICTP covered a wide range of the developments in the field of natural and socio-economic extreme hazards and disaster risks. During this advanced school, prominent experts delivered lectures and gave seminars to earlier career scientists, mostly from developing countries.

1.8 Focus on predictability

Natural and socio-economic disasters pose an intolerable threat to society. The most damaging and least understood of these are so-called extreme events. In different contexts these events are also called critical transitions, disasters, or catastrophes. The multidisciplinary research of the last decades reveals a surprising similarity and structural universality in the development of extreme events in systems of diverse origin. The ENHANS symposium 'Predicting Extreme Events in Natural and Socio-Economic Systems' was dedicated to V. Keilis-Borok (University of California, Los Angeles, USA), a distinguished scientist who greatly contributed to research on the predictability of extreme events in natural and socio-economic systems. The symposium was organised by A. Gabrielov (Purdue University, Lafayette, USA), I. Zaliapin, and A. Ismail-Zadeh. Invited speakers covered a wide spectrum of research on predictability: earthquake forecasting and prediction by D. Turcotte (University of California, Davis, USA), V. Kossobokov (Russian Academy of Sciences, Moscow, Russia), S. Uyeda (Japan Academy of Sciences, Tokyo, Japan), P. Varotsos (University of Athens, Greece), and T. Jordan (University of South California, Los Angeles, USA); scaling laws, chaos, and predictions by C. Allègre (Institut de Physique du Globe de Paris, France); prediction of human-induced events by H. Gupta (National Geophysical Research Institute, Hyderabad, India); prediction of extreme events in socio-economic systems by M. Ghil (École Normale Supérieure, Paris, France) and M. Intriligator (University of California, Los Angeles, USA).

1.9 Project results and conclusion

The ENHANS project concluded that a reduction of disaster risk could be achieved through in-depth scientific research on disaster risks and through risk assessments. The ENHANS project called for the following important actions (see Appendix 1.2 for the text of the Declaration):

- the promotion of comprehensive holistic inter- and transdisciplinary approaches to natural hazard and disaster risk research, which have to integrate knowledge from natural and social sciences, mathematics, engineering, disaster management, insurance sectors, and other stakeholders dealing with disaster risk;
- assistance in networking existing regional scientific and educational centres with the aim of establishing a regional centre of excellence in disaster risk research (e.g., in sub-Saharan Africa);
- negotiations to set up a process of assessing and synthesizing the policy-relevant results of peer-reviewed published research on: the understanding of natural phenomena and the social vulnerability associated with disasters; the capability of predictive systems to disseminate the timely and accurate information needed for policy and decision making; methodologies and approaches for reducing vulnerability and increasing resilience of societies; and the overall ability of societies to reduce risk (prevent, mitigate, and prepare for the increasing impact of natural events).

The major scientific results of the ENHANS project have been broadcast via the project homepage (http://www.ENHANS.org), newsletters of UN, international and national organizations, and the Internet (as video presentations or as report presentations). The ENHANS project results are summarized in this book *Extreme Natural Hazards, Disaster Risks and Societal Implications*.

Appendix 1.1

THE ENHANS INTERNATIONAL WORKSHOP RECOMMENDATION (PRETORIA, SOUTH AFRICA, 20 JANUARY 2011)

The following recommendation is made by attendees of the international workshop 'Extreme Natural Hazards and Disaster Risks in Africa'.

Whereas, natural hazards are an integral component of life in the African continent, and floods, droughts, earthquakes, tsunamis, landslides, and other extreme natural events hit Africa on a regular basis resulting in tragic loss of life and property due to tremendous vulnerability of the African countries to extreme hazards;

The ENHANS International Workshop "Extreme Natural Hazards and Disaster Risks in Africa"

Acknowledging the long-standing and ongoing contributions of

the American Geophysical Union (AGU);
the Global Oceanic Observing System (GOOS) of the Intergovernmental Oceanographic Commission (IOC) of UNESCO;

the International Council of Science (ICSU) including its Regional Office for Africa;
the ICSU/UNISDR/ISSC International Programme Integrated Research on Disaster Risks (IRDR);
the International Geographical Union (IGU);
the International Society for Photogrammetry and Remote Sensing (ISPRS);
the International Union of Geodesy and Geophysics (IUGG);
the International Union of Geological Sciences (IUGS);
the International Union of Theoretical and Applied Mechanics (IUTAM);
the Observing System Research and Predictability Experiment (THORPEX) of the World Meteorological Organization (WMO);
the United Nations Environment Programme (UNEP);
the United Nations International Strategy for Disaster Reduction (UNISDR);
the United Nations Platform for Space-based Information for Disaster Management and Emergency Response (UN-SPIDER); and
other United Nations, intergovernmental, international and national organizations dealing with natural hazards and disaster risks;

Recalling the relevant recommendations of the World Conference on Disaster Reduction in Kobe, Hyogo, Japan, 2005, in particular the Hyogo Framework for Action (2005–2015): Building the Resilience of Nations and Communities to Disasters (HFA);

Noting the valuable contributions to and intensive discussion during the Workshop;

Recognizing that:

(1) vulnerability of Africa is high and growing, and therefore, it is of crucial importance to reduce it for the safety and security of Africa;
(2) an extreme natural event can trigger major socio-economic disorder or even conflicts in the countries of the African continent that may have an impact of global significance;
(3) agricultural based African countries are highly dependent on climatic conditions, and there is a need to reduce dependence (e.g. on rainfall);
(4) marine and coastal hazard vulnerability reduction requires observational infrastructure;
(5) a lack of scientific data, their scattering and quality, access to data, and data exchange are major problems of research on natural hazards and disaster risks and in disaster risk reduction and management;
(6) better education of students in Africa going through higher education on integrated research (like the PeriPeriU project) as well as of policy makers on disaster risk reduction is required;
(7) better and more systematic use of the media would help promote risk reduction;
(8) enhanced collaboration among natural, social, and health scientists and engineers in Africa could contribute to reduction of adverse effects of natural hazards;
(9) closer and more active links among science, policy makers and end-users in operational disaster management bodies in Africa are essential for ensuring effective risk and vulnerability reduction;

Emphasizing that:

(1) The social and economic impacts of disasters usually exceed, by several orders of magnitude, the cost of risk reduction (prevention, mitigation and preparedness);
(2) Existing technology for satellite observations, real-time geophysical and environmental monitoring, and natural hazard forecasting models could prevent loss of life in Africa due to disastrous events if their predictions are timely prepared and delivered and warnings were heeded by the disaster management and other relevant authorities;
(3) For an improved management of disaster relief efforts, adequate scientific knowledge, existing technology and data could provide regional rescue agencies (national disaster management, civil defence or civil protection managers) immediate quantitative estimates of the occurrence, extent and severity of the disaster as well as data on impacts to be further utilized in recovery and risk reduction efforts;

Recommends to governments and funding institutions in cooperation with the relevant ICSU bodies, United Nations agencies and other international entities, that:

(1) the African Centre for Natural Hazards, Disaster Risk Analysis and Management, a network of already existing research and educational centres or new institutions in the African countries, coordinated by one of the network's partners, be set up in order to:
   (i) improve understanding of natural phenomena and human, social, economic and ecological vulnerabilities associated with disasters and develop predictive modeling capability;
   (ii) enhance the quality of the African PhD programmes to benefit students and earlier career specialists dealing with natural hazards and disaster risks through lecture and diploma courses;
   (iii) exchange knowledge and experience via research workshops, advanced schools, and networking of African scientists and experts in disaster risk management;
   (iv) integrate and provide free-access diverse data streams; and
   (v) promote international collaboration on natural hazards and disaster risks in Africa and with other regions of the world;
(2) Education and capacity building at all levels on natural hazards, disaster risk reduction (prevention, mitigation and preparedness) and post-disaster recovery should become a priority topic of the national disaster risk reduction policies of African countries;
(3) Disaster risk management and climate adaptation centres be established in African countries in order to:
   (i) catalogue, monitor and continuously update information on the population and infrastructure at risk and other hazard-relevant data;
   (ii) monitor land, water, sea and atmospheric processes, and their interaction, in relation to all kinds of natural hazards that can occur;
   (iii) assist emergency response agencies during disasters by providing timely information; and
   (iv) facilitate regional and international cooperation and coordination;

(4) As an essential part of a strategy for risk reduction, more infrastructure for reservoir holding and irrigation during dry seasons needs to be developed in the African regions affected by repeated threats of droughts and floods;
(5) Efficient collaboration be encouraged between natural and social scientists and engineers, health professionals as well as with mass-media and policy-makers;
(6) Insurance industry be more actively developed and involved in disaster reduction efforts;
(7) A partnership be promoted between the corporate industries and the scientific community towards establishing observation systems to enhance the application of geoscience in disaster risk reduction;
(8) A joint statement of the African national academies related to science education on natural hazards and disaster risks in Africa be issued;
(9) The ICSU Regional Office for Africa, with input from ICSU Scientific Unions and UN agencies, to facilitate and coordinate action on the science policy interface related to disaster risk mitigation in Africa; and

Resolves to promote fundamental research of the Earth systems, natural hazards, and natural and human-induced environmental disaster risks in Africa in order to improve monitoring and prediction of extreme natural hazards for the benefit of the society. Namely,

(1) To develop and to maintain an African seismic network combining local seismic networks, to monitor earthquake activity and deformation, and to intensify studies on quantitative predictions of large seismic events;
(2) To concentrate on studies of volcanoes combining the monitoring of gas flux, crustal deformation, volcano seismicity, gravity, electromagnetic and radioactivity anomalies with theoretical and numerical modeling;
(3) To improve research capabilities of African national meteorological services to forecast hydro-meteorological hazards;
(4) To develop specific research methodologies dealing with incomplete and scattered data sets (specifically important for Africa);
(5) To use the vast GOOS dataset operating in Africa to monitor and forecast natural hazards associated interaction of land, ocean and atmosphere (e.g., earthquakes, tsunami, landslides);
(6) In collaboration with the ICSU World Data System and ICSU Regional Office for Africa, to develop a network of African data centres and services to collect and index fundamental datasets, to document the datasets by creating metadata, to create data standards and data charters among many others.

Appendix 1.2

DECLARATION ON DISASTER RISK REDUCTION THROUGH RESEARCH AND ASSESSMENTS

The following Declaration, made by Partners of the Project EXTREME NATURAL HAZARDS AND SOCIETAL IMPLICATION (ENHANS), aims to reinforce the vital link between the scientific community, national governments and the public to mitigate disasters caused by extreme natural events and to contribute to sustainable development of society.

The first decade of the XXIst century has been marked by a significant number of disasters triggered by natural and human-induced hazards, such as devastating earthquakes (e.g. 2004 Sumatra–Andaman in the Indian Ocean, 2005 Kashmir, 2008 Sichuan in China, 2010 Haiti, and 2011 Great East Japan), which triggered tsunamis and landslides; floods (e.g. in western and central Europe in 2002; China in 2007; Taiwan and Philippines in 2009; Pakistan in 2010; and Australia in 2010); cyclones, hurricanes, and severe storms (e.g. Katrina in 2005; and Nargis in 2008); wildfire, and some other natural events resulting in tragic loss of life, property and a nuclear emergency.

The Partners of the Project 'Extreme Natural Hazards and Societal Implication – ENHANS' (*in alphabetical order*)

- American Geophysical Union (AGU);
- Global Oceanic Observing System (GOOS) of the Intergovernmental Oceanographic Commission (IOC);
- Integrated Research on Disaster Risks (IRDR) Programme;
- International Geographical Union (IGU);
- International Society for Photogrammetry and Remote Sensing (ISPRS);
- International Union of Geodesy and Geophysics (IUGG);
- International Union of Geological Sciences (IUGS); and
- International Union of Theoretical and Applied Mechanics (IUTAM);

Acknowledging the long-standing and ongoing contributions of

the International Council for Science (ICSU);
the International Social Sciences Council (ISSC);
the United Nations Educational, Scientific and Cultural Organization (UNESCO);
the United Nations Environment Programme (UNEP);
the United Nations International Strategy for Disaster Reduction (UNISDR);
the United Nations Platform for Space-based Information for Disaster Management and Emergency Response (UN-SPIDER);
the World Meteorological Organization (WMO); and
other United Nations, intergovernmental, international and national organizations dealing with natural hazards and disaster risks;

Recalling the relevant recommendations of the World Conference on Disaster Reduction in Kobe, Hyogo, Japan, 2005, in particular the Hyogo Framework for Action (2005–2015): Building the Resilience of Nations and Communities to Disasters (HFA); the 2011 IPCC Special Report on Managing the Risk of Extreme Events (SREX); the 2011 Beijing Declaration on Integrated Research on Disaster Risk; and the 2011 ESF-COST Declaration on Extreme Geohazards and the Reduction of Disaster Risks;

Considering the rapidly increasing vulnerability to natural and human-induced hazards at global, regional, and local levels; and

the number of people affected, and property damage caused by natural events;

Realizing, that disaster risk reduction (prevention, mitigation and preparedness) needs long term planning of scientific research and implementation; and

that reducing the impact of disasters should be addressed with higher priority from global background to local levels;

Noting, that the scientific knowledge and expertise in the relevant topics of extreme natural hazards and disaster risks exist in many hazard-prone regions;

that existing scientific knowledge and technology for disaster risk assessment and mitigation could provide impetus to more effective preventive measures and recovery processes; and

that the research on why, when and how natural hazards turn to be disasters and reduction of predictive uncertainty are the most important scientific agenda items in natural hazard and disaster risk reduction;

that the economic impact of disasters exceeds the cost of mitigation by orders of magnitude,

Call for a reduction of disaster risk through scientific research on disaster risks and through risk assessments; namely,

1. promotion of comprehensive holistic inter- and trans-disciplinary approaches to natural hazard and disaster risk research, which have to integrate knowledge from natural and social sciences, mathematics, engineering, disaster management, insurance, climate change, sustainable development sectors and other stakeholders dealing with disaster risk;
2. help in networking existing regional research and educational centres as well as new institutions with the aim to establish regional centres of excellence in disaster risk research and management (e.g., sub-Sahara Africa, Latin America and Caribbean); and
3. negotiation on setting up a process of assessing and synthesizing the policy-relevant results of peer-reviewed published research on (i) understanding of natural phenomena and the social vulnerability associated with disasters; (ii) the capability of predictive systems to disseminate timely and accurate information needed for policy and decision making; (iii) methodologies and approaches for reducing vulnerability and increasing resilience of societies; and (iv) the overall ability of societies to reduce risk (prevent, mitigate and prepare for the increasing impact of natural events). The assessment would contribute to enhance the knowledge of disaster risk at global, regional and local levels and the awareness of the people living with risk. A high-level intergovernmental body comprising of experts on natural hazards and disaster risk analysis should be the set up and undertake the assessment.

Endnotes

1 Available at: http://www.un-spider.org/news-en/4847/2011-02-04t123100/un-spider-enhans-international-workshop-%E2%80%9Cextreme-natural-hazards-and (accessed on 23 August 2013).

2 Available at: http://www.insurancejournal.com/news/international/2011/02/03/183169.htm (accessed on 23 August 2013).

3 Available at: http://web.up.ac.za/default.asp?ipkCategoryID=3523&ArticleID=6252 (accessed on 23 August 2013).

4 Available at: http://www.aussmc.org/2011/07/online-briefing-predicting-natural-disasters-from-a-shaking-earth/ (accessed on 23 August 2013).

2
The grand challenges of integrated research on disaster risk

Gordon McBean

2.1 Introduction

A sustainable society is one which progresses in its development while equitably meeting its present needs and not compromising the ability of future generations to develop and meet their own needs. With careful planning, progress can occur allowing a society to meet the challenges posed by disasters, technological changes, and other challenges and innovatively seize development opportunities along the way. Disasters can have very detrimental impacts on development and on efforts to eradicate global poverty and there is need for focus on the 'intrinsic relationship between disaster reduction, sustainable development and poverty eradication' (UNISDR, 2005). This is the grand challenge of integrated research on disaster risk.

While very large events are, fortunately, fairly rare, the frequency of recorded disasters has been rising rapidly. A disaster is defined as 'a serious disruption of the functioning of a community or a society involving widespread human, material, economic or environmental losses and impacts, which exceeds the ability of the affected community or society to cope using its own resources' (UNISDR, 2009). From a global average total of about 10 disasters per year in the period 1900–1940, to 200 per year in the 1980s, the number of disasters in the period 2001–2010 averaged 384 per year (Guha-Sapir *et al.*, 2011) with millions of people killed, injured, or displaced each year. The impacts of hazards have been increasing around the globe and the economic costs have increased to about \$110B (B = billion) per year averaged over the period 2001–2010.

The World Economic Forum's 2013 Global Risks Report (WEF, 2013) had as one of its three risk cases: Testing Economic and Environmental Resilience. Their conclusions on global economic and environmental systems include the statement: 'Future simultaneous shocks to both systems could trigger the "perfect global storm", with potentially insurmountable consequences. On the economic front, global resilience is being tested by bold monetary and austere fiscal policies. On the environmental front, the Earth's resilience is being tested by rising global temperatures and extreme weather events that are likely to become more frequent and severe. A sudden and massive collapse on one front is certain to doom the other's chance of developing an effective, long-term solution. Given the likelihood of future financial crises and natural catastrophes, are there ways to build resilience in our economic and environmental systems at the same time?' The Report urged policy leaders to step up efforts to tackle these grand challenges. 'These global risks are essentially a health warning regarding our most critical systems. National resilience to global risks needs to be a priority so that critical systems continue to function despite a major disturbance' (WEF, 2013). Resilience needs to include disaster risk management to reduce the impacts of extreme events as these costs add to the indebted state of governments. The World Bank President (Kim, 2013) stated that 'global warming imperils all of the development gains we have made.' Addressing resilience means addressing the grand challenges of disaster risk reduction.

In the fall of 2012, Hurricane Sandy caused extensive damage from the Caribbean along the eastern coastal areas of the USA with some impacts extending into Canada. The death toll is estimated at 253 lives, with about 131 being in the USA, at least 104 in Haiti and others in Cuba, the Bahamas, Canada, the Dominican Republic, and Jamaica. The estimated total economic costs are about \$75B, with more than \$71.4B in the USA and about \$2B in Cuba, \$700M (M = million) in both Haiti and the Bahamas and in the range \$100–300M in each of Canada and Jamaica (IBC, 2012a,b). Over 150 000 homes were damaged in Cuba (*Havana Times*, 2012). The costs of these events

Extreme Natural Hazards, Disaster Risks and Societal Implications, eds. Alik Ismail-Zadeh, Jaime Urrutia Fucugauchi, Andrzej Kijko, Kuniyoshi Takeuchi and Ilya Zaliapin. Published by Cambridge University Press.

are always estimates as it is difficult, if not impossible, to put dollar values on all impacts and usually not possible to identify all the impacts. An extra dimension of disasters as illustrated by Hurricane Sandy can be their role in national politics. *The Economist* (2012) noted that among the small list of factors contributing to the re-election of US President, Obama was 'maybe lucky even that Hurricane Sandy appeared when it did.'

On 11 March 2011 a devastating tsunami resulting from a *M*9.0 earthquake struck the northern Pacific coast of Japan, and completely destroyed many coastal communities, particularly in the Iwate, Miyagi, and Fukushima prefectures. The earthquake and tsunami resulted in the most expensive natural disaster ever recorded, with estimated economic damages of $210B. The death toll is estimated to be 19 846 and approximately 82 000 residents lost their houses (Guha-Sapir *et al.*, 2012).

In 2010, the Haitian earthquake caused over 220 000 fatalities with about 40% of Haitians being affected (Guha-Sapir *et al.*, 2011). For an already impoverished society, the earthquake was especially destructive with the costs of $8B, surpassing the country's gross domestic product (GDP). In comparison to Hurricane Sandy, the economic costs in dollars are small but in the context of Haiti, the economic costs are devastating. In the summer of the same year, extreme temperatures in Russia, coupled with floods and wildfires, led to a total of 55 800 deaths, while floods in Pakistan caused more than 2000 deaths and economic costs of about 6% of GDP. The varying effects of hazards on societies are one of the challenges that the disaster risk reduction community must address.

Although all countries have been impacted by natural disasters, the relative impacts usually are larger in human lives in developing countries and larger, in absolute terms, in economic costs in developed countries (Mileti, 1999). In highly developed countries, the average number of deaths per disaster is 23, while the number increases dramatically to about 150 deaths per disaster in medium developed countries and to over 1000 deaths per disaster in less developed countries (Mutter, 2005). While the absolute dollar costs of disasters in highly developed countries are large, the damage as a percentage of GDP is much larger in the developing countries (Handmer, 2003). Development is then not sustainable. Sustainable development needs to include decisions on investments and how they will alter societies' exposure to the risk and occurrence of natural hazards.

Hurricane Sandy and costly climate-related events in Canada (for example, the Slave Lake wildfire of 2011) demonstrate the potential for impacts on highly developed countries. Insured losses in Canada due to catastrophic events cost over $1B in 2012 (IBC, 2012b), about $1.6B in 2011, and almost $1B in each of the two previous years. The majority of these insured losses were caused by extreme weather events (McBean *et al.*, 2012). The Insurance Bureau of Canada's Chief Economist (IBC, 2012a) stated: 'We hope this research will act as a catalyst for governments, industry, communities and individuals to recognize the weather risks we are facing and to enter discussions about how to reduce their effects on Canadians' lives and communities.'

The World Risk Index (World Risk Report, 2012) seeks answers to the following questions: How probable is an extreme natural event, and will it affect people? How vulnerable are the people to the natural hazards? To what extent can societies cope with acute disasters? Is a society taking preventive measures to face natural hazards to be reckoned with in the future? Based on their assessment of the 173 countries that were ranked, the five countries most at risk are, in order: Vanuatu, Tonga, Philippines, Guatemala, and Bangladesh. All are countries with vulnerable populations exposed to hazards and with relatively small ability to cope. A grand challenge for the science community is to constructively assess these rankings, including their methodology, and to provide continually upgraded and enhanced scientific information for revised methodologies and as input to these rankings. The scientific community must also provide information that allows countries to improve their at-risk ranking in the World Risk Index.

In 2011, the Philippines, ranked as the third most at-risk country in the World Risk Index, experienced the highest number of natural disasters ever registered in its history (33), being affected by 18 floods and landslides, 12 storms, 2 volcanic eruptions, and 1 earthquake. Ethiopia (4.8 million victims), Kenya (4.3 million), and Somalia (4.0 million) were affected by droughts and consecutive famines with over 40% of Somalia's people being victims. Over the last decade, China, the USA, the Philippines, India, and Indonesia constitute together the five countries that are most frequently hit by disasters. In 2011, there were about 159M victims in China with a drought from January to May affecting 35 million people. Some progress is being made. For example, the global risk of being killed by a cyclone or flood is lower today than it was 20 years ago, except for those who live in a country with low GDP and weak governance (UNISDR, 2011).

This chapter introduces the issues of disaster risk reduction by first presenting the international context for disaster risk reduction as a framework for identifying issues and priorities. In the context of those reports, consideration is then given to grand research challenges. The chapter ends with this author's perceptions of the main issues – the 'grand challenges'; that is: foreseeing or predicting the hazards; scientific evidence-based decision making for reducing vulnerability and exposure; and understanding, influencing, and obtaining the necessary societal responses.

2.2 International context for disaster risk reduction

The 2000 Millennium Summit formally established a series of Millennium Development Goals (MDGs) as a comprehensive and multi-dimensional development framework with clear, quantifiable targets to be achieved in all countries by 2015. In 2002, the World Summit on Sustainable Development (MDG, 2002) drew connections between international development and natural hazards: 'An integrated, multi-hazard, inclusive approach to address vulnerability, risk assessment and disaster management, including prevention, mitigation, preparedness, response and recovery, is an essential element of a safer world in the twenty-first century.' The United Nations International Strategy for Disaster Reduction (UNISDR)[1] convened in 2005 at the World Conference on Disaster Reduction in Kobe, Japan. The resulting Hyogo Declaration (UNISDR, 2005) stated: 'The starting point for reducing disaster risk and for promoting a culture of disaster resilience lies in the knowledge of the hazards and the physical, social, economic and environmental vulnerabilities to disasters that most societies face, and of the ways in which hazards and vulnerabilities are changing in the short and long term . . .'. Note that the starting point relates to knowledge of hazards and vulnerabilities and ways in which they are changing. This succinctly describes the grand challenges of disaster risk reduction: hazards; vulnerabilities; knowledge and prediction. The UNISDR (2009) defines a hazard as a 'dangerous phenomenon, substance, human activity or condition that <u>may</u> cause loss of life, injury or other health impacts, property damage, loss of livelihoods and services, social and economic disruption, or environmental damage.' The word '<u>may</u>' is underlined to emphasize that a hazard alone does not cause a disaster. The objective of disaster risk reduction research is to provide the information upon which societies can make the 'may' non-applicable for their case. In the context of this chapter, hazards are usually natural but the disasters that follow are not.

Noting the definition of a disaster as given earlier, it is important to stress the aspect of exceeding 'the <u>ability of the affected community or society to cope</u> using its own resources'. The sense of the community's ability to cope emphasizes the human aspects of disasters. Disasters result when there is an intersection of a hazard and a vulnerable community or system that is exposed. Vulnerability is determined by the characteristics and circumstances of a community, system, or asset that make it susceptible to the damaging effects of a hazard. Exposure is determined by the people, property, systems, or other elements present in hazard zones that are thereby subject to potential losses. Thus, disasters relate to three aspects: the hazard; the vulnerability; and the exposure.

There are different approaches (Godschalk, 1991; Mileti, 1999) to disaster risk reduction. *Mitigation* is the adoption and implementation of structural and non-structural measures to limit the adverse impact of natural hazards, environmental degradation, and technological hazards and *preparedness* reflects the activities and measures taken in advance to ensure effective response to the impact of hazards, including the issuance of timely and effective early warnings, based on predictions. *Response* is the provision of assistance or intervention during or immediately after a disaster to meet the life preservation and basic subsistence needs of those people affected. *Recovery* comprises decisions and actions taken after a disaster with a view to restoring or improving the pre-disaster living conditions of the stricken community, while encouraging and facilitating necessary adjustments to reduce disaster risk. While recognizing that response and recovery are critically important, this chapter will focus on *mitigation* and *preparedness* as components of disaster risk reduction. In saying that, it needs to be noted that during recovery is the ideal time to initiate the mitigation aspects to reduce vulnerability and exposure to the next hazardous event.

Hazards are of many types, such as hydro-meteorological, geophysical, biological, health epidemics, and technological. This chapter is concerned with the first two types. The largest number of events are hydro-meteorological hazards (UNISDR, 2009), which in some analyses (Guha-Sapir *et al.*, 2012) are broken into hydrological (floods of all kinds and wet mass movements), meteorological (storms of all kinds including tropical cyclones (also known as typhoons and hurricanes), thunderstorms, hailstorms, tornadoes, blizzards, heavy snowfall, avalanches, coastal storm surges), and climatological (extreme temperature, drought, wildfire). Geophysical hazards include earthquakes, volcanic activity and emissions, and related geophysical processes such as dry mass movements, landslides, rockslides, surface collapses, and debris or mud. These types of hazards intersect; for example, a tsunami which is usually triggered by an undersea earthquake or a landslide into the ocean becomes an essentially oceanic process that is manifested as a coastal water-related hazard.

The enormity of the impacts of climate-related events was outlined by Wahlstrom who stated 'Over the last two decades (1988–2007), 76% of all disaster events were hydrological, meteorological or climatological in nature; these accounted for 45% of the deaths and 79% of the economic losses caused by natural hazards' (Wahlstrom, 2009). As the climate changes in the future, these events will increase in number and costs, in victims, deaths, and dollars. The Intergovernmental Panel on Climate Change Special Report on *Managing the Risks of Extreme Events and Disasters to Advance Climate Change Adaptation* has summarized the scientific knowledge to 2012 (IPCC,

2012). Wahlstrom (2009) concluded that: 'The real tragedy is that many of these deaths can be avoided.' This is the ultimate grand challenge for disaster risk reduction: how can scientific research and advances lead to actions that result in dramatic reductions in the number of deaths – so that many of these deaths are avoided?

Climate-related extreme events are already rolling back development gains with particularly evident impacts on low- and medium-income developing countries (McBean and Rodgers, 2010). The Global Framework for Climate Services[2] of the World Meteorological Organization has as its main goal to 'enable better management of the risks of climate variability and change and adaptation to climate change, through the development and incorporation of science-based climate information and prediction into planning, policy, and practice on the global, regional, and national scale'. The role of science is highlighted as science-based information and prediction and the need to address global, regional, and national scales. Sub-national scales will also be important for most countries. The Global Framework for Climate Services provides grand challenges for disaster risk reduction with respect to climate-related hazards.

In 2012, the United Nations convened the Rio+20 Conference (UN Secretary-General's High Level Panel on Global Sustainability, 2012; United Nations, 2012) on the '*future we want*'. On the topic of disaster risk reduction, the final report had four paragraphs (186–189) with the following key and relevant points. There is a 'call for disaster risk reduction and the building of resilience to disasters to be addressed with a renewed sense of urgency . . .' and recognition of the importance of: 'early warning systems as part of effective disaster risk reduction'; 'comprehensive hazard and risk assessments, and knowledge- and information-sharing, including reliable geospatial information'; and 'stronger interlinkages among disaster risk reduction, recovery and long-term development planning'. The '*future we want*' is one of greatly reduced impacts, especially deaths, of hazards around the world.

The report also reconfirmed a commitment to the Hyogo Framework for Action (UNISDR, 2005; see also Chapter 26), which had identified its Priorities for Action as:

1. Ensure that disaster risk reduction is a national and a local priority with a strong institutional basis for implementation.
2. Identify, assess, and monitor disaster risks and enhance early warning.
3. Use knowledge, innovation, and education to build a culture of safety and resilience at all levels.
4. Reduce the underlying risk factors.
5. Strengthen disaster preparedness for effective response at all levels.

In exploring the grand challenges of integrated research on disaster risk, these five priorities for action are used as a basis for development with appropriate order and focus.

2.3 Identify, assess, and monitor disaster risks and reduce the underlying risk factors

Mitigation and preparedness are ongoing actions to reduce risk factors and to be ready to act effectively before and have effective response to the impact of hazards. Both these actions need to be based on identification and assessment of disaster risks and their ongoing monitoring. Since disasters relate to the hazards that may occur in an area and the vulnerability of exposed people, infrastructure, and other aspects in the potentially impacted area, each of these aspects needs to be identified and assessed. Disaster risk reduction is formally defined as: 'the concept and practice of reducing disaster risks through systematic efforts to analyse and manage the causal factors of disasters, including through reduced exposure to hazards, lessened vulnerability of people and property, wise management of land and the environment, and improved preparedness for adverse events'. Note the words: reduced exposure; lessened vulnerability; wise management; and improved preparedness; all of these need to focus the scope of disaster risk reduction research. Reducing the underlying risk factors needs to be done through careful analysis of interconnections and interdependencies, looking for synergies and mutual benefits. There is need for ongoing monitoring to identify changes in any of these parameters coupled with analysis to see how the risks have changed or might change. Mitigation actions include standards and codes to protect infrastructure, people, etc., from extreme events, based on an analysis of probabilities of events, which implies being able to predict the probabilities of future events, and the costs of implementation.

A critical element of reducing disaster impacts in the future is the application of science and evidence to assess disaster risk in order to anticipate and prepare for future hazards. The IPCC Special Report on *Managing the Risks of Extreme Events and Disasters to Advance Climate Change Adaptation* (IPCC, 2012) and a United Kingdom report on the drivers of future disaster risk (Government Office for Science, UK, 2012) both concluded that the main determinants of disaster risk are taken to be the magnitude of the hazard, exposure, and vulnerability. These determinants, and hence disaster risk, will be influenced in the future by a wide range of drivers. The UK report summarised as 'Two drivers stand out in this analysis because of their potentially large and negative effects on disaster risk, and the low associated uncertainty: global environmental change and demographic change. Global environmental change and demographic shifts are likely to continue over the next three decades, leading to greater hazard exposure and vulnerability, as well as reduced resilience and increased uncertainties. The speed of urbanisation in developing countries is also an important driver of change: urban design and planning that both improves the quality of life for residents and makes expanding cities resilient to natural hazards is therefore a key priority.' Hence, to reduce risk there is need to reduce vulnerability, which means addressing poverty and exposure. A focus needs to be on planning and location of people and infrastructure, and infrastructure design and standards and their enforcement.

2.4 Enhance early warning

Effective disaster risk reduction requires the ability to predict the occurrence of hazards or in a sense to '*see the future*'. The new programme, Future Earth – Research for Global Sustainability, now launched by the Science and Technology Alliance for Global Sustainability, is focusing on the very broad aspects of 'seeing the future'. The 'future' may be later today, tomorrow, next week, month or year or beyond. Integrated, multi-disciplinary, science-based predictions of the future are needed and these are one of the grand challenges (McBean, 2008; Mesjasz, 2008). Prediction is the process of looking ahead on the basis of incomplete knowledge of the present and with incomplete understanding of how the system works. Sarewitz *et al.* (2000) have reviewed the role of science in decision-making. Early warning systems are the effective 'seeing the future' and acting on it. They are based on prediction of impending hazards and contribute to societal actions before (*mitigation and preparedness*) and during the events (*response*). *Preparedness* needs to include the issuance of timely and effective early warnings, based on predictions. Warnings (McBean 2000, 2002) advise people about impending events (e.g., tornado, cyclone, flood, volcano, and earthquake). Prediction can also make response and recovery more effective and lower the risk to the interveners. Prediction should also be seen as a key basis for disaster management through reducing the impacts of hazards by providing information for effective decision-making, for increasing awareness of the impacts of hazards, and by demonstrating how disasters are affected by collective societal activities. Early warning (UNEP, 2012) systems lead to 'the provision of timely and effective information, through identified institutions, that allows individuals exposed to hazard to take action to avoid or reduce their risk and prepare for effective response'. An early warning system integrates risk knowledge that provides information on which to base the warning and relies on monitoring and prediction systems. In order to be effective in disaster risk reduction, there needs to be effective response to the warning, which will depend on coordination, good governance, and appropriate action plans. Public awareness and education and communications systems are key elements in order for the system to be effective. If potentially affected locations are not alerted and/or if the local and regional governmental agencies do not understand the messages and know how to respond, the system will not be effective. For Hurricane Sandy, Newman (2012) noted that 'hurricane research and warning system saved lives' and reported how astoundingly accurate the forecast was and the positive return on investment in continued hurricane research.

Although early warning systems are usually thought of in terms of impending events, they need also be thought of in terms of longer periods and the assessment of risk. They can be used to identify anticipated climate events such as seasonal drought or long-term climate changes and the appropriate preparedness strategies can vary with the event. In the case of an impending event such as a tornado within the next 10 minutes, people should seek safe cover. For a river cresting within the next five days, preparations for evacuation and implementation of emergency actions can be taken. If a prediction or scenario predicts more tornadoes over the next year, decade, or century, building codes can be changed and educational programmes and warning systems improved. Early warning systems are the set of capacities needed to generate and disseminate timely and meaningful warning information to enable individuals, communities, and organizations threatened by a hazard to prepare and to act appropriately and in sufficient time to reduce the possibility of harm or loss. The UNEP *Early Warning Systems: A State of the Art Analysis and Future Directions* (2012) makes a useful distinction between types of hazard. One type is the ongoing and rapid/sudden-onset hazards, such as geological hazards and hydrometeorological hazards (except droughts), accidental oil spills, nuclear plant failures, and chemical plant accidents. The second are the slow-onset (or 'creeping') incremental but long-term and cumulative environmental changes. These include such issues as deteriorating air and water quality, acid rain, climate change, desertification and drought, ecosystems change and loss of biodiversity, deforestation, land cover/land changes, and environmental impacts of conflict, among others. It is noted that these usually receive little attention in their early phases but that, over time, they may cause serious crises.

Early warning systems need to differentiate between predictions of hazards and how human or social systems respond to the hazards and the predictions of their occurrence. Weather and weather-related hazards are responsible for the majority of disasters and, although there are difficulties in weather forecasting, the skill of weather forecasts on all scales has been improving (Nichols 2001; Jolliffe and Stephenson 2003; Government Office for Science, UK, 2012). Weather forecasting is now a major activity of all governments[3] and for a substantial private sector. The Government Office for Science (2012) assessment found that in general for forecasting of hydrometeorological hazards, more robust observation systems, higher resolution models, and enhanced computing power will lead in the next 20 years to improved predictions. For flood forecasting, they noted recent progress and projected significant improvements over the next 10–20 years. For drought forecasting, which is still in its infancy, improvements can be expected over the next 20 years. Forecasts of volcanic eruptions, where monitoring is done, have now been successful. For these events, satellite observations and improved models and understanding are important factors for the continued improvements in predictions. Their assessment of earthquake prediction was much more conservative and noted that 'the ability to forecast the timing of earthquakes remains a distant possibility, and whether it will ever be realised is uncertain' (Government Office for Science, UK, 2012). They then stated that the 'study of slow earthquakes and the modelling of complex seismic cycles offer potential routes forward'. Better observations from both satellite and seabed monitoring will be required. Predicting the risks of extreme seismic events in a broader sense of '*seeing the future*' will be discussed in Chapters 4 and 10. Tsunamis occur with different triggers, which are hard to predict. However, once the trigger has happened the time of landfall of the deep-water wave can be forecast. Progress in the improvement of these predictions is expected over the next 10–20 years.

To develop the optimum disaster management strategy requires predictions of: the influence of human systems on natural (both physical and biological) systems; the occurrence and impacts of hazards as they may occur fully naturally or as a result of human influences; how humans will respond to the information that the hazard might or will occur; and how their actions will have consequences for the resultant impacts. Hence, predictions in the realms of natural and social systems are needed. This has been well recognized for some time in the hazards field. The title of the book *Disasters by Design* (Mileti, 1999) stresses how much the impacts of natural hazards are due to choices or actions of humans. In its action plan, the World Conference on Disaster Reduction (UNISDR, 2005) called for 'people-centred early warning systems' to be a key part of the response to the tragedies of natural hazards. The Government Office for Science, UK (2012) in its assessment of the forecasting of the societal elements, vulnerability, and exposure concluded: 'The scientific advances in anticipating natural hazards discussed above can only be exploited for disaster risk forecasting if exposure and vulnerability of people and assets are also assessed. These are crucial components in forming accurate disaster risk projections. However, they are hard to measure because they depend on local circumstances and priorities. The quality and coverage of data on vulnerability and exposure are also generally very poor in developing countries. Developing methods of measurement that take into account local context and priorities would improve the situation.'

Accurate prediction, both of the hazard occurrence and of the response of humans, provides a new relationship with the future. A smog forecast for today allows individuals to respond in limited ways by adapting to reduce their exposure. A smog forecast for a few days allows individuals, industry, and governments to adapt by reducing emissions and, hence, reducing the smog level. These predictions of future states can or should lead to actions that change the outcome; fate can become a choice and choices can make the prediction wrong.

The conclusions and recommendations of the UNEP (2012) analysis of the state of the art and future directions in early warning systems noted that they have 'greatly benefited from recent advances in communication and information technologies and an improved knowledge on natural hazards and the underlying science'. At the same time, they note weaknesses and gaps in capacities and technologies and recommend that more be done to develop a global scale multi-hazard system. The report identifies early warning systems for landslides, droughts, desertification, dust storms, and fire that need to be established as well as full global coverage for earthquake and tsunami early warnings. Coordination across types of events and national boundaries is another area of concern. The Report recommends that an 'institutional mechanism to regularly monitor and communicate slow-onset changes is needed to keep changes under review and to enable rational and timely decisions to be taken based on improved information'.

Noting the technological divide between developed and developing countries, it is important that basic early warning infrastructures and capacities in parts of the developing world most affected by disasters be implemented. As noted above, poorer countries suffer relatively the most from hazard impacts and these impacts can reverse development. The key targets (UNEP, 2012) for capacity building include: (1) developing national research, monitoring, and assessment capacity, including training in assessment and early warning; (2) supporting national and regional institutions in data collection, analysis, and monitoring of natural and man-made hazards; (3) providing access to scientific and technological information, including information on state-of-the-art technologies; (4) education and awareness-raising, including networking among universities with programmes of excellence in the field of the emergency management; (5) organizing of training courses for local decision makers and communities; and (6) bridging the gap between emergency relief and long-term development. It is also important to bridge the gap between science and decision-making, and strengthen coordination and communication links.

2.5 Strengthen disaster preparedness

Since disaster preparedness reflects the activities and measures taken in advance to ensure effective response to the impact of hazards, it should result in reduced costs of disasters. Regarding decision-making and acting on risk information, the Government Office for Science, UK (2012) reported that while new science has considerable potential to improve the quality of information in the forecasting of many disasters, acting on that advice in a prudent and balanced way will be critical to reducing impacts. Decisions can be impeded by the very infrequent nature of some disasters as well as uncertainty in terms of severity, location, and precise timing. Also, while it may be unpalatable, in some cases there may be grounds for accepting the risk because the costs of implementing disaster risk reduction outweigh the benefits. There are no easy answers to such dilemmas and it will be for decision makers to consider when investment in enhanced resilience is justified.

The grand challenge for science is to provide the basis for evidence-based policy making regarding these issues, such as reducing the uncertainties and when does accepting the risks outweigh the benefits of implementing disaster risk reduction?

2.6 Priority for disaster risk reduction with a strong institutional basis for implementation

The science community can and should play a role through reports, presentations, and recommendations that provide the scientific evidence on the benefits of making disaster risk reduction a priority. In the USA, it has been recommended (Cutter *et al.*, 2012) that national resilience be a focal point of national strategies and a guiding principle to inform the mission and actions. An important part of this would be the establishment of a national resource of disaster-related data that documents injuries, loss of life, property loss, and impacts on economic activity. Measuring resilience is challenging but essential if communities want to track their progress toward resilience and prioritise their actions accordingly. Some organizations and communities in the USA have already developed tools to help them gauge their progress.

2.7 The international Integrated Research on Disaster Risk Programme

The International Council for Science[4] Planning Group on Natural and Human-induced Environmental Hazards and Disasters assessment was that 'despite all the existing or already planned activities on natural hazards, an integrated research programme on disaster risk reduction, sustained for a decade or more and integrated across the hazards, disciplines and geographical regions, is an imperative. The value-added nature of such a programme would rest with the close coupling of the natural, socio-economic, health and engineering sciences' (ICSU, 2008; McBean, 2009). In 2008, the International Council for Science, together with the International Social Sciences Council[5] and the UNISDR[6] created the new Integrated Research on Disaster Risk (IRDR)[7] Program, addressing the challenge of natural and human-induced environmental hazards (McBean, 2009). The IRDR is taking an international, multi-disciplinary (natural, health, engineering, and social sciences, including socio-economic analysis) approach and its objectives are: (1) characterization of hazards, vulnerability, and risk; (2) understanding decision making in complex and changing risk contexts; and (3) reducing risk and curbing losses through knowledge-based actions, with cross-cutting themes on capacity building, case studies, and demonstration projects, and assessment, data management, and monitoring.

The first objective on hazards, vulnerability, and risk takes an integrated approach. The second objective focuses on individual and collective choices and decisions: how these decisions are made and how that process can be influenced in ways that address the future more rather than responding to the past or just today (Leiss, 2001; Henstra and McBean, 2005; McBean, 2010). The 2011 Global Assessment Report on Disaster Risk Reduction (UNISDR, 2011) noted that 'Mortality risk for all weather-related hazards continues to be concentrated in countries with low GDP and weak governance, and mortality is still increasing in countries with weak risk governance capacities.' Governments must address complex issues and the challenge for IRDR and for all disaster risk reduction research is to contribute towards more effective decision-making processes. The aim of the third IRDR objective is to connect research with reality, i.e., how the lessons identified through the research studies can be used as the basis for disaster risk reduction in the real world.

The IRDR programme has partnered with the Global Change SysTem for Analysis, Research and Training (START)[8] for capacity enhancement. START, which has focused on global environmental change science and building regional research networks, has expanded its mandate into full consideration of environmental and related stresses including hazards and disasters. Partnerships with the ICSU Regional Offices for Asia and the Pacific, Africa, and Latin America[9] have been also established. Further, within the ICSU family, partnerships have been made with the International Union of Geodesy and Geophysics (IUGG)[10] to co-sponsor the Extreme Natural Hazards and Societal Implications (ENHANS) project. Since data and information are critical to the IRDR, a partnership agreement with the ICSU Committee on Data for Science and Technology (CODATA)[11] enables the effective exchange of information on projects on data and information. There are several centres in the world that compile disaster loss data and publish information. Unfortunately, their published numbers do not always agree (Gall *et al.*, 2009). The IRDR Data Working Group[12] has identified needs for: education of users regarding data bias and issues of social loss data; definition of a 'loss' and the creation of a methodology for assessing it; comparable and accessible human disaster loss data to support research and policy; and increased downscaling of loss data to sub-national geographies for policy makers.

A new set of international disaster case studies under the title Forensic Investigations of Disasters or FORIN[13] (Burton, 2011) is now underway. The goal behind the case studies is to probe more deeply into the complex and underlying causes of growing disaster losses and identify the fundamental cause of disasters. It is important to trace out and assign causal explanation of losses and to examine the intervening conditions that increase or reduce losses. This is being done through a series of case studies using a common template and methodology.[14]

The focus of the Risk Interpretation and Action project[15] is on the question of how people – both decision-makers and ordinary citizens – make decisions, individually and collectively, in the face of risk. There are many research questions that involve diverse fields (Eiser *et al.*, 2012).

The IRDR will be implemented through internationally planned and coordinated projects, such as those on FORINs. It will also build on national projects. An example is the Canadian-funded Coastal Cities at Risk (CCaR)[16]: Building Adaptive Capacity for Managing Climate Change in Coastal Megacities, a 5-year multidisciplinary project focusing on the large cities in the coastal zone and river deltas (Vancouver, Bangkok, Manila, and Lagos). The overall objective of the CCaR programme is to develop the knowledge base and enhance the capacity of megacities to successfully adapt to and when necessary cope with risks posed by the effects of climate change, including sea level rise, in the context of urban growth and development (McGranahan *et al.*, 2007; Nicholls *et al.*, 2007; Nicholls *et al.*, 2008; START, 2009; McBean and Ajibade, 2009).

2.8 Future Earth: Research for Global Sustainability

Future Earth: Research for Global Sustainability[17] is a ten-year international research programme, which will provide the knowledge required for societies to face the challenges posed by global environmental change and to identify opportunities for a transition to global sustainability. It will support science of the highest quality, integrate the natural and social sciences, as well as engineering, the humanities, and law. The Science and Technology Alliance for Global Sustainability is responsible for establishing Future Earth and will promote and support its development as its scientific sponsors. Its members consist of the International Council for Science (ICSU), the International Social Science Council (ISSC), the Belmont Forum, the United Nations Educational, Scientific, and Cultural Organization (UNESCO), the United Nations Environment Programme (UNEP), the United Nations University (UNU), and the World Meteorological Organization (WMO) as an observer.

2.9 Grand challenges

As Burton *et al.* (1993) stated, it is important to consider *The Environment as Hazard*. The natural environment is essential for human development but it is also essential that humans, in order to sustain their development, must take actions to reduce the growing impacts of hazards on societies. In 2001, the same authors published a paper entitled 'Knowing better and losing even more: the use of knowledge in hazards management' (White *et al.*, 2001). In the analysis they noted that knowledge was (and still is) improving and they recognized the need for continued advances. However, an overall concern was that there has been a lack of use of the knowledge – a failure to act. The situation is somewhat different in developing countries where the lack of use of knowledge can be attributed, in part, to the lack of resources for its implementation. In some cases the increasing vulnerabilities and exposure are overwhelming the implementation of disaster risk management. The grand challenge for the Risk Interpretation and Action component of the IRDR and for the scientific community in general is to understand why and how knowledge is used and why it is not used, by governments and as well as by individuals. In well-developed countries an additional concern is that the knowledge is not used effectively. Examples can be found in the realms of regulation and land-use planning and control. One of the challenges for disaster risk management is that, mainly due to the intervals of time between major hazards, the benefits only become evident over time – often beyond the next election.

It is important to learn from experience and to examine the underlying causes of disaster losses. This knowledge then needs to be used both in places where disasters have occurred and also in places where they are likely to occur in the future. An additional challenge to the scientific community is to identify the potential megadisasters of the future.

Following the motto of the International Council for Science, *science for the benefit of society*, science, natural, social, health, and engineering, and the law must address the grand challenges of disaster risk reduction. Society needs to be able to 'see the future', that future being today, tomorrow, and next week, months, years, and decades. There has been significant progress in the forecasting of some types of natural hazards, mainly meteorological ones, but challenges remain, especially for geological hazards. In addition to addressing the prediction of individual hazards, it is important to build the comprehensive, integrated capacity to predict all hazards and their interactions. The natural science community must work with the social science community, for example, to provide integrated predictions of risk, including the prospective changes in the hazards and the vulnerability and exposure of communities.

The scientific community must face the grand challenge of providing the basis for scientific evidence-based decision making on reducing vulnerability and exposure. In addition, a grand personal challenge for scientists is to undertake to be involved in the decision-making process, by not only providing evidence, but also explaining it and participating in the overall process. To do this, the scientific community must better understand the processes of influencing and obtaining the necessary societal responses. It is appropriate to note that for climate change, the international scientific community has effectively produced knowledge and the IPCC has assessed and presented it in policy-relevant ways. Yet there has been little international action to directly address the issue. Across many issues, such as climate change and disaster risk management, there is policy-relevant science. How can the scientific community more effectively promote the implementation of science-relevant policy – a grand challenge?

Endnotes

1 United Nations International Strategy for Disaster Reduction (UNISDR) (www.unisdr.org).
2 http://www.wmo.int/pages/gfcs/index_en.php.
3 The World Meteorological Organization, a specialized UN agency coordinates activities among national meteorological hydrological services. For further information and connections to all national services, visit www.wmo.int.
4 International Council for Science (www.icsu.org)
5 International Social Sciences Council (www.issc.org)
6 United Nations International Strategy for Disaster Reduction (www.unisdr.org).
7 Integrated Research on Disaster Risk (www.irdrinternational.org).
8 START (http://www.start.org).
9 ICSU Regional Offices – Africa (http://www.icsu.org/africa), Asia and the Pacific (http://www.icsu.org/asia-pacifc), Latin America and Caribbean (http://www.icsu.org/latin-america-caribbean).
10 International Union of Geodesy and Geophysics (http://www.iugg.org/).
11 International Council for Science: Committee on Data for Science and Technology (http://www.codata.org).
12 http://www.irdrinternational.org/about-irdr/scientific-committee/working-group/disaster-loss-data/
13 FORIN (http://www.irdrinternational.org/about-irdr/scientific-committee/working-group/forensic-investigations).
14 FORIN (http://www.irdrinternational.org/about-irdr/scientific-committee/working-group/forensic-investigations).
15 http://www.irdrinternational.org/about-irdr/scientific-committee/working-group/risk-interpretation-and-action/
16 Coastal Cities at Risk (http://www.coastalcitiesatrisk.com
17 http://www.icsu.org/future-earth).

References

Burton, I., Kates, R. W., and White, G. F. (1993). *The Environment as Hazard*. New York: The Guilford Press.

Burton, I. (2011). Forensic disaster investigations in depth: a new case study model. *Environment Magazine*, **52**(5), 36–41.

Cutter, S. L., Ahearn, J. A., Amadei, B. *et al.* (2012). *Disaster Resilience: A National Imperative*. Committee on Increasing National Resilience to Hazards and Disasters; Committee on Science, Engineering, and Public Policy; The National Academies. ISBN 978-0-309-26150-0, 260 pp. http://www.nap.edu/catalog.php?record_id=13457

Economist, The (2012). November 10–16, 405–8810, 11.

Eiser, J. R., Bostrom, A., Burton, I. *et al.* (2012). Risk interpretation and action: a conceptual framework for responses to natural hazards. *International Journal of Disaster Risk Reduction*, **1**, 5–16.

Gall, M., Borden, K. A., and Cutter, S. L. (2009). When do losses count? Six fallacies of natural hazards loss data. *Bulletin of the American Meteorological Society*, **90**, 799–809.

Godschalk, D. R. (1991). Disaster mitigation and hazard management In *Emergency Management: Principles and Practice for Local Government*, eds. Drabek, T. E. and Hoetmer, G. J. Washington, DC: International City Management Association, pp. 131–159.

Government Office for Science, UK (2012). *Foresight Reducing Risks of Future Disasters: Priorities for Decision Makers*. Final project report. London: The Government Office for Science.

Guha-Sapir, D., Vos. F., and Below, R., with Ponserre, S. (2011). *Annual Disaster Statistical Review 2010: The Numbers and Trends*. Brussels: CRED. Available at: http://www.cred.be/sites/default/files/ADSR_2010.pdf. Accessed on 7 August 2013.

Guha-Sapir, D., Vos. F., and Below, R., with Ponserre, S. (2012). *Annual Disaster Statistical Review 2011: The Numbers and Trends*. Brussels: CRED. Available at: http://www.cred.be/sites/default/files/ADSR_2011.pdf. Accessed on 7 August 2013.

Handmer, J. (2003). Adaptive capacity: what does it mean in the context of natural hazards. In *Climate Change, Adaptive Capacity and Development*, eds. Smith, J. B., Klein, R. J. T., and Huq, S. London: Imperial College Press, pp. 51–70.

Havana Times (2012). Sandy damaged over 150,000 homes in Cuba. http://www.havanatimes.org/?p=81060

Henstra, D., and McBean, G. (2005). Canadian disaster management policy: Moving toward a paradigm shift? *Canadian Public Policy*, **31**(3), 303–318.

IBC, Insurance Bureau of Canada (2012a). Media briefing – New IBC research on weather – Can Canada manage the storms ahead? Available at: http://www.newswire.ca/en/story/981597/media-briefing-new-ibc-research-on-weather-can-canada-manage-the-storms-ahead. Accessed on 7 August 2013.

IBC, Insurance Bureau of Canada (2012b). 2012 another billion-dollar year in severe weather damages for Canadian insurers, reports IBC. Available at: http://www.newswire.ca/en/story/1087193/it-s-another-billion-dollar-year-in-severe-weather-damages-for-canadian-insurers. Accessed on 7 August 2013.

ICSU, International Council for Science (2008). A science plan for integrated research on disaster risk: addressing the challenge of natural and human-induced environmental hazards. http://www.irdrinternational.org/reports/

IPCC (2012) Summary for Policymakers. In: *Managing the Risks of Extreme Events and Disasters to Advance Climate Change Adaptation*, eds. Field, C. B., Barros, V., Stocker, T. F. *et al*. A Special Report of Working Groups I and II of the Intergovernmental Panel on Climate Change. Cambridge, UK: Cambridge University Press, pp. 1–19.

Jolliffe, I. T., and Stephenson, D. B. (2003). *Forecast Verification: A Practitioner's Guide in Atmospheric Science*. Chichester: Wiley and Sons.

Kim, J. Y. (2013). Op-Ed – Make climate change a priority. Available at: http://www.worldbank.org/en/news/opinion/2013/01/25/oped_-_make_climatechangeapriority. Accessed on 7 August 2013.

Leiss, W. (2001). *Understanding Risk Controversies*. Montreal: McGill-Queen's University Press.

McBean, G. A. (2000). *Forecasting in the 21st Century*. Geneva: World Meteorological Organization. WMO-No. 916.

McBean, G. A. (2002). Prediction as a basis for planning and response. *Water International*, **7**(1), 70–76.

McBean, G. A. (2008). Role of prediction in sustainable development and disaster management. In *Globalization and Environmental Challenges: Reconceptualizing Security in the 21st Century*, eds. Brauch, H. G., Spring, Ú. O., Mesjasz, C. *et al*. Hexagon Series on Human and Environmental Security and Peace, vol. 3. Berlin: Springer, pp. 929–938.

McBean, G. A. (2009). Integrated research on disaster risk – the challenge of natural and human-induced environmental hazards. In *Geophysical Hazards, Minimizing Risk, Maximizing Awareness*, ed. Beer, T. London: Springer, pp. 59–70.

McBean, G. A. (2010). Coping with global environmental change – the need for an interdisciplinary, integrated approach. In *Coping with Global Environmental Change, Disasters and Security – Threats, Challenges, Vulnerabilities and Risks*, eds. Brauch, H. G., Spring, Ú. O., Mesjasz, C. *et al*. Hexagon Series on Human and Environmental Security and Peace, vol. 5. Berlin: Springer, pp. 1193–1204.

McBean, G. A., and Ajibade, I. (2009). Climate change, related hazards and human settlements. *Current Opinions in Environmental Sustainability*, **1**(2), 179–186.

McBean, G. A., and Rodgers, C. (2010). Climate hazards and disasters: the needs for capacity building. *Wiley Interdisciplinary Reviews: Climate Change*, **1**(6), 871–884.

McBean, G. A., Bruce, J. P., and Kovacs, P. (2012). *Telling the Weather Story*. Prepared by the Institute for Catastrophic Loss Reduction for Insurance Bureau of Canada, 50pp.

McGranahan, G., Balk, D., and Anderson, B. (2007). The rising tide: assessing the risks of climate change and human settlements in low-elevation coastal zones. *Environment and Urbanization*, **19**(1), 17–37.

MDG (2002). *Millennium Development Goals – Report of the World Summit for Sustainable Development*, Johannesburg, South Africa, 26 August–4 September 2002 (www.un.org).

Mesjasz, C. (2008). Prediction in security, theory and policy. In *Globalization and Environmental Challenges: Reconceptualizing Security in the 21st Century*, eds. Brauch, H. G., Spring, Ú. O., Mesjasz, C., *et al*. Hexagon Series on Human and Environmental Security and Peace, vol. 3. Berlin: Springer, pp. 889–900.

Mileti, D. S. (1999). *Disasters by Design: A Reassessment of Natural Hazards in the United States*. Washington, DC: Joseph Henry Press.

Mutter, J. C. (2005). The earth sciences, human well-being, and the reduction of global poverty. *EOS Transactions of the American Geophysical Union*, **86**(16), 157, 164–165.

Newman, A. (2012). Hurricane research and warning system saved lives. *The Miami Herald*, 8 November. Available at: http://www.miamiherald.com/2012/11/08/3088401/hurricane-research-and-warning.html. Accessed on 7 August 2013.

Nichols, N. (2001). Atmospheric and climatic hazards: improved monitoring and prediction for disaster mitigation. *Natural Hazards*, **23**, 137–155.

Nicholls, R. J., Wong, P. P., Burkett, V. R. *et al*. (2007). Coastal systems and low-lying areas. *Climate Change 7: Impacts, Adaptation and Vulnerability*. Contribution of Working Group II to the Fourth Assessment Report of the Intergovernmental Panel on Climate Change, eds. Parry, M. L., Canziani, O. F., Palutikof, J. P., van der Linden, P. J., and Hanson, C. E. Cambdirge, UK: Cambridge University Press, pp. 315–356.

Nicholls, R. J., Hanson, S., Herweijer, C. *et al*. (2008). *Ranking Port Cities with High Exposure and Vulnerability to Climate Change Extremes Exposure Estimates*. Organization for Economic Co-operation and Development: Environment working paper No.1.

Sarewitz, D., Pielke, R. A., Jr., and Byerly, R. (2000). *Prediction: Science, Decision Making and the Future of Nature*. Washington, DC: Island Press.

START: Cities at Risk: Developing Adaptive Capacity for Climate Change in Asia's Coastal Megacities. 2009. http://www.start.org

UNEP (2012). Early Warning Systems: A State of the Art Analysis and Future Directions. Division of Early Warning and Assessment (DEWA). Available at http://na.unep.net/siouxfalls/publications/Early_Warning.pdf. Accessed on 20 October 2013.

UNISDR (2005). Hyogo framework for action 2005–2015: building the resilience of nations and communities to disasters. World Conference on Disaster Reduction. http://www.unisdr.org/2005/wcdr/wcdr-index.htm

UNISDR (2009). Terminology. http://www.unisdr.org/we/inform/terminology

UNISDR (2011). *Global Assessment Report on Disaster Risk Reduction*. Geneva, Switzerland: United Nations. http://www.unisdr.org/we/inform/gar

United Nations (2012). *The Future We Want. Outcome Document of the United Nations Conference on Sustainable Development. Rio de Janeiro*, Brazil, 20–22 June 2012. http://sustainabledevelopment.un.org/index.php?menu=1559

UN Secretary-General's High Level Panel on Global Sustainability (2012). *Resilient People, Resilient Planet. A Future Worth Choosing*. New York. www.un.org/gsp

Wahlstrom, M. (2009). Quotation. In *Addressing the Challenge: Recommendations and Quality Criteria for Linking Disaster Risk Reduction and Adaptation to Climate Change*, eds. Birkmann, J., Tetzlaff, G., and Zentel, K. O. Bonn: DKKV Publication Series, 38:5.

WEF, World Economic Forum (2013). 2013 Global Risks Report. http://www3.weforum.org/docs/AM13/WEF_AM13_Report.pdf.

White, G. F., Kates, R. W., and Burton, I. (2001). Knowing better and losing even more: the use of knowledge in hazards management. *Environmental Hazards*, **3**(3–4), 81–92.

World Risk Report (2012). World Risk Report 2012: focus – environmental degradation and disasters. ISBN 978-3-9814495-0-3. www.WorldRiskReport.org.

PART II

Extreme hazards and disaster risks

3

Extreme volcanism: disaster risks and societal implications

Amy Donovan and Clive Oppenheimer

3.1 Introduction

Volcanic eruptions occur on Earth every day. Stromboli volcano in Sicily explodes roughly every eight minutes and has become an attraction for tourists. In a typical year, around 60 volcanoes erupt, though this figure includes continuously active volcanoes such as Stromboli, and excludes most eruptions on the sea floor (which are rarely observed). Some of the events may result in serious local impacts, or may disrupt aviation. There is a crude relationship between eruption magnitude and frequency: the larger the eruption, the rarer it is. While the largest magnitude eruptions are exceedingly rare, their impacts can be extreme. It has even been suggested that such eruptions provide a limit to the development of technologically advanced civilisations (Rampino, 2002), and that they have shaped the trajectory of human culture and society (e.g., Oppenheimer, 2011).

The threat from volcanic eruptions has never been higher than at the start of the twenty-first century: the global population has reached seven billion and is growing rapidly towards an estimated ten-billion peak. At the same time, Western democracies, in particular, have become increasingly conscious of 'risk' (Beck, 1992; Jasanoff, 1999). Risk has come to dominate many areas of life, not least in association with new and emerging technologies such as nanotechnology and biotechnology. This has, among other things, perpetuated the increasing politicisation of science, as scientists are asked to be expert advisors to governments, as well as required to make their research socially relevant. It is increasingly clear that there are some risks, generated in part by modernisation, that require a global response – the key example being climate change, but others include risks of infectious disease, which can be more readily spread in a globalised world. There are complex feedbacks between technological development and increased risk – demonstrated dramatically by the 2010 eruption of Eyjafjallajökull in Iceland, a relatively small volcanic event that nevertheless resulted in chaos for global aviation (Donovan and Oppenheimer, 2011; Adey *et al.*, 2011).

While medical science is allowing people to live longer, the resulting growth in populations also means that risk from natural hazards has increased because more people are living in dangerous areas. This chapter considers the risks associated with volcanic activity, particularly at the extreme end of the scale. A starting point is to define the size or magnitude, M, of an eruption. We do so using the scale introduced by Pyle (1995), which relates magnitude to the mass, m, of erupted materials (including lava and tephra), as follows: $M = \log_{10}(m) - 7.0$. This scale corresponds somewhat to the more qualitative *Volcanic Explosivity Index* (Newhall and Self, 1982) thanks to the constant in the right hand side of the equation. Of course, eruption magnitude does not scale in a straightforward way with societal impact – many other factors come into play, including: co-eruptive release of sulphur and other gases; proximity and vulnerability of populations and infrastructure; and eruption intensity.

Mason *et al.* (2004) compiled a list of the largest documented explosive volcanic eruptions, focusing on events of $M = 8$ or above. They identified 42 such eruptions that occurred in the last 36 My (million years), especially in two temporal clusters: 36–25 My ago and 17 My ago to the present. Only six of these took place in the Quaternary. The more recently published LaMEVE database (Crosweller *et al.*, 2012) identifies 28 eruptions of magnitude 8 or higher in the Quaternary period. The difference demonstrates the uncertainty that surrounds large eruptions: it is likely that not all such eruptions have yet been identified.

In a globalised world, the complexity of international trade, migration, development, and social mobility arguably increases vulnerability to extreme natural hazards.

Extreme Natural Hazards, Disaster Risks and Societal Implications, eds. Alik Ismail-Zadeh, Jaime Urrutia Fucugauchi, Andrzej Kijko, Kuniyoshi Takeuchi and Ilya Zaliapin. Published by Cambridge University Press.

The systems upon which many people rely are at risk in the event of a large magnitude eruption: electricity, clean water, food security, technology. In this chapter, we draw on evidence from past eruptions, models, and scientific data to describe potential scenarios for an 'extreme volcanic eruption'. We consider the likely impacts of such events, and discuss the mitigation measures that might be necessary. We do not consider flood-basalt eruptions, such as those that produced the Deccan Traps in India, though note that these appear to be closely associated with many of the mass extinction events identified in the palaeontological record (Courtillot and Olsen, 2007).

3.2 Facts and figures

This section draws on the United Nations (UN) population density estimates, the Smithsonian Institution Volcanoes of the World database (Simkin and Siebert, 1994), and the Large Magnitude Explosive Volcanic Eruptions (LaMEVE) database (Crosweller *et al.*, 2012).

There are estimated to be 1585 active volcanoes in the world (Simkin and Siebert, 1993), though this figure neglects most submarine eruptions. Of these, 315 are within 100 km of a major urban area – home to at least 161.5 million citizens (the figure neglects smaller towns and rural populations). When other settlements are considered, there are over half a billion people within 100 km of what might be considered 'a potentially active volcano'. The fact that over a fifth of them are in major urban areas increases the practical challenge of mitigation, because transport systems may not be able to cope with the evacuation of densely populated regions. Many of these areas are located in poorer nations, which may have limited resources to record and identify precursory activity (Fig. 3.1).

Taking the LaMEVE database, and extending the dataset to the 3132 volcanoes with documented large magnitude Quaternary eruptions ($M = 4$ and greater), then the 'threatened' population exceeds 0.75 billion, taking a conservative estimate, i.e., more than 10% of the global population. While it is statistically unlikely that many of these volcanoes will erupt catastrophically in the next century, the key challenge is identifying the few that are most likely to do so.

There are different approaches to understanding volcanic risk. These include:

- Frequency-based statistical analyses of long-term eruption rates, either at a particular volcano or over larger geographical areas (e.g. Mason *et al.*, 2004).
- Belief-based probabilistic methods that seek to combine multiple models and assessments of volcano behaviour (e.g. Marzocchi *et al.*, 2004).

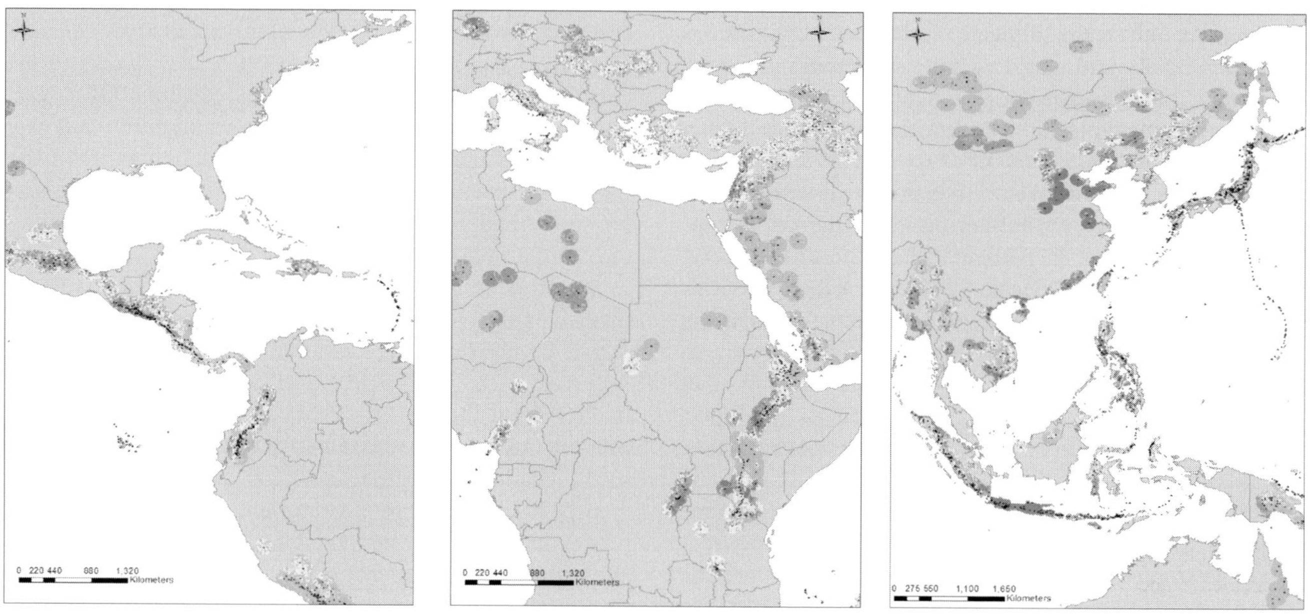

(a) (b) (c)

Figure 3.1. Population density within 100 km of volcanoes listed in the LaMEVE database (Crosweller *et al.*, 2012): (a) Americas; (b) Europe, Near-East and Africa; and (c) East Asia and West Pacific. Population data obtained from the Center for International Earth Science Information Network / Columbia University, and Centro Internacional de Agricultura Tropical, 2005. Population density is shown in five classes: 0 people per km^2 (dark green), 1–9 per km^2 (light green), 10–49 per km^2 (yellow), 50–199 per km^2 (orange), and more than 200 per km^2 (red). For larger colour version, see Plates section.

- Monitoring volcanoes to detect signals of unrest (e.g. Tilling, 2008), providing data that can be used to update belief-based event trees or Bayesian belief networks, or to parameterise deterministic models for eruption precursory activity (e.g. Voight, 1988; Kilburn, 2003).

While probabilistic methods can be applied to calculate global statistics, deterministic methods generally apply in specific situations. A holistic analysis would involve detailed data about all of the potentially active volcanoes in the world; this is currently not possible – not only is the status of many volcanoes unknown, but the task of merely documenting the potentially active volcanoes of the world remains incomplete. A few volcanoes have very closely studied eruption histories, but many more do not; dating old eruption deposits can be expensive and subject to numerous sources of uncertainty. Likewise, estimates of the magnitude of both past and present-day eruptions are subject to substantial errors.

The LaMEVE database lists a total of 235 eruptions in Africa, 14% of which have been dated and have eruptive volumes calculated. Other examples are given in Table 3.1, which shows the percentage of known large magnitude volcanic eruptions that have been dated and quantified. Note that these figures will underestimate the shortfall of knowledge, because some of the unstudied volcanoes may have had more than one large magnitude eruption and it is also possible that there are unknown volcanoes in these regions. The values in Table 3.1 are therefore maxima.

Volcanologists and seismologists have both referred to two main types of uncertainty – epistemic uncertainty, due to a lack of knowledge, and aleatory uncertainty, due to the inherent randomness of natural systems (Woo, 1999; Sparks, 2003; Marzocchi and Bebbington, 2012). There is clearly a high level of epistemic uncertainty in volcanology concerning the eruptive histories of many volcanoes. High epistemic uncertainty undermines frequency-based and belief-based probabilistic methods since both are ultimately knowledge-dependent. Monitoring data are clearly important, but without an awareness of a volcano's past activity, it is very challenging to exploit observations for forecasting. Given that the highest magnitude eruptions are more likely to occur at long-dormant volcanoes, two imperatives face volcanologists: extending the geographical spread of knowledge, and obtaining adequate monitoring networks. Both of these are ultimately dependent on the availability of funding.

The governance of low-probability, high-impact events is inevitably controversial (Taleb, 2007): like the frequency–magnitude distribution, cost–benefit analysis seems to break down for the highest magnitude events. Indeed, in some cases the cost may be unavoidable, suggesting that perhaps investment is not worthwhile. It may be that there are limits to the eventualities for which we can prepare. In this context, it is interesting to consider the combined history of humanity and volcanoes. We will do this in two parts, first examining the long- and short-term climate impacts of eruptions, and then examining the historical, archaeological, and geological evidence for past eruption impacts on humanity.

Table 3.1. *The global distribution of knowledge of Quaternary volcanism can be gauged crudely from the* LaMEVE *database*

Region	% of cited large eruptions with volumes
Japan	77
Iceland	57
Europe (including Azores, Canaries, Cape Verde)	57
Melanesia	35
USA (incl. Alaska and Hawaii)	33
Kamchatka and Asian mainland	32 (all but 3 volcanoes, 1% of this, in Kamchatka)
Philippines and SE Asia	17
Indonesia	16
Africa	14

3.3 Volcanoes and climate

Volcanic eruptions can have wide-ranging impacts on climate. There are two main mechanisms – on short timescales (years) via the generation of sulphate aerosol from sulphur dioxide (or hydrogen sulphide) emissions, and on long timescales (millennia) as a result of changes in volcanic outgassing of carbon dioxide. We will concentrate on the former, noting that the interest in carbon dioxide is relevant more to the case of flood-basalt eruptions in the geological past and possibly to periods of rapid deglaciation and associated increased rates of volcanism (e.g. Huybers and Langmuir, 2009).

Sulphur dioxide emitted during volcanic eruptions partially oxidises in the atmosphere to form sulphate aerosols. If the volcanic plume has penetrated the stratosphere, the residence time of such aerosols in the atmosphere can be several years, during which time the Earth's albedo is increased, direct shortwave radiation from the Sun is scattered, and longwave radiation emitted from the Earth's surface and lower atmosphere is absorbed by the aerosol layer. These processes cause a net warming of the stratosphere and cooling of the troposphere, averaged at global and annual scales. In the few years following the eruption of Mount Pinatubo in 1991, a boreal-summer cooling (of up to a few degrees Celsius) was observed in continental regions of the northern hemisphere (McCormick *et al.*, 1995; Oppenheimer *et al.*, 2011). A number of historical and prehistoric large and intense, sulphur-rich eruptions, notably of volcanoes located in the tropics, have been linked with climate change, associated poor harvests, and, in some historical cases, social unrest (e.g. Oppenheimer, 2011; Timmreck, 2012).

Evidence for the climatological impacts of volcanic eruptions is frequently difficult to interpret. Multiple datasets are combined, including evidence of sulphate and sometimes tephra deposition in ice cores, dendrochronological records, documentary sources, and archaeological studies. Several major eruptions have been associated with stratospheric aerosol veils and increased optical thickness of the atmosphere, as inferred from contemporary reports (Stothers, 2002). Regional- and hemispheric-scale multi-proxy studies can reveal short summer growing seasons or lower than average summer temperatures compatible with the summer-cooling patterns identified after Pinatubo and some other large, sulphur-rich, equatorial volcanic eruptions.

Ice cores allow the extension of climate records up to 800 000 years back in time, but with decreasing chronological and glaciochemical integrity further back in time. Additionally, it can be extremely difficult to attribute particular volcanoes (and eruptions) to ice core signatures where only sulphate is identified. Nevertheless, the presence of sulphate in the cores can be used to track periods of eruptive activity, some of which we discuss below.

In addition to evidence for climate impacts, several attempts have been made to model the climate forcing that results from eruptions. While the general finding of many scenarios is of surface cooling of continental regions of the northern hemisphere during the summer, patterns of winter warming have also been observed – as occurred after Pinatubo in some northern regions (Stenchikov *et al.*, 2002). Models of much larger eruptions and associated sulphur emissions (e.g. Timmreck and Graf, 2006; Robock *et al.*, 2009) suggest that the overall cooling effect, though short term geologically (on the order of a few tens of years), could be pronounced (up to 10–20 °C in some regions immediately after events releasing of order 100 times the amount of sulphur as the Pinatubo eruption). Such an event occurring in the future would be dramatic and would likely have consequences for society that outlast the climatic impact itself.

3.4 Societal impacts of major eruptions: prehistoric and historical perspectives

Humans have always lived with volcanoes and their eruptions. Oppenheimer (2011) argues that Quaternary volcanism probably had significant impacts on human evolution, migration, and demography. In this section, we outline just a few examples: the early Holocene eruption of Laacher See (Germany), and eruptions of Popocatépetl (Mexico; circa 2000 years ago) and Ilopango (El Salvador; circa sixth century CE). We then briefly discuss the more recent demographic change observed on Montserrat in the West Indies.

The eruption of the Laacher See volcano in the East Eiffel Volcanic Field of western Germany, approximately 12 900 years ago, has been studied by both volcanologists and archaeologists. Its impacts on the prehistoric societies of Scandinavia and northern Europe have been considered by Riede (2008), who argued that the eruption prompted migration to southwest Europe by the Federmesser people, and the simplification of technologies in southern Scandinavia, possibly due to increased isolation. The nature and impacts of the Laacher See eruption have been reconstructed from tephrastratigraphic and palaeoenvironmental records by Schmincke *et al.* (1999). Harms and Schmincke (2000) estimated that the eruption released approximately 15 Tg of sulphur dioxide into the stratosphere. This would have likely perturbed the northern hemisphere climate for several years (Graf and Timmeck, 2001). Schmincke *et al.* (1999) pointed out that a future eruption at this location should not be ruled out, judging by the history of other phonolitic volcanoes in the region.

A more recent example of inferred demographic change associated with volcanism concerns the eruptions of Popocatépetl in the first century CE, and Ilopango, possibly in the sixth century CE. Plunket and Uruñuela (2000) argue that the Popocatépetl eruption was a key event in the extensive demographic shifts that occurred in Mesoamerica during this period. Refugees from the eruption, mostly rural settlers, moved to the cities, resulting in the early rise of Teotihuacán – a city that came to specialise in the reworking of obsidian from the plentiful sources in the region. Archaeological evidence from the period demonstrates not only that the eruption did destroy major settlements, but also that there may have been some warnings, as sites in the vicinity of Popocatépetl showed signs of both abandonment and temporary occupation (i.e., people left but returned periodically to tend their crops).

The Ilopango eruption was a considerably larger caldera-forming eruption, and has been associated with the migration of highland Maya people into Guatemala. The eruption magnitude was about 7, but it has proved difficult to date using radiocarbon methods. The most recent attempts suggest that the eruption might be responsible for a major atmospheric aerosol veil suggested by mediaeval chronicles and ecclesiastical documents pertaining to 546 CE, but this has yet to be demonstrated conclusively (Dull *et al.*, 2010). As with the eruption of Popocatépetl, the Ilopango eruption may have aided the prosperity of Teotihuacán by severely limiting the regional influence of rivals in the Mayan highlands (Dull, 2004). In this way, we can view Teotihuacán's demographic, political, and economic authority arising partly as a consequence of two volcanic disasters.

The eruption of Soufrière Hills, Montserrat, which commenced in July 1995, provides a modern example of the complex impacts of volcanism on society. This eruption threatened the capital city, Plymouth, and much of island's infrastructure. Plymouth was evacuated for the final time in April 1996, and was destroyed in the summer of 1997. During the eruption, almost 10 000 people left the island. Even for a culture with a long history of migration (Fergus, 2004), this was an unprecedented event. It also raises questions about the nature of migration due to hazards in the modern world, where people are tracked across borders. In an ancient culture, for example, Montserratians might have had to fight the inhabitants of nearby islands in order to gain land to live on – and they could have attempted to colonise the tiny island of Redonda, or a similar uninhabited location in the region. In the modern world, such migrations are subject to political and legal approval. Many Montserratians moved to the UK or to the USA: the characteristics and laws of these countries dictated the migration patterns, as did the globalised networks of which they are a part. In the event of a large magnitude eruption in a very heavily populated region, however, today's governments might struggle to retain control.

3.5 Some catastrophe scenarios

Evidence from historical and archaeological studies of past eruptions provides important insights into human resilience and adaptability. The largest eruption that has been closely observed from a scientific perspective was that of Mount Pinatubo in 1991: there have been much larger eruptions that pre-date scientific monitoring, and there will be such eruptions in the future. Much of our knowledge of these eruptions is derived from studies of deposits – for example, at Yellowstone, Toba, and Campi Flegrei.

Detailed study of the deposits of past eruptions demonstrates that the key hazards with which they are associated are pyroclastic density currents, ash fallout, and volatile release. Pyroclastic density currents devastate the area around the volcano, ash fallout smothers forests, agricultural crops, and pasture and contaminates water supplies, and gas release into the atmosphere can perturb the global climate (Section 3.3). We consider the threats to humanity from 'extreme volcanoes' in the context of three contrasting scenarios:

- very large magnitude eruptions ($M = 8$ and greater, 'super-eruptions') occurring anywhere in the world, with substantial global climatic impacts and continental-scale deposition of ash;
- moderate-to-high magnitude eruptions ($M = 6$–7) occurring close to large populations;
- large-scale volcanogenic pollution events. These could be due to moderate-to-large magnitude eruptions (such as Laki, 1783–4; Thordarson and Self, 1993) releasing large quantities of volatiles, potentially episodically, for several months.

We consider each of these in turn, focusing on the evidence from past examples and the likely consequences of such events in the future.

3.5.1 Super-eruption!

The term 'super-eruption' has been used loosely to describe eruptions of magnitude 8 and greater or those ejecting 1000 km^3 or more tephra (see Mason *et al.*, 2004), though it is not a universally popular term amongst the volcanological community. Among oft-cited super-eruptions are several associated with the Yellowstone caldera in Wyoming, but there are numerous other volcanoes in diverse tectonic settings that are associated with such large events. Caldera-forming eruptions of this size are among the more poorly understood volcanic phenomena, but are associated with extreme intensities (mass eruption rates), leading to a collapse as the crustal magma reservoir is drained. Studies have suggested that the rheological properties of magma may affect the timing of collapse (syn- or post-eruption) (Karlstrom *et al.*, 2012). Evidence from historical large-magnitude eruptions, notably Tambora in 1815 ($M = 6.9$), suggests that caldera formation accompanied the paroxysmal phase of the eruption, with a very high mass discharge rate (Oppenheimer, 2003b; Self and Blake, 2008).

The largest known eruption of the Quaternary was that of Toba, Indonesia, approximately 74 000 years ago (Storey *et al.*, 2012), known as the Youngest Toba Tuff (YTT) eruption. The eruption ejected a crudely estimated 2800 km^3 (as a dense-rock equivalent volume) of tephra (i.e., $M = 8.8$) and contributed to a composite collapse caldera measuring 100 km by 30 km, and now largely occupied by Lake Toba. More than 1% of the Earth's surface was covered in ash to a depth of at least 10 cm (Rose and Chesner, 1987; 1990). Ash from the eruption has even been found in sediment cores taken from the bed of Lake Malawi, more than 7000 km from source (Lane *et al.*, 2013). Although petrological evidence suggests that this eruption was relatively poor in sulphur (Scaillet *et al.*, 1998; Oppenheimer, 2002), this has not deterred climate modellers from simulating putative YTT impacts for a 100 × Pinatubo scenario, i.e., of order 1000 Tg of sulphur (e.g., Jones *et al.*, 2005; Robock *et al.*, 2009; Timmreck et al., 2010). These models all suggest substantial impacts, varying in magnitude and geographical distribution according to treatments of microphysical processes occurring in stratospheric aerosol clouds, and the coupling of components of the Earth climate system, but none indicates that substantial global cooling would be prolonged for more than a decade or so, or that fully glacial conditions would develop (i.e., an ice age). A debate concerning the impacts of the YTT eruption on human populations, sparked by milestone papers by Ambrose (1998) and Rampino and Ambrose (2000), continues, though the extreme scenarios they argued for have been challenged on several grounds (e.g., Oppenheimer, 2002; Petraglia *et al.*, 2007; Haslam *et al.*, 2010; Lane *et al.*, 2013).

The probability of a YTT-sized eruption is extremely low in any one year. However, there is an important element of epistemic uncertainty in the calculation of the event probability. A glance at the database compiled by Mason *et al.* (2004) concerning the largest eruptions demonstrates that the vast majority occurred in wealthier countries, with many in the USA, where greater resources exist for research into these deposits. Only one eruption is noted in Africa (Ethiopia's Green Tuff). The implication of this may be that many very large magnitude eruptions are as yet undocumented and undated (see also Chapters 13 and 14). Indeed, the collapse calderas database notes a number of collapse structures in Africa for which the magnitudes of associated eruptions are unknown. The LaMEVE database lists 14 eruptions with an estimated dense-rock equivalent volume of 1 km^3 in Africa. Compare this with 24 in Alaska and 40 in the USA and Canada, and 170 in Japan. Given that the total number of such eruptions in the database is 635, this is a significant proportion. In terms of 'super-eruptions', the LaMEVE database lists 28 Quaternary eruptions from 15 volcanoes. Six of the volcanoes are in Japan; three are in the coterminous USA, and three are in New Zealand.

It is also clear that large caldera-forming events are relatively rare in the history of an individual volcano, and it cannot be assumed that a volcano with no history of caldera-forming lacks the potential for such an event in the future. An eruption of similar magnitude to that of Tambora ($M = 6.9$), depending on the location of the volcano responsible and magnitude of sulphur release, would likely have impacts around the world; a future eruption the size of the YTT event is difficult to imagine, especially if associated with a sulphur-rich magma. Should a super-eruption occur, its effects would be global. The duration of these events is difficult to determine. Estimates based on detailed analysis of the Bishop Tuff eruption of Long Valley, California (760 000 years ago), led Wilson and Hildreth (1998) to conclude that the event lasted only a few days. Other super-eruptions may have involved episodic events spanning several years. Regardless, a super-eruption would have major impacts on air quality across a large region, and the immediate area around the source volcano would be devastated by ash fallout and pyroclastic density currents. Agriculture and farming would be severely affected within the zone receiving more than a few centimetres of ash fallout – which in the case of past eruptions of Yellowstone has included much of North America – assuredly compromising food security. Ash would disrupt all forms of infrastructure, leaving hospitals without power and large areas without potable water due to contamination of water supplies by sulphate and halogens. This could lead to famine and disease at the continental scale, which would be difficult to mitigate because of the breakdown of transport links. Aviation could be grounded over much of the globe.

Outside of the immediate impact zone, a tropical super-eruption could cause significant climate perturbations worldwide of longer duration and greater magnitude than the $M = 6$ and 7 events known from the historical record. Jones *et al.* (2005) estimated that an eruption of Yellowstone could produce 5 °C of surface cooling throughout the year in North America, due to the high albedo of the ash deposit. Sulphate aerosols in the stratosphere would also cause significant cooling, and agriculture would be affected by precipitation deficits in regions dependent on monsoonal rains. Such widespread impacts would likely lead to high mortality rates around the world.

The resources required to mitigate this scale of eruption and associated impacts would need to be globally assembled. Winchester (2005) argued that the 1883 eruption of Krakatau was the first eruption to be communicated around the world. Currently, global networks, economies, and trade links render humanity vulnerable to large geophysical events in new ways. A very large eruption anywhere in the world would disrupt these links, cause economic collapses that could propagate across nations, and potentially lead to mass migrations (e.g., http://www.systemshock.org.uk/). The high global population would mean that living conditions in some regions would deteriorate significantly, particularly with crises of employment and food security. Rampino (2002) has argued that super-eruptions might present a limit on the natural progression of civilisation. Nevertheless, the traction of models and simulations for these events is limited by the small dataset available from past eruptions, let alone the complexities of the climate and societal responses. While a super-eruption would, like the impact of a large asteroid, have enormous repercussions for humankind, the record demonstrates that humanity has survived such eruptions in the past. The key questions concern how a future super-eruption would be forecast

Table 3.2. *Populations of large cities within 100 km of a volcano with recognised Quaternary activity, based on the* LaMEVE *database*

City	Population
Yerevan, Armenia	1.3 million
La Paz, Bolivia	1.6 million
Douala, Cameroon	1 million
Vancouver, Canada	1.4 million
Santiago, Chile	4.1 million
Changchun, China	1.7 million
Dalian, China	1.7 million
Nanjing, China	2.3 million
Qingdao, China	1.3 million
Shijiazhuang, China	1.2 million
Zhengzhou, China	1.2 million
Cali, Colombia	1.4 million
Santo Domingo, Dominican Rep.	1.3 million
Quito, Ecuador	1.1 million
Addis Ababa, Ethiopia	1.5 million
Tbilisi, Georgia	1.5 million
Cologne, Germany	1.8 million
Essen, Germany	3.9 million
Athens, Greece	3 million
Guatemala City, Guatemala	1.4 million
Bandung, Indonesia	1.8 million
Jakarta, Indonesia	8.6 million
Medan, Indonesia	1.4 million
Semarang, Indonesia	1 million
Surabaya, Indonesia	2 million
Tehran, Iran	6.4 million
Naples, Italy	2.9 million
Rome, Italy	3.2 million
Fukuoka, Japan	1.8 million
Hiroshima, Japan	1.6 million
Kawasaki, Japan	1.1 million
Kita Kyushu, Japan	1.5 million
Kobe, Japan	1.4 million
Kyoto, Japan	1.5 million
Nagoya, Japan	4.8 million
Osaka, Japan	15 million
Sapporo, Japan	1.9 million
Tokyo, Japan	23.6 million
Yokohama, Japan	3 million
Amman, Jordan	1.3 million
Nairobi, Kenya	1.3 million
Beirut, Lebanon	1.7 million
Guadalajara, Mexico	2.3 million
Mexico City, Mexico	14.1 million
Puebla de Zaragoza, Mexico	1.1 million
Manila, Philippines	5.5 million
Quezon City, Philippines	1.3 million
Taipei, Taiwan, Republic of China	6.1 million
Dakar, Senegal	1.4 million
Incheon, South Korea	1.6 million
Seoul, South Korea	15.9 million
Barcelona, Spain	3.2 million
Valencia, Spain	1.3 million
Aleppo, Syria	1.2 million
Damascus, Syria	1.9 million
El Paso, USA	1 million
Portland, USA	1.2 million
Seattle, USA	2 million
Ho Chi Minh City, Vietnam	3.1 million

with sufficient precision to enable the necessary political action to follow, and what would qualify as appropriate preparedness and mitigation measures.

3.5.2 Urban chaos

As noted at the start of the chapter, the exposure of urban populations to volcanic risk is substantial and increasing, particularly in the global south. Some population centres that are within 100 km of a volcano in the LaMEVE database are given in Table 3.2. Coastal cities are perhaps particularly vulnerable, because of the risk of volcanogenic tsunami, the high number of volcanoes that are close to the ocean, and the limited capacity for seaward escape. Indeed, the reason for the vulnerability of cities in general is the population density, rather than the population itself, particularly combined with the complexities and interdependencies of urban infrastructure, including power lines, water supplies, communications, transport, healthcare, and food supply chains.

Table 3.2 shows that there are striking vulnerabilities all over the world. Tokyo, for example, lies only 100 km from Mount Fuji, whose last eruption took place in 1707–8, with a magnitude of 5.3. Depending on prevailing wind patterns, an eruption of this magnitude today could result in significant ash fallout in the Tokyo metropolitan area, severely compromising urban resources.

Perhaps the most notorious volcano city in the world is Naples, sandwiched between Vesuvius and Campi Flegrei and close to Ischia. Many investigations have been carried out to develop impact models for a future eruption based on past events and probabilistic models (e.g. Baxter *et al.*, 2008; Neri *et al.*, 2008; Zuccaro *et al.*, 2008). However, the potential for hazards from Campi Flegrei has received less attention, and Ischia practically none. Figure 3.2 shows the Campi Flegrei caldera. It also shows the populated areas and schools. The caldera formed by the Campanian ignimbrite eruption, circa 39 300 years ago, is shown too: a sizeable population is located within the caldera itself.

The last eruption of Campi Flegrei was a relatively small event in 1538, forming the Monte Nuovo cone (Fig. 3.2). In this case, the vent opened some distance from what are

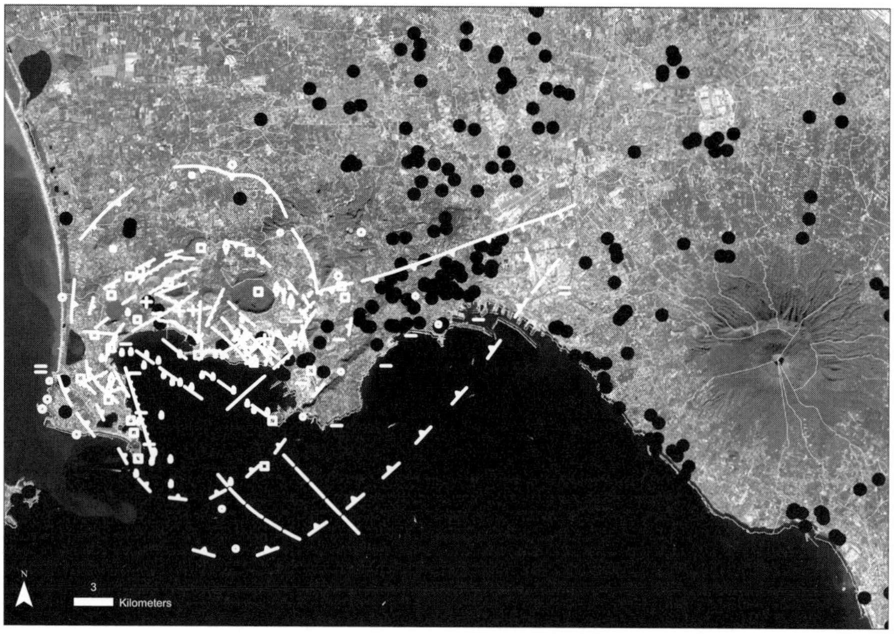

Figure 3.2. An ASTER image of the Campi Flegrei caldera system (from the ASTER Volcano Archive), overlain with the geology from past eruptions in white (Orsi *et al.*, 2004). Schools are shown as black circles. Thin white lines denote administrative districts.

now the most populated zones. Past vents have opened much closer to the urban areas, but activity seems to have moved towards the west with time. Campi Flegrei is a restless caldera associated with episodic uplift and subsidence (so-called *bradyseisms*). The Roman pillars at Serapis, for example, have exhibited over 10-m amplitude of vertical displacement since Roman times, and spent some time below sea level, as is evidenced by the boreholes of marine molluscs that decorate them (Bellucci *et al.*, 2006). In the event of a large eruption at Campi Flegrei, it is possible that some pyroclastic density currents would impact Naples, and even a small eruption could cause significant disruption through tephra fallout (Mastrolorenzo *et al.*, 2006). There are also a number of settlements to the northeast of the caldera, and these would be more exposed to pyroclastic density currents. The Campanian ignimbrite eruption resulted in deposition of tephra as far afield as Russia (Fig. 3.3) – extending across what are today highly populated areas.

In terms of eruption frequency, it would appear that Vesuvius is the greater threat to the Campanian region. Even a relatively small eruption could create chaos in Naples and the communities around the Bay, with ash fall and gas emissions. A large magnitude eruption could have catastrophic consequences, sending pyroclastic density currents into densely populated areas, and heavy ash fall further afield, rendering a wide area uninhabitable, possibly for some time. The EXPLORIS project, carried out with funding from the European Commission, estimated that such an eruption would, in the case where an evacuation of most of the population (98.5%) had taken place prior to the paroxysmal phase of the eruption, lead to the deaths of 8440 people (Baxter *et al.*, 2008). Around 130 000 would be homeless (Zuccaro *et al.*, 2008). Notwithstanding the uncertainties and limitations of such models, they demonstrate that the only effective mitigation measure for such an eruption would be a full evacuation.

There are a number of problems with evacuating urban areas due to high population density and limited routes. Models for evacuation do exist (e.g. Pidd *et al.*, 1996; Cova and Johnson, 2003) but are affected by a range of complex psychological factors as well as testing the limits of transportation systems. A further issue is raised concerning the application of mandatory evacuations: these are morally challenging because they appear to contradict the values of modern democratic societies. As has been witnessed in disasters such as Hurricane Katrina and the eruption of Mount Merapi (Java) in 2010, some people prefer to stay in their homes to protect their goods and chattels, livestock, etc. The counterargument is that those who resist mandatory evacuations place the emergency service personnel, and minors, in danger. The responsibility of the state to its inhabitants comes into conflict with the independence of those inhabitants in liberal democracies. We will discuss this further under 'mitigation', below.

3.5.3 'Yellow water-smoke'

Although this phrase was originally used in the myths of the Klamath people in stories thought to be linked to the caldera-forming eruption of Mount Mazama (resulting in

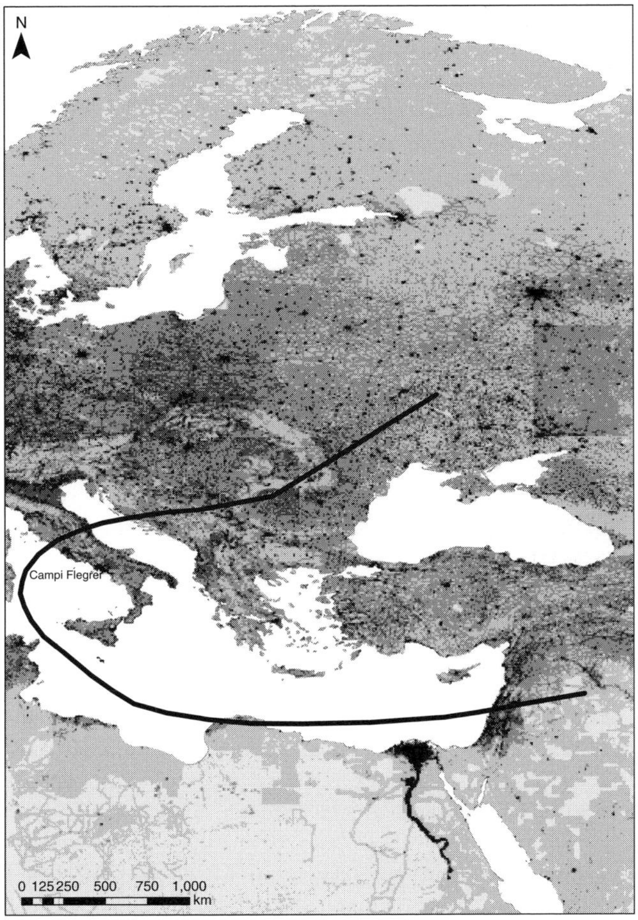

Figure 3.3. Map of Campi Flegrei fallout redrawn from Lowe *et al.* (2012). Population density (dataset as Fig. 3.1) is shown in five classes: 0 people per km^2 (very light), 1–9 per km^2 (light), 10–49 per km^2 (moderate), 50–199 per km^2 (dark), and more than 200 per km^2 (very dark).

Crater Lake, Oregon), it is a good description of the types of phenomenon mentioned in the literature describing volcanogenic pollution. Texts from the late eighteenth century, for example, describe the 'universal fog' produced during the basaltic fissure eruptions along the Lakagígar fissure (Laki) in Iceland in 1783–4:

> On this day and the six that followed there was an uncommon gloom in the air, with dead calm and very profuse dew. The sun was scarcely visible even at midday and entirely shorn of its beams, so as to be viewed with the naked eye without pain. It caused the corn to wither . . . (Henry Bryant, *Norfolk Chronicle*, 5 July 1783)

Other texts associated with a widespread stratospheric aerosol veil in 536 CE note:

> If the Sun becomes dim because the air is dense from rising moisture – as happened in the course of the recently passed fourteenth indiction [535/36] for nearly a whole year . . . (John the Lydian, De Ostentis IC (Arjava, 2005))

> And it came about during this year that a most dread portent took place. For the Sun gave forth its light without brightness, like the Moon, during this whole year, and it seemed exceedingly like the Sun in eclipse, for the beams it shed were not clear nor such as it is accustomed to shed. (Procopius, History of the wars 4.14.5-6 (Arjava, 2005))

Similarly, in 1257–9 CE, a widespread haze was observed that appears linked to one of the largest sulphate anomalies in Holocene glaciochemical records from polar ice cores (Oppenheimer, 2003a). In many cases, this is also associated with failed harvests.

The 1783–4 eruption of Laki in Iceland released an estimated 122 Tg of sulphur dioxide, 7 Tg of hydrogen chloride and 15 Tg of hydrogen fluoride. Much of this remained in the troposphere, contaminating air and water supplies in Iceland (Thordarson and Self, 1993). Around a fifth of the population of Iceland perished soon after the eruption. Witham and Oppenheimer (2005) documented increased mortality rates in England during the eruption, suggesting that both direct and indirect effects of associated climate anomalies and possibly the grounding of the volcanic plume might have been responsible. The eruption of Laki demonstrates that even dominantly effusive eruptions can produce significant gas-rich plumes whose impact may be widespread (Luterbacher *et al.*, 2004; Oman *et al.*, 2006). Given the impacts of the relatively small eruption of Eyjafjallajökull in 2010 (Sigmundsson *et al.*, 2010), a Laki-scale eruption or greater occurring today must be considered a threat to large areas of population (Schmidt *et al.*, 2011). A volcanogenic pollution crisis might involve:

- cardiovascular and respiratory impacts of sulphur dioxide, acidic sulphate aerosol, and ultrafine ash particles;
- fluoride contamination of grazing lands and water supply;
- acid contamination of soil and associated damage to vegetation;
- disruption of aviation due to ash clouds;
- perturbations of regional climate.

There have been other historical examples of large basaltic eruptions releasing prodigious quantities of volcanic volatiles into the atmosphere for protracted periods, including the Eldgjá eruption, also in Iceland, in circa 963 CE (Zielinski *et al.*, 1994). The mostly intrusive event beneath Miyakejima island in 2000 was likewise associated with prodigious sulphur emissions, peaking at around 0.1 Tg of sulphur dioxide per day and occasionally supplementing the burden of anthropogenic pollutants in the Tokyo metropolitan area (Fujita *et al.*, 2003; Kazahaya *et al.* 2004).

Variations in lifestyle, industry, and healthcare facilities make assessing the risk from this kind of event very challenging: local factors would play a key role. In particular, poor air quality aside from volcanogenic sources would exacerbate the situation, as could temperature stress. Schmidt *et al.* (2011) considered the atmospheric dispersion and chemistry of emissions from a future Laki-type event, along with present-day exposure in Europe and dose–response relationships for airborne particulate and human health impact. Their model predicted of the order of 100 000 cardiopulmonary fatalities across the continent. The potential long-range impacts of fine ash and sulphate aerosol in the atmosphere and deposited on the ground on farming and agriculture, and on utilities including power generation, are difficult to characterise and merit investigation. These would depend on synoptic meteorology, season, eruption duration, and topographic factors on a local and regional level.

The development of new technologies, and the increasing dependence on them, also raises vulnerability to this hazard. The most important example of this would be the impact of volcanic ash clouds on aviation. The 2010 eruption of Eyjafjallajökull highlighted the consequences for aviation, and quickly led to global transport disruption despite the comparatively limited dispersion of the ash plume. The eruption consisted of two phases: a fissure eruption on the flank of the volcano, and a summit eruption. The fissure eruption involved the emission of basaltic lava through flows and fountains, fed by a dyke, while the summit eruption was of dacitic magma that had been remobilised by the basaltic dyke (Sigmundsson *et al.*, 2010). The explosivity of the second phase of the eruption reflected the more evolved composition of the magma and its interaction with the ice sheet on the summit of the volcano. This, coupled with prevailing wind patterns in mid-April 2010, resulted in the spread of a cloud of fine ash towards Europe, resulting in closure of air space for nearly a week (Bolić and Sivčev, 2011).

Despite scientific awareness of the likelihood of such an event, such a scenario as the Eyjafjallajökull eruption had not been anticipated by national governments. The chaos quickly spread to airports around the world, with an estimated cost to the industry of several billion US dollars. In part, the situation was compounded by uncertainty concerning the ash cloud altitude, the varying intensity of the eruption, the atmospheric ash concentration, and the tolerance of jet engines to small amounts of ash. A Laki-type eruption in Iceland or elsewhere, involving episodic events spanning a period of months or even years, would likely have even greater impacts on the aviation industry. The Volcanic Ash Advisory Centres (VAACs) have been set up with the risk from volcanic eruptions in mind. Between them, the VAACs cover nearly all of global airspace and monitor it for volcanic plumes (Witham *et al.*, 2007).

While large magnitude lava eruptions like the eighteenth-century eruption of Laki differ in important ways from the eruption of Eyjafjallajökull (which involved magma–ice interaction and was dacitic in composition during the summit phase), the potential impact of a long-term eruption with even a relatively small degree of stratospheric injection would be significant and require long-term management strategies.

3.6 Mitigation and preparation

There are a number of different ways in which humanity could prepare itself for volcanic catastrophes. The first is simply early detection of signals from the volcano. This requires that volcanoes all over the world be appropriately monitored. It is in the interests of richer nations to subsidise this activity if necessary: a large eruption anywhere would have global consequences, and the global distribution of such eruptions is not geographically constrained (Fig. 3.4).

While detecting the signals is important, communicating the risk to governments is critical, as is communicating it internationally. This is perhaps the more challenging area because governments have very broad portfolios and may be affected by corruption, economic difficulties, and structural deficiencies. Nevertheless, they play an important role in preparation and disaster management. The communication of very low probability risks is not straightforward and may not even be meaningful when the uncertainty is very high. There may thus be space to circumvent quantitative risk assessment in these conditions and focus on mitigating the potential impacts of large magnitude events. The following sections discuss these issues in more detail.

3.6.1 Monitoring volcanoes and assessing risk

Ewert and Newhall (2004) estimated that less than 5% of volcanoes are 'adequately monitored'. Tilling (2008) considered the crucial role of volcano monitoring in volcanic risk reduction, arguing that establishing the baseline monitoring data prior to volcanic unrest is essential in identifying volcanic activity and managing the potential consequences. Volcanoes can be monitored in a variety of ways to track the motion of magma beneath the volcanic edifice. Seismic activity frequently accompanies magma movement, as it fractures surrounding rocks (volcano-tectonic earthquakes), or as it flows through cracks (long-period earthquakes). Ground deformation also precedes many eruptions, as magma rises towards the surface and expands in volume due to reduced pressure and volatile exsolution. There may also be changes in gas chemistry, such as increasing proportions of sulphur dioxide relative to carbon dioxide, as sulphur is increasingly exsolved from

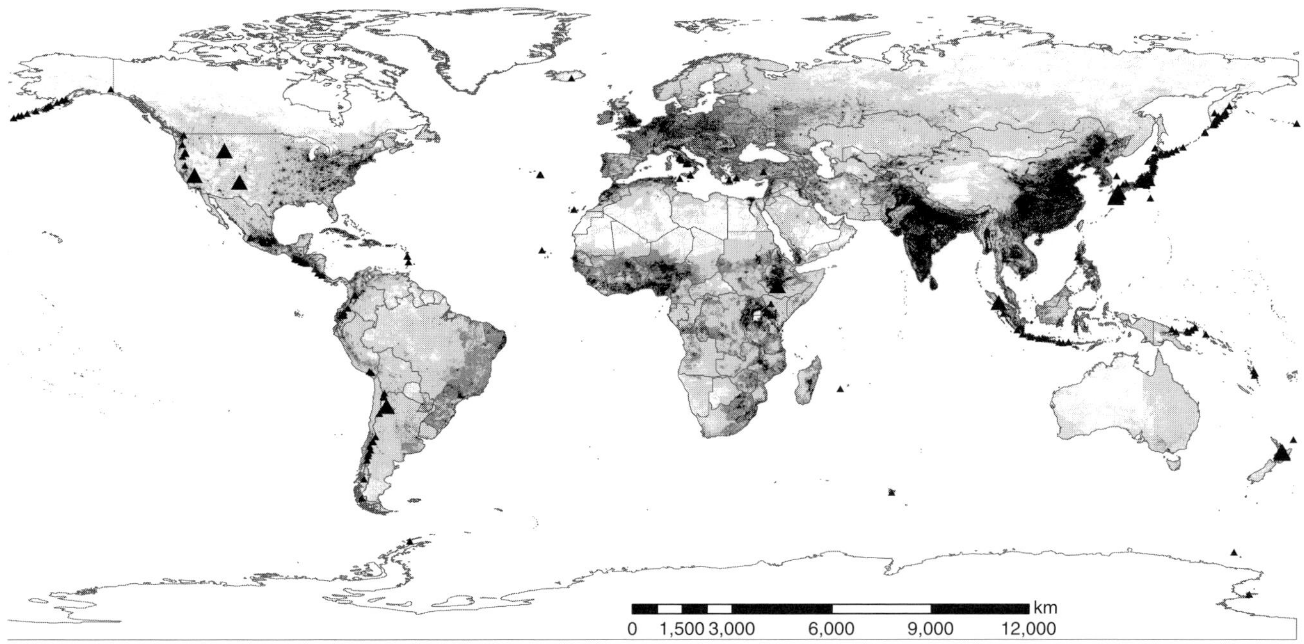

Figure 3.4. Global population density in relation to large magnitude eruptions. Large triangles show volcanoes with Quaternary $M = 8$ and $M = 9$ eruptions; small triangles are $M = 6$ or greater. Population density is shown in five classes: 0 people per km^2 (very light), 1–9 per km^2 (light), 10–49 per km^2 (moderate), 50–199 per km^2 (dark), and more than 200 per km^2 (very dark).

magmas at lower pressure (i.e., shallower depth). However, magma ascent need not culminate in eruption – and more often it does not. At Eyjafjallajökull, for example, there were two sill intrusions in 1994 and 1999, both of which showed very similar patterns of ground deformation to the pattern observed prior to the 2010 eruption (Sigmundsson *et al.*, 2010). Thus, the analysis of monitoring data can be complex and its application to forecasting the outcome of volcanic unrest fraught with uncertainty. Nevertheless, the movement of magma does suggest that the probability of eruption has increased from its baseline value (which could, for example, be calculated from the historical database). It is for this reason that more recent attempts at volcanic risk management have made use of belief-based probabilistic methods.

Belief-based methods, including Bayesian methods and expert elicitation, allow the inclusion of a range of diverse datasets in the analysis. These methods were used extensively during the EXPLORIS project (Baxter *et al.*, 2008), and have been employed operationally on Montserrat during the 1995–present eruptions (Aspinall *et al.*, 2002, 2003). They have also been used in New Zealand (Lindsay *et al.*, 2010; see also Marzocchi *et al.*, 2004, 2007). Notwithstanding their value, the application of quantitative probabilistic methods in volcanology raises a number of questions about the nature of events that can or should be quantified, and how uncertainty is represented in the process. Expert judgement is extremely important in assessing risk from volcanic eruptions: it provides a means by which different types of information can be combined (Donovan *et al.*, 2012b,c). These include data from monitoring networks, models of the magmatic plumbing system, models of hazards and impacts, geological and petrological information, and evidence of recurrence rates both for the volcano in question and for similar volcanoes around the world. Each of these datasets is associated with uncertainty, and requires careful analysis. In terms of 'extreme eruptions', this uncertainty is exacerbated by the inexperience of volcanologists with events of such magnitude: the largest well-observed and recorded eruption is the $M = 6.1$ paroxysm of Mount Pinatubo in 1991. The production of a risk assessment for high-impact events therefore relies heavily on interpretation of the tephrastratigraphic record, ice core data and other palaeoenvironmental proxies, and historical or archaeological information. It is hindered by the fact that the governance of risks that are very low probability tends to be dominated by the short time-span of political horizons (Donovan and Oppenheimer, 2012).

3.6.2 Managing uncertainty

As noted above, scientists distinguish between aleatory uncertainty and epistemic uncertainty. These have been further analysed by Donovan *et al.* (2012a), where a scheme was introduced based on Wynne (1992) and Stirling (2007) to analyse the types of epistemic uncertainty in accordance with information about likelihoods and outcomes. One of the largest areas of epistemic uncertainty in volcanology is the global eruption record: as we have discussed, there are large gaps in our knowledge of past eruptions. There are also gaps in the global monitoring of volcanoes, and in the experience of certain types of eruption in scientific times. A key example of the latter is caldera-forming activity. The largest eruption in the last several hundred years was that of Tambora in 1815, estimated to have been an $M = 6.9$ event. From eyewitness accounts, it seems that the paroxysmal phase of this eruption was accompanied by caldera formation and the passage of prodigious pyroclastic density currents on all sides of the volcano, probably coincident with erosion of the vent and a very high mass discharge rate. These are often the consequences of partial collapse of a Plinian eruption column rising above the central eruption vent. While studies of more recent eruptions (such as those of Pinatubo in 1991 and Mount Saint Helens in 1980) and of the tephra record, combined with theoretical considerations, have advanced our understanding of the dynamics of such plumes, there are some important lacunae in our knowledge of the dynamics of such eruptions. In particular, the timing and duration of caldera formation is not always readily reconstructed when past eruption histories are collated.

There are numerous collapse calderas around the world (Geyer and Marti, 2008). They occur in all tectonic settings and exhibit a range of behaviours. Most are associated with calc-alkaline silicic magmas (Geyer and Marti, 2008), but there are also a few examples involving less evolved magmas. Some volcanoes exhibit single calderas, while others consist of multiple, nested calderas. For many volcanoes, caldera formation is one of many potential styles of activity: many stratovolcanoes are capable of caldera-forming events even if they do not display evidence of such an eruption in the past.

There are very high uncertainties concerning the likelihood of such extreme events. Of the global inventory of potentially active volcanoes, few have been studied in any detail, and thus little is known about their eruptive histories. Likewise, many volcanoes are not monitored at all, or are monitored by the most basic equipment, such as a single regional seismometer. This creates a knowledge vacuum: epistemic uncertainty in terms of global volcanic history is very high. There are two ways in which scientists are working to reduce this uncertainty: the collation of available information into databases such as WOVOdat, Volcanoes of the World, and VOGRIPA; and the ongoing study of volcanic systems around the world. Putting numbers on the risk is, at this stage, very challenging, since the dataset is limited and managing very low probability estimates is complex. As Spiegelhalter and Riesch (2011) noted, the estimation of probabilities for highly uncertain events can be misleading if the uncertainty is not communicated properly. An estimate of the probability of a great eruption might be several orders of magnitude lower than the uncertainty bounds associated with it, rendering the estimate essentially meaningless and potentially counterproductive. Thus, the management and mitigation of very large magnitude eruptions is ultimately about the governance of low-probability, high-impact risks that may not be quantifiable but whose impacts merit preparation in any case.

In the event of a volcano showing signs of unrest, governments may depend heavily on scientific advice. An increasing number of countries have formal institutions that monitor volcanoes, such as observatories. Some observatories are focused on a single volcano, such as the Montserrat Volcano Observatory, while others are responsible for many, such as the Alaska Volcano Observatory. In these instances, the primary source for government advice is relatively clear. Frequently, some form of alert level system is employed (e.g. Leonard *et al.*, 2008; Fearnley *et al.*, 2012). While there is a global alert level system for volcanic ash hazard (applied by the various VAACs), there is no established global system for volcanic activity in general (other than those that are dependent upon temporary research funding) – and given the variation in volcanic activity between tectonic settings and individual volcanoes, such a system may not be appropriate. Geographical sensitivity, both to the social structure and culture and to the physical nature of particular volcanoes, may be required. However, volcanic impacts can affect very large areas, as demonstrated above, and the translation of volcanic risk management between cultures may become necessary. What is currently lacking is a global strategy for the management of large magnitude hazards, including volcanic eruptions. There are a number of issues that may have to be dealt with, including refugees, financial collapse, infrastructure collapse, famine, and disease.

One approach to modelling the consequences of an event of the magnitude of a super-eruption could be to use the kinds of 'integrated assessment models' that are used to forecast the impacts of greenhouse gas emissions and climate change. These combine both economic and Earth system sectors. The economic part of such models considers capital, labour, gross domestic product, and so on, with associated carbon energy costs and greenhouse emission projections. The geophysical part captures relationships between greenhouse gas emissions, radiative forcing, and climate-change induced damage such as flooding due to sea-level rise. The integration of economic and Earth system components allows modelling of different climate-change policies to find optimal balances between damage limitation and its associated costs in terms of restricted carbon emissions.

An important limitation to such investigations in the case of climate change and carbon policy is that they rely on information that is currently unavailable, and which may only become available when it is too late. Two central uncertainties, in particular, pose challenges to quantitative economic analysis. First, how bad will the climate get, and second, how bad will the worsening climate be for the economy, i.e., how much economic damage will be caused by increased temperatures and associated physical impacts of climate change? Notwithstanding these difficulties, such models remain useful for examining the effects of different policies such as market-based emissions trading or carbon taxes, and also for exploring the circumstances that can lead to catastrophic economic outcomes (Ackerman *et al.*, 2010). Given the existing capacity to model the climate change wrought by super-eruptions (Section 3.5.1), it might be feasible to adapt the integrated assessment model approach to consider impact scenarios for future YTT- or Yellowstone-like events. At the very least, we propose that there are parallels between aspects of climate economics and the global risk of a super-eruption that merit exploration.

3.6.3 Disaster management

Volcanic eruptions, even relatively small ones like that of the Soufrière Hills, Montserrat, can impact human migration significantly, redefining nations. As Figure 3.1 shows, there are numerous locations where significant migration as a result of a volcanic eruption could lead to extensive problems with housing refugees. One potential mitigation strategy for this is that already used in some locations: pairing up individual city districts with towns at some distance from the volcano. This ensures that people have places to go to, and prevents the inundation of a particular town or city. Nevertheless, there are problems with this approach, particularly in the event of a large magnitude eruption in a small country. Similarly, economic and infrastructure impacts of such an eruption would cripple a nation and, critically, reverberate well beyond its borders. The global economy was deeply affected by the Great East Japan Earthquake and Tsunami. This occurred on a linear feature that affected a single wealthy nation very heavily. Volcanoes are generally point features with the capacity to affect large areas in all directions. The global impact of a large eruption could prevent or diminish the current practice of unaffected nations providing aid. Foreign aid, particularly to poorer countries (for example, following the 2004 'Boxing Day' tsunami) is an important aspect of global disaster management networks. The capacity of individual countries to cope with large-scale disruption to their economy, infrastructure, and living conditions varies enormously with their wealth and governance. In the event of a large eruption with global effects, the poorest would indubitably suffer the most, in general. Examining the global distribution of volcanoes in relation to population density and the gross domestic product (GDP) suggests that the East African rift system, South East Asia and South and Central America are particularly vulnerable.

A key difference between the impact of eruptions and that of other hazards is duration. As the eruptions of the Soufrière Hills since 1995 and of Laki in the eighteenth century demonstrate, eruptions can continue episodically for months, years, or decades. Very large sulphur releases to the middle atmosphere can likewise result in prolonged climate anomalies that could impact food security for years (Robock *et al.*, 2009). Severely limited harvests could promote the spread of famine and disease over large parts of the globe. Tinned goods would be at a premium in these circumstances, and again this would affect the poorer nations first.

3.6.4 Outreach and communication

The communication of volcanic risk to communities in danger is a critical part of risk management (e.g., Bignami *et al.*, 2012). It is ultimately the responsibility of governments, but is frequently delegated to volcano observatories or similar institutions. A critical challenge is the communication of uncertainty, since the low probability of volcanic eruptions can lead to apathy among politicians and members of the public. This can hinder the provision of funding for outreach during periods of volcanic inactivity. The 2010 eruption of Eyjafjallajökull demonstrated the potential for volcanoes to take over the news cycle in countries that had not even realised that they were at risk (Donovan and Oppenheimer, 2011). The communication of risk to such communities varies significantly around the globe in accordance with scientific knowledge and resources. However, numerous studies have provided important insights into the practice and impact of risk communication at active volcanoes.

Public awareness of volcanic risk tends to be much greater at volcanoes with a history of eruptive activity (e.g. Perry *et al.*, 1982; Paton *et al.*, 2001, 2008; Tobin and Whiteford, 2002; Gregg *et al.*, 2004; Dominey-Howes and Minos-Minopoulos, 2004; Barberi *et al.*, 2008; Haynes *et al.*, 2008; Perry and Lindell, 2008; Bird *et al.*, 2010). The presence of other risks, particularly those encountered on a daily basis, may also decrease popular concern about volcanic risk (e.g. Gaillard, 2008): in poorly developed areas, for example, daily anxieties about food and amenities may detract from concern about low-probability volcanic eruptions. Studies have also demonstrated a number of important insights into the effectiveness of risk communication. On Montserrat, it was found that direct communication between scientists and the public aided public trust and understanding (Haynes *et al.*, 2008). Independent experts may enjoy much higher levels of trust than government representatives, for example. Other studies have demonstrated the importance of participatory methods in engaging with local communities (e.g. Cronin *et al.*, 2004).

Communication between scientists and policymakers is also critical, both before eruptions and during them. Interaction between scientists and officials prior to crises aids trust and awareness of uncertainty (Donovan and Oppenheimer, 2012). Furthermore, studies have demonstrated that even local officials on recently-active volcanoes can be ill informed about the possible impacts of an eruption (e.g. Solana *et al.*, 2008). A key issue facing volcanologists in communicating both with the public and with local officials is the typically low probability of eruptions at any individual volcano. The comments of Louisiana Governor Bobby Jindal in 2009 criticising US President Barack Obama for funding 'something called volcano monitoring' summed up the feeling that can be prevalent in political circles. Communicating risks from volcanic eruptions to policymakers and the public during long periods of volcanic quiescence may be most effectively done through ongoing visibility. This can include regular reports, social media pages, outreach in schools, and public lectures. Many volcano observatories and research institutions participate in such work. However, the impact of outreach on policymaking is dependent in large part on the governmental structures in particular countries and the existence of organisations such as civil protection departments. Such institutions can play a key role in maintaining communication links with scientists (Donovan and Oppenheimer, 2012).

3.7 Concluding remarks

Large magnitude volcanic eruptions have occurred in the past with significant impacts on climate, societies, and landscapes. Such eruptions will occur in the future, and the probability of a magnitude 7 (or even 8) eruption occurring in the next century is finite (Oppenheimer, 2011). We have identified and discussed three potential scenarios for such eruptions, and have indicated a number of global hotspots for high impacts. A key challenge for scientists is intensive data gathering and collation concerning the many unstudied volcanic eruptions around the world in addition to impact assessment and disaster planning.

This brings us to the crux of the challenge of volcano catastrophe risk management: what international institutional models and organisations are relevant to addressing the threat? In fact, there are numerous existing platforms to consider and prior experience on which to draw. They include UNESCO's Intergovernmental Oceanographic Commission (International Oceanographic Commission / UNESCO, 2009), which promotes (among other things) regional tsunami warning networks, development of national disaster plans, community awareness programmes, and evacuation drills. Another is the World Meteorological Organization (WMO, also an agency of the United Nations). Reading its 'vision and mission' statement it is easy to replace 'meteorological' by 'volcanological', and instructive, therefore, to reproduce it here (http://www.wmo.int/pages/about/mission_en.html):

> The vision of WMO is to provide world leadership in expertise and international cooperation in weather, climate, hydrology and water resources and related environmental issues and thereby contribute to the safety and well-being of people throughout the world and to the economic benefit of all nations.

We might borrow further from the mission statement of the WMO, substituting 'meteorology' with 'volcanology', so as to propose the following remit, of a global volcanological organisation, which would be to:

> Facilitate worldwide cooperation in the establishment of networks of stations for the making of volcanological observations as well as hydrological and other geophysical observations related to volcanology, and to promote the establishment and maintenance of centres charged with the provision of volcanological and related services;
>
> Promote the establishment and maintenance of systems for the rapid exchange of volcanological and related information;
>
> Promote standardization of volcanological and related observations and to ensure the uniform publication of observations and statistics;
>
> Further the application of volcanology to aviation, shipping, water problems, agriculture and other human activities;
>
> Promote activities in operational volcanology and to further close cooperation between Meteorological and Hydrological Services;
>
> Encourage research and training in volcanology and, as appropriate, in related fields, and to assist in coordinating the international aspects of such research and training.

The primary international body in volcanology is the International Association of Volcanology and Chemistry of the Earth's Interior (IAVCEI; http://www.iavcei.org), and while it does provide an umbrella for a number of commissions (including the World Organization of Volcano Observatories (WOVO; http://www.wovo.org)), it does not take a coordinating role in volcano monitoring and eruption warning. IAVCEI is a constituent association of the International Union of Geodesy and Geophysics (IUGG; http:www.iugg.org; a non-governmental organisation founded in 1919), which is a member of the International Council for Science (ICSU; http://www.icsu.org, a non-governmental organisation founded in 1931). Then there are the VAACs, set up under the umbrella of the

International Civil Aviation Organization (ICAO, also a UN offshoot). A joint WMO-IUGG Volcanic Ash Scientific Advisory Group (VASAG) was created in 2011 after the Eyjafjallajökull crisis to work closely with ICAO, national and regional civil aviation regulatory bodies with a goal to define such levels of ash concentration which create problems in safely and air worthiness of aircraft.

We suggest that there are existing pertinent agencies from which to build a global vision 'to provide world leadership in expertise and international cooperation in *volcanology* and related environmental issues and thereby contribute to the safety and well-being of people throughout the world and to the economic benefit of all nations'.

Many communities today tolerate proximity to geophysical hazards: the perceived benefits outweigh often incalculable longer-term risks, and uprooting an individual family let alone a whole community or megacity is traumatic. Thus, the many co-locations of people, flood plains, fault lines, and/or volcanoes (Tokyo, Los Angeles, Naples, Istanbul, and Port-au-Prince) pose considerable challenges. As the global human population heads towards ten billion, an increasing proportion of whom will likely live in poverty in cities and near coasts, it is undeniable that future large and very large volcanic eruptions pose vital management challenges for national governments and the global community. A major problem in risk management is that most extreme possible scenarios become a kind of science fiction. On the other hand, considering only the 'worst' probable eruptions, earthquakes, tsunami, and so on may fail to prepare us for more frequent events.

Volcanic catastrophe risk should not be reduced to the product of the probability of the event and the magnitude of the associated losses. Nor should its prospect be ignored or viewed simply with fear. The human track record demonstrates that we have the capacity to manage volcanic threats with resolve, flexibility, and creativity. To quote Adger (2006), 'the diversity of individuals and institutions that draw on reservoirs of practices, knowledge, values, and worldviews . . . is crucial for preparing . . . for change, building resilience, and for coping with surprises'.

References

Ackerman, F., Stanton, E. A. and Bueno, R. (2010). Fat tails, exponents, extreme uncertainty: Simulating catastrophe in DICE. *Ecological Economics*, **69**, 1657–1665.

Adey, P., Anderson, B. and Guerrero, L. L. (2011). An ash cloud, airspace and environmental threat. *Transactions of the Institute of British Geographers*, **36**, 338–343.

Adger, W. N. (2006). Vulnerability. *Global Environmental Change*, **16**, 268–281.

Ambrose, S. H. (1998). Late Pleistocene human population bottlenecks, volcanic winter, and differentiation of modern humans. *Journal of Human Evolution*, **34**, 623–651.

Arjava, A. (2005). *The Mystery Cloud of 536 CE in the Mediterranean*. Dumbarton Oaks Papers, vol. 59. Washington, DC: Dumbarton Oaks, pp. 73–94.

Aspinall, W. P., Loughlin, S. C., Michael, F. V. *et al.* (2002). The Montserrat Volcano Observatory: its evolution, organization, role and activities. *Geological Society of London Memoirs*, **21**, 71–91.

Aspinall, W. P., Woo, G., Voight, B. and Baxter, P. J. (2003). Evidence-based volcanology: Application to eruption crises. *Journal of Volcanology and Geothermal Research*, **128**, 273–285.

Barberi, F., Davis, M. S., Isaia, R., Nave, R. and Ricci, T. (2008). Volcanic risk perception in the Vesuvius population. *Journal of Volcanology and Geothermal Research*, **172**, 244–258.

Baxter, P. J., Blong, R. and Neri, A., eds. (2008). Evaluating explosive eruption risk at European volcanoes – contributions from the EXPLORIS project. *Journal of Volcanology and Geothermal Research*, **178**, 331–592.

Beck, U. (1992). *Risk Society: Towards a New Modernity*. New Delhi: Sage (German original, 1986).

Bellucci, F., Woo, J., Kilburn, C. and Rolandi, G. (2006). Ground deformation at Campi Flegrei, Italy: Implications for hazard assessment. *Geological Society of London Special Publications*, **269**, 141–157.

Bignami, C., Bosi, V., Costantini, L. *et al.*, eds. (2012). *Handbook for Volcanic Risk Management: Prevention, Crisis Management, Resilience*. Available at: http://miavita.brgm.fr (accessed on 20 August 2013).

Bird, D. K., Gisladottir, G. and Dominey-Howes, D. (2010). Volcanic risk and tourism in southern Iceland: Implications for hazard, risk and emergency response education and training. *Journal of Volcanology and Geothermal Research*, **189**, 33–48.

Bolić, T. and Sivčev, Z. (2011). Eyjafjallajökull eruption: European air traffic management experience. *Transportation Research Record: Journal of the Transportation Research Board*, **2214**, 136–143, doi: 10.3141/2214-17.

Courtillot, V., and Olsen, P. (2007). Mantle plumes link magnetic superchrons to Phanerozoic mass depletion events. *Earth and Planetary Science Letters*, **260**, 495–504.

Cova, T. J. and Johnson, J. P. (2003). A network flow model for lane-based evacuation routing. *Transportation Research Part A: Policy and Practice*, **37**, 579–604.

Cronin, S., Gaylord, D., Charley, D. *et al.* (2004). Participatory methods of incorporating scientific with traditional knowledge for volcanic hazard management on Ambae Island, Vanuatu. *Bulletin of Volcanology*, **66**, 652–668.

Crosweller, H., Arora, B., Brown, S. *et al.* (2012). Global database on large magnitude explosive volcanic eruptions (LaMEVE). *Journal of Applied Volcanology*, **C7**–4, 1–13.

Dominey-Howes, D. and Minos-Minopoulos, D. (2004). Perceptions of hazard and risk on Santorini. *Journal of Volcanology and Geothermal Research*, **137**, 285–310.

Donovan, A. R. and Oppenheimer, C. (2011). The 2010 Eyjafjallajökull eruption and the reconstruction of geography. *The*

Geographical Journal, **177**, 4–11, doi: 10.1111/j.1475-4959.2010.00379.x.

Donovan, A. and Oppenheimer, C. (2012). Governing the lithosphere: Insights from Eyjafjallajökull concerning the role of scientists in supporting decision-making on active volcanoes. *Journal of Geophysical Research*, **117**, B03214.

Donovan, A., Oppenheimer, C. and Bravo, M. (2012a). Science at the policy interface: volcano-monitoring technologies and volcanic hazard management. *Bulletin of Volcanology*, **74**, 1005–1022.

Donovan, A., Oppenheimer, C. and Bravo, M. (2012b). Reply to comment from W. P. Aspinall on "Social studies of volcanology: knowledge generation and expert advice on active volcanoes" by Amy Donovan, Clive Oppenheimer and Michael Bravo [*Bull Volcanol* (2012) 74:677–689]. *Bulletin of Volcanology*, **74**, 1571–1574.

Donovan, A. R., Oppenheimer, C. and Bravo, M. (2012c). The use of belief-based probabilistic methods in volcanology: scientists' views and implications for risk assessments. *Journal of Volcanology and Geothermal Research*, **247–248**, 168–180.

Dull, R. A. (2004). An 8000-year record of vegetation, climate, and human disturbance from the Sierra de Apaneca, El Salvador. *Quaternary Research*, **61**, 159–167.

Dull, R., Southon, J. R., Kutterolf, S. *et al.* (2010). Did the TBJ Ilopango eruption cause the AD 536 event? American Geophysical Union, Fall Meeting 2010, abstract #V13C-2370.

Ewert, J. W. and Newhall, C. G. (2004). Status and challenges of volcano monitoring worldwide. In: *Proceedings of the 2nd International Conference on Volcanic Ash and Aviation Safety, 21–24 June 2004, Alexandria, Virginia, USA*, pp. 9–14. Available at: http://www.ofcm.gov/ICVAAS/Proceedings2004/pdf/11-session2.pdf (accessed on 22 August 2013).

Fearnley, C. J., McGuire, W. J., Davies, G. and Twigg, J. (2012). Standardisation of the USGS Volcano Alert Level System (VALS): analysis and ramifications. *Bulletin of Volcanology*, **74**, 2023–2036.

Fergus, H. A. (2004). *Montserrat: History of a Caribbean Colony*. Oxford: MacMillan.

Fujita, S., Sakurai, T. and Matsuda, K. (2003). Wet and dry deposition of sulfur associated with the eruption of Miyakejima volcano, Japan. *Journal of Geophysical Research*, **108**, 4444, doi:101029/2002JD003064.

Gaillard, J.-C. (2008). Alternative paradigms of volcanic risk perception: The case of Mt. Pinatubo in the Philippines. *Journal of Volcanology and Geothermal Research* **172**, 315–328.

Geyer, A. and Marti, J. (2008). The new worldwide collapse caldera database (CCDB): A tool for studying and understanding caldera processes. *Journal of Volcanology and Geothermal Research*, **175**, 334–354.

Graf, H.-F. and Timmreck, C. (2001). A general climate model simulation of the aerosol radiative effects of the Laacher See eruption. *Journal of Geophysical Research*, **106**, 14747–14756, doi:10.1029/2001JD900152.

Gregg, C. E., Houghton, B. F., Johnston, D. M., Paton, D. and Swanson, D. A. (2004). The perception of volcanic risk in Kona communities from Mauna Loa and Hualālai volcanoes, Hawaii. *Journal of Volcanology and Geothermal Research*, **130**, 179–196.

Harms, E. and Schmincke, H. U. (2000). Volatile composition of the phonolitic Laacher See magma (12,900 yr BP): implications for syn-eruptive degassing of S, F, Cl and H_2O. *Contributions to Mineralogy and Petrology*, **138**, 84–98.

Haslam, M., Clarkson, C., Petraglia, M. *et al.* (2010). The 74 ka Toba super-eruption and southern Indian hominins: archaeology, lithic technology and environments at Jwalapuram Locality 3. *Journal of Archaeological Science*, **37**, 3370–3384, doi:10.1016/j.jas.2010.07.034.

Haynes, K., Barclay, J. and Pidgeon, N. (2008). The issue of trust and its influence on risk communication during a volcanic crisis. *Bulletin of Volcanology*, **70**, 605–621.

Huybers, P. and Langmuir, C. (2009). Feedback between deglaciation, volcanism, and atmospheric CO_2. *Earth and Planetary Science Letters*, **286**, 479–491.

International Oceanographic Commission / UNESCO (2009). Five years after the tsunami in the Indian Ocean: from strategy to implementation, advancements in global early warning systems for tsunamis and other ocean hazards 2004–2009, available at: http://ioc-unesco.org/ (accessed on 20 August 2013).

Jasanoff, S. (1999). The songlines of risk. *Environmental Values*, **8**, 135–152.

Jones, G. S., Gregory, J. M., Stott, P. A., Tett, S. F. and Thorpe, R. B. (2005). An AOGCM simulation of the climate response to a volcanic super-eruption. *Climate Dynamics*, **25**, 725–738, doi:10.1007/s00382-005-0066-8.

Karlstrom, L., Rudolph, M. L. and Manga, M. (2012). Caldera size modulated by the yield stress within a crystal-rich magma reservoir. *Nature Geoscience*, **5**, 402–405.

Kazahaya, K., Shinohara, H., Uto, K. *et al.* (2004). Gigantic SO_2 emission from Miyakejima volcano, Japan, caused by caldera collapse. *Geology*, **32**, 425–428.

Kilburn, C. R. J. (2003). Multiscale fracturing as a key to forecasting volcanic eruptions. *Journal of Volcanology and Geothermal Research*, **125**, 271.

Lane, C. S., Chorn, B. T. and Johnson, T. C. (2013). Ash from the Toba supereruption in Lake Malawi shows no volcanic winter in East Africa at 75 ka. *Proceedings of the National Academy of Sciences*, **110**, 8025–8029.

Leonard, G. S., Johnston, D. M., Paton, D. *et al.* (2008). Developing effective warning systems: Ongoing research at Ruapehu volcano, New Zealand. *Journal of Volcanology and Geothermal Research*, **172**, 199–215.

Lindsay, J., Marzocchi, W., Jolly, G. *et al.* (2010). Towards real-time eruption forecasting in the Auckland Volcanic Field: application of BET_EF during the New Zealand National Disaster Exercise 'Ruaumoko', *Bulletin of Volcanology*, **72**, 185–204.

Lowe, J., Barton, N., Blockley, S. *et al.* (2012). Volcanic ash layers illuminate the resilience of Neanderthals and early modern humans to natural hazards. *Proceedings of the National Academy of Sciences*, **109**, 13532–13537.

Luterbacher, J., Dietrich, D., Xoplaki, E., Grosjean, M. and Wanner, H. (2004). European seasonal and annual temperature variability, trends, and extremes since 1500. *Science*, **303**, 1499–1503.

Marzocchi, W. and Bebbington, M. (2012). Probabilistic eruption forecasting at short and long time scales. *Bulletin of Volcanology*, **74**, 1777–1805.

Marzocchi, W., Sandri, L., Gasparini, P., Newhall, C. and Boschi, E. (2004). Quantifying probabilities of volcanic events: The example of volcanic hazard at Mount Vesuvius. *Journal of Geophysical Research*, **109**, B11201, doi:10.1029/2004JB003155.

Marzocchi, W., Sandri, L. and Selva, J. (2007). BET_EF: A probabilistic tool for long- and short-term eruption forecasting. *Bulletin of Volcanology*, **70**, 623–632.

Mason, B. G., Pyle, D. M. and Oppenheimer, C. (2004). The size and frequency of the largest explosive eruptions on Earth. *Bulletin of Volcanology*, **66**, 735–748, doi:10.1007/s00445-004-0355-9.

Mastrolorenzo, G., Pappalardo, L., Troise, C. *et al.* (2006). Volcanic hazard assessment at the Campi Flegrei caldera. *Geological Society of London, Special Publications*, **269**(1), 159–171.

McCormick, M. P., Thomason, L. W. and Trepte, C. R. (1995). Atmospheric effects of the Mt Pinatubo eruption. *Nature*, **373**, 399–404, doi:10.1038/373399a0.

Neri, A., Aspinall, W. P., Cioni, R. *et al.* (2008). Developing an event tree for probabilistic hazard and risk assessment at Vesuvius. *Journal of Volcanology and Geothermal Research*, **178**, 397–415.

Newhall, C. G. and Self, S. (1982). The volcanic explosivity index (VEI) an estimate of explosive magnitude for historical volcanism. *Journal of Geophysical Research*, **87**, 1231–1238.

Oman, L., Robock, A., Stenchikov, G. L. *et al.* (2006). Modelling the distribution of the volcanic aerosol cloud from the 1783–1784 Laki eruption. *Journal of Geophysical Research*, **111**, D12209.

Oppenheimer, C. (2002). Limited global change due to largest known Quaternary eruption, Toba ≈ 74 kyr BP? *Quaternary Science Reviews*, **21**, 1593–1609.

Oppenheimer, C. (2003a). Ice core and palaeoclimatic evidence for the timing and nature of the great mid-13th century volcanic eruption. *International Journal of Climatology*, **23**, 417–426.

Oppenheimer, C. (2003b). Climatic, environmental and human consequences of the largest known historic eruption: Tambora volcano (Indonesia) 1815. *Progress in Physical Geography*, **27**, 230–259.

Oppenheimer, C. (2011). *Eruptions That Shook the World.* Cambridge: Cambridge University Press.

Oppenheimer, C., Scaillet, B. and Martin, R. S. (2011). Sulfur degassing from volcanoes: source conditions, surveillance, plume chemistry and impacts. *Reviews in Mineralogy and Geochemistry*, **73**, 363–421.

Orsi, G., Di Vito, M. A. and Isaia, R. (2004). Volcanic hazard assessment at the restless Campi Flegrei caldera. *Bulletin of Volcanology*, **66**, 514–530.

Paton, D., Millar, M. and Johnston, D. (2001). Community resilience to volcanic hazard consequences. *Natural Hazards*, **24**, 157–169.

Paton, D., Smith, L., Daly, M. and Johnston, D. (2008). Risk perception and volcanic hazard mitigation: Individual and social perspectives. *Journal of Volcanology and Geothermal Research*, **172**, 179–188.

Perry, R. W. and Lindell, M. K. (2008). Volcanic risk perception and adjustment in a multi-hazard environment. *Journal of Volcanology and Geothermal Research*, **172**, 170–178.

Perry, R. W., Lindell, M. K. and Greene, M. R. (1982). Threat perception and public response to volcano hazard. *Journal of Social Psychology*, **116**, 199–204.

Petraglia, M., Korisettar, R., Boivin, N. *et al.* (2007). Middle Paleolithic assemblages from the Indian subcontinent before and after the Toba super-eruption. *Science*, **317**, 114–116.

Pidd, M., De Silva, F. N. and Eglese, R. W. (1996). A simulation model for emergency evacuation. *European Journal of Operational Research*, **90**, 413–419.

Plunket, P. and Uruñuela, G. (2000). The quick and the dead: decision making in the abandonment of Tetimpa. *Mayab*, **13**, 78–87.

Procopius (2007). *The Secret History.* Translated by G. A. Williamson and P. Sarris. London: Penguin Classics.

Pyle, D. M. (1995). Mass and energy budgets of explosive volcanic eruptions. *Geophysical Research Letters*, **22**, 563–566.

Rampino, M. R. (2002). Supereruptions as a threat to civilizations on Earth-like planets. *Icarus*, **156**, 562–569.

Rampino, M. R. and Ambrose, S. H. (2000). Volcanic winter in the Garden of Eden: the Toba super-eruption and the Late Pleistocene human population crash. *Geological Society of America Special Paper*, **345**, 71–82.

Riede, F. (2008). The Laacher See-eruption (12,920 BP) and material culture change at the end of the Allerød in Northern Europe. *Journal of Archaeological Science*, **35**, 591–599, doi:10.1016/j.jas.2007.05.007.

Robock, A., Ammann, C. M., Oman, L. *et al.* (2009). Did the Toba volcanic eruption of ~74 ka B.P. produce widespread glaciation? *Journal of Geophysical Research*, **114**, D10107, doi:10.1029/2008JD011652.

Rose, W. I. and Chesner, C. A. (1987). Dispersal of ash in the great Toba eruption, 75 kyr. *Geology*, **15**, 913–917.

Rose, W. I. and Chesner, C. A. (1990). Worldwide dispersal of ash and gases from earth's largest known eruption: Toba, Sumatra, 75 kyr. *Palaeogeography, Palaeoclimatology, Palaeoecology*, **89**, 269–275.

Scaillet, B., Clemente, B., Evans, B. W. and Pichavant, M. (1998). Redox control of sulphur degassing in silicic magmas. *Journal of Geophysical Research*, **103**, 23937–23949.

Schmidt, A., Ostro, B., Carslaw, K. S. *et al.* (2011). Excess mortality in Europe following a future Laki-style Icelandic eruption. *Proceedings of the National Academy of Sciences*, **108**, 15710–15715.

Schmincke, H. U., Park, C. and Harms, E. (1999). Evolution and environmental impacts of the eruption of Laacher See Volcano (Germany) 12,900 a BP. *Quaternary International*, **61**(1), 61–72.

Self, S. and Blake, S. (2008). Consequences of explosive supereruptions. *Elements*, **4**, 41–46.

Sigmundsson, F., Hreinsdóttir, S., Hooper, A. *et al.* (2010). Intrusion triggering of the 2010 Eyjafjallajökull explosive eruption. *Nature*, **468**, 426–430.

Simkin, T. and Siebert, L. (1994). *Volcanoes of the World: A Regional Directory, Gazetteer, and Chronology of Volcanism during the Last 10,000 Years*. Tucson, Arizona: Geoscience Press.

Solana, M. C., Kilburn, C. R. J. and Rolandi, G. (2008). Communicating eruption and hazard forecasts on Vesuvius, Southern Italy. *Journal of Volcanology and Geothermal Research*, **172**, 308–314.

Sparks, R. S. J. (2003). Forecasting volcanic eruptions. *Earth and Planetary Science Letters*, **210**, 1–15.

Spiegelhalter, D. J. and Riesch, H. (2011). Don't know, can't know: embracing deeper uncertainties when analysing risks. *Philosophical Transactions of the Royal Society A: Mathematical, Physical and Engineering Sciences*, **369**, 4730–4750.

Stenchikov, G., Robock, A., Ramaswamy, V. *et al*. (2002). Arctic Oscillation response to the 1991 Mount Pinatubo eruption: effects of volcanic aerosols and ozone depletion. *Journal of Geophysical Research*, **107**, 4803, doi:10.1029/2002JD002090.

Stirling, A. (2007). Risk, precaution and science: Towards a more constructive policy debate. *EMBO Reports*, **8**, 309–315.

Storey, M., Roberts, R. G. and Saidin, M. (2012). Astronomically calibrated 40Ar/39Ar age for the Toba supereruption and global synchronization of late Quaternary records. *Proceedings of the National Academy of Sciences*, **109**, 18684–18688.

Stothers, R. B. (2002). Cloudy and clear stratospheres before A. D. 1000 inferred from written sources. *Journal of Geophysical Research*, **107**, 4718, doi:10.1029/2002JD002105.

Taleb, N. N. (2007). *The Black Swan: The Impact of the Highly Improbable*. London: Allen Lane.

Thordarson, T. and Self, S. (1993). The Laki (Skaftár Fires) and Grímsvötn eruptions in 1783–1785. *Bulletin of Volcanology*, **55**, 233–263.

Tilling, R. I. (2008). The critical role of volcano monitoring in risk reduction. *Advances in Geosciences*, **14**, 3–11.

Timmreck, C. (2012). Modeling the climatic effects of large explosive volcanic eruptions. *Wiley Interdisciplinary Reviews: Climate Change*, **3**, 545–564.

Timmreck, C. and Graf, H.-F. (2006). The initial dispersal and radiative forcing of a Northern Hemisphere mid-latitude super volcano: a model study. *Atmospheric Chemistry and Physics*, **6**, 35–49, doi:10.5194/acp-6-35-2006.

Timmreck, C., Graf, H.-F., Lorenz, S. J. *et al*. (2010). Aerosol size confines climate response to volcanic super-eruptions. *Geophysical Research Letters*, **37**, 24.

Tobin, G. A. and Whiteford, L. M. (2002). Community resilience and volcano hazard: The eruption of Tungurahua and evacuation of the Faldas in Ecuador. *Disasters*, **26**, 28–48.

Voight, B. (1988). A method for prediction of volcanic eruptions. *Nature*, **332**, 125–130.

Wilson, C. J. and Hildreth, W. (1998). Hybrid fall deposits in the Bishop Tuff, California: A novel pyroclastic depositional mechanism. *Geology*, **26**, 7–10.

Winchester, S. (2005). *Krakatoa: The Day the World Exploded*. New York: Harper Perennial, ISBN-13: 978-0060838591.

Witham, C. S., Hort, M. C., Potts, R. *et al*. (2007). Comparison of VAAC atmospheric dispersion models using the 1 November 2004 Grimsvötn eruption. *Meteorological Applications*, **14**, 27–38.

Witham, C. S. and Oppenheimer, C. (2005). Mortality in England during the 1783–4 Laki Craters eruption. *Bulletin of Volcanology*, **67**, 15–26, doi:10.1007/s00445-004-0357-7.

Woo, G. (1999). *The Mathematics of Natural Catastrophes*. London: Imperial College Press.

Wynne, B. (1992). Misunderstood misunderstanding: Social identities and the public uptake of science. *Public Understanding of Science*, **1**, 281–304.

Zielinski, G. A., Mayewski, P. A., Meeker, L. D. *et al*. (1994). Record of volcanism since 7000 B.C. from the GISP2 Greenland ice core and implications for the volcano-climate system, *Science*, **264**, 948–952, doi:10.1126/science.264.5161.948.

Zuccaro, G., Cacace, F., Spence, R. J. S. and Baxter, P. J. (2008). Impact of explosive eruption scenarios at Vesuvius. *Journal of Volcanology and Geothermal Research*, **178**, 416–453.

4

Extreme seismic events: from basic science to disaster risk mitigation

Alik Ismail-Zadeh

> An earthquake is an evil that spreads everywhere;
> it is inevitable and damaging for everybody.
> Besides the destruction of people, houses and entire towns,
> it can make disappear nations and even entire nations
> without leaving a trace of what has been once
> (Lucius A. Seneca, *Quaestiones Naturales*, 1 BC–65 AD)

4.1 Introduction

Although the extreme nature of earthquakes has been known for millennia due to the resultant devastation from many of them, the vulnerability of our civilization to extreme seismic events is still growing. This is partly because of the increase in the number of high-risk objects and clustering of populations and infrastructure in the areas prone to seismic hazards. Today an earthquake may affect several hundred thousand lives and cause damage costing hundreds of billions of dollars. A large earthquake can trigger an ecological catastrophe if it occurs in close proximily to a nuclear power plant. For instance, the 2011 Great East Japan (Tohoku) earthquake and subsequent tsunami damaged the cooling system of the Fukushima Dai-ichi nuclear power plant and resulted in nuclear radiation leaks.

About a million earthquakes with a magnitude greater than 2 are registered annually, of which about a thousand are strong enough to be felt, and about 10–100 earthquakes cause damage. On average, an extreme seismic event occurs once in a few decades. In a broad sense, an extreme seismic event can be defined as the occurrence of an earthquake that, with respect to other earthquakes, is notable, rare, unique, and profound or otherwise significant in terms of its impacts, effects, and outcomes (Ismail-Zadeh *et al.*, 2012a).

We distinguish two types of extreme seismic events: (i) large magnitude earthquakes, and (ii) earthquakes leading to disasters. Extreme seismic events, like the 1755 Lisbon, 1906 San Francisco, 1960 Chile, 2004 Sumatra-Andaman, 2008 Sichuan, and the 2011 Tohoku earthquakes, belong to both types of extreme seismic events (high magnitude events and great disasters at the same time). Some great earthquakes around the Pacific seismic belt, fortunately, do not lead to disasters, but nevertheless they are considered to belong to the first type of extreme seismic events. The 2003 Bam $M = 6.6$ and 2010 Haiti $M = 7.0$ catastrophic earthquakes (just to mention the more recent events) belong to the second type of extreme, events in which despite, the magnitude of the earthquakes being only moderate, the resulting human and economic losses are considerable.

In this chapter I deal with the first type of extreme events, that is, the events, which are of large magnitude and occur infrequently, that is, speaking statistically, they are located in the tail of the frequency–magnitude distribution curve. Meanwhile the importance of studies on the second type of the extreme seismic events should not be underestimated; this type of extreme event shows that an earthquake disaster is a social phenomenon and depends not only on the magnitude of the natural event but also on the vulnerability and resilience of societies to seismic hazards.

Great earthquakes, when they occur, come as a surprise, and society is poorly prepared to deal with them. Protecting human life and property against earthquake disasters requires an uninterrupted chain of geoscience, social science, engineering, and civil protection tasks (Ismail-Zadeh, 2010). The geophysical tasks range from (i) monitoring earthquake-prone regions using geodetic, seismic, electromagnetic, and other observations, and multiple earthquake precursors, (ii) analysis and physical understanding of the preparation processes of large earthquakes, (iii) studies of past seismic events using geology,

Extreme Natural Hazards, Disaster Risks and Societal Implications, eds. Alik Ismail-Zadeh, Jaime Urrutia Fucugauchi, Andrzej Kijko, Kuniyoshi Takeuchi and Ilya Zaliapin. Published by Cambridge University Press.

palaeoseismology, and geochemistry, (iv) analysis of the vast number of data through modelling and interpretation to (v) comprehensive seismic hazard analysis and earthquake forecasting/prediction. The social science and engineering tasks could be concentrated on social and physical vulnerabilities, resilience, and sustainability of society, and studies of exposures. The civil protection tasks are designed to implement significant scientific and engineering knowledge and range from (1) communication of knowledge on natural hazards and disaster risks to local authorities and the public using 'non-expert language', (2) an increase in awareness of society with respect to natural hazards and risks, (3) an enhancement of disaster preparedness, to (4) assistance in preventive disaster management to improve the resilience of nations to earthquake disasters.

4.2 Plate tectonics and earthquakes

Observations of high seismic activity and large gravity anomalies, associated with oceanic trenches, resulted in the hypothesis that trenches are the sites of crustal convergence and consumption compared to surface augmentation at seafloor spreading zones (e.g., Hess, 1962). As the lithosphere moves away from an oceanic ridge, it cools, densifies, and thickens. Once the lithosphere becomes sufficiently dense compared to the underlying mantle rocks, it bends, founders, and begins sinking into the hot mantle due to gravitational instability. The downward buoyancy forces promote the sinking of the lithosphere, but elastic, viscous, and frictional forces resist the descent. The combination of these forces produces tectonic stresses high enough to cause earthquakes. Other processes contributing to stress generation in the descending lithosphere and its release in earthquakes include plastic instability at high temperature (e.g. Griggs and Baker, 1969), faulting due to metamorphic phase transitions (e.g. Green and Burnley, 1989), and dehydration-induced embrittlement (e.g. Hacker *et al.*, 2003). Earthquakes occur as a sudden release of stresses. When an earthquake occurs, part of the released energy generates elastic waves propagating through the Earth. These waves generate ground motions and shaking, which result sometimes in the damage or collapse of buildings, landslides, tsunami wave generation, etc.

Oceanic trenches are the sites of the world's largest earthquakes. The earthquakes in oceanic trench zones can occur along the descending lithosphere to depths of about 700 km depending on the thermal state in the mantle. Large earthquakes occur not only in subduction zones, but also within continents (so-called 'intraplate earthquakes') especially in regions of continental collisions, rifts, and grabens. For example, the 1811–1812 New Madrid earthquakes of magnitude $M = 7.5–7.7$ were typical intraplate earthquake series associated with the reactivated faults of the Late Precambrian New Madrid Rift complex (Braile *et al.*, 1986). Large intermediate-depth earthquakes in the southeastern Carpathians (Vrancea) are associated with the relic continental/oceanic collision zone (e.g., Ismail-Zadeh *et al.*, 2008, 2012b).

An extreme (large magnitude) earthquake is a key manifestation of the lithosphere dynamics exhibiting a nonlinear system behaviour and evolving from stability to a catastrophe over space and time (e.g., Keilis-Borok, 1990; Sornette and Sammis, 1995; Turcotte, 1999). The lithosphere presents a hierarchy of blocks where the largest blocks are the major tectonic plates. They are divided into smaller blocks: from shields or mountain chains to the grains of rock. The blocks are separated by less rigid boundary zones, which present a similar hierarchical structure: from fault zones to cracks and microcracks. The blocks and faults interact along and across the hierarchy and move relative to each other under the control of lithosphere dynamics. In large part, these movements are realized through the formation and subsequent healing of failures on surfaces where displacements are discontinuous resulting in defects, slips, fractures, or earthquakes (e.g., Rice and Ben-Zion, 1996; Keilis-Borok *et al.*, 2001).

4.3 Seismic hazard and earthquake surprises

A seismic hazard can be defined as a potentially damaging earthquake, which may cause injury or the loss of life, property damage, social and economic disruption, or environmental degradation. Seismic hazard assessment (SHA) in terms of the engineering parameters of strong ground motion (namely, peak ground acceleration, and seismic intensity) is based on information about the features of excitation of seismic waves at the source, seismic wave propagation (attenuation), and site effects in the region under consideration and combines the results of seismological, geomorphological, geological, and tectonic investigations and modelling (e.g., Ismail-Zadeh *et al.* 2007a).

Two principal methods are intensively used in SHA: probabilistic and deterministic SHA. Probabilistic SHA determines the probability rate of exceeding various levels of ground motion estimated over a specified period of time (Cornell, 1968). Probabilistic assessment considers uncertainties in earthquake source, path, and site conditions. The uncertainties are classified as epistemic and aleatory. Epistemic uncertainties reflect the incomplete knowledge about input model parameters to the assessment and the variability of interpretations of available data. Aleatory uncertainties consider the inevitable unpredictability of the parameters: the uncertainties are mainly quantified through the use of the standard deviation of the scatter around the mean values. Although the probabilistic SHA method is dominant in hazard assessment, there have been several criticisms of it, e.g., the validity of a point source model, ground motion uncertainties in the mathematical formulation of the method, and its inability to correctly model the dependencies between large numbers of uncertain random parameters (Klügel *et al.*, 2006; Panza *et al.*, 2010).

Deterministic SHA is an alternative method for hazard analysis and is based on a specified earthquake scenario. For a given earthquake, the deterministic assessment model analyses the attenuation of seismic energy with distance to determine the level of ground motion at a particular site. Ground motion calculations capture the effects of local site conditions and use the available knowledge on earthquake sources and wave propagation processes. Namely, attenuation relationships are used for a given earthquake magnitude to calculate ground shaking for hard rock, which is then amplified by factors based on local soil conditions. Although the occurrence frequency of the ground motion is usually not addressed in deterministic SHA, the method is robust for an assessment of seismic hazard due to specified events and remains a useful approach for decision-making (Babayev *et al.*, 2010).

The parade of extreme seismic events (2004 Sumatra-Andaman, 2005 Kashmir, 2008 Wenchuan, 2010 Chile, and 2011 Tohoku earthquakes) demonstrated that big magnitude events come as a surprise, even though scientific knowledge of historical large earthquakes was available in the most of these areas. Moreover, existing seismic hazard maps did not mark these regions as those of significant hazards. This reveals a weakness in current hazard assessment methods: expected ground shaking due to the earthquakes was significantly underestimated compared to that from observed events (e.g. Stein *et al.*, 2011; Stirling, 2011; Hanks *et al.*, 2012; Kossobokov and Nekrasova, 2012). And hence, the numbers of fatalities in these earthquake disasters were underestimated by approximately 2–3 orders of magnitude (Wyss *et al.*, 2012).

In the most cases large magnitude earthquakes are not accounted for in probabilistic SHA due to the lack of information about them and the unknown reoccurrence time of extreme events. Our present knowledge about the characteristics of seismicity is based on observed (recorded) data and sometimes on historical data (from palaeoseismological and archaeological studies, written stories about intensities of large earthquakes and some other sources). Information about extreme events in a particular region is incomplete as large events are rare. Modelling of extreme seismic events (using so called 'earthquake simulators') is a way to overcome the difficulties in seismic hazard assessment by combining observations, historic data, and modelled results (see Section 4.4).

Natural hazard analysis should be considered from a holistic point of view (from the whole to details). According to Ismail-Zadeh (2010), a holistic comprehensive analysis of seismic hazard should be based on multidisciplinary research (Fig. 4.1) in: (i) geology, geodynamics, and geodesy (to identify zones of tectonic stress and strain localization); (ii) seismology (to understand the physics of significant ruptures and to localise areas prone to large events); (iii) nonlinear dynamics (to model and analyse statistical properties of the earthquake sequences, large earthquake occurrences, their clustering, and other properties of earthquakes); (iv) soil sciences (to assist in analysis of liquefaction and seismic shaking); (v) SHA (to determine peak ground acceleration, response spectra amplitude, and seismic intensity); and (vi) hydrology and electromagnetic (EM) studies (to search for and monitor precursors for extreme seismic events). This approach to seismic hazard should be accompanied by a holistic approach to earthquake prediction (e.g., Keilis-Borok and Soloviev, 2003) and by a holistic approach to earthquake disaster risk, when a convolution of hazard, vulnerability, and exposure (as functions of space and time) should be

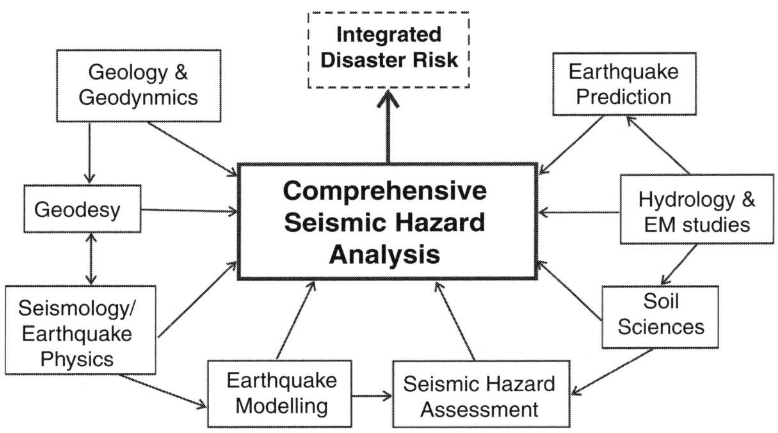

Figure 4.1. Essential parts of comprehensive SHA.

viewed also from socio-psychological (e.g., resilience of community to extreme seismic events) and legislation (e.g., role of law in risk reduction) points of view (Beer and Ismail-Zadeh, 2003).

4.4 Modelling extreme seismicity

Stress accumulation and its release in earthquakes are governed by non-linear systems, which have a number of degrees of freedom and, therefore, cannot be understood by studying them piece-by-piece. Studying seismicity using the statistical and phenomenological analysis of earthquake catalogues has the disadvantage that instrumental observations cover a short time interval compared to the duration of the tectonic processes responsible for earthquakes. The patterns of earthquake occurrence identifiable in a catalogue may be apparent and yet may not be repeated in the future. Moreover, the historical data on seismicity are usually incomplete. Numerical modelling of seismogenic processes allows the generation of synthetic earthquake catalogues covering very long time intervals and, therefore, provides a basis for reliable estimates of the parameters of the earthquake occurrences (Soloviev and Ismail-Zadeh, 2003; Ismail-Zadeh *et al.*, 2012a).

It is challenging to detect the impact of a single process on the dynamics of seismicity as well as to single out the impact of an isolated process by using seismic observations only. This challenge may be resolved by modelling processes that generate seismicity and by studying the synthetic earthquake catalogues so obtained. An adequate model should indicate the physical basis of premonitory patterns determined prior to large events. The model can also be used to suggest new premonitory patterns that might exist in catalogues of seismic events (Soloviev and Ismail-Zadeh, 2003).

Three types of quantitative models of stress localisation and stress drop ('earthquakes') have so far been developed. The first type is related to models of tectonic stress generation, localisation, and transfer in a fault zone or its surroundings (e.g., King *et al.*, 1994; Ismail-Zadeh *et al.*, 2005, 2010). These models provide results on high-shear-stress localisations in a region and/or static, viscoelastic, and dynamic Coulomb stress changes after an earthquake to determine the potential sites of impending large events.

Models of the second type consider dynamic systems reproducing 'universal' features of seismicity common to a wide class of nonlinear systems (e.g., Burridge and Knopoff, 1967; Ogata, 1988; Zaliapin *et al.*, 2003; Turcotte *et al.*, 2007). These models provide earthquake statistics, and assist in analysis of earthquake patterns, occurrences, clustering, and frequencies. Among others, a widely used model to simulate earthquakes and aftershocks is the epidemic type aftershock sequence (ETAS) model by Ogata (1988), which employs the concept of aftershock cascades, when a main shock generates aftershocks, the aftershocks generate their own aftershocks, etc.

Despite the fact that the models of the second type describe many features of earthquake occurrences, only the Earth-specific models (of the third type) simulate earthquakes in a system of faults (e.g., Gabrielov *et al.*, 1990; Soloviev and Ismail-Zadeh, 2003; Rundle *et al.*, 2006; Ismail-Zadeh *et al.*, 2007b, 2012a). In particular, the block-and-fault dynamics (BAFD) model by Gabrielov *et al.* (1990) can answer the following questions. How do upper crustal (or lithospheric) blocks react to the plate motions and to a flow of the lower ductile crust (or highly viscous asthenosphere)? How do earthquakes cluster in the system of major regional faults? At which part of a fault system can large (extreme) events occur, and what is the occurrence time of the extreme events? How do the properties of the frequency–magnitude relationship change prior to extreme events? How do fault zones properties influence earthquake clustering, its magnitude, and fault slip rates? The BAFD model was applied to several earthquake-prone areas (see Ismail-Zadeh *et al.* (2012a) for a review of the model's applications).

Using a BAFD model Ismail-Zadeh *et al.* (2007b) developed numerical experiments to analyse earthquake clustering, frequency-to-magnitude relationships, earthquake focal mechanisms, fault slip rates, and the frequency of large events in the Tibet-Himalayan region. Synthetic seismicity spanned 4000 years, a considerably longer interval of time than the duration of the recorded regional seismicity. The biggest magnitude of the synthetic events was found to be $M = 8.9$. These events occurred five times at the same fault. It was shown that the recurrence time of the largest events varied from about 1200 to 50 years. It is remarkable that the BAFD model identified a cluster of large synthetic events (magnitude about 8) along the Longmen Shan fault before the 2008 Wenchuan $M = 7.9$ earthquake occurred at this fault (see Fig. 6c in Ismail-Zadeh *et al.* (2007b)).

Quantitative models of lithosphere dynamics are not only tools for studying earthquake preparation processes, but also can be useful in SHA by incorporating observed and modelled extreme events into the ground motion estimation and earthquake forecasting. Figure 4.2 compares the results of the probabilistic SHA for the Eastern Sichuan region (China) obtained by two methods: the standard probabilistic assessment and the new assessment accounting for large synthetic events in the region (Sokolov and Ismail-Zadeh, in preparation, 2013). The peak ground acceleration in the new hazard model is greater by a factor of 2–3 than that in the standard hazard model, and hence better explains the observed shaking in the Wenchuan

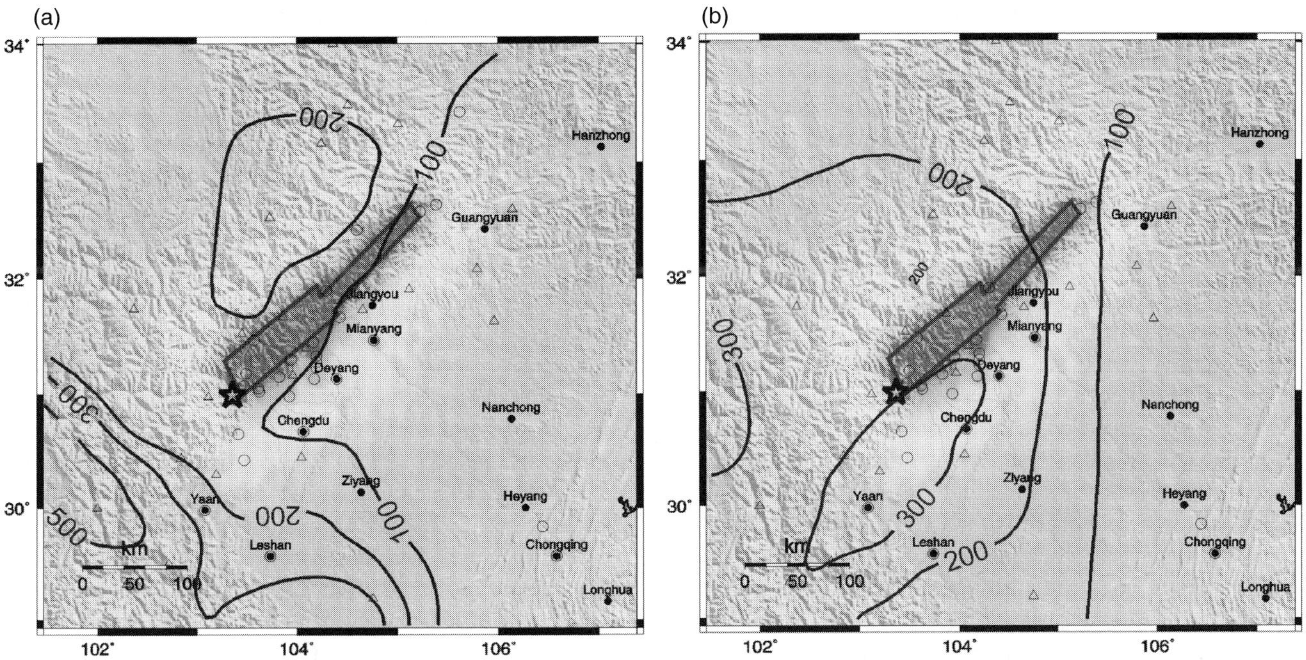

PERCEIVED SHAKING	Not felt	Weak	Light	Moderate	Strong	Very strong	Severe	Violent	Extreme
POTENTIAL DAMAGE	none	none	none	Very light	Light	Moderate	Moderate/Heavy	Heavy	Very Heavy
PEAK ACC.(%g)	<.17	.17-1.4	1.4-3.9	3.9-9.2	9.2-18	18-34	34-65	65-124	>124
PEAK VEL.(cm/s)	<0.1	0.1-1.1	1.1-3.4	3.4-8.1	8.1-16	16-31	31-60	60-116	>116
INSTRUMENTAL INTENSITY	I	II-III	IV	V	VI	VII	VIII	IX	X+

Figure 4.2. Probabilistic SHA for the Eastern Sichuan region, China, for the return period of 475 y. Peak ground acceleration obtained using (a) the standard methodology (GSHAP map, http://www.seismo.ethz.ch/static/gshap/) and (b) the methodology accounting for modelled extreme seismic events. The background map is the United States Geological Survey (USGS) shake map for the 12 May 2008 Wenchuan $M_W = 7.9$ earthquake. Numbers at the contours are scaled in centimetres per second. For colour version, see Plates section.

region. The proposed approach to seismic hazard modelling may provide a better understanding of ground shaking and could be useful for earthquake risk assessment, engineering purposes, and emergency planning.

4.5 Predicting earthquakes

Scientific research aimed at predicting earthquakes started in the second half of nineteenth century, when seismology reached the level of a recognised scientific discipline. For more than 100 years earthquake prediction research experienced alternating periods of great enthusiasm and scepticism. The abruptness along with apparent irregularity and infrequency of large earthquake occurrences perpetuated the perception that earthquakes are random unpredictable phenomena. Earthquake prediction research has been widely debated, and opinions on the possibilities of prediction vary from the statement that earthquake prediction is intrinsically impossible (Geller *et al.*, 1997) to the statement that prediction is possible, but difficult (Knopoff, 1999). Although many observations reveal unusual changes of geophysical fields before a large earthquake (e.g. animal behaviour, ground elevation, water level in boreholes, radon emission), most of them report a unique case history and lack a systematic description (Wyss, 1991).

To predict an earthquake, one must 'specify the expected magnitude range, the geographical area within which it will occur, and the time interval within which it will happen with sufficient precision so that the ultimate success or failure of the prediction can readily be judged. Only by careful recording and analysis of failures as well as successes can the eventual success of the total effort be evaluated and future directions charted. Moreover, scientists should also assign a confidence level to each prediction' (Allen *et al.*, 1976). A prediction of an earthquake of certain magnitude range can be identified by temporal or spatial extents. Based on the temporal extent, the following stages can be distinguished in earthquake prediction: long-term (a decadal time scale), intermediate-term (one to several years), short-term (weeks and months), and immediate (seconds to hours). Based the rupture size (L) of the incipient earthquake, the area covered by predictions could be up to 100*L* (long-range prediction), 5*L*–10*L* (middle-range), 2*L*–3*L* (narrow), and exact (indicating the fault segment to rupture).

Long-term prediction of earthquakes (also called long-term earthquake forecasting (Kanamori, 2003)) is based on the elastic rebound theory, a physical theory about increasing the shear stress on the fault that forms the boundary between crustal blocks until the fault reaches its yield stress and ruptures (Reid, 1911). The use of characteristic earthquake (Schwartz and Coppersmith, 1984) and seismic gap (Fedotov, 1965; Kelleher *et al.*, 1973) hypotheses allowed the development of a long-term prediction of large earthquakes on specified faults or fault segments. Based on fault slip rates, geodetic measurements, and paleoseismological studies, it is possible to determine the probabilities of active fault segments rupturing in the next few decades (e.g., Working Group on California Earthquake Probabilities, 2007). A long-term earthquake forecast is complicated by the irregular occurrence of large earthquakes. Long-term earthquake forecasts can guide engineering and emergency planning measures to mitigate the impact of pending large earthquakes.

Earthquake prediction is based on monitoring of precursor(s) of earthquakes (physical, chemical, or biological signals, which indicate that a large earthquake can be anticipated), and issuing an alarm at the time of the abnormal behaviour of the precursor (the so-called time of increased probability of large-event occurrence). Several intermediate-term earthquake prediction methods have been developed since 1990 (e.g., Keilis-Borok and Kossobokov, 1990; Rundle *et al.*, 2002; Shebalin *et al.*, 2006). An intermediate-term earthquake prediction method, the *M*8 algorithm (Keilis-Borok and Kossobokov, 1990), aims to forecast large (magnitude 8 and greater) earthquakes by monitoring and analysing several parameters of the seismic activity in a particular region. This prediction algorithm has received a fair amount of attention due to on-going real-time experimental testing that is unprecedented in rigour and global coverage (http://www.mitp.ru/en/predictions.html). The accumulated statistical data from this experiment confirm the intermediate-term predictability of large earthquakes with a middle- to exact-range of location (e.g., Ismail-Zadeh and Kossobokov, 2011). Independent assessments of the *M*8 algorithm performance confirm that the method is reliable for predicting large earthquakes (Zechar and Jordan, 2008; Molchan and Romashkova, 2011). Today the *M*8 algorithm is the only continuous (more than two decades), operational, alarm-based method of predicting large earthquakes.

The short-term prediction of the devastating 1975 Haicheng (China) $M_s = 7.0$ earthquake by Chinese seismologists was based on monitoring anomalies in land elevation, in ground water level, and in seismicity prior to the large event and on the observations of the peculiar behaviour of animals (Zhang-li *et al.*, 1984). The success of this prediction stimulated the design of further methods for the diagnosis of the approach of a large earthquake. Unfortunately, most of the prediction methods suggested at that time were not confirmed in the following years. The catastrophic 1976 Tangshan (China) $M_s = 7.4$ earthquake, which caused hundreds of thousands of fatalities, was not predicted.

Stress-induced variations in the electric conductivity of rocks have been studied in laboratory experiments (e.g., Morat and Le Mouël, 1987). In the 1980s P. Varotsos, K. Alexopoulos and K. Nomicos proposed a method of short-term earthquake prediction, the VAN method (named after the researchers' initials), which is based on detection of characteristic changes in the geoelectric potential (so called 'seismic electric signals', SESs) via a telemetric network of conductive metal rods inserted in the ground (e.g., Varotsos *et al.*, 1986). The anomaly pattern is continually refined as to the manner of differentiating SESs from the abundant electric noise the VAN sensors pick up. Despite many years of investigations and some progress since the 1980s (Lazaridou-Varotsos, 2013), short-term prediction by the VAN method is still controversial.

Another type of short-term prediction is based on calculating the probabilities of target events within future space-time domains. An example of this type of prediction is the short-term earthquake probability (STEP) method developed by Gerstenberger *et al.* (2005) and employed by the United States Geological Survey (USGS) for operational forecasting in California. STEP uses aftershock statistics to make hourly revisions of the probabilities of strong ground motion. The probability-based forecasts are the means for transmitting information about probabilities of earthquakes in the particular region being monitored. While the probability gains of short-term forecasts can be high, the probabilities of potential destructive earthquakes remain much smaller than 0.1 as the forecasting intervals are much shorter than the recurrence intervals of large earthquakes.

At present only long-term and intermediate-term prediction methods can give some reliable (in a probabilistic sense) information about seismic hazard: (1) probability-based long-term (about 30-year interval) earthquake forecasts; and (2) alarm-based intermediate-term (about 3–5 years interval), long- to middle-range earthquake predictions. Global efforts in studies of earthquake predictability resulted in the multi national CSEP programme (http://www.cseptesting.org). The principal goal of the CSEP program is to 'support a wide range of scientific prediction experiments in multiple regional or global natural laboratories'. The programme aims to answer the questions: How should scientific prediction experiments be conducted and evaluated? What is the intrinsic predictability of the earthquake rupture process? Can knowledge about the predictability of large earthquakes be used to give reliable advance warning of potentially destructive events? The establishment of CSEP testing centres, developing computational infrastructure and running actual forecasts, is a great advance in the area of earthquake predictability for the last decade.

Earthquake forecasts/predictions have a large economic impact, and as prediction methods improve along with seismological data analysis, these economic effects will increase. Meanwhile, compared to the accuracy of weather forecasting, the current accuracy of earthquake prediction might appear low. Our knowledge of earthquake physics and earthquake dynamics is limited to predicting large earthquakes with a relatively high accuracy. There is no quantitative description of earthquake physics, namely, no mathematical equations to describe the non-linear dynamics of fault systems and earthquake 'flow'. This situation is unlike meteorology, where the Navier–Stokes equations describe atmospheric flow and hence allow weather forecasting with a relatively high accuracy for time scales ranging from a few hours to a few days. Nevertheless, even the current level of earthquake prediction capacity can be useful for seismic risk assessment and disaster preparedness.

4.6 Seismic risk

Seismic risk can be defined as a measure that combines, over a given time, the likelihoods and the consequences of a set of earthquake scenarios (Beer and Ismail-Zadeh, 2003). The risk can be estimated as the probability of harmful consequences or expected losses (of lives and property) and damage (e.g., people injured, economic activity disrupted, environment damaged) due to an earthquake resulting from interactions between seismic hazards (H), vulnerability (V), and exposed values or exposure (E). Conventionally, seismic risk (R) is expressed quantitatively by the convolution of these three parameters: $R = H \otimes V \otimes E$ (e.g., Kantorovich *et al.*, 1973). The vulnerability to earthquakes depends on several factors including the quality of building structures, ground conditions, and the distribution of population. An exposure accounts normally for infrastructure at risk.

A rapid growth of population, intensive civil and industrial building, land and water instabilities, and the lack of public awareness regarding seismic hazard contribute to the increase of vulnerability of big cities. For example, Babayev *et al.* (2010) assessed earthquake risk in Baku (Azerbaijan) based on the convolution of scenario-based seismic hazards, vulnerability (due to building construction fragility, population features, the gross domestic product per capita, and landslide's occurrence), and exposure of infrastructure and critical facilities. One of the remarkable results of this assessment was that the western-central part of the city is exposed to the highest risk independent of the magnitude and epicentral distance of modelled large earthquakes. This is because of the high density of population, low quality of building construction, and significant exposure values in this part of the city.

With economic and technological development more and more structural elements (and sometimes the entire infrastructure) become exposed to risks due to natural hazards. Risk assessment allows the elaboration of strategic countermeasure plans for earthquake risk mitigation. An estimation of seismic risk may facilitate a proper choice in a wide variety of seismic safety measures, ranging from building codes and insurance to establishment of rescue-and-relief resources. Different representations of seismic risk require different safety measures. Most of the practical problems require the estimation of seismic risk for a territory as a whole, and within this territory separately for the objects of each type: areas, lifelines, sites of vulnerable constructions, etc. The choice of the territory and the objects is determined by the jurisdiction and responsibility of a decision-maker. Each specific representation of seismic risk is derived from the models of seismic hazards, the territorial distribution of population, property, and vulnerable objects, and the damage caused by an episode of strong motion.

Problems of estimating seismic risks have become highly important. In the last few decades a number of concepts on disaster risks have been suggested and a number of international projects on risk assessment and management have been conducted. For example, the ongoing Global Earthquake Model project, an international initiative of the OECD Global Science Forum supported financially by Munich Re and other organisations, is designed to provide a global framework for assessment of earthquake risk on a local and global scale.

Significant difficulties in decision-making in the field of seismic risk are related to uncertainties in data (especially those related to social and physical vulnerabilities and exposure), imperfect methods of hazard assessment (which are sometimes biased against scenario-based hazard assessments), and limitations in using mathematical tools to carry out the historical analysis and forecasting.

4.7 Preventive measures and disaster mitigation

Several extreme seismic events have struck the Earth and affected our society in recent times. Such natural events are rare, but not unexpected. The 2004 earthquake and tsunami in the Indian Ocean and the earthquakes in Kashmir (2005), Sichuan (2008), and Haiti (2010) resulted in large humanitarian disasters because of weak preventive disaster management. The 2011 Tohoku earthquake is a special case in the sense of disaster preparedness. Although this extreme event generated significant human and economic losses, the nation was prepared to withstand a great earthquake, but not a great tsunami in this region of Japan. Considerable losses were associated with the extreme tsunami inundations, and not with the damage to buildings and infrastructure due to the great earthquake.

Following Ismail-Zadeh and Takeuchi (2007), I consider here the economic and political factors as well as those of awareness and preparedness, which brought about the humanitarian tragedies of the early twenty-first century.

Economic factors. The decision on whether to adopt preventive mitigation measures to reduce losses from catastrophic natural disasters is based on the cost–benefit ratio. This decision may not be easily justified if a standard (5%) annual discount rate (the annual interest divided by the capital including that interest) is employed to estimate the benefits of preventive mitigation. The discount rate is determined by societal preference and the opportunity cost (the relationship between scarcity and choice) of investment in the market. It is decided by a number of factors and reflects the overall societal as well as economic reality. The rate for public investment should be different from that of commercial goods, especially if it concerns the future generation and sustainable development. Let us explain the latter statement using a simple example. Assume that a T-year large earthquake can cause a disaster in a big city and that the probability of occurrence of such an earthquake (and the resulting losses) in a year is constant over time and equals $1/T$. Also assume that if an earthquake disaster occurs, the city will be restored to its pre-disaster condition and be functional again prior to the next disaster whenever it comes. The decision as to whether or not to invest in disaster mitigation measures is determined by comparing the cost of the preventive mitigation with the expected benefits $B_e \propto \sum_{t=1}^{N}(1/T)L/(1+r)^t$, where L is the average loss reduction from mitigation measures (preventive works) against T-year or more severe event (assume that $L =$ US\$ 20×10^9 for $T = 100$ y), r is the annual discount rate, and N is the time horizon to be considered for the assessment. Ismail-Zadeh and Takeuchi (2007) showed that the expected benefit, calculated over 10–100 years, ranges from about US\$1.6 billion to about US\$4.2 billion for a 100-year extreme seismic event at the annual discount rate of 5%, and the benefit increases with the reduction of the annual discount rate (US\$6.5 billion and US\$12.7 billion at the discount rates of 3% and 1%, respectively). This should be taken into account by the governments of hazard-prone areas.

A large investment is made at the stage of a response to an earthquake disaster, and the investment decreases until the next large earthquake, especially at the final stage of risk management related to prediction and preparedness (Fig. 4.3). The reaction of media and the societal attention to disasters follow the same cycle. I call this cycle the 'seismo-illogical cycle' (Ismail-Zadeh, 2006). If a seismic cycle is marked by an *increase* of tectonic stresses towards an earthquake (stress drop), the 'seismo-illogical cycle' is characterized by a *decrease* of funding toward the next big earthquake.

The disaster policy of many nations, including developed countries, continues to be primarily focused on response after the earthquake or other natural hazard events (e.g., Rubin, 2012). 'If about 5 to 10% of the funds, necessary for recovery and rehabilitation after a disaster, would be spent to mitigate an anticipated earthquake, it could in effect save lives, constructions, and other resources ... The tendency to reduce the funding for preventive disaster management of natural catastrophes rarely follows the rules of responsible stewardship for future

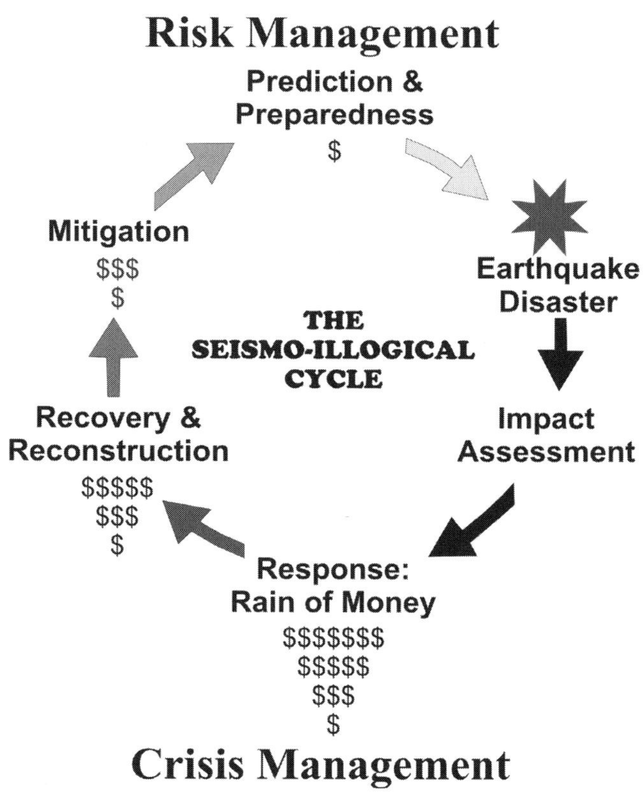

Figure 4.3. The 'seismo-illogical cycle' of earthquake disaster management.

generations neither in developing countries nor in highly developed economies' (Ismail-Zadeh and Takeuchi, 2007).

A new approach should be developed to: (i) mitigate disasters in advance of an extreme seismic event by investing in preventive measures; (ii) be prepared for the next large earthquakes and subsequent environmental, economic, and financial events associated with the earthquakes; and (iii) recover easily from a disaster. We need stronger efforts to develop a society resilient to natural hazards and disasters.

Political factor. The investment to avoid potential earthquake losses tends not to be easily accepted in political decision-making (compared to investments to gain positive benefits). This is because the cost of preventive measures may outweigh the benefit during the typical terms served by presidents and governments (usually between 4 and 8 years), while the long-term positive benefit is obvious to the people as a whole. Investment in disaster preparedness is a matter of political decision. Politicians assisted by scientists and engineers could inform the public on major (vital but costly) regional, national, or international disaster mitigation project proposals and learn their opinion on the proposals. For example, Ismail-Zadeh and Takeuchi (2007) suggested drawing the public opinion via the contingent valuation method by Cummings *et al.* (1986) to quantify the benefits of non-marketed goods so that they could be entered directly into cost–benefit calculations.

It is the poor who live in the least-safe areas as was borne out by the 2010 Haiti earthquake disaster. Many of the buildings that collapsed in Port-au-Prince (the capital city of Haiti) during the earthquake were occupied by the poor. The political influence of these people is much lower than that of the rich as they have little power in the economy. Although they have an equal vote, they do not have an equal voice in their parliament. One year after the Haiti earthquake, Oxfam International published a report on the recovery status. The report mentions 'only five percent of the rubble has been cleared and only 15 percent of the required basic and temporary houses have been built . . . Relief and recovery were at a standstill due to government inaction and indecision on the part of the donor countries' (http://www.oxfam.org/en/policy/relief-recovery). Political decisions reflect the views of a majority of the ruling population. It seems that fairness and equity are not necessarily unquestionable even in the most democratic countries.

Factor of preparedness and awareness. The level of preparedness for the 2004 Indian Ocean tsunami disaster in many countries turned out to be extremely low, partly because it happened without any warning. But even if a warning had been sent to the appropriate local authorities, it is unlikely that it would have been delivered to the public in a timely manner. It is also unlikely that people would have responded to it in an appropriate manner, since most of them did not believe that such a disaster could affect them (Ismail-Zadeh and Takeuchi, 2007).

In Japan, where the preparedness and awareness of people regarding earthquakes and tsunamis are the highest in the world, the initial data on the earthquake of 11 March 2011 sent to the Tohoku coastal region underestimated the height of the tsunami waves. In response, some people did not evacuate their buildings and move to safer places, because they considered that the available sea walls (in some places up to 10 m) would protect their houses and lives. Unfortunately, the tsunami waves were much higher than expected, and the sea walls were incapable of preventing the great inundation. Another important issue revealed by a sociological study is that frequent warnings that overestimate the heights of tsunami waves play a negative role, as the coastal community then begins not to believe tsunami alarms (Ando *et al.*, 2011).

Emergency management practices should always be planned and rehearsed well before a disaster occurs. The measures taken to help humans adjust to living with risk are important. Such measures include hazard maps, education and awareness raising, warning, evacuation, sea-wall patrols, emergency protection, periodic exercises of evacuation, etc. Such measures apply not only to earthquakes, tsunamis, and floods but also to other geological and hydrometeorological hazards and associated disasters. Scientists can help with the design of emergency management plans and practices. For example, earthquake hazards and risk maps are available in many countries. The maps change with time, specifically in big cities and coastal regions, and there is a need to proceed to georisk monitoring. Therefore, research on time-dependent hazard, vulnerability, and risk is essential to periodically update earthquake risk maps.

Scientists know about historical devastating earthquakes and tsunamis, such as those that occurred in the Indian and Pacific Ocean regions or the earthquakes in Himalayas-Tibet, in Haiti, and in the Japanese islands. Moreover, scientists have issued alerts and proposals regarding hazard-prone areas, but it seems that their warnings have been buried under other matters such as political and financial affairs, environmental issues, and other social affairs. In particular, geophysicists and geodesists warned of a potential strong earthquake near the capital city of Haiti. They wrote '. . . the Enriquillo fault in Haiti is currently capable of a Mw7.2 earthquake if the entire elastic strain accumulated since the last major earthquake was released in a single event today' (Manaker *et al.*, 2008). Seismologists and tsunami researchers knew about the 869 Jogan earthquake (off Miyagi, Japan), which produced unusually large tsunamis and killed 1000 people on the Sendai plain (Satake *et al.*, 2007). Large magnitude (>9) seismic events produced by the BAFD earthquake simulator have been clustered in the north Sumatran region, where the 2004 Aceh-Sumatra *M*9.3 earthquake occurred (Soloviev and Ismail-Zadeh, 2003). 'Without having the scientific awareness raised, no political and governmental actions are possible. Here there is a large room for geoscientists to take responsibility' (Ismail-Zadeh and Takeuchi, 2007).

It is known that it is not earthquakes that kill people, but buildings and corruption (Ambraseys and Bilham, 2011), irresponsibility, and ignorance. After the 1755 Great Lisbon earthquake and tsunami, the German philosopher Immanuel Kant wrote: 'If humans are building on inflammable material, over a short time the whole splendor of their edifices will be falling down by shaking' (Kant, 1756; Fuchs, 2005). To clarify this sentence, I would mention that in the eighteenth century scientists believed that an earthquake generates fires, and hence Kant's remark about 'inflammable material'. But the meaning of the message is clear: 'Build safer houses to withstand an earthquake!' More than 250 years later, the words by Kant are still relevant, especially when we think about the L'Aquila (2009), Haiti (2010), and other disasters caused by strong (but not extreme in magnitude) earthquakes.

Disaster preparedness is vital for society. But unfortunately, often disaster managers postpone preparedness for extreme events, considering them rare. How can disaster-management authorities be convinced to pay attention to preventive measures in advance of rare extreme events and hence mitigate or even in some cases prevent humanitarian disasters?

Scientists could improve knowledge of extreme seismic event occurrences (including great earthquakes and potential earthquake disasters). Seismologists should improve the current seismic hazard assessment methods to tackle extreme events and should work in close cooperation with earthquake engineers to implement the new knowledge in future building codes. Geoscientists could assist disaster managers in terms of timely delivery of reliable forecasts of potentially large earthquakes and their possible aftermath. Studies of accurate temporal and spatial prediction of such events could be enhanced with the aim of forecasting two important parameters for preventive disaster management: 'where' and 'when'. Scientists should establish a strong link with disaster management authorities and scientists of governmental departments responsible for disaster mitigation to bring to their attention comprehensive seismic hazard assessments and earthquake forecasts. Predictions of earthquakes can assist the authorities in choosing adequate strategies related to earthquakes losses and cost–benefit analysis used for disaster preparedness. A methodology of optimal preparedness, which considers the concept of unpreventable damage in extreme events and the level of damage prevented by taking action in response to an earthquake prediction, opens a new way of utilising scientific knowledge about earthquake occurrences for the benefit of disaster mitigation (Davis, 2012).

4.8 Challenging problems in seismology and earthquake disaster risk

Although the origin of seismology can be dated back to the Eastern Han Dynasty (AD 25–220) in China and great progress has been achieved in this branch of science, several challenging problems still exist (Forsyth *et al.*, 2009), and among them an important question: *how do faults interact and slip to generate a big earthquake?* (Lay and Kanamori, 2011) The fundamental difficulty in answering this question comes from the fact that no earthquake (process of rupture initiation at depth) has ever been observed directly and just a few of them have been subject to an in-situ verification of their physical parameters.

Another important question is: *how can the available basic scientific knowledge and earthquake-forecasting strategies be linked to risk reduction strategies to produce cost effective mitigation?* 'Optimized earthquake prediction algorithms will greatly aid disaster managers and decision makers in their preparations once a prediction is made. The loss functions help to develop a greater understanding between earthquake prediction research and disaster preparedness implementation, allowing for future improvements in earthquake disaster prevention' (Davis, 2012).

Natural hazards were recognized as a grand challenge a long time ago, and the scientific community has concentrated its efforts on solving the challenging problem of hazard analysis and assessment and significant progress has been achieved. Meanwhile with time it has been recognized that there are much more challenging problems, including extreme natural and concatenated hazards and disaster risks. The problems affect society directly and require truly trans disciplinary studies and a systems analysis approach.

Many of the extreme seismic, tsunami, and volcanic events of the beginning of this century are linked in a chain of subsequent events that produce a disaster (the chain of events is called also '*concatenated events*'). The 2004 great Sumatra-Andaman earthquake followed the Indian Ocean tsunami, which affected the coastal regions by flooding, contaminated water resources, and influenced the tourism business in the entire region. The 2008 Wenchuan earthquake was followed by a number of big landslides with significant damage and human/economic losses. An eruption of the Eyjafjallajökull volcano in Iceland in 2010, although relatively small in magnitude, affected civil aviation causing enormous disruption to air traffic across northern and western Europe over a week, with accompanying travel chaos and huge financial losses. The 2011 Great East Japan earthquake was followed by extreme tsunami waves, significant inundation, a technological accident at a nuclear power plant, and environmental pollution with food and health security problems in the country and abroad. How to tackle concatenated extreme events is a new challenging problem.

One of the ways to understand extreme events is integrated research on disaster risk, which combines knowledge from natural and social sciences, engineering, law, policy making and management, the insurance industry, and other stakeholders dealing with disaster risks. This research would be truly transdisciplinary (Stokols, 2006) aimed at in-depth investigations of a disaster problem using a systems analysis approach and with recommendations for actions to reduce risks and to improve the resilience of society. This requires arduous efforts to tackle the challenging problems of extreme natural hazards and disaster risks. One of the complications is that the scientific community and other stakeholders still live on isolated islands in the 'sea of ignorance'. 'So oft in *academic* wars, The disputants, I ween, Rail on in utter ignorance Of what each other mean, And prate about an *extreme* Not one of them has seen!' (from the poem *The Blind Men and the Elephant* by John G. Saxe, modified by the author).

Consider the geoscience community: seismologists have only recently started to communicate with geodesists on problems of impending large earthquakes; a link between seismologists and earthquake engineers exists but it is weak; there is little communication between the electromagnetic and seismological communities (there is a high barrier between theoreticians/statisticians and experimentalists); the list of examples can be continued. When it concerns communication between natural and social sciences, scientists and policy makers, scientists and insurance representatives, the situation becomes even worse. This could be changed (and, in fact, the situation is changing, but quite slowly). Strong links should be established between all stakeholders who have to deal with disaster risk, involving both joint research and improved communications. An entire great network should be developed that connects the isolated islands in a 'sea of mutual understanding' (Fig. 4.4).

Forensic investigations of disasters (FORIN) and assessments of disaster risks (AIRDR) are two new major scientific challenges dealing with integrated, transdisciplinary research, which have been formulated by the International Scientific Program 'Integrated Research on Disaster Risk' of the International Council for Science (ICSU), the International Social Sciences Council (ISSC), and the UN International Strategy for Disaster Reduction (UNISDR). FORIN will concentrate among other topics on: (i) investigation of the circumstances, causes, and consequences of losses in disasters: (ii) identification of the conditions that have limited or prevented loss; (iii) testing a series of hypotheses of damage causality (including primary and concatenated hazards, settlement, land use, the built environment, development paths, values and perceptions, lack of accountability, and others); (iv) identification of key factors in the expanding losses in disasters since 2000 and showing how they enter into risk and disaster; and (v) investigation of the use of existing scientific knowledge in disaster risk assessment and management (IRDR, 2011). AIRDR aims to: (i) provide the state-of-the-art of the science on integrated research on disaster risk; (ii) measure the effectiveness of multiple programmes; (iii) identify and support a long-term science agenda for the research community and funding entities; and (iv) create a mechanism for substantiating advances in the scientific evidentiary basis for supporting policy and practice (IRDR, 2012).

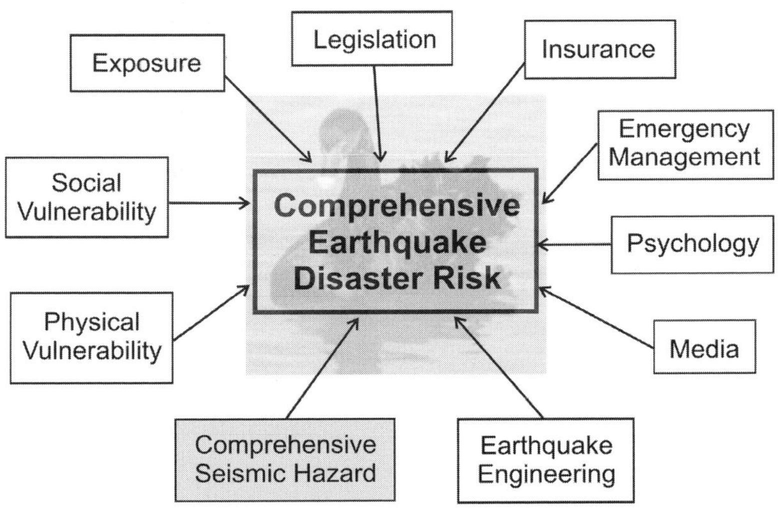

Figure 4.4. Basic components of integrated research for disaster risk.

4.9 Concluding remarks

'A Dream is the World without Disasters. Make the Dream a Reality' should become a leitmotif to the great opera called 'Disaster Risk Reduction'. To contribute to this statement I summarize below some basic scientific tasks still to be tackled:

- Significant new knowledge should be gained on extreme seismicity through observations, monitoring, analysis, modelling, and interpretations.
- Comprehensive seismic hazard and risk assessment tools should be developed and/or enhanced.
- Epistemic uncertainties in earthquake forecasting should be reduced to improve the link between the disaster mitigation community and the public. Multiple and concatenated extreme events need to be considered.
- Social vulnerability should be reduced. Earthquake hazards cannot be reduced, as an earthquake is a natural event and an abrupt response to complex tectonic stress localization. But vulnerability to extreme seismic events can be reduced significantly by monitoring human systems at local, regional, and global scales regarding demographic and economic changes, e.g., household mitigation, daily commuting, neighbourhood change, etc. (Peacock *et al.*, 2008; Cutter and Zoback, 2013).
- Stronger links to policy-makers, media, and the insurance industry, should be developed.
- Science education should be improved and awareness of extreme seismic hazards and earthquake disaster risk should be enhanced.

Unless these issues are addressed and the necessary social and political action is taken, escalating disaster losses (both human and economic) due to extreme seismic events presage a great challenge to sustainable development of society.

References

Allen, C. R., Edwards, W., Hall, W. J. *et al.* (1976). *Predicting Earthquakes: A Scientific and Technical Evaluation – With Implications for Society.* Panel on Earthquake Prediction of the Committee on Seismology, National Research Council. Washington, DC: US National Academy of Sciences.

Ambraseys, N. and Bilham, R. (2011). Corruption kills. *Nature*, **469**, 153–155.

Ando, M., Ishida, M., Hayashi, Y. and Mizuki, C. (2011). Interviews with survivors of Tohoku earthquake provide insights into fatality rate. *EOS Transactions, American Geophysical Union*, **92**, 411, doi:10.1029/2011EO460005.

Babayev, G., Ismail-Zadeh, A. and Le Mouël, J.-L. (2010). Scenario-based earthquake hazard and risk assessment for Baku (Azerbaijan). *Natural Hazard and Earth System Sciences*, **10**, 2697–2712.

Beer, T. and Ismail-Zadeh, A. T., eds. (2003). *Risk Science and Sustainability.* Dordrecht: Kluwer Academic Publishers.

Braile, L. W., Hinze, W. J., Keller, G. R., Lidiak, E. G. and Sexton, J. L. (1986). Tectonic development of the New Madrid Rift Complex, Mississippi Embayment, North America. *Tectonophysics*, **131**, 1–21.

Burridge, R. and Knopoff, L. (1967). Model and theoretical seismicity. *Bulletin of the Seismological Society of America*, **57**, 341–371.

Cornell, C. A. (1968). Engineering seismic risk analysis. *Bulletin of the Seismological Society of America*, **58**, 1583–1606.

Cummings, R. G. Brookshire, D. S., Bishop, R. C., and Arrow, K. J. (1986). *Valuing Environmental Goods: An Assessment of the Contingent Valuation Method.* Totowa: Rowman & Allanheld.

Cutter, S. L. and Zoback, M. L. (2013). Improving the nation's resilience to disasters, *EOS Transactions, American Geophysical Union*, **94**, 89, doi:10.1002/2013EO090007.

Davis, C. A. (2012). Loss functions for temporal and spatial optimizing of earthquake prediction and disaster preparedness. *Pure and Applied Geophysics*, **169**, 1989–2010.

Fedotov, S. A. (1965). Regularities of the distribution of strong earthquakes in Kamchatka, the Kurile islands, and northeastern Japan. *Trudy Inst. Fiziki Zemli Akad. Nauk SSSR (Proceedings of the Institute Physics of the Earth of the USSR Academy of Sciences)* **36**, 66–93.

Forsyth, D. W., Lay, T., Aster, R. C. and Romanowicz, B. (2009). Grand challenges for seismology. *Eos Transactions American Geophysical Union*, **90**, doi:10.1029/2009EO410001.

Fuchs, K. (2005). The great earthquakes of Lisbon 1755 and Aceh 2004 shook the world – seismologists' societal responsibility. In: *Proceedings of the 250th Anniversary of the 1755 Lisbon Earthquake International Conference*, Lisbon, Portugal, 1–4 November.

Gabrielov, A. M., Levshina, T. A. and Rotwain, I. M. (1990). Block model of earthquake sequence. *Physics of the Earth and Planetary Interiors*, **61**, 18–28.

Geller, R. J., Jackson, D. D., Kagan, Y. Y. and Mulargia, F. (1997). Earthquakes cannot be predicted. *Science*, **275**, 1616–1617.

Gerstenberger, M. C., Wiemer, S., Jones, L. M. and Reasenberg, P. A. (2005). Real-time forecasts of tomorrow's earthquakes in California. *Nature*, **435**, 328–331.

Green, H. W. II and Burnley, P. C. (1989). A new self-organizing mechanism for deep-focus earthquakes. *Nature*, **341**, 733–737.

Griggs, D. T. and Baker, D. W. (1969). The origin of deep-focus earthquakes. In *Properties of Matter Under Unusual Conditions*, eds. Mark, H. and Fernbach, S. New York: Wiley, pp. 23–42.

Hacker, B. R., Peacock, S. M., Abers, G. A. and Holloway, S. D. (2003). Subduction factory. 2. Are intermediate-depth earthquakes in subducting slabs linked to metamorphic dehydration reactions? *Journal of Geophysical Research* **108**, 2030, doi:10.1029/ 2001JB001129.

Hanks, T. C., Beroza, G. C. and Toda, S. (2012). Have recent earthquakes exposed flaws in or misunderstandings of probabilistic seismic hazard analysis? *Seismological Research Letters* **83**, 759–764.

Hess, H. (1962). History of ocean basins. In *Petrologic Studies*, eds. Engeln, A., James, H. L., and Leonard, B. F. New York: Geological Society of America, pp. 599–620.

IRDR (2011). *Forensic Investigations of Disasters: The FORIN Project (IRDR FORIN Publication No. 1)*. Beijing: Integrated Research on Disaster Risk.

IRDR (2012). *Assessment of Integrated Research on Disaster Risk: The AIRDR Project*. Beijing: Integrated Research on Disaster Risk.

Ismail-Zadeh, A. (2006). Earthquake science for society. Keynote lecture, *OECD Global Science Forum Workshop on Earthquake Science and its Contribution to Society*, 1–2 June 2006, Potsdam, Germany. Paris: Organisation for Economic Co-operation and Development. Available on CD-ROM.

Ismail-Zadeh, A. (2010). Computational geodynamics as a component of comprehensive seismic hazards analysis. In *Geophysical Hazards: Minimizing Risk and Maximizing Awareness*, ed. Beer, T. Amsterdam: Springer, pp. 161–178.

Ismail-Zadeh, A. T. and Kossobokov, V. G. (2011). Earthquake prediction M8 algorithm. In *Encyclopaedia of Solid Earth Geophysics*, ed. Gupta, H. Heidelberg: Springer, pp. 178–182.

Ismail-Zadeh, A. and Takeuchi, K. (2007). Preventive disaster management of extreme natural events. *Natural Hazards*, **42**, 459–467.

Ismail-Zadeh, A., Mueller, B. and Schubert, G. (2005). Three-dimensional modeling of present-day tectonic stress beneath the earthquake-prone southeastern Carpathians based on integrated analysis of seismic, heat flow, and gravity observations. *Physics of the Earth and Planetary Interiors*, **149**, 81–98.

Ismail-Zadeh, A. T., Sokolov, V. and Bonier, K. (2007a). Geodynamics, seismicity and seismic hazard of the south-eastern Carpathians. *Natural Hazards*, **42**, 493–514.

Ismail-Zadeh, A. T., Le Mouël, J. L., Soloviev, A., Tapponnier, P. and Vorobieva, I. (2007b). Numerical modeling of crustal block-and-fault dynamics, earthquakes and slip rates in the Tibet-Himalayan region. *Earth and Planetary Science Letters*, **258**, 465–485.

Ismail-Zadeh, A. T., Schubert, G., Tsepelev, I. A. and Korotkii, A. I. (2008). Thermal evolution and geometry of the descending lithosphere beneath the SE-Carpathians: An insight from the past. *Earth and Planetary Science Letters*, **273**, 68–79.

Ismail-Zadeh, A., Aoudia, A., and Panza, G. F. (2010). Three-dimensional numerical modeling of contemporary mantle flow and tectonic stress beneath the Central Mediterranean. *Tectonophysics*, **482**, 226–236.

Ismail-Zadeh, A., Le Mouël, J.-L. and Soloviev, A. (2012a). Modeling of extreme seismic events. In *Extreme Events and Natural Hazards: The Complexity Perspective* eds. Sharma, S. A., Bunde, A., Dimri, V. P. and Baker, D. N. Geophysical Monograph 196. Washington, DC: American Geophysical Union, pp. 75–97.

Ismail-Zadeh, A., Matenco, L., Radulian, M., Cloetingh, S. and Panza, G. (2012b). Geodynamic and intermediate-depth seismicity in Vrancea (the south-eastern Carpathians): current state-of-the-art. *Tectonophysics*, **530–531**, 50–79.

Kanamori, H. (2003). Earthquake prediction; An overview. In: *International Handbook of Earthquake and Engineering Seismology*, Part B, eds. Lee, W. H. K., Kanamori, H., Jennings, P. C. and Kisslinger, C. London: Academic Press, pp. 1205–1216.

Kant, I. (1756). *Geschichte und Naturbeschreibung der merkwürdigsten Vorfälle des Erdbebens welches an dem Ende des 1755sten Jahres einen großen Theil der Erde erschüttert hat*. Königsberg.

Kantorovich, L., Keilis-Borok, V. I., and Molchan, G. (1973). Seismic risk and principles of seismic zoning. In *Computational and Statistical Methods for Interpretation of Seismic Data*, ed. Keilis-Borok, V. I. Moscow: Nauka, pp. 3–20 (in Russian).

Keilis-Borok, V. I. (1990). The lithosphere of the Earth as a nonlinear system with implications for earthquake prediction. *Reviews of Geophysics*, **28**, 19–34.

Keilis-Borok, V. I., and Kossobokov, V. G. (1990). Premonitory activation of earthquake flow: algorithm M8. *Physics of the Earth and Planetary Interiors*, **61**, 73–83.

Keilis-Borok, V. I. and Soloviev, A. A. (2003). *Nonlinear Dynamics of the Lithosphere and Earthquake Prediction*. Heidelberg: Springer.

Keilis-Borok, V. I., Ismail-Zadeh, A. T., Kossobokov, V. G. and Shebalin, P. N. (2001). Non-linear dynamics of the lithosphere and intermediate-term earthquake prediction. *Tectonophysics*, **338**, 247–259.

Kelleher, J., Sykes, L. and Oliver, J. (1973). Possible criteria for predicting earthquake locations and their application to major plate boundaries of Pacific and Caribbean. *Journal of Geophysical Research*, **78**, 2547–2585.

King, G. C. P., Stein, R. S. and Lin, J. (1994). Static stress changes and the triggering of earthquakes. *Bulletin of the Seismological Society of America*, **84**, 935–953.

Klügel, J. U., Mualchin, L. and Panza, G. F. (2006). A scenario-based procedure for seismic risk analysis. *Engineering Geology*, **88**, 1–22.

Knopoff, L. (1999). Earthquake prediction is difficult but not impossible. *Nature Debates*. http://www.nature.com/nature/debates/earthquake. Accessed 15 March 2013.

Kossobokov, V. G., and Nekrasova, A. K. (2012). Global Seismic Hazard Assessment Program maps are erroneous. *Seismic Instruments*, **48**, 162–170.

Lay, T. and Kanamori, H. (2011). Insights from the great 2011 Japan earthquake. *Physics Today*, **64**, 33–39.

Lazaridou-Varotsos, M. S. (2013). *Earthquake Prediction by Seismic Electric Signals: The Success of the VAN Method over Thirty Years*. Springer, Heidelberg, 252 pp.

Manaker, D. M., Calais, E., Freed, A. M. *et al.* (2008). Interseismic plate coupling and strain partitioning in the Northeastern Caribbean. *Geophysical Journal International*, **174**, 889–903.

Molchan, G., and Romashkova, L. (2011). Gambling score in earthquake prediction analysis. *Geophysical Journal International*, **184**, 1445–1454.

Morat, P., and Le Mouël, J.-L. (1987). Variation of the electrical resistivity of large rock samples with stress. *Geophysics*, **52**, 1424–1430.

Ogata, Y. (1988). Statistical models for earthquake occurrences and residual analysis for point processes. *Journal of the American Statistical Association*, **83**, 9–27.

Panza, G. F., Irikura, K., Kouteva, M. *et al.* (2010). Advanced seismic hazard assessment. *Pure and Applied Geophysics*, **168**, DOI 10.1007/s00024-010-0179-9.

Peacock, W. G., Kunreuther, H., Hooke, W. H. *et al.* (2008). *Toward a Resiliency and Vulnerability Observatory Network: RAVON*. HRRC reports: 08-02R, http://www.nehrp.gov/pdf/RAVON.pdf.

Reid, H. F. (1911). *The Elastic-Rebound Theory of Earthquakes*. University of California Publications in Geological Sciences. Berkeley, CA: University of California Press, pp. 413–444.

Rice, J. R. and Ben-Zion, Y. (1996). Slip complexity in earthquake fault models. *Proceedings of the National Academy of Sciences of the United States of America*, **93**, 3811–3818.

Rubin, C. B., ed. (2012). *Emergency Management: The American Experience 1900–2010*. Boca Raton: CRC Press, Taylor & Francis Group.

Rundle, J. B., Tiampo, K. F., Klein, W. and Martins, J. S. S. (2002). Self-organization in leaky threshold systems: The influence of near-mean field dynamics and its implications for earthquakes, neurobiology, and forecasting. *Proceedings of the National Academy of Sciences of the United States of America*, **99**, 2514–2521.

Rundle, P. B., Rundle, J. B., Tiampo, K. F., Donnellan, A. and Turcotte D. L. (2006). Virtual California: Fault model, frictional parameters, application. *Pure and Applied Geophysics*, **163**, 1819–1846.

Satake, K., Sawai, Y., Shishikura, M. *et al.* (2007). Tsunami source of the unusual AD 869 earthquake off Miyagi, Japan, inferred from tsunami deposits and numerical simulation of inundation. American Geophysical Union (AGU), Fall Meeting 2007, abstract T31G-03. Washington, DC: AGU.

Schwartz, D. P. and Coppersmith, K. J. (1984). Fault behavior and characteristic earthquakes – examples from the Wasatch and San Andreas fault zones. *Journal of Geophysical Research*, **89**, 5681–5698.

Shebalin, P., Keilis-Borok, V., Gabrielov, A., Zaliapin, I. and Turcotte, D. (2006). Short-term earthquake prediction by reverse analysis of lithosphere dynamics. *Tectonophysics*, **413**, 63–75.

Soloviev, A. A. and Ismail-Zadeh, A. T. (2003). Models of dynamics of block-and-fault systems. In *Nonlinear Dynamics of the Lithosphere and Earthquake Prediction*, eds. Keilis-Borok, V. I. and Soloviev, A. A. Heidelberg: Springer, pp. 69–138.

Sornette, D. and Sammis, C. G. (1995). Complex critical exponents from renormalization group theory of earthquakes: Implication for earthquake prediction. *Journal of Physics I*, **5**, 607–619.

Stein, S., Geller, R. and Liu, M. (2011). Bad assumptions or bad luck: why earthquake hazard maps need objective testing. *Seismological Research Letters*, **82**, 623–626.

Stirling, M. W. (2012). Earthquake hazard maps and objective testing: the hazard mapper's point of view. *Seismological Research Letters*, **83**, 231–232.

Stokols, D. (2006). Toward a science of transdisciplinary action research. *American Journal of Community Psychology*, **38**, 63–77.

Turcotte, D. L. (1999). Seismicity and self-organized criticality. *Physics of the Earth and Planetary Interiors*, **111**, 275–294.

Turcotte, D. L., Holliday, J. R. and Rundle, J. B. (2007). BASS, an alternative to ETAS. *Geophysical Research Letters*, **34**, L12303, doi:10.1029/2007GL029696.

Varotsos, P., Alexopoulos, K., Nomicos, K. and Lazaridou, M. (1986). Earthquake predictions and electric signals. *Nature*, **322**, 120.

Working Group on California Earthquake Probabilities (2007). The Uniform California Earthquake Rupture Forecast, Version 2 (UCERF 2). US Geological Survey Open-file Report 2007–1437 (also California Geological Survey Special Report 203).

Wyss, M., ed. (1991). *Evaluation of Proposed Earthquake Precursors*. Special Publication No. 32. Washington, DC: American Geophysical Union.

Wyss, M., Nekraskova, A. and Kossobokov V. (2012). Errors in expected human losses due to incorrect seismic hazard estimates. *Natural Hazards*, **62**, 927–935.

Zaliapin, I., Keilis-Borok, V. and Ghil M. (2003). A Boolean delay model of colliding cascades. II: Prediction of critical transitions. *Journal of Statistical Physics*, **111**, 839–861.

Zechar, J. D. and Jordan, T. H. (2008). Testing alarm-based earthquake predictions. *Geophysical Journal International*, **172**, 715–724.

Zhang-li, C., Pu-xiong, L., De-yu, H. *et al.* (1984). Characteristics of regional seismicity before major earthquakes. In *Earthquake Prediction*. Paris: UNESCO, pp. 505–521.

5
The spatial-temporal dimensions of landslide disasters

Irasema Alcántara-Ayala

5.1 Introduction

5.1.1 The significance of disasters in the world

Disasters continue to be a major issue on the international agenda as human, economic, and environmental consequences have certainly increased in recent decades. According to the UNISDR (2012) global estimates indicate that since 1992 1.3 million casualties have occurred, in addition to 4.4 billion other people have been affected and there have been economic losses of US$2 trillion: for instance, 2.2 billion inhabitants in China were affected by disasters, economic losses in the USA reached US$560 billion, and in Haiti, one of the most vulnerable nations on the planet, 230 675 people were killed.

The office of US Foreign Disaster (OFDA) disaster figures (EM-DAT database, http://www.emdat.be/database) bring to light the fact that, of the total events for the period 1900–2011 reported worldwide as disasters, 36.54% were floods, followed by storms (31.71%), earthquakes (10.95%), landslides (5.85%), droughts (5.63%), extreme temperature (3.93%), wildfires (3.31%), and volcanic eruptions (2.05%). Nonetheless, in spite of the number of events, consequences vary in terms of social and economic costs as the vulnerability of exposed communities differs from hazard to hazard. For instance, the physical susceptibility of infrastructure and the vulnerability of population groups have proved to be significant in the effects of disasters associated with earthquakes. Some 2.5 million people have died and more than 170 million others have been affected by the consequences of living in non-resistant buildings and infrastructure in both developed and developing countries (Alcántara-Ayala, 2008). 'Silent hazards', specifically droughts (Sheffield and Wood, 2011), are a major concern since their impact has accounted for more than 11.7 million casualties, with 2.1 billion people affected; these represent 51% and 31.47% of the effects of natural hazards worldwide. Flood disasters are the deadliest hazard, accounting for 30% of all disaster-related deaths, and they have affected the largest number of people, in other words, the equivalent of 51% of the total population (Table 5.1).

Disasters associated with landslides are not the most important in terms of number of events, people killed, population affected, or estimated economic damage; they rank fourth (Table 5.1). However, patterns of landslide distribution in space and time have not yet been defined with great accuracy. Therefore, human, economic, and environmental impacts are still difficult to assess with precision since landslides are triggered by a multifaceted interaction of factors. Bearing in mind the forthcoming increasing landslide occurrence derived from climate change (Borgatti and Soldati, 2010) and the high vulnerability levels of exposed populations, data analysis and some considerations on the impact of landslide disasters at different scales in space and time are presented in this chapter.

5.1.2 Magnitude and frequency

The comprehension of magnitude and frequency of forces in geomorphic processes dates back to the pioneering research of Wolman and Miller (1960). Their work was centred on understanding the effective force of extreme or catastrophic events and those of relatively lower significance in landscape development. Thus, the importance of extreme events of high magnitude and low frequency, as well as those of low magnitude and high frequency, was understood in terms of measuring the relative amount of work done on the landscape and by the physical and morphological features created on it. Accordingly, considerations regarding low-frequency events of high magnitude

Extreme Natural Hazards, Disaster Risks and Societal Implications, eds. Alik Ismail-Zadeh, Jaime Urrutia Fucugauchi, Andrzej Kijko, Kuniyoshi Takeuchi and Ilya Zaliapin. Published by Cambridge University Press.

Table 5.1. *Disaster impacts throughout the world during 1900–2011 (source: EM-DAT database)*

	Number of events	People killed	People affected	Estimated damage (US$,000)
Drought	616	11 708 271	2 139 534 240	98 147 906
Earthquake	1198	2 563 226	170 648 422	735 935 084
Extreme temperature	430	169 211	97 021 992	57 374 542
Flood	3998	6 929 305	3 461 636 168	559 237 000
Landslides	640	64 089	13 656 983	8 653 598
Storm	3469	1 380 151	902 827 403	844 179 677
Volcano	225	96 312	5 271 981	3 040 348
Wildfire	363	3 622	5 942 174	52 107 055
TOTAL	10 939	22 914 187	6 796 539 363	2 358 675 210

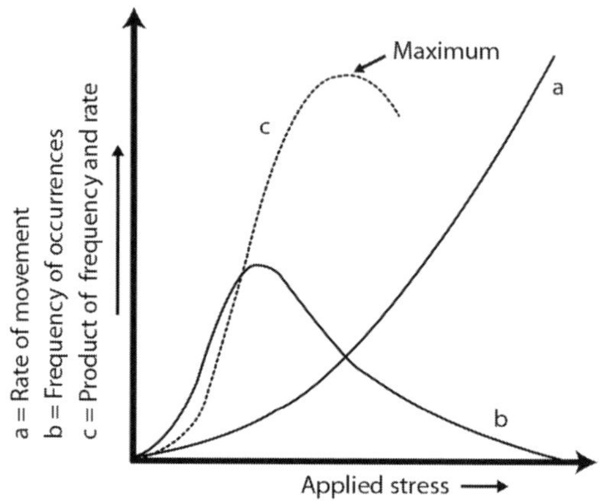

Figure 5.1. Relationships between the rate of transport, applied stress, and the frequency of stress application in landform development. (Source: Wolman and Miller, 1960.)

as responsible for the most effective processes for progressive denudation of the landscape were questioned, because catastrophic events are not necessarily the crucial elements in landform development (Fig. 5.1). The effectiveness of specific mechanisms and the relative significance of geomorphological processes in the generation of landforms involve both the frequency of occurrence and the magnitude of single events (Wolman and Miller, 1960).

From a hazards perspective, the concept of magnitude and frequency was applied to landslides by Guthrie and Evans (2007), who on the basis of a systematic analysis of rapid landsliding suggested that most work transporting material on hillslopes is done by moderate-sized landslides, and the moderate magnitude is defined by the work peak, with magnitude being determined mostly by the physiographic, climatic, and geotechnical settings of the landscape. The same reasoning needs to be applied to the occurrence of disasters, not only to hazards, because low-magnitude and high-frequency events are also responsible for creating and increasing vulnerable conditions, decreasing resilience, and interfering with development. Although huge disasters (mainly caused by high-magnitude, low-frequency events) are brought to the attention of the international community by the mass media, it is often the more frequent natural events of lower magnitudes that cause widespread disruption of the economy and the environment on a local scale.

The definition of large, moderate, or small disasters has been greatly associated with quantitative factors rather than with qualitative characterisations. The largest disasters occur under circumstances, where the highest numbers of casualties, affected people, and economic impacts occur. A failure to consider prevailing social and economic conditions, people's perceptions, capabilities, and needs has led to the misinterpretation and underestimation of the impact of small and moderate disasters on daily life, resilience construction, development, and adequate disaster risk management.

5.2 Impact of landslide disasters in space and time

The human and economic damage associated with landslides remains unquantified (Alcántara-Ayala, 2008; Kirschbaum *et al.*, 2010; Petley, 2012). However, their occurrence has caught the attention of the scientific community and civil protection authorities given that their negative impact on wellbeing, the environment, and economic activities has increased in recent decades. This situation may worsen in forthcoming years as a result of climate change (Borgatti and Soldati, 2010; IPCC, 2012).

Of particular importance is the fact that since landslides are triggered by natural mechanisms such as seismicity, precipitation, and, to a lesser extent, volcanic activity, it is difficult to measure their impact with great accuracy. Statistics thus regularly express the effects of multihazard events (e.g. earthquake–landslides, floods–landslides) and there is very limited specific account of landslide consequences.

It is clear that databases are very useful since they provide specific information on events (Pelling, 2005). However, it must be borne in mind that data represent a general picture of what really happens when disasters take place. Data on disasters generally also embody a series of inconsistencies derived from data collection, definition of losses, and availability and reliability. What is more, despite of having been natural processes of the Earth's surface even before the appearance of humankind, during recent decades, landslides have been influenced by anthropogenic activities, and in many cases it is the interaction between natural processes and human action that controls their impact.

5.2.1 Large landslide disasters

The Emergency Events Database (EM-DAT database) is one of the major global disaster databases available and is used by the scientific community. For EM-DAT, a devastating situation or event that overwhelms local capacity, involving a request to a national or an international agency for external assistance, is regarded as a disaster. For a disaster to be included in the database, at least one of the following criteria must be met: (1) 10 or more people reported killed; (2) 100 people reported affected; (3) the declaration of a state of emergency; and (4) a call for international assistance.

According to the EM-DAT database (Fig. 5.2), during the period 1900–2011 there have been 640 landslide disaster events which resulted in the loss of 64 089 human lives and more than 13.6 million people have been affected worldwide (Table 5.2). Asia is the region where the most catastrophic consequences have occurred because their aftermaths have involved 53% of all registered landslide events, 37% of all casualties, and 59% of all affected people. The American continent is the second most affected region, more than 22 000 deaths, 35% of all human deaths and people affected, in other words, 40% of the entire analysed record.

In terms of decades since 1900, there have been two major periods of landslide disasters (Fig. 5.3). The first was in the 1940s, when in both Europe and the Americas there was a sharp increase in casualties, owing to single landslide events that were associated with the losses of over 12 000 and 5000 human lives, respectively. The second period has been 1980–2011, when the cumulative effects of several landslide episodes around the world have given rise to one-third of the total casualties registered. During the period 1980–2011, the population affected by landslides worldwide reached >9.5 million (70% of the total since 1900), and since 2000, some 4 million people have been affected, one-third of the total recorded since 1900.

In terms of casualties, the most disastrous landslide event was in Tajikistan in 1949. Hundreds of landslides constituting rock-slope failures and flowslides in loess were triggered by the *M*7.4 Khait earthquake near the Tien Shan Mountains. One of the most tragic movements was a rockslide that transformed into a very rapid flow that travelled ~7.5 km with an estimated average velocity of ~30 m/s over a vertical distance of 1421 m (Evans *et al.*, 2009) in which 12 000 people died.

Another striking event occurred in December 1941, in Huaraz, Peru. Approximately 25% of the city was destroyed by a debris flow caused by an unexpected failure of a moraine dam at Laguna Cohup, Peru (Schuster *et al.*, 2002); 5000 casualties were reported. The third most catastrophic event took place in Honduras on 20 September 1973; rainfall from hurricane Fifi triggered numerous flash floods and mudslides involving about 2800 casualties (Table 5.3).

One crucial landslide disaster directly linked to human activities took place in Italy in 1963 (Fig. 5.4). After a series of instability episodes derived from the filling of the Vaiont reservoir, a mass of ~270 million m^3 was detached into the water and produced a wave that overtopped the reservoir and swept away the towns of Longarone, Pirago, Rivalta, Villanova, and Faè (Hendron and Patton, 1987; Semenza and Ghiritti, 2000; Kilburn and Petley, 2003) killing about 2000 people. Fifty years after the tragedy, controversy still remains about the landslide mechanism involved in the Vaiont failure, attracting the attention of the Earth's science community (Paronuzzi and Bolla, 2012; Spagna, 2012; Da-Chao *et al.*, 2013). Investigation and understanding of catastrophic flows resulting from dams and the potential generation of waves linked to hillslope instability in different regions of the world are important for landslide risk assessment (Costa and Schuster, 1991; Schuster, 2006; Evans and Alcántara-Ayala, 2007; Alcántara-Ayala and Domínguez Morales, 2008; Evans *et al.*, 2011; Strom, 2013).

Landslide disasters take place in both developed and developing countries (Alcántara-Ayala, 2002), and not surprisingly, their impact is greater in economically developing nations because of their greater vulnerability (Fig. 5.5). Considering the total number of people affected, the three worst landslide disasters registered in the database have taken place in three of the five most populated countries in the world (Population Reference Bureau, 2012).

More than 4 million people were affected in Brazil in January 1966 as a result of a series of landslides triggered by rainfall. In July of the same year, a landslide episode associated with seasonal rainfall caused severe damage in India, where 2.5 million inhabitants suffered the consequences. In May 2010, the area of the Huang Gu Po district of Badong on the lower reaches of the Three Gorges Dam in China was severely damaged by landslides, with 2.1 million people affected (see Table 5.4 for more examples of severe landslides).

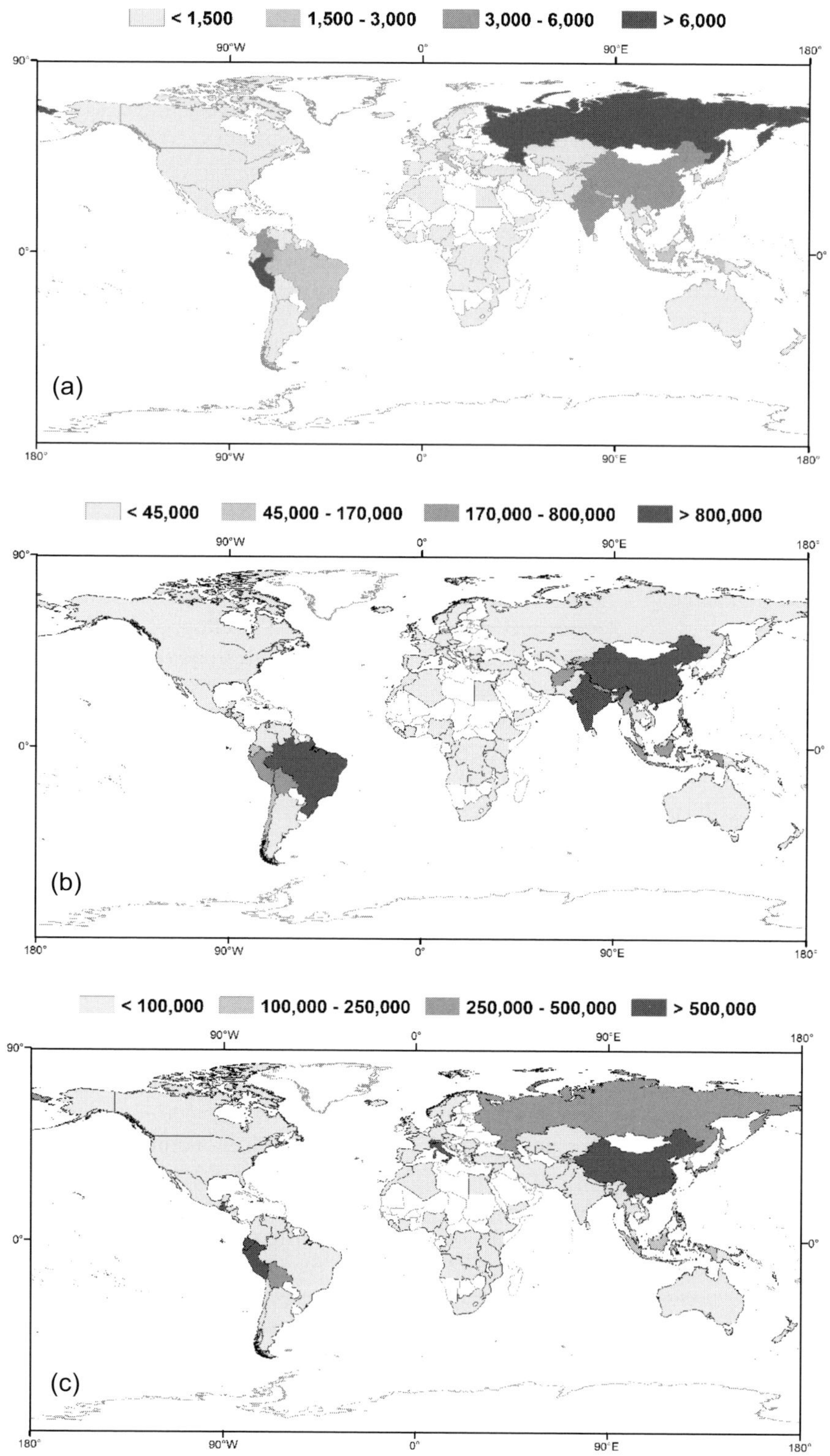

Figure 5.2. Impact of landslide disasters in terms of (a) people killed, (b) total people affected, and (c) estimated damage in US dollars (thousands) in the world from 1900 to 2011. (Source: EM-DAT database.)

Short- and long-term economic consequences of landslides have been assessed through various investigations since the 1980s, particularly for developing countries (Brabb and Harrod, 1989; Schuster, 1996; Schuster and Highland, 2001, 2007; Petley *et al.*, 2005; Kjekstad and Highland, 2008). Nonetheless, as a result of the Great East Japan Earthquake of March 2011 and its associated landslides (Miyagi *et al.*, 2011; Has *et al.*, 2012), developed countries with low levels of socio-economic vulnerability, high standards of disaster risk management, and an extensive culture of disaster prevention have also attracted attention.

In the case of landslide disasters, the limitations of economic impact assessments are greater than those linked to earthquakes, floods, and volcanic activity. Impediments to a realistic evaluation of landslide economic damage include: (i) a lack of constant, realistic, and homogenous information; (ii) the nature landslides as an ingredient of a multihazard; and (iii) budget reporting that favours political *momenta*, social engagements and/or economic needs (i.e. international debts, donor seeking). In this regard, information provided by OFDA/CRED (EM-DAT Database) represents one of the major sources of data on large landslide disasters. According to this, Peru and China are the countries where the highest economic impact of landslide disasters has occurred (Table 5.5).

In Peru, during January 1983, a series of debris and mud flows triggered by precipitation associated with El Niño affected Santa Eulalia and partially destroyed San José de Palle, Las Kiskas, and La Trinchera (Chuquisengo and Ferradas, 2007); economic damage amounted to US$1 billion. The Gangkouzhen landslide triggered by rainfall in the areas of Wulong and Chongqing, China, in May 1998 caused approximately US$0.9 billion of damage and 79 deaths (Huang, 2009; Huang and Li, 2011). In 2010, also in China, the devastating Zhouqu debris flow in the Gansu province, triggered by heavy rainfall, involved 1675 casualties and US$760 million of economic damage (Huang and Li, 2011).

Table 5.2. *Impact of landslide disasters in the world from 1900 to 2011 (source: EM-DAT database)*

Region	Number of events	People killed	Total people affected
Africa	31	1334	45 032
America	177	22 176	5 518 459
Asia	337	23 252	8 023 397
Europe	76	16 755	48 780
Oceania	19	572	21 315
Total	640	64 089	13 656 983

5.2.2 Small- and medium-scale landslide disasters

Investigations carried out by La RED (Network for Social Studies on Disasters Prevention in Latin-America) have indicated the need to look at the impact of small- and medium-size disasters, because their accumulated effects are at least equal in significance to those of large events. According to an analysis focused on disasters and urban development in Africa, which can be applied to other regions of the world, the number, territorial extent, and impact of small disasters is rising quickly and, as population and vulnerability increase in areas where hazards are intensified, events that once would have been regarded as small may acquire greater significance (Table 5.6). It is also suggested that the ability to prevent larger disasters might be enhanced by experiences gained through developing the ability to prevent small disasters (Bull-Kamanga *et al.*, 2003). Furthermore, it is evident that small- and large-scale disasters contribute extensively to the persistence of, or increase in, poverty (Lavell, 2008).

As part of the focus on the accumulated impact of small- and medium-size disasters on society and the environment, Marulanda *et al.* (2010) carried out an in-depth evaluation of disasters at the national level in Colombia, on the basis of two main databases: EM-DAT (OFDA/CRED) and

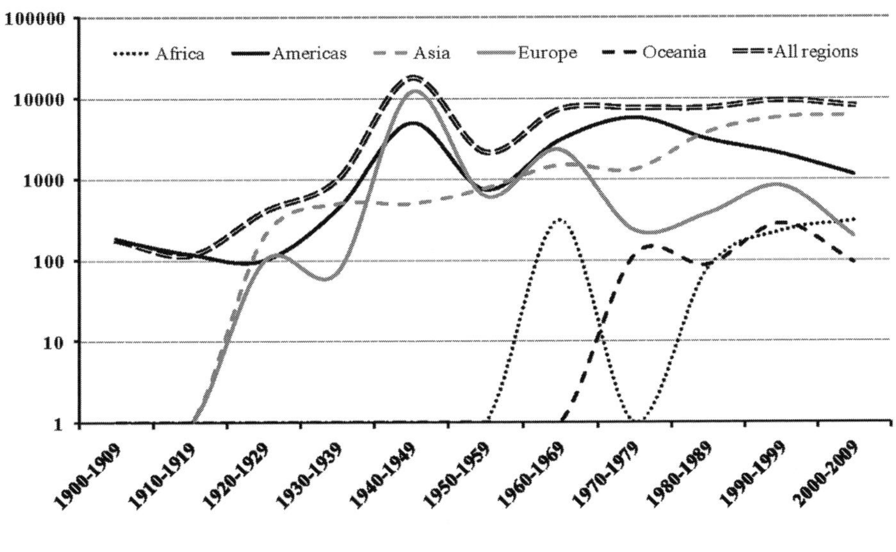

Figure 5.3. Decadal impact of landslide disasters in the different regions of the world in terms of people killed.

Table 5.3. *Ten largest landslide disasters for the period 1900–2011 in terms of the number of deaths at country level (source: EM-DAT database)*

	Country	Date	People killed	Triggering mechanism
1	Tajikistan RSS	1949	12,000	Earthquake
2	Peru, Huaraz	Dec. 1941	5,000	Moraine dam failure
3	Honduras	20 Sep. 1973	2,800	Hurricane Fifi
4	Peru	10 Jan. 1962	2,000	Snow melting
5	Italy	9 Oct. 1963	1,917	Landslide at Vaiont dam
6	China	7 Aug. 2010	1,765	Rainfall
7	Philippines	17 Feb. 2006	1,126	Rainfall
8	India	1 Oct. 1968	1,000	Rainfall
9	Colombia	27 Sep. 1987	640	Rainfall
10	Peru	18 Mar. 1971	600	Earthquake

Figure 5.4. General view of the Vaiont landslide from Casso village, Italy. (Photo courtesy: A. P. Dykes, September 2011.) For colour version, see Plates section.

DesInventar (La RED, 2012). For the period 1971–2002, EM-DAT recorded 97 events, whereas DesInventar recorded about 19 000. This analysis indicated that the spatial variability and dispersion of risk in Colombia comprises small events (derived from socio-natural processes linked to environmental deterioration) that pose a significant development problem at local and national level, but are not considered in the international databases.

Given their nature and the triggering mechanisms involved in the failure, landslide disasters are among the most difficult to evaluate. Tables 5.7 and 5.8 present data on landslide disasters in nine Latin-American countries from the DesInventar and EM-DAT databases, respectively.

Whereas DesInventar recorded about 34 500 deaths, the EM-DAT database recorded 23% of that amount. DesInventar recorded 19 253 injured people, whereas EM-DAT again recorded only 23% of them. Overall, EM-DAT accounted for only 15% of the total number of people registered as affected in DesInventar. The difference in recording homelessness was particularly significant; DesInventar recorded 880 000 people and EM-DAT recognized only 4% of that figure.

At the national level, the largest difference was in the case of Peru: DesInventar reported more than 21 000 deaths, and EM-DAT only 3341. Dissimilar information is also found in the case of Mexico, with 2497 versus 214 deaths, respectively. For Colombia, DesInventar reported 5551 deaths, and EM-DAT registered only 2892. For Costa Rica, DesInventar reported 209 deaths and EM-DAT recorded only 7. For El Salvador, DesInventar reported 1419 deaths and EM-DAT recorded 44.

Table 5.4. *Ten most important landslide disasters for the period 1900–2011 in terms of total number of people affected at country level (source: EM-DAT database)*

	Country	Date	Total affected people	Triggering mechanism
1	Brazil	11 Jan. 1966	About 4 000 000	Rainfall
2	India	Jul. 1966	About 2 500 000	Rainfall
3	China	30 May 2010	About 2 100 000	Rainfall/seismicity
4	India	12 Sep. 1995	About 1 100 000	Rainfall
5	Peru	Jan. 1983	About 700 000	Rainfall
6	Afghanistan	13 Jan. 2006	About 300 000	Rainfall/snow
7	Nepal	15 July 2002	265 865	Rainfall
8	Indonesia	31 Mar. 2003	229 548	Rainfall
9	Philippines	19 Dec. 2003	217 988	Rainfall
10	India	17 Aug. 1998	About 200 000	Rainfall

Figure 5.5. Landslides in the city of Teresopolis, state of Rio de Janeiro, Brazil. (Photo courtesy: Renato De Lima, January 2011.)

Comparison of the two databases also reveals that information on the number of evacuees and the people resettled is likely to be found only in the DesInventar database. Should the impact of landslide disasters be analysed in depth, detailed information of events at local level is undoubtedly required. Quite commonly, not only human losses are involved in landslide episodes: housing, agriculture, cattle, tourism, and other economic activities in addition to environmental effects frequently occur as well. The challenge then is to incorporate such ingredients in the evaluation of disasters in order to understand their real impact on society.

5.3 Behind the scenes: on the infinite dimensions of landslide disasters

5.3.1 The notion of nescience and uncertainty. Landslide storytelling: the Vargas Tragedy, Venezuela, 1999

Tens of thousands people died due to the occurrence of landslides, flash-floods, debris, and mudflows triggered by rainfall in the state of Vargas, Venezuela, in December 1999 (Fig. 5.6). Several sources indicated widely contrasting numbers of casualties ranging from 100 (17 December 1999) to 50 000 (29 December) deaths. The numbers most widely publicised and considered (USAID, 2000) were those provided at the end of December by the Civil Defence (30 000 fatalities) and by the Red Cross (50 000 fatalities) (Altez and Revet, 2005).

According to the United States Geological Survey (USGS), an extraordinarly wet period resulted from the interaction of a cold front with a moist southwestward flow from the Pacific Ocean towards the Caribbean Sea; rainfall accumulation of 200 mm on 2–3 December, along with a further 911 mm from 14 to 16 December, led to a series of debris flows and flash floods that affected the Vargas region. Massive debris flows moved very quickly through steep narrow canyons of metamorphic rocks and onto the alluvial fans causing considerable devastation (Wieczorek *et al.*, 2001).

The aftermath included 240 000 people affected (the equivalent to 70% of the total population), 100 000

Table 5.5. *Ten top landslide disasters for the period 1900–2011 in terms of economic cost at country level (source: EM-DAT database)*

	Country	Date	Damage (000 US$)
1	Peru	Jan. 83	988,800
2	China P. Republic	1 May 1998	890,000
3	China P. Republic	7 Aug. 2010	759,000
4	Italy	14 Dec. 1982	700,000
5	Switzerland	21 Feb. 1999	685,000
6	Italy	28 Jul. 1987	625,000
7	Ecuador	28 Mar. 1993	500,000
8	Guatemala	4 Sep. 2010	500,000
9	Soviet Union	10 Mar. 1989	423,000
10	Bolivia	8 Dec. 1992	400,000

evacuees, 8000 houses destroyed, 5 hospitals damaged, the collapse of water sewage systems and 85% of the infrastructure destroyed. Harbour, airport, and tourism activities were paralysed, 30% of the infrastructure for education was affected, between 10 000 and 15 000 casualties were reported, and economic damage reached US$5 billion (Genatios, 2012).

The physical susceptibility of the Vargas region to hillslope instability is beyond question; it is due to several factors including geological setting, morphological features, weathering and erosion processes, the presence of shallow soils, intense precipitation, and antecedent rainfall (Audemard *et al.*, 2000; Cartaya *et al.*, 2006). Nevertheless, according to an extensive investigation published in a forensic medicine periodical, the consequences of the

Table 5.6. *Comparison of disasters, small-size disasters, and everyday hazards in urban areas (source: Bull-Kamanga* et al., *2003)*

	Disasters	Small-size disasters	Everyday hazards
Frequency	Generally infrequent	Frequent (often seasonal)	Every day
Scale	Large, or potential to be large (e.g. 10 or more killed, 100 or more seriously injured, need for external assistance)	3–9 persons killed, 10 or more injured	1–2 persons killed, up to 9 injured
Total impact	Can be catastrophic for particular places and times in most low- and middle-income nations, but generally a low overall contribution to premature death and serious injury	Probably a significant and considerably underestimated contribution to premature death and serious illness or injury	In most African urban areas, these remain the main cause of premature death and serious injury
An integrated framework incorporating risk from disasters and 'non-disaster' events	*Very large impact for city, low frequency*	*Continuum of risk*	*Small impact for city, very high frequency*

Table 5.7. *Impact of landslide disasters in several countries of Latin America for the period of 1970–2011 (source: DesInventar database)*

Country	Number of events registered	Deaths	Missing people	Injured	Homeless	Affected	Evacuees	Resettled people
Bolivia	440	410	164	289	42 739	41 082	1600	0
Chile	382	357	123	909	75 914	650 921	3753	180
Colombia	7534	5551	794	2720	61 099	2 592 810	41 334	3009
Costa Rica	3097	209	18	21	15 542	8478	3802	441
Ecuador	1481	1467	584	612	10 532	152 348	10 947	799
El Salvador	1005	1419	119	430	7807	65 813	10 371	0
Mexico	1529	2497	1419	1093	354 639	3 094 616	30 626	72 116
Peru	3302	21 768	5042	12 857	258 977	588 878	18 042	11 118
Venezuela	1339	847	101	322	52 619	250 400	13 856	2601
TOTAL	20 109	34 525	8364	19 253	879 868	7 445 346	134 331	90 264

Vargas tragedy, particularly those related to casualties, were overestimated by the media (Altez, 2007).

The analysis carried out by Altez (2007) concluded that only 521 bodies were found, which with 331 missing persons accounts for 852 people; as 290 of the bodies recovered could not be indentified and this figure was lower than that of the missing people reported, the author pointed out the likelihood that unidentified bodies were those of missing persons. On the basis of that argument, the study then suggested that the total number of deaths did not exceed 700, a figure quite different to that widely reported by mass media and therefore also in some of the scientific literature.

Table 5.8. *Impact of landslide disasters in some countries of Latin America 1970–2011 (source: EM-DAT database)*

Country	Deaths	Injured	Homeless	Affected
Bolivia	218	134	800	169 690
Chile	203	170	17 671	65 000
Colombia	2892	3565	6188	22 520
Costa Rica	7	0	0	200
Ecuador	866	120	180	81 306
El Salvador	44	0	0	0
Mexico	214	0	120	200
Peru	3341	162	9776	780 740
Venezuela	134	318	2200	19 000
TOTAL	7919	4469	36 935	1 138 656

5.3.2 Scientific uncertainties and decision making: the 2006 Southern Leyte Landslide, Philippines

After a two-week period of rainfall and the occurrence of two small earthquakes of *M*2.3 and *M*2.7 on 16 February 2006, a complex mass movement in the form of a rockslide–debris avalanche buried Guinsaugon village on Leyte Island in the Philippines (Fig. 5.7). The initial source of the movement was the failure of a 450 m, mostly vegetated, 45° rocky slope; the rock mass sheared, and 15 million cubic metres of brecciated volcanic, sedimentary, and volcano-clastic rocks formed an avalanche that covered a 3800 m horizontal distance over a vertical distance of 810 m (Evans *et al.*, 2007).

The failure occurred on the ridge, Mount Canabag, a scarp that is the morphological expression of the active Leyte segment of the Philippine Fault Zone (PFZ) (Allen, 1962; Aurelio, 1992). Activity of the PFZ comprises a left-lateral strike-slip creep movement, with rates between 0.55 and 3.5 cm y^{-1} (Cole *et al.*, 1989; Barrier *et al.*, 1991; Duquesnoy *et al.*, 1994; Aurelio *et al.*, 1997; Duquesnoy, 1997), and offsetting at the northeast and southwest sectors of the ancestral Mount Bao (Duquesnoy *et al.*, 1994; Lagmay *et al.*, 2003). Investigations suggest that across the PFZ at the location of the Guinsaugon landslide, there is significant vertical displacement and a structure analogous to a pull-apart basin (Makino *et al.*, 2007). The disaster involved 1221 casualties, including 6 teachers and 248 primary students. More than 19 000 people were displaced and the economic impact was centred particularly on the agriculture sector, reaching estimated losses of about US$460 billion

1-2
DOMINGO 19 DE DICIEMBRE DE 1999
NACIONAL
EL UNIVERSAL CARACAS, VENEZUELA
http://politica.eud.com

CATASTROFE NATURAL

Operación rescate en Vargas

Los Corales desapareció bajo rocas

EN EL LITORAL todo es desolación y persisten aún los saqueos porque no ha llegado aun la anunciada ayuda oficial en alimentos y agua potable

Gustavo Rodríguez
El Universal

Caracas.- La mayoría de los damnificados del litoral central exigen que se les reponga a la mayor brevedad los servicios públicos y aún sobreviven en palafitos anclados en un mar de piedras, lodo y arena.

Desde Maiquetía hasta Caraballeda la desolación es casi absoluta. Sólo se escucha el ruido de los helicópteros en los diversos puntos helitácticos donde las víctimas llegan por sus propios medios para huir de la devastada zona. El tarantín utilizado para promocionar la Feria de la Hallaca Patriótica es ocupado por los peregrinos cuando la lluvia arrecia. Luego continúan su marcha hasta Maiquetía.

"Nosotros estamos vivos, pero dame razón de tu familia", suelen decir los damnificados cuando se topan con algún vecino que dejaron de ver desde el día del referéndum. Las bombonas de gas son parte del obligado equipaje aunque si en el camino consiguen

La dimensión de los aludes puede apreciarse desde el aire. Los Corales fue una de las zonas más afectadas por los deslizamientos

FELIX GERARDI

Figure 5.6. Image of *El Universal* newspaper headlines on 19 December 1999 (Caracas, Venezuela): Los Corales overwhelmed by rocks.

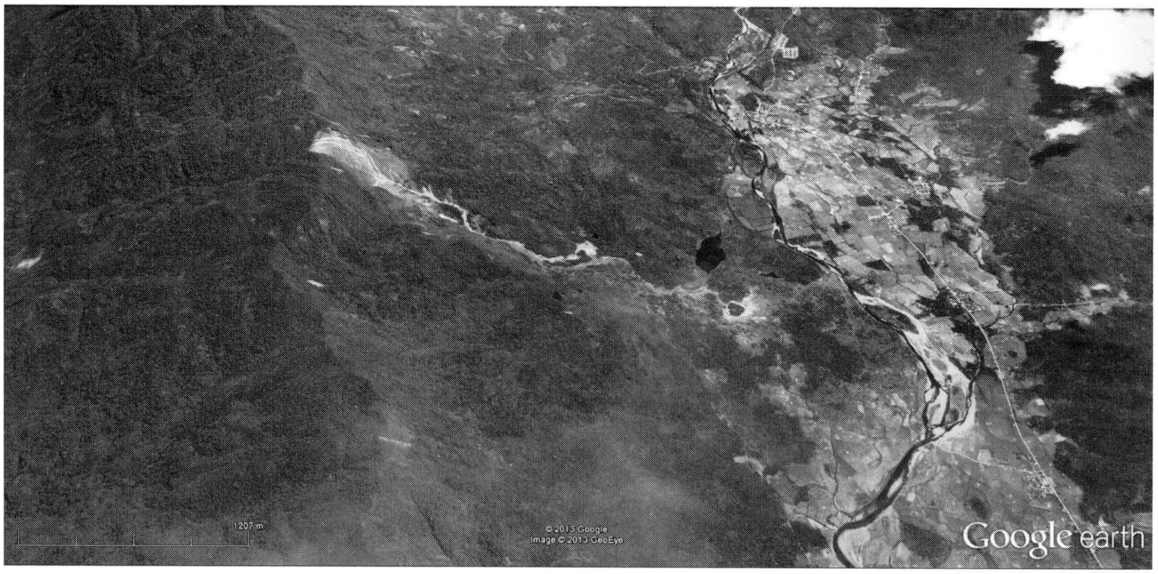

Figure 5.7. Image in 2013 of the area affected by the 2006 landslide on Leyte Island, Philippines (Google Earth).

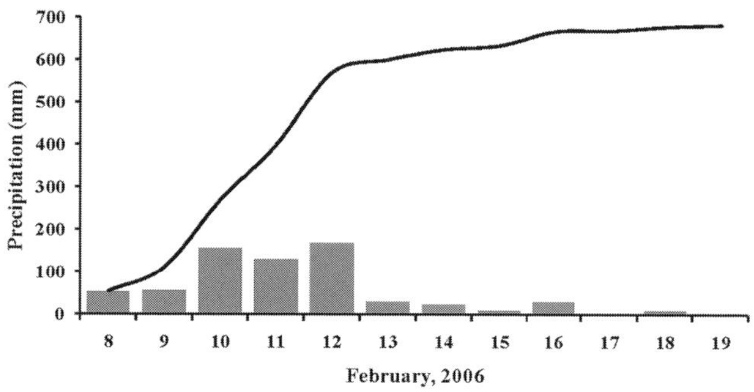

Figure 5.8. Daily and cumulative rainfall data at Libagon rain gauge near Guinsaugon, Leyte. (Source: PAGASA, http://kidlat.pagasa.dost.gov.ph/index.shtml.)

(Luna *et al.*, 2011). No particular direct triggering mechanism has yet been identified for the Leyte landslide (Evans *et al.*, 2007). However, there are three major ingredients that might have influenced landslide occurrence: precursor unstable conditions, seismicity, and precipitation.

The tectonic setting of the area, particularly of the fault scarp, can be linked to pre-failure instability conditions such as weathering of materials, and the existence of a discontinuity that acted as a sliding plane, a series of tension cracks (Evans *et al.*, 2007; Guthrie *et al.*, 2009), including one that appeared in May 2005, a small landslide on the 2006 landslide headscarp, and the downslope tilting of trees since the end of 2005 (Suwa, 2006).

Three seismic events were reported on the day of the Leyte landslide episode. According to the Philippine Institute of Volcanology and Seismology (PHIVOLCS), on 17 February 2006, an *M*2.6 earthquake occurred 21 km west of Guinsaugon (PHIVOLCS, 2006). One minute afterwards, USGS registered a further earthquake of *M*4.3 at a depth of 35 km, just 2 km north of the Guinsaugon landslide (USGS, 2006). The National Research Institute for Earth Science and Disaster Prevention in Japan recorded another, subsequent, ground movement from which, as Yamanaka (2006) suggested, the generation of the Rayleigh wave produced by the actual failure could be deduced.

Rainfall conditions had been severe since the beginning of February 2006. According to the data registered at the Libagon rain gauge about 7 km southwest of Guinsaugon, a 24 h precipitation of 157.2 mm was registered on 10 February, followed by 131 mm and 171 mm in the two subsequent days. From 8 to 12 February, the cumulative rainfall was 571.2 mm, and for the whole period between 8 and 19 February, it was 683.6 mm (Fig. 5.8), the equivalent of 235% of the mean monthly rainfall.

Although the triggering mechanism of the Leyte landslide is still under discussion, Sassa *et al.* (2009) have suggested that it was a combination of the *M*2.6 earthquake and rainfall. On the basis of field investigations and laboratory tests, they estimated that although the peak of groundwater level in the slope was reached before the earthquake, the 'sliding-surface liquefaction phenomenon' produced the subsequent rapid motion of the landslide owing to the generation of high pore-water pressure within the shear zone, derived crushing by grains of the volcano-clastic debris at the site subjected to shearing under a high normal stress.

Approximately 45 ha were covered by 25–40 m of mud and rocks from Mount Canabag (CDRC and LCDE, 2006) that had travelled at an estimated mean velocity of 27–38 m s^{-1} (Lagmay *et al.*, 2006, 2008). Earlier in the same month, on around 10–12 February, seven people had been killed in another landslide due to intense rainfall in a nearby municipality. Residents of 26 villages including Guinsaugon had been evacuated, but because weather conditions had improved they returned home the day before the disaster (Luna *et al.*, 2011).

The Recovery Status Report prepared for the Leyte Landslide indicates that lessons learned from the event have been translated into the establishment of measures for disaster risk reduction (DRR) at municipal level by means of creating an office for disaster and emergency management and the incorporation of DRR into its plans. Moreover, specific activities were linked to the priority actions of the Hyogo Framework for Action endorsed by the United Nations International Strategy for Disaster Reduction. Such activities have included hazard assessment, land-use planning, relocation to safer places, community training, partnerships among non-governmental organisations (NGOs), and the academic and private sectors (Luna *et al.*, 2011).

5.3.3 Disasters within disasters: landsliding associated with the 2010 Haiti Earthquake

On 12 January 2010, an *M*7.0 earthquake at a depth of 10 km struck Haiti, 17 km southwest of Port-au-Prince. Considerable damage occurred in the Port-au-Prince metropolitan area, and 80% of the town of Léogâne was destroyed. Specific consequences included 222 570 human losses, 3.7 million people were affected, and the damage cost US$8 billion (based on EM-DAT database).

This was a nation already devastated by social inequity and inequality closely associated with the situation inherited from slavery (Oliver-Smith, 2010), and with the contemporary complexity characterised by poverty, lack of development and access to education, high levels of corruption, environmental deterioration, and deforestation.

A number of landslides were induced by the seismicity (Fig. 5.9), and other mass movement processes took place soon afterwards as a result of the precipitation that occurred in subsequent days. The landslides exacerbated the pre-existing need for shelter and required the evacuation of people. Additionally, the area of the seaport at Port-au-Prince was considerably damaged, since the north wharf collapsed owing to liquefaction and lateral spreading; hence, aid supplies and relief personnel could not easily reach the affected areas.

Landslides along the roads generally occurred where they crossed swampy ground and at the distal ends of small alluvial valleys; local liquefaction may have contributed to the road failures, but was not the basic controlling factor (DesRoches *et al.*, 2011). In the epicentral zone, a series of rock-slides and rock-falls occurred on steep limestone slopes and road cuts. Shallow and deep-seated failures took place: the shallow ones were mainly concentrated on colluvial soil and fractured rock within the 1–3 m depth weathered zone; the deeper ones occurred in less-weathered, deeper bedrock in steep mountainous slopes (Carydis and Lekkas, 2011).

In the Rivière Froide drainage basin, for example, by using remote sensing images produced before and after the event, Zhao *et al.* (2012) identified 165 earthquake-induced

Figure 5.9. Earthquake-triggered landslides in Rivière Froide, Haiti. (Photo: courtesy of Raúl Carreño.)

landslides. Landslides were classified as new, extended, unaffected, and old, which cover areas of 54 132, 40 315, 186 962, and 5618 m^2, respectively.

Based on field recognitions and the magnitude of the 2010 Haiti earthquake (Keefer, 2002; Malamud *et al.*, 2004), USGS estimated that the earthquake triggered between 4000 and 5000 landslides, affecting an approximate area spreading 60 km east and west from the epicentre and across the entire north–south extent of the peninsula west of Port-au-Prince. Landslides were concentrated mainly in the south of the Enriquillo-Plantain Garden fault zone, particularly in deeply weathered, sheared, fractured, and/or altered limestone. Additionally, stream drainage and impounded lakes were blocked by numerous landslides. Potential further effects derived from existing hazards include the detachment of landslide masses, reactivation of landslides in areas where cracks have developed, and increasing downstream sedimentation (Jibson and Harp, 2011).

5.3.4 The 2011 Tohoku (Japan) global multidisaster

On 11 March 2011, the Tohoku *M*9 earthquake, also known as the Great East Japan Earthquake, triggered a devastating tsunami (MLIT, 2011). The main earthquake was followed by five aftershocks of $M > 7$ and hundreds of $M > 5$. A chain reaction was produced by the seismic activity: the tsunami, floods, land subsidence, landslides, and, particularly, the impact on Fukushima Dai-ichi Nuclear Power Station of the Tokyo Electric Power Company (TEPCO); these were the ingredients of one of the most unsettling and challenging contemporary disastrous events. The aftermath of the multidisaster associated with the Tohoku earthquake involved 19 846 human losses, a total of 368 820 people affected, economic damage in the order of US$210 billion (EM-DAT database), and the evacuation of 468 000 people (MLIT, 2011).

Subsequent to the major earthquake, and resulting from several stages of field reconnaissance, the Japan Landslide Society identified a range of mass movement processes. One of the most significant movements took place in Hanoki-daira, Shirakawa City, Fukushima Prefecture, where a rapid and long runout fluidised landslide destroyed a local community and 13 human lives were lost (Fukuoka *et al.*, 2012). Additionally, the Tohoku Geographical Association reported a distribution of landslides in other zones of Japan, such as the earth-fill residential areas in the mountainous sector of Sendai city, the volcanic deposits in the inland area of southern Fukushima, the Neogene soft rocks of the coastal region around Matsushima Bay, and in the vicinity of the displaced faults derived from the large aftershock in Iwaki city (Higaki *et al.*, 2011).

The effects of landslides and dam failures were not comparable with those that resulted from the major earthquake and the tsunami. However, they involved about 40 casualties and economic losses. Field observations and analysis showed landsliding in the mesoseismal region. In particular, detailed inspections of dams indicated from minor to moderate damage, except for the Fujinuma earth-dam that started breaching 20 minutes after the earthquake. In this case, modelling performed by Pradel *et al.* (2012) suggested that seismic shaking generated large lateral displacements of the downstream face and a major drop in the crest elevation, producing the conditions needed for dam overtopping and failure.

A day after the Tohoku earthquake and tsunami, an *M*6.2 earthquake occurred in the Northern Nagano Prefecture; 73 houses were destroyed, 66 people were injured, and there was damage to infrastructure including the railways. More than a hundred landslides were triggered by the Nagano earthquake, and its effects extended over a wide mountainous area, where heavy snowfall is quite common. According to the investigations carried out by Has *et al.* (2012), most landslides were shallow and occurred on the slopes of the left bank of the Chikuma River and on the active Miyanohara Fault, with very few on artificial slopes along the local railway. The shallow mass failures had a very sharp slip surface and were translational movements approximately 90–50 m long and 80–40 m wide. To a lesser extent, larger deep-seated landslides also occurred and had long runout distances as material travelled along the heavy snowfall courses.

Major effects of the landsliding included damage to buildings and transport networks and the blocking of rivers. Disrupted failures were rather common; nonetheless, lateral spreads were also very significant because they contributed greatly to erosion. Landslides quite frequently occurred in sedimentary rocks and were initiated at the crest of the slope, suggesting a topographic control due to amplification of the ground motions. Contrastingly, lateral spreading took place typically in Quaternary units, but also in sedimentary and volcanic terrains (Wartman *et al.*, 2012). In general terms and according to the field assessment carried out by the Structural Engineers Association of Washington (2011), the landslides triggered by the Tohoku earthquake took place in unstable areas, whereas liquefaction was concentrated in zones with loose to medium-dense, saturated sand and silt, principally on reclaimed land.

5.3.5 El Vergel disaster, Candelaria Loxicha, Oaxaca, Mexico 2012: where on Google Earth is that?

On 9 September 2012, a rainfall-triggered landslide buried a house in the village of El Vergel, in the municipality of Candelaria Loxicha, a mountainous zone within the Sierra del Sur physiographic province and 120 km to the south of Oaxaca City, in the state of Oaxaca, Mexico. The aftermath included four deaths (aged 40, 20, 5 years, and a 6 months old baby) and three injured people from a single family. Forty-four houses were destroyed, and coffee plantations, one of the major sources of employment in the area, were severely damaged.

The region is within the Oaxacan Complex, a lithological high-grade metamorphic terrain formed mainly by gneiss and extending over 10 000 km^2 (Mora and Valley, 1985). Mean annual precipitation reaches 2692.3 mm, although less than 60 mm of this falls during the dry season. The Cuajinicuil River runs across the municipality in a north–south direction; floods and hillslope instability due to river erosion are not infrequent in the area.

Certainly, the landslide event of El Vergel would not have been regarded as a disaster to be included in a major international database, since its consequences were neither 'considerable', nor very attractive to mass media. Nonetheless, the obliteration of a family and the economic losses were indeed a major disaster at individual and local levels. Countless examples of this sort take place on a daily basis, particularly in developing countries.

Not surprisingly, 89% of the total population (7948 inhabitants) of Candelaria Loxicha municipality live in poverty, and 63% of these are in extreme poverty; 96% of the population have no social security and only 46% receive any education. In the whole region, the presence of indigenous groups is significant, with local *zapotecas* living in 32 rural communities. Of all Mexican states, Oaxaca is second only to Guerrero in social deprivation (CONEVAL, 2010). Such socio-economic vulnerability, superimposed on the physiographic complexity of the territory, has played a significant role in the construction of risk.

5.4 Concluding remarks

In general, scientists, mass media, and governmental authorities usually pay attention to disasters of high magnitude and low frequency. However, events of low magnitude and high frequency, including landslide disasters, also contribute to the increase in vulnerability, decreasing resilience, and interfering with development, as life, economic assets, and the environment are affected, and greater needs arise.

Traditional definitions of large-, moderate-, or small-size disasters have been derived usually from quantitative factors without taking into account qualitative characterisations. The largest disasters are characterised by the highest numbers of casualties and affected people, and the greatest economic impact. Failure to consider factors associated with vulnerability and resilience, such as prevailing social and economic conditions, people's perceptions, capabilities, and needs, has led to the misinterpretation and underestimation of small and moderate disasters.

In the particular case of landslides, assessment of both hazard and risk is complicated; there are diverse natural triggering mechanisms including seismicity, precipitation, and, to a lesser extent, volcanic activity, but also the combination of some of them (i.e. earthquake and rainfall) impedes accurate measurement of landslide impact. Moreover, quite frequently data cover the effects of multi-hazards events, for instance earthquake–landslides or floods–landslides, with no specific account of landslide consequences.

According to world landslide datasets, from 1900 to 2011 landslide disasters have impacted most on countries in Asia, followed by the Americas, Europe, Africa, and Oceania. There have been 640 landslide disasters worldwide in the period, resulting in more than 60 000 deaths, and with more than 13 million people affected.

What is more, particular examples of historical and contemporary landslide disasters, one of the key points made in this chapter, provide challenging pictures of the diversity of dimensions of landslide disasters in space and time. For example, the hyped tragedy of Vargas, Venezuela, in 1999 was not only an example of the alarmist mode adopted by mass media for reporting disasters, but was also a disaster that is still believed by a significant sector of academia to have involved a high number of casualties. The notion of nescience and uncertainty and the natural scientific uncertainties of hazard and risk assessment are major ingredients in decision making, because they play a significant role in disaster risk reduction and in cases such as the Leyte landslide in the Philippines, the consequences can be devastating.

An important component contributing to disasters is time-dependent socio-economic vulnerability. The high impact of landslide disasters in economically less-developed countries is associated not only with their high level of vulnerability but is also partly due to their lack of preparedness and of a disaster prevention culture.

There has certainly been no interest in analysing the significance of small and moderate disasters at the local scale, such as the case of El Vergel landslide, yet their impact is without doubt one of the major drivers for the construction of the different spheres of vulnerability and resilience, and above and beyond all, for shaping impending development and achieving sustainability.

Herein lie challenges, as it is evident that the spatial-temporal dimensions of landslide disasters have not been correctly understood on a global scale, and are understood even less at the local scale; little is known about their impact on daily life, vulnerability conditions, resilience construction, development, and adequate disaster risk management.

Acknowledgements. The author acknowledges the support kindly provided by CONACyT and UNAM through the research projects 156242 and IN307410 respectively. Also thanks are due to Ricardo J. Garnica-Peña, Miguel A. Ramírez-Beltrán, and Laura Diana López-Ascencio for their collaboration in the preparation of some figures. Dr Alan P. Dykes (Kingston University London), Dr Renato de Lima (CENACID, Federal University of Parana) and Dr Raúl Carreño (Grudec Ayar, Peru) kindly provided photographs of the Vaiont, Teresopolis, and Haiti landslides respectively. Disaster data were kindly provided by and downloaded from the ODFA/CRED International Disaster Database Centre for Research into the Epidemiology of Disasters (CRED), Catholic University of Louvain, and DesInventar. Thanks to Ann Grant for the English revision and editing of the manuscript.

References

Alcántara-Ayala, I. (2002). Geomorphology, natural hazards, vulnerability and prevention of natural disasters in developing countries. *Geomorphology*, **47**, 107–124.

Alcántara-Ayala, I. (2008). On the historical account of disastrous landslides in Mexico: the challenge of risk management and disaster prevention. *Advances in Geociences-ADGEO*, **14**, 159–164. Tokyo: UNU.

Alcántara-Ayala, I. and Domínguez Morales, L. (2008). The San Juan de Grijalva catastrophic landslide, Chiapas, Mexico: lessons learnt. In *Proceedings of The First Landslide World Forum*, United Nations University, International Programme on Landslides, 18–21 November, Tokyo, Japan, pp. 53–60.

Allen, C. R. (1962). Circum-Pacific faulting in the Philippines-Taiwan region. *Journal of Geophysical Research*, **67**, 4795–4812.

Altez, R. (2007). Muertes bajo sospecha: Investigación sobre el número de fallecidos en el desastre del estado Vargas, Venezuela, en 1999 (Deaths under suspicion: investigation on the number of deaths in the Vargas state disaster, Venezuela, in 1999), *Cuadernos de Medicina Forense*, **13**(50), 255–268.

Altez, R. and Revet, S. (2005). Contar los muer tos para contar la muer te: discusión en torno al número de fallecidos en la tragedia de 1999 en el estado Vargas-Venezuela (Counting the diseases to tell the death: discussion about the number of deaths in Vargas'99 tragedy, Vargas state, Venezuela), *Revista Geográfica Venezolana*, Número especial 2005, 21–43.

Audemard, F. A., Rodriguez, J. A. and Stockhausen, H. W. (2000). La vertiente Caracas del Macizo Avila en el espejo del desastre de Vargas de diciembre 99. XVI Seminario Venezolano de Geotecnia. *Calamidades Geotécnicas Urbanas con Visión al Siglo XXI, Caracas.* Caracas: Corporación Andina de Fomento/Sociedad Venezolana de Geotecnia.

Aurelio, M. A. (1992). Tectonique du segment central de al faille Philippine (étude structurale, cinématique et evolution geodynamique). Doctoral thesis University Pierre et Marie Curie, Paris, France.

Aurelio, M. A., Barrier, E., Gaulon, R. and Rangin, C. (1997). Deformation and stress states along the central segment of the Philippine fault: implications to wrench fault tectonics. *Journal of Asian Earth Sciences*, **15**, 107–119.

Barrier, E., Huchon, P. and Aurelio, M. A. (1991). Philippine fault: a key for Philippine kinematics. *Geology*, **19**, 32–35.

Borgatti, L. and Soldati, M. (2010). Landslides and climate change. In *Geomorphological Hazards and Disaster Prevention*, eds. Alcántara-Ayala, I. and Goudie, A. S. Cambridge, UK: Cambridge University Press, pp. 87–95.

Brabb, E. E. and Harrod, B. L. (1989). Landslides: extent and economic significance. In *Proceedings of the International Geological Congress*, volume 28.

Bull-Kamanga, L., Diagne, K., Lavell, A. *et al.* (2003). From everyday hazards to disasters: the accumulation of risk in urban areas. *Environment & Urbanization*, **15**(1), 193–204.

Cartaya, S., Méndez, W. and Pacheco, H. (2006). Modelo de zonificación de la susceptibilidad a los procesos de remoción en masa a través de un sistema de información geográfica. *Interciencia*, **31**(9), 638–646.

Carydis, P. G. and Lekkas, E. L. (2011) The Haiti Earthquake $M_w = 7.0$ of January 10th 2010: structural and geotechnical engineering field observations, near-field ground motion estimation and interpretation of the damage to buildings and infrastructure in the Port-au-Prince area. *Ingegneria sismica*, **XXVIII**(3), 24–42.

CDRC and LCDE (2006). A preliminary report on the landslide in St. Bernard, Southern Leyte. Citizen Disaster Response Center and Leyte Center for Development.

Chuquisengo, O. and Ferradas, P. (2007). *Gestión de riesgos en Ancash, Lima*. Soluciones Prácticas – ITDG.

Cole, J., McCabe, R., Moriarty, T. *et al.* (1989). A preliminary Neogene palaeomagnetic dataset from Leyte and its relation to motion on the Philippine fault. *Tectonophysics*, **168**, 205–221.

CONEVAL (2010). *Indicadores de pobreza por municipio*, Medición de la pobreza, http://www.coneval.gob.mx/cmsconeval/rw/pages/medicion/medicion_pobreza_municipal_2010.es.do (retrieved on 15 December 2012).

Costa, J. E. and Schuster, R. L. (1991). *Documented Historical Landslide Dams from Around the World*. United States Geological Survey Open-File Report, 91–239:486.

Da-Chao, L., Jing-Luo, C., Zhang-Lin, G. *et al.* (2013). Evaluation of landslide risk based on synchronization of nonlinear motions in observed data. *Natural Hazards*, **65**, 581–603.

DesRoches, R., Comerio, M., Eberhard, M., Mooney, W. and Rix, G. J. (2011). Overview of the 2010 Haiti Earthquake. *Earthquake Spectra*, **2**, 1–21.

Duquesnoy, T. (1997). Contributions de la géodésie à l'étude de grands décrochements actifs associés à des zones de subduction à convergence oblique. D.Sc. thesis University of Paris XI, Orsay.

Duquesnoy, T., Barrier, E., Kasser, M. *et al.*, the French–Philippine Cooperation team (1994). Detection of creep along the Philippine fault: first results of geodetic measurements on Leyte Island, central Philippines. *Geophysical Research Letters*, **21**, 975–978.

Evans, S. G. and Alcántara-Ayala, I. (2007). Disasters resulting from landslides, snow avalanches, and geotechnical failures in North America (Canada, United States, and Mexico) 1841–2006: A first assessment. In *Landslides and Society*, eds. Turner, K. and Schuster, R. L. AEG Special Pub. No. 22. Denver, CO: Association of Environmental & Engineering Geologists, pp. 3–23.

Evans, S. G., Guthrie, R. H., Roberts, N. J. and Bishop, N. F. (2007). The disastrous February 17, 2006 rock avalanche on Leyte Island, Philippines: a catastrophic landslide in tropical mountain terrain. *Natural Hazards and Earth System Sciences*, **7**, 89–101.

Evans, S., Roberts, N., Ischuk, A. *et al.* (2009). Landslides triggered by the 1949 Khait Earthquake, Tajikistan, and associated loss of life. *Engineering Geology*, pp. 195–212.

Evans, S. G., Hermanns, R., Strom, A. L. and Scarascia-Mugnozza, G., eds. (2011). *Natural and Artificial Rockslide Dams*. Lecture Notes in Earth Sciences, vol. 133. Berlin: Springer.

Fukuoka, H., Higaki, D. and Ugai, K. (2012). *Landslides Induced by the 11 March 2011 Tohoku Earthquake, Japan*. Geophysical Research Abstracts, 14, EGU2012-14337, EGU General Assembly 2012.

Genatios, C. (2012). *Vargas, del riesgo al desastre: Los aludes torrenciales de 1999. Gestión de riesgo: entre lo planificado y lo ejecutado* (Vargas, from risk to disaster: the flash-floods of 1999. Risk management: between the planned and the executed), Instituto Latinoamericano de Investigaciones Sociales (ILDIS), Oficina en Venezuela de la Fundación Friedrich Ebert, 8 pp.

Guthrie, R. H. and Evans, S. G. (2007). Work, persistence, and formative events: the geomorphic impact of landslides. *Geomorphology*, **88**, 266–275.

Guthrie, R. H., Evans, S. G., Catane, S. G., Zarco, M. A. H. and Saturay, R. M. (2009). The 17 February 2006 rock slide–debris avalanche at Guinsaugon Philippines: a synthesis. *Bulletin of Engineering Geology and the Environment*, **68**, 201–213.

Has, B., Noro, T., Maruyama, K. *et al.* (2012). Characteristics of earthquake-induced landslides in a heavy snowfall region – landslides triggered by the Northern Nagano Prefecture earthquake, March 12, 2011, Japan. *Landslides*, **9**, 539–546.

Hendron, A. J. and Patton, F. D. (1987). The Vaiont slide. A geotechnical analysis based on new geologic observations of the failure surface. *Engineering Geology*, **24**, 475–491.

Higaki, D., Tsuchiya, T. and Tushima, M. (2011). *Landslide Disasters Induced by the 2011 off the Pacific Coast of Tōhoku Earthquake*. The 2011 East Japan Earthquake Bulletin of the Tōhoku Geographical Association. Available at: http://tohokugeo.jp/disaster/disaster-e.html (retrieved on 20 January, 2013).

Huang, R. Q. (2009). Some catastrophic landslides since the twentieth century in the southwest of China. *Landslides*, **6**, 69–81.

Huang, R. Q. and Li, W. (2011). Formation, distribution and risk control of landslides in China. *Journal of Rock Mechanics and Geotechnical Engineering*, **3**(2), 97–116.

IPCC (2012). *Managing the Risks of Extreme Events and Disasters to Advance Climate Change Adaptation. A Special Report of Working Groups I and II of the Intergovernmental Panel on Climate Change*, eds. Field, C. B., Barros, V., Stocker, T. F., *et al.* Cambridge, UK: Cambridge University Press.

Jibson, R. W. and Harp, E. L. (2011) *Field Reconnaissance Report of Landslides Triggered by the January 12, 2010, Haiti Earthquake*. US Geological Survey Open-File Report 2011–1023.

Keefer, D. K. (2002). Investigating landslides caused by earthquakes – a historical review. *Surveys in Geophysics*, **23**(6), 473–510.

Kilburn, C. R. J. and Petley, D. N. (2003). Forecasting giant, catastrophic slope collapse: lessons from Vajont, Northern Italy. *Geomorphology*, **54**, 21–32.

Kirschbaum, D. B., Adler, R., Hong, Y., Hill, S., and Lerner-Lam, A. L. (2010). A global landslide catalog for hazard applications – method, results and limitations. *Natural Hazards*, **52**, 561–575.

Kjekstad, O. and Highland, L. M. (2008). Economic and social impacts of landslides. In *Plenary Session Volume, First World Landslide Forum*, eds. Sassa, D. and Canuti, P., 18–21 November 2008, Tokyo, Japan. Tokyo: UNU, p. 573.

LA RED (2012). DesInventar. http://www.desinventar.org/

Lagmay, A. M. F., Tengonciang, A. M. P., Marcos, H. V. and Pascua, C. S. (2003). A structural model guide for geothermal exploration in ancestral Mount Bao, Leyte, Philippines. *Journal of Volcanology and Geothermal Research*, **122**, 133–141.

Lagmay, A. M. F., Ong, J. B. T., Fernandez, D. F. D. *et al.* (2006). Scientists investigate recent Philippine landslide. *EOS Transactions of the American Geophysical Union*, **87**, 121–124.

Lagmay, A. M. F., Rodolfo, R. S., Tengonciang, A. M. P. *et al.* (2008). Science guides search and rescue after the 2006 Philippine landslide. *Journal of Disasters*, **32**(3), 416–433.

Lavell, A. (2008). *Community and Local Level Disaster Risk Management: Considerations as Regards Relations with Poverty Alleviation. A Contribution to the 2009 ISDR Global Assessment Report on Disaster Risk Reduction*. Available at: http://www.desenredando.org/public/varios/2008/2008_ISDR_GAR_Lavell_CBRMPoverty.pdf (retrieved on 22 August 2013).

Luna, E. M., Luna, G. D. A., Molina, J. G. J. and Molina, F. G. J. (2011). *The Southern Leyte Landslide 2006, Recovery Status Report*, International Recovery Platform, 86 pp.

Makino, M., Mandanas, A. A. and Catane, S. G. (2007). Gravity basement of the Guinsaugon landslide along the Philippine fault zone. *Earth Planets Space*, **59**, 1067–1071.

Malamud, B. D., Turcotte, D. L., Guzzetti, F. and Reichenbach, P. (2004). Landslides, earthquakes and erosion. *Earth and Planetary Science Letters*, **229**(1–2), 45–59.

Marulanda, M. C., Cardona, O. D. and Barbat, A. H. (2010). Revealing the socioeconomic impact of small disasters in

Colombia using the DesInventar database. *Disasters*, **34**, 552–570.

Miyagi, T., Higaki, D., Yagi, H. *et al.* (2011). Reconnaissance report on landslide disasters in northeast Japan following the M 9 Tohoku earthquake. *Landslides*, **8**(3), 339–342.

MILT (2011). *White Paper on Land, Infrastructure, Transport, and Tourism in Japan, 2011*, Ministry of Land, Infrastructure, Transport, and Tourism, http://www.mlit.go.jp/report/press/toshi07_hh_000056.html.

Mora, C. I. and Valley, J. W. (1985). Ternary feldspar thermometry in granulites from the Oaxacan complex, Mexico. *Contributions to Mineralogy and Petrology*, **89**, 215–225.

Oliver-Smith, A. (2010). *Haiti and the Historical Construction of Disasters* NACLA Report on the Americas, July–August, pp. 32–36.

Paronuzzi, P. and Bolla, A. (2012). The prehistoric Vajont rockslide: An updated geological model. *Geomorphology*, **169–170**, 165–191.

Pelling, M. (2005). *Disaster Data: Building a Foundation for Disaster Risk Reduction*. International Federation of the Red Cross and Red Crescent Societies, World Disasters Report 2005, pp.173–180.

Petley, D. N. (2012). Global patterns of loss of life from landslides. *Geology*, **40**, 927–930.

Petley, D. N., Dunning, S. A. and Rosser, N. J. (2005). The analysis of global landslide risk through the creation of a database of worldwide landslide fatalities. In *Landslide Risk Management*, eds. Hungr, O., Fell, R., Counture, R. and Ebergardt, E. Amsterdam: Balkema, pp. 367–374.

PHIVOLCS (2006). Earthquake wave data. Accessed online: http://www.phivolcs.dost.gov.ph/Earthquake/LatestEQ/2006/20060217_0236.htm.

Population Reference Bureau (2012). World Population Data Sheet. Available at: http://www.prb.org/pdf12/2012-population-data-sheet_eng.pdf (retrieved on 9 December 2012).

Pradel, D., Wartman, J. and Tiwari, B. (2012). *Failure of Fujinuma Dam during the 2011 Tohoku Earthquake. Joint Conference Proceedings 9th International Conference on Urban Earthquake Engineering / 4th Asia Conference on Earthquake Engineering, March 6–8, 2012*. Tokyo: Tokyo Institute of Technology.

Sassa, K., Fukuoka, H., Soridum, R. *et al.* (2009). Mechanism of the initiation and motion of the 2006 Leyte landslide, Philippines. *Bulletin of Engineering Geology and the Environment*, **68**, 193–207.

Schuster, R. L. (1996). Landslides: investigation and mitigation. Chapter 2 – Socioeconomic significance of landslides. *Transportation Research Board Special Report* No. 247.

Schuster, R. L. (2006). *Interaction of Dams and Landslides – Case Studies and Mitigation*. US Geological Survey Professional Paper, 1723, 107 pp.

Schuster, R. L. and Highland, L. M. (2001). *Socioeconomic and Environmental Impacts of Landslides in the Western Hemisphere*. US Geological Survey Open-File Report, 1–276.

Schuster, R. L. and Highland, L. M. (2007). The Third Hans Cloos Lecture. Urban landslides: socioeconomic impacts and overview of mitigative strategies. *Bulletin of Engineering Geology and the Environment*, **66**, 1–27.

Schuster, R. L., Salcedo, D. A. and Valenzuela, L. (2002). Overview of catastrophic landslides of South America in the twentieth century. In *Catastrophic Landslides: Effects, Occurrence, and Mechanisms*, eds. Evans, S. G. and Degraff, J. V., Reviews in Engineering Geology, vol. 15. Boulder: Geological Society of America, pp. 1–34.

Semenza, E. and Ghirotti, M. (2000). History of the 1963 Vaiont slide: the importance of geological factors. *Bulletin of Engineering Geology and the Environment*, **59**, 87–97.

Sheffield, J., and Wood, E. F. (2011). *Drought: Past Problems and Future Scenarios*. London: Earthscan.

Spagna, V. (2012). Fifty years ago: the Vaiont night. *Annals of Geophysics*, **55**(3), 389–392.

Strom, A. (2013). Geological prerequisites for landslide dams' disaster assessment and mitigation in Central Asia. In *Progress of Geo-Disaster Mitigation Technology in Asia*, eds. Wang, F., Miyajima, M., Li, T., Shan, W., Fathani, T. F. Berlin: Springer, pp. 17–53.

Structural Engineers Association of Washington (2011). *The 2011 Great East Japan (Tohoku) Earthquake & Tsunami, Earthquake Reconnaissance Observations*.

Suwa, H., ed. (2006). *Catastrophe caused by the 17 February 2006 southern Leyte landslide in the Philippines. Journal of the Japanese Society Natural Disaster Science*, **25**, 83–97.

UNISDR (2012). *Impacts of Disasters Since the 1992 Rio de Janeiro Earth Summit*, http://www.unisdr.org/files/27162_infographic.pdf (retrieved on 8 December 2012).

USAID (2000). *Venezuela Factsheet, February 2000*. USAID-Office of Foreign Disaster Assistance.

USGS (2006). *NEIC Earthquake Search Catalog*, United States Geological Survey. Available at: http://neic.usgs.gov/.

Wartman, J., Dunham, L., Tiwari, B. and Pradel, D. (2012). Landslides triggered by the great Tohoku, Japan earthquake. AGU Fall Meeting, NH13A-1585, San Francisco, 3–7 December 2012. Washington, DC: American Geophysical Union.

Wieczorek, G. F., Larsen, M. C., Eaton, L. S., Morgan, B. A. and Blair, J. L. (2001). *Debris-flow and Flooding Hazards Associated with the December 1999 Storm in Coastal Venezuela and Strategies for Mitigation*. US Geological Survey, Open File Report, 01–0144, http://pubs.usgs.gov/of/2001/ofr-01-0144/ (retrieved on 7 December 2012).

Wolman, M. G. and Miller, W. P. (1960). Magnitude and frequency of forces in geomorphic processes. *Journal of Geology*, **68**, 54–74.

Yamanaka, Y. (2006). Website EIC note of seismology, Earthquake Research Institute, University of Tokyo, 23 February 2006. In *Catastrophe caused by the 17 February 2006 southern Leyte landslide in the Philippines*, ed. Suwa, H. *Journal of the Japanese Society Natural Disaster Science*, **25**, 83–97.

Zhao, J., Liu, C. and Dai, L. T. (2012). Identification of landslide spatial distribution and their types along the Riviere Frorse Drainage Basin triggered by the earthquake in Haiti on 12 January 2010. *Disaster Advances*, **5**(1), 5–13.

6

Global climate model and projected hydro-meteorological extremes in the future

Akio Kitoh

High-resolution climate models are vital to project quantifiable and reliable future changes in weather extremes, such as tropical cyclones and rain systems that cause heavy rainfall and strong winds. Global climate model simulations are performed for the present day (1979–2003) and the future (2075–2099) under the Special Report on Emission Scenario A1B scenario using 20-km and 60-km mesh atmospheric models that have not been used before in global warming research. The 20-km mesh MRI-AGCM3 enabled the representation of the global spatial distribution of tropical cyclones and the frequency distribution of intense tropical cyclones. The projected number of geneses of tropical cyclones in the future is found to be statistically significantly decreasing in the globe and both hemispheres. On the other hand, the genesis frequency of intense tropical cyclones will increase. In order to assess the uncertainty of climate change projections, twelve ensemble projections combining three different cumulus schemes and four different sea surface temperature patterns were conducted using the 60-km mesh MRI-AGCM3. The occurrence frequency of tropical cyclones decreases in the West Pacific, South Pacific, and South Indian Ocean and increases in the Central Pacific around Hawaii, regardless of the experimental settings. Further dynamical downscaling using a regional climate model, focusing on Japan and the surrounding region, is also presented.

6.1 Introduction

In the Fourth Assessment Report of the Intergovernmental Panel on Climate Change (IPCC), a dataset of more than 20 global coupled atmosphere–ocean general circulation models (AOGCMs) under the Coupled Modeling Intercomparison Project phase 3 (CMIP3) are utilised to project future climate changes for various scenarios (IPCC, 2007). These results give good information on future climate changes, but the low resolution of the models has an undesirable impact on the results particularly for extreme weather events, because the horizontal resolution of these models is about 100–400 km. The resolution of the new CMIP5 models is finer (60–200 km), but it is still insufficient to study extreme weather events and to project their modification by climate changes.

Higher horizontal resolution models are thus needed to better reproduce the extreme rainfall associated with tropical cyclones and precipitation systems such as the Meiyu/Baiu rain band. To this end, dynamical downscaling approaches have been developed, including regional climate models (RCM), variable-resolution atmospheric general circulation models (AGCM), and global high-resolution AGCMs. At the Meteorological Research Institute (MRI), Tsukuba, Japan, both a global high-resolution AGCM and an RCM have been developed (Kitoh *et al.*, 2009).

A new global model (MRI-AGCM3.2) has been developed by Mizuta *et al.* (2012). This model shows improvements in simulating heavy precipitation around the tropical western Pacific, the global distribution of tropical cyclones, the seasonal march of the East Asian summer monsoon, and blockings in the North Pacific. Climate projection is not deterministic due to the imperfect representation of various processes such as cumulus convection in climate models and the atmospheric nature of chaos. Thus quantifying uncertainty in projections of future climate changes is a critical issue. Differences in the physical parameterisation employed in the models as well as the differences in sea surface temperature (SST) changes projected by AOGCMs could lead to uncertainty in future

Extreme Natural Hazards, Disaster Risks and Societal Implications, eds. Alik Ismail-Zadeh, Jaime Urrutia Fucugauchi, Andrzej Kijko, Kuniyoshi Takeuchi and Ilya Zaliapin. Published by Cambridge University Press.

precipitation projections. Using the MRI-AGCM3.2, climate projections have been performed for the end of the twenty-first century using the 60-km-mesh model under a multi-physics multi-SST framework (twelve ensemble experiments) as well as using the 20-km-mesh model. Using this approach differences of impacts due to the model physics and prescribed future SST changes on projected future changes can be addressed (Endo *et al.*, 2012; Murakami *et al.*, 2012b). In this chapter state-of-the-art high-resolution climate models and results on hydro-meteorological extremes that are projected for the end of the twenty-first century, are presented.

6.2 Methods

6.2.1 Global Climate Model (GCM)

There are two versions of MRI-AGCM (versions 3.1 and 3.2). Version 3.1 (v3.1) is based on the model developed by the Japan Meteorological Agency (JMA) and MRI (Mizuta *et al.*, 2006). Mizuta *et al.* (2006) showed that simulated global distributions of the seasonal mean atmospheric circulation fields, surface air temperature, and precipitation agree well with the observations. Moreover, the model improves the representation of regional-scale phenomena and local climate by increasing horizontal resolution to better express topographical effects and physical processes without degrading the representation of the global climate. This model was used for the 20-km climate projections, addressing future changes in extreme events such as precipitation extremes (Kamiguchi *et al.*, 2006), tropical cyclones (Oouchi *et al.*, 2006; Murakami *et al.*, 2011), extratropical cyclones (Mizuta *et al.*, 2011), blocking (Matsueda *et al.*, 2009), the Baiu rain band (Kusunoki and Mizuta, 2008; Kusunoki *et al.*, 2011), and monsoons (Rajendran and Kitoh, 2008).

The new version 3.2 (v3.2) is the atmospheric component of MRI-ESM1 (Yukimoto *et al.*, 2011), the ESM of MRI. The dynamical framework has not been changed from that used for v3.1, which is a hydrostatic primitive equation system employing a spectral transform method of spherical harmonics. Vertical levels have been changed from 60 layers (top at 0.1 hPa) in the previous version to 64 layers (top at 0.01 hPa) in the new version. Only levels above the tropopause have been changed. Some physical processes are replaced with new schemes in v3.2. Details are described in Mizuta *et al.* (2012). Among them, a new cumulus parameterisation scheme largely changes tropical precipitation and tropical cyclone climatology. The v3.1 is equipped with a prognostic Arakawa and Schubert (AS) cumulus convection scheme (Randall and Pan, 1993), in which multiple convective updraughts with different heights are explicitly calculated within a single grid cell, although each updraught is a simplified entraining plume. In the new version, v3.2, a Tiedtke (1989)-type scheme has been developed and introduced. In the original Tiedtke scheme, only a single convective updraught is calculated within a single grid cell, but is represented as a more detailed entraining and detraining plume than in the AS scheme. However, in v3.2, a new cumulus convection scheme, called the 'Yoshimura scheme (YS)' (Yukimoto *et al.*, 2011; Mizuta *et al.*, 2012), is introduced. The YS scheme is a modified Tiedtke-type scheme, in which detailed entraining and detraining plumes are considered for multiple plumes with different entraining rates. In other words, the new scheme makes use of the benefits of both the AS and the original Tiedtke-type schemes. Mizuta *et al.* (2012) verified the present day climate projections, showing improvements in heavy monthly-mean precipitation around the tropical western Pacific, the seasonal march of the East Asian summer monsoon, and blockings in the Pacific.

In this study, 20-km mesh and 60-km mesh versions of the MRI-AGCM3 are used. The 20-km mesh model employs a triangular truncation of 959 with a linear Gaussian grid (T_L959) in the horizontal. The transform grid uses 1920 × 960 grid cells. The 60-km mesh model has 640 × 320 grid cells (T_L159). The 60-km mesh model is equipped with multiple cumulus convection schemes. Three cumulus convection schemes are used to develop the multi-physics ensemble simulations: the AS scheme; the YS scheme; and the Kain–Fritsch (KF) convection scheme (Kain and Fritsch, 1993).

6.2.2 Regional Climate Model (RCM)

A non-hydrostatic model with a horizontal resolution of 5 km (NHM-5km), which is a climate model version of the operational model of JMA (Saito *et al.*, 2007) is employed. The model uses a cloud microphysics scheme, but also uses the KF scheme. The experimental domain is 669 × 594 horizontal grids and 50 vertical layers (top height at 22 km), covering a wide region from the eastern part of the Asian Continent to the Japanese Islands. The spectral boundary coupling method (Kida *et al.*, 1991) is employed to reduce phase errors between the outer model and NHM-5km. Kanada *et al.* (2008) and Nakano *et al.* (2012) gave the model performance in the reproducibility of extreme precipitation over Japan by the NHM-5km.

6.2.3 Experiment

A schematic of the experiment is shown in Fig. 6.1. The present day realisation is a so-called 'AMIP'-style simulation, where the observed monthly SST and sea ice concentration during 1979–2003 (HadISST1; Rayner *et al.*, 2003) are prescribed as the lower boundary

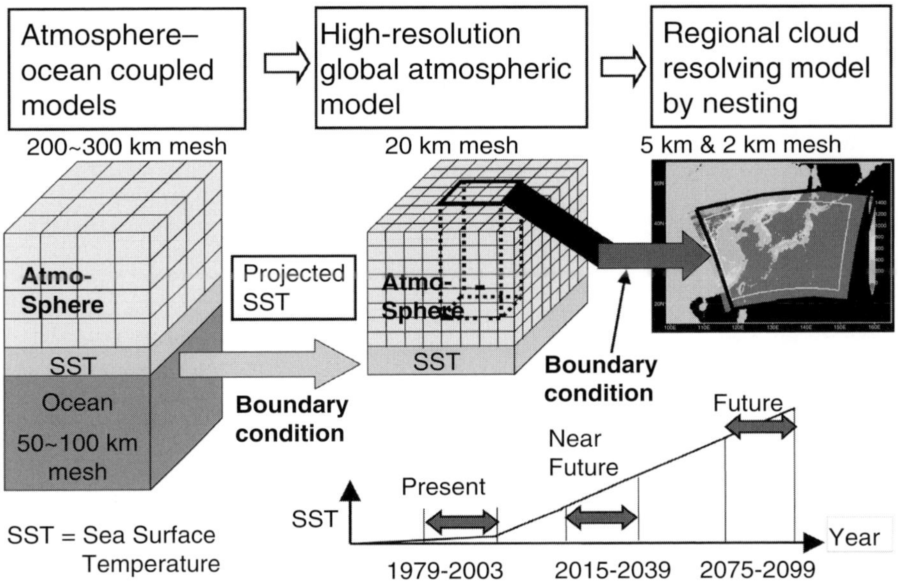

Figure 6.1. A schematic of the experiment.

conditions. For the future globally-warmed simulation, the targeted projection period is the last quarter of the twenty-first century (2075–2099). For the lower boundary conditions in the global warming simulation, future changes in SST and sea ice were estimated from the CMIP3 multi-model ensemble means under the Special Report on Emission Scenario (SRES) A1B scenario to which detrended interannual variations in observed SST were added (Mizuta *et al.*, 2008), assuming that interannual variations in SST and sea ice in the global warming simulation are similar to those of the present day. This assumption was made because there is large uncertainty in future changes in the SST spatial pattern. The design retains observed year-to-year variability and El Niño and Southern Oscillation events in the future climate, but with a higher mean and a clearly increasing trend in SST. The changes in global annual mean 25-year averaged SST prescribed are 2.17 °C. Note that SST warming is inhomogeneous spatially (Xie *et al.*, 2010): a greater warming occurs in the northern than in the southern tropics. The SST also increases more in the central Pacific and eastern Pacific, which is similar to the El Niño events.

Integration of NHM-5km was conducted from June to October by nesting within the results of the 25-year periods using the 20-km AGCM for the present and future climates. To assess the uncertainty of climate change projections, ensemble simulations were performed with the 60-km resolution model: multi-physics (AS, YS, and KF) and multi-SST ensemble simulations. For the multi-SST simulations, in addition to the CMIP3 multi-model ensemble mean SST pattern, three different SST patterns were used. They were created using a cluster analysis, where normalised tropical SST anomalies derived from the 18 CMIP3 models were grouped to avoid the subjective selection of a single model (Murakami *et al.*, 2012a).

6.2.4 Observed precipitation data for validation

For model verification three precipitation datasets were used. They were Climate Prediction Center Merged Analysis of Precipitation (CMAP: Xie and Arkin, 1997), GPCP_1DD_v1.1 by the Global Precipitation Climatology Project (GPCP: Huffman and Bolvin, 2009), and Tropical Rainfall Measuring Mission (TRMM) PR3A25 V6 dataset (TRMM 3A25: Iguchi *et al.*, 2000). For the observed climatology, a 25-year (1979–2003) mean for CMAP on a 2.5° latitude–longitude grid, a 12-year (1997–2008) mean for GPCP1dd on a 1.0° grid, and an 8-year (1998–2005) mean for TRMM 3A25 on a 0.5° grid were calculated.

6.3 Results

6.3.1 Present day climate

The present day climate simulated by the 20-km mesh MRI-AGCM3.2 was evaluated in Mizuta *et al.* (2012). The global distribution of precipitation compared better with observations in MRI-AGCM3.2 than those in MRI-AGCM3.1. The seasonal cycle of the East Asian summer monsoon including the Meiyu-Baiu characteristics was well simulated. The model also shows improvements in the global distribution and intensity of tropical cyclones (Murakami *et al.*, 2012b).

Figure 6.2 compares the geographical distributions of annual precipitation climatology (25-year average of the present day simulation) for the 180-, 60- and 20-km

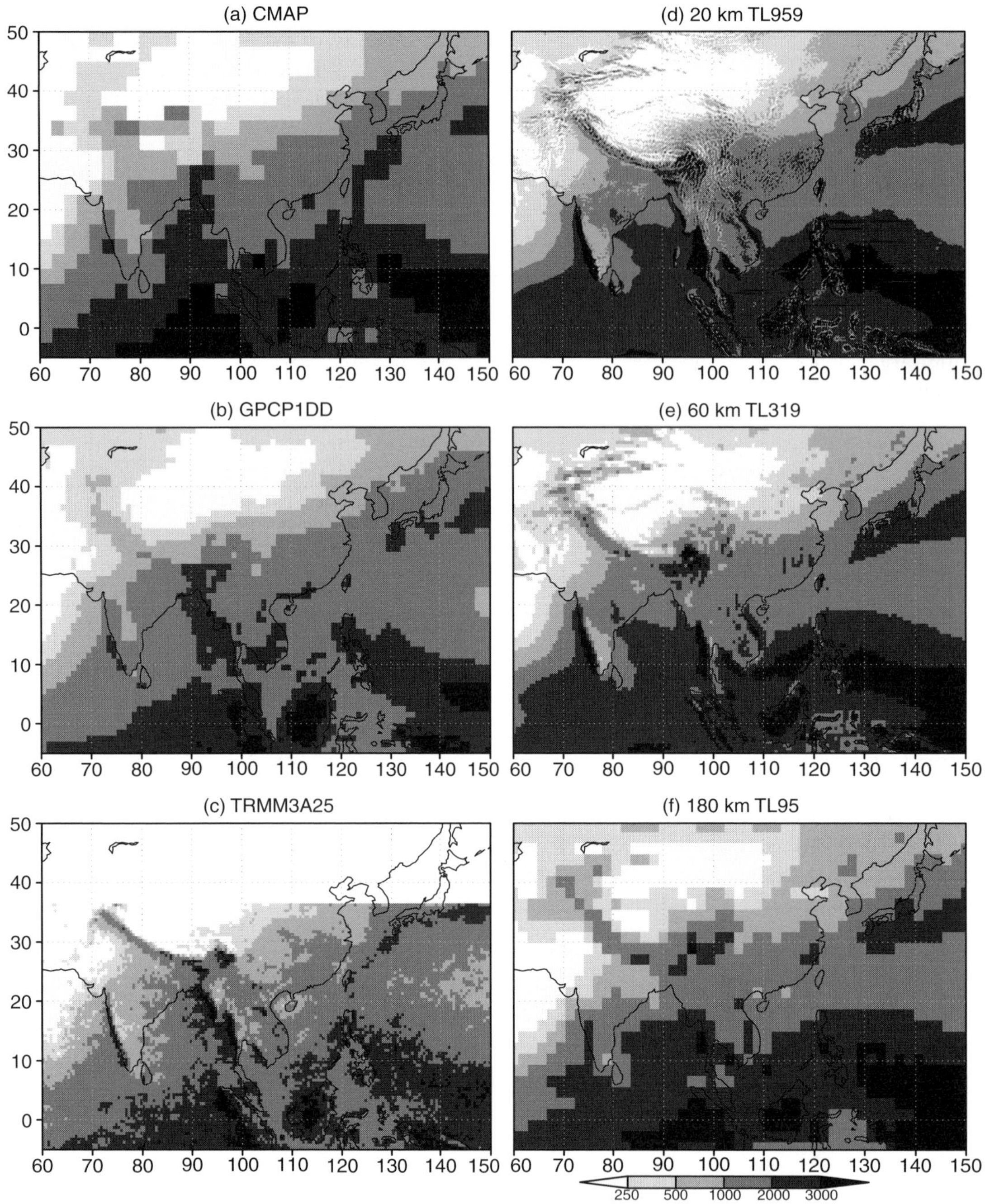

Figure 6.2. Global distribution of annual precipitation climatology (mm) for (a)–(c) observations and (d)–(f) model simulations. (a) CMAP°, (b) GPCP 1° grid data, (c) TRMM 3A25, (d) 20-km model, (e) 60-km model, (f) 180-km model.

models with three observed climatologies: CMAP, GPCP, and TRMM 3A25. Observations show major convection centres over the Philippines Sea and the South China Sea as well as over the equatorial Indian Ocean off Sumatra. Due to the dominant westerly winds in the summer monsoon season, there are precipitation maxima along the western side of southern India and Myanmar. This feature becomes clearer as the resolution of data improves from CMAP (2.5°) through GPCP (1.0°) to TRMM (0.5°).

All three versions of the MRI-AGCM3.2 reproduced the general features of annual mean precipitation in this region. There are high rainfalls over the western side of mountains of the Western Ghats, the Malay Peninsula and the Philippines together with lee-wind side precipitation minima. Compared with the lower resolution versions, the 20-km model more successfully simulated narrow precipitation bands windward of the mountains and contrasting low precipitation areas in the lee-wind side. Orographic precipitation along the Himalayas is also seen (Yatagai *et al.*, 2005). The coarser-resolution datasets (CMAP and GPCP) did not show such a sharp horizontal contrast in precipitation. It is noted that the 180-km model overestimated precipitation around 100° E, 30° N, which was remedied in the 20-km model. The rain band associated with Meiyu-Baiu is seen over southeast China and to the south of Japan.

6.3.2 Asian monsoon precipitation in the future

Asia is one of the most water-rich regions of the world where billions of people depend on the water for their life. The region is sometimes inundated with too much water, which results in disastrous floods particularly in the more vulnerable countries. The monsoon brings about the rainfall in this region. Therefore, reliable projections of future precipitation changes associated with the monsoon are required for both mean precipitation and the degree of variability. Past assessments reported a very likely increase in the frequency of intense precipitation events in East Asia and in parts of South Asia (Seneviratne *et al.*, 2012). Kusunoki and Arakawa (2012) obtained an increase in precipitation intensity in June–July over almost all regions of East Asia, based on the AOGCM projections of the CMIP3 dataset. However, the horizontal resolution of these models might be too coarse to fully represent orographic rainfall and precipitation extremes.

Endo *et al.* (2012) investigated future changes in mean and extreme precipitation over the Asian monsoon region. They used 20-km mesh MRI-AGCM3.2 simulations for the present day (1979–2003) and the future (2075–2099). To assess the uncertainty of the projections, 12 ensemble projections (i.e., combinations of three different cumulus schemes and four different SST change patterns) were conducted using 60-km-mesh AGCMs. For the future climate simulations, an overall agreement was found between the 20-km and 60-km model projections. In south Asia and southeastern Asia, mean and extreme precipitation generally increased, but their changes showed marked differences among the projections, suggesting these trends include some uncertainty over these regions. In east Asia, northwestern China, and Bangladesh, in contrast, mean and extreme precipitation showed consistent increases in the projections, suggesting these increases are reliable for this model framework over these regions.

Figure 6.3 shows the changes in the simple daily precipitation index (SDII) for the 60- and 20-km mesh models. The SDII is defined as the annual total precipitation divided by the number of rainy days, where 'a rainy day' is defined as a day with precipitation greater than or equal to 1 mm day^{-1}. Compared with the case for mean precipitation, areas with significant positive changes were more extensive, particularly over land (Endo *et al.*, 2012). At the end of the twenty-first century, the SDII became larger and increased more significantly over land areas, increasing almost everywhere with the exception of some oceanic areas. Large and significant increases in SDII were projected in Bangladesh, southern China, and western Japan. The last corresponds to heavier rainfall in the Meiyu-Baiu region in the future. In southeast Asia, most land areas will experience increasing intensity of

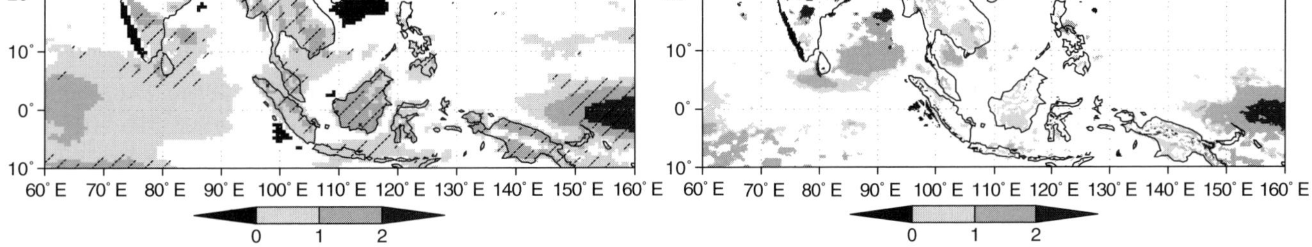

Figure 6.3. Changes in SDII (mm day^{-1}) between the present and the end of the twenty-first century for (a) the 60-km model and (b) the 20-km model. For the 60-km model, areas where more than 10 of the 12- experiment project the same sign are hatched. For the 20-km model, statistically significant changes are shaded.

precipitation, which is contrasted with adjacent oceans where changes are not significant or are sometimes negative. Endo *et al.* (2012) showed that the percentage increase of extreme precipitation differs between indices. It is larger for the R5d (maximum 5-day precipitation intensity index) and more for the R95T (fraction of annual precipitation due to very wet days exceeding the ninety-fifth percentile in the present climate) than the mean precipitation. Therefore, the more intense the extremes, the larger were the percentage increases in the warming world.

There are some exceptional regions where SDII decreases. The Western Ghats in southwestern India is such a region. Rajendran *et al.* (2012) investigated the issue of likely future changes in the summer monsoon over the Western Ghats orographic region of India in response to global warming. The 20-km mesh MRI-AGCM resolved orographic features on fine spatial scales leading to a quasi-realistic simulation of the seasonal variation and spatial distribution of the present day summer monsoon rainfall over India. The seasonal mean summer monsoon rainfall, the intensity of the Indian summer monsoon circulation, and the vertical velocities over the Western Ghats orographic region have weakened significantly during the recent past and the model simulated these features realistically. The reduced orographic precipitation over the narrow Western Ghats mountains south of 16° N by the end of the twenty-first century as projected by the 20-km model was found to be associated with a drastic reduction in the southwesterly winds and moisture transport into the region, and weakening of the summer monsoon Hadley circulation and vertical velocities in response to global warming. The drastic reduction of vertical ascent and weakening of large-scale monsoon circulation due to the 'upper tropospheric warming effect' predominates over the moisture build-up effect in reducing the rainfall over this narrow orographic region.

6.3.3 Tropical cyclones

In order to investigate future changes of tropical cyclone activity under a warming world, a model with a horizontally fine grid is necessary and a 20-km-mesh AGCM has been developed for this purpose. Figure 6.4(a) shows the observed distribution of tropical cyclone tracks during the period 1979–2003. The average annual number of observed tropical cyclones during this period was 84.8. More than 30% of tropical cyclones were observed in the western North Pacific. The 20-km mesh MRI-AGCM3.1 (Fig. 6.4(b)) underestimated the number of tropical cyclone geneses in the western North Pacific. This model also underestimated tropical cyclone strength. The improved model (MRI-AGCM3.2) enabled the representation of the global distribution of observed tropical cyclones and extremely strong tropical cyclones, such as those of Categories 4 and 5 (Figure 6.4(c)), which were impossible to simulate by conventional GCMs. With this model, extremely strong tropical cyclones were simulated for research using a global climate model.

As a result of the future warming projection experiment using the improved model, the projected number of geneses of tropical cyclones was found to be statistically significantly decreasing around the globe and in both Southern and Northern Hemispheres (Murakami *et al.*, 2012b; Sugi *et al.*, 2012). When viewed by region, the projected genesis frequency will decrease statistically significantly in the northwest Pacific and South Pacific and will increase in the tropical central Pacific. On the other hand, the genesis frequency of intense tropical cyclones will increase. Although these results are consistent with previous studies (Oouchi *et al.*, 2006; Sugi *et al.*, 2009; Murakami *et al.*, 2011), the increase in the genesis frequency in the improved model is limited to extremely strong tropical cyclones with a maximum wind speed of 70 m s^{-1} or more, i.e., Category 5 cyclones. The projected future changes in the occurrence frequency distribution of tropical cyclones of Category 5 increase from Taiwan to the southern coasts of the Japanese Archipelago. Although this suggests the possibility of an increase in the frequency of tropical cyclones of extremely strong intensity approaching Japan in the future, it is possible that the frequency of extremely strong typhoons in the vicinity of Japan has been overestimated due to a bias in that the occurrence frequency distribution of tropical cyclones in the present day simulation shows a northward shift relative to observations, probably related to a missing cooling process of SST by air–sea interaction in this experimental setting.

The uncertainty in the future projection of tropical cyclones has been investigated by multi-physics and multi-SST ensemble simulations using the 60-km mesh MRI-AGCM3.2 (Murakami *et al.*, 2012a). As a tropical cyclone is a convective system in which individual cumulus convections gather together, a factor in the uncertainty lies in the cumulus convection scheme among the physical processes. Also, as tropical cyclones appear over oceans with high SSTs, a difference in future changes in SSTs would affect future changes in tropical cyclone activity.

The means of the twelve ensemble future changes in the occurrence frequency of tropical cyclones are shown in Fig. 6.5. The crosses and open circles in Fig. 6.5 indicate consistent future changes in which more than 10 of 12 experiments have the same sign in changes (crosses indicate negative signs, and open circles positive ones). The occurrence frequency of tropical cyclones decreased in the west Pacific, South Pacific, and south Indian Ocean and

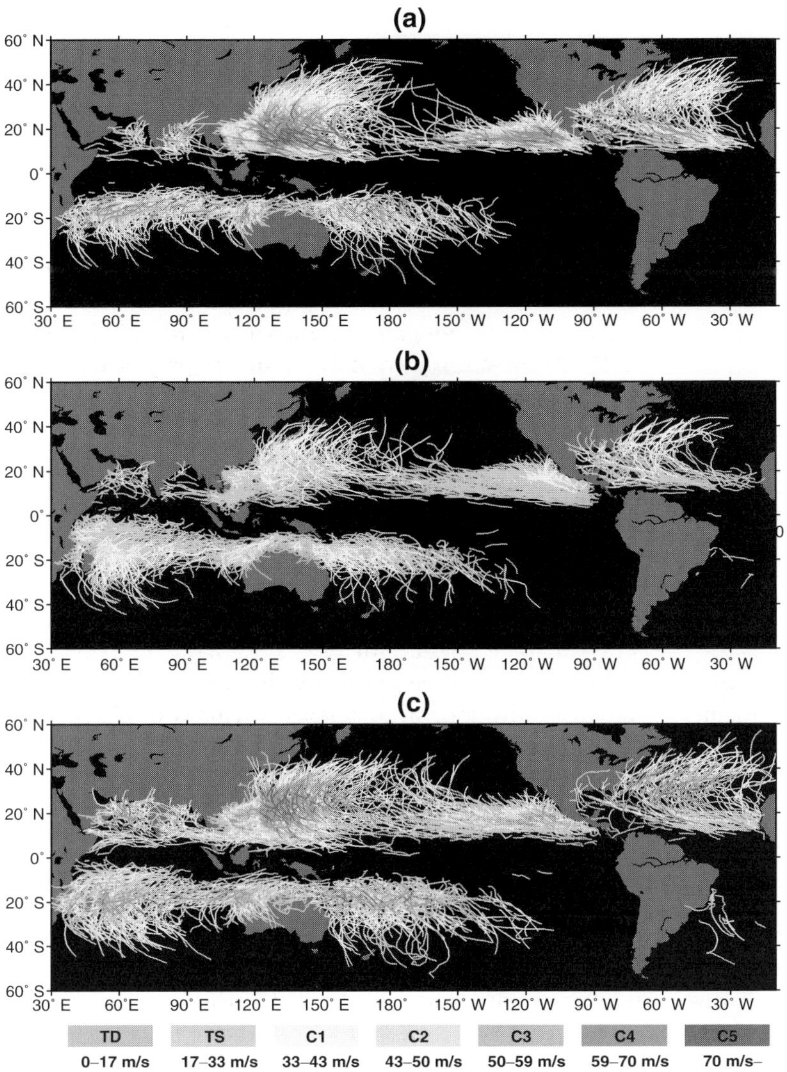

Figure 6.4. Global distributions of tropical cyclones for (a) observations, (b) the present day simulation using the MRI-AGCM3.1S, and (c) using the MRI-AGCM3.2S. Tropical cyclone tracks are coloured according to their intensity. For colour version, see the Plates section.

increased in the Central Pacific around Hawaii, regardless of the experimental settings. It was also found that the differences in the SSTs contributed to the differences between the experiments, expanding the range of the uncertainty (Murakami *et al.*, 2012a). This suggests that a projection of SSTs with a high degree of accuracy is important to reduce the uncertainty in future projections of tropical cyclones.

6.3.4 Further downscaling with the use of RCM

The global 20-km mesh AGCM can be used as the lateral boundary conditions for further dynamical downscaling of regions of interest. The global 20-km mesh AGCM itself is not yet sufficient to reproduce very intense precipitation. An NHM using a cloud microphysics scheme with a horizontal resolution on the order of a few kilometres would be needed for such a purpose (Kanada *et al.*, 2008). A cloud-system resolving RCM with a horizontal resolution of 5 km (NHM-5km) has performed dramatically well in reproducing extreme indices of precipitation, compared with the global atmospheric model, and enables discussion on more realistic precipitation extremes (Kanada *et al.*, 2010a). Changes in the quality of precipitation phenomena in the vicinity of Japan have been examined, focusing particularly on heavy rainfall and intense precipitation, through nesting the cloud-system resolving RCM into the results of the warming projection experiment with the high-resolution global atmospheric model.

Kanada *et al.* (2010b, 2012) investigated future changes in the characteristics of the Baiu frontal zone and precipitation. First, NHM-5km very realistically simulated the

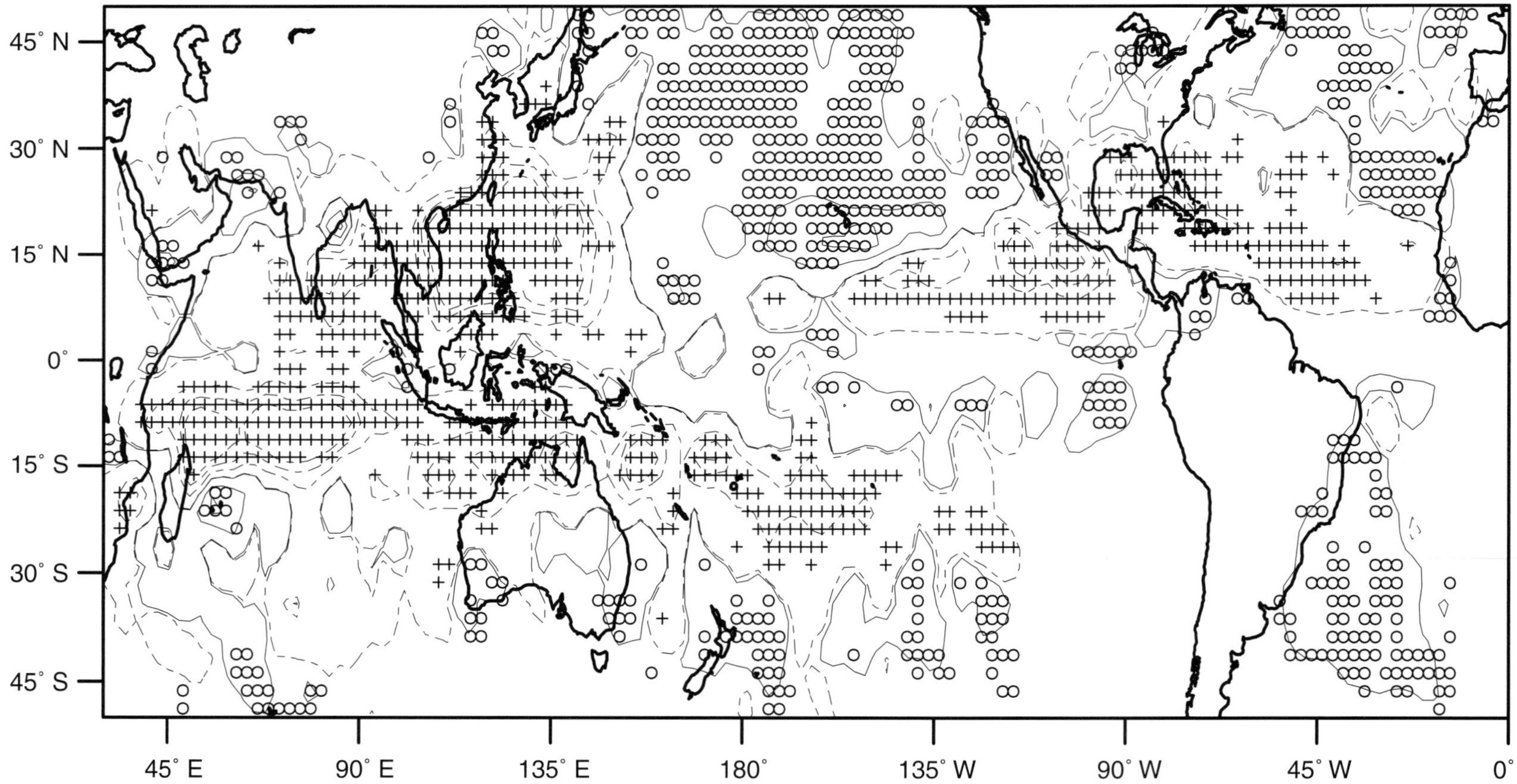

Figure 6.5. Ensemble mean future changes in tropical cyclone frequency. The contour interval is the number per 25 years. An open circle indicates that more than 10 out of 12 experiments project a positive sign, while a cross indicates that more than 10 out of the 12 experiments project a negative sign.

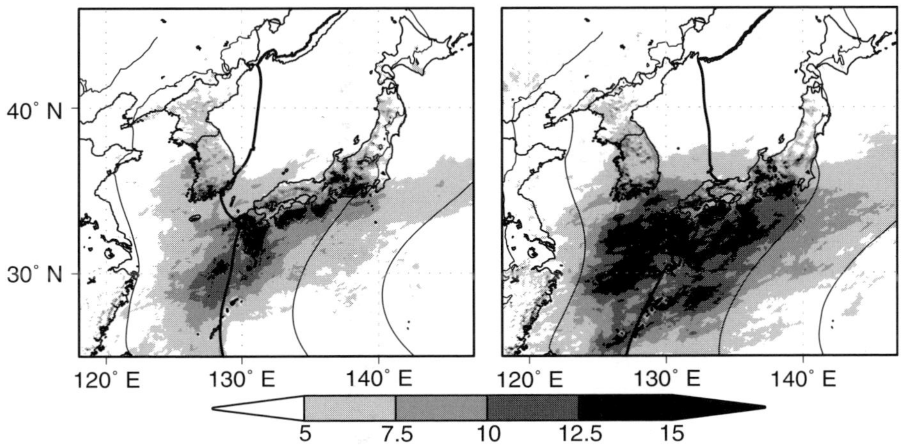

Figure 6.6. Horizontal distributions of mean precipitation (mm day^{-1}) and sea level pressure (hPa) between 1–10 July in (a) the present day (1979–2003) and (b) future climates (2075–2099) projected by NHM-5km.

seasonal variation of precipitation around western Japan, such as the peak during the Baiu period from mid-June to mid-July, the low rainfall in August, and the peak due to autumnal rain and typhoons from September to October. In the future climate, a significant increasing tendency in precipitation, compared with the present day climate, was projected around western Japan in early July, which corresponds to the late Baiu season (Fig. 6.6). The Baiu front in the future climate was characterized by the tall and intense vertical velocity along an intensified frontal zone with intense mean updraughts and large amounts of rainwater. The increases in the supply of water vapour from the southern side of the front and the intensified convective activity along the Baiu front were supposed to contribute to the projected increase of intense precipitation. During the late Baiu period, not only did the precipitation increase quantitatively, but also the percentage of the precipitation caused by heavy rainfall with daily precipitation above 100 mm of the total amount increased from around 10% to around 15%. In other words, it suggests the possibility that while the precipitation caused around western Japan in the latter period of the Baiu period increases, the frequency of intense precipitation will increase.

6.4 Concluding remarks

This study used the 20-km mesh global model and the 5-km mesh cloud-resolving regional model for the climate change projections. These models were developed based on the JMA operational global numerical forecast model for the short- and medium-range forecast and the regional meso-scale model for the very short-range forecast, respectively. The global and regional super-high-resolution models, which are able to simulate present weather realistically, project a more frequent occurrence of weather extremes in the future. These results together with information on the uncertainty are significant for decision-making processes. These data have been used for various application studies on disaster prevention, such as changes of disaster environment regarding landslide, debris flow, flood, drought, storm surge and strong wind, in collaboration with other institutes, nationally and internationally.

Acknowledgements. This work was done under the framework of the KAKUSHIN and SOUSEI programmes funded by MEXT. The calculations were performed on the Earth Simulator.

6.5 Abbreviations

AGCM	Atmospheric general circulation model
AOGCM	Atmosphere–ocean general circulation model
AS	Arakawa and Schubert
CMAP	NOAA NCEP Climate Prediction Center Merged Analysis of Precipitation
CMIP	Coupled Modelling Intercomparison Project
ESM	Earth system model
GCM	Global climate model
GPCP	Global Precipitation Climatology Project
IPCC	Intergovernmental Panel on Climate Change
JMA	Japan Meteorological Agency
KF	Kain–Fritsch
MRI	Meteorological Research Institute
NHM	Non-hydrostatic model
RCM	Regional climate model
SDII	Simple daily precipitation intensity index
SRES	Special Report on Emission Scenario
SST	Sea surface temperature
TRMM	Tropical Rainfall Measuring Mission
YS	Yoshimura scheme

References

Endo, H., Kitoh, A., Ose, T., Mizuta, R. and Kusunoki, S. (2012). Future changes and uncertainties in Asian precipitation simulated by multi-physics and multi-sea-surface temperature ensemble experiments with high-resolution Meteorological Research Institute atmospheric general circulation models (MRI-AGCMs). *Journal of Geophysical Research*, **117**, D16118, doi:10.1029/2012JD017874.

Huffman, G. J. and Bolvin, D. T. (2009). GPCP one-degree daily precipitation data set documentation. ftp://precip.gsfc.nasa.gov/pub/1dd-v1.1/1DD_v1.1_doc.pdf

Iguchi, T., Kozu, T., Meneghini, R., Awaka, J. and Okamoto, K. (2000) Rain-profiling algorithm for the TRMM Precipitation Radar. *Journal of Applied Meteorology*, **39**, 2038–2052.

IPCC (2007). *Climate Change 2007: The Physical Science Basis. Contribution of Working Group I to the Fourth Assessment Report of the Intergovernmental Panel on Climate Change*, eds. Solomon, S, Qin, O., Manning, M. *et al.* Cambridge, UK: Cambridge University Press.

Kain, J. and Fritsch, J. (1993). Convective parameterization for meso-scale models: The Kain–Fritsch scheme. In: *The Representation of Cumulus Convection in Numerical Models.* Meteorological Monographs, No 24, American Meteorological Society, pp. 165–170.

Kamiguchi, K., Kitoh, A., Uchiyama, T., Mizuta, R. and Noda, A. (2006). Changes in precipitation-based extremes indices due to global warming projected by a global 20-km-mesh atmospheric model. *Scientific Online Letters on the Atmosphere*, **2**, 64–67.

Kanada, S., Nakano, M., Hayashi, S. *et al.* (2008). Reproducibility of maximum daily precipitation amount over Japan by a high-resolution non-hydrostatic model. *Scientific Online Letters on the Atmosphere*, **4**, 105–108.

Kanada, S., Nakano, M. and Kato, T. (2010a). Climatological characteristics of daily precipitation over Japan in the Kakushin regional climate experiments using a non-hydrostatic 5-km-mesh model: comparison with an outer global 20-km-mesh atmospheric climate model. *Scientific Online Letters on the Atmosphere*, **6**, 117–120.

Kanada, S., Nakano, M. and Kato, T. (2010b). Changes in mean atmospheric structures around Japan during July due to global warming in regional climate experiments using a cloud-system resolving model. *Hydrological Research Letters*, **4**, 11–14.

Kanada, S., Nakano, M. and Kato, T. (2012). Projections of future changes in precipitation and the vertical structure of the frontal zone during the Baiu season in the vicinity of Japan using a 5-km-mesh regional climate model. *Journal of the Meteorological Society of Japan*, **90A**, 65–86.

Kida, H., Koide, T., Sasaki, H. and Chiba, M. (1991). A new approach to coupling a limited area model to a GCM for regional climate simulations. *Journal of the Meteorological Society of Japan*, **69**, 723–728.

Kitoh, A., Ose, T., Kurihara, K. *et al.* (2009). Projection of changes in future weather extremes using super-high-resolution global and regional atmospheric models in the KAKUSHIN Program: results of preliminary experiments. *Hydrological Research Letters*, **3**, 49–53.

Kusunoki, S. and Arakawa, O. (2012). Change in the precipitation intensity of the East Asian summer monsoon projected by CMIP3 models. *Climate Dynamics*, **38**, 2055–2072.

Kusunoki, S. and Mizuta, R. (2008). Future changes in the Baiu rainband projected by a 20-km mesh global atmospheric model: sea surface temperature dependence. *Scientific Online Letters on the Atmosphere*, **4**, 85–88.

Kusunoki, S., Mizuta, R. and Matsueda, M. (2011). Future changes in the East Asian rainband projected by global atmospheric models with 20-km and 60-km grid size. *Climate Dynamics*, **37**, 2481–2493.

Matsueda, M., Mizuta, R. and Kusunoki, S. (2009). Future change in wintertime atmospheric blocking simulated using a 20-km-mesh atmospheric global circulation model. *Journal of Geophysical Research*, **114**, D12114.

Mizuta, R, Oouchi, K., Yoshimura, H. *et al.* (2006). 20-km-mesh global climate simulations using JMA-GSM model – Mean climate states. *Journal of the Meteorological Society of Japan*, **84**, 165–185.

Mizuta, R., Adachi, Y., Yukimoto, S. and Kusunoki, S. (2008). *Estimation of the Future Distribution of Sea Surface Temperature and Sea Ice using the CMIP3 Multi-Model Ensemble Mean. Technical Report, Meteorological Research Institute*, No 56, 28 pp.

Mizuta, R., Matsueda, M., Endo, H. and Yukimoto, S. (2011). Future change in extratropical cyclones associated with change in the upper troposphere. *Journal of Climate*, **24**, 6456–6470.

Mizuta, R., Yoshimura, H., Murakami, H. *et al.* (2012). Climate simulations using MRI-AGCM3.2 with 20-km grid. *Journal of the Meteorological Society of Japan*, **90A**, 233–258.

Murakami, H., Wang, B. and Kitoh, A. (2011). Future change of western North Pacific typhoons: Projections by a 20-km-mesh global atmospheric model. *Journal of Climate*, **24**, 1154–1169.

Murakami, H., Mizuta, R. and Shindo, E. (2012a). Future changes in tropical cyclone activity projected by multi-physics and multi-SST ensemble experiments using the 60-km-mesh MRI-AGCM. *Climate Dynamics*, **39**, 2569–2584.

Murakami, H., Wang, Y., Yoshimura, H. *et al.* (2012b). Future changes in tropical cyclone activity projected by the new high-resolution MRI-AGCM. *Journal of Climate*, **25**, 3237–3260.

Nakano, M., Kato, T., Hayashi, S. *et al.* (2012). Development of a 5-km-mesh cloud-system-resolving regional climate model at the Meteorological Research Institute. *Journal of the Meteorological Society of Japan*, **90A**, 339–350.

Oouchi, K., Yoshimura, J., Yoshimura, H. *et al.* (2006). Tropical cyclone climatology in a global-warming climate as simulated in a 20 km-mesh global atmospheric model: frequency and wind intensity analyses. *Journal of the Meteorological Society of Japan*, **84**, 259–276.

Rajendran, K. and Kitoh, A. (2008). Indian summer monsoon in future climate projection by a super high-resolution global model. *Current Science*, **95**, 1560–1569.

Rajendran, K., Kitoh, A., Srinivasan, J., Mizuta, R. and Krishnan, R. (2012). Monsoon circulation interaction with Western Ghats orography under changing climate – Projection by a 20-km mesh AGCM. *Theoretical and Applied Climatology*, **110**, 555–571.

Randall, D. and Pan, D.-M. (1993). Implementation of the Arakawa-Schubert cumulus parameterization with a prognostic closure. In: *The Representation of Cumulus Convection in Numerical Models.* Meteorological Monograph, No 46. Boston, MA: American Meteorological Society, pp. 145–150.

Rayner, N. A., Parker, D. E., Horton, E. B. *et al.* (2003). Global analyses of sea surface temperature, sea ice, and night marine air temperature since the late nineteenth century. *Journal of Geophysical Research*, **108**, 4407, doi:10.1029/2002JD002670.

Saito, K., Ishida, J., Aranami, K. *et al.* (2007). Nonhydrostatic atmospheric models and operational development at JMA. *Journal of the Meteorological Society of Japan*, **85B**, 271–304.

Seneviratne, S. I., Nicholis, N., Easterling, D. *et al.* (2012). Changes in climate extremes and their impacts on the natural physical environment. In: *Managing the Risks of Extreme Events and Disasters to Advance Climate Change Adaptation*. Eds. Field, C.B., Barros, V., Stocker, T.F. *et al.* A Special Report of Working Groups I and II of the Intergovernmental Panel on Climate Change (IPCC). Cambridge, UK: Cambridge University Press, pp. 109–230.

Sugi, M., Murakami, H. and Yoshimura, J. (2009). A reduction in global tropical cyclone frequency due to global warming. *Scientific Online Letters on the Atmosphere*, **5**, 164–167.

Sugi, M., Murakami, H. and Yoshimura, J. (2012). On the mechanism of tropical cyclone frequency changes due to global warming. *Journal of the Meteorological Society of Japan*, **90A**, 397–408.

Tiedtke, M. (1989). A comprehensive mass flux scheme for cumulus parameterization in large-scale models. *Monthly Weather Review*, **117**, 1779–1800.

Xie, P. and Arkin, P. A. (1997). Global precipitation: A 17-year monthly analysis based on gauge observations, satellite estimates, and numerical model outputs. *Bulletin of the American Meteorological Society*, **78**, 2539–2558.

Xie, S.-P., Deser, C., Vecchi, G. A. *et al.* (2010). Global warming pattern formation: Sea surface temperature and rainfall. *Journal of Climate*, **23**, 966–986.

Yatagai, A, Xie, P. and Kitoh, A. (2005). Utilization of a new gauge-based daily precipitation dataset over monsoon Asia for validation of the daily precipitation climatology simulated by the MRI/JMA 20-km-mesh AGCM. *Scientific Online Letters on the Atmosphere*, **1**, 193–196.

Yukimoto, S., Yoshimura, H., Hosaka, M. *et al.* (2011). *Meteorological Research Institute Earth System Model Version 1 (MRI-ESM1) – Model description. Technical Report, Meteorological Research Institute*, No. 64, 83 pp.

7

Physically-based hurricane risk analysis

Ning Lin, Kerry Emanuel, and Erik Vanmarcke

7.1 Introduction

Tropical cyclones (TCs) cause tremendous damage worldwide due to the associated strong winds, heavy rainfall, and storm surge. As the climate warms, these TC hazards may intensify (e.g., Emanuel *et al.*, 2008; Knutson *et al.*, 2010; Lin *et al.*, 2012), making their societal impacts an increasing concern (Mendelsohn et al., 2012). Events have revealed the vulnerability of the USA to severe TCs. In 2011, Hurricane Irene produced more than US$10 billion in damage in the northeastern USA. In 2012, Hurricane Sandy struck the Northeastern Seaboard, killing over 200 people, causing more than US$65 billion in damage, and leaving millions without electric service. Hurricane Katrina of 2005 resulted in more than 1800 fatalities and caused more than US$80 billion in losses (the costliest natural disaster in the US history). To prevent such TC disasters in the future, major advances in TC risk management are urgently needed.

Due to uncertainties in future climate and demographic conditions, including TC activity, sea-level rise (SLR), and exposure and vulnerability, effective TC risk management should be based on probabilistic risk assessment. Currently, increases in exposure appear to dominate over the effects of anthropogenic climate change as the leading cause of change in TC damage (Pielke, 2007, Handmer *et al.*, 2012). However, potential changes in storm characteristics and hazards due to climate forcing and related SLR also affect risk (Mendelsohn *et al.*, 2012). Accordingly, changes in the physical state of the atmosphere and ocean also must be taken into account. Thus reliable TC risk assessment, unlike traditional assessments, cannot rely solely on direct statistical analyses of the (quite limited) historical TC records and damage data. Rather, it requires a new physically-based approach that incorporates information on current and projected future climates as well as exposure and vulnerability.

Developing a physically-based risk assessment methodology is challenging, as multi-scale features of the phenomena and gaps in current research limit our knowledge of TC hazards and damage. For instance, TC research, especially that related to the influence of climate change, has focused on storm occurrence and intensification over oceans (e.g., Bender *et al.*, 2010; Knutson *et al.*, 2010, 2013), which may not directly and sufficiently inform the hazards at landfall (Lin *et al.*, 2012). Dynamic multi-model analyses by Bender *et al.* (2010) and by Knutson *et al.* (2013) projected a significant reduction in the Atlantic basin-wide hurricane frequency (−20% to −30%) but an increase in lifetime-maximum hurricane intensity (+4% to +6%) over the twenty-first century. Hurricane rainfall rates are projected to increase substantially (+10% to +30%). The frequency of very intense (Category 4–5) hurricanes is also projected to increase substantially (+40% to +90%). Questions remain about how such projected basin-wide changes translate into changes in landfalling TC features and hazards at more regional scales (i.e., along part of the US coastline).

In this chapter, we review studies that take a physically-based approach to better assess and manage hurricane risk. Such a methodology includes three components: modelling the storm climatology (which defines TC risk in terms of the upper tail of the storm statistics); modelling landfalling hazards; and characterising damage and losses. These three components are discussed in the next three sections, respectively, followed by a summary.

7.2 TC risk assessment

A challenge in TC risk assessment is that historical records of TCs, especially those making landfall in a local area, are

Extreme Natural Hazards, Disaster Risks and Societal Implications, eds. Alik Ismail-Zadeh, Jaime Urrutia Fucugauchi, Andrzej Kijko, Kuniyoshi Takeuchi and Ilya Zaliapin. Published by Cambridge University Press.

very limited. Statistical methods based solely on local TC landfall history are unreliable because the estimated frequency of high-intensity events is highly sensitive to exactly how the tail of the probability distribution is modelled. Hurricane return levels may be estimated from geological information and/or the historical data about hurricanes landfalling in neighbouring regions (Elsner *et al.*, 2008). More often, TC risk assessment makes use of Monte Carlo simulation to generate synthetic basin-wide storms to obtain landfall statistics; two principal approaches have been developed. One approach, pioneered by Vickery *et al.* (2000), uses the statistics of the storm parameters (i.e., track, intensity, and size) to construct synthetic storm sets (e.g., Powell *et al.*, 2005; Rumpf *et al.*, 2007; Hall and Jewson, 2007). A drawback of this method is that it relies entirely on historical TC data and cannot readily incorporate or predict future changes. An extension of this method is to develop statistical relationships between the storm parameters and environmental parameters (e.g., Yonekura and Hall, 2011) so that future storms may be simulated using projected future environments, assuming that the statistical relationships remain the same under different climate conditions. The other approach, developed by Emanuel *et al.* (2006, 2008), applies a deterministic, coupled ocean–atmosphere TC model to simulate storm development driven by large-scale environmental conditions whose statistics are derived from observations or global climate model data. This statistical/deterministic TC risk model does not rely on the limited historical TC data but generates large samples of synthetic storms that are in statistical agreement with observations (Emanuel *et al.*, 2006), and compares well with other methods used to study the effects of climate change on TCs (Emanuel *et al.*, 2008, 2010; Knutson *et al.*, 2010). This approach has been used to investigate TC wind and surge risk (e.g. Lin *et al.*, 2010a, 2012; Klima *et al.*, 2011; Mendelsohn *et al.*, 2012).

Here we briefly summarize the Emanuel and coworkers method for generating synthetic storms. Storm genesis points are randomly selected from a distribution constructed from the best-track historical dataset or generated by a random seeding technique. Once initiated, storm displacements are calculated using the 'beta and advection model' in which 850 and 250 mb environmental flows vary randomly but in accordance with the monthly mean, variance, and covariances of the observations or the climate model prediction. Figure 7.1 shows an example of simulated synthetic storms affecting Boston. Along each simulated track, the coupled hurricane intensity prediction system (CHIPS), a deterministic ocean–atmospheric coupled model, is used to simulate the intensity evolution of the storm (Emanuel, 2004). TC risk for a certain area is estimated by examining the frequency and characteristics

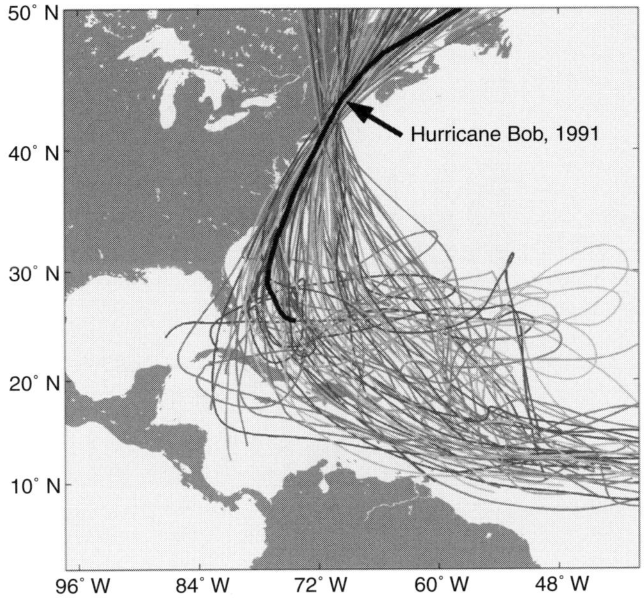

Figure 7.1. Tracks of the 100 most intense among 3000 synthetic storms affecting downtown Boston. Shown for comparison (in black) is the observed track of Hurricane Bob of 1991 (after Emanuel *et al.*, 2006). For colour version, see Plates section.

of storms passing by the area (e.g., crossing a boundary of the area or passing within a certain radius of the centre of the area). Figure 7.2 shows a comparison between the simulated and observed storm intensity distributions for the Atlantic Basin. The CHIPS model also estimates the storm radius of maximum wind, conditional on the storm outer radius (where the storm wind vanishes). The storm outer radius was found to obey a lognormal distribution (Chavas and Emanuel, 2010). Theoretically, the storm outer radius scales with the storm potential intensity (Emanuel, 1986). However, the scaling relationship remains to be evaluated against observations; at the same time, further investigation is needed to determine how storm outer radii will change with climate. Storm size (as described by the storm radius of maximum wind, the outer radius, and other related parameters) is an important factor, in addition to storm intensity and track, in determining TC hazards, as amply demonstrated by Hurricanes Katrina, Irene, and Sandy.

7.3 TC hazards modelling

TCs induce multiple hazards during and after landfall, including strong winds, heavy rainfall, and storm surge. TC wind and rainfall may be simulated using high-resolution regional dynamic models, such as the weather

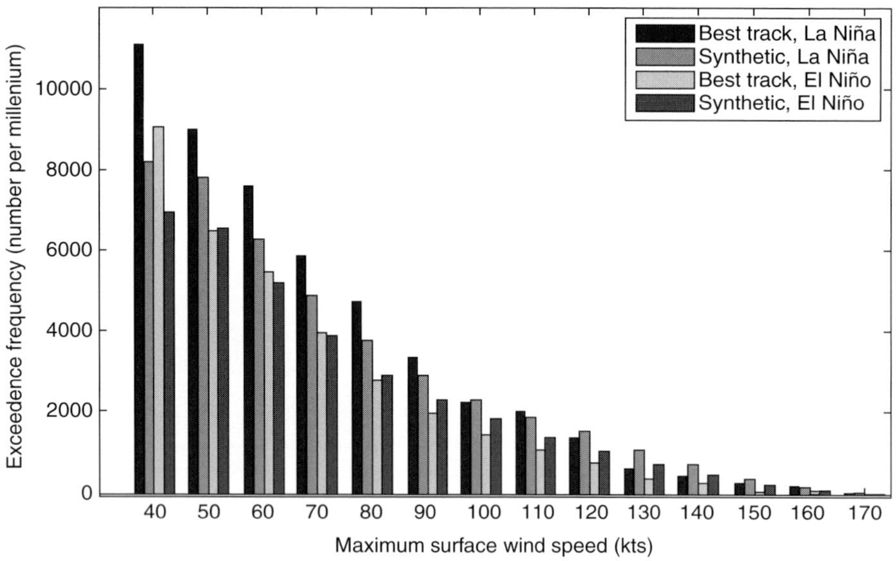

Figure 7.2. Exceedance frequency distributions by intensity for North Atlantic storms for La Niña years (black and light grey bars), and El Niño years (dark grey and white bars). In each pair of bars, the best-track data pertain to the left member, and the synthetic data to the right. Data from the two phases of ENSO during the period 1949–2004 are used for both the best-track and reanalysis data (after Emanuel *et al.*, 2008).

research and forecasting (WRF; Skamarock *et al.*, 2005) model. For example, Lin *et al.* (2010b) carried out a case study for Hurricane Isabel (2003), applying the WRF model to examine storm properties and wind and rainfall hazards down to 1-km resolution at landfall. They further coupled the WRF model with a hydrodynamic model to simulate the storm surge in Chesapeake Bay. This approach may be applied to studying the impact of climate change on TC hazards; in particular, high-resolution dynamic models can be used to downscale relatively lower-resolution TC simulations, such as Knutson *et al.*'s (2013) projection of storms in different climates, to local scales. However, this approach is, at present, computationally too expensive to be applied directly to risk assessment, which should involve very large numbers of simulations to cover nearly all possible scenarios. Consequently, computationally much more efficient parametric models are often used in risk analysis. Parametric wind and pressure models can also be applied to storm surge analysis. Such parametric modelling and analysis can be readily coupled with TC risk models (see above) to estimate TC hazard risk. In this section, we review parametric wind modelling and storm surge modelling as well as statistical methods to estimate TC hazard risk from the model results. A physically-based parametric rainfall model, which depends on the wind model, is currently (2013) under development by the authors.

7.3.1 Wind field modelling

In the parametric approach, the TC surface wind field may be estimated as the sum of an axisymmetric wind field associated with the storm itself and a background wind field representing the local environment (Lin and Chavas, 2012), with effects of sea state and terrain accounted for (see Vickery *et al.*, 2009a). Given the storm characteristics (i.e., track, intensity, and size), the axisymmetric component of the surface wind may be estimated by calculating the wind velocity at gradient level with a TC gradient wind profile and translating the gradient wind to the surface level with an empirical surface wind reduction factor (SWRF) (e.g., Powell *et al.*, 2003) and inflow angle (e.g., Bretschneider, 1972) to account for the effect of surface friction on the storm. A boundary layer model (e.g., Thompson and Cardone, 1996; Vickery *et al.*, 2009b; Kepert, 2010) may be applied to more accurately calculate the surface wind from the gradient condition, but it is nonparametric and more computationally demanding. A number of gradient wind profiles (e.g., Holland, 1980; Jelesnianski *et al.*, 1992; Emanuel, 2004; Emanuel and Rotunno, 2011) have been used in wind and surge analysis; it may be difficult at this point to identify the 'best' wind profile, as each profile has its own strengths and limitations. Figure 7.3 shows comparisons of these profiles for two storms. Lin and Chavas (2012) investigated the sensitivities of simulated wind and associated surge fields to these gradient wind profiles and the other above-mentioned factors, which may then also be used to quantify the uncertainties in parametric wind modelling.

The surface background wind is often related to the storm's translation velocity, based on two assumptions: the storm's movement is mainly due to advection by (some vertically integrated measure of) the background wind in the free troposphere near the storm, and surface friction causes the surface background wind to deviate from the free tropospheric wind in magnitude and direction. However, previous applications disagreed on the nature of this deviation. Some assumed that the surface background

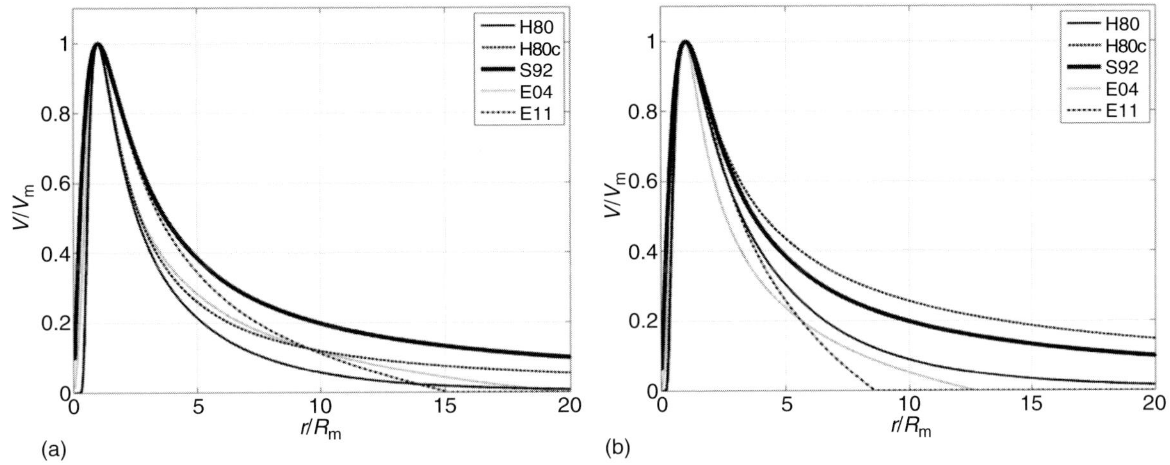

Figure 7.3. Comparison of parametric wind profiles for (a) a relatively intense storm and (b) a relatively weak storm, making landfall near Tampa, FL. H80, H80c, S92, E04, and E11 denote the parametric wind profiles of Holland (1980; with gradient wind balance assumption), Holland (1980; with cyclostrophic approximation), Jelesnianski *et al.* (1992), Emanuel (2004), and Emanuel and Rotunno (2011), respectively. V is the wind speed at a certain radius, V_m is the maximum wind speed of the storm, r is the radius, and R_m is the radius of maximum wind of the storm. (after Lin and Chavas, 2012).

wind is approximately equal in direction to the storm translation velocity and reduced in magnitude by a variously-valued factor (e.g., 0–0.5 used by Jelesnianski *et al.* (1992) and Phadke *et al.* (2003); 0.6 by Emanuel *et al.* (2006); and 0.5 by Lin *et al.* (2012)). In many other applications, the full velocity of the storm's translation is added to the storm's wind field (e.g., Powell *et al.*, 2005; Mattocks and Forbes, 2008; Vickery *et al.*, 2009b), neglecting the velocity difference between the free tropospheric wind and surface background wind. Lin and Chavas (2012) performed an observational analysis to estimate the magnitude and orientation of the surface background wind relative to the storm translation velocity and found that, with relatively small spatial variations, the surface background wind on average was reduced in magnitude from the storm translation velocity by a factor of about 0.55 and was rotated in the counter-clockwise direction by about 20 degrees. They also showed that the wind fields are very sensitive to the representation of the surface background wind, so that previous methods may have significantly under- or over-estimated wind speeds.

It is noted, however, that Lin and Chavas's (2012) parametric representation of the surface background wind assumed a uniform surface background wind to be added to the symmetric storm wind, which may only be valid for storms developing in a relatively uniform wind environment. TCs moving to higher latitudes may undergo extratropical transition and become 'hybrid' storms (Hart and Evans, 2001) – a critical aspect of Hurricane Sandy's devastating impact – about which our current knowledge is limited (Emanuel, 2005). These hybrid events, having a partially baroclinic structure, are often highly asymmetric. To account for this baroclinic effect, the surface background wind estimation may be improved by applying a representation of the interaction between the highly localised potential vorticity anomaly of the TC and its environmental baroclinic fields. In particular, Hart's (2003) TC phase space technique may be applied to study the characteristics of hybrid storms and facilitate these developments.

7.3.2 Surge modelling

The storm surge is a rise of coastal shallow water driven by a storm's surface wind and pressure gradient forces; its magnitude is determined, in a complex way, by the characteristics of the storm plus the geometry and bathymetry of the coast. When observed wind and pressure fields are used, state-of-the-art storm surge models can often produce successful hindcasts of surges (e.g., Houston *et al.*, 1999; Westerink *et al.*, 2008; Bunya *et al.*, 2010). In real-time forecasting, predictions of winds, and thus surges, can be performed using advanced numerical weather forecasting models (e.g., Colle *et al.*, 2008; Lin *et al.*, 2010b). In risk analysis, parametric wind and pressure field models may be used to drive the storm surge simulations. Although high-resolution numerical grids can better capture the spatial variation of the storm characteristics and coastal features, in order to conduct surge estimates for large numbers of storm scenarios in risk analysis a trade-off between efficiency and accuracy is often required.

Here we compare two widely used storm surge models, which represent the hydrodynamics of the storm surge by means of shallow water equations. One is the sea, lake, and overland surges from hurricanes (SLOSH; Jelesnianski *et al.*, 1992) model, used by the Natural Hurricane Center of the National Oceanic and Atmospheric Administration (USA) for real-time forecasting of hurricane storm surge. The performance of the SLOSH model has been evaluated using observations of storm surge heights from past hurricanes (Jarvinen and Lawrence, 1985; Jarvinen and Gebert, 1986); the accuracy of surge heights predicted by the model is $\pm 20\%$ when the hurricane is adequately described (Jelesnianski *et al.*, 1992). The SLOSH model applies finite difference methods to solve the equations and uses a polar grid, which allows for a fine mesh in primary coastal regions of interest and a coarse mesh in the open ocean. For example, the New York basin grid is a polar coordinate system with 75 arcs and 82 radials, with resolution of about 1 km near New York City (NYC). With some simplification in the physics represented (Jelesnianski *et al.*, 1992) and with relatively coarse grids, the SLOSH simulation runs relatively fast. Another storm surge model is the Advanced Circulation model (ADCIRC; Luettich *et al.*, 1992; Westerink *et al.*, 1992). It has been evaluated and applied to simulate storm surges and make forecasts for various coastal regions (e.g., Westerink *et al.*, 2008; Colle *et al.*, 2008; Dietrich *et al.*, 2011; Lin *et al.*, 2010b, 2012). The ADCIRC model fully describes the complex physical process associated with storm surge and can also simulate astronomical tides and wind-induced waves during the surge event (when coupled with a wave model; see Dietrich *et al.* (2011)). It allows the use of an unstructured grid over a relatively large domain, with very fine resolution near the coast and much coarser resolution in the deep ocean. The high-resolution ADCIRC model is computationally expensive, compared to the SLOSH model, and thus is not feasible for very large numbers of simulations.

Lin *et al.* (2010a) compared nine synthetic storm surge estimates for NYC using the SLOSH model (with resolution of $\sim$1 km around NYC) and the ADCIRC model (with resolution as high as 10 m around NYC). When driven by the same wind fields, the SLOSH model performed well (judging by the ADCIRC model) in simulating the maximum storm surge at a location with relatively simple coastal features, although subgrid-scale variations in the local surge were averaged out. Lin *et al.* (2012) further carried out comparisons of over 1000 synthetic storm surge estimates from the SLOSH (with a resolution of $\sim$1 km) and ADCIRC (with a resolution of $\sim$100 m) simulations for NYC. They also found that the results from SLOSH simulations were not biased relative to those of the ADCIRC simulations when the same wind fields were applied, although the SLOSH simulations were less sensitive to storm characteristics and often predicted similar surges for a range of different storms, compared to the ADCIRC simulations. However, it should be noted that the SLOSH model has an internal parametric wind model that may underestimate the surface background wind and thus may underestimate wind-field intensity. Thus, when the SLOSH wind fields are used, the storm surge may be underestimated; nevertheless, the estimated surges from the two model simulations are highly correlated (see Lin *et al.* 2012, supplementary material).

The SLOSH and ADCIRC models may be applied together in surge risk analysis. To make it possible to simulate surges with reasonable accuracy for the large synthetic storm sets for a risk assessment for NYC, Lin *et al.* (2012) applied the two hydrodynamic models with numerical grids of various resolutions in such a way that the main computational effort was concentrated on the storms that determine the risk of concern. First, the SLOSH model with a resolution of $\sim$1 km around NYC was applied as a filter to select the storms that have return periods, in terms of the surge height at the Battery, greater than 10 y. Second, the ADCIRC model with a resolution of $\sim$100 m around NYC was applied to each of the selected storms. To determine whether the resolution of the ADCIRC simulation is sufficient, another ADCIRC mesh with a resolution as high as $\sim$10 m around NYC was used to simulate over 200 most-extreme synthetic events under present climate conditions. The differences between the results from the two grids were very small. Thus, the $\sim$100 m ADCIRC simulations were used, with a 2.5% reduction of the surge magnitude motivated by the $\sim$10-m simulations, to estimate the surge levels at the Battery for return periods of 10 y and longer. Figure 7.4 shows the estimated return period curve for storm surge, along with the return period curve for the storm tide, which includes the contribution of astronomical tides. Applying the two storm surge models in this way, large numbers of surge events can be efficiently simulated to estimate the risk. In addition to the analysis for the current climate as shown in Fig. 7.4, Lin *et al.* (2012) applied this approach, together with the Emanuel *et al.*'s (2008) TC risk model, to study the impact of climate change on surge risk through simulating 40 000 synthetic storm surge events under current and future climate conditions projected by four climate models. Their results indicate a greatly increased storm surge threat in a future climate due to the change in storm climatology, and the combination of the change of storm climatology and the rise of sea level will greatly reduce coastal-flood return periods for NYC (Fig. 7.5).

7.3.3 TC hazard risk assessment

The risk of TC hazards, such as wind speed and surge height, may be estimated from a physically simulated database of the hazards. One may assume the annual storm

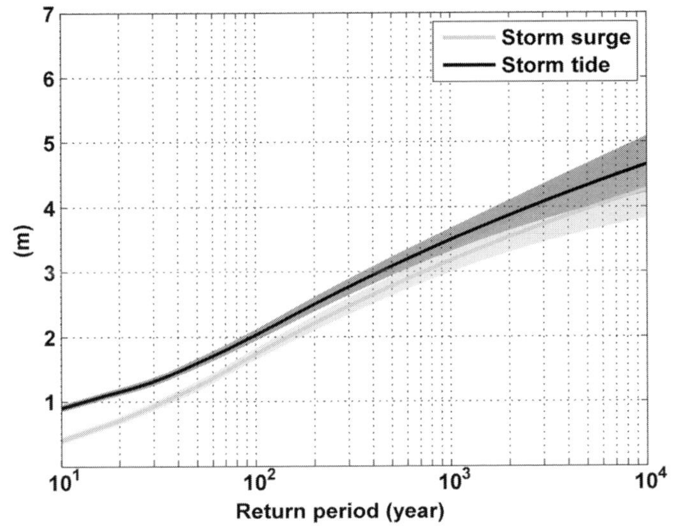

Figure 7.4. Estimated return periods for the Battery, NYC, of the storm surge and storm tide for the climate of 1981–2000, estimated from observation (based on the NCEP/NCAR reanalysis). The shading shows the 90% confidence interval. The estimation was based on 5000 synthetic storm surge events for the NYC area (after Lin *et al.*, 2012).

counts for a region to be Poisson-distributed (Elsner and Bossak, 2001; Lin *et al.*, 2012), with the mean estimated by the TC risk model. The probability density function (PDF) of TC hazards appears to be characterized by a thick upper tail, which tends to dominate the risk (Lin *et al.*, 2012). For the (marginal) PDFs of wind speed and surge height, one may apply a peaks-over-threshold (POT) method to model the upper tail with a generalized Pareto distribution (GPD) and the rest of the distribution through nonparametric density estimation. The PDF of coastal-flood levels will be a convolution of the PDFs of surge, astronomical tide, wave setup, and sea-level rise (see Lin *et al.*, 2012). These PDFs and the storm frequency model can then be combined to estimate hazard-specific return-period curves and associated statistical confidence intervals. Return-period maps can also be generated, and one can proceed to estimate joint distributions for wind and coastal flooding and for wind and inland flooding.

7.4 TC damage modelling

TCs cause damage to the natural and built environment via a multiplicity of hazards. Residential buildings, in particular, are often heavily damaged by TC winds, surge/wave, and flooding caused by severe rainfall, as seen

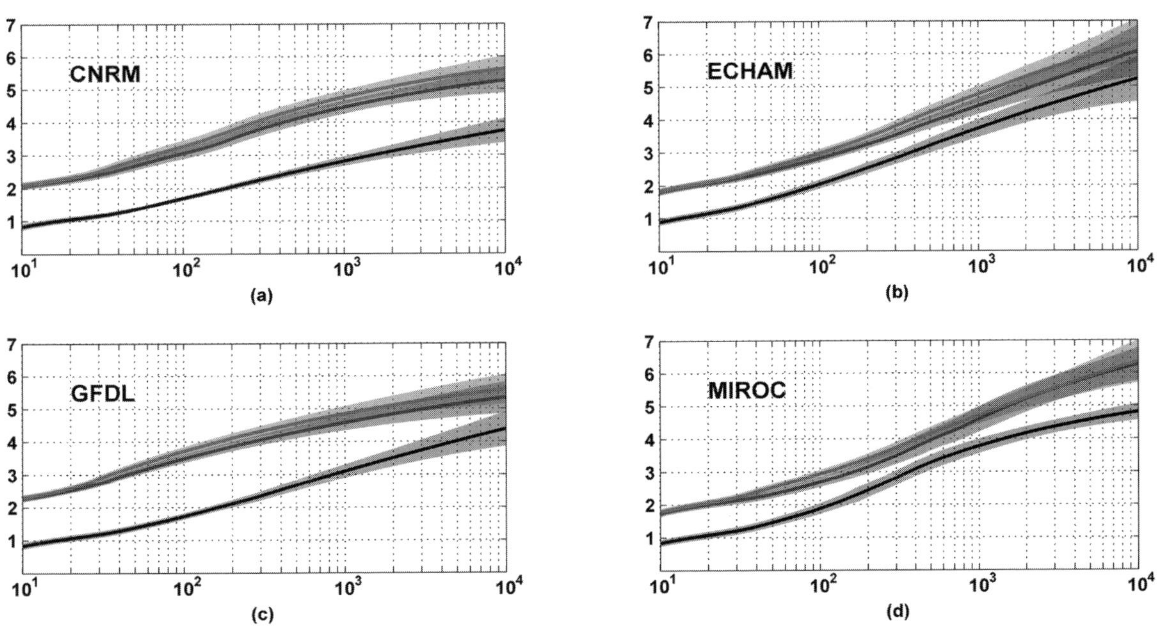

Figure 7.5. Estimated flood return levels for the Battery, NYC, predicted with each of the four climate models (CNRM-CM3, ECHAM5, GFDL-CM2.0 and MIROC3.2). The black is for the climate of 1981–2000 (current climate), the blue is for the IPCC A1B climate and the red is for the IPCC A1B climate with the storm outer radius increased by 10% and radius of maximum wind increased by 21% for 2081–2100 (future climate). The shading shows the 90% confidence interval. The estimation for each climate model was based on 5000 synthetic storm surge events for the current climate condition and 5000 synthetic storm surge events for the future climate condition. 1-m SLR was included for the future flood (after Lin *et al.*, 2012). For colour version, see Plates section.

in storms such as Hurricane Sandy. Hurricane wind damage to residential buildings is mainly due to direct wind pressure effect, windborne debris impact, and their interaction over time. Typical debris sources are roof materials, such as roof covers, sheathings, and timbers. These roof materials may become windborne, due to high wind pressure on the roof, and fly at high speeds so as to damage surfaces of neighbouring buildings. When the building envelopes are penetrated, in addition to causing wind and rain damage to building contents, internal pressurisation increases the net loading in suction zones, possibly leading to failure of roofing and wall cladding, generating new debris and thus more pressure damage, and so on until the end of the hurricane wind. Lin and Vanmarcke (2010) developed a debris risk model, based on the Poisson theory, large numbers of wind-tunnel experiments (Lin *et al.*, 2006, 2007), and available post-damage survey data (Twisdale *et al.*, 1996). Lin *et al.* (2012) further integrated this debris model with a component-based pressure damage model to estimate the wind damage to a residential area during the passage of a storm. This wind damage model is an improvement over other methods, such as the FEMA HAZUS-MH hurricane model (Vickery *et al.*, 2006, 2009a) and the Florida public hurricane loss projection model (Pinelli *et al.*, 2004; Gurley *et al.*, 2005), in that it explicitly models, for the first time, the (two-way) interaction between pressure damage and debris damage over time. The flow chart of this debris-pressure coupled vulnerability model is shown in Fig. 7.6. Results from a case study involving a residential area in Florida are shown in Fig. 7.7, indicating that wind damage to residential areas may be greatly underestimated if the effects of windborne debris are not included. Yau *et al.* (2011) made use of this wind damage model to estimate economic losses from TCs.

This wind damage model can be extended to account for the effects of the surge/wave and rainfall. Damage due to rainfall may be modelled by converting the amounts of water that penetrated (through openings created by pressure and debris) into an interior damage ratio (Pita *et al.*, 2012). The surge/wave damage component can be developed using damage survey data. Based on data from Hurricane Ike (2008), Kennedy *et al.* (2011) developed empirical relationships between surge/wave damage and the freeboard height and significant wave height. Such empirical damage functions can be tested and further improved using damage data, for example, from Hurricane Sandy. Surge/wave-generated debris is another important damage factor that should be considered in future research.

TC vulnerability models can be combined with the TC risk assessment and hazard modelling (see above) to

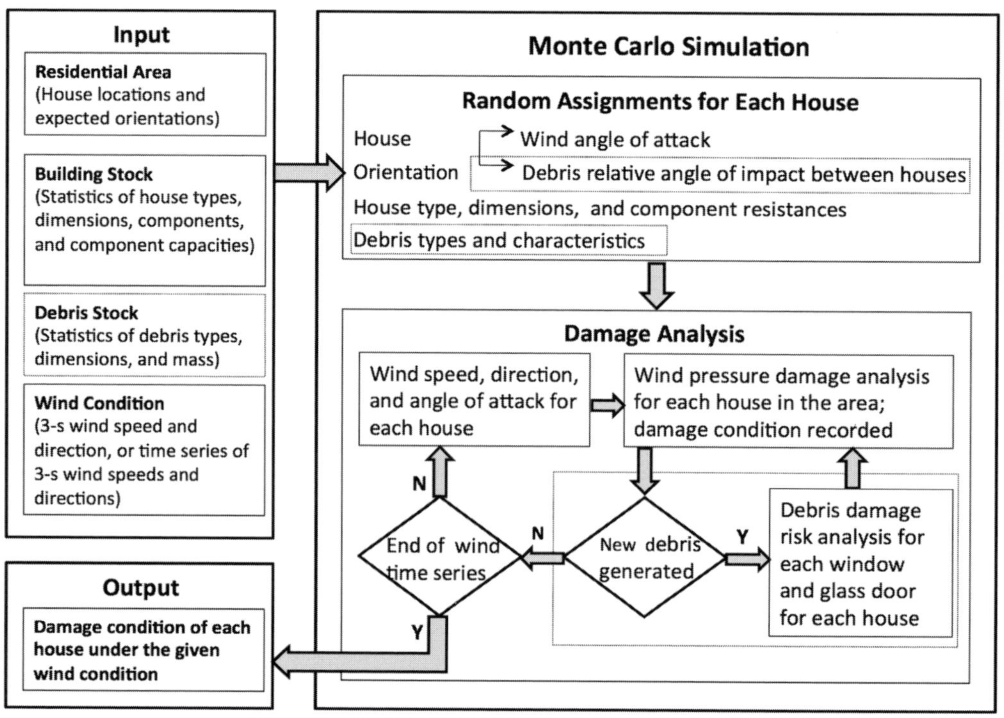

Figure 7.6. Flowchart of the integrated vulnerability model for residential developments. Dashed line boxes highlight the components related to debris risk analysis; the functions of these components may be turned off if the debris effect is ignored or if the openings are fully protected from debris damage (after Lin *et al.*, 2010c).

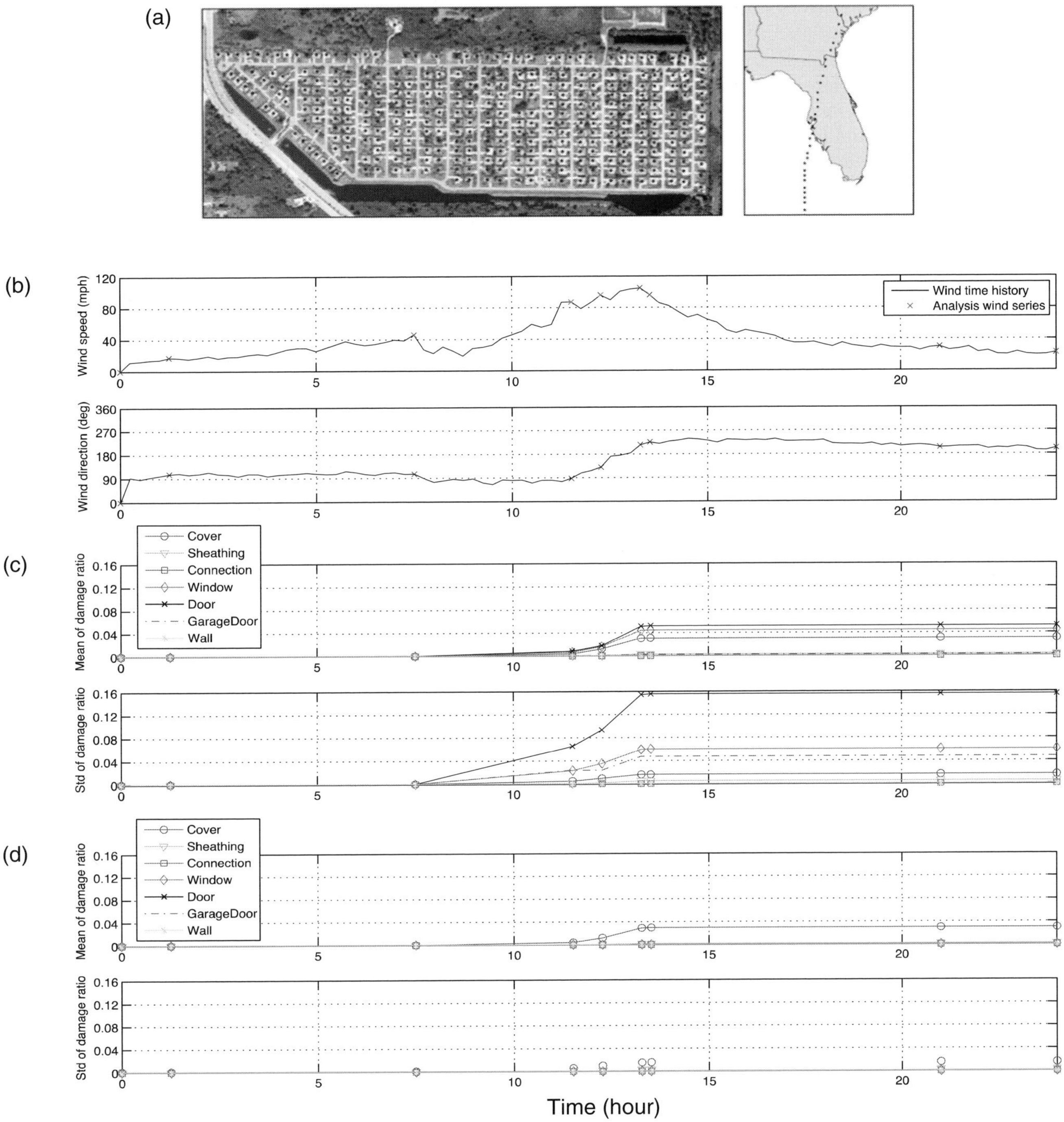

Figure 7.7. Case study of wind damage modelling for residential areas. (a) Study area of a residential development of 358 houses in Sarasota County, Florida (left; the dots represent the centres of the houses) and simulated storm track of Hurricane Charley of 2004 (right). The star (in both panels) represents the recording location of the wind time history. (b) Wind speed and direction time history during the passage of the simulated storm. (c) Estimated progressive damage in the study area, using the debris–pressure damage model. (d) Estimated progressive damage in the study area, using only the pressure damage model (after Lin *et al.*, 2010c). For colour version, see Plates section.

predict the aggregate impact of TCs in coastal regions. Aerts *et al.* (2013) predicted the storm surge damage risk, in terms of expected annual losses, for NYC. Klima *et al.* (2011) estimated and compared expected damage/losses with the costs of alternative mitigation measures (and their combinations) for Miami-Dade County, Florida. Further improvements may include accounting for the correlation of the TC hazards. Also, dynamic economic models may be applied to derive near-optimal mitigation strategies and implementation time paths.

7.5 Concluding comments

We sought to assess the status of quantifying risks associated with hurricanes, including the effect on TC risk of various climate-change scenarios. The principal hazards, entailing various combinations of catastrophic loss potential, are storm surge, strong wind pressures, high-impact debris, and heavy rainfall and flooding. Approaches to modelling in each of these areas were reviewed and illustrated with (references to and summaries of) case studies. The challenge remains to tackle the multiplicity of hazards in an integrated way, aiming at improved risk estimation and quantification of uncertainties across a range of spatial scales. The results are of considerable interest to those seeking to quantify the effectiveness of risk mitigation measures and perform risk-based scientific and policy analysis pertaining to the resilience of the natural and built environments.

References

Aerts, J., Lin, N., Moel, H., Emanuel, K. and Botzen, W. (2013). Low probability flood-risk modeling for New York City. *Risk Analysis*, in press.

Bender, M. A., Knutson, T. R., Tuleya, R. E. *et al.* (2010). Modeled impact of anthropogenic warming on the frequency of intense Atlantic hurricanes. *Science*, **327**(5964), doi:10.1126/science.1180568.

Bretschneider, C. L (1972). A non-dimensional stationary hurricane wave model. In *Proceedings of the Offshore Technology Conference, 1–3 May 1972, Houston, Texas*. Offshore Technology Conference, Paper No. OTC 1517.

Bunya, S., Dietrich, J. C., Westerink, J. J. *et al.* (2010). A high-resolution coupled riverine flow, tide, wind, wind wave, and storm surge model for southern Louisiana and Mississippi. Part I: Model development and validation. *Monthly Weather Review*, **138**, 345–377.

Chavas, D. R. and Emanuel, K. A. (2010). A QuikSCAT climatology of tropical cyclone size, *Geophysical Research Letters*, **37**, L18816, doi:10.1029/2010GL044558.

Colle, B. A., Buonaiuto, F., Bowman, M. J. *et al.* (2008). New York City's vulnerability to coastal flooding. *Bulletin of the American Meteorological Society*, **89**, 829–841.

Dietrich, J. C., Zijlema, M., Westerink, J. J., *et al.* (2011). Modeling hurricane waves and storm surge using integrally-coupled, scalable computations. *Coastal Engineering*, **58**, 45–65.

Elsner, J. B. and Bossak, B. H. (2001). Bayesian analysis of U.S. hurricane climate. *Journal of Climate*, **14**, 4341–4350.

Elsner, J. B., Jagger, T. H. and Liu, K. B. (2008). Comparison of hurricane return levels using historical and geological records. *Journal of Applied Meteorology*, **47**, 368–374.

Emanuel, K. A. (1986). An air–sea interaction theory for tropical cyclones. Part I: Steady-state maintenance. *Journal of Atmospheric Sciences*, **43**, 585–605.

Emanuel, K. A. (2004). Tropical cyclone energetics and structure. In *Atmospheric Turbulence and Mesoscale Meteorology*, eds. Fedorovich, E. Rotunno, R. and Stevens, B. Cambridge: Cambridge University Press, pp. 165–192.

Emanuel, K. A. (2005). *Divine Wind: The History and Science of Hurricanes*. Oxford: Oxford University Press.

Emanuel, K. and Rotunno, R. (2011). Self-stratification of tropical cyclone outflow. Part I: Implications for storm structure. *Journal of Atmospheric Sciences*, **68**, 2236–2249.

Emanuel, K., Ravela, S., Vivant, E. and Risi, C. (2006). A statistical deterministic approach to hurricane risk assessment. *Bulletin of the American Meteorological Society*, **87**, 299–314.

Emanuel, K., Sundararajan, R. and Williams, R. (2008). Hurricanes and global warming: results from downscaling IPCC AR4 simulations. *Bulletin of the American Meteorological Society*, **89**, 347–367.

Emanuel, K., Oouchi, K., Satoh, M., Hirofumi, T. and Yamada, Y. (2010). Comparison of explicitly simulated and downscaled tropical cyclone activity in a high-resolution global climate model. *Journal of Advances in Modeling Earth Systems*, **2**, doi:10.3894/JAMES.2010.2.9.

Gurley, K., Pinelli, J. P., Subramanian, C. *et al.* (2005). *Florida Public Hurricane Loss Projection Model Engineering Team Final Report*, Technical report, International Hurricane Research Center, Florida International University.

Hall, T. and Jewson, S. (2007). Statistical modelling of North Atlantic tropical cyclone tracks. *Tellus A*, **59A**, 486–498.

Handmer, J., Honda, Y., Kundzewicz, Z. W. *et al.* (2012). Changes in impacts of climate extremes: human systems and ecosystems. In *Managing the Risks of Extreme Events and Disasters to Advance Climate Change Adaptation*, eds. Field, C. B., Barros, V., Stocker, T. F. A Special Report of Working Groups I and II of the Intergovernmental Panel on Climate Change (IPCC). Cambridge, UK: Cambridge University Press, pp. 231–290.

Hart, R. E. (2003). A cyclone phase space derived from thermal wind and thermal asymmetry. *Monthly Weather Review*, **131**, 585–616.

Hart, R. E. and Evans, J. L. (2001). A climatology of the extra-tropical transition of Atlantic tropical cyclones. *Journal of Climate*, **14**, 546–564.

Holland, G. J. (1980). An analytical model of the wind and pressure profiles in hurricanes. *Monthly Weather Review*, **108**, 1212–1218.

Holland, G. J., Belanger, J. I. and Fritz, A. (2010). A revised model for radial profiles of hurricane winds. *Monthly Weather Review*, **138**, 4393–4401.

Houston, S. H., Shaffer, W. A., Powell, M. D. and Chen, J. (1999). Comparisons of HRD and SLOSH surface wind fields in hurricanes: Implications for storm surge modeling, *Weather and Forecasting*, **14**, 671–686.

Jarvinen, B., and Gebert, J. (1986). Comparison of observed versus SLOSH model computed storm surge hydrographs along the Delaware and New Jersey shorelines for Hurricane Gloria, September 1985, NOAA Tech. Memo NWS NHC 32.

Jarvinen, B. R. and Lawrence, M. B. (1985). An evaluation of the SLOSH storm surge model, *Bulletin of the American Meteorological Society*, **66**, 1408–1411.

Jelesnianski, C. P., Chen, J. and Shaffer, W. A. (1992). *SLOSH: Sea, Lake, and Overland Surges from Hurricanes*, NOAA Tech. Rep. NWS 48, NOAA/AOML Library, Miami, Florida.

Kennedy, A., Rogers, S., Sallenger, A., Gravois, U., Zachry, B., Dosa, M. and Zarama, F. (2011). Building destruction from waves and surge on the Bolivar Peninsula during Hurricane Ike. *Journal of Waterway, Port, Coastal, Ocean Engineering*, **137**(3), 132–141.

Kepert, J. D. (2010). Slab- and height-resolving models of the tropical cyclone boundary layer. Part I: Comparing the simulations. *Quarterly Journal of the Royal Meteorological Society*, **136**, 1686–1699.

Klima, K., Lin, N., Emanuel, K. A., Morgan, M. G. and Grossmann, I. (2011). Hurricane modification and adaptation in Miami-Dade County, Florida. *Environmental Science and Technology*, **45**, 636–642.

Knutson, T. R., Sirutis, J. J., Garner, S. T., Held, I. M. and Tuleya, R. E. (2007). Simulation of the recent multi-decadal increase of Atlantic hurricane activity using an 18-km-grid regional model. *Bulletin of the American Meteorological Society*, **88**, 1549–1565.

Knutson, T. R., McBride, J., Chan, J. *et al.* (2010). Tropical cyclones and climate change. *Nature Geoscience*, **3**, 157–163.

Knutson, T. R., Sirutis, J. J., Vecchi, G. A. *et al.* (2013). Dynamical downscaling projections of 21st century Atlantic hurricane activity: CMIP3 and CMIP5 model-based scenario. *Journal of Climate*, submitted for publication.

Lin, N. and Chavas, D. (2012). On hurricane parametric wind and applications in storm surge modeling. *Journal of Geophysical Research*, **117**, D09120, doi:10.1029/2011JD017126.

Lin, N. and Vanmarcke, E. (2008). Windborne debris risk assessment. *Probabilistic Engineering Mechanics*, **23**(4), 523–530.

Lin, N. and Vanmarcke, E. (2010). Windborne debris risk analysis. Part I: Introduction and methodology. *Wind and Structures*, **13**(2), 191–206.

Lin, N., Letchford, C. W. and Holmes, J. D. (2006). Investigations of plate-type windborne debris, I. Experiments in wind tunnel and full-scale. *Journal of Wind Engineering and Industrial Aerodynamics*, **94**, 51–76.

Lin, N., Holmes J. D. and Letchford, C. W. (2007). Trajectories of windborne debris and applications to impact testing. *ASCE Journal of Structural Engineering*, **133**(2), 274–282.

Lin, N., Emanuel, K. A., Smith, J. A. and Vanmarcke, E. (2010a). Risk assessment of hurricane storm surge for New York City. *Journal of Geophysical Research*, **115**, D18121, doi:10.1029/2009JD013630.

Lin, N., Smith, J. A., Villarini, G., Marchok, T. and Baeck, M. L. (2010b). Modeling extreme rainfall, winds, and surge from Hurricane Isabel (2003). *Weather and Forecasting*, **25**, 1342–1361.

Lin, N., Vanmarcke, E. and Yau, S. C. (2010c). Windborne debris risk analysis: Part II. Application in structural vulnerability modeling. *Wind and Structures*, **13**(2), 207–220.

Lin, N., Emanuel, K. A., Oppenheimer, M. and Vanmarcke, E. (2012). Physically based assessment of hurricane surge threat under climate change. *Nature Climate Change*, **2**, 462–467.

Luettich, R. A., Westerink, J. J. and Scheffner, N. W. (1992). *ADCIRC: An Advanced Three-dimensional Circulation Model for Shelves, Coasts and Estuaries, Report 1: Theory and Methodology of ADCIRC-2DDI and ADCIRC-3DL. DRP Technical Report DRP-92-6.* (Department of the Army, US Army Corps of Engineers, Waterways Experiment Station, Vicksburg, MS, 1992).

Mattocks, C. and Forbes, C. (2008). A real-time, event-triggered storm surge forecasting system for the state of North Carolina. *Ocean Modelling*, **25**, 95–119.

Mendelsohn, R., Emanuel, K., Chonabayashi, S. and Bakkensen, L. (2012). The impact of climate change on global tropical cyclone damage. *Nature Climate Change*, **2**, 205–209.

Phadke, A. C., Martino, C. D., Cheung, K. F. and Houston, S. H. (2003). Modeling of tropical cyclone winds and waves for emergency management. *Ocean Engineering*, **30**, 553–578.

Pielke, Jr, R. A. (2007). Future economic damage from tropical cyclones: sensitivities to societal and climate changes. *Philosophical Transactions of the Royal Society A*, **365**, 2717–2729.

Pinelli, J.-P., Simiu, E., Gurley, K., Subramanian, C., Zhang, L., Cope, A., Filiben, J. J. and Hamid, S. (2004). Hurricane damage prediction model for residential structures. *Journal of Structural Engineering*, **130**, 1685–1691.

Pita, G., Pinelli, J.-P., Cocke, S. *et al.* (2012). Assessment of hurricane-induced internal damage to low-rise buildings in the Florida Public Hurricane Loss Model, *Journal of Wind Engineering and Industrial Aerodynamics*, **104–106**, 76–87.

Powell, M. D., Houston, S. H., Amat, L. R. and Morisseau-Leroy, N. (1998). The HRD real-time hurricane wind analysis system. *Journal of Wind Engineering and Industrial Aerodynamics*, **77–78**, 53–64.

Powell, M. D., Vickery, P. J. and Reinhold T. A. (2003). Reduced drag coefficient for high wind speeds in tropical cyclones, *Nature*, **422**, 279–283.

Powell, M., Soukup, G., Cocke, S. *et al.* (2005). State of Florida hurricane loss projection model: Atmospheric science component, *Journal of Wind Engineering and Industrial Aerodynamics*, **93**, 651–674.

Resio D. T. and Westerink, J. J. (2008). Modeling the physics of storm surges. *Physics Today*, **61**(9), 33–38.

Rumpf, J., Weindl, H., Höppe, P., Rauch, E. and Schmidt, V. (2007). Stochastic modeling of tropical cyclone tracks. *Mathematical Methods of Operations Research*, **66**, 475–490.

Skamarock, W. C., Klemp, J. B., Dudhia, J. *et al.* (2005). A description of the Advanced Research WRF version 2. NCAR Tech. Note *NCAR/ TN-4681STR*, 88 pp.

Thompson, E. F. and Cardone, V. J. (1996). Practical modeling of hurricane surface wind fields, *Journal of Waterway, Port, Coastal Ocean Engineering*, **122**, 195–205.

Twisdale, L. A., Vickery, P. J. and Steckley, A. C. (1996). *Analysis of hurricane windborne debris risk for residential structures*, Technical report. Raleigh, NC: Applied Research Associates, Inc.

Vickery, P. J., Skerjl, P. F. and Twisdale, L. A. (2000). Simulation of hurricane risk in the U.S. using empirical track model. *Journal of Structural Engineering*, **126**, 1222–1237.

Vickery, P. J., Skerlj, P. F., Lin, J. *et al.* (2006). HAZUS-MH Hurricane Model methodology. II: Damage and loss estimation. *Natural Hazards Review*, **7**(2), 94–103.

Vickery, P. J., Masters, F. J., Powell, M. D. and Wadhera, D. (2009a) Hurricane hazard modeling: The past, present, and future, *Journal of Wind Engineering and Industrial Aerodynamics*, **97**, 392–405.

Vickery, P. J., Wadhera, D., Powell, M. D. and Chen, Y. (2009b). A hurricane boundary layer and wind field model for use in engineering applications. *Journal of Applied Meteorology and Climatology*, **48**, 381–405.

Westerink, J. J., Luettich, R. A., Blain, C. A. and Scheffer, N. W. (1992). ADCIRC: An advanced three-dimensional circulation model for shelves, coasts and estuaries. Report 2: *User's Manual for ADCIRC-2DDI, Dredging Research Program Technical Report DRP-92-6*. Vicksburg, MA: Coastal Engineering Research Center.

Westerink, J. J., Luettich, R. A., Feyen, J. C. *et al.* (2008). A basin to channel scale unstructured grid hurricane storm surge model applied to southern Louisiana. *Monthly Weather Review*, **136**, 833–864.

Yau, S. C., Lin, N. and Vanmarcke, E. (2011). Hurricane damage and loss estimation using an integrated vulnerability model. *Natural Hazards Review*, **12**(4), 184–189.

Yonekura, E., and Hall, T. M. (2011). A statistical model of tropical cyclone tracks in the Western North Pacific with ENSO-dependent cyclogenesis. *Journal of Applied Meteorology and Climatology*, **50**, 1725–1739.

8

Satellite-based remote sensing estimation of precipitation for early warning systems

Soroosh Sorooshian, Phu Nguyen, Scott Sellars, Dan Braithwaite, Amir AghaKouchak, and Kuolin Hsu

8.1 Introduction

Co-evolution between the earth system and human systems presents many difficult and growing challenges to society. Natural hazards have direct impacts on human systems, stressing the preparedness and resilience of society. Understanding and observing complex earth system phenomena, as well as alerting the public to an imminent threat, such as an extreme precipitation event, is a major priority for emergency managers, regional planners, decision makers, and leaders around the world. Advances in science, technology, and engineering allow near real-time satellite-based information to estimate a wide range of earth variables and phenomena. Communication networks between international leaders and organisations combined with near real-time satellite-based information can work in unison to warn society of imminent natural hazards and prepare human systems for the impact.

Extremes in the hydrologic cycle can cause massive flooding, flash flooding, long-term drought, infrastructure damage, agricultural disasters and collapse, and stress on water resources. From a hydrologic perspective, global early warning systems play a pivotal role in observing, monitoring, and forecasting imminent extreme precipitation events and flooding, or drought. Precipitation observations can be seen as vital information needed to prepare society for potential extreme precipitation events. Precipitation events are highly complex, with some precipitation events lasting a few hours causing flash floods and catastrophic infrastructure failure. Other precipitation events can last for weeks or months, with monsoon flows causing widespread regional flooding. At the opposite extreme, long-term droughts, the lack of sufficient precipitation, can last for years, decades, or even centuries, initiating agricultural collapse and stresses on water resources, and heightening tensions between countries. Many nations around the world have seen long-term drought as a hindrance to societal advancement and regional stability. One region can experience widespread flooding, while another region experiences widespread drought, all of which promote the need for dynamic global early warning systems able to alert for both widespread drought and imminent flash flooding. Both extreme precipitation events are intertwined in a complex web of short-, medium-, and long-term variations in weather and climate, making the development of any global early warning system challenging. Accurate precipitation measurements are extremely challenging, with in situ (ground-based) observations being sparse and not representing the true nature of precipitation, which has unique spatial patterns and varying intensities that point source ground-based observations often do not capture.

Extreme precipitation events are considered rare events that occur on varying spatial and temporal scales across the Earth. Due to limited global observational data, determining what is considered an extreme precipitation event can be very difficult, especially at point-scale representation, or in other words, a single point on the Earth. Remotely sensed precipitation estimates provide enhanced observations and information on the location, intensity, and the duration of extreme events for regions with very few ground-based rain gauges. These observations can be used in observing and monitoring extreme precipitation events in global early warning systems. Global early warning systems must be able to detect extreme events as they develop and, given the complex nature of precipitation events as described above, the early warning system must operate near real-time for emergency managers to be able to provide critical alerts and information to the public. Few systems allow this real-time monitoring, especially in remote regions of the world. Technology and communication platforms have advanced the capability of

Extreme Natural Hazards, Disaster Risks and Societal Implications, eds. Alik Ismail-Zadeh, Jaime Urrutia Fucugauchi, Andrzej Kijko, Kuniyoshi Takeuchi and Ilya Zaliapin. Published by Cambridge University Press.

alerting and informing the public of imminent natural hazards. With quickly developing extreme precipitation or the slow evolution of a widespread drought, more can be done to prevent loss of life and property caused by these hazards. This includes advancing remote sensing sensors and capabilities, improving precipitation and drought monitoring algorithms, and designing systems to address the needs of society for the preparation and advance warning of meteorological threats.

This chapter will focus on observations of precipitation employing remote sensing capabilities for use in advanced global early warning systems. The remote sensing of precipitation is reviewed in Section 8.2 including an in-depth look at the remote sensing precipitation estimation algorithm called Precipitation Estimation from Remotely Sensed Information using Artificial Neural Networks (PERSIANN and PERSIANN-CCS) and a near real-time platform for displaying precipitation estimates (G-WADI PERSIANN-CCS GeoServer). The G-WADI PERSIANN-CCS GeoServer is one example of a state-of-the-art system that can be used for precipitation monitoring and it will be used as an example of how high spatial and temporal resolution precipitation estimates can be provided, via the Internet, for use in near global precipitation monitoring. Next, Section 8.3 describes the current state-of-the-art global early warning systems for use in flood and drought monitoring and the applications of remotely sensed precipitation estimates for global early warning systems. Subsection 8.3.3 will discuss the strengths and limitations of remote sensing precipitation estimates for use in global early warning systems. Finally, we conclude in Section 8.4 with a summary of the chapter and a discussion of the future of remote sensing capabilities, with a few caveats.

8.2 Remote sensing precipitation estimation

As a whole, satellites are automated, near global, and near real-time observing networks providing data and observations of meteorological, geological, and oceanic features and phenomena. Developments and proposed enhancements in satellite remote sensing techniques provide a unique opportunity for improving the quality of precipitation observation, at the global scale, for most applications. The task of obtaining accurate measurements from satellites still poses significant challenges in terms of the sensor and the sampling frequency associated with the orbit of each platform. Geostationary earth orbiting (GEO) satellites are capable of providing images every 15–30 min in multiple spectral bands, but their spectral coverage is limited to visible (VIS) and infrared (IR) wavelengths that resolve cloud patterns rather than the hydrometeors directly relevant to surface precipitation rates. Therefore, as reported in numerous studies, GEO-IR-based rainfall algorithms are most effective at identifying tropical convective systems whose cold high tops show prominently in GEO-IR imagery, but perform less well in the presence of warm low-lying rain clouds and cold high cirrus, non-raining, clouds (Arkin and Meisner, 1987; Adler and Negri, 1988). Some improvement may be gained by employing cloud classification approaches using texture measures and cloud-patch identification, and further improvements may result from combining information from multi-spectral imagery (Hsu *et al.*, 1999; Bellerby *et al.*, 2000; Ba and Gruber, 2001; Bellerby, 2004; Hong *et al.*, 2004; Capacci and Conway, 2005; Turk and Miller, 2005). Some improvements in precipitation retrieval have been achieved by locally adjusting GEO-IR retrievals using low earth orbit (LEO) passive microwave (PMW)-based or radar rainfall estimations (Hsu *et al.*, 1997; Vicente *et al.*, 1998; Bellerby *et al.*, 2000; Sorooshian *et al.*, 2000; Todd *et al.*, 2001; Ba and Gruber, 2001; Kidd *et al.*, 2003; Nicholson *et al.*, 2003a,b; Bellerby, 2004; Marzano *et al.*, 2004; Turk and Miller, 2005; Huffman *et al.*, 2007).

PMW imagery from LEO satellites provides more direct information on the hydrometeor distribution in rain clouds than does GEO imagery. The low sampling frequency of LEO satellites, however, limits the effectiveness of PMW rainfall data retrieval at short time scales. By integrating data from multiple LEO satellites, considerable improvement in short-time-scale rainfall retrieval has been achieved (Huffman *et al.*, 2007). The Tropical Rainfall Measurement Mission (TRMM), launched in 1997, carried the first orbital rainfall radar, which may be used to calibrate passive microwave sensors on other satellites, resulting in significant improvements in rainfall retrievals over the tropical regions of the globe (Simpson *et al.*, 1988; Kummerow *et al.*, 1998, 2000). The follow-up to the TRMM mission is the Global Precipitation Measurement (GPM), planned for launch in 2014. The GPM mission will deploy an enhanced dual-frequency radar sensor. The GPM programme aims to combine observations from multiple passive microwave sensors mounted on both pre-existing and newly-deployed satellites. The planned GPM satellite constellation will be able to sample more than 90% of the globe with a return interval of 3 hours or less (Hou *et al.*, 2008). Such intensive precipitation sampling will provide unprecedented information on global water and energy distributions at fine time scales.

Because precipitation is highly variable in space and rapidly evolves in time, precipitation processes are very difficult to measure effectively. A thunderstorm from a convective process can generate several inches of precipitation over a small catchment area within a 30-minute period, causing a severe flash flood. The combined temporal sampling frequency provided by LEO satellites has varied significantly over time due to the number of functioning satellites available in earth orbits. This poses challenges to researchers for providing reliable operational precipitation measurements and generating a consistent long-term database for hydro-meteorological and hydro-climatological studies. Therefore, continuing to develop effective retrieval algorithms capable of integrating multiple sensors and platforms with flexible and efficient modern data assimilation techniques is essential.

Developments have demonstrated the effectiveness of using GEO-derived cloud motion vectors to 'morph' PMW rainfall estimates between sensor overpasses. The Climate Prediction Center morphing (CMORPH) algorithm applied GEO-IR cloud motion and then linearly interpolated PMW rainfall estimates along advection streamlines (Joyce *et al.*, 2004). Using a Kalman filter approach to update the advected rainfall values of CMORPH is under development (Okamoto *et al.*, 2005; Joyce and Xie, 2011). Studies also have been done on the relationships between developing cloud cells in GEO imagery and evolving rainstorms (Machado *et al.*, 1998; Horsfield, 2006; Bellerby *et al.*, 2009; Hsu *et al.*, 2009; Behrangi *et al.*, 2010, 2011). These developments suggest the possibility of running a simple storm development model between PMW overpasses rather than interpolating or updating PMW rainfall estimates directly. Such an approach would use the relationship between changes in cloud patterns and in rainfall processes, rather than attempting to determine a static relationship between GEO cloud imagery and rainfall, and would use data assimilation techniques to update model representations of rainfall processes, rather than attempting to adjust the rainfall values themselves.

In the 1990s, artificial neural networks (ANNs) became a popular tool for representing complex relationships between different phenomena and they have been applied in many fields of science and engineering. The ANNs are built on 'artificial network neurons', inspired by neurobiology. ANNs are considered artificial intelligence algorithms, which are essentially flexible mathematical models capable of defining a function that represents the complex relationship between input signals or variables and output data sets. A more detailed description can be found in Hsu *et al.* (1995). Two data fusion methods using ANNs for near real-time precipitation estimation have been developed at the Center for Hydrometeorology and Remote Sensing, University of California, Irvine. They are: (1) precipitation estimation from remotely sensed information using artificial neural networks (PERSIANN) and (2) PERSIANN cloud-patch classification system (PERSIANN-CCS). PERSIANN is a multiple-satellite-based precipitation estimation algorithm, which uses local GEO-IR image texture and LEO PMW rainfall rates to estimate precipitation. PERSIANN-CCS extends PERSIANN from local IR pixel and texture features to using patch-based segmentation, image features extraction and classification.

8.2.1 PERSIANN

The development of PERSIANN is based on using the more reliable, but less frequently sampled instantaneous precipitation rate from microwave sensors to adjust the IR to rain rate mapping function of the infrared image of GEO satellites and rainfall rate. The input features of PERSIANN are extracted from the local image texture of the long-wave IR imagery (10.2–11.2 μm) of GEO satellites. The mapping function of ANNs converting a longwave image to the rainfall map is implemented by (1) extracting the local image texture by calculating the mean and variance of IR brightness temperatures near each calculation pixel, (2) classifying the extracted features, and (3) mapping classified features to the surface rainfall rate for that pixel. An adaptive training feature facilitates updating the network parameters whenever independent estimates of rainfall are available (Hsu *et al.*, 1997; Sorooshian *et al.*, 2000). The parameters of PERSIANN are continually updated when PMW-based rainfall is available (Ferraro and Marks, 1995; Hsu *et al.*, 1997; Janowiak *et al.*, 2001). The PERSIANN system first used only GEO-IR imagery. It was later extended to use both GEO-IR and VIS imagery and the rainfall estimates improved (Hsu *et al.*, 1999). The PERSIANN algorithm estimates the rainfall rate at each 0.25×0.25 pixel every 30 min. The estimated rainfall is then integrated to various temporal scales, such as six-hourly, daily, monthly, etc. A description of the current PERSIANN operation is shown in Fig. 8.1. IR imagery is provided by GEO satellites, such as GOES (covering the Americas), MeteoSat (covering Europe and Africa), MTSAT (covering the west Pacific and Asia), while PMW rainfall calculated from the information of LEO satellites, including TRMM of NASA, Defense Meteorological Satellite Program (DMSP) satellites of Department of Defense (DoD), and Polar Orbiting Environmental Satellites (POES) of NOAA, is used to train the mapping parameters of PERSIANN (Ferraro and Marks, 1995; Kummerow *et al.*, 1998; Janowiak *et al.*, 2001).

8.2.2 PERSIANN-CCS

The continuing development of the PERSIANN system has extended the classification features of the GEO IR image from local pixel neighbourhood texture-based to cloud-patch-based features and improved the resolution of the retrieved product from $0.25° \times 0.25°$ latitude–longitude scale to a much finer resolution of

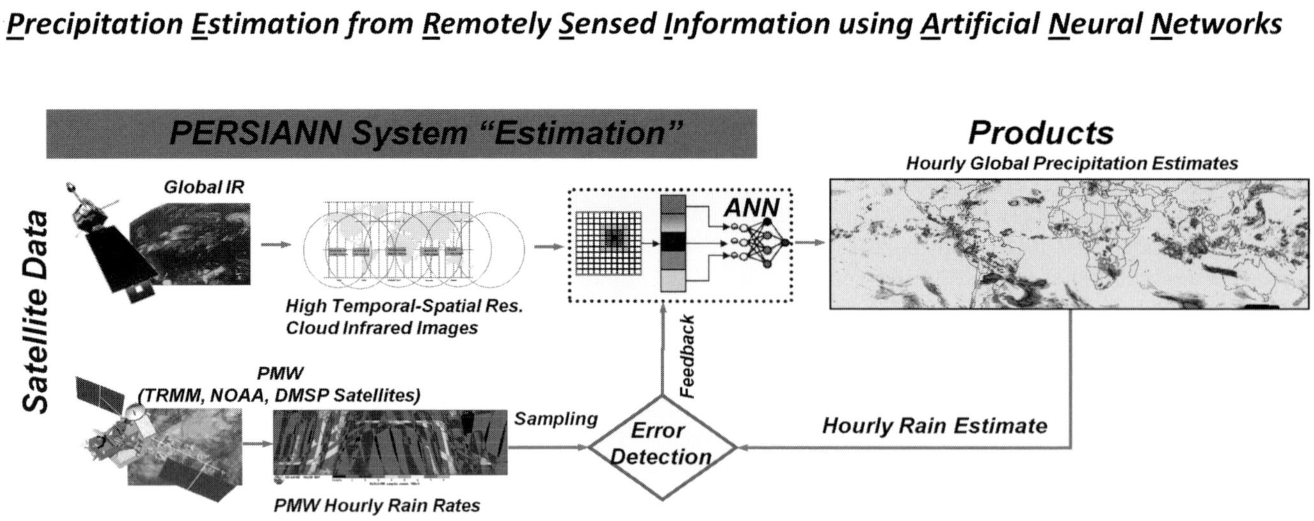

Figure 8.1. Rainfall estimation from the PERSIANN system using GEO and LEO satellite information.

$0.04° \times 0.04°$. This patch-based algorithm is named PERSIANN-CCS. Similar to PERSIANN, PERSIANN-CCS uses an ANN (self-organising feature map) algorithm for image classification and rainfall estimation. It is considered as a different approach from PERSIANN in its unique object-based image segmentation and classification. The PERSIANN-CCS patch-based cloud classification and rainfall estimation process is depicted in Fig. 8.2. This algorithm is implemented in four major steps: (1) GEO-IR cloud image segmentation, (2) GEO-IR cloud patch feature extraction, (3) patch classification, and (4) rainfall estimation. A detailed description of PERSIANN-CCS can be found from Hong *et al.* (2004) and Hsu *et al.*, (2007). These four major processing steps are described below.

(1) GEO-IR cloud image segmentation Each GEO-IR cloud image is segmented to divide it into separable patches. This is performed via a watershed-based segmentation algorithm (Vincent and Soille, 1991). The algorithm starts by finding the local minima of a GEO-IR temperature image, followed by raising the temperature threshold gradually and connecting the neighbourhood pixels attracted to each local minimum pixel until all the local minimum basins are separated into distinct patches.
(2) GEO-IR cloud patch feature extraction The features to be extracted from the segmented cloud patches cover three categories: coldness, geometry, and texture. From these categories, representative features, such as the cloud height (coldest temperature), cloud size and shape, surface textures, and surface gradients are estimated. All of these relevant features are calculated for three GEO-IR temperature thresholds (220 K, 235 K, and 253 K) for each patch.
(3) Patch classification Each GEO-IR cloud patch is classified based on similarities of the extracted features, as one of an array of 20×20 (400 total) categorized groups. This array was created via an unsupervised clustering method, self-organizing feature map (SOFM), and is used to classify patch features into one of a number of cloud patch categories (Kohonen, 1995; Hsu *et al.*, 1999).
(4) Rainfall estimation The GEO-IR temperature (T_b) for each pixel in each cloud patch is converted to an hourly rainfall rate *(R)* using the T_b–R relationship defined for that cloud classification. Initially, a large number of GEO-IR values and surface rainfall values are collected from concurrent GEO IR pixels and LEO-PMW rainfall pixels. The probability matching method (PMM) is used to match the relationship between the GEO-IR temperature (T_b) and the hourly rainfall rate *(R)* in each classified IR patch group, assuming that higher rainfall rates associate with lower IR temperatures (Atlas *et al.*, 1990; Rosenfeld *et al.*, 1994). Stage IV radar, gauge, and LEO satellite PMW rainfall data are used for training the T_b–R relationship. The T_b–R relationship is fitted with an exponential function of five parameters, where those parameters are estimated from the shuffled complex evolution UA (SCEUA) optimisation algorithm (Duan *et al.*, 1992; Hong *et al.*, 2004; Hsu *et al.*, 2007).

8.2.3 G-WADI PERSIANN-CCS GeoServer

With global early warning systems needing near real-time precipitation data, the G-WADI PERSIANN-CCS GeoServer provides one example of how remote sensing

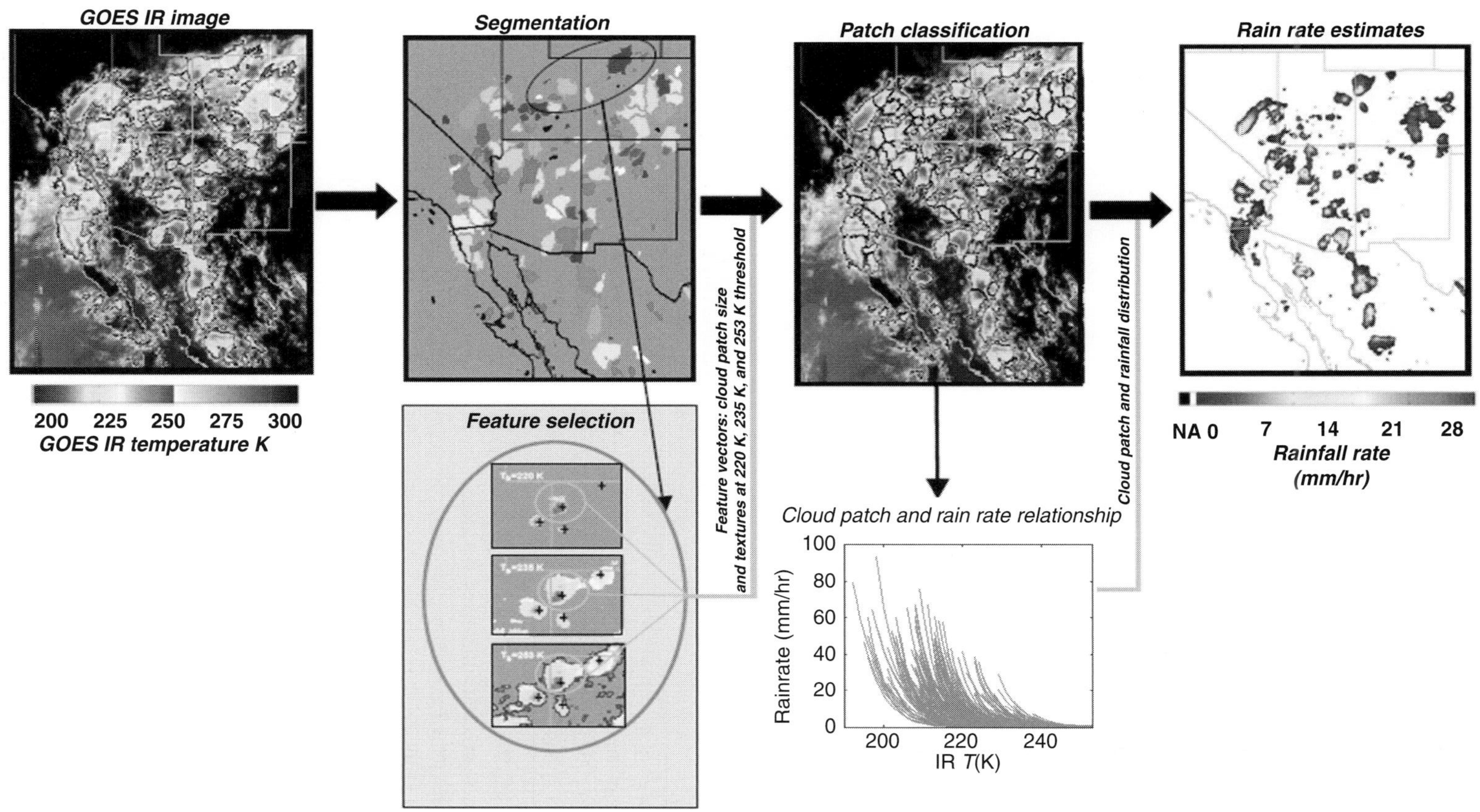

Figure 8.2. Image segmentation, extraction, classification, and rainfall estimation steps of the PERSIANN-CCS algorithm.

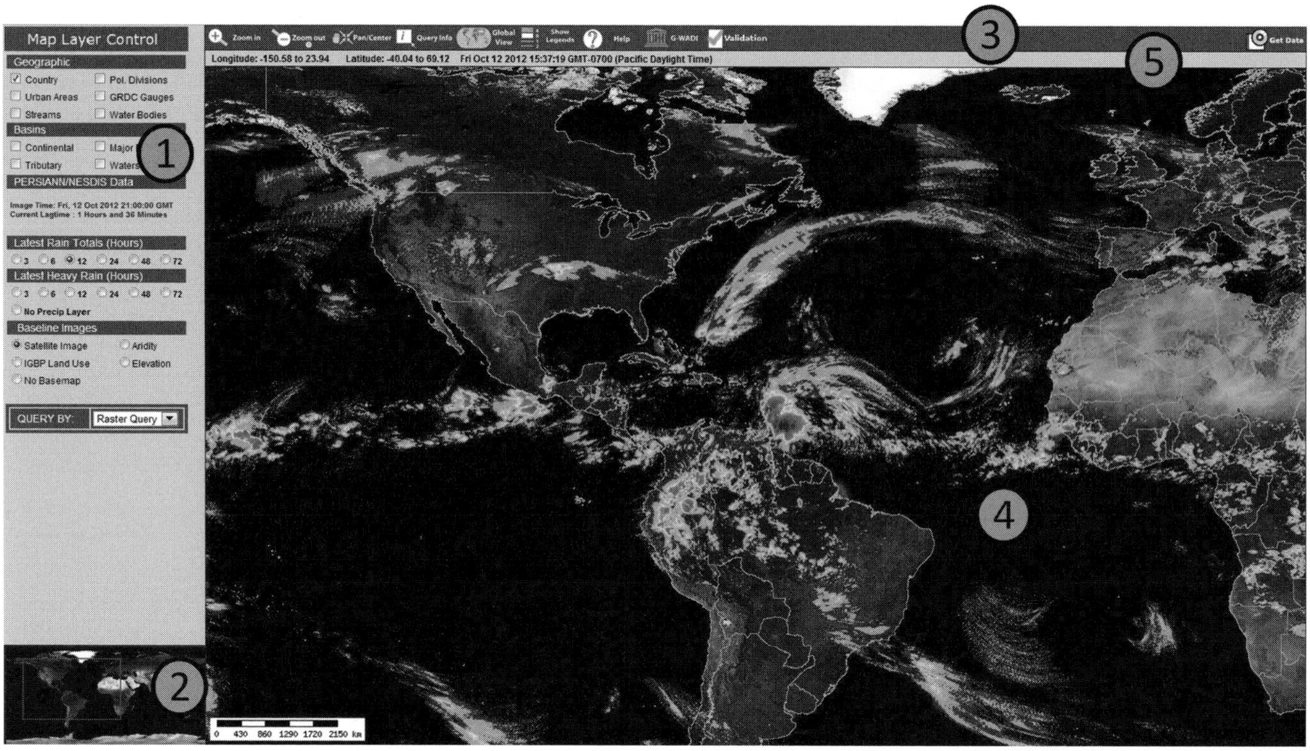

Figure 8.3. Components of the G-WADI GeoServer web-interface include: (1) map layer control menu, (2) interactive reference map, (3) map navigation and query menu, (4) map canvas, and (5) location bar. For colour version, see Plates section.

precipitation estimates can be accessed quickly and efficiently for use in monitoring near global precipitation. The G-WADI PERSIANN-CCS GeoServer is designed to provide user-friendly web-based visualisation and mapping of PERSIANN-CCS precipitation estimates with a time latency of around 1 hour and should be thought of as a good example of how satellite-based remotely sensed precipitation can be estimated, displayed, and used by decision makers. Any global early warning system will need to ensure a similar process in accessing and displaying precipitation information. The G-WADI PERSIANN-CCS GeoServer utilises the open-source MapServer technology (http://mapserver.org//) from the University of Minnesota to provide near real-time visualisation of precipitation for the most recent 3-, 6-, 12-, 24-, 48-, and 72-hour accumulation periods, mapped over various basemaps and overlaid with vector information, such as country, state, and watershed boundaries. The server also provides updated precipitation summaries for each watershed, including minimum, mean, maximum, and average precipitation within the rain area, while updating and populating database tables with this data. Figure 8.3 shows the G-WADI PERSIANN-CCS GeoServer interface, which consists of (1) a map layer control menu, (2) an interactive reference map, (3) a map navigation and query menu, (4) a map canvas, and (5) a location bar. The layer control menu includes four sections that allow the user to add one or more vector overlays, choose the PERSIANN-CCS data accumulated period, and select a base image map.

From the initial global view, a user can zoom and pan to an area or event of interest. An event can be tracked by showing successive accumulation periods up to 72 hours. The *3hr* will show the most recent 3 hours of accumulated precipitation. Clicking on the *6hr* will show you what happened for the 3 hours before the event and how the weather system has moved. The high spatial resolution ($0.04^\circ \times 0.04^\circ$) of the PERSIANN-CCS data allows many levels of zoom, down to a very fine scale. Overlaying the basin boundaries and streams will help a user to visualise how runoff will contribute to potential flooding. Another helpful feature of the G-WADI PERSIANN-CCS GeoServer is the ability to query by pixel, basin, political division, or country. Each query produces a report page showing not only the recent precipitation accumulations, but also including areas with the heaviest precipitation, climatological information, basemap data such as aridity and elevation, and even a link to download the precipitation data. Once a desired area is defined, the data for all the various accumulation totals can be easily downloaded using the '*get data*' button. The user can manually adjust the bounding coordinates before submitting the selection, and the server will provide links for each accumulation total for the chosen subset area.

One important aspect of the G-WADI PERSIANN-CCS GeoServer is that all of the software and technical work is done on the server and the user, whether she/he is a GIS expert using a workstation or a farmer using a simple personal computer, only needs to have a modern browser and Internet access to take advantage of all of the tools available on the GeoServer. With quick and easy access to precipitation data over remote regions, extreme weather events can be observed. These observations provide the user with critical information for planning purposes and potential integration into global early warning systems. A tutorial describing the functions of the G-WADI PERSIANN-CCS GeoServer has been made available in several languages (English, Farsi, Chinese, Spanish, and Vietnamese) and can be accessed online: http://persiann.eng.uci.edu/gwadi_tutorial_videos.html.

8.3 Remote sensing precipitation measurement for flood and drought warnings

The United Nations (2006) defines early warning as 'the provision of timely and effective information, through identified institutions, that allows individuals exposed to hazard to take action to avoid or reduce their risk and prepare for effective response'. An effective early warning system consists of four key elements: risk knowledge, monitoring and predicting, disseminating information, and response (UNEP, 2012). All parts of the system are important and interact with each other. Monitoring and predicting hazardous impacts plays a crucial role in producing warnings to the public. Any data collected and forecast should be accurate, timely, and reliable, and use appropriate technologies.

8.3.1 Flood

Thanks to the developments in earth observation technologies since the 1960s, more data of better quality have become available from various satellites by organisations such as NASA, NOAA, and ESA (Europe). In a flood early warning system, precipitation data might come from ground-based gauges, radars, and satellites. However, floods, especially flash floods, often occur in rural areas, where no rain gauge data are available, and areas out of radar coverage; satellite data are often the only option. Radars provide accurate data in areas under their coverage. However, radars are not effective in mountainous areas where natural hazards (e.g. flash floods and landslides) occur most often. In addition, radars require high skill levels and are expensive to operate and maintain. For example, in Pakistan, there are currently seven meteorological radars but most of them are only in continuous operation during the monsoon season (WMO, 2010). Satellites have been considered a good option for providing input to early warning systems because they offer large areal coverage, continuous data at short time intervals, and good quality.

Scientists have made significant efforts in developing early warning systems to alert the public to water-related hazards before they occur. Among such systems, the All-hazard Warning, Analysis, and Risk Evaluation system (Disaster AWARE) has been under development by the Pacific Disaster Center (PDC) in Hawaii since 1996. The system automatically retrieves hazard bulletins from various agencies such as NOAA National Weather Service, National Hurricane Center, National Earthquake Information Center, and Pacific Tsunami Warning Center. Data are processed to produce various warnings (e.g. flash floods, hurricanes, earthquakes, tsunami, and landslides) on the web. Alerts can also be sent to registered users via email and text messaging. The Global Flash Flood Guidance System (GFFGS) with global coverage has been developed by the Hydrologic Research Center (HRC) at the University of California, San Diego. The system is fed input data from satellites and/or in situ gauge data. GFFGS can be set to run in real-time mode with observed data or forecast mode with forecast precipitation data. The system can also be tailored to a local scale as a regional flash flood system. It was successfully implemented for seven countries in Central America (Panama, Costa Rica, Nicaragua, El Salvador, Honduras, Guatemala, and Belize) and other parts of the world such as Cambodia, Laos, Thailand, and Vietnam. The Flood Early Warning System has been developed by Deltares in the Netherlands (Delft-FEWS). Delft-FEWS is a state-of-the-art hydrologic forecast and warning system, which can retrieve, store, and process inputs from various external models to produce forecasts and warnings through an open-source interface (Deltares, 2010). The system has been successfully applied operationally in more than 20 countries, among them the USA (Community Hydrological Prediction System – CHPS), several southeastern Asian countries (FEWS Mekong), England and Wales (National Flood Forecasting System – NFFS), and the Netherlands (Water Information System).

Last but not least, an excellent effort has been made by the International Centre for Water Hazard and Risk Management under the auspices of UNESCO (ICHARM) for effective flood management. The Integrated Flood Analysis System (IFAS) has been developed by ICHARM for local flood management of areas prone to floods. With rainfall input automatically retrieved from various sources (satellites, radars, gauges), IFAS estimates river flow to produce forecasts and analyses for floods in the basins. IFAS has been provided to developing countries through training programmes for operational meteorologists, hydrologists, and disaster management officers (ICHARM, 2012). It is hoped that IFAS will be an effective system for the prevention/mitigation of water-related disasters. Flooding disasters can be caused by a number of complex phenomena. In Asia, one source of abundant moisture is typhoons. To provide an example of the need for global early warning systems to use satellite-based remote sensing precipitation products in regions of sparse observations, we describe the observations of Tropical Storm Nock-Ten using the G-WADI PERSIANN-CCS GeoServer.

Tropical Storm Nock-Ten caused severe flooding across Thailand starting in late July 2011. It is considered one of the most devastating natural hazards in modern history in terms of the number of people affected (over 8 million, reported by MRC (2011)) and economic damage (totalled US$45.7 billion, estimated by the World Bank 2011). One application of a platform that can potentially assist global early warning systems is the G-WADI PERSIANN-CCS GeoServer (G-WADI GeoServer) hosted by the Center of Hydrometeorology and Remote Sensing at the University of California, Irvine. Figure 8.4 shows the 72-hour accumulation over Thailand of the PERSIANN-CCS rainfall estimates during 29–31 July 2011 displayed on the G-WADI GeoServer. Emergency managers and decision makers can employ the G-WADI GeoServer to monitor the development of extreme events. Near real-time precipitation data from the system can also be fed to global early warning systems such as GFFGS and Delft-FEWS to produce public alerts.

Other monitoring and warning systems provide similar precipitation estimates and regional warnings for associated flooding and other natural disasters. Figure 8.5 shows an example of an early warning system on PDC's web page (http://www.pdc.org/atlas/) at 00:00 5 October 2012 which captures various active natural hazards including volcanoes, earthquakes, tropical storm Gaemi and its floods. Gaemi developed on 29 September 2012 at 17.2° N, 114.7° E in the northwestern Pacific Ocean. The 3-day average rainfall (inches/hour) data for 2–4 October 2012 (UTC) was retrieved from the TRMM satellite. Heavy rain during the storm caused floods across the region. Observed data from water level stations show floods occurring in Thailand and the south of Vietnam. This early warning system offers valuable tools for emergency managers and decision makers in preventing/mitigating impacts of the hazards.

8.3.2 Drought

Droughts are among the most costly natural disasters. Real-time and reliable precipitation measurement plays a significant role in drought monitoring and prediction. Droughts are typically categorised into three major classes: (a) meteorological, (b) agricultural, and (c) hydrological (Wilhite and Glantz, 1985). All types of droughts are consequences of a sustained lack of precipitation for a period of time with respect to the climatology. For example, agricultural drought represents a deficit in soil moisture due to lack of precipitation. Similarly, hydrological drought is defined as abnormally below average stream flow, which is directly proportional to precipitation. In fact, regardless of the type of drought, reliable

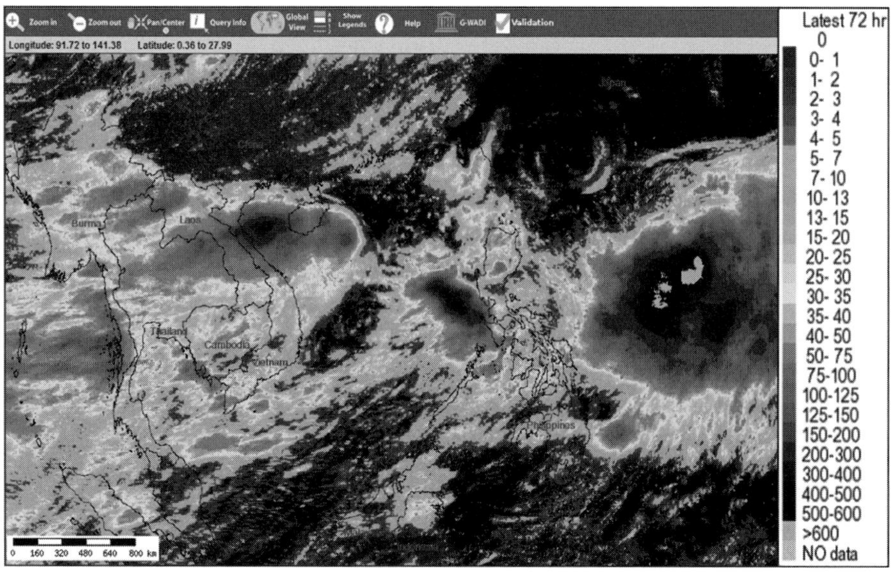

Figure 8.4. G-WADI GeoServer visualisation of 72-hour accumulated PERSIANN-CCS precipitation totals in millimeters for the downpour rainfall caused by the Nock-Ten depression in the Thailand 2011 flood from 00:00 UTC on 29 July 2011 up to 00:00 UTC on 1 August 2011 displaying the widespread heavy precipitation across Thailand. For colour version, see Plates section.

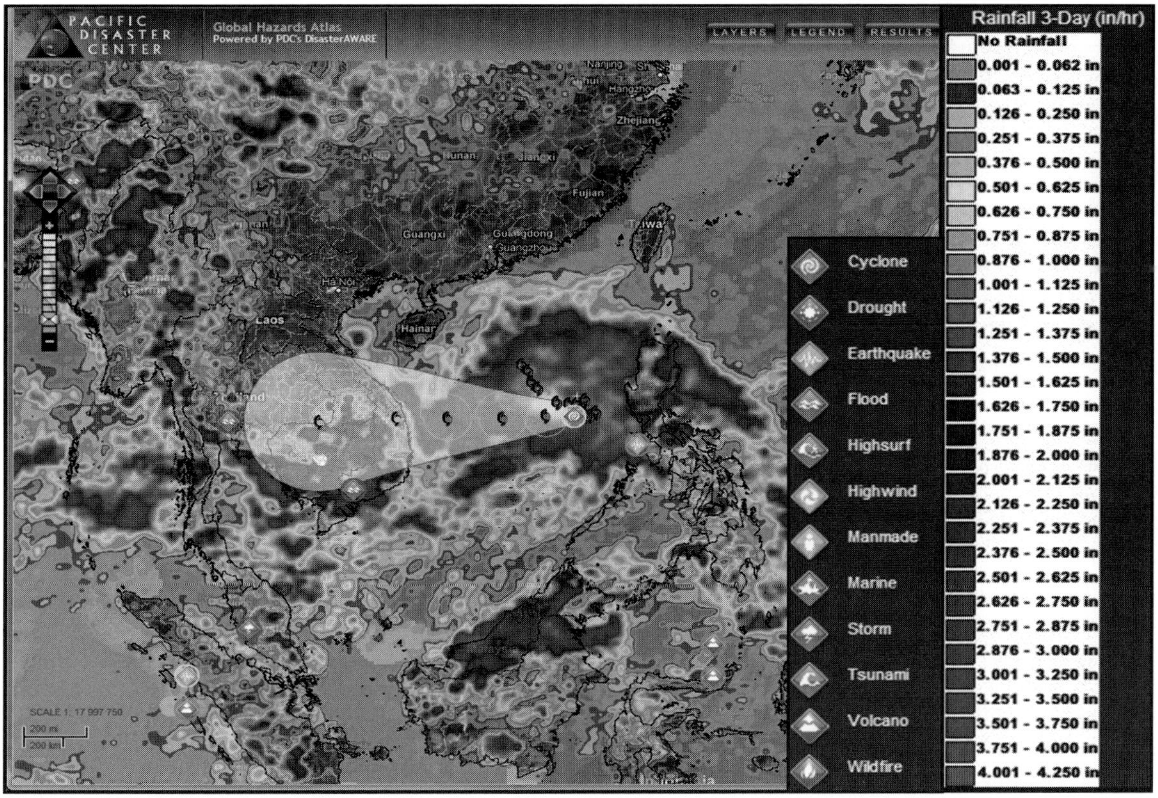

Figure 8.5. PDC Global Hazards Atlas showing rainfall (inches/hour) during 2–4 October 2012 (UTC) in the northwestern Pacific Ocean, with the Joint Typhoon Warning Center (JTWC) forecast positions and error cone for active storm Gaemi (source from http://www.pdc.org/atlas/). For colour version, see Plates section.

precipitation information can be used to better understand the behaviour and patterns of droughts.

There are indices such as the standardized precipitation index (SPI) (McKee *et al.*, 1993) that can be used to describe droughts based on precipitation. One attractive feature of SPI is that with availability of precipitation in real-time, one can provide timely information. The SPI has been used for drought monitoring based on long-term rain gauge measurements in numerous studies (e.g. see Hayes *et al.*, (1999) and references therein). Rain gauges, however, are unequally distributed across the world, and in most parts of the globe gauge measurements are either not available or not representative (Easterling, 2012). Limitations of gauge networks lead to shortcomings in drought monitoring especially over remote regions. Satellite precipitation products, on the other hand, provide the opportunity to capture the spatial variability of precipitation beyond the traditional interpolated rain gauge data (Sorooshian *et al.*, 2011).

Satellite precipitation data have been already used in several drought monitoring studies (e.g. Sheffield *et al.*, 2006; AghaKouchak and Nakjiri, 2012). The experimental African Drought Monitor, for example, provides drought information using a land-surface model and satellite precipitation data as input (Sheffield *et al.*, 2006). Anderson *et al.* (2007) used TRMM data, along with other remotely sensed products, to study vegetation response to droughts. Anderson *et al.* (2007) concluded that remote sensing data sets can improve drought monitoring and analysis. Using CMORPH precipitation data, Zhang *et al.* (2008) investigated droughts in northern China. The potential application of satellite data for drought monitoring is also highlighted in Paridal *et al.* (2008) after assessing drought over the Philippines using remote sensing data.

While the advantages of remotely sensed data over the traditional ground-based measurements (e.g. high spatial resolution and global coverage) have been argued in several studies they have not been fully integrated into operational drought monitoring. Drought monitoring requires long-term observations since most indices are defined relative to the climatological records (e.g. precipitation condition relative to long-term precipitation climatology). However, real-time high-resolution satellite observations do not provide long-term data to derive a reliable climatology. There are other long-term low-resolution satellite data products, such as the Global Precipitation

Climatology Project (GPCP, Adler *et al.*, 2003). GPCP provides satellite-based gauge-adjusted precipitation estimates on a global scale that can be used in drought monitoring. However, the GPCP data are not available in real-time. In their study, AghaKouchak and Nakhjiri (2012) have developed a Bayesian algorithm, based on Tian *et al.* (2010), that creates consistent climatology by combining near real-time PERSIANN data and long-term GPCP measurements. In this approach, based on the overlap between the two data sets, the near real-time satellite data are adjusted to have the same climatology as that of the reference data (here, GPCP). The objective is to resolve the limitations in the number of data, which is fundamental to drought monitoring.

Figure 8.6 displays an example application of using PERSIANN precipitation data for drought monitoring and prediction. Figure 8.6(a) displays GPCP as reference only (not used to derive the rest of the panels). Figure 8.6(b) shows combined GPCP-PERSIANN with no correction to climatology. Given the biases in real-time satellite data (AghaKouchak *et al.*, 2011, 2012), one can see Fig. 8.6(a) and Fig. 8.6(b) are not in agreement. Figure 8.6(c) demonstrates the combined GPCP (excluding 2010) and PERSIANN data, after the Bayesian correction for creating a consistent climatology for drought monitoring. As shown the Bayesian correction in Fig. 8.6(c) reasonably captures the features in the reference data (Fig. 8.6(a)). Having such a model, one can apply near real-time satellite observations for timely drought monitoring.

Figure 8.7 displays an example application of the proposed methodology for drought monitoring. One can see that some of the major droughts such as Texas 2011 and Ethiopia 2011 can be seen in the satellite-based drought time series. These example applications highlight the potential of real-time satellite precipitation data for real-time drought monitoring and for developing early warning systems.

8.3.3 Remote sensing precipitation estimation: strengths and limitations

Precipitation is a very challenging variable to observe via remote sensing instruments. Using the current suite of GEO and LEO satellites and combining the information and data from these remote sensing sources has resulted in many approaches and algorithms for precipitation estimation at high spatial and temporal resolutions. High spatial and temporal resolution and near global coverage are considered the main advantage of using remote sensing approaches for precipitation estimation. GEO satellites providing images every 15–30 minutes and remaining permanently in the same location provide almost constant

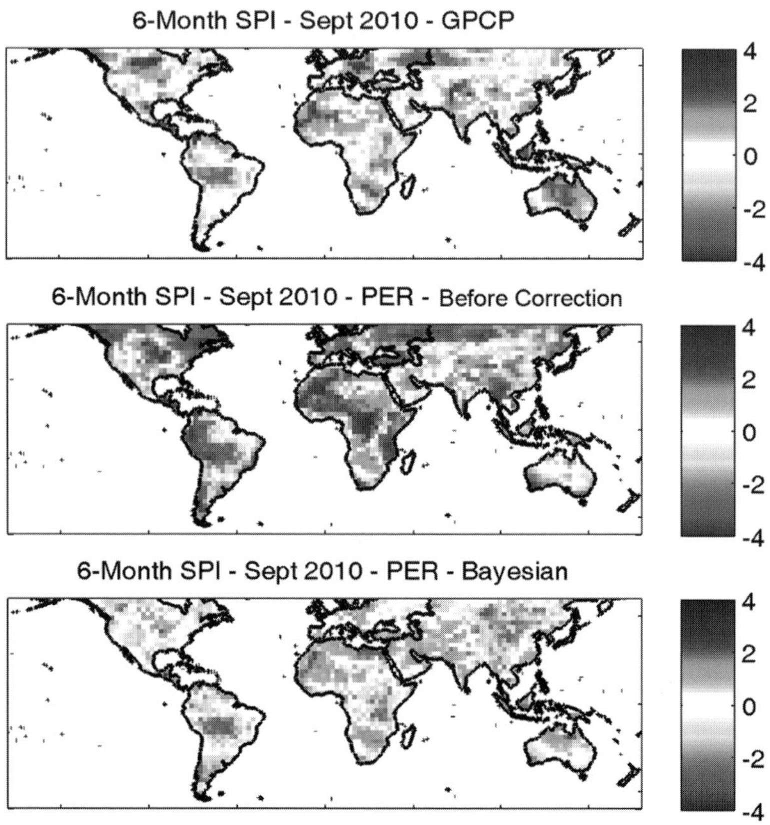

Figure 8.6. (a) Reference GPCP data, not used to derive the other two panels; (b) PERSIANN data before applying the Bayesian algorithm for making the climatology consistent; (c) combined GPCP (excluding 2010) and PERSIANN data, after the Bayesian correction for creating consistent climatology (AghaKouchak and Nakhjiri, 2012). For colour version, see Plates section.

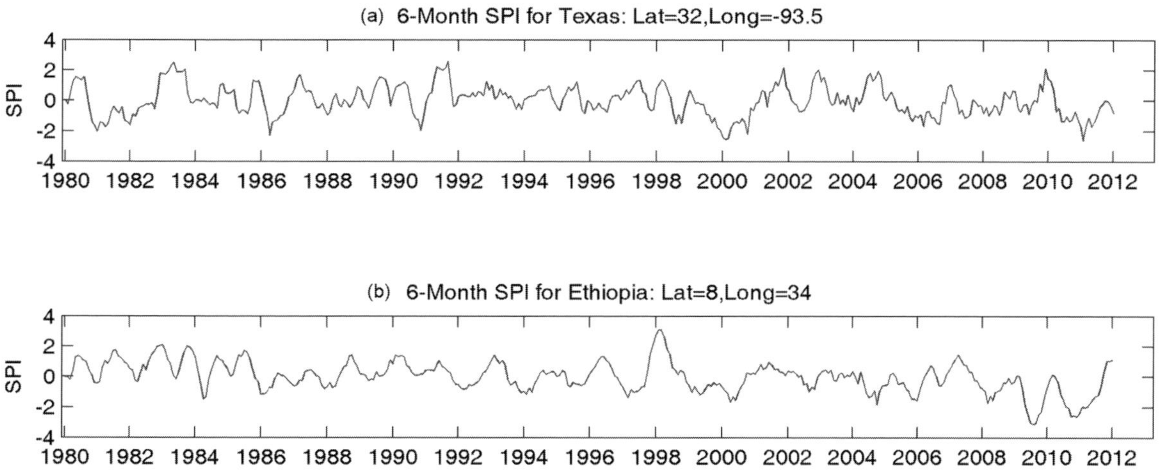

Figure 8.7. Six-month SPI data over (a) Texas and (b) Ethiopia derived using satellite precipitation data (AghaKouchak and Nakhjiri, 2012).

observations of the viewable earth from their respective positions. LEO satellites, having a lower altitude and including PMW sensors, provide more direct information regarding precipitation, with the main constraint being the temporal resolution. Sampling limitations from LEO satellites in temporal and spatial resolution prevent accurate predictions of flooding and landslide potential. The upcoming launch of GPM satellites will increase sampling and improve precipitation measurement and coverage. Using effective sensors and combining estimates from LEO and GEO satellites, nearly the entire earth can be observed at regular intervals. With many research groups continuing to progress towards finer and finer resolution data, global early warning systems will likely be the largest user of this high-resolution precipitation data. Considering the scarcity of ground-based observations globally, high spatial and temporal resolution precipitation data can be seen as the most useful to global early warning systems. A number of limitations have been addressed and require further improvement. Satellite imagery can provide measurement of precipitation at regional to global scales without referencing the ground measurement, but those estimates may be biased relative to ground-based estimation. Effective use of measurements from ground-based gauges/radars can reduce bias from satellite measurement. Some studies have shown improvements from merging satellite and other sources, such as gauge and radar measurements, as numerical model estimation as well (Xie and Arkin, 1997; Huffman *et al.* 2001; Ebert *et al.*, 2007; Boushaki *et al.*, 2009; Tian *et al.*, 2010; Behrangi *et al.*, 2011). Overall, remotely sensed precipitation estimates can be improved by bias correction and by adding other GEO channels containing additional information affecting precipitation estimates. The accuracy of remotely sensed precipitation estimates is dependent on algorithms used to interpret the data provided by the remote sensing platforms. Uncertainties in precipitation estimations can depend on season, storm type, elevation, and climate region, which all greatly impact the precipitation estimated from retrieval algorithms (Sorooshian *et al.*, 2011). Long-term ground observations exist only for a few regions around the world. With the scarcity of observations, developing training and validation data sets for retrieval algorithm development is difficult. In addition, local knowledge of landslide potential factors and flood prone areas, which is lacking in remote areas, is very important, and manual assessments need to be made of the data.

8.4 Conclusions

Global early warning systems must provide timely and accurate information, which can be used to adapt and/or mitigate potential impacts of natural hazards on society. Reducing society's risk to natural hazards must be the goal of any global early warning system. With floods and droughts being the most devastating natural hazards to society, satellite information for precipitation estimation will continue to play an important role in global early warning systems. With the proper investment in technological capacity (e.g. data storage and processing power), communication infrastructure, and early warning system design, advanced remote sensing capabilities can be harnessed to mitigate the loss of life and property caused by flooding and landslides due to heavy precipitation. Drought monitoring also takes advantage of the high spatial and temporal resolution of remotely sensed precipitation data, which can be used for drought mitigation and adaptation. Advanced precipitation estimation algorithms continue to evolve and show promise with reliable high spatial and temporal resolution estimates. Global early warning systems are well positioned to take full advantage of remotely sensed precipitation estimates with global coverage, high spatial resolution, and near real-time data.

Remote sensing precipitation estimates provide a powerful view into global precipitation. Considering all the strengths and limitations of remotely sensed precipitation estimates, there is opportunity for significant improvement in the near future. With the strengths of near global coverage, high temporal resolution, and near real-time capabilities, satellite derived precipitation estimates must be seen as a large contributor to any global early warning system. Challenging limitations still remain, such as precipitation errors, uncertainties, and biases, lack of regional training and validation data, sensor consistency, and future funding. Sound research and investment in improved satellite platforms can seek to minimise these limitations. The remotely sensed precipitation products available have numerous differences as described above; all have their own biases and errors, which cause some difficulties in deciding which products are the most accurate. The accuracy of the individual products can be very dependent on region, elevation, and season. Minimising these errors and improving the fundamental understanding of the root causes of biases is a major priority for remotely sensed precipitation estimates. With continued development of precipitation algorithms, the accuracy and availability of precipitation data will only continue to improve. From major floods to widespread droughts, remote sensing precipitation data provide vital information on a spatial and temporal resolution necessary for observing, monitoring, and preparing for precipitation-related natural hazards.

Acknowledgments. The authors acknowledge Bisher Imam for his development of CHRS G-WADI. Partial financial support was provided by the NASA-PMM (Grant# NNX10AK07G), the US Army Research Office project (Grant# W911NF-11-1-0422), NOAA-CCDD (Grant# NA10OAR4310122), and NOAA-CICS (Prime award NA09NES4400006, NCSU CICS subaward# 2009–1380–01).

References

Adler, R. F. and Negri, A. J. (1988). A satellite infrared technique to estimate tropical convective and stratiform rainfall. *Journal of Applied Meteorology*, **27**, 30–38.

Adler, R. F., Huffman, G. J., Chang, A. T. C. *et al.* (2003). The Version 2 Global Precipitation Climatology Project (GPCP) monthly precipitation analysis (1979–present). *Journal of Hydrometeorology*, **4**, 1147–1167.

AghaKouchak, A. and Nakhjiri, N. (2012). A near real-time satellite-based global drought climate data record. *Environmental Research Letters*, **7**(4), 044037, doi:10.1088/1748-9326/7/4/044037.

AghaKouchak, A., Behrangi, A., Sorooshian, S., Hsu, K. and Amitai, E. (2011). Evaluation of satellite-retrieved extreme precipitation rates across the Central United States. *Journal of Geophysical Research*, **116**, D02115, doi:10.1029/2010JD014741.

AghaKouchak, A., Mehran, A., Norouzi, H. and Behrangi, A. (2012). Systematic and random error components in satellite precipitation data sets. *Geophysical Research Letters*, **39**, L09406, doi:10.1029/2012GL051592.

Anderson, L., Malhi, Y., Aragao, L. and Saatchi, S. (2007). Spatial patterns of the canopy stress during 2005 drought in Amazonia. In *2007 IEEE International Geoscience and Remote Sensing Symposium, IGARSS 2007*, Barcelona, Spain, 23–28 July 2007. Piscataway, NJ: Institute of Electrical and Electronic Engineers, pp. 2294–2297.

Arkin, P. A. and Meisner, B. N. (1987). The relationship between large-scale convective rainfall and cold cloud over the western hemisphere during 1982–84. *Monthly Weather Review*, **115**, 51–74.

Atlas, D., Rosenfeld, D. and Short, D. (1990). The estimation of convective rainfall by area integrals. 1. The theoretical and empirical basis. *Journal of Geophysical Research*, **95**, No D3, 2153–2160.

Ba, M. B. and Gruber, A. (2001). GOES Multispectral Rainfall Algorithm (GMSRA). *Journal of Applied Meteorology*, **40**, 1500–1514.

Behrangi, A., Hsu, K., Imam, B. *et al.* (2010). PERSIANN-MSA: a precipitation estimation method from satellite-based multispectral Analysis. *Journal of Hydrometeorology*, **10**(6), 1414–1429.

Behrangi, A., Imam, B., Hsu, K. *et al.* (2011). Rain estimation using forward adjusted-advection of microwave estimates (REFAME). *Journal of Hydrometeorology*, **11**, 1305–1321.

Bellerby, T. (2004). A feature-based approach to satellite precipitation monitoring using geostationary IR imagery. *Journal of Hydrometeorology*, **5**, 910–921.

Bellerby, T., Todd, M., Kniveton, D. and Kidd, C. (2000). Rainfall estimation from a combination of TRMM Precipitation Radar and GOES multi-spectral satellite imagery through the use of an artificial neural network. *Journal of Applied Meteorology*, **39**, 2115–2128.

Bellerby, T., Hsu, K. and Sorooshian, S. (2009). LMODEL: A satellite precipitation algorithm using cloud development modeling and model updating. Part I: Model development and calibration. *Journal of Hydrometeorology*, **10**(5), 1081–1095. doi: 10.1175/2009JHM1091.1.

Boushaki, F. I., Hsu, K., Sorooshian, S. *et al.* (2009). Bias adjustment of satellite precipitation estimation using ground-based measurement. *Journal of Hydrometeorology*, **10**, 1231–1242.

Capacci, D. and Conway, B. J. (2005). Delineation of precipitation areas from MODIS visible and infrared imagery with artificial neural networks. *Meteorological Applications*, **12**, 291–305.

Deltares. (2010). Software Delft-FEWS. Innovative solutions for water and subsurface issues. Available at http://www.deltares.nl/en/software/479962/delft-fews/479964.

Duan, Q., Sorooshian, S. and Gupta, V. K. (1992). Effective and efficient global optimization for conceptual rainfall-runoff models. *Water Resources Research*, **28**(4), 1015–1031.

Easterling, D. (2012). Global data sets for analysis of climate extremes. In *Extremes in a Changing Climate: Detection, Analysis and Uncertainty*, eds. AghaKouchak, A., Easterling, D., Hsu, K., Schubert, S. S., and Sorooshian, S., eds., Springer, doi: 10.1007/978-94-007-4479-012.

Ebert, E. E., Janowiak, J. E. and Kidd, C. (2007). Comparison of near-real-time precipitation estimates from satellite observations and numerical models. *Bulletin of the American Meteorological Society*, **88**, 47–64.

Ferraro, R. R. and Marks, G. F. (1995). The development of SSM/I rain-rate retrieval algorithms using ground-based radar measurements. *Journal of Atmospheric and Oceanic Technology*, **12**, 755–770.

Hayes, M., Svoboda, M., Wilhite, D. and Vanyarkho, O. (1999). Monitoring the 1996 drought using the standardized precipitation index. *Bulletin of the American Meteorological Society*, **80**, 429–438.

Hong, Y., Hsu, K., Sorooshian, S. and Gao, X. (2004). Precipitation estimation from remotely sensed imagery using an artificial neural network cloud classification system. *Journal of Applied Meteorology*, **43**, 1834–1852.

Horsfield, N. (2006). Development of a mass balance approach to modeling cloud lifecycles and rainfall using satellite observations. Ph.D. Thesis. University of Hull, UK. 353pp.

Hou, A., Jackson, G. S., Kummerow, C. and Shepherd, J. M. (2008). Global precipitation measurement. In *Precipitation: Advances in Measurement, Estimation, and Prediction*, ed. Michaelides, S. Berlin: Springer, pp. 131–170.

Hsu, K., Gupta, H. V. and Sorooshian, S. (1995). Artificial Neural Network Modeling of the Rainfall-Runoff Process. *Water Resources Research*, **31**(10), 2517–2530, doi:10.1029/95WR01955.

Hsu, K., Gao, X., Sorooshian, S. and Gupta, H. V., (1997). Precipitation estimation from remotely sensed information using artificial neural networks. *Journal of Applied Meteorology*, **36**(9), 1176–11.

Hsu, K., Gupta, H. V., Gao, X. and Sorooshian, S. (1999). Estimation of physical variables from multiple channel remotely sensed imagery using a neural network: application to rainfall estimation. *Water Resources Research*, **35**(5), 1605–1618.

Hsu, K., Hong, Y. and Sorooshian, S. (2007). Rainfall estimation using a cloud patch classification map. In *Measuring Precipitation from Space: EURAINSAT and Future*, eds. Levizzani, V., Bauer, P. and Turk, F.J. Dordrecht: Springer Publishing Company, pp. 329–242.

Hsu, K., Bellerby, T. and Sorooshian, S. (2009). LMODEL: A satellite precipitation algorithm using cloud development modeling and model updating. Part II: Model updating. *Journal of Hydrometeorology*, **10**(5), 1096–1108. doi: 10.1175/2009JHM1092.1.

Huffman, G. J., Adler, R. F., Morrissey, M. *et al.* (2001). Global precipitation at one-degree daily resolution from multisatellite observations. *Journal of Hydrometeorology*, **2**, 36–50.

Huffman, G. J., Adler, R. F., Bolvin, D. T. *et al.* (2007). The TRMM Multisatellite Precipitation Analysis (TMPA): Quasi-global, multiyear, combined-sensor precipitation estimates at fine scales. *Journal of Hydrometeorology*, **8**, 38–55.

ICHARM. (2012). *ICHARM Newsletter*, **7**, No. 2, July 2012.

Janowiak, J. E., Joyce, R. J. and Yarosh, Y. (2001). A real-time global half-hourly pixel resolution infrared dataset and its applications. *Bulletin of the American Meteorological Society*, **82**, 205–217.

Joyce, R. J., Janowiak, J. E., Arkin, P. A. and Xie, P. (2004). CMORPH: A method that produces global precipitation estimates from passive microwave and infrared data at high spatial and temporal resolution. *Journal of Hydrometeorology*, **5**(3), 487–503.

Joyce, R. J. and Xie, P. (2011). Kalman filter based CMORPH. *Journal of Hydrometeorology*, **12**(6), 1547–1563.

Kidd, C., Kniveton, D. R., Todd, M. C. and Bellerby, T. J. (2003). Satellite rainfall estimation using combined passive microwave and infrared algorithms. *Journal of Hydrometeorology*, **4**, 1088–1104.

Kohonen, T. (1995). *Self-Organizing Map*, 1st edn. Berlin: Springer-Verlag.

Kummerow, C., Barnes, W., Kozu, T., Shiue, J. and Simpson, J. (1998). The Tropical Rainfall Measuring Mission (TRMM) sensor package. *Journal of Atmospheric and Oceanic Technology*, **15**(3), 809–817.

Kummerow, C., Simpson, J., Thiele, O. *et al.* (2000). The status of the Tropical Rainfall Measuring Mission (TRMM) after two years in orbit. *Journal of Applied Meteorology*, **39**(12), 1965–1982.

Machado, L. A. T., Rossow, W. B., Guedes, R. L. and Walker, A. W. (1998). Life cycle variations of mesoscale convective systems over the Americas. *Monthly Weather Review*, **126**, 1630–1654.

Marzano, F. S., Palmacci, M., Cimini, Giuliani, G. and Turk, F. J. (2004). Multivariate statistical integration of satellite infrared and microwave radiometric measurements for rainfall retrieval at the geostationary scale. *IEEE Transactions on Geoscience and Remote Sensing*, **42**, 1018–1032.

McKee, T., Doesken, N. and Kleist, J. (1993). The relationship of drought frequency and duration to time scales. In *Proceedings of the 8th Conference of Applied Climatology, 17–22 January 1993*. Anaheim, CA: American Meteorological Society, pp. 179–184.

MRC (2011). Flood Situation Report 2011. MRC Technical Paper No. 36, November 2011.

Nicholson, S. E., Some, B., McCollum, J. *et al.* (2003a). Validation of TRMM and other rainfall estimates with a high-density gauge dataset for West Africa. Part I: Validation of GPCC rainfall product and pre-TRMM satellite and blended products. *Journal of Applied Meteorology*, **42**, 1337–1354.

Nicholson, S. E., Some, B., McCollum, J. *et al.* (2003b) Validation of TRMM and other rainfall estimates with a high-density gauge dataset for West Africa. Part II: Validation of TRMM rainfall products. *Journal of Applied Meteorology*, **42**, 1355–1368.

Okamoto, K., Iguchi, T., Takahashi, N., Iwanami, K. and Ushio, T. (2005). The global satellite mapping of precipitation (GSMaP) project. In *Proceedings of the International Geoscience and Remote Sensing Symposium, IGARSS 2005, Seoul, South Korea, 25–29 July 2005*. Piscataway, NJ: Institute of Electrical and Electronic Engineers, pp. 3414–3416.

Paridal, B. R., Collado, W. B., Borah, R., Hazarika, M. K. and Sarnarakoon, L. (2008). Detecting drought-prone areas of rice agriculture using a modis-derived soil moisture index. *GIScience & Remote Sensing*, **45**(1), 109–129.

Rosenfeld, D.Wolff, D. B. and Amitai, E. (1994). The window probability matching method for rainfall measurements with radar. *Journal of Applied Meteorology*, **33**, 683–693.

Sheffield, J., Goteti, G. and Wood, E. (2006). Development of a 50-yr, high resolution global dataset of meteorological forcings for land surface modeling. *Journal of Climate*, **13**, 3088–3111.

Simpson, J., Adler, R. F. and North, G. R. (1988). A proposed Tropical Rainfall Measurement Mission (TRMM) satellite. *Bulletin of the American Meteorological Society*, **69**(3), 278–295.

Sorooshian, S., Hsu, K., Gao, X. *et al.* (2000). Evaluation of PERSIANN system satellite-based estimates of tropical rainfall. *Bulletin of the American Meteorological Society*, **81**, 2035–2046.

Sorooshian, S., AghaKouchak, A., Arkin, P. *et al.* (2011). Advanced concepts on remote sensing of precipitation at multiple scales. *Bulletin of the American Meteorological Society*, **92** (10), 1353–1357, doi: 10.1175/2011BAMS3158.1.

The World Bank. (2011). The World Bank supports Thailand's post-floods recovery effort. Feature story, 13 December 2011.

Tian, Y., Peters-Lidard, C. D. and Eylander, J. B. (2010). Real-time bias reduction for satellite-based precipitation estimates. *Journal of Hydrometeorology*, **11**, 1275–85.

Todd, M., Kidd, C., Kniveton, D. R. and Bellerby, T. J. (2001). A combined satellite infrared and passive microwave technique for the estimation of small scale rainfall. *Journal of Atmospheric and Oceanic Technology*, **18**, 742–755.

Turk, F. J. and Miller, S. D. (2005). Toward improving estimates of remotely-sensed precipitation with MODIS/AMSR-E blended data techniques. *IEEE Transactions on Geoscience and Remote Sensing*, **43**, 1059–1069.

UCAR (2010). Flash Flood Early Warning System Reference Guide 2010. Available at http://www.meted.ucar.edu/communities/hazwarnsys/haz_fflood.php.

UNEP (2012). *Early Warning Systems: A State of the Art Analysis and Future Directions. Division of Early Warning and Assessment (DEWA)*, United Nations Environment Programme (UNEP), Nairobi.

United Nations (2006). *Global Survey of Early Warning Systems*. United Nations report.

Vincent, L. and Soille, P. (1991). Watersheds in digital spaces: an efficient algorithm based on immersion simulations. *IEEE Transactions on Pattern Analysis and Machine Intelligence*, **13**(6), 583–598.

Vicente, G., Scofield, R. A. and Mensel, W. P. (1998). The operational GOES infrared rainfall estimation technique. *Bulletin of the American Meteorological Society*, **79**, 1881–1898.

Wilhite, D. and Glantz, M. (1985). Understanding the drought phenomenon: The role of definitions. *Water International*, **10**, 111120.

WMO (2010). *WMO Fact-finding and Needs-assessment Mission to Pakistan*. Mission Report, November 2010.

Xie, P., and Arkin, P. A. (1997). Global precipitation: A 17-year monthly analysis based on gauge observations, satellite estimates, and numerical model outputs. *Bulletin of the American Meteorological Society*, **78**, 2539–2558.

Zhang, Q., Xu, C.-Y., Zhang, Z. *et al.* (2008). Spatial and temporal variability of precipitation maxima during 1960–2005 in the Yangtze river basin and possible association with large-scale circulation. *Journal of Hydrology*, **353**, 215–227.

9

Predicting and mitigating socio-economic impacts of extreme space weather: benefits of improved forecasts

Daniel N. Baker, Jamie M. Jackson, and Lauren K. Thompson

9.1 Introduction

Modern societies face many natural hazards and potential threats to their technical systems and fundamental infrastructure. These hazards can be geographically localised or they can extend to broader regional, or even national, scales. One of the hazards with the widest possible reach is what is often termed 'space weather'. This concept has been defined as '... variable conditions on the Sun, throughout space, and in the Earth's magnetic field and upper atmosphere that can influence the performance and reliability of space-borne and ground-based technological systems and endanger human life and health' (NSWP, 2010).

Unlike many other natural hazards such as floods, hurricanes, and earthquakes, space weather can be more subtle and difficult for human senses to perceive directly. However, as societies evolve to highly sophisticated states, they become ever more dependent on interconnected and increasingly susceptible technologies. The National Space Weather Program (NSWP) strategic plan (NSWP, 2010) notes that 'adverse conditions in the space environment can cause disruption of satellite operations, communications, navigation, and electric power distribution grids, leading to a variety of socio-economic losses and impacts on security'. It has rightly been observed that most of the developed world is encased in a cyber-electric 'cocoon' that is multi-layered and is highly interconnected (NRC, 2008). As in most complex networks, a failure in one node or limited part of the network can quickly propagate throughout the system causing widespread collapse (Buldyrev *et al.*, 2010).

Our goal here is to describe in general terms the connected Sun–Earth system and to help the reader understand several of the presently defined space weather threats. The intent is to indicate the day-to-day effects of moderate space weather features as well as to assess the most severe space weather events. These latter extreme episodes are quite rare, but when they do occur they can be amongst the most widespread and challenging hazards to be faced by developed nations (Baker and Green, 2011). Such unusual, but potentially highly damaging, phenomena have been termed 'low-frequency/high consequence' events (NRC, 2008). As such, extreme space weather events fall into categories that may also include the largest hurricanes, widespread floods, and the strongest earthquakes (Taleb, 2010). Like these other extreme natural events, the societal costs for severe space weather can be staggering and every possible measure must be taken by vulnerable societies to mitigate possible consequences.

In this chapter we will briefly describe some of the many documented impacts of space weather events. We will also discuss the clear benefits of observing the Sun and detecting the near-Earth environmental drivers of space weather. We will particularly attempt to make evident the potential benefits of accurate, timely space weather forecasts. One of the most clearly understood and possibly seriously damaging threats of space weather is to the electric power distribution system. Consequently, we will focus here on mitigation strategies and present-day policy imperatives related to the national power grid. We conclude with suggestions on next key steps to deal with the evolving space weather threat.

9.2 The connected Sun–Earth system

A key to evaluating the evolving space weather threat is to understand the Sun, its variability, and its many modes of connecting to the near-Earth space environment. Figure 9.1 presents a stylised visualisation of the Sun with a magnified portrayal of the region surrounding our home planet. Superimposed upon this rather arresting graphic of the

Extreme Natural Hazards, Disaster Risks and Societal Implications, eds. Alik Ismail-Zadeh, Jaime Urrutia Fucugauchi, Andrzej Kijko, Kuniyoshi Takeuchi and Ilya Zaliapin. Published by Cambridge University Press.

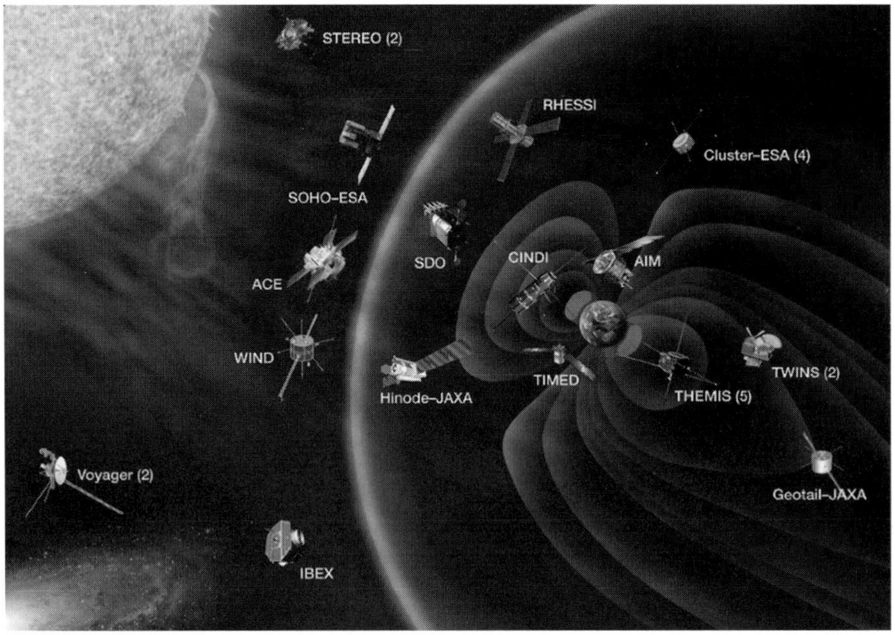

Figure 9.1. The Heliophysics System Observatory constellation of operating and planned NASA spacecraft. Together, these missions observe the Sun, the interplanetary medium, and the near-Earth space environment (courtesy of NASA Headquarters). For colour version, see Plates section.

Sun–Earth 'connection' are iconic images of many of the National Aeronautics and Space Administration (NASA) spacecraft that presently are observing the space environment (or else soon will be doing so). It is beyond the scope of this chapter to describe all the spacecraft missions portrayed in Fig. 9.1, but these are well described in the solar and space physics Decadal Survey (NRC, 2012).

An essential fact is that our Sun is a highly active magnetic star. It undergoes a broad range of cyclic changes over an approximately 11-year activity recurrence period. A measure of the solar disturbance level is the number of sunspots or sunspot groups visible on the Sun's surface (photosphere). During solar activity minimum conditions, the Sun can be virtually devoid of sunspot regions and the Sun's outer atmosphere (corona) can be relatively stable and quiescent. In contrast, as the Sun approaches its period of sunspot maximum, the surface and corona are a roiling mass of gnarled magnetic field structures, bright X-ray flaring regions, and frequent outbursts of plasma structures called *coronal mass ejections* (CMEs). Accompanying solar flares and CME releases can be intense bursts of high-energy charged particles (called *solar energetic particles*, or SEPs). Underlying all of these transient solar outbursts is a highly variable outflow of magnetised hot plasma flow called the solar wind. This solar wind streaming outward away from the Sun distorts and confines the Earth's magnetic field (see Fig. 9.1) to form a region called the magnetosphere (NRC, 2012). Over the more than five decades since the dawn of the Space Age, research has revealed a great deal about the near-Earth space environment. Confined within the magnetospheric volume portrayed in Fig. 9.1 are intense bands, or 'belts', of high-energy charged particles comprising the Van Allen radiation zones. Commingled with these dangerous energetic particle populations are lower-energy particles that can cause surface ablation of space vehicles or electrostatic charging and discharging that may adversely affect space systems. In the upper parts of Earth's atmosphere are layers of charged gases that constitute the ionosphere. In this region above 100 km altitude or so, powerful currents can flow in global-scale electrical patterns. When these intensify sufficiently, they often couple downward near Earth's surface to long-haul power lines, oil pipelines, and undersea communication cables. During the most intense solar-storm-induced magnetospheric disturbances, the '*geomagnetically-induced currents*' (GICs) flowing in the electric power grid can cause catastrophic upsets and electrical system failures (NRC, 2008).

Figure 9.1 makes clear that there are a substantial number of present and planned scientific satellites operating in near-Earth space and in the interplanetary medium. These are capable of making in situ measurements of key magnetospheric and solar wind properties. The spacecraft also have sensors that can remotely determine the changes induced by solar disturbances in Earth's ionosphere and atmosphere. Perhaps one of the most fundamental roles that the scientific satellite fleet performs is to provide a nearly constant vigil on the Sun. Such observing of the Sun often provides the first hint of developing major space weather episodes. Taken together, the constellation of NASA science satellites is called the Heliophysics System Observatory (NRC, 2012).

Crucial, absolutely indispensable additional space weather observation tools come from other agencies of the US government. These include the Geostationary Orbit Environmental Satellite (GOES) series of weather spacecraft operated by the National Oceanic and Atmospheric Administration (NOAA), ground-based magnetometer facilities operated by the US Geological Survey (USGS), and numerous ground- and space-based systems operated by the US Department of Defense (DoD). When merged with ground-based atmospheric, ionospheric, magnetospheric, and solar observing tools from the National Science Foundation (NSF), there presently are powerful systems in place to follow space weather disturbances from their origins on the Sun all the way to their ultimate impacts deep within the geospace (near-Earth) environment.

An important point made in the recent *Decadal Survey* in solar and space physics (NRC, 2012) is that the world desperately needs to establish a truly operational space weather observational system. In the sense used here, 'operational' means key observing platforms returning real-time data 24 hours per day, 7 days per week. The fleet of spacecraft portrayed in Fig. 9.1, for example, is contributing greatly to studying the connected Sun–Earth system, but they are *scientific* missions, not operational space weather monitors. We will return later to this fundamental policy point.

9.3 Key space weather threats

It is common practice to focus attention on the most extreme space weather events and try to estimate the societal and economic consequences of such intervals (NRC, 2008). This is the equivalent of policy makers asking about the 100-year flood or the largest probable earthquake in a given region and for a given time interval. Scaling the resiliency of human technological systems to the likely 'worst-case' scenario makes reasonable sense from a public policy standpoint.

One of the problems with policy formation regarding the space weather threat is that the effects are so diverse and range so widely over different systems and technologies. Figure 9.2 makes the point that solar disturbances (such as X-ray flares) can have prompt effects on aircraft and ground-based communications through radio-frequency blackouts. But in short order after a solar flare, there can be prompt solar protons impacting space-based systems. Ionospheric scintillations can then almost immediately affect global positioning system (GPS) satellite navigation and timing signals. On only slightly longer time scales, the radiation belts can be pumped up to high levels, thereby increasing greatly the chances of deep-dielectric charging and damage to spacecraft electronics. Enhancement of ionospheric currents – as was discussed in the previous section – can couple into power lines to produce GICs in the electrical grid. And historical evidence well demonstrates that induced currents in oil and gas pipelines – or in submarine cables – can cause huge operational headaches as well (NRC, 2008).

We must very much bear in mind that the solar and solar wind drivers of the near-Earth space environment can produce changes ranging from mild disturbances to the most profound and devastating (Schrijver and Mitchell, 2013). And yet, the entire range of space weather effects is integrated into the long-term performance of technical systems. A good example is the accumulated radiation dose of energetic electrons and protons that affects spacecraft flying through the radiation zones around the Earth (Baker, 2002). While a huge solar storm can cause acute changes in spaceborne electronics, the gradual accumulation of radiation doses can eventually lead to space hardware failure.

For the purposes of this discussion, let us focus on one main aspect of the severe space weather threat, viz., the powerful coronal mass ejection-driven development of a geomagnetic storm. The breadth and depth of a major solar-driven storm can be appreciated by recounting a historically important set of events that occurred over 150 years ago. These effects were documented in an earlier NRC report (NRC, 2008):

> From August 28 through September 4, auroral displays of extraordinary brilliance were observed throughout North and South America, Europe, Asia, and Australia, and were seen as far south as Hawaii, the Caribbean, and Central America in the Northern Hemisphere and in the Southern Hemisphere as far north as Santiago, Chile. Even after daybreak, when the aurora was no longer visible, its presence continued to be felt through the effect of the auroral currents. Magnetic observatories recorded disturbances in Earth's field so extreme that magnetometer traces were driven off scale, and telegraph networks around the world experienced major disruptions and outages. "The electricity which attended this beautiful phenomenon took possession of the magnetic wires throughout the country," the Philadelphia Evening Bulletin reported, "and there were numerous side displays in the telegraph offices where fantastical and unreadable messages came through the instruments, and where the atmospheric fireworks assumed shape and substance in brilliant sparks." In several locations, operators disconnected their systems from the batteries and sent messages using only current induced by the aurora.
>
> The auroras were the visible manifestation of two intense magnetic storms that occurred near the peak of the sunspot cycle. On September 1, the day before the onset of the second storm, Richard Carrington, a British amateur astronomer, observed an outburst of "two patches of intensely bright and white light" from a large and complex group of sunspots near the center of the Sun's disk. The outburst lasted 5 minutes and was also observed, independently, by Richard Hodgson from his home observatory near London. Carrington noted that the solar outburst – a white-light flare – was followed the next day by a magnetic storm, but he cautioned against inferring a causal connection between the two events.

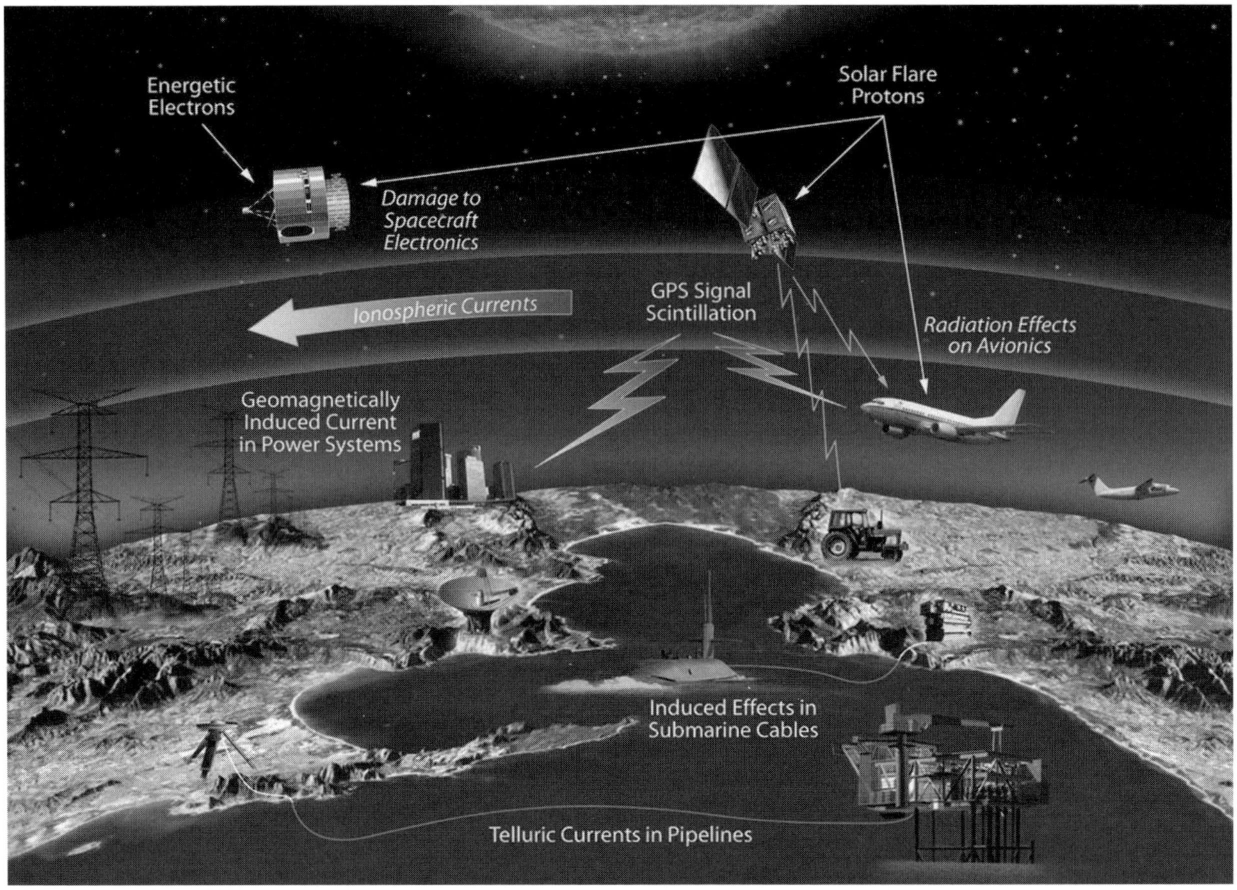

Figure 9.2. Space weather effects on critical infrastructure (adapted from Baker *et al.* (2013a)).

The cause of the magnetic storm that began on 2 September 1859 was not the highly energetic flare that Carrington and Hodgson had observed the previous morning. It was a fast CME launched from near the same huge sunspot region just northwest of the Sun's centre that had produced the flare. Had the Heliophysics System Observatory (Fig. 9.1) been in operation in 1859, the coronograph instruments would have observed the CME, probably 20 minutes or more after the flare's peak emission time. The CME would have appeared as a bright 'halo' of material surrounding the coronograph-occulted solar disc. This would indicate that the CME was headed directly towards Earth. Between the time of the flare/CME eruption on 1 September and the onset of the magnetic storm the next morning, over 17 hours had elapsed. Dividing the mean distance between Earth and the Sun by the 17.5-hour propagation time yields a speed of approximately 2300 kilometres per second, making the CME of 1 September 1859 among the fastest CMEs on record (NRC, 2008).

Moving substantially faster than the surrounding solar wind, fast CMEs create a shock wave that can accelerate coronal and solar wind ions (mostly protons) and electrons to relativistic energies. Particles are accelerated by solar flares as well; and large SEP events, although dominated by shock-accelerated particles, generally include flare-accelerated particles as well. Travelling near the speed of light, SEPs begin arriving at Earth less than an hour after the CME lift-off/flare eruption and are channelled along geomagnetic field lines into the upper atmosphere above the Earth's north and south poles. There they enhance the ionisation of the lower ionosphere over the entire polar regions and can initiate ozone-depleting chemistry in the middle atmosphere. Solar radiation storms can last several days and can cause intense versions of all the space weather phenomena portrayed above in Fig. 9.2. Strong SEP event occurrence can be an early indication that a powerful solar-generated CME is on the way, but there is not a perfect CME–SEP correlation.

The substantial concern for all involved in space weather research and emergency preparedness is that an event of the 1859 sort could occur today (NRC, 2008). If this did happen, it would not produce the relatively subtle and obscured effects on society that were the case in the Victorian Age of 150 years ago. Rather than the inconvenience of effects on the telegraph system – or, indeed, the awe of incredibly bright and spectacular auroral displays – we would today expect that a whole host of technological systems could (and probably would) be brought to their knees. Modern society has both vastly increased challenges when confronting severe solar storms, but also remarkable opportunities through the observing, modelling, and forecasting tools that have emerged through heliophysical research (NRC, 2012). The next section of this chapter describes how we can use the combination of real-time observations and advanced models to aid greatly the forecasting and alerting of space weather events.

9.4 The role of modelling and forecasting

A combination of empirical and numerical modelling methods has produced remarkable strides in the comprehensive understanding of solar wind conditions throughout the inner heliosphere (Arge *et al.*, 2004; Odstrcil *et al.*, 2004). This work was motivated by the goal of providing improved forecasts of ambient solar wind conditions at Earth's location for space weather purposes. The modelling methods that have been developed rely on remote observations of the Sun from Earth-based and space-based instruments combined with empirical and physics-based numerical modelling tools. The operational use of such modelling has demonstrated dramatically improving capabilities for space weather prediction for both quiet and disturbed solar conditions.

As a first model component, the Wang–Sheeley–Arge (WSA) model involves combined empirical and physics-based forecasts of the global solar wind flow (Arge *et al.*, 2004). It is used to predict the solar wind speed and the interplanetary magnetic field (IMF) strength at Earth (as well as at other points in the inner heliosphere). The model uses ground-based observations of the solar surface magnetic field as input to a magnetostatic potential-field source surface model to estimate the current sheet properties between $2.5R_s$ and $5R_s$, where R_s is the solar radius. For the work commonly used for forecasts, solar magnetic field maps for the portion of the Sun's surface visible from Earth-based observatories updated once per day (24 h cadence) are used. Outward flows in the corona are approximated by the imposition of radial field boundary conditions at the source surface and by empirical relationships of expansion factors to initialise solar wind speeds at this point. This surface is a Sun-centred sphere of radius $21.5R_s$.

Photospheric magnetic field observations from Earth-based observatories are the basic properties used to drive the modelling. They serve as a key input to the coronal and solar wind models. Updated photospheric magnetic maps, incremented regularly with new data, are constructed with magnetograms from the National Solar Observatory's Global Oscillation Network Group (GONG) data set. The line-of-sight field measurements from these data sources are converted to the radial field component (Arge *et al.*, 2004). The radial magnetic field is the basis of the standard 'forecast' version of the models.

Building upon WSA results for the region near the Sun, an ideal magnetohydrodynamic (MHD) simulation, ENLIL, is subsequently used to model the solar wind flow outward to distances beyond 1 astronomical unit (AU) (Odstrcil *et al.*, 2004). The computational domain of ENLIL (note that '*Enlil*' was the Sumerian god of the wind) is a uniform grid in latitude and longitude occupying the sector of a sphere. The position of the inner boundary is set at 0.1 AU ($\approx 21.5R_s$), and the outer boundary was commonly set at 1.7 AU. The meridional and azimuthal ranges spanned 30–150° in heliospheric colatitude and 0–360° in longitude, respectively. The inner boundary lies in the supersonic flow region. The outer boundary at 1.7 AU allows comparison of simulated temporal profiles of solar wind properties at and near the Earth's position with spacecraft measurements by the Advanced Composition Explorer (ACE) or other space platforms (Odstrcil *et al.*, 2004). More recent modelling efforts have gone to great lengths to use spacecraft information to include the effects of CME features along with the ambient solar wind flows (see Baker *et al.* (2013b) and references therein).

The combined WSA–ENLIL modeling provides specification of the solar wind flow speed, plasma density, solar wind mean plasma temperature, and magnetic field strength throughout the inner heliosphere. As an example, a greyscale representation of the radial flow speed (V_r) of the solar wind plasma in the heliospheric equatorial plane computed for the entire inner heliosphere on 24 July 2012 is shown in Fig. 9.3. The model results in this snapshot demonstrate that a powerful CME region was present near the ecliptic plane during this time. The modelled solar wind speed enhancements (up to speeds of $>1600\,\mathrm{km\,s^{-1}}$) at ~1 AU were primarily in the longitude sector encompassing the azimuthal location far from the Earth (i.e., 90–180° longitude). According to the model, the stream had enveloped the STEREO-A spacecraft at the time of the snapshot but passed far from the Earth, Mercury, Mars, and the STEREO-B spacecraft.

The case illustrated in Fig. 9.3 for 24 July 2012 is an immensely important one. The active region on the Sun that ultimately gave rise to the emitted CME had passed over the part of the Sun facing Earth about a week prior to 24 July. In that passage there were flares and smaller transient disturbances from the active region. However, late on 23 July, a massive X-ray flare occurred followed by the release of the huge CME that initially was observed

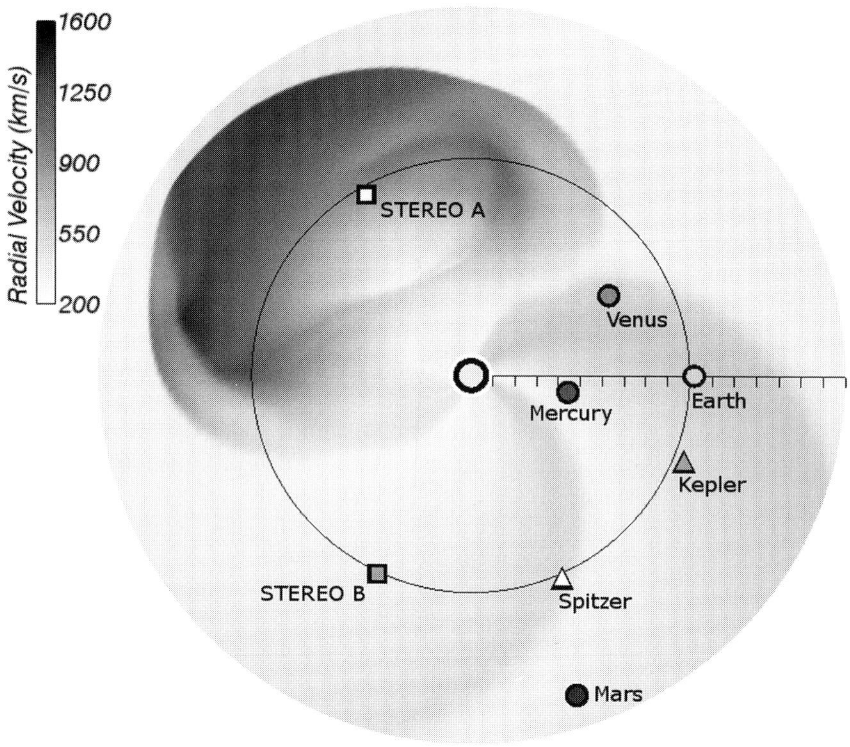

Figure 9.3. The inner heliospheric map of solar wind speed at ~1200 UT on 24 July 2012 as modelled using the WSA-ENLIL space weather tool as described in the text (courtesy of G. Millward).

only by STEREO-A (which was in the perfect viewing location). The developing CME quickly expanded so it could be detected by SOHO, STEREO-B, and other observing platforms.

Triangulated observations of the 23–24 July CME showed that it initially was moving at over 3500 km s^{-1} outward from the Sun. This is the fastest CME motion on record. The WSA-ENLIL modelling quickly alerted space weather users that a massive solar storm was in progress, but fortunately for Earth's residents, the solar blast was headed almost directly away from the Earth (see Fig. 9.3).

About 17 hours after the CME's release from the Sun, the mass ejection reached the STEREO-A spacecraft location at 1 AU (Fig. 9.3). Note that this transit time of 17 h is comparable to (or slightly faster than) the 17.5 h estimated (see above) for the 1859 Carrington event. Indeed, as the modelling would suggest (Fig. 9.3), the CME scored a direct hit upon the STEREO-A spacecraft. Using measurements of magnetic field data, plasma flow speeds, and solar wind densities from STEREO (augmented by WSA-ENLIL modelling), it is possible to characterise the solar wind forcing strength for this immensely powerful CME. Using these solar wind forcing values to compute what the geomagnetic response would have been at Earth for such an event, we find that the results would have been remarkable. Using reasonable and prudent timing assumptions, we find that the 24 July 2012 event would have produced a storm at Earth comparable to the largest observed geomagnetic storms of the twentieth century. Thus, we do not have to go back 100 or 150 years in solar history to find a 'Carrington' type event. We as Earth citizens missed such an event by just a few days in July 2012.

A well-observed case such as 24 July 2012 gives us an immensely useful object lesson for space weather analysis. Very fortunately, Earth 'dodged a bullet' in that the CME moved in a relatively harmless direction. Yet, the STEREO-A observations allowed us to characterise the CME properties to such a point that we can now have plausible, directly measured 'severe case' knowledge of an extreme space weather event. Using our space environmental models we can make clear, concrete estimates of what this extreme event would have done to Earth-based technological systems.

9.5 Space weather mitigation strategies

If a solar storm of the size and strength observed in July 2012 had scored a direct hit upon the Earth and its magnetosphere, what would have been the likely consequences? It is quite probable that every one of the space weather 'impacts' portrayed in Fig. 9.2 would have been encountered, starting with immediate solar flare effects on radio communication, followed by solar energetic particle impacts on spacecraft a short time later, and very likely SEP threats to airline polar routes over a period of hours to

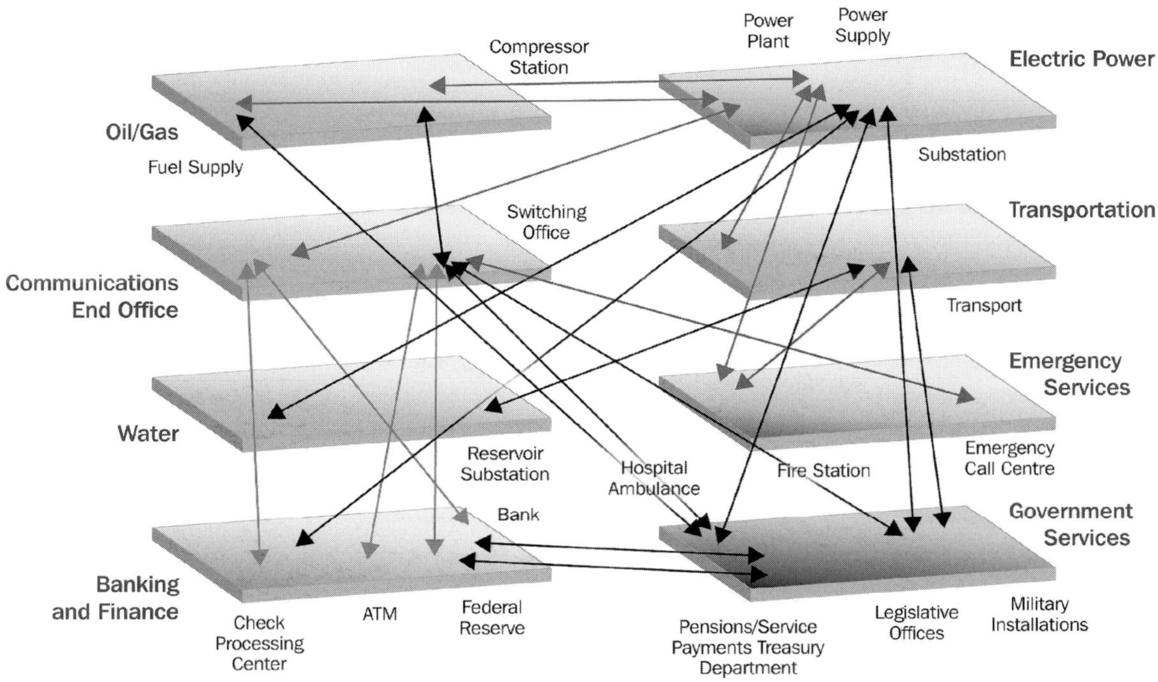

Figure 9.4. Connections and key interdependencies of various components across the economy. This schematic shows the highly interconnected infrastructures of modern societies and how effects in one sector might impact other sectors (adapted from Baker (2012a)).

days (NRC, 2008). Overall, a top-tier solar storm hitting Earth today would cause significant concerns for communication, navigation, remote sensing, precision agriculture, and a plethora of other modern activities. It is only mildly speculative to say that had the 24 July 2012 CME event occurred a week earlier, we would have quite a different perspective today on space weather threats.

Clearly the biggest concern for developed countries and advanced societies is the space weather impact on the electric power grid. Figure 9.4 (from the report by NRC (2008)) shows the immense interdependencies of modern society. The arrows in the diagram are meant to illustrate some of the main connections between different aspects of society. Electric power, for example, is a 'cornerstone' technology in every sense. Without abundant, reliable electrical energy, the availability of oil and gas quickly ends. Communication systems, provision of potable water, carrying out of banking and financing, and execution of government services would quickly grind to a halt without electricity. Transportation systems of all sorts and emergency services require electrical power to be at all effective. The arrows in the diagram show this immense dependence on the electrical power supply.

Some may say: 'But we have backup generators.' Indeed, this can stave off immediate effects in some cases, but prolonged power failures start to trigger all the node 'linkage' problems portrayed (or hinted at) in Fig. 9.4: the failure of one system or sector can propagate in complex (and not fully predictable) ways leading to complete network collapse (see Buldyrev *et al.*, 2010). A very real concern is that evolving societies are building in more interdependencies every day, often without fully thinking through the possible consequences.

What are possible mitigation strategies for severe space weather events? A place to start goes back to earlier suggestions made in this chapter. We, as a modern society and a modern world, must have continuous real-time monitoring of the connected Sun–Earth system (NRC, 2012). We must endeavour to convert the rather 'organically' grown science observing network of satellites (Fig. 9.1) and related ground facilities into an efficient, effective, operational, real-time, space weather observing system. We must also support and supplement the observing network with the data analysis and modelling tools to ingest information about the space environment and convert this into actionable alerts and warnings. As remarked by Arthur C. Clarke, we must convert information into knowledge, and knowledge into wisdom.

A further set of mitigative steps was summarised as an outgrowth of the 2012 Electric Infrastructure Security (EIS) Summit meeting held in London, UK (Baker, 2012b):

- Establish a severe space weather working group to identify and define the most reasonable extreme space weather event(s) that would serve as the threat analysis baseline for bulk power operators and system engineers to measure against.

- Identify the critical infrastructures and facilities that must continue to have electric power during extreme space weather event scenarios.
- Establish detailed modelling of the effects and interconnections within a national power grid under the influence of severe space weather (see recommendation 1 in Baker (2012b)). Modelling would be undertaken by companies, system engineers, and operators with the protection of key assets (see recommendation 2, *ibid.*) a foremost priority in the scenarios analysed.
- Identify techniques and engineering solutions that would keep ground-induced currents from space weather events and from EMP events isolated and away from key infrastructure (blocking solutions). This would require efforts by engineers, transformer experts, and other professionals.

Several of these recommendations are in progress on a national level in the USA, but are lacking international coordination, a critical step in this globalised and intensely interconnected international society. The next section discusses policy steps that are currently taking place in the USA.

9.6 Policy imperative: present steps and future prospects

As the discussion above outlines, space weather can pose significant societal hazards due to its potential for disruption of critical infrastructure, in particular the electric power grid. Strong solar magnetic storms such as the 1859 Carrington event, discussed above, which disrupted telegraphs worldwide and in some cases generated fires (JASON, 2011), are low-frequency/high-consequence events, as described above. Such events occur at wide intervals but cause catastrophic damage when they do occur, particularly if critical infrastructure is not designed properly to manage their impact. In addition to the need for establishment of a truly operational space weather observation system, mitigation efforts must accompany the information-gathering and observation efforts. Activities are underway, as described below, in the USA to make use of GIC data collection to determine a path forward to protect such infrastructure.

In the case of the electric power grid, the components most vulnerable to damage due to GICs caused by geomagnetic disturbances (GMDs) are high-voltage transformers. The impact of a geomagnetic storm on the grid, and the resultant impact on society, will vary with the number of transformers affected. The most extreme scenario, which could result from an event on the order of magnitude of the Carrington event, could bring about substantial loss of access to electricity for up to several years (JASON, 2011). Because of the dependence of many segments of the economy on electricity, the effects of such a storm could be severe.

Transformers connected to the grid are subject to a plethora of factors that can increase their vulnerability to GICs (Molinski, 2000). The length and direction of transmission lines, characteristics of the surrounding geography and geology, and the level of demand for electricity at the time of a GIC can heighten a transformer's vulnerability (Molinski, 2000). Figure 9.5 shows the modelling and analysis approach presently being applied to power transformers. Currently, the average age of US transformers is about 40 years; older transformers, due to cycling and repeated exposure to excess current, are far more likely to fail after exposure to GICs than some newer transformers (Kappenman, 2012). Moreover, some of the newest and most efficient transformers may be even more vulnerable to failure after GIC exposure (McConnell, 2012). While loss of only a few transformers may not have widespread effect, events on the high-voltage transmission system cannot be isolated: a 'physical event anywhere on a grid – a downed transmission line, a failed relay switch, a large generator coming on line – affects every other point on the network' (Amici, 2002). Consequently, much of the policy discussion has centred on a national plan for mitigation and response to GMD.

Two disasters resulting from natural weather and geophysical phenomena underscore the peril of inadequate mitigation planning. First, in March 2011, an earthquake and tsunami struck Japan and affected several reactors at the Fukushima Daiichi nuclear power plant (US Nuclear

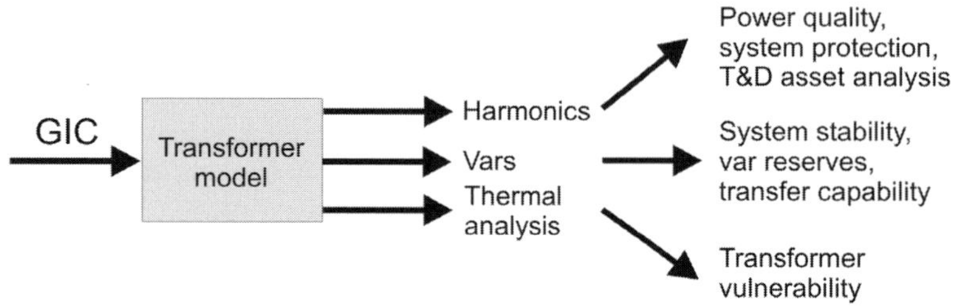

Figure 9.5. Modelling and analysis approach for GIC impact on transformers.

Regulatory Commission, 2012). The Fukushima plant was designed and constructed to withstand an earthquake of magnitude 8.2, and a tsunami wave of 5.7 m (Buongiorno *et al.*, 2011). However, the actual event was an earthquake of magnitude 9.0 and a tsunami wave of 14 m (Buongiorno *et al.*, 2011), later assessed to be a '1,000 year event' (Rincon, 2011). Although the probability of such an event was low, the result of failure to design for worst-case events caused catastrophic damage that will haunt the surrounding areas for years to come and broadly affect nations' stance on nuclear power. For this reason, design bases for nuclear facilities are quite stringent; for example, many nuclear plants are designed to withstand earthquakes twice as strong as the 1000-year event for each site (Nöggerath *et al.*, 2011; World Nuclear Association, 2012). This event provides an important object lesson when it comes to severe space weather events and their effects on the power system.

Similarly, in 2005, Hurricane Katrina made landfall on the coast of Florida travelling over the Gulf of Mexico and eventually striking land again in Mississippi and Louisiana, causing tremendous devastation as the resulting tidal surges overtopped the levees east of New Orleans (Knabb *et al.*, 2005). The levees were designed for a 200- to 300-year event. 'But at the time that these levees were designed and constructed, it was felt that that was an adequate level, given the probability of an event like this occurring,' said Lieutenant General Carl Strock, Commander, US Army Corps of Engineers, and Chief of Engineers at the Defense Department Special Briefing on 2 September 2005 (DoD, 2005). Other nations have adopted a longer-term perspective in engineering for such events. As a point of reference, the Dutch use a 1000–10 000 year flood design basis for major towns and cities (Independent Levee Investigation Team, 2006). The case of Katrina drives home the point that being prepared for worst-case scenarios due either to tropospheric weather or to space weather is crucial in a policy sense.

The fact that the magnitude of a geomagnetic disturbance is a result of a confluence of factors makes it difficult to predict a CME's magnitude until all of the variables have revealed themselves. For this reason, notice of an oncoming event will be very short, on the order of tens of minutes rather than hours or days (NERC, 2012a). While relatively few CME events will cause a geomagnetic disturbance on the order of magnitude of the Carrington event, sizable CMEs at the Sun's surface are a regular occurrence (JASON, 2011) and by one estimate the probability of a Carrington event within the next decade is approximately 12% (Riley, 2013). In the past 150 years, the Carrington event was certainly the largest GMD event experienced on Earth, but it has had several near-competitors – and one event on 24 July 2012 that the Earth missed by mere days, as discussed previously – suggesting that the Carrington event may provide a basis for national emergency protection and response planning in order to avoid the type of societal impact seen in the wakes of the Fukushima and Katrina disasters (JASON, 2011). The return period for a severe, Carrington-style event, which could cause wide-scale disruption and prolonged outages to the US grid network, could be around 150 years (Lloyd, 2013).

This risk has not gone unappreciated by regulators and industry. As the insurance industry – expert in assessment and quantification of risk – has noted, '[t]he power blackout risk is generally underestimated. Due to the increasing interconnectedness in combination with rather old infrastructure we expect this risk to increase in both frequency and severity' (Bruch *et al.*, 2011). In the past decade, this issue has been the focus of increased attention from the US Congress, the US Federal Energy Regulatory Commission (FERC), the North American Electric Reliability Corporation (NERC), the insurance industry, and the public utilities. The policy debate at the federal level centres on what magnitude of GMD should be a benchmark for mitigating action. Classifying smaller GMD events as the potential threat will require fewer and less expensive mitigating measures, while preparing for larger events would require more. This, paired with uncertainties surrounding prediction and the specific effects of a large event on the grid, makes legislators and agencies hesitant to adopt the Carrington event as the threat level to assess.

FERC is the US federal agency responsible for the reliability of the bulk power system (BPS). NERC is the entity required to develop mandatory and enforceable reliability standards, subject to FERC's review and approval (Federal Power Act § 215, 2006). While NERC typically promulgates standards *sua sponte* based on its own experience and recognition of necessity, FERC also has the authority to order NERC to promulgate reliability guidelines on specific matters (Federal Power Act §215(d)(5), 2012). The stakeholders and their relationships for GMD assessment are illustrated in Fig. 9.6.

On 16 May 2013, FERC issued directions with respect to reliability guidelines for GMD and the BPS. FERC's rule directed NERC to develop and submit new GMD standards in a two-stage process (FERC, 2013). FERC did not require NERC to include any specific requirements in the GMD reliability standards; it only identified issues to be considered and addressed in the standards development process (FERC, 2013).

In the first stage, NERC must file, within six months of the rule, reliability standards requiring owners and operators of the BPS to develop and implement operational procedures to mitigate GMD effects. NERC must also conduct a GMD vulnerability assessment and identify facilities most at risk from a severe disturbance (FERC, 2013).

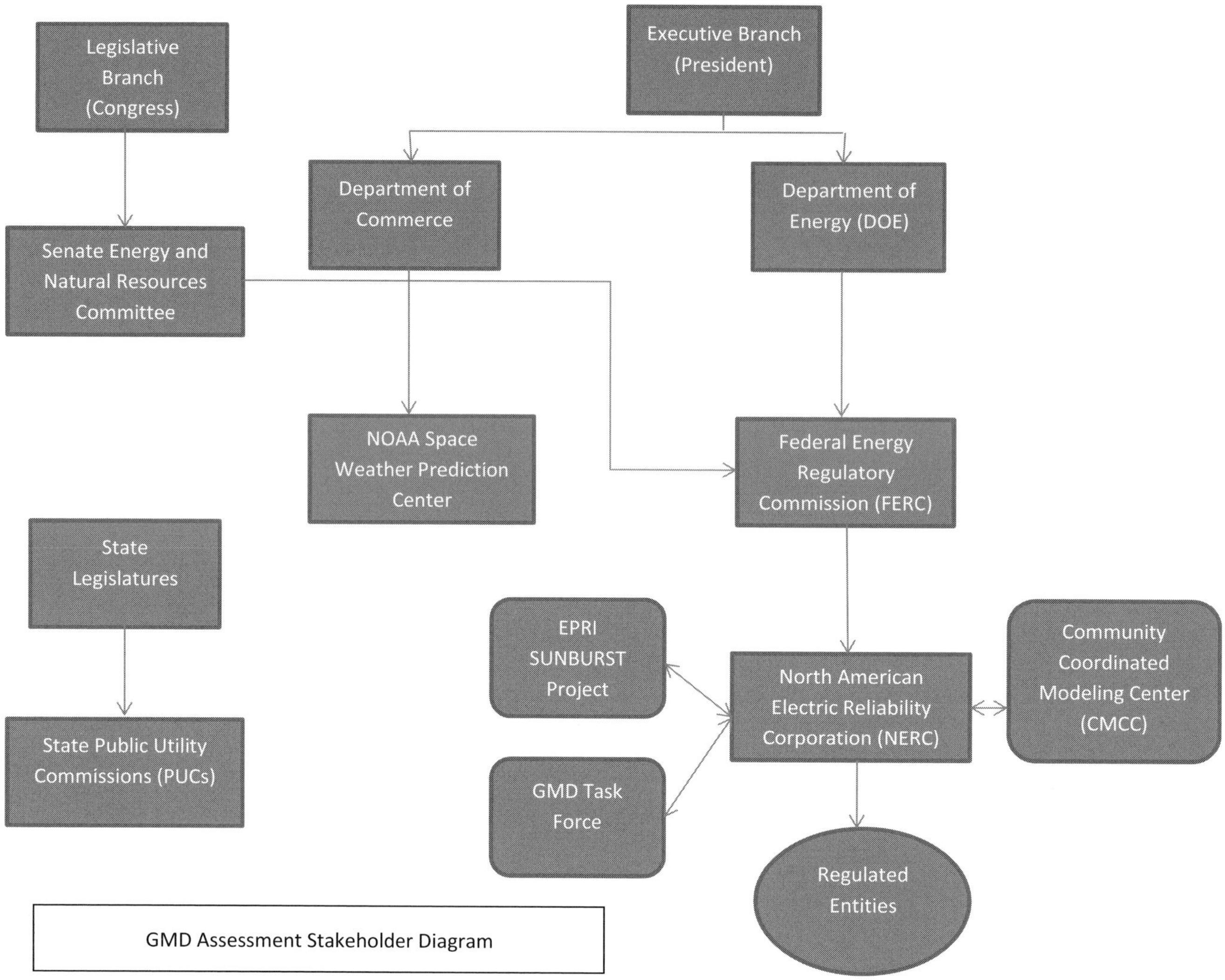

Figure 9.6. Stakeholders in GMD assessment and policy analysis.

In the second stage, NERC must file standards within 18 months of the rule identifying 'benchmark GMD events', which define the severity of GMD events a responsible entity must assess for potential impacts on the BPS. NERC's standards must require owners and operators to conduct initial and continuing assessments of the potential effects of specified 'benchmark GMD events' on equipment and the BPS. If the assessments identify potential effects from such events, the reliability standards should require a responsible entity to develop and implement plans to protect against instability, uncontrolled separation, or cascading failures of the system (FERC, 2013).

The stance taken by FERC leaves in the hands of NERC the definition of the magnitude of GMD event for which utilities should plan, and states that '[g]iven that the scientific understanding of GMDs is still evolving, we recognize that Reliability Standards cannot be expected to protect against all GMD-induced outages' (FERC, 2013). NERC itself has taken a somewhat hands-off approach, stating in a landmark 2012 reliability assessment report that the most likely consequence of a GMD would be system voltage collapse (NERC, 2012a,b). Nevertheless, NERC has also been occupied with this issue, establishing a GMD task force (GMDTF) following the publication of its initial report on the subject of GMD and the BPS in 2009 (NERC, 2010). Within the task force, NERC-commissioned research teams are tackling issues necessary to assess the threat and implement appropriate measures (NERC, 2012a,b). These teams are creating models to use in calculating the GIC levels entering the BPS due to a given magnitude of GMD event, and are modelling the functioning of various types of transformers in order to design and carry out assessments of vulnerability to thermal variation and fluctuating current stresses (NERC, 2012a,b).

The US electric power industry itself has demonstrated interest in the issue, although its general consensus is that '[m]any aspects of the science surrounding GMDs are still immature, the methods of grid impact analysis remain crude and unrefined, and necessary assessments of the impact of methods of remediation are unproven' (American Public Power Association, 2012). The industry has been involved in development of societal understanding of the impact of GMD through NERC's GMDTF, the Electric Power Research Institute (EPRI) SUNBURST project, the NOAA, and the Community Coordinated Modeling Center (CCMC) (Pulkkinen, 2012). The EPRI SUNBURST project, of particular importance in understanding the effects of GMD on high voltage transformers, was initiated shortly after the 13 March 1989 solar event that precipitated the 'Quebec Hydro Blackout' (American Public Power Association, 2012). The SUNBURST network consists of a consortium of utilities continuously monitoring GICs on various large power transformers across North America. This collection of information will be important in developing further understanding of the technological mitigation required to protect transformers from GMD.

At a US congressional level, no legislation has addressed the issue to date. Members of the US Congress have introduced legislation attempting to address the potential for GMDs to impact the BPS. However, this legislation has stalled, due in part to the tendency of proposals to group GMDs together with other issues, and to overlap existing legislation directing FERC to defer to NERC to create reliability standards. The SHIELD Act, proposed by R. Trent Franks in 2011, would have authorised FERC to issue orders to NERC, and users, owners, and operators of the BPS to take emergency measures to protect the reliability of the BPS from GMD or electromagnetic pulses (EMPs) from nuclear weapons (Franks, 2011). The SHIELD Act also would have authorized FERC to provide for cost recovery of 'substantial costs' incurred in compliance with such emergency orders and would have required NERC to propose reliability standards addressing 'reasonably foreseeable' GMDs, based upon FERC's specification of the 'nature and magnitude' of such threats. Finally, the Act would have required that NERC promulgate reliability standards addressing the availability of 'large transformers' capable of restoring reliable operations after an EMP or GMD event. Because predictions of an imminent GMD would come minutes before the effects, the emergency authority in the SHIELD Act would have been limited in its scope, but would also have had the potential to extend FERC authority in unanticipated ways (Franks, 2011). A US Senate Bill also introduced in February of 2011 (Cyber Security and Internet Freedom Act of 2011, 2011) addressed GMD along with not only EMP, but also cyber security. How to ensure cyber security is a highly contested issue, and so legislation is likely not forthcoming on the topic. A very important policy issue that is beyond the scope of this present chapter is to clarify further the relationships between EMP mitigation techniques and those for GMD. While it is not uncommon to discuss these issues together, we urge that technical and policy leaders make concerted efforts to resolve better the differences between these distinct power grid threats.

At a state level, Maine has become the first state in the USA, led by R. Andrea Boland, to launch its own project to protect its state grid from GMDs (Maine Legislature, 2013). Because GMDs enter the Earth's atmosphere near the poles and spread toward the equator (Jason, 2011), Maine could be more vulnerable than states to the south. The state's bill requires the state Public Utilities Commission (PUC) to identify vulnerabilities in the state's transmission infrastructure and take cost-effective steps to minimize that vulnerability (Maine Legislature, 2013). However, because the BPS extends beyond state lines, grid failures outside Maine could also affect the delivery of power within the state (Amici, 2002). Although state-level mitigation may not be effective in protecting the national grid, establishing state collaboration with FERC will both simplify monitoring and reporting of GIC levels, and ease the process of hardening the grid later. Additionally, state governments can help by coordinating emergency plans to ensure availability of infrastructure such as hospitals and water supplies in the event the grid is affected by GMD.

9.7 Summary and conclusion

Overall, the impact of space weather on critical infrastructure is an evolving, multi-faceted policy issue, involving many stakeholders and governing entities (Fig. 9.6). Going forward, key steps for policy makers should focus on both emergency response planning – particularly at the state level where policies can be implemented to ensure the continuous functioning of critical facilities such as hospitals and water supply – and on mitigation measures, including the identification of a design basis event and development of reliability and engineering standards to prepare the ageing electricity infrastructure for such an event, through assessment and modelling as pictured in Fig. 9.6. The magnitude of event that should form the basis of mitigation planning has yet to be determined by policy makers. In the USA this decision is to be made by NERC under the new FERC rule issued in May 2013, but should bear in mind the potential for disruption that a Carrington-magnitude event carries.

Looking to the more distant future, the ultimate goal would be to make technological systems that are effectively immune to severe space weather effects. This will require continued efforts by the space research community to define and characterise plausible worst-case scenarios. It will also require that critical societal infrastructure be clearly identified and designated. Thereafter, advanced engineering and mitigation methods can be sought to provide the hoped-for immunity to disruption. However, until such a desired state is reached, accurate space weather forecasts and alerts represent an important first line of defence. US federal agencies such as NASA, NOAA, and DoD have important observational tools that can provide immense benefits for space weather predictive capabilities. It should be clearly directed national policy that these agencies and other agencies (such as USGS, DOE, NSF, etc.) with space- or ground-based assets or modelling tools join more fully in the efforts of a reinvigorated NSWP. It should be a US national mandate and presidential imperative that the relatively modest marginal commitment of agency resources be made by each organisation to use its capabilities to produce a truly operational space weather system as described in the 2013–2022 *Decadal Survey* (NRC, 2012).

Acknowledgments. DB was supported by the National Science Foundation Center for Integrated Space Weather Modeling. JJ and LT thank Brad Bernthal and Roger Koenig for their invaluable guidance and support, and Patton Boggs LLP for its support of clinical education, as well as the many other individuals who have provided generous assistance in their academic and professional capacities through advice, discussions, and education. The authors thank A. Pulkkinen and G. Millward for useful discussions and for graphics support.

References

American Public Power Association (2012). Comments of the American Public Power Association, the Edison Electric Institute, the Large Public Power Council, and the National Rural Electric Cooperative Association. FERC Docket No. RM12-22-000. Available at: http://www.lppc.morphatic.com/wp-content/uploads/2013/02/LPPC-FERC-Comments-Docket-No.-RM12-22-000-12-26-2012.pdf. Accessed on 7 August 2013.

Amici (2002). Amici Brief for Electrical Engineers, Energy Economists and Physicists as Amici Curiae Supporting Respondents (2002). New York v. Federal Energy Regulatory Comm'n, 535 US 1 (No. 00-568).

Arge, C. N., Luhmann, J. G., Odstrcil, D., Schrijver, C. J. and Li, Y. (2004). Stream structure and coronal sources of the solar wind during the May 12th, 1997 CME. *Journal of Atmospheric and Solar-Terrestrial Physics*, **66**, 1295–1309.

Baker, D. N. (2002). How to cope with space weather. *Science*, **297**, 1486.

Baker, D. N. (2012a). Extreme space weather: Forecasting behavior of a nonlinear dynamical system. In *Extreme Events and Natural Hazards: The Complexity Perspective*, eds. Sharma, A. S., Bunde, A., Dimri, V. P. and Baker, D. N. AGU Geophysical Monograph Series, 196. Washington, DC: American Geophysical Union, pp. 255–266.

Baker, D. N. (2012b). The Third Electric Infrastructure Security World Summit meeting. *Space Weather*, **10**, S07002, doi:10.1029/2012SW000820.

Baker, D. N. and Green, J. L. (2011). The perfect solar superstorm. *Sky and Telescope*, February issue, 28–34.

Baker, D. N., Charo, A. and Zurbuchen, T. (2013a). Science for a technological society: The 2013–2022 survey in solar and space physics. *Space Weather*, **11**, 50–51, doi:10.1002/swe.20022.

Baker, D. N., Poh, G., Odstrcil, D. *et al.* (2013b). Solar wind forcing at Mercury: WSA-ENLIL model results. *Journal of Geophysical Research*, **118**, 1–13, doi:10.1029/2012JA018064.

Bruch, M., Kuhn, M. and Schmid, G. (2011). *Power Blackout Risks: Risk Management Options, Emerging Risk Initiative – Position Paper.* Munich: CRO Forum. Available at: https://www.allianz.com/v_1339677769000/media/responsibility/documents/position_paper_power_blackout_risks.pdf. Accessed on 7 August 2013.

Buldyrev, S. V., Parshani, R., Paul, G., Stanley, H. E. and Havlin, S. (2010). Catastrophic cascade of failures in interdependent networks. *Nature*, **464**, 1025–1028.

Buongiorno, J., Ballinger, R., Driscoll, M. *et al.* (2011). *Technical Lessons Learned from the Fukushima-Daichii Accident and Possible Corrective Actions for the Nuclear Industry: An Initial Evaluation.* Center for Advanced Nuclear Energy Systems Publications No. MIT-NSP-TR-025. Cambridge: Massachusetts Institute of Technology. Available at: http://web.mit.edu/nse/pdf/news/2011/Fukushima_Lessons_Learned_MIT-NSP-025.pdf. Accessed on 6 August 2013.

Cybersecurity and Internet Freedom Act of 2011 (2011). 112th Cong. 1st Sess. Available at: http://www.govtrack.us/congress/bills/112/s413/text. Accessed on 7 August 2013.

DoD (2005). *Defense Department Special Briefing on Efforts to Mitigate Infrastructure Damage from Hurricane Katrina.* Office of the Assistant Secretary of Defense (Public Affairs). US Department of Defense. Available at: http://www.defense.gov/transcripts/transcript.aspx?transcriptid=2070. Accessed on 7 August 2013.

Federal Power Act § 215 (2006). 16 USC § 824.

Federal Power Act § 215(d)(5) (2012). 16 U.S.C. § 824o(d)(5); 18 CFR § 39.5(f).

FERC, Federal Energy Regulatory Commission (2013). Order No. 779, Reliability Standards for Geomagnetic Disturbances (Docket No. RM12-22-000). 143 FERC ¶61,147 (to be codified at 18 C.F.R. Pt. 40). Available at: http://www.ferc.

gov/whats-new/comm-meet/2013/051613/E-5.pdf. Accessed on 7 August 2013.

Franks, T. (2011). SHIELD Act. HR 668, The Secure High-voltage Infrastructure for Electricity from Lethal Damage (or SHIELD) Act, 112th Cong., 1st Sess. Available at: http://www.govtrack.us/congress/bills/112/hr668/text. Accessed on 7 August 2013.

Independent Levee Investigation Team (2006). *Final Report on Investigation of the Performance of the New Orleans Flood Protection Systems in Hurricane Katrina on August 29, 2005*. Available at: http://www.ce.berkeley.edu/projects/neworleans/report/CH_10.pdf. Accessed on 6 August 2013.

JASON (2011). *The MITRE Corporation, Impacts of Severe Weather on the Electric Grid*. Available at: http://www.fas.org/irp/agency/dod/jason/spaceweather.pdf. Accessed on 6 August 2013.

Kappenman, J. (2012). A perfect storm of planetary proportions. *IEEE Spectrum*, February. Available at: http://spectrum.ieee.org/energy/the-smarter-grid/a-perfect-storm-of-planetary-proportions/0. Accessed on 6 August 2013.

Knabb, R. D., Rhome, J. R., and Brown, D. P. (2005). *Tropical Cyclone Report: Hurricane Katrina*. National Hurricane Center. Available at: http://www.nhc.noaa.gov/pdf/TCR-AL122005_Katrina.pdf. Accessed on 6 August 2013.

Lloyd's (2013). *Solar Storm Risk to the North American Electric Grid*. Report by Lloyd's and the Atmospheric and Environmental Research, Inc. Available at: http://www.lloyds.com/?/media/lloyds/reports/emerging%20risk%20reports/solar%20storm%20risk%20to%20the%20north%20american%20electric%20grid.pdf. Accessed on 7 August 2013.

Maine Legislature (2013). Session – 126th Maine Legislature. H-121. Available at: http://www.mainelegislature.org/legis/bills/bills_126th/billtexts/HP010602.asp. Accessed on 6 August 2013.

McConnell, D. B. (2012). Testimony of Dr. Ben McConnell, technical witness at the FERC Staff Technical Conference on Geomagnetic Disturbances to the Bulk-Power System. Available at: http://www.nerc.com/pa/Stand/Geomagnetic%20Disturbance%20Resources%20DL/Transcript_GMD_112012.pdf. Accessed on 7 August 2013.

Molinski, T. S. (2000). Shielding grids from solar storms. *IEEE Spectrum*, **37**, 55–60.

NERC, North American Electric Reliability Corporation (2010). *High-Impact, Low Frequency Event Risk to the North American Bulk Power System*. A jointly-commissioned summary report of the North American Electric Reliability Corporation and the US Department of Energy, November 2009 Workshop. Available at: http://www.nerc.com/files/HILF.pdf. Accessed on 7 August 2013.

NERC, North American Electric Reliability Corporation (2012a). *Effects of Geomagnetic Disturbances on the Bulk Power System*. 2012 Special Reliability Assessment Interim Report. Available at: https://www.frcc.com/Public%20Awareness/Lists/Announcements/Attachments/105/GMD%20Interim%20Report.pdf. Accessed on 7 August 2013.

NERC, North American Electric Reliability Corporation (2012b). NERC Meeting Minutes, Geomagnetic Disturbance Task Force Phase 2. Available at: http://www.nerc.com/docs/pc/gmdtf/GMDTF_Leadership_Meeting_Minutes.pdf. Accessed on 6 August 2013.

Nöggerath, J., Geller, R. J. and Gusiakov, V. K. (2011). Fukushima: The myth of safety, the reality of geoscience. *Bulletin of the Atomic Scientists*, **67**, 37–46.

NRC, National Research Council (2008). *Severe space weather events – Understanding societal and economic impacts*, A workshop report, National Research Council, Space Studies Board. Washington, DC: National Academies Press.

NRC, National Research Council (2012). *Decadal Survey, Solar and Space Physics: A Science for a Technological Society*. Washington, DC: National Academies Press.

NSWP, National Space Weather Program (2010). Strategic Plan, FCM-P30–2010. Washington, DC: NSWP.

Odstrcil D., Pizzo, V. J., Linker, J. A. *et al.* (2004). Initial coupling of coronal and heliospheric numerical magnetohydrodynamic codes. *Journal of Atmospheric and Solar-Tereestrial Physics*, **66**, 1311–1320.

Pulkkinen, A. (2012). *Electric Fields, GIC Flow, and Transformer Response*. *CCMC 2012 Workshop*, Key Largo, Florida, 16–20 January 2012. Available at: http://ccmc.gsfc.nasa.gov/RoR_WWW/workshops/2012/Thursday_pdf/lordan_ccmc2012.pdf. Accessed on 6 August 2013.

Riley, P. (2013). On the probability of occurrence of extreme space weather events. *Space Weather*, **11**(5), 214–218.

Rincon, P. (2011). Japan tsunami 'could be 1,000-year event'. *BBC News: Science & Environment*. Available at: http://www.bbc.co.uk/news/science-environment-12740649. Accessed on 7 August 2013.

Schrijver, C. J. and Mitchell, S. D. (2013). Disturbances in the U.S. electric grid associated with geomagnetic activity. *Journal of Space Weather and Space Climate*, **3**, A19, doi: 10.1051/swsc/2013041.

Taleb, N. N. (2010). *The Black Swan: The Impact of the Highly Improbable*. New York: Random House.

US Nuclear Regulatory Commission (2012). Japan Nuclear Accident – NRC Actions. Available at: http://www.nrc.gov/japan/japan-info.html. Accessed on 6 August 2013.

World Nuclear Association. (2012). Retrieved from Nuclear Power Plants And Earthquakes. Available at: http://www.world-nuclear.org/info/inf18.html. Accessed on 6 August 2013.

10

Predictability of extreme events in a branching diffusion model

Andrei Gabrielov, Vladimir Keilis-Borok, Sayaka Olsen, and Ilya Zaliapin

10.1 Introduction

Extreme events are the most important yet least understood feature of natural and socioeconomic complex systems. In different contexts these events are also called *critical transitions*, *disasters*, *catastrophes*, or *crises*. Among examples are destructive earthquakes, El Niños, heat waves, electric power blackouts, economic recessions, stock-market crashes, pandemics, armed conflicts, and terrorism surges. Extreme events are rare, but consequential: they inflict the lion's share of the damage on population, economy, and environment. This chapter is focused on the theoretical foundations of predicting individual extreme events. The prediction problem is pivotal both for the fundamental understanding of complex systems and for disaster preparedness (e.g., Keilis-Borok and Soloviev, 2003; Sornette, 2004; Albeverio *et al.*, 2005).

There exist several well-developed approaches to prediction of extreme events including classical Kolmogorov–Wiener extrapolation of time series (Kolmogorov, 1941a,b; Wiener, 1949), linear (Kalman–Bucy) (Kalman and Bucy, 1961) and non-linear (Kushner–Zakai) (Kushner, 1964; Zakai, 1969; Chow, 2007) filtering, sequential Monte-Carlo methods (Doucet *et al.*, 2001), and the extreme-value theory (Embrechts *et al.*, 2008). The approach to prediction described in this chapter is complementary to the above methods. The need for an alternative approach is dictated by a non-standard formulation of the prediction problem, where one is particularly interested in the future occurrence times of rare events rather than the complete unobserved state of the system in continuous time. We notice, accordingly, that often the easily observed extreme events cannot be defined as the instants of threshold exceedance by the observed physical or economical fields, like air temperature or asset price. A paradigmatic example is an earthquake initiation time, which is determined by the complex interplay of stress and strength fields in the heterogeneous Earth lithosphere. The physical theory for spatio-temporal evolution of these fields is still in its infancy, their values can hardly be measured with the existing instruments, or predicted using the available statistical methods. At the same time, earthquakes are readily defined, measured, and studied.

Prediction here is based on analysis of the observable permanent background activity of the complex system. We look for *premonitory patterns*, i.e., particular deviations from long-term averages that emerge more frequently as an extreme event approaches. These patterns might be either *perpetrators* contributing to triggering an extreme event, or *witnesses* merely signalling that the system became unstable, ripe for a disaster. An example of a witness is the proverbial 'straws in the wind' preceding a hurricane.

The following types of premonitory patterns have been established by exploratory data analysis and numerical modelling: (i) increase of background activity; (ii) deviations from self-similarity, i.e. a change of the size distribution of events in favour of relatively strong yet subextreme events; (iii) increase of event clustering; and (iv) emergence of long-range correlations. Solid empirical evidence for the existence of these patterns in seismology and other forms of multiple fracturing has been accumulated since the 1970s (Haberman, 1981; Mogi, 1981; Aki, 1985; Keilis-Borok and Shebalin, 1999; Sykes *et al.*, 1999; Keilis-Borok and Soloviev, 2003). Importantly, these patterns are universal, common for complex systems of distinctly different origin. Similar premonitory patterns have been observed in socio-economic systems (Keilis-Borok *et al.*, 2000, 2005), dynamic clustering in elastic billiards (Gabrielov *et al.*, 2008), hydrodynamics, and hierarchical models of extreme event development (Narkunskaya and

Extreme Natural Hazards, Disaster Risks and Societal Implications, eds. Alik Ismail-Zadeh, Jaime Urrutia Fucugauchi, Andrzej Kijko, Kuniyoshi Takeuchi and Ilya Zaliapin. Published by Cambridge University Press.

Shnirman, 1990; Newman *et al.*, 1995; Blanter *et al.*, 1997; Gabrielov *et al.*, 2000; Zaliapin *et al.*, 2003a,b). We discuss here a general mechanism that reproduces these universal premonitory patterns.

We focus in particular on *premonitory deviations from self-similarity*. Self-similarity is one of the most prominent features of complex systems. A canonical example is a power-law (self-similar) distribution of system observables, whose remarkable feature is the inevitability of extremely large events that dwarf numerous smaller events. The power-law distribution is well known under different names in such diverse phenomena as inertial-range self-similarity in turbulence (Kolmogorov–Obukhov laws) (Kolmogorov, 1941a; Obukhov, 1941; McWilliams, 1990; Frisch, 1996), energy released in an earthquake (Gutenberg–Richter law) (Gutenberg and Richter, 1944; Ben-Zion, 2003), word usage frequency in a language (Zipf law) (Zipf, 1965), allocation of wealth in a society (Pareto law) (Pareto, 1897; Klass *et al.*, 2006), war casualties (Richardson law) (Richardson, 1960), number of papers published by a given scientist (Lotka law) (Lotka, 1926), mass of a landslide (Malamud *et al.*, 2004; Brunetti *et al.*, 2009), stock price returns (Mandelbrot and Taylor, 1967; Gabaix *et al.*, 2003; Plerou and Stanley, 2008), number of species per genus (Burlando, 1990), and many others (Mandelbrot, 1983; Newman, 1997, 2005; Turcotte, 1997; Albert and Barabasi, 2002). An important paradigm of *self-organised criticality* (Bak *et al.*, 1988; Turcotte, 1999) that is demonstrated by sand-pile (Dhar, 1990), forest-fire (Drossel and Schwabl, 1992), and slider-block (Burridge and Knopoff, 1967; Olami *et al.*, 1992; Rundle and Klein, 1993) models and their numerous ramifications has been introduced in order to understand dynamic processes whose only attractor corresponds to self-similarity (criticality) of the size distribution of appropriately defined *events*.

Exact self-similarity, as well as many other universal properties, however, is only an approximation to (or a mean-field property of) the observed and modelled systems; at each particular time moment the distribution of event sizes deviates from a pure power-law form. We show in this chapter how to use such deviations for understanding the dynamics of a complex system in general and the occurrence of extreme events in particular.

The rest of the chapter is organised as follows. We informally outline our model and the corresponding prediction problem in Section 10.2. A formal model description is given in Section 10.3. Section 10.4 summarises the study's results most relevant to the prediction problem. Section 10.5 derives the spatio-temporal model distribution as a function of the control parameter. Section 10.6 uses these results to find spatio-temporal deviations of the event size distribution from its mean-field form. Results of numerical experiments are illustrated in Section 10.7. In Section 10.8 we further discuss the relation of our results to prediction of extreme events. Proofs and necessary technical information are collected in Appendices.

10.2 Model outline and prediction problem

The model combines *external driving* ultimately responsible for the occurrence of events, including the extreme ones, a *cascade process* responsible for redistribution of energy (or another appropriate physical quantity such as mass, moment, stress, etc.) within the system, and *spatial dynamics*. We first outline the process of populating a system space Ω with *particles* of discrete ranks and then proceed with the definition of the *observation space* and *events*. We assume that Ω is an n-dimensional Euclidean space.

A direct cascade (branching) within a system starts with consecutive injection (immigration) of *particles* of the largest possible rank, $r_{\max}$, into the origin $0 \in \Omega$, which we call the *source*. After injection, each particle diffuses freely and independently of the others across the space Ω. Eventually, it splits into a random number of particles of smaller rank, $r_{\max} - 1$, each of which continues to diffuse from the location of the parent and independently of the other particles. These particles split in their turn into even smaller particles, and so on.

At each time instant $t \geq 0$, *observations* can be done on a subspace $\mathcal{R}_t \subset \Omega$. In this chapter we assume that $\mathcal{R}_t$ is an affine subspace of dimension $d < n$. An observed *event* corresponds to an instant when a particle crosses the subspace of observations. Each event is characterised by its occurrence time t, spatial location $\mathbf{x} \in \mathcal{R}_t$ within the observation space, and rank r. Model observations at instant t thus consist of a collection of events $\mathcal{C}_t = (t_i \leq t, \mathbf{x}_i, r_i)$, $i \geq 1$, referred to as the *catalogue*. An *extreme event* is defined as a sufficiently large, although not necessarily the largest, event, $r \geq r_0$, where r_0 is a rank threshold.

Importantly, the location of $\mathcal{R}_t$ within Ω is (a) unknown to an observer, and (b) time-dependent. One can interpret this as movement of the observation space relative to the source, movement of the source relative to the observation space, or a combination of the two. A principal goal of an observer is to assess the likelihood of the occurrence of an extreme event using the catalogue $\mathcal{C}_t$. It is readily seen that the probability of an extreme event increases as the observation space approaches the source and achieves its maximal value when the source belongs to the observation space, $0 \in \mathcal{R}_t$. The distance between the observation subspace and the source thus becomes a natural control parameter and allows one to reduce the prediction problem to estimating the distance to the source. This latter problem is the focus of the chapter.

As the observation subspace approaches the source, the intensity of the observed events increases, larger events become relatively more frequent, and clustering and long-range correlations become more prominent (see Fig. 10.1 and Sections 10.4, 10.8). The emergence of these patterns, each individually and all together, can be therefore used to forecast an approach of a large event; indeed, such a prediction should be understood in a statistical sense. We focus on quantitative description of two of these patterns, intensity increase and deviations from self-similarity,

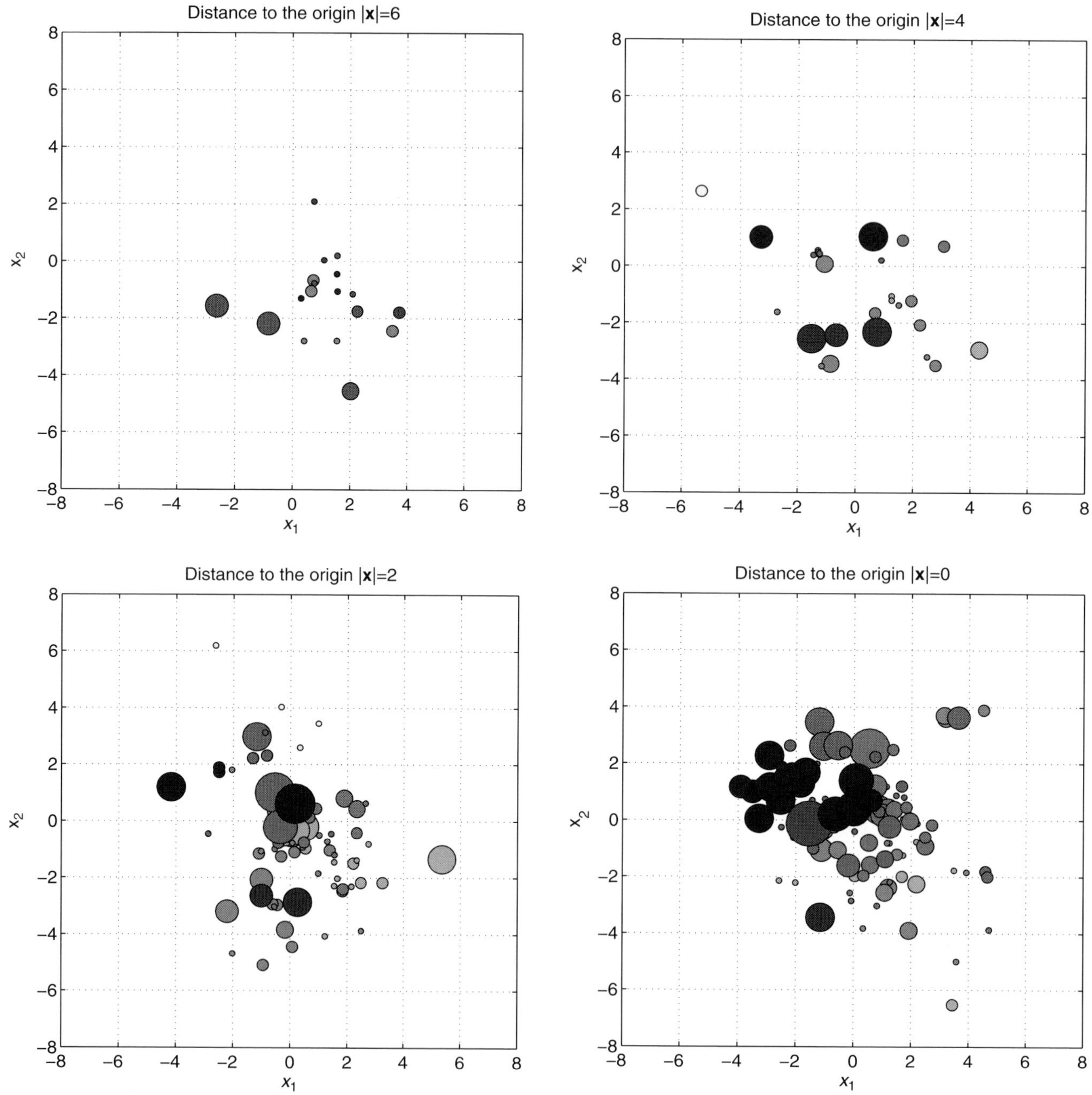

Figure 10.1. Example of a three-dimensional model population. Different panels show two-dimensional subspaces of the model three-dimensional space at different distances $|\mathbf{x}|$ to the origin. Model parameters are $\mu = \lambda = 1$, $D = 1$, $B = 2$. Circle size is proportional to the particle rank. Different shades correspond to populations from different immigrants; the descendants of earlier immigrants are a lighter shade. The clustering of particles is explained by the splitting histories. Note that, as the origin approaches, the particle activity significantly changes, indicating the increased probability of an extreme event.

for a classical branching process formally introduced in the next section.

We emphasise that the location and dynamics of the observation space $\mathcal{R}_t$ within Ω depend on details of a particular system of interest and may be hard to estimate or model. Important points of this chapter are that (i) the information about this unknown dynamics can be summarised by a scalar value of the control parameter (distance between the observational subspace and the origin); and (ii) knowledge of the control parameter is sufficient to solve the prediction problem.

Finally, it is important to mention that we do not use direct cascade as a dynamical model of event formation, which would imply that large events *cause* smaller ones. We merely use this analytically tractable approach to create a *hierarchical network* of spatially distributed particles. A dynamic interpretation of the latter will depend on a particular application, and may include inverse cascading or other physically relevant processes.

10.3 Model formulation

We consider an age-dependent multi-type branching diffusion process with immigration in $\mathbb{R}^n$. The system consists of particles, each of which belongs to a *generation* $k = 0, 1, \ldots$. Particles of zero generation (the largest ones) appear in a system as a result of external driving (forcing); we will refer to them as *immigrants*. Particles of any other generation $k > 0$ are produced as a result of the splitting of the particles of generation $k - 1$. Immigrants ($k = 0$) are born at the origin $\mathbf{x}:=(x_1,\ldots,x_n) = \mathbf{0}$ at a constant rate μ; that is, the probability for a new immigrant to appear within the time interval of length Δt is $\mu\Delta t + o(\Delta t)$ as $\Delta t \to 0$. Accordingly, the birth instants form a homogeneous Poisson process with intensity μ. Each particle lives for some random time τ and then transforms (splits) into a random number β of particles of the next generation. The probability laws of the lifetime τ and branching β are generation-, time-, and space-independent. We assume that new particles are born at the location of their parent at the moment of splitting.

The particle lifetime has an exponential distribution:

$$G(t) := \mathrm{P}\{\tau < t\} = 1 - \mathrm{e}^{-\lambda t}, \ \lambda > 0. \tag{10.1}$$

The conditional probability that a particle transforms into $k \geq 0$ new particles (0 means that it disappears) given that the transformation took place is denoted by p_k. The probability generating function (pgf) for the number β of new particles is thus

$$h(s) = \sum_k p_k\, s^k. \tag{10.2}$$

The expected number of offspring (also called the *branching number*) is $B := E(\beta) = h'(1)$ (e.g., Athreya and Ney (2004) Chapter 1).

Each particle diffuses in $\mathbb{R}^n$ independently of other particles. This means that the density $p(\mathbf{x}, \mathbf{y}, t)$ of a particle that was born at instant 0 at point $\mathbf{y}$ solves the equation

$$\frac{\partial p}{\partial t} = D\left(\sum_i \frac{\partial^2}{\partial x_i^2}\right)p \equiv D\Delta_{\mathbf{x}}p \tag{10.3}$$

with the initial condition $p(\mathbf{x}, \mathbf{y}, 0) = \delta(\mathbf{x} - \mathbf{y})$. The solution of (10.3) is given by Evans (1998):

$$p(\mathbf{x},\mathbf{y},t) = (4\,\pi\,D\,t)^{-n/2}\exp\left\{-\frac{|\mathbf{x}-\mathbf{y}|^2}{4\,D\,t}\right\}, \ |\mathbf{x}|^2 = \sum_i x_i^2. \tag{10.4}$$

Accordingly, the density of each particle, given that it is alive at the instant t, is $\phi(\mathbf{x}, t) := p(\mathbf{x}, 0, t)$. Naturally, the positions of the particles produced by the same immigrant are correlated. This can be reflected by the joint distribution of pairs, triplets, etc.

The model is specified by the following parameters: immigration intensity $\mu > 0$, branching intensity $\lambda > 0$, diffusion constant $D > 0$, and branching distribution $\{p_k\}$, which will be often represented by its pgf $h(z)$ or simply by the branching number B. An appropriate choice of the temporal and spatial scales allows one to assume $\mu = 1$ and $D = 1$.

It is convenient to introduce particle *rank* $r := r_{\max} - k$ for an arbitrary integer $r_{\max}$ and thus consider particles of ranks $r \leq r_{\max}$. Particle rank can be considered a logarithmic measure of the size. Similar to the analysis of the real-world systems, we sometimes only consider particles of the first several generations $0 \leq k \leq r_{\max} - 1$, which corresponds to the largest ranks $1 \leq r \leq r_{\max}$. Figure 10.1 illustrates the model population.

10.4 Prediction: summary

This section summarises important findings that are most relevant to the prediction problem. Recall that the prediction problem consists of assessing the likelihood of an extreme event; the latter corresponds to an instant when a sufficiently large particle crosses the observation space. The likelihood of an extreme event is thus directly related to the distance between the space of observations and the origin. Accordingly, the prediction problem is reduced to the estimation of this distance from available data. For that, one should look for increase in the intensity of medium-to-large-sized events, as well as upward deviations in the event size distribution. We believe that this general idea can be useful in a wide range of models and observed systems, not necessarily based on a branching diffusion mechanism. Statistical assessment of particular prediction schemes based on this idea is left for a future study.

All statements below refer to a steady state of the model (dynamics after a transient). All asymptotic statements have been confirmed numerically in finite models.

(1) *Mean-field self-similarity* Particle ranks, averaged over time and space, have an exponential distribution; this is equivalent to a power-law distribution of particle sizes; see (10.33) and Fig. 10.2.
(2) *Small-size self-similarity* The particle rank distribution at *any* spatial point is asymptotically exponential as rank decreases, with the exponent index $-B$; see (10.38) and Figs. 10.3 and 10.4. This is equivalent to a power-law distribution of particle sizes with power-law index $-B$. Furthermore, this implies that deviations from self-similarity, if any, can be only seen at large ranks (large particle sizes).
(3) *Upward deviations close to the origin* At any point sufficiently close to the origin, the particle size distribution deviates from a self-similar power-law form so as to have a larger number of medium-to-large-sized events. The magnitude of this deviation increases with the event size, as well as with dimension of the model space; see (10.36) and the upper lines in Figs. 10.3 and 10.4.
(4) *Downward deviations away from the origin* At any point sufficiently far from the origin, the particle size distribution deviates from a self-similar power-law form so as to have a smaller number of medium-to-large-sized events. The magnitude of this deviation increases with the event size and is independent of the model's dimension; see (10.37) and the lower lines in Figs. 10.3 and 10.4.
(5) *Exponential decay of event intensity* The intensity of events of any fixed size is exponentially decaying away from the origin; see (10.28).
(6) *Divergence of event intensity at the origin* For models with a spatial dimension larger than 1, the intensity of sufficiently large events diverges at the origin in a power-law fashion; see (10.28), (10.30), and Fig. 10.3(b)–(d).

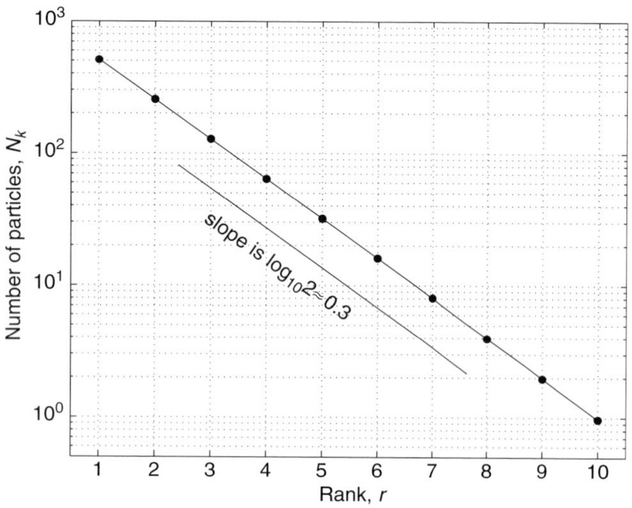

Figure 10.2. Spatially averaged particle rank distribution at $t = 30$. The distribution is averaged over 4000 independent realisations of a three-dimensional model with parameters $\mu = \lambda = 1$, $D = 1$, $B = 2$, $r_{\max} = 10$. One can clearly see the exponential rank distribution of Eq. (10.33).

10.5 Model solution: moment generating functions

The model introduced in Section 10.3 is a superposition of independent branching processes generated by individual immigrants. Sections 10.5.1 and 10.5.2 analyse, respectively, the one-point and two-point moments of a particle distribution produced by a single immigrant. Then we expand these results to the case of multiple immigrants in Section 10.5.3.

10.5.1 Single immigrant: one-point properties

10.5.1.1 *Moment generating functions*

Let $p_{k,i}(G, \mathbf{y}, t)$ be the conditional probability that at time $t \geq 0$ there exist $i \geq 0$ particles of generation $k \geq 0$ within spatial region $G \subset \mathbb{R}^n$ given that at time 0 a single immigrant was injected at point $\mathbf{y}$. The corresponding *moment generating function* is

$$M_k(G, \mathbf{y}, t; s) = \sum_i p_{k,i}(G, \mathbf{y}, t)e^{si}. \tag{10.5}$$

Proposition 5.1 *The moment generating functions M_k $(G, \mathbf{y}, t; s)$ solve the following recursive system of non-linear partial differential equations:*

$$\frac{\partial}{\partial t} M_k(G, \mathbf{y}, t; s) = D\Delta_{\mathbf{y}} M_k - \lambda M_k + \lambda\, h(M_{k-1}), \quad k \geq 1, \tag{10.6}$$

with initial conditions $M_k\,(G, \mathbf{y}, 0; s) \equiv 1$, $k \geq 1$, and

$$M_0(G, \mathbf{y}, t; s) = (1 - P) + Pe^s, \;\; P := e^{-\lambda t} \int_G p(\mathbf{x}, \mathbf{y}, t)d\mathbf{x}. \tag{10.7}$$

Here $h(s)$ is defined by (10.2) *and* $\Delta_{\mathbf{y}} = \sum_i \partial^2/\partial y_i^2$.

The proof is given in Appendix 10.A.

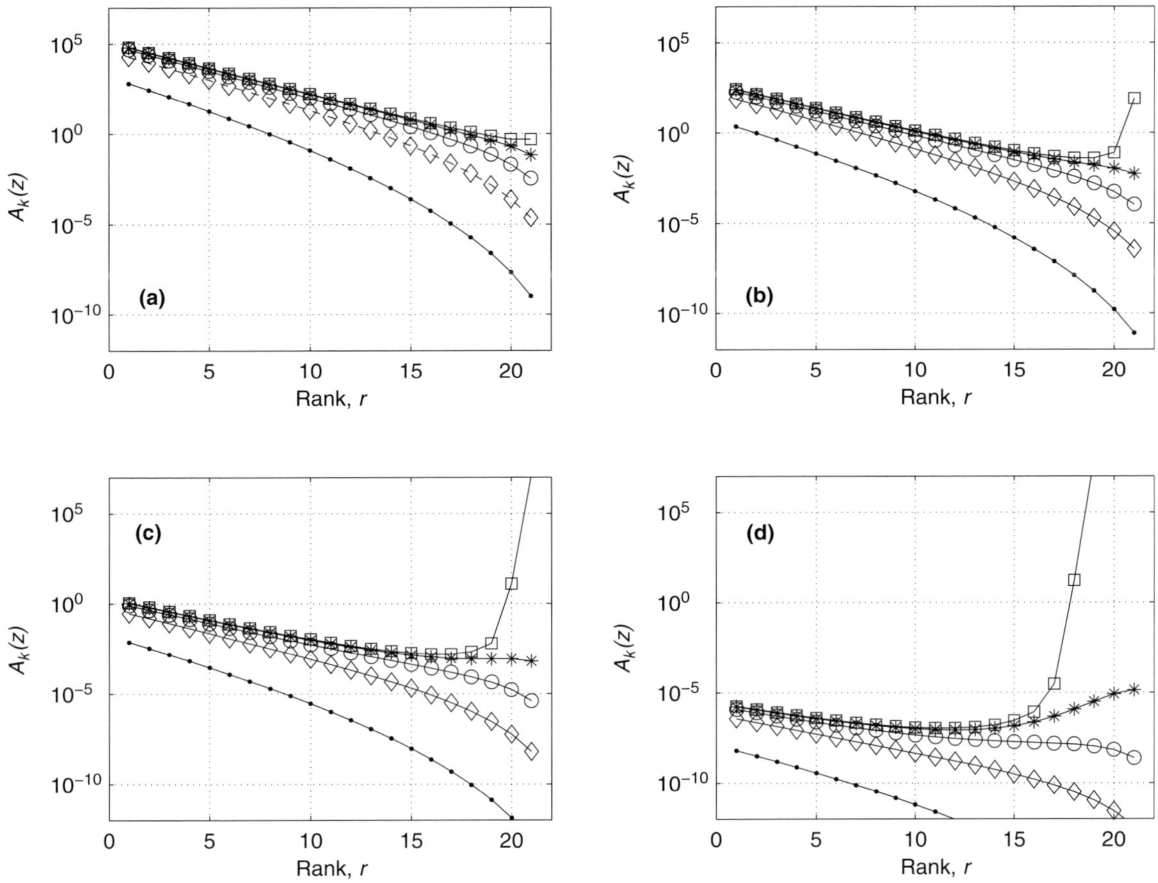

Figure 10.3. Expected number A_k (z) of generation-k particles at distance z from the origin (cf. Proposition 10.6.2). The distance z is increasing (from top to bottom line in each panel) as $z = 10^{-3}$, 2, 5, 10, 20. Model dimension is (a) $n = 1$, (b) $n = 3$, (c) $n = 5$, and (d) $n = 10$. Other model parameters: $\mu = \lambda = 1$, $D = 1$, $B = 2$, $r_{\max} = 21$.

10.5.1.2 *The first moment densities*

Let $\overline{A}_k(G, \mathbf{y}, t)$ be the expected number of generation-k particles at instant t within the region G, produced by a single immigrant injected at point $\mathbf{y}$ at time $t = 0$. It is given by the following partial derivative (e.g., Athreya and Ney, 2004, Chapter 1):

$$\overline{A}_k(G, \mathbf{y}, t) := \left. \frac{\partial M_k(G, \mathbf{y}, t; s)}{\partial s} \right|_{s=0}. \tag{10.8}$$

Consider also the expectation density $A_k(\mathbf{x}, \mathbf{y}, t)$ that satisfies, for any $G \subset \mathbb{R}^n$,

$$\overline{A}_k(G, \mathbf{y}, t) = \int_G A_k(\mathbf{x}, \mathbf{y}, t) d\mathbf{x}. \tag{10.9}$$

Corollary 10.5.2. *The first moment densities $A_k(\mathbf{x}, \mathbf{y}, t)$ solve the following recursive system of partial differential equations:*

$$\frac{\partial A_k(\mathbf{x}, \mathbf{y}, t)}{\partial t} = D\Delta_{\mathbf{x}} A_k - \lambda A_k + \lambda B A_{k-1}, \; k \geq 1, \tag{10.10}$$

with the initial conditions $A_k(\mathbf{x}, \mathbf{y}, 0) \equiv 0$, $k \geq 1$,

$$A_0(\mathbf{x}, \mathbf{y}, 0) = \delta(\mathbf{y} - \mathbf{x}), \; A_0(\mathbf{x}, \mathbf{y}, t) = e^{-\lambda t} p(\mathbf{x}, \mathbf{y}, t), \; t > 0. \tag{10.11}$$

The solution to this system is given by

$$\begin{aligned} A_k(\mathbf{x}, \mathbf{y}, t) &= \frac{(\lambda B t)^k}{k!} A_0(\mathbf{x}, \mathbf{y}, t) \\ &= \frac{(\lambda B)^k}{k!(4\pi D)^{n/2}} t^{k-n/2} \exp\left\{ -\lambda t - \frac{|\mathbf{x} - \mathbf{y}|^2}{4Dt} \right\}. \end{aligned} \tag{10.12}$$

The proof is given in Appendix 10.C. It follows from a general result for the higher moments obtained in Appendix 10.B.

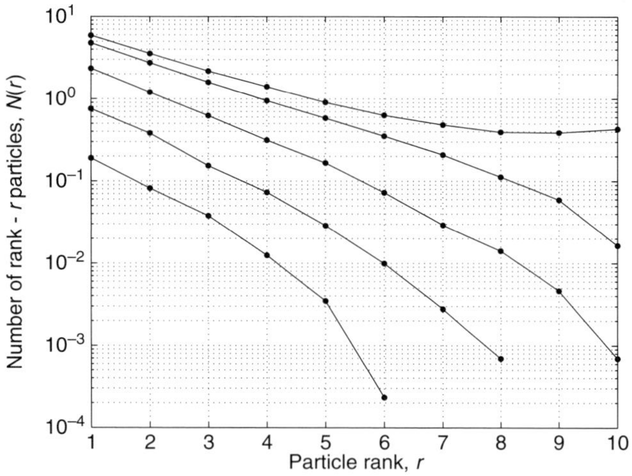

Figure 10.4. Particle rank distribution at $t = 30$ and fixed distance z from the origin (cf. Proposition 10.6.2). The distribution is averaged over 4000 independent realisations of a three-dimensional model with parameters $\mu = \lambda = 1$, $D = 1$, $B = 2$, $r_{\max} = 10$. Different lines correspond to different distances (from top to bottom): $z = 0, 2, 4, 6, 8$. One can clearly see that the rank distribution deviates from the pure exponential form, which corresponds to a straight line in the semilogarithmic scale used here. One observes *downward deviations* at large distances from the origin, and *upward deviations* close to the origin.

The system (10.10) has a transparent intuitive meaning. The rate of change of the expectation density $A_k(\mathbf{x}, \mathbf{y}, t)$ is affected by three processes: diffusion of the existing particles of generation k (the first term in the right hand side of (10.10)), splitting of the existing particles of generation k at the rate λ (the second term), and splitting of the generation $k - 1$ particles that produce on average B new particles of generation k (the third term).

To obtain the solution for the entire population, we sum up the contributions from all generations:

$$A(\mathbf{x}, 0, t) = \sum_{k=0}^{\infty} A_k(\mathbf{x}, 0, t) = e^{-\lambda t(1-B)}\, p(\mathbf{x}, 0, t) = \frac{e^{-\lambda t(1-B)}}{(4\,\pi\,D\,t)^{n/2}} \exp\left(-\frac{|\mathbf{x}|^2}{4\,D\,t}\right). \tag{10.13}$$

This formula emphasises the role of the branching parameter B: in the subcritical case, $B < 1$, the population goes extinct exponentially; in the supercritical case, $B > 1$, the population grows exponentially; in the critical case, $B = 1$, the expected number of particles remains the same (steady state) and is given by the diffusion density $p(\mathbf{x}, 0, t)$.

10.5.2 Single immigrant: two-point properties

10.5.2.1 *Moment generating functions*

Let $p_{k_1,k_2,i,j}(G_1, G_2, \mathbf{y}, t)$ be the conditional probability that at instant $t \geq 0$ there exist $i \geq 0$ particles of generation $k_1 \geq 0$ within region $G_1 \subset \mathbb{R}^n$ and $j \geq 0$ particles of generation $k_2 \geq 0$ within region $G_2 \subset \mathbb{R}^n$ given that at time 0 a single immigrant was injected at point $\mathbf{y}$. Assume that G_1 and G_2 do not overlap. The corresponding moment generating function is

$$M_{k_1,k_2}(G_1, G_2, \mathbf{y}, t; s_1, s_2) = \sum_{i,j \geq 0} p_{k_1,k_2,i,j}(G_1, G_2, \mathbf{y}, t) e^{i\,s_1 + j\,s_2}. \tag{10.14}$$

Proposition 10.5.3 *The moment generating functions $M_{k_1,k_2}(G_1, G_2, \mathbf{y}, t; s_1, s_2)$ solve the following recursive system of non-linear partial differential equations:*

$$\frac{\partial}{\partial t} M_{k_1,k_2} = D\Delta_{\mathbf{y}} M_{k_1,k_2} - \lambda M_{k_1,k_2} + \lambda h(M_{k_1-1,k_2-1}), k_1,k_2 \geq 1, \tag{10.15}$$

with the initial conditions

$$M_{k_1,k_2}(G_1, G_2, \mathbf{y}, 0; s_1, s_2) \equiv 1,\ k_1, k_2 \geq 1, \tag{10.16}$$

$$M_{0,0}(G_1, G_2, \mathbf{y}, t; s_1, s_2) = P_1 e^{s_1} + P_2 e^{s_2} + 1 - P_1 - P_2, \tag{10.17}$$

$$M_{0,k}(G_1, G_2, \mathbf{y}, t; s_1, s_2) = (M_k(G_2, \mathbf{y}, t; s_2) - e^{-\lambda t}) + (e^{-\lambda t} - P_1) + P_1 e^{s_1}, \tag{10.18}$$

where $P_i := e^{-\lambda t} \int_{G_i} p(\mathbf{x}, \mathbf{y}, t) d\mathbf{x}$, $i = 1,2$. Here, as before, $h(s)$ is defined by (10.2) and $\Delta_{\mathbf{y}} = \sum_i \partial^2/\partial y_i^2$.

The proof is given in Appendix 10.D.

10.5.2.2 *Moments*

Consider the expected value $\overline{A}_{k_1,k_2}(G_1, G_2, \mathbf{y}, t)$ of the product of the number of generation-k_1 particles in region G_1 and the number of generation-k_2 particles in region G_2 at instant t, produced by a single immigrant injected at point $\mathbf{y}$ at time $t = 0$. It is given by the following partial derivative:

$$\overline{A}_{k_1,k_2}(G_1, G_2, \mathbf{y}, t) := \left.\frac{\partial^2 M_{k_1,k_2}(G_1, G_2, \mathbf{y}, t; s_1, s_2)}{\partial s_1 \partial s_2}\right|_{s_1=s_2=0}. \tag{10.19}$$

We notice that the expectations $\overline{A}_{k_1}(G_1, \mathbf{y}, t)$ and $\overline{A}_{k_2}(G_2, \mathbf{y}, t)$ of (10.8) can be represented as

$$\overline{A}_{k_1}(G_1,\mathbf{y},t) := \left.\frac{\partial M_{k_1,k_2}(G_1,G_2,\mathbf{y},t;s_1,s_2)}{\partial s_1}\right|_{s_1=s_2=0} \tag{10.20}$$

and

$$\overline{A}_{k_2}(G_2,\mathbf{y},t) := \left.\frac{\partial M_{k_1,k_2}(G_1,G_2,\mathbf{y},t;s_1,s_2)}{\partial s_2}\right|_{s_1=s_2=0}. \tag{10.21}$$

Consider also the expectation density $A_{k_1,k_2}(\mathbf{x}_1,\mathbf{x}_2,\mathbf{y},t)$ that satisfies, for any non-overlapping $G_1, G_2 \subset \mathbb{R}^n$,

$$\overline{A}_{k_1,k_2}(G_1,G_2,\mathbf{y},t) = \int_{G_2}\int_{G_1} A_{k_1,k_2}(\mathbf{x}_1,\mathbf{x}_2,\mathbf{y},t)\mathrm{d}\mathbf{x}_1\mathrm{d}\mathbf{x}_2. \tag{10.22}$$

Corollary 10.5.4 *The moment densities* $A_{k_1,k_2} \equiv A_{k_1,k_2}(\mathbf{x}_1,\mathbf{x}_2,\mathbf{y},t)$ *solve the following recursive system of partial differential equations:*

$$\frac{\partial A_{k_1,k_2}}{\partial t} = D\,\Delta_{\mathbf{y}}A_{k_1,k_2} - \lambda\,A_{k_1,k_2} + \lambda\,B\,A_{k_1-1,k_2-1}$$
$$+\lambda\,h''(1)\,A_{k_1-1}(\mathbf{x}_1)\,A_{k_2-1}(\mathbf{x}_2),\;\; k_1,k_2 \ge 1, \tag{10.23}$$

with the initial conditions

$$A_{k_1,k_2}(\mathbf{x}_1,\mathbf{x}_2,\mathbf{y},0) \equiv 0,\;\; k_1,k_2 \ge 1, \tag{10.24}$$

$$A_{0,k}(\mathbf{x}_1,\mathbf{x}_2,\mathbf{y},t) \equiv 0,\;\; k \ge 0,\;\; t \ge 0, \tag{10.25}$$

and $A_k(\mathbf{x}) \equiv A_k(\mathbf{x},\mathbf{y},t)$ *given by* (10.12).

The proof is given in Appendix 10.E.

10.5.3 Multiple immigrants

Here we expand the results of Section 10.5.1 to the case of multiple immigrants that appear at the origin according to a homogeneous Poisson process with intensity μ. The expectation $\mathcal{A}_k$ of the number of particles of generation k is given by

$$\mathcal{A}_k(\mathbf{x},t) = \int_0^t A_k(\mathbf{x},0,s)\,\mu\,\mathrm{d}s$$
$$= \frac{\mu\,(\lambda\,B)^k}{k!(4\,\pi\,D)^{n/2}}\int_0^t s^{k-n/2}\exp\left\{-\lambda\,s - \frac{|\mathbf{x}|^2}{4\,D\,s}\right\}\mathrm{d}s. \tag{10.26}$$

The steady-state spatial distribution corresponds to the limit $t \to \infty$:

$$\mathcal{A}_k(\mathbf{x}) := \mathcal{A}_k(\mathbf{x},\infty) = \frac{2\,\mu\,(\lambda\,B)^k}{k!(4\,\pi\,D)^{n/2}}\left(\frac{|\mathbf{x}|^2}{4\,D\,\lambda}\right)^{\nu/2} K_\nu\left(|\mathbf{x}|\sqrt{\frac{\lambda}{D}}\right). \tag{10.27}$$

Here $\nu = k - n/2 + 1$ and K_ν is the modified Bessel function of the second kind (see Appendix 10.G). Introducing the normalised distance from the origin $z := |\mathbf{x}|\sqrt{\lambda/D}$ we obtain

$$\mathcal{A}_k(z) = \frac{\mu}{\lambda\,k!}\left(\frac{B}{2}\right)^k\left(\frac{2\,\pi\,D}{\lambda}\right)^{-n/2} z^\nu\,K_\nu(z). \tag{10.28}$$

For odd n, there are explicit expressions for $K_\nu(z)$ (Appendix 10.G, (10.G2), (10.G3)). In particular, we have

$$\mathcal{A}_0(z) = \frac{\mu}{\sqrt{4\,D\,\lambda}}\mathrm{e}^{-z},\text{ for } n=1,$$
$$\mathcal{A}_0(z) = \sqrt{\frac{\lambda}{D^3}}\,\frac{\mu}{4\,\pi\,z}\mathrm{e}^{-z},\text{ for } n=3. \tag{10.29}$$

From (10.28) and the asymptotic behaviour of $K_\nu(z)$ as $z \to 0$ (Appendix 10.G, (10.G5)) it follows that

$$\lim_{z\to 0}\mathcal{A}_k(z) = \begin{cases} \infty, & \text{for } \nu \le 0,\text{ i.e., } k \le n/2-1 \\ \text{const} < \infty, & \text{for } \nu > 0,\text{ i.e., } k > n/2-1. \end{cases} \tag{10.30}$$

Thus, in a model with spatial dimension $n \ge 2$, the elements of several of the lowest generations ($k \le n/2 - 1$) have an infinite concentration at the origin.

10.5.4 Alternative model representation

In this section we derive a system of equations for the steady-state expectations $\mathcal{A}_k(\mathbf{x})$ using the radial symmetry of the problem. By integrating (10.10) from $t = 0$ to ∞, we obtain

$$D\Delta_{\mathbf{x}}\mathcal{A}_k(\mathbf{x}) - \lambda\mathcal{A}_k(\mathbf{x}) + \lambda B\mathcal{A}_{k-1}(\mathbf{x}) = 0,$$

since $A_k(\mathbf{x},\mathbf{y},\infty) = 0$. We now rewrite this equation in terms of the normalised distance from the origin, $z := |\mathbf{x}|\sqrt{\lambda/D}$, using the fact that $\mathcal{A}_k(\mathbf{x}) \equiv \mathcal{A}_k(z)$ as soon as $|\mathbf{x}|=|z|$:

$$\mathcal{A}_k''(z) + \frac{n-1}{z}\,\mathcal{A}_k'(z) - \mathcal{A}_k(z) + B\mathcal{A}_{k-1}(z) = 0. \tag{10.31}$$

We notice, furthermore, that one can rewrite the expectation densities (10.12) as a function of z, which results in $A_k(z) \equiv A_k(\mathbf{x}, 0, t)$. It is then readily seen that

$$A_k'(z) = -\frac{B}{2k}\,z\,A_{k-1}(z). \tag{10.32}$$

The same recursive system holds for $\mathcal{A}_k(z)$, which is shown by integrating the last equation with respect to time.

10.6 Particle rank distribution

We analyse here the particle rank distribution; recall that the rank is defined as $r = r_{\max} - k$, where k is the particle's generation. A self-similar branching mechanism that governs the model suggests an exponential distribution of particle ranks. Indeed, the spatially averaged steady-state rank distribution is a pure exponential law with index B:

$$A_k := \int_{\mathbb{R}^n} \int_0^\infty A_k(\mathbf{x}, 0, t)\mu \, dt \, d\mathbf{x} = \frac{\mu \, B^k}{k!} \int_0^\infty (\lambda t)^k e^{-\lambda t} \, dt = \frac{\mu}{\lambda} \, B^k \propto B^{-r}. \tag{10.33}$$

Remark 10.6.1 *The use of the term 'self-similar' with respect to the exponential distribution, often seen in physical literature, requires some explanation. As we mentioned earlier, the particle rank serves as a logarithmic measure of its size. Thus, the exponential distribution of ranks corresponds to the power-law distribution of sizes; hence the term 'self-similarity'.*

To analyse rank- and space-dependent deviations from the pure exponential distribution, we consider the ratio $\gamma_k(\mathbf{x})$ between the number of particles of two consecutive generations:

$$\gamma_k(\mathbf{x}) := \frac{\mathcal{A}_k(\mathbf{x})}{\mathcal{A}_{k+1}(\mathbf{x})}. \tag{10.34}$$

For the purely exponential rank distribution, $A_k(\mathbf{x}) = c\, B^k$, the value of $\gamma_k(\mathbf{x}) = 1/B$ is independent of k and $\mathbf{x}$; while deviations from the pure exponential distribution cause γ_k to vary as a function of k and/or $\mathbf{x}$. Plugging (10.28) into (10.34) we find

$$\gamma_k(\mathbf{x}) = \frac{2(k+1)}{B\, z} \frac{K_\nu(z)}{K_{\nu+1}(z)}, \tag{10.35}$$

where, as before, $z := |\mathbf{x}| \sqrt{\lambda/D}$ and $\nu = k - n/2+1$.

Proposition 10.6.2 *The asymptotic behaviour of the function $\gamma_k(z)$ is given by*

$$\lim_{z\to 0} \gamma_k(z) = \begin{cases} \infty, & \nu \le 0, \\ \frac{1}{B}\left(1 + \frac{n}{2\,\nu}\right), & \nu > 0, \end{cases} \tag{10.36}$$

$$\gamma_k(z) \sim \frac{2(k+1)}{B\, z}, \quad z \to \infty, \text{ fixed } k, \tag{10.37}$$

$$\gamma_k(z) \sim \frac{1}{B}\left(1 + \frac{n}{2\,\nu}\right), \quad k \to \infty, \text{ fixed } z. \tag{10.38}$$

The proof and the explicit rates of divergence in (10.36) are given in Appendix 10.F.

Proposition 10.6.2 describes the spatio-temporal deviations of the particle rank distribution from the pure exponential law (10.33). We interpret below each of (10.36)–(10.38) in some detail. Equation (10.38) implies that at any spatial point, the distribution asymptotically approaches the exponential form as generation k increases (rank r decreases). In other words, the distribution of small ranks (large generation numbers) is close to the exponential with index $-B$; thus the deviations can only be observed at the largest ranks (small generation numbers). Analysis of the large-rank distribution is done using (10.36) and (10.37). Near the origin, where the immigrants enter the system, (10.36) implies that $\gamma_k(z) > \gamma_{k+1}(z) > 1/B$ for $\nu > 0$. Hence, one observes the *upward deviations* from the pure exponential distribution: for the same number of rank r particles, the number of rank $r + 1$ particles is larger than predicted by the exponential law. The same behaviour is in fact observed for $\nu \le 0$ (see Appendix 10.F, (10.F5)). In addition, for $\nu \le 0$ the ratios $\gamma_k(z)$ do not merely deviate from $1/B$, but diverge to infinity at the origin. Away from the origin, according to (10.37), we have $\gamma_k(z) < \gamma_{k+1}(z) < 1/B$, which implies *downward deviations* from the pure exponent: for the same number of rank r particles, the number of rank $r + 1$ particles is smaller than predicted by the exponential law.

Figure 10.3 illustrates the above findings; it shows the distribution of particles for the largest ranks at different distances from the origin. One can clearly see the transition from downward to upward deviation of the rank distributions from the pure exponential form as we approach the origin. Notably, the magnitude of the upward deviation close to the origin (the upper line in all panels) strongly increases with the model dimension n.

10.7 Numerical analysis

Our analytical results and asymptotics are closely reproduced in numerical experiments with a finite number of generations, limited spatial extent, and spatial averaging (unavoidable when working with observations). Here, to mimic the ensemble averaging, the numerical results have been averaged over 4000 independent realisations of a three-dimensional model with parameters $\mu = \lambda = 1$, $D = 1$, and $B = 2$.

First, we check the exponential rank distribution of (10.33). Figure 10.2 shows the observed spatially averaged particle rank distribution. The exponential form (10.33) is indeed well reproduced.

Next, we see how the spatial averaging affects the rank distribution. Figure 10.4 shows the rank distribution at $t = 30$ at various distances to the origin. The spatial averaging has been done within spherical shells (space between two concentric spheres) of a constant volume $V = 5$ with inner radius z. Thus, here we see an observable counterpart of the theoretical distributions shown in Fig. 10.3(b). Although the spatial averaging somewhat tapers off the upward bend at the largest ranks close to the origin, the predicted transition from the downward to upward bend is clearly seen.

Figure 10.5 illustrates in more detail how the spatial averaging affects the upward bend in a three-dimensional model. It shows the particle rank distributions at $t = 30$ spatially averaged over spheres of different volumes centred at the origin. The upward bend is prominent for the spheres with volumes $V \leq 5$; and it gradually disappears within larger spheres in favour of an exponential distribution observed after a complete spatial averaging. Notably, the pure exponential distribution can only be achieved by averaging over *all* events in the model ($V = \infty$).

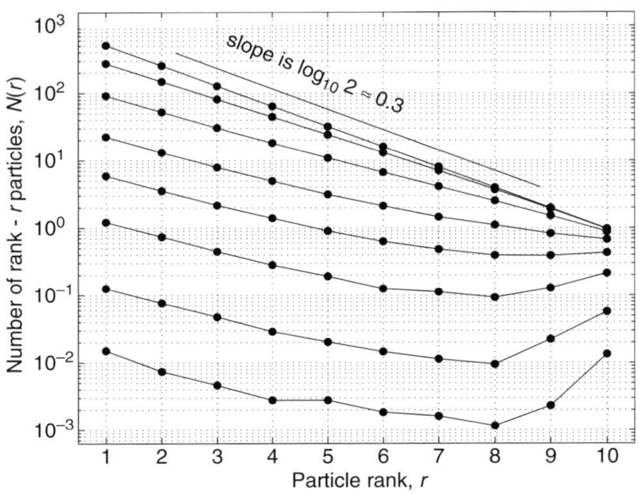

Figure 10.5. Particle rank distribution at $t = 30$ in a three-dimensional model. The distribution is spatially averaged over spheres of volume V centred at the origin with (from top to bottom): $V = \infty$, 500, 100, 200, 5, 1, 0.01. Model parameters: $\mu = \lambda = 1$, $D = 1$, $B = 2$, $r_{\max} = 10$. The upward deviations from the exponential distribution (a straight line) are fading away with the extent of the spatial averaging.

10.8 Discussion

The problem of predicting extreme events in complex systems is challenging. This chapter proposes a simple mechanism and a single control parameter for premonitory patterns that has been reported in the literature.

Quantitative analysis is performed here for a classical model of a spatially distributed population of particles of different sizes governed by direct cascade of branching and external driving (see Section 10.3). In probability theory this model is known as the age-dependent multi-type branching diffusion process with immigration (Athreya and Ney, 2004). This chapter discusses a new scope of problems for this model. We assume that observations (detection of particles) are only possible on a subspace of the system space while the source of external driving (origin) remains unobservable, as is the case in many real-world systems. The natural question under this approach concerns the dependence of the process statistics on the distance to the source. A complete analytical solution to this problem, in terms of the moments with respect to the particle density, is given by Proposition 10.5.1. In addition, the correlation structure of the particle field can be found using Proposition 10.5.3.

It is natural to consider rank as a logarithmic measure of the particle size. The exponential rank distribution derived in (10.33) corresponds to a self-similar, power-law distribution of particle sizes, characteristic for many complex systems. The self-similarity in the considered model, as well as in the real-world systems, is only observed after global spatial averaging in a steady state. Proposition 10.6.2 and Fig. 10.3 describe space-dependent deviations from the self-similarity. Recall that an extreme event in the examined system is defined as an observation of a particle of sufficiently large size. As the source approaches the observation subspace, the probability of an extreme event increases. These results are thus directly connected to prediction: when the location of the source changes in time and approaches the subspace of observation (or vice versa), the increase of event intensity and the downward bend in the event size distribution becomes premonitory to an extreme event. The numerical experiments confirm the validity of the analytical results and asymptotics in a finite model.

The examined model exhibits very rich and intriguing premonitory behaviour. Figure 10.1 shows several two-dimensional snapshots of a three-dimensional model at different distances from the source. One can see that, as the source approaches, the following changes in the background activity emerge: (a) the intensity (total number of particles) increases; (b) particles of larger size become relatively more numerous; (c) particle clustering becomes more prominent; (d) the correlation radius increases. All these premonitory changes have been independently observed in natural and socioeconomic systems. Here they are all determined by a single control parameter – the distance between the source and the observation space.

The above-mentioned premonitory patterns closely resemble universal properties of models of statistical physics in the vicinity of second order phase transition (Stanley, 1971; Ma, 2000; Kadanoff, 2000), percolation models near the percolation threshold (Stauffer and Aharony, 1994; Grimmett, 1999), and random graphs prior to the emergence of a giant cluster (Bollobás, 2001; Durrett, 2006; Newman *et al.*, 2006). In these models, the approach of an extreme event, usually referred to as a critical point, and the emergence of premonitory patterns, called critical phenomena, correspond to an instant when a control parameter crosses its critical value. In statistical physics a typical control parameter is temperature or magnetisation; in percolation it is the site or bond occupation density; in a random graph it is the probability for two vertices to be connected. The theory of critical phenomena (Ma, 2000) quantifies a system's behaviour at the critical value of the corresponding control parameter. The remarkable power of this theory is connected to the fact that very different systems demonstrate similar behaviour near to criticality. More precisely, when the control parameter is close to its critical value, the system sticks to one of just a few types of possible limit behaviours, each being described by an appropriate scale-invariant statistical field theory. In particular, each limit behaviour corresponds to the asymptotic power-law size distribution of system observables with a characteristic value of critical exponent.

We have focused in this chapter on a problem inverse to that considered by the critical phenomena theory: estimating the deviation of a control parameter from the critical value using the observed system behaviour. The motivation for this comes from environmental, geophysical, and other applied fields where one faces the problem of assessing the likelihood of the occurrence of an extreme event associated with a critical point. We have formulated and solved such a prediction problem for a spatially embedded cascade process, which enjoys both mean-field self-similarity and realistic premonitory time- and space-dependent deviations from the latter. The methods presented in this chapter may provide a framework for studying predictability of extreme events in complex systems of arbitrary nature.

Acknowledgment. This research was partly supported by NSF grants ATM-0620838, DMS-0934871 (to IZ), DMS-0801050 and DMS-1161629 (to AG).

Appendix 10.A

PROOF OF PROPOSITION 10.5.1

We will need the following calculus lemma that is readily proven by using the definition of derivative:

Lemma 10.A.1 *Let f(z), g(z), $z \in \mathbb{R}$ be continuous functions such that the definite integral* $G(t) = \int_0^t f(z)\, g(t-z)\mathrm{d}z$ *exists. We also assume that g(z) is differentiable. Then,*

$$\frac{\mathrm{d}}{\mathrm{d}t}G(t) = \int_0^t f(z)\, g'(t-z)\mathrm{d}z + f(t)\, g(0).$$

There are two possible scenarios for the model development up to time t. In the first one, the initial immigrant will not split; the probability for this is $P = e^{-\lambda t}$. In the second one, the initial immigrant will split at instant $0 \le u \le t$; the probability of the first split within the time interval $[u, u + \mathrm{d}u]$ is $\lambda \mathrm{e}^{-\lambda t}\, du + o(du)$ as $\mathrm{d}u \to 0$. The spatial position of the split is given by the diffusion density $p(\mathbf{x}, \mathbf{y}, u)$. If the immigrant splits, the composition property of generating functions gives $M_k = h[M_{k-1}]$. Integrating over all possible split instants and locations, we obtain

$$M_k(G, \mathbf{y}, t; s) = \mathrm{e}^{-\lambda t} + \int_{\mathbb{R}^n} \mathrm{d}\mathbf{y}' \int_0^t \mathrm{d}u\, \lambda \mathrm{e}^{-\lambda u} p(\mathbf{y}', \mathbf{y}, u)\, h[M_{k-1}(G, \mathbf{y}', t-u; s)]. \tag{10.A1}$$

Here the first and the second terms correspond to the first and second scenarios, respectively. Using the new integration variable $z = t - u$, we write

$$\begin{aligned} M_k(G, \mathbf{y}, t; s) &= \mathrm{e}^{-\lambda t} + \mathrm{e}^{-\lambda t} \int_{\mathbb{R}^n} \mathrm{d}y' \int_0^t \mathrm{d}u\, \lambda \mathrm{e}^{\lambda(t-u)} p(y', \mathbf{y}, u)\, h[M_{k-1}(G, y', t-u; s)] \\ &= \mathrm{e}^{-\lambda t}\left(1 + \int_{\mathbb{R}^n} \mathrm{d}y' \int_0^t \mathrm{d}z\, \lambda \mathrm{e}^{\lambda z} p(y', \mathbf{y}, t-z)\, h[M_{k-1}(G, y', z; s)]\right). \end{aligned}$$

Now we take the derivative with respect to t of both sides and apply Lemma 10.A.1 using the fact that $p(\mathbf{y}', \mathbf{y}, 0) = \delta(\mathbf{y}' - \mathbf{y})$ and $(\partial/\partial t - D\Delta_{\mathbf{y}})p = 0$:

$$\begin{aligned} \frac{\partial}{\partial t} M_k(G, \mathbf{y}, t; s) &= -\lambda M_k(G, \mathbf{y}, t; s) \\ &+ \mathrm{e}^{-\lambda t}\left[\left[\int_{\mathbb{R}^n} \mathrm{d}y' \int_0^t \mathrm{d}z\, \lambda \mathrm{e}^{\lambda z}\, h[M_{k-1}(G, y', z; s)]\, D\Delta_{\mathbf{y}} p(y', \mathbf{y}, t-z) + \lambda \mathrm{e}^{\lambda t}\, h[M_{k-1}(G, \mathbf{y}, t; s)]\right]\right]. \end{aligned}$$

Taking the operator $\Delta_{\mathbf{y}}$ out of the integration signs, we find

$$\frac{\partial}{\partial t} M_k(G, \mathbf{y}, t; s) = D\Delta_{\mathbf{y}} M_k - \lambda M_k + \lambda\, h[M_{k-1}].$$

It is left to establish the initial conditions. Since we start the model with a particle of generation $k = 0$ and the distribution of splitting is continuous, at $t = 0$ there are no other particles with probability 1. Hence, $M_k(G, \mathbf{y}, 0; s) = 1$ for all $k \geq 1$. For generation $k = 0$, we can only have 1 or no particles at time $t > 0$. The probability to have 1 particle is given by the product of probabilities that there was no split up to time t and that the particle happens to be within region G at time t: $P = \mathrm{e}^{-\lambda t} \int_G p(\mathbf{x}, 0, t) d\mathbf{x}$. The probability to have no particles is then $(1 - P)$. This implies (10.7).

Appendix 10.B

MOMENTS IN ONE-POINT SYSTEM

For any natural number j, consider the jth moment $\overline{A}_k^{(j)}(G, \mathbf{y}, t)$ of the number of generation-k particles at instant t within the region G, produced by a single immigrant injected at point $\mathbf{y}$ at time $t = 0$. It is given by the following partial derivative (e.g., Athreya and Ney, 2004, Chapter 1):

$$\overline{A}_k^{(j)}(G, \mathbf{y}, t) := \left. \frac{\partial^j M_k(G, \mathbf{y}, t; s)}{\partial s^j} \right|_{s=0}. \tag{10.B1}$$

Corollary 10.B.1 *The moments $\overline{A}_k^{(j)}(G, \mathbf{y}, t)$ solve the following recursive system of partial differential equations:*

$$\frac{\partial}{\partial t} \overline{A}_k^{(j)}(G, \mathbf{y}, t) = D\Delta_{\mathbf{y}} \overline{A}_k^{(j)} - \lambda \overline{A}_k^{(j)} + \lambda \left[\sum \frac{j!}{m_1! m_2! \ldots m_j!} h^{(m)}(1) \prod_{i=1}^{j} \left(\frac{\overline{A}_{k-1}^{(i)}}{i!} \right)^{m_i} \right], \tag{10.B2}$$

where $m = m_1 + \cdots + m_j$ and the sum is over all partitions of j, i.e., values of $m_1, \ldots, m_j$ such that $m_1 + 2m_2 + \cdots + jm_j = j$, with the initial conditions

$$\overline{A}_k^{(j)}(G, \mathbf{y}, 0) \equiv 0, k \geq 1, \tag{10.B3}$$

$$\overline{A}_0^{(j)}(G, \mathbf{y}, 0) = \int_G \delta(\mathbf{y} - \mathbf{x}) d\mathbf{x}, \tag{10.B4}$$

$$\overline{A}_0^{(j)}(G, \mathbf{y}, t) = \mathrm{e}^{-\lambda t} \int_G p(\mathbf{x}, \mathbf{y}, t) d\mathbf{x}, \; t > 0, \tag{10.B5}$$

and

$$h^{(i)}(1) := \left. \frac{d^i}{ds^i} h(s) \right|_{s=1} = \sum_{n=i}^{\infty} \frac{n!}{(n-i)!} p_n.$$

Proof The validity of (10.B2) follows from Proposition 10.5.1. Namely, applying the operator $\partial^j / \partial s^j (\cdot)|_{s=0}$ to both sides of (10.6), changing the order of differentiation, and using Faà di Bruno's formula (Faá di Bruno, 1855) for the jth derivative of a composition function, one finds, for each $k \geq 1$,

$$\left. \frac{\partial^j}{\partial s^j} \left[\frac{\partial M_k(G, \mathbf{y}, t; s)}{\partial t} \right] \right|_{s=0} = \left. \frac{\partial^j}{\partial s^j} \left[D\Delta_{\mathbf{y}} M_k - \lambda M_k + \lambda\, h(M_{k-1}) \right] \right|_{s=0},$$

$$\frac{\partial}{\partial t} \left[\left. \frac{\partial^j M_k(G, \mathbf{y}, t; s)}{\partial s^j} \right|_{s=0} \right] = \left. \left[D\Delta_{\mathbf{y}} \frac{\partial^j M_k}{\partial s^j} - \lambda \frac{\partial^j M_k}{\partial s^j} + \lambda \frac{\partial^j}{\partial s^j} h(M_{k-1}) \right] \right|_{s=0},$$

$$\frac{\partial}{\partial t} \overline{A}_k^{(j)}(G, \mathbf{y}, t) = \left[D\Delta_{\mathbf{y}} \frac{\partial^j M_k}{\partial s^j} - \lambda \frac{\partial^j M_k}{\partial s^j} + \lambda \left(\sum \frac{j!}{m_1! m_2! \cdots m_j!} h^{(m)}(M_{k-1}) \prod_{i=1}^{j} \left(\frac{M_{k-1}^{(i)}}{i!} \right)^{m_i} \right) \right] \Bigg|_{s=0}$$

$$= D\Delta_{\mathbf{y}} \overline{A}_k^{(j)} - \lambda \overline{A}_k^{(j)} + \lambda \left[\sum \frac{j!}{m_1! m_2! \cdots m_j!} h^{(m)}(1) \prod_{i=1}^{j} \left(\frac{\overline{A}_{k-1}^{(i)}}{i!} \right)^{m_i} \right],$$

where $m = m_1 + \cdots + m_j$ and the sum is over all partitions of j, i.e., values of $m_1, \ldots, m_j$ such that $m_1 + 2m_2 + \cdots + jm_j = j$. The initial conditions are established by applying the operator $\partial^j / \partial s^j (\cdot)|_{s=0}$ to both sides of (10.7) and using the definition of $\overline{A}_k^{(j)}(G, \mathbf{y}, t)$ in (10.8).

Appendix 10.C

PROOF OF COROLLARY 10.5.2

For $j = 1$, the equation in Corollary 10.B.1 simplifies to

$$\frac{\partial}{\partial t}\overline{A}_k(G,\mathbf{y},t) = D\Delta_{\mathbf{y}}\overline{A}_k - \lambda\overline{A}_k + \lambda B\overline{A}_{k-1}.$$

Using the definition of A_k (**x**, **y**, t) given in (10.9), one obtains for each $k \geq 1$,

$$\frac{\partial}{\partial t}A_k(\mathbf{x},\mathbf{y},t) = D\Delta_{\mathbf{y}}A_k - \lambda A_k + \lambda BA_{k-1}.$$

It is left to use the translation property A_k (**x**, **y**, t) = A_k (**x** − **y**, 0, t) to change $\Delta_{\mathbf{y}}$ to $\Delta_{\mathbf{x}}$.

The validity of general solution (10.12) is proven by induction using the fact that

$$\left[\frac{\partial}{\partial t} - D\Delta_{\mathbf{x}} + \lambda\right]A_0 = 0.$$

The last equality in (10.12) follows from (10.B5) and (10.4).

Appendix 10.D

PROOF OF PROPOSITION 10.5.3

The proof of Proposition 10.5.3 follows the line of the proof of Proposition 10.5.1. There are two possible scenarios for the model development up to time t. In the first one, the initial immigrant will not split; the probability for this is $P = e^{-\lambda t}$. In the second one, the initial immigrant will split at instant $0 \leq u \leq t$; the probability of the first split within the time interval $[u, u + du]$ is $\lambda e^{-\lambda t}\, du + o(du)$ as $du \to 0$. The spatial position of the split is given by the diffusion density p(**x**, **y**, u). If the immigrant splits, the composition property of generating functions gives $M_{k_1,k_2} = h[M_{k_1-1,k_2-1}]$. Integrating over all possible split instants and locations, we obtain

$$\begin{aligned}M_{k_1,k_2}&(G_1,G_2,\mathbf{y},t;s_1,s_2)\\&= e^{-\lambda t} + \int_{\mathbb{R}^n} d\mathbf{y}'\int_0^t du\, \lambda e^{-\lambda u}p(\mathbf{y}',\mathbf{y},u)\\&\quad h[M_{k_1-1,k_2-1}(G_1,G_2,\mathbf{y}',t-u;s_1,s_2)].\end{aligned}$$

Here the first and second terms correspond to the first and second scenarios, respectively. Using the new integration variable $z = t - u$, we write

$$\begin{aligned}M_{k_1,k_2}&(G_1,G_2,\mathbf{y},t;s_1,s_2)\\&= e^{-\lambda t} + e^{-\lambda t}\int_{\mathbb{R}^n} d\mathbf{y}'\int_0^t du\, \lambda e^{\lambda(t-u)}p(\mathbf{y}',\mathbf{y},u)\\&\quad h[M_{k_1-1,k_2-1}(G_1,G_2,\mathbf{y}',t-u;s_1,s_2)]\\&= e^{-\lambda t}\left(1 + \int_{\mathbb{R}^n} d\mathbf{y}'\int_0^t dz\, \lambda e^{\lambda z}p(\mathbf{y}',\mathbf{y},t-z)\right.\\&\quad \left. h[M_{k_1-1,k_2-1}(G_1,G_2,\mathbf{y}',z;s_1,s_2)]\right).\end{aligned}$$

Now we take the derivative with respect to t of both sides and apply Lemma 10.A.1 using the fact that $p(\mathbf{y}',\mathbf{y},0) = \delta(\mathbf{y}'-\mathbf{y})$ and $(\partial/\partial t - D\Delta_{\mathbf{y}})p = 0$:

$$\begin{aligned}\frac{\partial}{\partial t}&M_{k_1,k_2}(G_1,G_2,\mathbf{y},t;s_1,s_2)\\&= -\lambda M_{k_1,k_2}(G_1,G_2,\mathbf{y},t;s_1,s_2)\\&\quad + e^{-\lambda t}\left[\int_{\mathbb{R}^n} d\mathbf{y}'\int_0^t dz\,\lambda e^{\lambda z}\right.\\&\quad h[M_{k_1-1,k_2-1}(G_1,G_2,\mathbf{y}',z;s_1,s_2)]D\Delta_{\mathbf{y}}p(\mathbf{y}',\mathbf{y},t-z)\\&\quad \left. + \lambda e^{\lambda t}\, h[M_{k_1-1,k_2-1}(G_1,G_2,\mathbf{y},t;s_1,s_2)]\right].\end{aligned}$$

Taking the operator $\Delta_{\mathbf{y}}$ out of the integration signs, we find

$$\begin{aligned}\frac{\partial}{\partial t}M_{k_1,k_2}(G_1,G_2,\mathbf{y},t;s_1,s_2) &= D\Delta_{\mathbf{y}}M_{k_1,k_2} - \lambda M_{k_1,k_2}\\&\quad + \lambda\, h[M_{k_1-1,k_2-1}].\end{aligned}$$

It is left to establish the initial conditions. Since we start the model with a particle of generation $k = 0$ and the distribution of splitting is continuous, at $t = 0$ there are no other particles with probability 1. Hence, $M_{k_1,k_2}(G_1,G_2,\mathbf{y},0;s_1,s_2) = 1$ for all k_1, $k_2 \geq 1$. For generation $k_1 = k_2 = 0$, we have three possibilities: the initial immigrant has not split and is in G_1 ($i = 1, j = 0$), the initial immigrant has not split and is in G_2 ($i = 0$, $j = 1$), and neither ($i = 0$, $j = 0$), with corresponding probabilities of P_1, P_2, and $1 - P_1 - P_2$, respectively. This implies (10.17).

For generation $k_1 = 0$ and $k_2 = k \geq 1$, we again have three possibilities: the initial immigrant has not split and is in G_1 ($i = 1, j = 0$), the initial immigrant has not split and is not in G_1 ($i = 0$, $j = 0$), and the initial immigrant has split ($i = 0, j \geq 0$), with corresponding probabilities of P_1, $e^{-\lambda t} - P_1$, and $1 - e^{-\lambda t}$, respectively. In the last case, the number of the zeroth generation particles in G_1 is 0 with probability 1 while the information on the kth generation particles in G_2 is given by

$$\int_{R^n} d\mathbf{y}'\int_0^t du\, \lambda e^{-\lambda u}p(\mathbf{y}',\mathbf{y},u)\, h[M_{k-1}(G_2,\mathbf{y}',t-u;s_2)].$$

From (10.A1), we see that the above expression equals $M_k(G_2,\ \mathbf{y},\ t;\ s_2) - e^{-\lambda t}$. This implies (10.18). We notice that setting $s_2 = 0$ in (10.17) and (10.18) each yields $(1 - P_1) + P_1 e^{s_1}$ as it should (cf. (10.7)).

Appendix 10.E

PROOF OF COROLLARY 10.5.4

The validity of (10.13) follows from Proposition 10.5.3 and the definition of $\overline{A}_{k_1,k_2}(G_1, G_2, \mathbf{y}, t)$, $A_{k_1,k_2}(\mathbf{x}_1, \mathbf{x}_2, \mathbf{y}, t)$. Formally, applying the operator $\partial^2/\partial s_1 \partial s_2(\cdot)|_{s_1=s_2=0}$ to both sides of (10.15) and changing the order of differentiation, one finds, for each k_1, $k_2 \geq 1$,

$$\frac{\partial^2}{\partial s_1 \partial s_2}\left[\frac{\partial M_{k_1,k_2}}{\partial t}\right]\Bigg|_{s_1=s_2=0}$$

$$= \frac{\partial^2}{\partial s_1 \partial s_2}[D\Delta_{\mathbf{y}} M_{k_1,k_2} - \lambda M_{k_1,k_2} + \lambda h(M_{k_1-1,k_2-1})]|_{s_1=s_2=0},$$

$$\frac{\partial}{\partial t}\left[\frac{\partial^2 M_{k_1,k_2}}{\partial s_1 \partial s_2}|_{s_1=s_2=0}\right]$$

$$= \left[D\Delta_{\mathbf{y}} \frac{\partial^2 M_{k_1,k_2}}{\partial s_1 \partial s_2} - \lambda \frac{\partial^2 M_{k_1,k_2}}{\partial s_1 \partial s_2} + \lambda h'(M_{k_1-1,k_2-1}) \frac{\partial^2 M_{k_1-1,k_2-1}}{\partial s_1 \partial s_2} + \lambda h''(M_{k_1-1,k_2-1}) \frac{\partial M_{k_1-1,k_2-1}}{\partial s_1} \frac{\partial M_{k_1-1,k_2-1}}{\partial s_2}\right]\Bigg|_{s_1=s_2=0},$$

$$\frac{\partial}{\partial t} = D\Delta_{\mathbf{y}} \overline{A}_{k_1,k_2} - \lambda \overline{A}_{k_1,k_2} + \lambda B \overline{A}_{k_1-1,k_2-1} + \lambda h''(1) \overline{A}_{k_1-1}(G_1) \overline{A}_{k_2-1}(G_2).$$

The system (10.23) readily follows now from the definition of $A_{k_1,k_2}(\mathbf{x}_1, \mathbf{x}_2, \mathbf{y}, t)$. The initial conditions (10.24)–(10.25) are established by applying the operator $\partial^2/\partial s_1 \partial s_2(\cdot)|_{s_1=s_2=0}$ to both sides of (10.16)–(10.18) and using again the definition of $A_{k_1,k_2}(\mathbf{x}_1, \mathbf{x}_2, \mathbf{y}, t)$.

Appendix 10.F

PROOF OF PROPOSITION 10.6.2

The asymptotic (10.37) readily follows from (10.G4). To prove (10.38), let $r_\nu(z) := K_\nu(z)/K_{\nu+1}(z)$. From (10.G1) one finds that

$$\frac{K_{\nu+1}(z)}{K_\nu(z)} = \frac{K_{\nu-1}(z)}{K_\nu(z)} + \frac{2\nu}{z} \tag{10.F1}$$

and furthermore

$$\frac{z}{2\nu} \frac{1}{r_\nu(z)} = \frac{z}{2\nu} r_{\nu-1}(z) + 1. \tag{10.F2}$$

From the monotonicity of $K_\nu(z)$ with respect to the index $\nu > 0$ it follows that $r_\nu(z) < 1$ for $\nu > 0$. Accordingly, the first term on the right hand side of (10.F2) goes to zero as $k \to \infty$. Hence,

$$\lim_{k\to\infty} \frac{z}{2\nu} \frac{1}{r_\nu(z)} = 1, \text{ or } r_\nu(z) \sim \frac{z}{2\nu}, \quad k \to \infty. \tag{10.F3}$$

To complete the proof of (10.38), we use this asymptotic in (10.35). Finally, we prove (10.36). In fact, we will derive a stronger result showing the asymptotics of $r_\nu(z)$ and $\gamma_\nu(z)$ as $z \to 0$. To find the asymptotics for $r_\nu(z)$, we use (10.G5) for all possible combinations of signs for ν and $\nu + 1$. We take into account that by definition ν can only take values $(i, i + 1/2\}_{i\in\mathbf{Z}}$.

$$r_\nu(z) = \begin{cases} \dfrac{K_\nu(z)}{K_{\nu+1}(z)} \sim \dfrac{\Gamma(-\nu)}{\Gamma(-\nu-1)}\left(\dfrac{2}{z}\right)^{-\nu-(-\nu-1)} \sim 2(-\nu-1)/z, & \nu \leq -3/2, \\ \dfrac{K_{-1}(z)}{K_0(z)} \sim [z(\ln(2/z)-\gamma)]^{-1} \sim -(z\ln z)^{-1}, & \nu = -1, \\ \dfrac{K_{-1/2}(z)}{K_{1/2}(z)} = 1, & \nu = -1/2, \\ \dfrac{K_0(z)}{K_1(z)} \sim z(\ln(2/z)-\gamma) \sim -z\ln z, & \nu = 0, \\ \dfrac{K_\nu(z)}{K_{\nu+1}(z)} \sim \dfrac{\Gamma(\nu)}{\Gamma(\nu+1)}\left(\dfrac{2}{z}\right)^{\nu-(\nu+1)} = z/(2\nu), & \nu > 0. \end{cases} \tag{10.F4}$$

Combining this with (10.35) we find

$$\gamma_\nu(z) = \frac{2(k+1)}{Bz} r_\nu(z) \sim \begin{cases} \dfrac{4}{Bz^2}(\nu + n/2)(-\nu - 1), & \nu \leq -3/2, \\ -\dfrac{(n-2)}{Bz^2 \ln z}, & \nu = -1, \\ \dfrac{n-1}{Bz}, & \nu = -1/2, \\ -\dfrac{n \ln z}{B}, & \nu = 0, \\ \dfrac{1}{B}\left(1 + \dfrac{n}{2\nu}\right), & \nu > 0. \end{cases} \tag{10.F5}$$

One can see that for $\nu \leq 0$ the ratio $\gamma_\nu(z)$ diverges at the origin. The rate of divergence increases monotonously from $\ln z$ to z^{-2} with the absolute value of ν.

Appendix 10.G

PROPERTIES OF K_N

Here we summarise some essential facts about the modified Bessel function of the second kind $K_\nu(z)$. The sources of this as well as further information about $K_\nu(z)$ are handbooks (Abramowitz and Stegun, 1965, Chapters 9, 10; Gradshtein and Ryzhik, 2007, Section 8.4). The function K_ν can be defined as a decreasing solution of the modified Bessel differential equation

$$x^2\, y'' + x\, y' - \left(x^2 + \nu^2\right) y = 0.$$

The function $K_\nu(z)$ exponentially decreases as $z \to \infty$ and diverges at $z = 0$. In addition, $K_{-\nu}(z) = K_\nu(z)$ and

$$K_{\nu+1}(z) = K_{\nu-1}(z) + \frac{2\,\nu}{z}\, K_\nu(z). \tag{10.G1}$$

For integer $k \geq 0$ we have

$$K_{k+1/2}(z) = \sqrt{\frac{\pi}{2z}}\mathrm{e}^{-z}\sum_{m=0}^{k} \frac{(k+m)!}{m!(k-m)!(2z)^m}, \tag{10.G2}$$

and in particular

$$K_{\pm 1/2}(z) = \sqrt{\frac{\pi}{2\,z}}\,\mathrm{e}^{-z};\; K_{3/2}(z) = \sqrt{\frac{\pi}{2\,z^3}}\,\mathrm{e}^{-z}. \tag{10.G3}$$

For arbitrary fixed ν and $z \gg \nu$

$$K_\nu(z) \sim \sqrt{\frac{\pi}{2\,z}}\,\mathrm{e}^{-z},\; z \to \infty. \tag{10.G4}$$

The asymptotic behaviour at $z = 0$ is given by

$$K_\nu(z) \sim \begin{cases} \dfrac{\Gamma(|\nu|)}{2}\left(\dfrac{2}{z}\right)^{|\nu|}, & |\nu| \neq 0, \\ \log\left(\dfrac{2}{z}\right) - \gamma, & \nu = 0, \end{cases} \tag{10.G5}$$

where $\gamma \approx 0.577$ is the Euler–Mascheroni constant (Euler, 1734).

References

Abramowitz, M. and Stegun, I. A. eds. (1965). *Handbook of Mathematical Functions with Formulas, Graphs, and Mathematical Tables*. New York, NY: Dover.

Aki, A. (1985). Theory of earthquake prediction with special reference to monitoring of the quality factor of lithosphere by the coda method. *Earthquake Prediction Research*, **3**, 219–230.

Albert, R. and Barabasi, A. L. (2002). Statistical mechanics of complex networks. *Reviews of Modern Physics*, **74** (1), 47–97.

Albeverio, S., Jentsch, V. and Kantz, H., eds, (2005). *Extreme Events in Nature and Society*. Heidelberg: Springer.

Allegre, C. J., Lemouel, J. L. and Provost, A. (1982). Scaling rules in rock fracture and possible implications for earthquake prediction. *Nature*, **297**, 5861, 47–49.

Athreya, K. B. and Ney, P. E. (2004). *Branching Processes*. New York: Dover Publications.

Bak, P., Tang, C. and Wiesenfeld, K. (1988). Self-organized criticality. *Physical Review A*, **38**, 1, 364–374.

Ben-Zion, Y. (2003). Appendix 2: Key formulas in earthquake seismology, In *International Handbook of Earthquake and Engineering Seismology*, Part B, 1857–1875. Academic Press.

Blanter, E. M., Shnirman, M. G. and LeMouel, J. L. (1997). Scaling laws in blocks dynamics and dynamic self-organized criticality. *Physics of the Earth and Planetary Interiors*, **103** (1–2), 135–150.

Bollobás, B. (2001). *Random Graphs*. Cambridge: Cambridge University Press, second edn.

Brunetti, M. T., Guzzetti, F. and Rossi, M. (2009). Probability distributions of landslide volumes. *Nonlinear Processes in Geophysics*, **16**, 2, 179–188.

Burlando, B. (1990). The fractal dimension of taxonomic systems. *Journal of Theoretical Biology*, **146**, 99–114.

Burridge, R. and Knopoff, L. (1967). Model and theoretical seismicity. *Bulletin of the Seismological Society of America*, **57**, 3, 341–371.

Chow, P. L. (2007). *Stochastic Partial Differential Equations* Boca Raton, FL: Chapman Hall/CRC Press.

Dhar, D. (1990). Self-organized critical state of sandpile automaton models. *Physical Review Letters*, **64**, 14, 1613–1616.

Doucet, A., de Freitas, N. and Gordon, N., eds. (2001). *Sequential Monte Carlo Methods in Practice*. Springer.

Drossel, B. and Schwabl, F. (1992). Self-organized critical forest-fire model. *Physical Review Letters*, **69**, 11, 1629–1632.

Durrett, R. (2006). *Random Graph Dynamics*. Cambridge: Cambridge University Press.

Embrechts, P., Klüppelberg, C. and Mikosch, T. (2008). *Modelling Extremal Events for Insurance and Finance (Stochastic Modelling and Applied Probability)*. Springer.

Euler, L. (1734). De Progressionibus harmonicis observationes, *Commentorii: Academiae Scientarium Imperialis Petroplitanae*, **7** (1734/35), 150–161.

Evans, L. C. (1998). *Partial Differential Equations*. Providence, RI: American Mathematical Society.

Faá di Bruno, F. (1855). Sullo sviluppo delle Funzioni, *Annali di Scienze Matematiche e Fisiche* (in Italian), **6**: 479–480.

Frisch, U. (1996). *Turbulence: The Legacy of A. M. Kolmogorov*, Cambridge: Cambridge University Press.

Gabaix, X., Gopikrishnan, P., Plerou, V. and Stanley, H. E. (2003). A theory of power-law distributions in financial market fluctuations. *Nature*, **423**, (6937), 267–270.

Gabrielov, A., Keilis-Borok, V., Sinai, Y. and Zaliapin, I. (2008). Statistical properties of the cluster dynamics of the systems of statistical mechanics. In *Boltzmann's Legacy*, ESI Lecture Notes in Mathematics and Physics: G. Gallavotti, W. Reiter and J. Yngvason. Zurich: European Mathematical Society, pp. 203–216.

Gabrielov, A. M., Zaliapin, I. V., Keilis-Borok, V. L. and Newman W. I. (2000). Colliding cascades model for earthquake prediction. *Geophysical Journal International*, **143**, 427–437.

Gradshtein, I. S. and Ryzhik, I. M., (2007). *Tables of Integrals, Series and Products*, eds., Jeffrey, A. and Zwillinger, D. Academic Press, seventh edn.

Grimmett, G. R. (1999). *Percolation*. Springer, second edn.

Gutenberg, B. and Richter, C. F. (1944). Frequency of earthquakes in California. *Bulletin of the Seismological Society of America*, **34**, 185–188.

Haberman, R. E. (1981). Precursory seismicity pattern: Stalking the mature seismic gap. In *Earthquake Prediction: An International Review*, Maurice Ewing Series, 4. Washington, DC: American Geophysical Union, pp. 29–42.

Kadanoff L. P. (2000). *Statistical Physics: Statics, Dynamics and Renormalization*. World Scientific Publishing Company.

Kalman, R. E. and Bucy, R. S. (1961). New results in linear filtering and prediction theory. *Transactions of the ASME, Series D Journal of Basic Engineering*, **83**, 95–108.

Keilis-Borok, V. I. and Shebalin, P. N., eds. (1999). Dynamics of lithosphere and earthquake prediction (special issue). *Physics of the Earth and Planetary Interiors*, **111**.

Keilis-Borok, V. I. and Soloviev, A. A., eds (2003). *Nonlinear Dynamics of the Lithosphere and Earthquake Prediction*. Heidelberg: Springer.

Keilis-Borok, V. I., Soloviev, A. A., Allegre, C. B. *et al.* (2005). Patterns of macroeconomic indicators preceding the unemployment rise in Western Europe and the USA. *Pattern Recognition*, **38**, 3, 423–435.

Keilis-Borok, V. I., Stock, J. H., Soloviev, A. and Mikhalev, P. (2000). Pre-recession pattern of six economic indicators in the USA. *Journal of Forecasting*, **19**, 65–80.

Klass, O. S., Biham, O., Levy, M., Malcai, O. and Soloman, S. (2006). The Forbes 400 and the Pareto wealth distribution. *Economics Letters*, **90**, 2, 290–295.

Kolmogorov, A. N. (1941a). The local structure of turbulence in incompressible viscous fluid for very large Reynolds numbers. *Doklady Akademii Nauk. USSR*, **30**, 299–303 (in Russian). English translation: *Proceedings of the Royal Society of London. Series A*, **434**, 9–13 (1991).

Kolmogorov, A. N. (1941b). Interpolation, extrapolation of stationary random sequences. *Izv. Akad. Nauk. SSSR, Ser. Mat.*, **5**, 3–14. (English translation: W. Doyle, RAND corporation memorandum RM–3090–PR, April 1962).

Kushner H. J. (1964). On the differential equations satisfied by conditional probability densities of Markov processes, with applications. *SIAM Journal on Control, Ser. A*, **2**(1), 106–119.

Lotka, A. J. (1926). The frequency distribution of scientific productivity, *Journal of the Washington Academy of Sciences*, **16**, 12, 317–324.

Ma, S.-K. (2000). *Modern Theory of Critical Phenomena*. Boulder, CO: Westview Press.

Malamud, B. D., Turcotte, D. L., Guzzetti, F. and Reichenbach, P. (2004). Landslide inventories and their statistical properties. *Earth Surface Processes and Landforms*, **29**, 6, 687–711.

Mandelbrot, B. (1983). *The Fractal Geomery of Nature*, W. H. Freeman.

Mandelbrot, B. and Taylor, H. M. (1967). On the distribution of stock price differences. *Operations Research*, **15**, 6, 1057–1062.

McWilliams, J. C. (1990). The vortices of two-dimensional turbulence. *Journal of Fluid Mechanics*, **219**, 361–385.

Mogi, K. (1981) Seismicity in Western Japan and long-term forecasting. In *Earthquake Prediction: An International Review*, Maurice Ewing Series, 4. Washington, DC: American Geophysical Union, pp. 43–51.

Narkunskaya, G. S. and Shnirman, M. G. (1990). Hierarchical model of defect development and seismicity. *Physics of the Earth and Planetary Interiors*, **61**, 29–35.

Newman, M. E. J. (1997). Evidence for self-organized criticality in evolution. *Physica D*, **107**, 293–296.

Newman, M. E. J. (2005). Power laws, Pareto distributions and Zipf's law. *Contemporary Physics*, **46** (5), 323–351.

Newman, M. E. J., Barabasi, A.-L. and Watts, D. J. (2006). *The Structure and Dynamics of Networks*. Princeton, NJ: Princeton University Press.

Newman, W. I., Turcotte, D. L. and Gabrielov, A. M. (1995). Log-periodic behavior of a hierarchical failure model with applications to precursory seismic activation. *Physical Review E*, **52**, 4827–4835.

Obukhov, A. M. (1941). Spectral energy distribution in a turbulent flow. *Doklady Akademii Nauk. USSR*, **1**, 22–24.

Olami, Z., Feder, H. J. S. and Christensen, K. (1992) Self-organized criticality in a continuous, nonconservative cellular automaton modeling earthquakes. *Physical Review Letters*, **68**, 8, 1244–1247.

Pareto, V. (1897). *Cours d'Economie Politique*. Lausanne: F. Rouge.

Plerou, V. and Stanley, H. E. (2008) Stock return distributions: Tests of scaling and universality from three distinct stock markets. *Physical Review E*, **77**, 3, Art. No. 037101.

Richardson, L. F. (1960). *Statistics of Deadly Quarrels*. Pacific Grove, CA: Boxwood Press.

Rundle, J. B. and Klein, W. (1993). Scaling and critical phenomena in a cellular automaton slider-block model for earthquakes. *Journal of Statistical Physics*, **72**, (1–2), 405–412.

Rundle, J., Turcotte, D. and Klein, W., eds. (2000). *Geocomplexity and the Physics of Earthquakes*. Washington, DC: American Geophysical Union.

Sornette, D. (2004). *Critical Phenomena in Natural Sciences*. Heidelberg: Springer-Verlag, second edn.

Stanley, H. E. (1971). *Introduction to Phase Transitions and Critical Phenomena*. Oxford: Oxford University Press.

Stauffer, D. and A. Aharony (1994). *Introduction to Percolation Theory*. Boca Raton, FL: CRC.

Sykes, L. R., Shaw, B. E. and Scholz, C. H. (1999). Rethinking earthquake prediction. *Pure and Applied Geophysics*, **155**, 2–4, 207–232.

Turcotte, D. L. (1997). *Fractals and Chaos in Geology and Geophysics*. Cambridge: Cambridge University Press, second edn.

Turcotte, D. L. (1999). Self-organized criticality. *Reports on Progress in Physics*, **62**, 10, 1377–1429.

Wiener, N. (1949). *Extrapolation, Interpolation and Smoothing of Stationary Time Series with Engineering Applications*. Wiley.

Zakai, M. (1969). On the optimal filtering of diffusion processes. *Zeitschrift für Wahrscheinlichkeitstheorie und Verwandte Gebiete* **11**, 3, 230–243.

Zaliapin, I., Keilis-Borok, V. and Ghil, M. (2003a). A Boolean delay equation model of colliding cascades. Part I: Multiple seismic regimes. *Journal of Statistical Physics*, **111**, 3–4, 815–837.

Zaliapin, I., Keilis-Borok, V. and Ghil, M. (2003b). A Boolean delay equation model of colliding cascades. Part II: Prediction of critical transitions. *Journal of Statistical Physics*, **111**, 3–4, 839–861.

Zipf, G. K. (1965). *Psycho-Biology of Languages*, Houghton-Mifflin, MIT Press.

PART III

Case studies: Latin America and the Caribbean region

11

Earthquakes, tsunamis, and the related vulnerability in South America and the Caribbean – an overview

Omar J. Pérez, Carlos Rodríguez, and José L. Alonso

11.1 Introduction

The main purposes of this chapter are: to provide a brief review of the earthquake activity during the period of instrumental seismology in South America and the Caribbean; to highlight the socio-economic impact of severe earthquakes in the region and the vulnerability of the countries located along the coast of western South America (WSA) and those surrounding the Caribbean Sea; and to review the historical occurrences of tsunamis along these coastal areas.

We show here that much of the population of WSA, from Patagonia to Colombia, is located along one of the most active seismic belts in terms of seismic moment release. The eastward subduction of the Nazca plate with the rate of about 7–8 cm y^{-1} (e.g., DeMets *et al.*, 1990) results in frequent large earthquakes. These have included several of the largest and giant (of moment magnitude, $M_w \geq 8.6$) seismic events recorded, among them the Chilean earthquakes of 1960 (M_w9.5) and 2010 (M_w8.8). For the Caribbean, most of the surrounding countries are embedded within a seismicity belt that defines the boundaries of the Caribbean plate, which slips easterly at about 2 cm y^{-1} relative to the North American and South American plates (e.g., Dixon *et al.*, 1998; Pérez *et al.*, 2001). Although the seismic activity of this belt is smaller than that for other earthquake-prone regions of the world, strong earthquakes ($M_w \leq 7.8$) have resulted in several catastrophes at national levels, e.g. the 2010 Haiti M_w7 earthquake, which occurred in the poorest country of the Western Hemisphere, in a region of a very-low-quality building construction.

In addition to the high intensities, ground accelerations, and shaking induced by major seismic events, particularly within and near rupture zones, tsunamis are another dangerous earthquake-induced hazard in WSA, that have produced catastrophic events (e.g., the tsunamis generated by the 1960 and 2010 giant earthquakes). They occur due to the shallow-dip subduction of the Nazca plate beneath continental South America (e.g., Barazangi and Isacks, 1976), which results in interplate shallow-depth (less than 70 km) thrust events, whose ruptures occasionally reach the marine floor. Tsunamis are less frequent in the Caribbean because of the strike-slip nature of more than 50% of the plate boundary and its low seismic productivity, although the subduction parts of this boundary including the Greater and Lesser Antilles have generated tsunamis in the past, for example, in 1946 when a destructive tsunami seriously affected the Dominican Republic with an estimated death toll of 1600–1800 people.

Statistical analysis of tsunami occurrence in WSA using the database provided by the National Oceanic and Atmospheric Administration (NOAA) shows that every single earthquake rupture with a calculated or estimated magnitude $M_w \geq 8$ known to have occurred along the interplate boundary between the Nazca and the Pacific plates since colonial times (i.e. since about 1560) generated, beyond any reasonable doubt, a tsunami with a total death toll of a few tens of thousands people. Knowledge of this fact by the Chilean navy and its Oceanography Service (SHOA) on 27 February 2010 and an appropriate interaction with the US Pacific Tsunami Warning Center would have led the Chilean President to issue and maintain a tsunami warning for the country, thus a significant number of the human casualties caused by the giant M_w8.8 event on that date, would have been avoided.

First, we look at the last 30 years of earthquake activity in South America and the Caribbean and compare the corresponding seismicity belts with the location of the majority of the population in those regions. Then we

Extreme Natural Hazards, Disaster Risks and Societal Implications, eds. Alik Ismail-Zadeh, Jaime Urrutia Fucugauchi, Andrzej Kijko, Kuniyoshi Takeuchi and Ilya Zaliapin. Published by Cambridge University Press.

analyse the location and extent of great ($M_w \geq 8$) seismic ruptures that have occurred in South America during the period of instrumental seismology (from about 1900 onwards) and point out various aspects of their socio-economic impact. We do a similar analysis for the Caribbean. Then we present our analysis of the tsunamis known to have occurred in these regions in both historical and recent times, followed by a final discussion.

11.2 Earthquakes

Figures 11.1 and 11.2 are seismicity maps showing the epicentral locations (dots) of earthquakes with body-wave magnitude $m_b \geq 5$ that occurred in South America and the Caribbean, respectively, during the period 1980–2010. All locations are taken from the International Seismological Centre (ISC, 2011) world seismicity catalogue. These figures also show the tectonic features of both regions. Figure 11.3 is a NASA light map of these regions illustrating in white the corresponding plate boundary zones, whose actual width is provided by the corresponding belts of shallow-depth seismicity (black dots in Figs. 11.1 and 11.2). It is clear from these three figures that much of the populations of the WSA and Caribbean countries (as shown by lights in Fig. 11.3) is concentrated along seismicity belts whose shallow-depth zones ($h < 70$ km) have generated many destructive earthquakes and tsunamis in the past. Unfortunately, the majority of these countries and their populations are poor, and the civil and governmental building constructions are highly vulnerable with respect to an occurrence of nearby strong and shallow-depth earthquakes.

Figure 11.4 shows the main epicentre locations (stars) and aftershock distribution (dots, $m_b \geq 5$) for all the shallow-depth great shocks of $M_w \geq 8$ that have occurred in South America during the period of instrumental seismology. The main epicentre and aftershock locations of each event define their rupture zones (Fedotov, 1965), which, in general, straddle the coast line of WSA crossing the complete interplate boundary zone (Scholz, 1990), that is, the zone of contact between the subducted Nazca plate and the overriding South American plate (Fig. 11.5(a)). All these great events show thrust motion on shallow dipping faults (Fig. 11.5(b)). The vast majority of these shocks occurred at depths $h < 70$ km. For the earlier events which lacked aftershock data, we estimated their rupture areas (white circles in Fig. 11.4) from their scalar seismic moment (Pacheco and Sykes, 1992) using a circular crack elastic model (Kanamori and Anderson, 1975). The South American events account for about 50% of the total seismic moment released by the Earth between 1900 and 2010, making WSA the most seismically active region in the world with a total death toll estimated at about 75 000 people. In Fig. 11.4, we can see the segments of plate

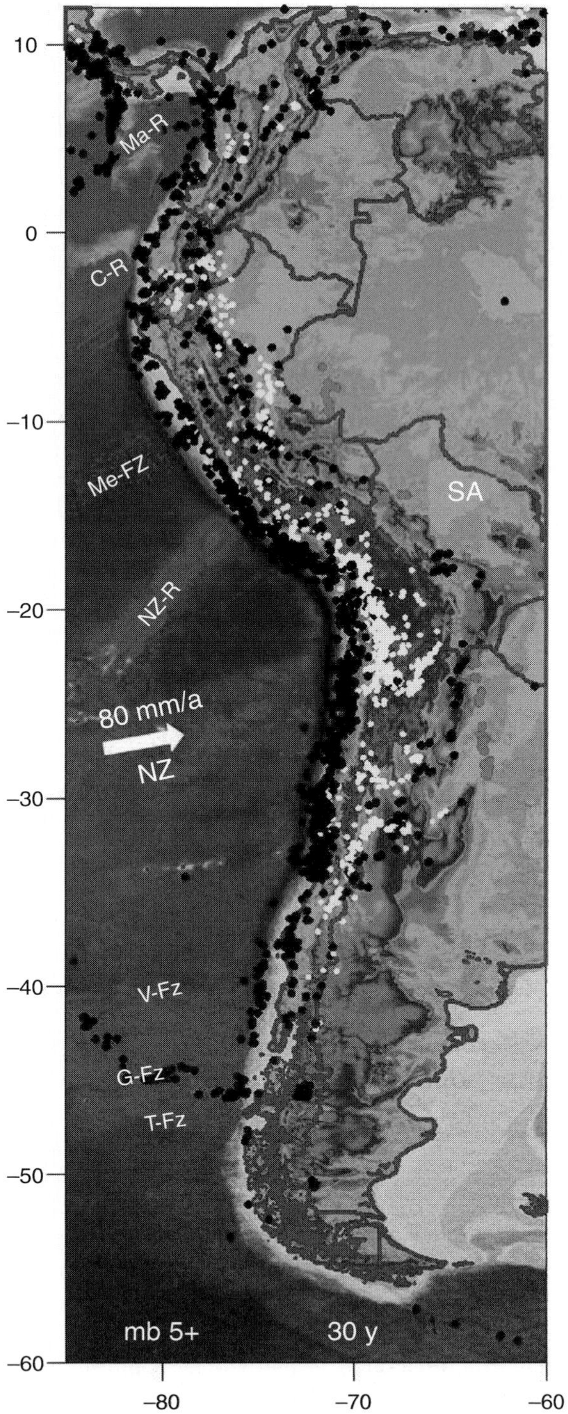

Figure 11.1. Epicentral locations (dots) of earthquakes with $m_b \geq 5$ in WSA during the period 1980–2010. Focal depths are indicated in black for $h < 70$ km, in yellow for 70 km $< h \leq 300$ km, and in red for $h > 300$ km. Tectonic plates: NZ Nazca, SA South American. Blue contours delineate the countries of the region. The depth distribution of these seismic events clearly defines a Benioff zone associated with the east-northeast subduction of the Nazca plate beneath South America that starts along the trench. Ridges (R): Ma, Malpelo; C, Carnegie; Nz, Nazca. Fracture zones (Fz): Me, Mendaña; V, Valdivia; G, Guafo; T, Taitao. For colour version, see Plates section.

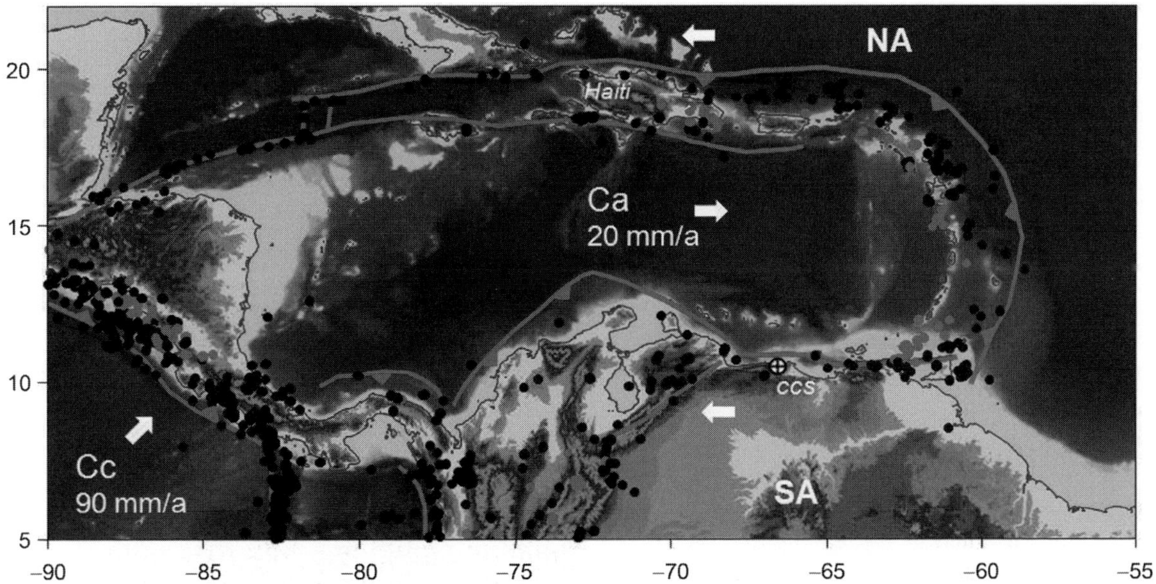

Figure 11.2. Epicentral locations (dots) of earthquakes with $m_b \geq 5$ in the Caribbean during the period 1980–2010. Focal depths are indicated in black for $h < 70$ km; and red for $h \geq 70$ km. Maximum focal depth reported for the region is 233 km. Tectonic plates: Ca, Caribbean; SA, South American; Cc, Cocos; NA, North American. CCS: City of Caracas. Red lines delineate the main plate boundaries of the Caribbean plate; triangles as the line show the areas where subduction takes place; they are strike-slip otherwise, left-lateral in the north and right-lateral in the south. The depth distributions of seismic events along the eastern and western Caribbean plate boundaries clearly define Benioff zones associated with the westerly and easterly subductions of the North Atlantic and Cocos plates, respectively, beneath the Caribbean plate. For colour version, see Plates section.

boundary that have not broken in great earthquakes for over a century, and thus are likely candidates to break in future.

Figure 11.6 shows the shallow-depth earthquake activity ($6 \leq M_w \leq 7.8$) that has occurred in the Caribbean surrounding areas since 1900. Although the regional seismic and tsunami activity is small compared to other regions of the world, several large earthquakes have occurred in the region in historic and recent times including the 1976 Guatemalan M_w7.6 and the 2010 Haiti M_w7.0 shocks. The Guatemalan earthquake resulted in the loss of 23 000 lives and caused 74 000 reported injuries (Plafker, 1976). The Haiti earthquake resulted in a great number of human casualties (the loss of 200 000–250 000 lives and many thousands of injuries), and over US$14 billion in economic losses including the collapse of or severe damage to over a quarter of a million residences and commercial buildings (e.g., Cavallo *et al.*, 2010). The conjunction of poverty and the low quality of building construction in many Caribbean countries significantly increases the social vulnerability to seismic events of lower magnitudes (compared to WSA).

11.3 Tsunamis

Figure 11.7 shows the tsunamis (black dots) known to have occurred in South America since 1560 (Soloviev's scale; Iida, 1963); tsunami data are compiled by the NOAA National Geophysical Data Center (http://www.ngdc.noaa.gov/hazard/tsu.shtml). The figure also presents the epicentres (white stars) of the earthquakes of magnitude $M_w \geq 8$ for the same time span (e.g., Pacheco and Sykes, 1992). Although this integrated catalogue is clearly incomplete for the years prior to 1900 (i.e., not all events of $M_w \geq 8$ or tsunamis that occurred prior to 1900 are listed), we can see that the vast majority of these tsunamis were generated by events of $M_w \geq 7.5$, but only about 50% of the events in the range $7.5 \leq M_w \leq 7.9$ generated a tsunami. On the other hand, every single listed event of $M_w \geq 8$ (white stars) did generate a tsunami (black dots) and their corresponding ruptures occurred along the shallow-depth earthquake belt shown in Fig. 11.1. This provides useful information to scientists and authorities of the WSA countries and to the Pacific Tsunami Warning Center when deciding whether to issue a tsunami alarm. We point this out to avoid cases like that in 2010, when the Chilean navy's SHOA failed to recommend to the authorities that a tsunami alarm should be issued after the giant (M_w8.8) earthquake.

Tsunamis are less frequent in the Caribbean (those since 1900 are shown as squares in Fig. 11.6) than in WSA and relatively smaller (tsunami magnitudes < 3), and many of them have a volcanic origin along the Lesser Antilles. Yet the subduction tectonic regimes along the western, southwestern and eastern Caribbean plate boundary regions, including the Greater Antilles east of Haiti

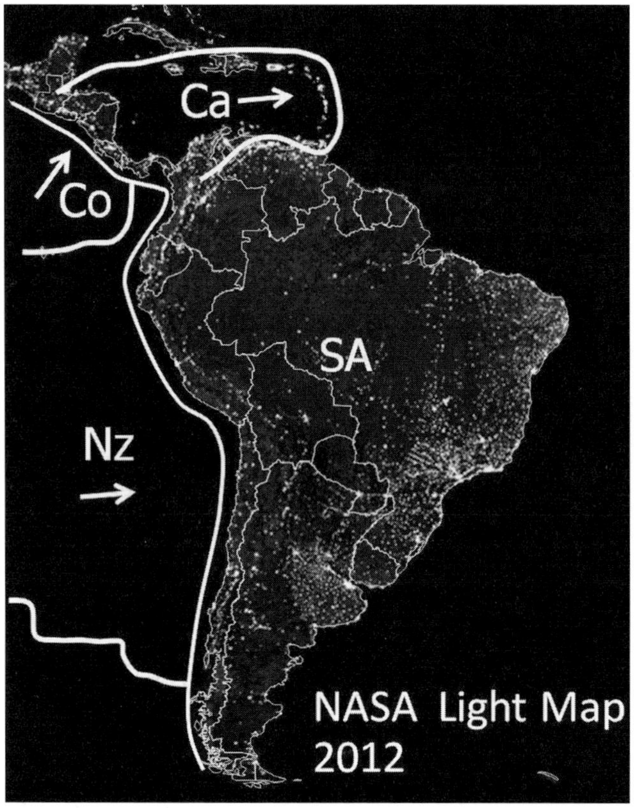

Figure 11.3. NASA 2012 light map and main plate boundaries (white lines) along western South America and the Caribbean. Note that much of the population of WSA and the Caribbean countries is selectively concentrated along plate boundary zones and their associated seismicity belts shown in Figs. 11.1 and 11.2. White contours delineate the countries of the region.

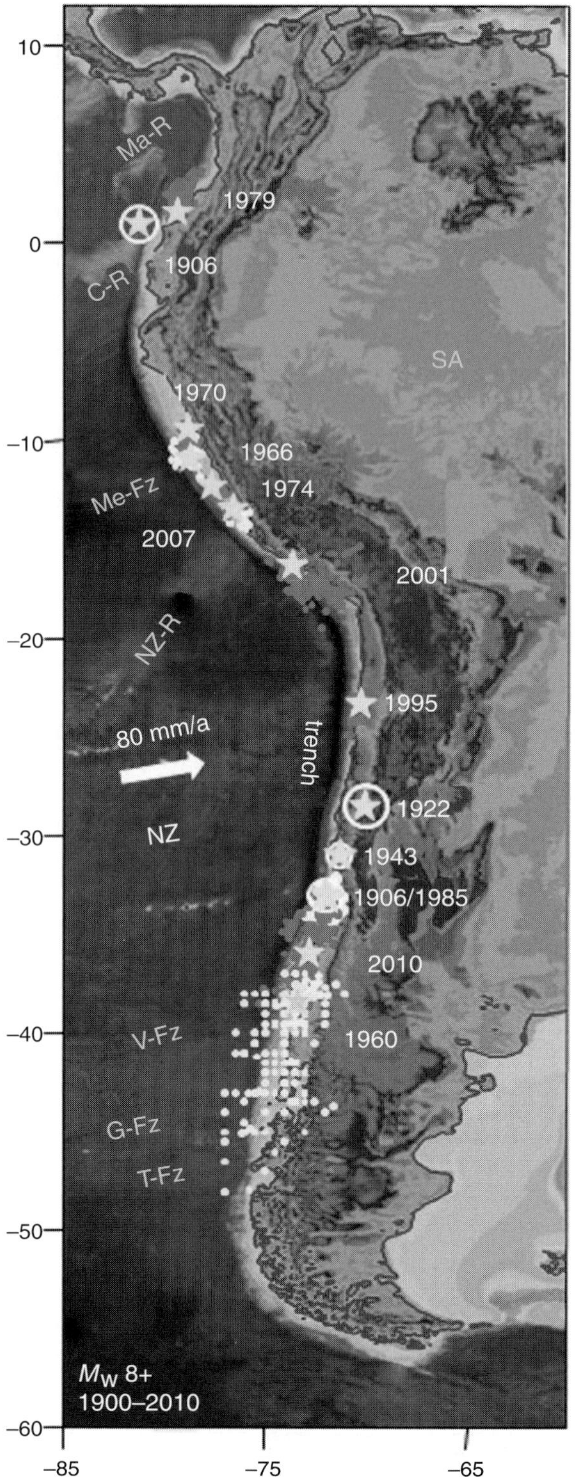

Figure 11.4. Main epicentres (stars) and aftershocks (dots, 5–8 days, $m_b \geq 5$) of great events ($M_w \geq 8$) that occurred in South America during the period 1900–2010. Other symbols as in Fig. 11.1. For colour version, see Plates section.

(e.g., Molnar and Sykes, 1969; Malavé and Suárez, 1995; Pérez et al., 1997), make these areas potentially capable of tsunami generation. For instance, we mentioned earlier that a severe tsunami struck the Dominican Republic in 1946, and the southeast subduction of the Caribbean plate beneath South America (Fig. 11.6) may have a great enough interplate contact area to generate a great shock ($M_w > 8$) and a subsequent large tsunami.

11.4 Discussion and conclusion

We have briefly pointed out the severe damage, large numbers of human casualties, and economic losses generated by large (extreme) earthquakes and tsunamis in WSA and the Caribbean. The earthquake and tsunami disasters negatively impact the sustainable development of the countries in the region. Unfortunately, most of the

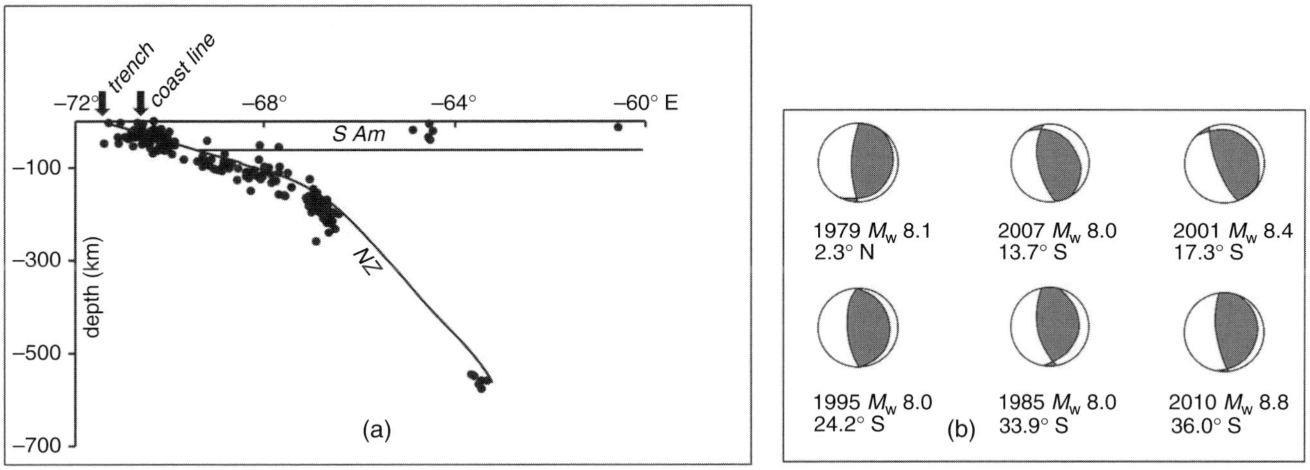

Figure 11.5. (a) Depth distribution of earthquakes within the latitudes 23° S to 25° S along a west–east profile, showing the easterly oriented Benioff zone that defines the subduction of the Nazca (NZ) plate beneath South America (S-Am). Black lines are the schematic representations of the top of the NZ plate and the bottom of the S-Am plates. (b) Harvard's Centroid Moment Tensor Solutions in lower hemisphere equal area projections for six of the great shallow depth (h < 70 km) shocks presented in Fig. 11.4 along the South American arch.

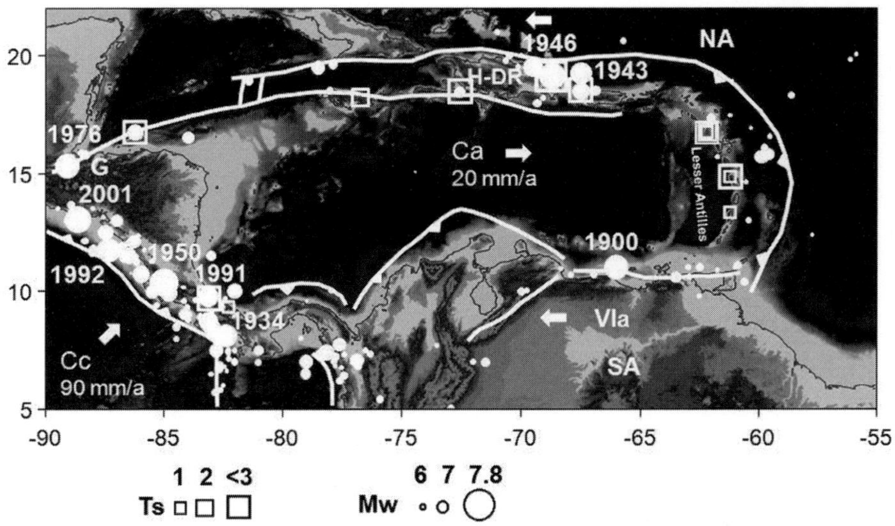

Figure 11.6. Shallow depth (h < 70 km) earthquakes (dots, $M_w \geq 6$) and tsunamis (squares) occurred in the surrounding area of the Caribbean since 1900. Other symbols as in Fig. 11.2.

population in South America and the Caribbean is urban with belts of misery and poverty characterised by poor construction and facilities surrounding the rich portions of large cities and towns. The city of Caracas, the Venezuelan capital (Fig. 11.8), is a notable example.

Caracas is embedded within the deformation zone of the southern Caribbean plate boundary (Fig. 11.2) and was hit by severe shocks in 1812 (M_w7+), 1900 (M_w7.6), and 1967 (M_w6.5) (e.g., Fielder, 1961), all of which caused severe damage and significant numbers of human casualties. The city has a population over 2.3 million people, half of which live in a ring of poverty, misery, and consequent poor facilities and constructions, surrounding the inner rich portions of the city. There are notable social, economic, and construction quality differences between the eastern (right) and central (left) sides of the city (Fig. 11.8). The poor side is clearly very vulnerable to earthquakes.

This increases significantly the vulnerability of the population in the region and indicates not only the need for research in seismology, geodesy, tectonics, earthquake engineering, and some other branches of natural sciences as well as social and economic sciences, but also the need to define and implement responsible policies in each country that include an inter-American system of mutual assistance, designed to mitigate the societal impact of earthquakes and other natural hazards. For instance, an

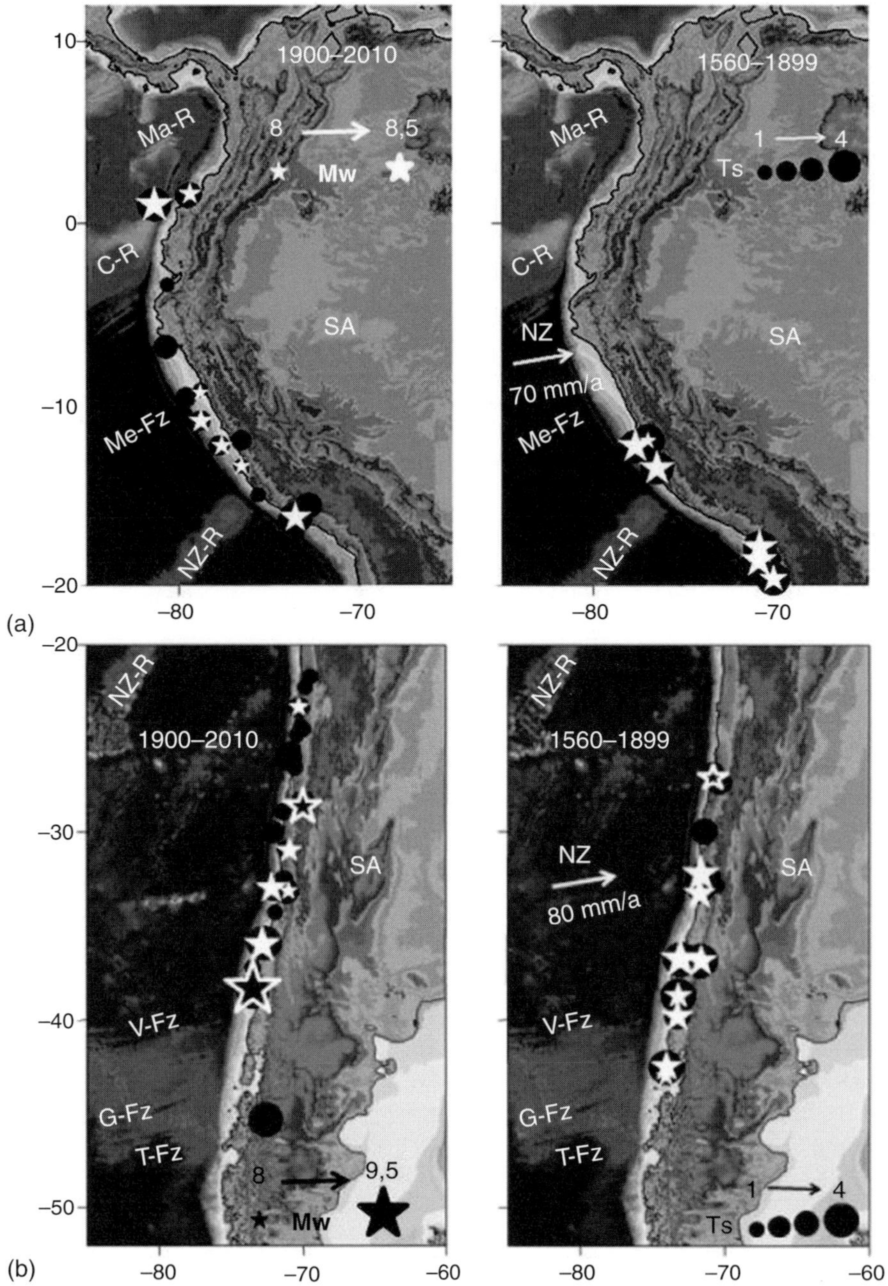

Figure 11.7. Tsunamis (NOAA, dots) and great earthquakes (stars, $M_w \geq 8$) during the periods 1900–2010 (left) and 1560–1899 (right) in (a) the northern and (b) the southern parts of South America. The size of each symbol is proportional to the M_w magnitude (8–9.5) of the event. Other symbols are as in Fig. 11.1. See text for further explanations.

appropriate interaction of local scientists and authorities with the US Warning System, the US Coast and Geodetic Survey, and the Preparatory Commission of the Comprehensive Nuclear-Test-Ban Treaty Organization (CTBTO) would have led the Chilean government to issue a tsunami warning in time after the 2010 giant event hit the region, and thus to avoid the large number of casualties generated by the tsunami. These organisations may provide in a matter of ~10 minutes or less (e.g., Duputel *et al.*, 2011; Singh *et al.*, 2012) crucial earthquake and tsunami information on the hypocentral location, size, focal mechanism, and rupture area of strong seismic events, and on the generation and development of tsunamis, including the size and expected arrival times of tsunami waves at nearby and remote localities. We hope this review will help the authorities and scientists of the Americas to mitigate disasters associated with natural hazards in South America and the Caribbean.

Figure 11.8. Photographic collage of eastern Caracas.

Acknowledgments. We thank Jaime Urrutia Fucugauchi, Raúl Castro, and Alik Ismail-Zadeh for their critical review of the initial manuscript.

References

Barazangi M. and Isacks, B. L. (1976). Spatial distribution of earthquakes and subduction of the Nazca plate beneath South America. *Geology*, **4**, 686–692.

Cavallo, E., Powell, A. and Becerra, O. (2010). Estimating the direct economic damages of the earthquake in Haiti. *The Economic Journal*, **120**, F298–F312.

DeMets, C., Gordon, R. G., Argus, D. F. and Stein, S. (1990). Current plate motions. *Geophysical Journal International*, **101**, 425–478.

Dixon, T. H., Farina, F., DeMets, C. *et al.* (1998). Relative motion between the Caribbean and North American plates and related boundary zone of deformation from a decade of GPS observations. *Journal of Geophysical Research*, **103**, 15157–15182.

Duputel, Z., Rivera, L., Kanamori, H. *et al.* (2011). Real-time W phase inversion during the 2011 Tohoku-oki earthquake. *Earth, Planets and Space*, **63**, 535–539.

Fedotov, S. A. (1965). Regularities of the distribution of strong earthquakes in Kamchatka, the Kurile Islands and northeastern Japan (in Russian). *Trudy Inst. Fiz. Zemli Akad. Nauk SSSR*, 66–93.

Fielder G. (1961). Areas affected by earthquakes in Venezuela (in Spanish). *Bol. Soc. Geol. Ven.*, **4**, 1791–1810.

Iida, K. (1963). Magnitude, energy and generation of tsunamis, and catalogue of earthquakes associated with tsunamis. *IUGG Monograph*, **24**, 7–18.

ISC (2011). *On-line bulletin*, http://www.isc.ac.uk, International Seismological Centre, Thatcham, United Kingdom.

Kanamori, H. and Anderson, D. L. (1975). Theoretical basis of some empirical relations in seismology. *Bulletin of the Seismological Society of America*, **65**, 1073–1095.

Malavé, G. and Suárez, G. (1995). Intermediate-depth seismicity in northern Colombia and western Venezuela and its relationship to Caribbean plate subduction. *Tectonics*, **14**, 617–628.

Molnar, P. and Sykes, L. R. (1969). Tectonics of the Caribbean and Middle America regions from focal mechanisms and seismicity. *Geological Society of America Bulletin*, **80**, 1639–1684.

Pacheco, J. F. and Sykes, L. R. (1992). Seismic moment catalog of large earthquakes. *Bulletin of the Seismological Society of America*, **82**, 1306–1349.

Pérez, O. J., Jaimes, M. and Garciacaro, E. (1997). Microseismicity evidence for subduction of the Caribbean plate beneath the South-American plate in northwestern Venezuela. *Journal of Geophysical Research*, **102**, 17875–17882.

Pérez, O. J., Bilham, R., Bendick, R., Velandia, R., Hernández, N., Moncayo, C., Hoyer, M. and Kosuch, M. (2001). Velocity field across the southern Caribbean plate boundary and estimates of Caribbean/South-American plate motion using GPS geodesy 1994–2000. *Geophysical Research Letters*, **28**, 2987–2990.

Plafker, G. (1976). Tectonic aspects of the Guatemala earthquake of 4 February 1976. *Science*, **193**, 1201–1208.

Scholz, C. H. (1990). *The Mechanics of Earthquake and Faulting*. New York, NY: Cambridge University Press.

Singh, S. K., Pérez-Campos, X., Iglesias, A. and Melgar, D. (2012). A method for rapid estimation of moment-magnitude for early tsunami warning based on coastal GPS networks. *Seismological Research Letters*, **83**, 516–530.

12

Magnetic studies of active volcanoes in Mexico: implications for volcanic hazards and volcano monitoring

Jaime Urrutia Fucugauchi

12.1 Introduction

Since the 1990s studies of volcanic hazards and volcano monitoring have intensified, with the development of improved observational techniques and mitigation programmes. The study of volcanoes undergoing eruptive activity phases poses major challenges related to access restrictions, security issues, measurement precision, identification of eruption precursors, documentation of volcano structure, deep processes, and the installation of monitoring systems. Advances in methods, instrumentation, computing, data storage, telecommunications, satellite observation systems, interferometric synthetic aperture radar (InSAR), global positioning systems (GPS), the use of a large number high density monitoring arrays, borehole instrumentation, gas emission analyses, and aerogeophysical surveys have provided increased understanding of magmatic and eruption processes (e.g. Francis and Rothery, 2000; Dzurisin, 2003; McNutt, 2005; Smith *et al.*, 2009).

Hazard and risk analyses of stratovolcanoes, which could produce large explosive eruptions, present particular challenges. Stratovolcanoes are characterised by long periods of inactivity, or relatively low level fumarolic, seismic, and deformation activity, making it difficult to forecast a time before reawakening episodes that could result in large explosive eruptions (Smith *et al.*, 2009). This was the case with the eruptions of Mount St Helens in the Cascades, western USA, in 1980, El Chichon volcano in southern Mexico in 1982, and Pinatubo volcano in the Philippines in 1991. Explosive eruptive episodes present major hazards associated with flank collapses, debris avalanches, and eruption-triggered lahars such as those generated in the 1985 Nevado del Ruiz eruption in Colombia. Generally, many population centres are located in young volcanic provinces, close to historically active volcanoes. Therefore, analysis of volcanic hazards and implementation of volcano monitoring programmes constitute a major component of hazard prevention and mitigation programmes. Explosive volcanic eruptions pose hazards even when they occur in remote sparsely populated areas, because high and dense volcanic ash plumes may present a potential danger to planes.

Progress in studying active volcanoes, with the implementation of systematic monitoring programmes, has been slow, partly because it is difficult to investigate the deep structure of volcanoes, where magmatic processes take place, and to detect reliably the unrest which would lead to an eruptive event. High resolution imaging of the underground structure and the conduit system as well as understanding magma–rock interactions, magma ascent, degassing, eruption triggering, and eruption dynamics are major areas of research. Aerogeophysical surveys have been implemented and successfully applied to investigate the underground structure of volcanoes and volcanic terrains (Fedi *et al.*, 1998; Finn *et al.*, 2001, 2007; De Ritis *et al.*, 2005, 2010; Nakatsuka *et al.*, 2009). They offer several advantages compared to ground based techniques in the way that they provide information rapidly on the structure and geological discontinuities. Since the 1990s, potential field and electromagnetic methods have been developed and used from helicopter and aircraft platforms; they are now able to provide accurate and useful information for analysing hazards and advancing techniques to use for ground based monitoring networks.

In this chapter, we focus on magnetic methods applied to active volcanoes, hydrothermally altered zones, underground structure, faults and fractures, and thermal and magma movement effects. They encompass ground magnetic and aeromagnetic surveys, continuous ground based magnetic observations, repeat magnetic surveys, and palaeomagnetic and rock magnetic analyses.

Extreme Natural Hazards, Disaster Risks and Societal Implications, eds. Alik Ismail-Zadeh, Jaime Urrutia Fucugauchi, Andrzej Kijko, Kuniyoshi Takeuchi and Ilya Zaliapin. Published by Cambridge University Press.

12.2 Magnetic surveys of active volcanoes

Magnetic, electromagnetic, and gravity surveys have been used to investigate the structure of volcanoes and volcanic terrains in a wide range of tectonic settings (e.g., Gudmundsson and Milsom, 1997; Gibson and Milligan, 1998; Kauahikaua *et al.*, 2000; Finn *et al.*, 2001, 2007; Ramillien and Wright, 2002; De Ritis *et al.*, 2005, 2010; Gottsmann *et al.*, 2008; Okuma *et al.*, 2009; Blanco-Montenegro *et al.*, 2011). Gravity surveys measure mass distributions characterised by density contrasts in volcanic edifices, magma conduits, and deep structures. Magnetic surveys detect magnetisation and magnetic susceptibility contrasts of the underground structure. These potential field methods have also been applied with success on island arcs, mid-ocean ridges, seamounts, and volcanic islands, which present challenges related to inaccessibility of the underwater structure and bathymetric and coast effects.

Airborne magnetic and electromagnetic methods have been improved with high resolution data acquisition systems, GPS positioning, accurate processing, topographic corrections, and forward and inverse modelling techniques, which define more precisely the geometry and location of the magnetic sources (Blakely, 1995). Magnetic surveys have been applied to a range of geological and geophysical problems, including the modelling of underground structure and stratigraphy, documenting folded and faulted structures, geologic mapping, searching for alteration zones, etc. (e.g. Gibson and Milligan, 1998). In volcanic terrains, analyses of magnetic anomalies allow buried structures to be located. De Ritis *et al.* (2010) used high resolution aeromagnetic data in the southern Tyrrhenian Sea to locate a buried volcano of the Calabrian arc. Modelling of the positive magnetic anomaly shows a 20 km × (3–5) km body about 3 km below sea level, which is interpreted in terms of a buried sill associated with a Pleistocene volcano. Finn *et al.* (2007) analysed hydrothermally altered zones in the Mt Adams volcano using magnetic and electromagnetic low altitude helicopter surveys. They found that hydrothermally altered rocks exhibit low resistivity and magnetic susceptibility, which can be distinguished from fresh massive rocks. Mapping rock alteration and water saturation provides information on weakened zones prone to slope collapse and debris avalanches. Since the 1950s, volcanomagnetic signals have been observed by repeat and continuous magnetic monitoring using mainly total field magnetic stations. New approaches have been developed to use repeat airborne and ground magnetic surveys. The volcanomagnetic signals have been related to magma ascent, fracturing and deformation processes, and temperature changes (Stacey *et al.*, 1965; Zlotnicki *et al.*, 1993; Martin del Pozzo *et al.*, 2002). Other interpretations proposed involve electrokinetic effects, underground fluid flows, piezomagnetic signals related to magnetization changes under stress field changes and changes in electrical resistivity (Zlotnicki *et al.*, 1993; Johnston, 1997). Laboratory experiments involving palaeomagnetic and rock magnetic studies have been implemented to investigate volcanic stratigraphy, thermal demagnetisation effects, hydrothermal alteration, magma crystallisation, magma mixing, gas contents, oxygen fugacity, and conditions in magma conduits and eruptive processes (e.g., Conte *et al.*, 2004, 2006; Speranza *et al.*, 2006; Kosterov *et al.*, 2009).

12.3 Studies in the Trans-Mexican Volcanic Belt

The active Popocatepetl and Colima volcanoes and other volcanoes in the Trans-Mexican Volcanic Belt (TMVB) in central Mexico have been intensively studied (Fig. 12.1). Volcanic activity in the central and southern parts of Mexico is associated with plate subduction of the Rivera and Cocos plates at the Middle America trench, with development of the TMVB and Chiapas continental magmatic arcs (Robin and Boudal, 1987; Luhr and Carmichael, 1990; Macias, 2007). Volcanic activity is characterised by the large stratovolcanoes forming the arc front, monogenetic cinder cone and maar fields, shield volcanoes, and large silicic calderas. There are several historically active volcanoes, including the Popocatepetl, Pico de Orizaba, Ceboruco, Colima, San Martin, and Tacana volcanoes (Fig. 12.1). Historic eruptions include the birth of the Jorullo (1759–1774) and Paricutin (1943–1952) monogenetic cinder cones in the Michoacan–Guanajuato field in the western-central TMVB sector and the 1952–1953 eruption of the Barcena volcano in the Revillagigedo Islands in the Equatorial Pacific Ocean.

The Colima volcano is historically the most active volcano in Mexico and has been extensively studied as part of the International Decade of Volcanic Hazard Reduction. It has an elevation of 3820 m above sea level and is located in western Mexico (Fig. 12.1) in a large north–south oriented extensional tectonic structure of the Colima rift. The volcanic activity in the Colima volcanic complex migrated southward towards the trench during the Pleistocene and Holocene. The Colima stratovolcano developed at the southern end of the range formed by Cantaro and Nevado de Colima (Luhr and Carmichael, 1990). The explosive activity of Colima volcano has been marked by partial edifice collapses, large debris avalanche deposits, and lava and pyroclastic flow emissions. The historical record documents explosive eruptive activity with partial flank collapses, lava, and pyroclastic flows in 1585, 1606, 1622, 1818, 1890, and 1913. More recently, activity has been characterised by a summit dome and block lava and

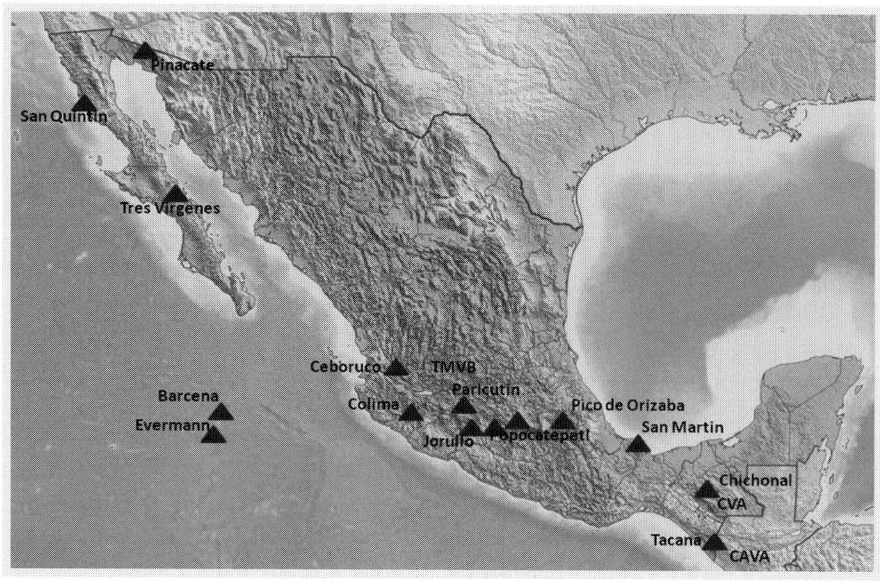

Figure 12.1. Map showing distribution of the active volcanoes in Mexico. TMVB, CVA, and CAVA stand for the continental magmatic arcs of the Trans-Mexican volcanic belt, the Chiapas volcanic arc, and the Central America volcanic arc, respectively.

pyroclastic eruptions in 1961–1962, 1975–1976, 1981–1982, 1991, 1998–1999, 2001–2003, 2004, and 2005 (Luhr and Carmichael, 1990; Macias, 2007). Current phases of activity include growth of the summit dome, explosions, small pyroclastic flows, and flank avalanches.

The Popocatepetl stratovolcano (Fig. 12.1) is 5452 m high, and is located in a densely populated zone in central Mexico, about 70 km southeast of the Mexico City metropolitan area and about 40 km west-southwest of Puebla City. It is similarly located at the southern end of a north–south trending volcanic range of the Sierra Nevada formed by the Tlaloc, the Iztaccihuatl, and the active Popocatepetl volcanoes. This region is characterised by intense Plio-Quaternary magmatic and tectonic activity. It developed through alternate periods of cone formation and destruction, with major cataclysmic events that destroyed ancient cones followed by the formation of dome complexes and new summit domes (Carrasco-Nuñez *et al.*, 1986; Robin and Boudal, 1987; Siebe *et al.*, 1996).

Popocatepetl is formed by two structures: the summit Popocatepetl volcano, with a truncated conical shape, and the basal Nexpayantla volcano. The structures were constructed by lava flows intercalated with breccias and pyroclastic flows. Extensive pyroclastic deposits of ash fall, ash flows, and debris avalanches are distributed around Popocatepetl and constitute part of the basal cone. Carrasco-Nuñez *et al.* (1986) and Robin and Boudal (1987) distinguish two major stages in the history of the volcano: an old Pleistocene age volcano, Nexpayantla, destroyed by a huge eruption 23 ka ago; and the construction of the modern Popocatepetl cone. The northwest facing 4200 m high Ventorillo represents the last external remnant of the old Nexpayantla cone. Since the construction of the modern cone, there have been several major Plinian eruptions. Seven pumice and ash flow deposits have been documented, with the last three occurring within the historical period at about 5, 2.1, and 1.1 ka ago (Siebe *et al.*, 1996). Stratigraphic and mineralogical studies document a succession, from the older Nexpayantla rocks to the more recent ones, of olivine-bearing basaltic andesites and andesites at the base of the Nexpayantla formation, to pyroxene-rich dacites and rhyodacites towards the top, followed by a discontinuity corresponding to the major eruption 23 ka ago. The younger Popocatepetl rocks are pyroxene-rich andesites, with a small increase in acidity with time (Carrasco-Núñez *et al.*, 1986).

After decades of quiescence Popocatepetl volcano reawakened in 1994. Background seismicity showed a sharp increase in October 1994, followed a few weeks later by a swarm of volcano-tectonic events and tremors accompanying a large ash plume on 21 December 1994 (Martin del Pozzo *et al.*, 2002). The activity gave rise to the emission of ash-rich non-juvenile material and volatiles, with tremor, volcano-tectonic, and long-period earthquakes. The ash and gas columns from some events distributed ash over Mexico City and surrounding areas, while lapilli and blocks fell over the volcano flanks. In March 1996, a dacitic dome built within the summit crater. In the past there have been several eruptive episodes, with construction and destruction of the lava dome. More recently, there have been several eruptive events, including those in June 1997, December 1998, December 2000–January 2001.

The increase in eruptive activity in December 2000 prompted emergency measures, including evacuation of about 40 000 inhabitants from the volcano area.

12.4 Aeromagnetic and ground magnetic surveys

Magnetic surveys are well suited for investigating volcanic structures and understanding volcanic processes. Magnetic susceptibility and remanent magnetisation contrasts are key tools in such studies. Here we illustrate potential applications to investigate active volcanoes, with reference to studies of Popocatepetl and Colima volcanoes (Urrutia-Fucugauchi *et al*., 2002a,b; Lopez-Loera *et al*., 2011).

Popocatepetl volcano shows a high amplitude dipolar anomaly centred over the summit crater with the high to the south and the low to the north. The anomaly is slightly elongated along a southwest–northeast trend following the topography and major fault zones (Fig. 12.2). The magnetic signal contrasts with that over the Iztaccihuatl volcanic complex, which is defined by a composite dipolar anomaly. Superimposed lower amplitude anomalies are present over the volcano flanks (Fig. 12.2).

The magnetic signature is also affected by the shift in trend corresponding to the Nexpayantla canyon, with a change in the anomaly shape to the northwest of the dipole low. The other large feature is a southwest–northeast trend passing through the centre of the magnetic dipole, which is reflected in an elongation of the dipole low and may represent a major fault zone.

The older Iztaccihuatl volcano is characterised by a complex composite elongated dipolar anomaly, which is related to the structure of the summit peaks. Regional fault zones are recognised from the shapes of anomalies and trends. The regional field is approximated with a first order horizontal surface of 41 900 nT. The residual anomaly shows a maximum for the magnetic high of about 250 nT and a minimum of about –250 nT.

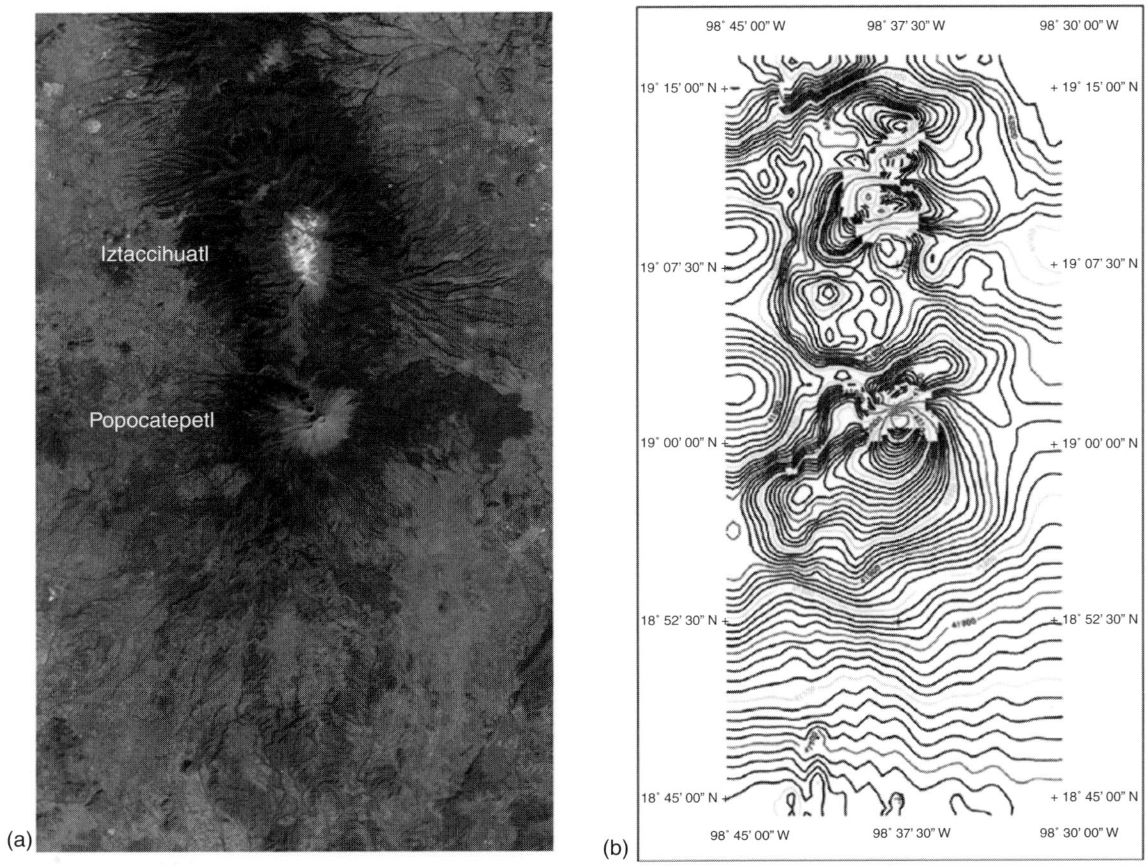

Figure 12.2. The Sierra Nevada volcanic complex, southern basin of Mexico. (a) Landsat 7-Enhanced Thematic Mapper Plus image of the basin of Mexico, showing the Iztaccihuatl and Popocatepetl volcanoes (courtesy NASA Earth Observatory, EROS Data Center; image 4, January 1999). (b) Aeromagnetic anomaly over the Iztaccihuatl and Popocatepetl volcanoes (Urrutia-Fucugauchi *et al*., 2002a,b). For colour version, see Plates section.

Detailed modelling of the structure of stratovolcanoes is difficult because of abrupt topography, high elevation, complex non-horizontal lithologic contacts, heterogeneous physical properties, and alteration zones. Forward and inverse methods have been used for modelling the magnetic sources and to constrain the underground structure (Urrutia-Fucugauchi *et al.*, 2002a,b).

Spectral analysis is employed to estimate depths to magnetic source bodies (Bhattacharyya and Leu, 1975; Spector and Grant, 1979). Results for the transect crossing the summit give average depth estimates of 690 m and 1000 m below the 4.7 km reference survey flight line (Fig. 12.3). Results for neighbour profiles across the summit cone give depths ranging from 580 m to 710 m and 1000 m to 1270 m, respectively, for two shallow and deep source assemblages. The depths to sources are estimated from the slope of the semi-logarithmic relation of the power spectrum as a function of the wavenumber in the radial power spectrum. Depths correlate with locations of volcano-tectonic earthquake hypocentres, which show a diffuse cluster between about 1 km above sea level and 6 km below sea level. This correlation suggests a possible connection to features in the local basement, at the carbonate–volcanic contact and brittle ductile deep transition.

Over the Iztaccihuatl and Popocatepetl volcanoes, upward analytical continuations of the magnetic anomalies to the reference plane show a simple pattern of a large dipolar anomaly slightly oriented north-northeast–south-southwest, suggesting the presence of a deep intrusive body that has no surface expression. The dipole is also normally polarised with a peak to peak amplitude of some 160 nT. The volcanoclastic deposits, including thick lahars from the Popocatepetl volcano, are marked by long wavelength, low amplitude anomalies. Examination of the low elevation data shows a more complex pattern with high frequency anomalies related to sources at various depths, whose effects are integrated in the upward continuation field.

The surrounding volcanic plains are at around 1800 m above sea level, with elevation decreasing to the south. Aeromagnetic anomalies over the adjacent volcanic and sedimentary terrains to the east and south of the Iztaccihuatl and Popocatepetl volcanoes map the surface and deep geological features, with high amplitude, high frequency anomalies over the volcanic units and small volcanoes and low amplitude, long wavelength anomalies over the sedimentary deposits.

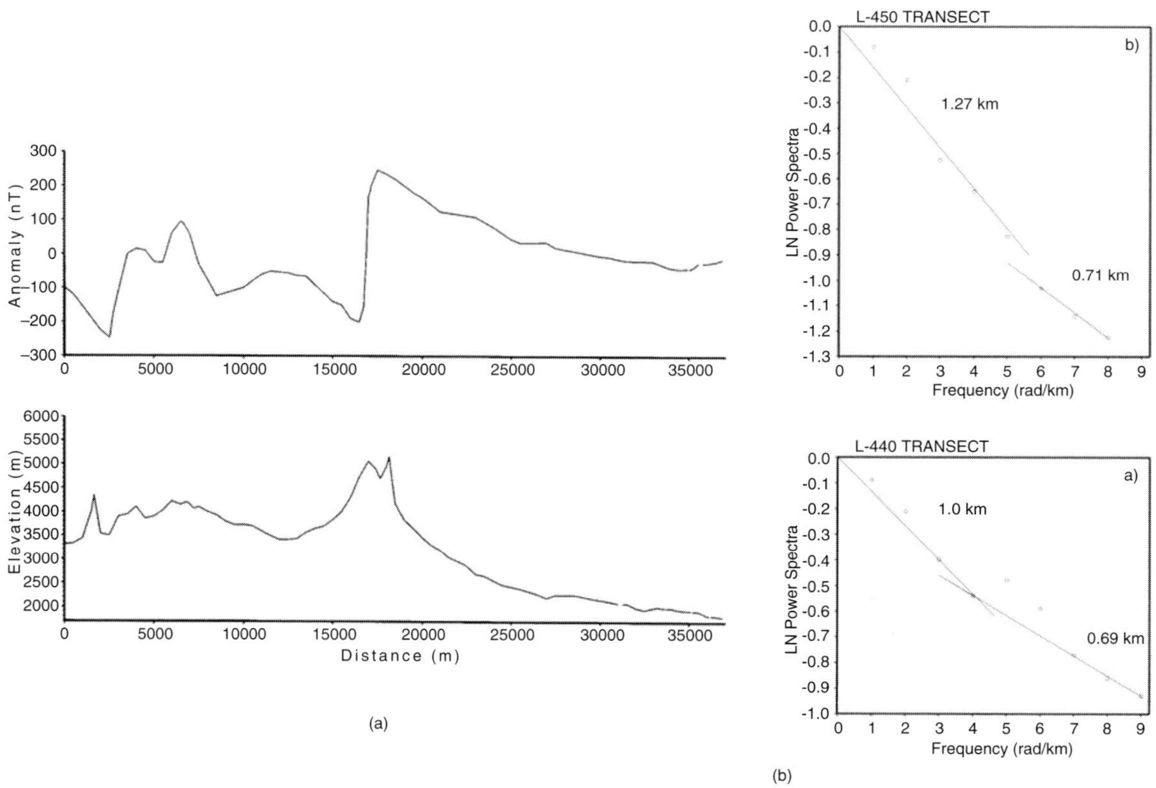

Figure 12.3. Aeromagnetic anomaly over the Iztaccihuatl and Popocatepetl volcanoes. (a) Aeromagnetic anomaly and topographic profiles over the volcanic complex; observe the large amplitude anomaly over the Popocatepetl summit. (b) Examples of spectral analysis of aeromagnetic profile data over the Popocatepetl volcano.

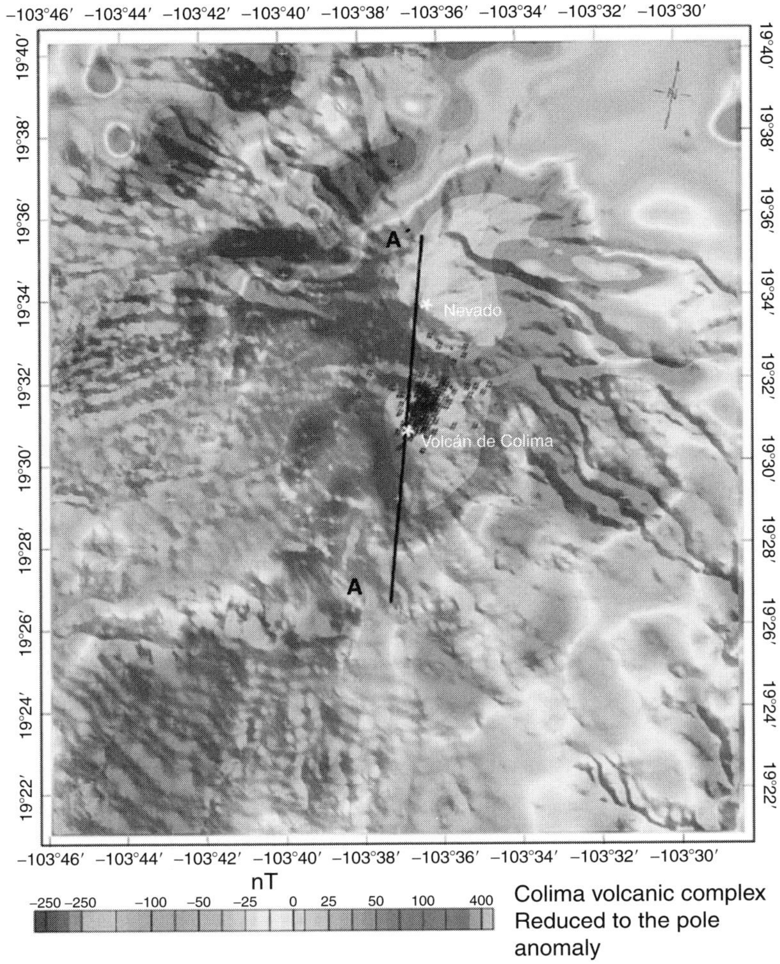

Figure 12.4. Reduced to the pole aeromagnetic anomaly over the Colima rift and volcanic complex. Observe the large amplitude anomaly over the Colima and Nevado de Colima volcanoes. The profile A–A′ shown in Fig. 12.5 is indicated by the line. Earthquake epicentres from the local RESCO 1997–1998 database of Colima volcano are indicated by crosses (Lopez-Loera *et al.*, 2011). For colour version, see Plates section.

The magnetic anomaly over the Colima rift is characterised by a series of highs and lows over the volcanic terrains and the Colima volcanic complex. First and second vertical derivatives, reduction to the pole, and upward analytical continuations of the residual aeromagnetic field have been used to localise the anomaly sources (Lopez-Loera *et al.*, 2011). Figure 12.4 shows the upward continued field to a reference level of 5 km of the reduced to the pole analytical signal. The magnetic response of the Colima complex is separated from the other anomalies of the Colima rift. The tectonic limits and fault systems in the rift are marked by high gradient zones and lineaments. The anomaly reduced to the pole over the Colima and Nevado de Colima volcanoes is shown in Figure 12.5. Forward and inversion models are affected by the non-uniqueness in the solutions. Therefore, different models are analysed, depending on different geometries and magnetic contrasts. The model shown here assumes deep magnetic sources possibly representing the magma chamber and conduit systems beneath the Colima and Nevado de Colima volcanoes. The source body has an elongated shape about 7 km in length at a depth of around 4–5 km below sea level. Beneath the Colima volcano, the source body has larger dimensions. In the modelling, part of the magnetic signal comes from the base of the volcanic sequences (marked by AEV-S-VC, AEV-N-VN, and LAN), which indicates a high contrast corresponding to the assumption of the existence of a carbonate sequence forming the local basement. Use of a flat contact surface for the model requires larger dimensions for the deep source bodies. Modelling of source beneath the Nevado de Colima volcano, suggests a smaller body than that beneath the Colima volcano, with a flat top. Further constraints from geological and geophysical observations are required to improve the model resolution and localise the deep structure.

12.5 Magnetic mapping

Aeromagnetic and ground magnetic surveys have been used to map faults, folded structures in sedimentary basins, and volcanic terrains (e.g., Gibson and Milligan, 1998; Grauch, 2001; Finn and Morgan, 2002; Saltus *et al.*, 2005; Lopez-Loera *et al.*, 2010, 2011). Over the Colima rift and volcanic

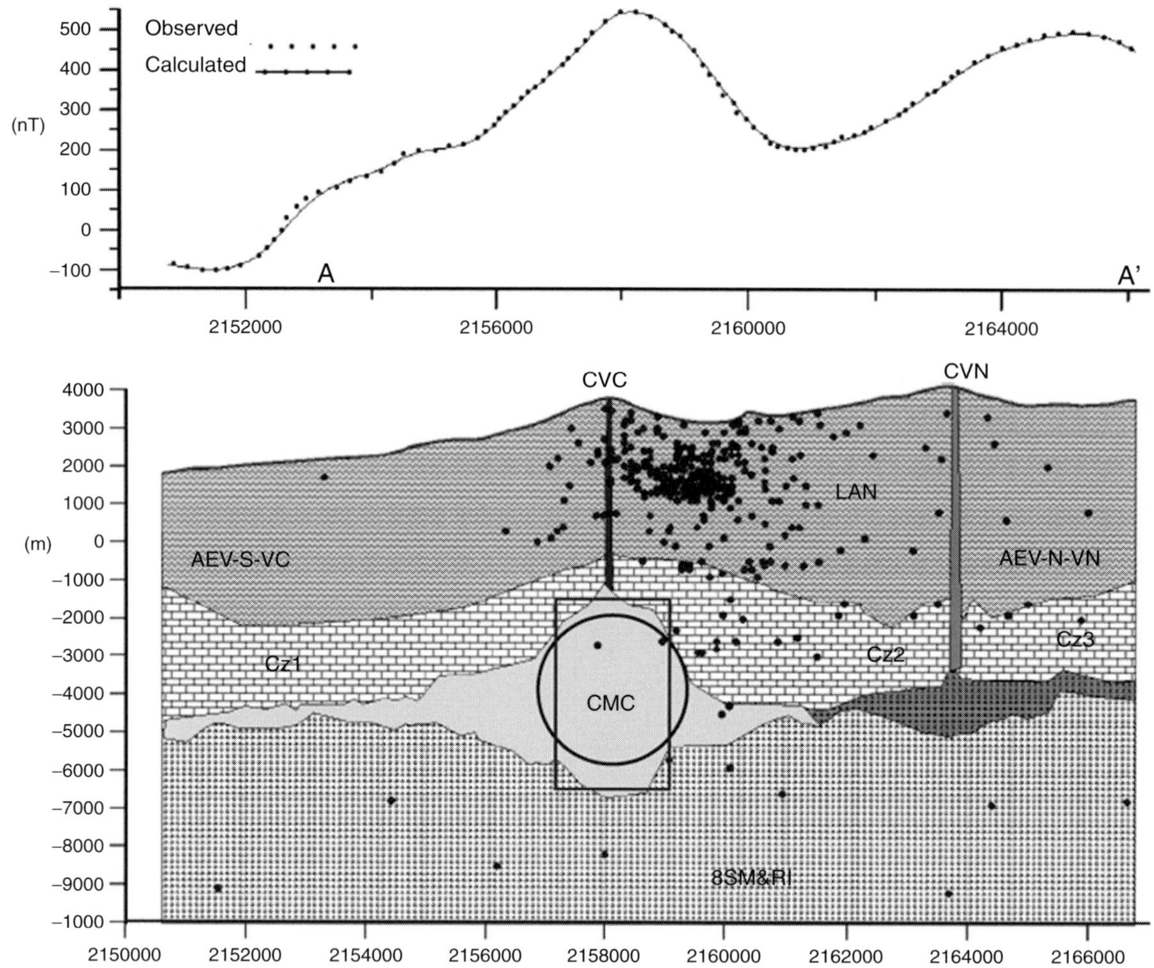

Figure 12.5. Geophysical model for the Colima volcano complex. The magnetic anomaly modelled corresponds to the reduced to the pole anomaly. The model assumes deep sources corresponding to the upper volcanic sequences, the carbonate local basement, and a large elongated body located 4–5 km below sea level (Lopez-Loera *et al.*, 2011).

complex (Figs. 12.4 and 12.6) the aeromagnetic anomalies correlate with the tectonic limits of the rift, which are marked by topographic lineaments and with the faults in the volcanic terrains (Lopez-Loera *et al.*, 2011).

Structural analyses and fault mapping on the Colima volcanic complex are difficult because young volcanic and pyroclastic rocks obscure structural features and stratigraphy. Most of the southern Colima rift is covered by volcanic avalanches and volcanoclastic units, which have resurfaced the volcanic terrains. Ground magnetic surveys have highlighted the identification of faults and mapping of volcano-sedimentary and volcanic units. Results are based on total magnetic field measurements spaced every 0.5 km along profiles, with an overall length of 284.5 km, covering the Colima volcanic complex sector of the Colima rift. Fractures and fault zones of local and regional character are mapped from their characteristic magnetic anomaly response (Fig. 12.7). Large mapped structures include the north–south Montitlan, northeast–southwest La Lumbre, and east–west La Escondida faults, which can be traced across the area from the magnetic profiles. Fault systems are modelled by lateral contrasts in terms of step models assuming thin dipping elongated zones along the fault planes.

12.6 Palaeomagnetism and rock magnetism

Magnetostratigraphic and rock magnetic studies in the Sierra Nevada and the Colima volcanic complex (e.g., Carrasco-Nuñez *et al.*, 1986; Urrutia-Fucugauchi, 1995; Conte *et al.*, 2004; Lopez-Loera *et al.*, 2010) have constrained physical property contrasts and remanent magnetisation directions. Rock magnetic data correlate with hydrothermal effects in alteration zones.

Conte *et al.* (2004) have studied 16 lava flows on the Popocatepetl volcano, mainly andesites and one trachyandesite. Coercivity and unblocking temperature spectra

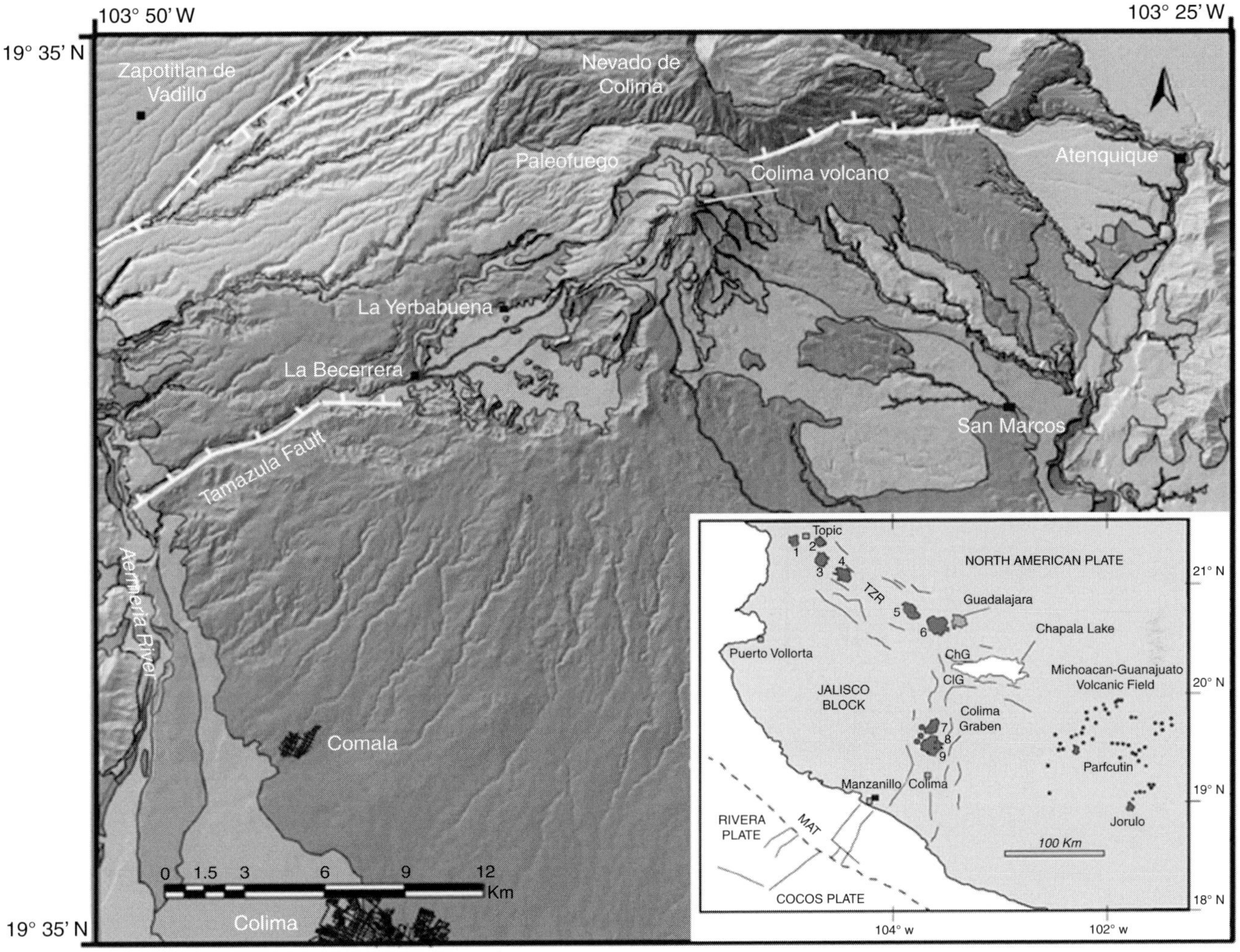

Figure 12.6. Geologic map of the Colima rift in the Colima volcanic complex area, showing the major structures and volcanic units. The inset shows the location of the Colima rift and major tectonic features in western Mexico (after Macias 2007).

suggest that magnetic carriers are in most cases titanium-poor titanomagnetites and magnetite, which likely result from oxy-exolution of iron-titanium oxides during magma cooling. Hysteresis loops and rock magnetic data point to small magnetic particles with pseudo-single domain states. Measurements of magnetic susceptibility versus increasing temperatures up to 700 °C mainly show reversible behaviour with high temperature Curie points. Remanent magnetisations are stable under alternating field and thermal demagnetisation, showing single component demagnetisation plots, pointing to thermal remanent magnetisations, which are well grouped for individual flows. The overall mean direction for characteristic magnetisations presents a northward declination and downward inclination with small angular dispersion.

To further constrain the volcanic stratigraphy, palaeointensity determinations using the double-heating Thellier technique have been carried out (Conte *et al.*, 2004). Paleointensity estimates were obtained for 14 samples from 4 cooling units, with flow-mean virtual dipole moments (VDM) ranging from 5.9 × 10 A m to 9.2 × 10 A m. The overall mean VDM is 7.2 × 10 A m, which is lower than the present-day dipolar value. Further palaeointensity data from several lava flows of different ages are still necessary for correlation with the reference palaeointensity curve, and to derive stratigraphic constraints for the volcanic sequence.

Petrographic and geochemical analyses have been carried out as part of the rock magnetic studies (Carrasco-Nuñez *et al.*, 1986; Conte *et al.*, 2004). Chemical compositions range from basic basaltic andesites and dacites to acidic rocks rhyodacites, with andesites and dacites most abundant. Flow units correspond to andesites and some trachyandesites. The samples studied were rich in plagioclase crystals and

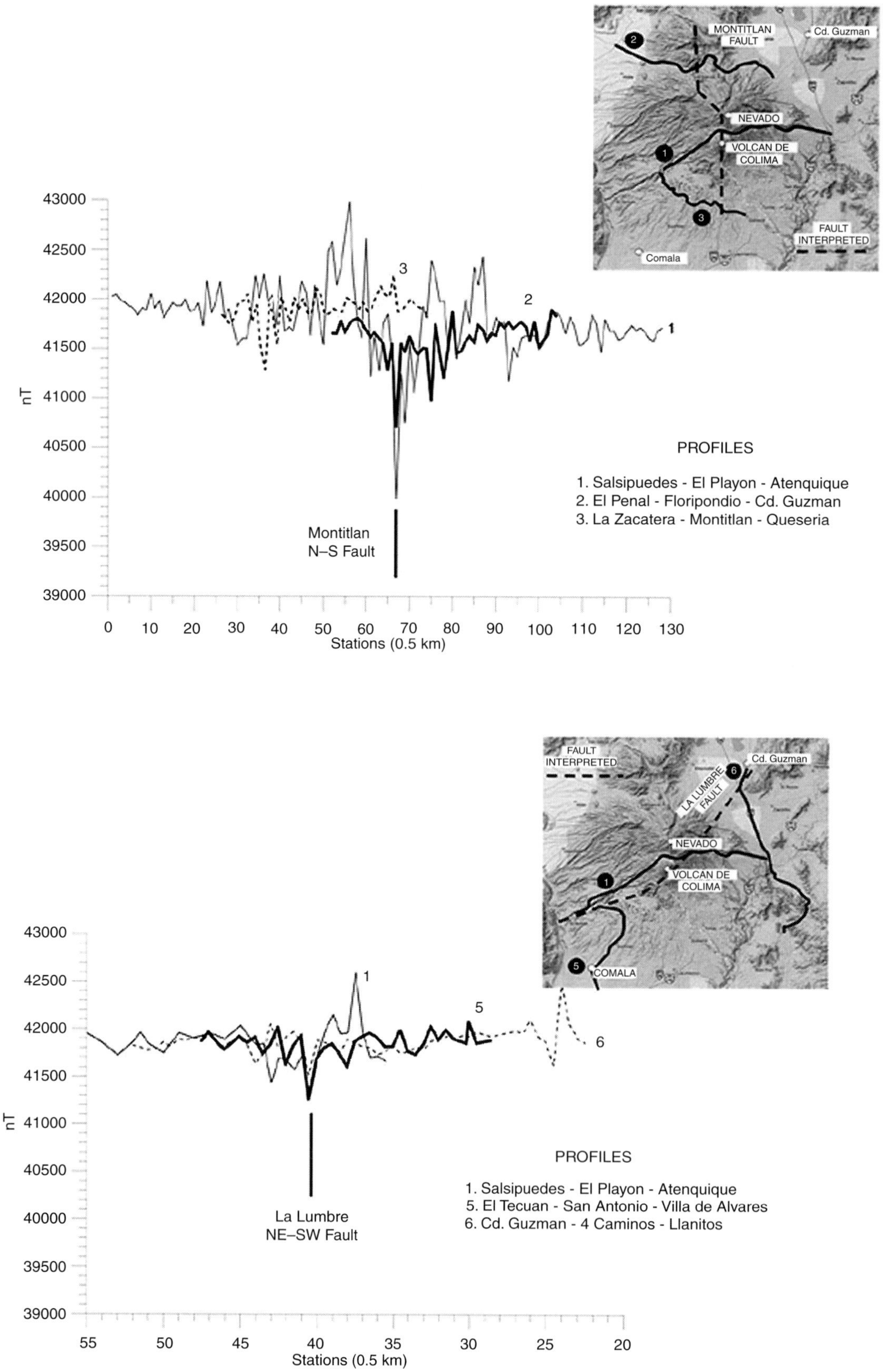

Figure 12.7. Examples of the magnetic response in ground magnetic profiles across fault zones in the volcanic terrain of the Colima rift. Observe the large amplitude sharp anomalies over the fault zone (Lopez-Loera *et al.*, 2010).

were microcrystalline. The samples were olivine-bearing calcalkaline and magnesian andesites, with similar major-oxide and trace-element compositions; 57–63% SiO_2; > 6% MgO; low CaO, Al_2O_3, and FeO; and high Na_2O. Popocatepetl andesites contain magnesian, cromium-spinel fosteritic olivine, which has been taken as evidence of magma mixing with a mafic component still recognizable (e.g., Robin and Boudal, 1987). Chemical data for Popocatepetl units plotted in alkalis–FeO–MgO (AFM) ternary diagrams fall within the calc-alkaline field. Chemical and petrographic data show between 1% and 3% of oxides in the mineralogy, mainly titanium-poor titanomagnetites.

The magnetic polarity of Popocatepetl units is normal, supporting the geological and radiometric data for development of the volcano. The stratigraphic relations indicate an older age, consistent with southward migration of activity along the north–south Sierra Nevada range (Urrutia–Fucugauchi, 1995). Palaeomagnetic polarity measurements indicate that volcanic activity developed during the normal polarity Brunhes chron. Radiometric dating and geological mapping suggest that volcanic activity in the Sierra Nevada Range is not older than 1 Ma, with reported potassium/argon (K/Ar) dates for Iztaccihuatl of ~200 ka in lavas covering the Nexpayantla units. Considering the Neogene activity of Popocatepetl, the Nexpayantla structure is probably no older than ~780 ka and possibly younger than ~200 ka. Palaeomagnetic data for Iztaccihuatl flows indicate dominant normal polarity, within the Brunhes chron (Urrutia-Fucugauchi, 1995). Ages of volcanic units that yield reliable palaeodirectional and palaeointensity estimates lie between about 200 ka and 23 ka, according to magnetic polarity stratigraphy and radiometric data (Carrasco-Nuñez *et al.*, 1986; Robin and Boudal, 1987; Siebe *et al.*, 1996; Conte *et al.*, 2004).

Rock magnetic data for samples from the summit area of Colima stratovolcano have been reported by Connor *et al.* (1993) and Urrutia-Fucugauchi *et al.* (1997). Lopez-Loera *et al.* (2010) reported rock magnetic properties for samples collected along transects crossing the volcano summit and the Colima volcanic complex. Magnetic properties of volcanic fragments and matrix in the debris avalanche deposits from Nevado de Colima and Colima have wide ranges of variation. The younger andesitic flows show higher magnetic susceptibility and remanence intensity values. The avalanche deposits show scattered magnetisation directions and variable magnetic properties. The data on magnetic susceptibility and remanent magnetisation have been correlated with the magnetic responses along profiles crossing the units.

A vectorial composition and the directional stability of samples investigated by stepwise demagnetisation show stable directions. In general, samples show high magnetic stability with mainly univectorial or in some cases multicomponent magnetisations. Unblocking temperatures for most of the samples are in the range 500–575 °C, characteristic of magnetite and titanium-poor titanomagnetites. Magnetic carriers are members of the titanomagnetite and titanohaematite series, with low and intermediate coercivity, and scattered and unblocking temperature spectra. The magnetic polarity of the andesitic lava flows of Nevado de Colima and Colima is normal, indicating an age within the Brunhes chron. An air-fall ash deposit showed scattered directions. Remanent magnetisation directions for volcanic clasts of the old and young debris avalanche deposits show scattered directional angular distributions (Urrutia-Fucugauchi *et al.*, 1997), suggesting emplacement at intermediate to low temperatures.

12.7 Magmatic processes and volcanomagnetic effects

Magnetic, electric, and electromagnetic field variations have been associated with thermal, stress, electrokinetic, and crystal dislocation effects of volcanic activity and tectonic deformation. Early studies identified volcanomagnetic and tectonomagnetic effects (e.g., Stacey *et al.*, 1965). Volcanomagnetic signals have been reported in relation to magmatic processes and eruptive events (e.g., Zlotnicki *et al.*, 1993; Johnston, 1997; Zlotnicki and Bof, 1998).

Geomagnetic stations have been installed on the Popocatepetl and Colima volcanoes (Cifuentes-Nava *et al.*, 1998; Hernandez-Quintero *et al.*, 2000) to record data on diurnal and secular variation of the geomagnetic field (and magnetic storms), as well as any short-term anomaly variations, which could be associated with volcanomagnetic effects. Spatio-temporal correlations of geomagnetic signals and weighed difference techniques have been used for detecting magnetic signals (Rikitake, 1968). The Teoloyucan Geomagnetic Observatory (TEO) in northern Mexico City (Urrutia Fucugauchi and Campos-Enriquez, 1993) has been used as a reference station for monitoring the magnetic field free of volcano-magnetic signals.

Volcano-magnetic signals have been recorded during the last phases of the ongoing eruptive phase of Popocatepetl volcano (Fig. 12.8). The signals have been associated with piezomagnetic and thermomagnetic effects related to ash and gas plume events and magma injection episodes (Cifuentes-Nava *et al.*, 1998; Martin del Pozzo *et al.*, 2002). Detailed modelling of the magnetic anomalies,

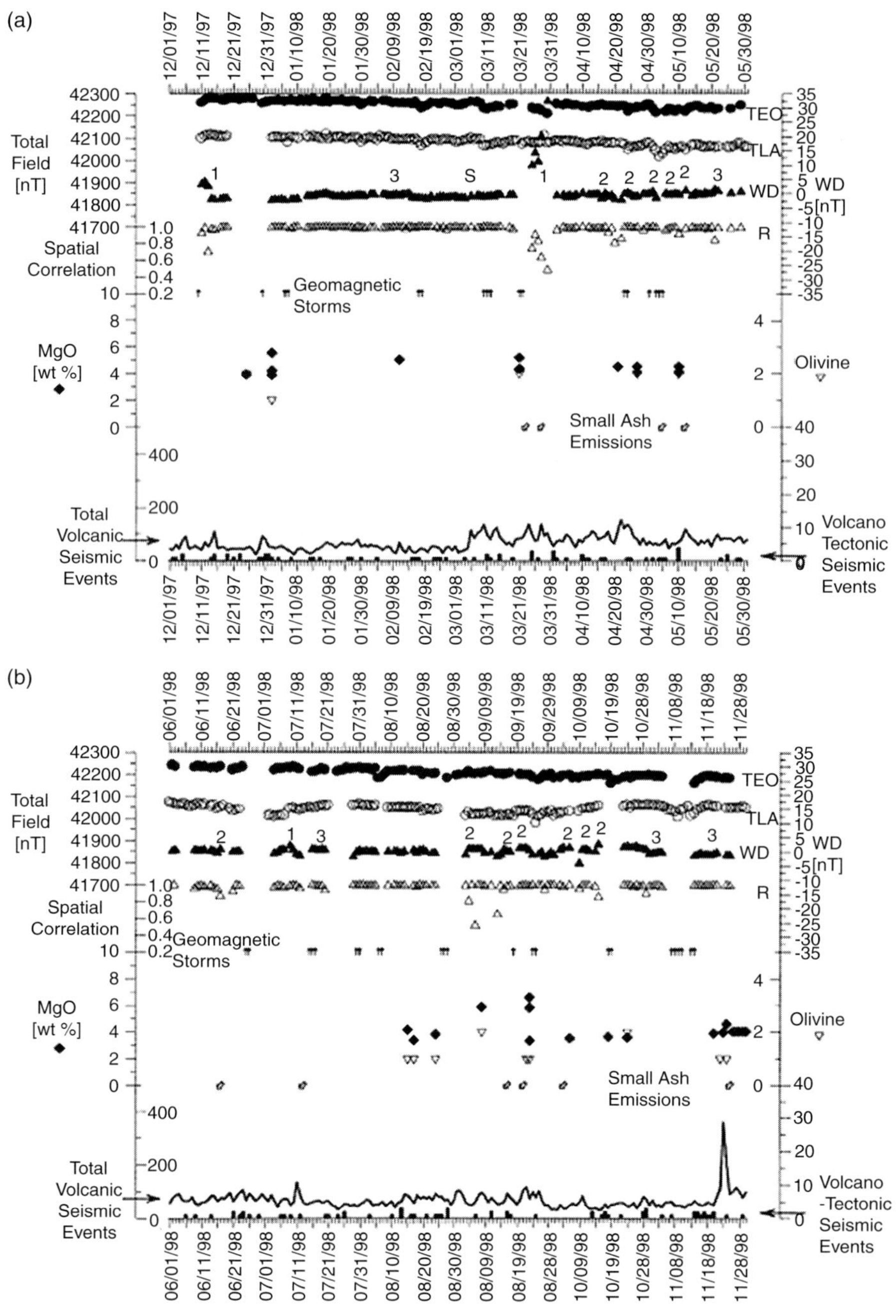

Figure 12.8. Examples of volcano-magnetic signals observed in 1997–1998 during the current eruptive activity phase of the Popocatepetl volcano. Magnetic total field records on Tlamacas (TLA) station and Teoloyucan Observatory (TEO) showing weighted differences (WD) and spatial correlation (R) and the magnesium oxide (MgO) weight per cent (wt%) and olivine in the ejecta. Seismic activity is shown as total volcanic and volcano-tectonic seismic events. Eruption dates are identified by arrows (small eruptions) and diamonds (larger eruptions). (a), (b) The diagrams show the correlations of geomagnetic signals (TEO, TLA), olivine geochemical data, ash eruptions, and the volcano-tectonic seismic events (Martin del Pozzo *et al.*, 2002).

however, requires further and longer observations based on several recording stations. The Tlamacas station (TLA) is located about 5 km away from the summit crater of Popocatepetl on the northern volcano flank. The station has been operational since December 1997. The volcano-magnetic signals were correlated with the eruptive activity in a two-year period between 1997 and 1999 by Martin del Pozzo *et al.* (2002).

Magnetic signals were interpreted in terms of gas pressure build-ups, deformation, and magma injection episodes, recorded by a range of proxies from monitoring ash and gas plumes, seismic tremor and events, and geochemical and petrographic analyses of pumice and ash (Fig. 12.8). Martin del Pozzo *et al.* (2002) interpreted the magnetic signals as due to changes in vent morphology and emission events, with some related to thermal and stress changes. Most magnetic events observed are transient signals reverting to baseline levels. Martin del Pozzo *et al.* (2002) used the weighted differences method in terms of overall peak-to-peak amplitude and shape

(Fig. 12.8). The anomalies were considered in relative terms, not modelled quantitatively, and interpreted from the correlation with geochemistry and seismic observations. Using magnetic differences between total fields might be useful. Further observations appear necessary to confirm the relatively large amplitudes of 27 nT; there were only few cases for which volcanic activity induced such large amplitudes. For their study, magnetic signals were divided into four signal types: type 1 are sharp peaks up to 0.7–33 nT or −0.3 to −27 nT; type 2 are groups of smaller increasing or decreasing variations between 0.5 and 2.5 nT or −0.5 to 7 nT; type 3 are smoother variations lasting several weeks; and type 4 are stepwise 8 nT increase signals in the baseline (only one example recorded). Type 1 magnetic anomalies are characterised by small amplitude, short-time signals above baseline levels, with decreasing anomalies up to −0.5 to −6.8 nT, preceding eruption events by 1–8 days.

On long time scales magma injection and ascent, associated with eruptive behaviour, result in thermal anomalies due to demagnetisation/magnetisation effects. In some volcanoes, the thermal anomalies result in pronounced low anomalies over the active areas. Magmatic processes have been related to magnetic anomalies over volcanic domes, e.g., Mt St Helens (Dzurisin *et al.*, 1990) and the Colima volcano (Connor *et al.*, 1993; Lopez-Loera and Urrutia-Fucugauchi, 1999).

Repeat ground based and aeromagnetic surveys using helicopter-borne surveys have been applied to study magma processes and eruptive activity (e.g., Tanaka, 1993; Lopez-Loera and Urrutia Fucugauchi, 1999; Sasai *et al.*, 2001; Ueda, 2007; Nakatsuka *et al.*, 2009). The resolution of airborne surveys is not high enough for monitoring magmatic activity before an eruption, though they can provide information on volcanic structure, deformation, and temperature changes. Thermal magnetic signals can reach up to few tens of nanoteslas, while piezomagnetic and electrokinetic signals have amplitudes of a few nanoteslas. Nakatsuka *et al.* (2009) studied the magma processes in the Asama volcano following the eruption in 2004, investigating magnetic anomalies associated with thermal demagnetisation effects and volcano deformation. Their study included analyses of methodological uncertainties arising from positioning and measurement, data reproducibility, noise, and anomaly detection.

Ground magnetic surveys and permanent stations have been used to record temporal variations associated with the eruptive activity at the Popocatepetl and Colima volcanoes. At the Colima volcano, high resolution ground magnetic repeat surveys along selected profiles have shown temporal variations linked to the volcanic activity in the summit area and the growing dome (Lopez-Loera and Urrutia-Fucugauchi, 1999).

12.8 Discussion

Few volcanoes in Mexico are monitored with enough spatial and temporal resolution to investigate long-term processes involved in magma generation and ascent. This has been a major limitation in accurate volcanic risk analyses and long-term forecasting. Explosive volcanic eruptions present hazards associated with pyroclastic flows, ground surges, and ash and pumice fallout. They may trigger flank and summit collapses with debris avalanches and lahars. Volcanic hazards are also associated with effusive eruptions, lava flow emplacement, ash plumes, and volcanic gases.

Eruptive activity is triggered and controlled by magmatic processes at depth, with ascent of magma, fracturing, pressure–temperature changes, degassing, and magma interactions within the conduit system. Monitoring of seismic activity and surface deformation associated with deep magma processes is being used to forecast impending eruptions (Swanson *et al.*, 1983; Dzurisin, 2003; McNutt, 2005).

Geophysical surveys provide high resolution data on underground structure, plumbing systems, and deep magma chambers to study active volcanoes. Geophysical studies encompass airborne and ground surveys, remote sensing, continuous-ground base observations, repeat surveys, and laboratory measurements of physical properties. In this chapter we have briefly reviewed magnetic surveys, which are applied to study the underground structure of volcanoes in Mexico, and more recently successfully applied to document hydrothermally altered zones, faults and fractures, and thermal and magma movement effects. Continuous monitoring and repeat surveys have identified volcano-magnetic signals related to magma ascent, fracturing and deformation processes, and temperature changes. Palaeomagnetic and rock magnetic studies provide information on volcanic stratigraphy, thermal demagnetisation effects, hydrothermal alteration, magma crystallisation, magma mixing, gas contents, oxygen fugacity, and conditions in magma conduits and eruptive processes.

Modelling of magnetic anomalies can provide information on the thermal state at depth. The relationship between magnetisation and temperature, which has been studied experimentally in the laboratory, provides simple ways to relate the blocking temperature spectra of rocks with the temperature at depth in the crust. The thermal regime can be inferred from magnetic anomaly models of crustal structure, which can be related to heat flow, variation of petrophysical properties, and crustal structure. Model estimates based on spectral analysis for the bottom of the magnetised crust have been interpreted in terms of the Curie point isotherm deep in the crust (Spector and Grant,

1970; Bhattacharyya and Leu, 1975). Spectral analyses of regional magnetic anomaly data have been used to estimate Curie point isotherms and crustal structure in a variety of tectonic settings (e.g., Tanaka and Ishikawa, 2005). The technique has been applied in the TMVB, and this has allowed the estimation of the estimate depths to the Curie point isotherm in the magmatic arc (Campos-Enríquez *et al.*, 1989), which have been correlated with gravity models of crustal structure (Urrutia Fucugauchi and Flores-Ruiz, 1996).

Basaltic and andesitic volcanoes are characterised by a wide range of sizes and morphologies (Valentine and Gregg, 2008). Conduit geometry is one of the important controlling factors in the eruptive dynamics of volcanoes (Gonnermann and Manga, 2012). Field evidence is limited since volcanic conduits are not, or are only partly, exposed. Keating *et al.* (2008) reported data on feeder dykes for cinder cones in Nevada and New Mexico. Development of complex systems results from dyke bifurcations, brecciation, and incorporation of country rock. Deep in the volcano, feeder dykes generally become narrow. For the Unzen volcano, Goto *et al.* (2008) documented a composite conduit system with sub-vertical individual dykes and a brecciated zone cut by numerous dykes and various dimensions. Urrutia Fucugauchi *et al.* (2012) imaged the conduit system of a monogenetic volcano using modelling of magnetic anomalies constrained by palaeomagnetic data, providing a tool for studying conduit volcanic systems. Individual dykes may be formed by a single magma injection and composite dykes by multiple injections. Once an eruptive phase stops, the magma-filled conduit will solidify on a short time scale depending on the characteristics of the magma feeder system and fluid circulation beneath the volcano.

Numerous studies are being conducted on magma ascent and eruption characteristics of active volcanoes in different tectonic settings. Among the geophysical methods used both in structural imaging and in volcano monitoring, magnetic studies might highlight key aspects concerning deep structures and local and regional heterogeneities, as well as the dynamics of eruptive processes. However, major questions on genesis, magma ascent, eruption controls, and trigger mechanisms remain for further integrated studies using geological and geophysical methods (Dzurisin, 2003; McNutt, 2005; Valentine and Gregg, 2008; Smith *et al.*, 2009).

Acknowledgments. Useful comments on a first version of the chapter by two anonymous reviewers are acknowledged. Partial financial support for the studies has been provided by grants from CONACYT and the National University of Mexico. This study forms part of the UNAM-AVGP Active Volcanoes Geophysical Program.

References

Bhattacharyya, B. K. and Leu, L. K. (1975). Analysis of magnetic anomalies over Yellowstone National Park: Mapping of Curie point isothermal surface for geothermal reconnaissance. *Journal of Geophysical Research*, **80**, 4461–4465.

Blakely, R. J. (1995). *Potential Theory in Gravity and Magnetic Applications*. Cambridge: Cambridge University Press.

Blanco-Montenegro, I., Nicolasi, I., Pignatelli, A., García, A. and Chiappini, M. (2011). New evidence about the structure and growth of ocean island volcanoes from aeromagnetic data: The case of Tenerife, Canary Islands. *Journal of Geophysical Research*, **116**, B03102, doi:10.1029/2010JB007646.

Campos-Enríquez, J. O., Urrutia Fucugauchi, J. and Arroyo-Esquivel, M. A. (1989). Depth to the Curie isotherm from aeromagnetic data and geothermal considerations for the western sector of the Trans-Mexican volcanic belt. *Geofisica Internacionale*, **28**, 993–1005.

Carrasco-Nuñez, G., Silva, L., Delgado-Granados, H. and Urrutia Fucugauchi, J. (1986). *Geologia y paleomagnetismo del Popocatepetl*. Mexico: Publ. Inst. Geofis. Ser. Inv., no. 33.

Cifuentes-Nava, G., Hernandez, E., Cabral, E., *et al.* (1998). Simultaneous magnetic observations at Teoloyucan Magnetic Observatory and Popocatépetl volcano. *Geofisica*, **48**, 149–159.

Connor, C., Lane, S. and Clement, B. M. (1993). Structure and thermal characteristics of the summit dome, March 1990 March 1991: Volcan Colima. *Geofisica Internacionale*, **32**, 643–657.

Conte, G., Urrutia Fucugauchi, J., Goguitchaichvili, A., *et al.* (2004). Paleomagnetic study of lavas from the Popocatepetl volcanic region, central Mexico. *International Geology Review*, **46**, 210–225.

Conte, G., Urrutia Fucugauchi, J., Goguitchaichvili, A., Incoronato, A. and Tiano, P. (2006). Lava identification by paleomagnetism: a case study and some problems surrounding the 1631 eruption of Mount Vesuvius, Italy. *Earth, Planets and Space*, **58**, 1061–1069.

De Ritis, R., Blanco-Montenegro, I., Ventura, G. and Chiappini, M. (2005). Aeromagnetic data provide new insights on the volcanism and tectonics of Vulcano Island and offshore areas (southern Tyrrhenian Sea, Italy). *Geophysical Research Letters*, **32**, L15305.

De Ritis, R., Dominici, R., Ventura, G. *et al.* (2010). A buried volcano in the Calabria arc (Italy) revealed by high-resolution aeromagnetic data. *Journal of Geophysical Research*, **115**, B11101.

Dzurisin, D. (2003). A comprehensive approach to monitoring volcano deformation as a window on the eruption cycle. *Reviews of Geophysics*, **41**, 1001, doi:10.1029/2001RG000107.

Dzurisin, D., Denlinger, R. P. and Rosenbaum, J. G. (1990). Cooling rate and thermal structure determined from progressive magnetization of the dacite dome at Mount St. Helens. *Journal of Geophysical Research*, **95**, 2763–2780.

Fedi, M., Florio, G. and Rapolla, A. (1998). 2.5D modelling of Somma-Vesuvius structure by aeromagnetic data. *Journal of Volcanology and Geothermal Research*, **82**, 239–247.

Finn, C. A. and Morgan, L. A. (2002). High-resolution aeromagnetic mapping of volcanic terrain, Yellowstone National Park. *Journal of Volcanology and Geothermal Research*, **115**, 207–231.

Finn, C., Sisson, T. W. and Deszcz-Pan, M. (2001). Aerogeophysical measurements of collapse-prone hydrothermally altered zones at Mount Rainier volcano. *Nature*, **409**, 600–603.

Finn, C., Deszcz-Pan, M., Anderson, E. D. and John, D. A. (2007). Three-dimensional geophysical mapping of rock alteration and water content at Mount Adams, Washington: Implications for lahar hazards. *Journal of Geophysical Research*, **112**, B1024, doi:10.1029/2006JB004783.

Francis, P. and Rothery, D. (2000). Remote sensing of active volcanoes. *Annual Review of Earth and Planetary Science*, **28**, 81–106.

Gibson, R. I. and Milligan, P. S., eds. (1998). *Geologic Applications of Gravity and Magnetics: Case Histories*, Geophysical Reference Series, vol. 8. Tulsa, OK: Society of Exploration Geophysicists.

Gonnermann, H. M. and Manga, M. (2012). Dynamics of magma ascent in the volcanic conduit. In *Modeling Volcanic Processes: The Physics and Mathematics of Volcanism*, eds. Fagents, S. A., Gregg, T. K. P. and Lopes, R. M. C. Cambridge: Cambridge University Press, pp. 55–84.

Goto, Y., Nakada, S., Kurokawa, M. *et al.* (2008). Character and origin of lithofacies in the conduit of Unzen volcano, Japan. *Journal of Volcanology and Geothermal Research*, **175**, 45–59.

Gottsmann, J., Camacho, A., Wooller, L. *et al.* (2008). Shallow structure beneath the central volcanic complex of Tenerife from new gravity data. *Physics of the Earth and Planetary Interiors*, **168**(3–4), 212–230.

Grauch, V. J. S. (2001). High-resolution aeromagnetic data, a new tool for mapping intrabasinal faults: Example from Albuquerque basin, New Mexico. *Geology*, **29**, 367–370.

Gudmundsson, M. T. and Milsom, J. (1997). Gravity and magnetic studies of the subglacial Grimsötn volcano, Iceland: Implications for crustal and thermal structure. *Journal of Geophysical Research*, **102**, 7691–7704.

Hernandez-Quintero, E., Cifuentes-Nava, G., Cabral, E. *et al.* (2000). A new permanent geomagnetic station at Colima volcano observatory, Mexico. *Geofisika Internationale*, **39**, 267–275.

Johnston, M. J. S. (1997). Review of electric and magnetic fields accompanying seismic and volcanic activity. *Surveys in Geophysics*, **18**, 441–475.

Kauahikaua, J., Hildenbrand, T. and Webring, M. (2000). Deep magmatic structures of Hawaiian volcanoes, imaged by three-dimensional gravity models. *Geology*, **28**, 883–886.

Keating, G. N., Valentine, G. A., Krier, D. J. and Perry, F. V. (2008). Shallow plumbing systems for small-volume basaltic volcanoes. *Bulletin of Volcanology*, **70**, 563–582.

Kosterov, A., Conte, G., Goguitchaichvili, A. and Urrutia-Fucugauchi, J. (2009). Low-temperature magnetic properties of andesitic rocks from Popocatepetl stratovolcano, Mexico. *Earth, Planets and Space*, **61**, 133–142.

Lopez-Loera, H. and Urrutia-Fucugauchi, J. (1999). Spatial and temporal magnetic anomalies of Colima volcano, western México. *Geofisica Internacionale*, **39**, 3–16.

Lopez-Loera, H., Urrutia-Fucugauchi, J. and Alva, L. M. (2010). Magnetic characteristics of fracture zones and constraints on the subsurface structure of the Colima Volcanic Complex, western Mexico. *Geosphere*, **6**, 35–46.

Lopez-Loera, H., Urrutia Fucugauchi, J. and Alva, L. M. (2011). Estudio aeromagnético del complejo volcánico de Colima, occidente de México – implicaciones tectónicas y estructurales. *Revista Mexicana de Ciencias Geológicas*, **28**, 349–370.

Luhr, J. F. and Carmichael, I. S. E. (1990). Geology of Volcan de Colima. *Boletin del Instituto Geologica de Mexico*, UNAM, 107.

Macias, J. L. (2007). Geology and eruptive history of some active volcanoes of México. In *Geology of México: Celebrating the Centenary of the Geological Society of México*, eds. Alaniz-Alvarez, S. A. and Nieto-Samaniego, A. F., Geological Society of America Special Paper 422. Boulder, CO: Geological Society of America, pp. 183–232.

Martin del Pozzo, A. L., Cifuentes, G., Cabral, E., *et al.* (2002). Volcanomagnetic signals during the recent Popocatepetl (Mexico) eruptions and their relation to eruptive activity. *Journal of Volcanology and Geothermal Research*, **113**, 415–428.

McNutt, S. R. (2005). Volcano seismology. *Annual Review of Earth and Planetary Sciences*, **33**, 461–491.

Nakatsuka, T., Utsugi, M., Okuma, S., Tanaka, Y. and Hashimoto, T. (2009). Detection of aeromagnetic anomaly change associated with volcanic activity: an application to the generalized mis-tie control method. *Tectonophysics*, **478**, 3–18.

Okuma, S., Stotter, C., Supper, R. *et al.* (2009). Aeromagnetic constraints on the subsurface structure of Stromboli volcano, Aeolian Islands, Italy. *Tectonophysics*, **478**, 19–33.

Ramillien, G. and Wright, I. C. (2002). Seamount gravity anomaly modelling with variably thick sediment cover. *Marine Geophysical Research*, **23**, 13–23.

Rikitake, T. (1968). Geomagnetism and earthquake prediction. *Tectonophysics*, **6**, 59–68.

Robin, C. and Boudal, C. (1987). A gigantic Bezymianitype event at the beginning of modern Volcan Popocatepetl. *Journal of Volcanology and Geothermal Research*, **31**, 115–130.

Saltus, R. W., Blakely, R. J., Haeussler, P. J. and Wells, R. E. (2005). Utility of aeromagnetic studies for mapping of potentially active faults in two forearc basins: Puget Sound, Washington, and Cook Inlet, Alaska. *Earth, Planets and Space*, **57**, 781–793.

Sasai, Y., Zlotnicki, J., Nishida, Y. *et al.* (2001). Evaluation of electric and magnetic field monitoring of Miyake-jima volcano (central Japan): 1995–1999. *Annali di Geofisica*, **44**, 239–260.

Siebe, C., Abrahms, M., Macias, J. L. and Obenholzner, J. (1996). Repeated volcanic disasters in prehispanic time at Popocatepetl, central Mexico: Past key to the future? *Geology*, **24**, 399–402.

Smith, J. G., Dehn, J., Hoblitt, R. *et al.* (2009). Volcano monitoring. In: *Geological Monitoring*, eds. Young, R. and Norby, L., Boulder, CO: Geological Society of America, pp. 273–305.

Spector, A. and Grant, F. S. (1970). Statistical models for interpreting aeromagnetic data. *Geophysics*, **35**, 293–302.

Speranza, F., Branca, S., Coltelli, M., Caracciolo, F. D. and Vigliotti, L. (2006). How accurate is "paleomagnetic dating"? New evidence from historical lavas from Mount Etna. *Journal of Geophysical Research – Solid Earth*, **111**, B12 Art B12533.doi: 10.1029/2006JB0004496.

Stacey, F. D., Barr, K. G. and Robson, G. R. (1965). The volcano-magnetic effect. *Pure and Applied Geophysics*, **62**, 96–104.

Swanson, D. A., Casadevall, T. J., Dzurisin, D., *et al.* (1983). Predicting eruptions at Mount St. Helens, June 1980 through December 1982. *Science*, **221**, 1369–1376.

Tanaka, A. and Ishikawa, Y. (2005). Crustal thermal regime inferred from magnetic anomaly data and its relationship to seismogenic layer thickness: The Japanese islands case study. *Earth and Planetary Interiors*, **152**, 257–266.

Tanaka, Y. (1993). Eruption mechanism as inferred from geomagnetic changes with special attention to the 1989–1990 activity of Aso Volcano. *Journal of Volcanology and Geothermal Research*, **56**, 319–338.

Ueda, Y. (2007). A 3D magnetic structure of Izu-Oshima Volcano and their changes after the eruption in 1986 as estimated from repeated airborne magnetic surveys. *Journal of Volcanology and Geothermal Research*, **164**, 176–192.

Urrutia Fucugauchi, J. (1995). Constraints on Brunhes low-latitude paleosecular variation-Iztaccihuatl stratovolcano. *Geofisica Internacionale*, **34**, 253–262.

Urrutia Fucugauchi, J. and Campos-Enriquez, J. O. (1993). Geomagnetic secular variation in central Mexico since 1923 AD and comparison with 1943–1990 IGRF models. *Journal of Geomagnetism and Geoelectricity*, **45**, 243–249.

Urrutia Fucugauchi, J. and Flores-Ruiz, J. H. (1996). Bouguer gravity anomalies and regional crustal structure in central Mexico. *International Geology Review*, **38**, 176–194.

Urrutia Fucugauchi, J., Martín del Pozzo, A. L., Alva, L., Lopez-Loera, L. and Ponce, R. (1997). Paleomagnetic estimates of emplacement temperatures of pyroclastic deposits from the Colima volcanic complex. In Abstracts of the IAVCEI Scientific Assembly, January 1997, Puerto Vallarta, Mexico. Puerto Vallarta, Mexico: International Association of Volcanology and Chemistry of the Earth's Interior, p. 12.

Urrutia Fucugauchi, J., Flores-Ruiz, J. H., Hernandez, I. and Anaya, C. (2002a). Aeromagnetic survey over an active stratovolcano in central Mexico. *Leading Edge*, **21**, 560–563.

Urrutia Fucugauchi, J., Martínez-Pepin, N., Hernández, I. *et al.* (2002b). Aeromagnetic anomalies and structure of the Iztaccihuatl-Popocatepetl volcanic region in Central Mexico. *Geofisica Internacionale*, **41**, 121–131.

Urrutia Fucugauchi, J., Trigo, A. and Pérez-Cruz, L. (2012). Magnetic links among lava flows, tuffs and the underground plumbing system in a monogenetic volcano, derived from magnetics and paleomagnetic studies. *Physics of the Earth and Planetary Interiors*, **212**, 10–18.

Valentine, G. A. and Gregg, T. K. P. (2008). Continental basaltic volcanoes – Processes and problems. *Journal of Volcanology and Geothermal Research*, **177**, 857–873.

Zlotnicki, J. and Bof, M. (1998). Volcanomagnetic signals associated with the quasi-continuous activity of andesitic Merapi volcano, Indonesia: 1990–1995. *Physics of the Earth and Planetary Interiors*, **105**, 119–130.

Zlotnicki, J., Le Mouel, J. L., Delmond, J. C., Pambrun, C. and Delorme, H. (1993). Magnetic variations on Piton de la Fournaise volcano. Volcanomagnetic signals associated with the November 6 and 30, 1987, eruptions. *Journal of Volcanology and Geothermal Research*, **56**, 281–296.

PART IV

Case studies: Africa

13

Volcanism in Africa: geological perspectives, hazards, and societal implications

Nils Lenhardt and Clive Oppenheimer

13.1 Introduction

Africa faces many and diverse risks associated with natural hazards. Phenomena such as floods, hurricanes, earthquakes, tsunamis, droughts, wildfires, and volcanic eruptions cause extensive losses to livelihoods and property, and claim many lives (ICSU, 2007). Currently, disasters associated with geological hazards have a far smaller impact on sub-Saharan Africa than those due to hydro-meteorological hazards. Earthquakes account for 2% and landslides and volcanic hazards account for 1% of the number of hazards occurring on the continent (OFDA-CRED, 2012). However, the impact of these hazards may change in the future. Although eruptions of Africa's many volcanoes rarely make it to the headline news, the high population density in the fertile plains surrounding some of the volcanoes increases the potential for severe impacts. Furthermore, as the 2011 eruption of Nabro volcano in the Southern Red Sea Region of Eritrea showed, eruptions of African volcanoes can pose a risk to international aviation. This was the first eruption of Nabro in recorded history, and Eritrea had no monitoring network for volcano observations at the time (see Chapter 14). Very little research had ever been carried out on the volcano. The same is true of many other late Quaternary/Holocene active volcanoes in Africa, and implies that there is currently considerable uncertainty as to the nature of the hazards posed.

In 2010 the United Nations (UN) estimated Africa's population to be 1 022 234 000 (UN estimate for 2012: 1 070 096 000), 20% more than in 2000 and more than double the figure in 1980 (United Nations, 2011). Most countries on the continent have a population growth in excess of 2% per year. According to the US Population Reference Bureau (2012) the population on the continent will double again by 2050 to reach 2.1 billion, so the numbers of people exposed to combinations of natural and human-induced hazards and disasters will continue to increase. The fact that most of the African countries are income-poor arguably compromises their capacity to cope with the impacts of hazards and disasters. Reducing disaster risk through preventive measures is therefore a central concern for Africa's sustainable development (e.g., ICSU, 2007).

Fournier d'Albe (1979) defined a hazard as the probability of occurrence, within a specified period of time, within a given area, of a potentially damaging natural phenomenon. According to the UN International Strategy for Disaster Reduction (UNISDR, 2009) a hazard is 'a dangerous phenomenon [...] that may cause loss of life, injury or other health impacts, property damage, loss of livelihoods and services, social and economic disruption, or environmental damage'. In contrast, risk is the measure that characterises the potential losses (to life, economy, the natural and built environment, etc.) that can result from the occurrence of a hazard event (Alberico *et al.*, 2002). Hazardous phenomena that are related to volcanic eruptions that are discussed in this chapter include earthquakes, lava flows, volcanic gas emissions, tephra falls, pyroclastic density currents, lahars, volcanic debris avalanches, and even tsunamis. Risk mitigation aims to lessen the effects of these volcanic hazards by planning, ideally in an integrated three-pronged approach, involving prevention, preparedness, and emergency response measures. Risk mitigation plans define the roles of governmental and non-governmental organisations, district councils, and the community in managing the impacts of potential natural hazard events. The most effective preventive measure is that of land-use planning, which limits or prohibits the development of hazardous areas. Preparedness measures include the installation of monitoring systems

Extreme Natural Hazards, Disaster Risks and Societal Implications, eds. Alik Ismail-Zadeh, Jaime Urrutia Fucugauchi, Andrzej Kijko, Kuniyoshi Takeuchi and Ilya Zaliapin. Published by Cambridge University Press.

(in conjunction with public awareness programmes), and the establishment of emergency response plans for warning and evacuation (where necessary), search and rescue, and rehabilitation and reconstruction.

Including the islands of the Comoros Archipelago, Madagascar, and two uninhabited (apart from a research station) islands belonging to South Africa (Marion and Prince Edward), Africa has around 150 volcanoes that are catalogued as either active or having erupted in the past 10 ka (Table 13.1). However, even among these, very few have been the focus of systematic geochronological studies, and many other volcanoes have yet to be studied at all. Nabro provides a good example: the volcano was not monitored nor was it the focus of any scientific attention when it reawakened in June 2011.

The volcanoes are related to hotspots (e.g., the Cameroon Volcanic Line volcanoes, the volcanoes of the Sahara, and the volcanic islands of the Indian Ocean), the rifting in East Africa, or a combination of the two, and are primarily characterised by central (stratovolcanoes) and fissure vents (shield volcanoes). Eruptive styles range from the fluid lava lakes (as seen today at Nyiragongo, DR Congo, and Erta 'Ale, Ethiopia) to Plinian-style explosive eruptions of silicic magma associated with widespread fall deposits and pyroclastic density current deposits (e.g., Dubbi, Eritrea; Rungwe, Tanzania). The silicic magmas include trachytic, phonolitic, and rhyolitic compositions, and in many cases reflect rift affinities in their chemical compositions. Of special interest are the many caldera complexes found in the East African rift system (EARS). The continent has recorded relatively few large (volcanic explosivity index (VEI $\geq$ 4; Newhall and Self (1982)) eruptions in the Holocene (and three of these four, for which quantitative data are available, were more than 2000 years ago). Nevertheless, at least seven larger Holocene eruptions can be inferred from indications of caldera collapses or Plinian-fall deposits (Table 13.2), and once more research is carried out on the volcanic record it is extremely likely that further evidence will emerge.

The first written account of African volcanic activity may be attributed to a Carthaginian, Hanno the Navigator, who described sulphurous fumes and streams of lava descending down a mountain, probably Mount Cameroon, in the fifth century BC (Blomquist, 1979; Burton, 2011). By the fifteenth century AD, however, when Portuguese exploration of Africa began and Vasco da Gama sailed to India via the Cape of Good Hope, only two more eruptions had been recorded, Fentale in Ethiopia (circa 1250 AD) and Dubbi in Eritrea (1400 AD). In the next 350 years, another 20 or so eruptions were recorded, but it is not until the opening of the Suez Canal at the end of 1869 that the written historical records of volcanic activity in Africa become more complete (Siebert *et al.*, 2010). In the period between 1900 and 2012, 17 major volcanic crises occurred on the African continent. In the course of these events 2218 people were killed, and more than 500 000 people were either injured, temporarily displaced, and/or lost their properties. The damage caused by these events has been estimated at around US$9 million (OFDA-CRED, 2012).

The largest known historical eruption in Africa is that of Dubbi (Eritrea) in 1861. This erupted several cubic kilometres of tephra and lava flows, and claimed 175 lives according to one contemporary account, almost certainly as a result of pyroclastic density currents reaching two villages (Wiart and Oppenheimer, 2000).

Another volcano that has claimed many lives is Nyiragongo (DR Congo). Eruptions in 1977 and 2002 both involved highly mobile lava flows sourced from a magma chamber high within the volcanic edifice (Demant *et al.*, 1994; Platz *et al.*, 2004). The 2002 eruption resulted in destruction of a large part of Goma, a town of approximately 400 000 inhabitants, up to 300 000 of whom spontaneously fled as the flows approached. The 2002 eruption disgorged a modest quantity of lava, $\sim(13–34) \times 10^6$ m^3, but its high temperature and low viscosity facilitated long and rapid run-out, resulting in the destruction of the housing of an estimated 120 000 people (Allard *et al.*, 2002; Tedesco *et al.*, 2007); 147 lives were lost during the eruption, more than half of them as a result of the explosion of a petrol station. A smaller eruption of Nyiragongo in 1977 claimed an estimated 700–1500 lives, though reliable figures are not available. Nyiragongo and its neighbour, Nyamuragira, are responsible for nearly two-fifths of Africa's recorded historical eruptions.

In 1999 and 2000, eruptions of Mt Cameroon precipitated the evacuation of several thousand people and polluted drinking water, affecting about 2600 people living in the area. Subsequent lahars in 2001 killed an estimated 23 people. In 2005 the eruption of Mt Karthala (Comoros) caused more than 10 000 villagers to flee their homes.

Each of these volcanoes and their eruptions is associated with a range of contrasting hazard types and specific human factors (including risk perception, socio-economic circumstances, and conflict), and will be discussed in more detail in the course of this chapter. Although many of the earlier Holocene eruptions (see Tables 13.1 and 13.2) were much larger in terms of eruption intensity and magnitude, they were not necessarily the most devastating eruptions due to their remoteness and the small contemporary populations. We know little, however, of the archaeological record that might be associated with prehistoric volcanism in Africa. With increasing population numbers and density on the slopes of active volcanoes, even smaller volcanic eruptions can have a high impact and pose a serious threat to life and property of local communities and national economies.

Table 13.1. *List of identified Holocene volcanoes (modified from data from the Global Volcanism Program of the Smithsonian Institute). It is important to recognise that the record is undoubtedly incomplete due to the limited geochronological studies that have been carried out on the volcanic provinces of the continent*

Name	Country	Type	Latitude	Longitude	Elevation above sea level (m)	Last known eruption
Atakor Volc. Field	Algeria	Scoria cones	23.33° N	5.83° E	2918 m	?
Manzaz Volc. Field	Algeria	Scoria cones	23.92° N	5.83° E	1672 m	?
Tahalra Volc. Field	Algeria	Pyroclastic cones	22.67° N	5.00° E	1467 m	?
In Ezzane Volc. Field	Algeria/Niger	Volcanic field	23.00° N	10.83° E	Unknown	?
Cameroon	Cameroon	Stratovolcano	4.20° N	9.17° E	4095 m	2000
Manengouba	Cameroon	Stratovolcano	5.03° N	9.83° E	2411 m	?
Ngaoundere Plateau	Cameroon	Volcanic field	7.25° N	13.67° E	Unknown	?
Oku Volc. Field	Cameroon	Stratovolcano	6.25° N	10.50° E	3011 m	?
Tombel Graben	Cameroon	Cinder cones	4.75° N	9.67° E	500? m	?
Emi Koussi	Chad	Pyroclastic shield	19.80° N	18.53° E	3415 m	?
Tarso Tôh	Chad	Volcanic field	21.33° N	16.33° E	2000 m	?
Tarso Toussidé	Chad	Stratovolcano	21.03° N	16.45° E	3265 m	?
Tarso Voon	Chad	Stratovolcano	20.92° N	17.28° E	3100 m	?
Karthala	Comoros	Shield volcano	11.75° S	43.38° E	2361 m	2007
La Grille	Comoros	Shield volcano	11.47° S	43.33° E	1087 m	?
Ardoukôba	Djibouti	Fissure vents	11.58° N	42.47° E	298 m	1978
Nyamuragira	DR Congo	Shield volcano	1.408° S	29.20° E	3058 m	2010
Nyiragongo	DR Congo	Stratovolcano	1.52° S	29.25° E	3470 m	2011 (in or after)
Tshibinda	DR Congo	Cinder cones	2.32° S	28.75° E	1460+ m	?
Karisimbi	DR Congo/Rwanda	Stratovolcano	1.50° S	29.45° E	4507 m	8050 BC (?)
Visoke	DR Congo/Rwanda	Stratovolcano	1.47° S	29.492° E	3711 m	1957
San Carlos	Equatorial Guinea (Bioko Island)	Shield volcano	3.35° N	8.52° E	2260 m	?
San Joaquin	Equatorial Guinea (Bioko Island)	Shield volcano	3.35° N	8.63° E	2009 m	?
Santa Isabel	Equatorial Guinea (Bioko Island)	Shield volcano	3.58° N	8.75° E	3007 m	1923
Alid	Eritrea	Stratovolcano	14.88° N	39.92° E	904 m	?
Assab Volc. Field	Eritrea	Volcanic field	12.95° N	42.43° E	987 m	?
Dubbi	Eritrea	Stratovolcano	13.58° N	41.808° E	1625 m	1861
Jalua	Eritrea	Stratovolcano	15.04° N	39.82° E	713 m	?
Nabro	Eritrea	Stratovolcano	13.37° N	41.70° E	2218 m	2011
Adwa	Ethiopia	Stratovolcano	10.070° N	40.840° E	1733 m	?
Afderà	Ethiopia	Stratovolcano	13.08° N	40.85° E	1295 m	?
Alayta	Ethiopia	Shield volcano	12.88° N	40.57° E	1501 m	1915
Ale Bagu	Ethiopia	Stratovolcano	13.52° N	40.63° E	1031 m	?
Alu	Ethiopia	Fissure vents	13.825° N	40.508° E	429 m	?
Alutu	Ethiopia	Stratovolcano	7.77° N	38.78° E	2335 m	50 BC (?)
Asavyo	Ethiopia	Shield volcano	13.07° N	41.60° E	1200? m	?
Ayelu	Ethiopia	Stratovolcano	10.082° N	40.702° E	2145 m	?
Beru	Ethiopia	Volcanic field	8.95° N	39.75° E	1100 m	?
Bilate River Field	Ethiopia	Maars	7.07° N	38.10° E	1700? m	?
Bishoftu Volcano Field	Ethiopia	Fissure vents	8.78° N	38.98° E	1850+ m	?
Bora-Bericcio	Ethiopia	Pumice cones	8.27° N	39.03° E	2285 m	?
Borale Ale	Ethiopia	Stratovolcano	13.725° N	40.60° E	668 m	?
Borawli	Ethiopia	Stratovolcano	13.30° N	40.98° E	812 m	?
Borawli	Ethiopia	Lava domes	11.63° N	41.45° E	875 m	?
Boset-Bericha	Ethiopia	Stratovolcanoes	8.558° N	39.475° E	2447 m	?
Butajiri-Silti Field	Ethiopia	Fissure vents	8.05° N	38.35° E	2281 m	?
Chiracha	Ethiopia	Stratovolcano	6.65° N	38.12° E	1650 m	?

Table 13.1. (*cont.*)

Name	Country	Type	Latitude	Longitude	Elevation above sea level (m)	Last known eruption
Corbetti Caldera	Ethiopia	Caldera	7.18° N	38.43° E	2320 m	?
Dabbahu	Ethiopia	Stratovolcano	12.60° N	40.48° E	1442 m	2005
Dabbayra	Ethiopia	Shield volcano	12.38° N	40.07° E	1302 m	?
Dalaffilla	Ethiopia	Stratovolcano	13.792° N	40.55° E	613 m	2008
Dallol	Ethiopia	Explosion craters	14.242° N	40.30° E	−48 m	1926
Dama Ali	Ethiopia	Shield volcano	11.28° N	41.63° E	1068 m	1631
Dofen	Ethiopia	Stratovolcano	9.35° N	40.13° E	1151 m	?
East Zway	Ethiopia	Fissure vents	7.95° N	38.93° E	1889 m	?
Erta Ale	Ethiopia	Shield volcano	13.60° N	40.67° E	613 m	2011 (in or after)
Fentale	Ethiopia	Stratovolcano	8.975° N	39.93° E	2007 m	1820 (?)
Gabillema	Ethiopia	Stratovolcano	11.08° N	41.27° E	1459 m	?
Gada Ale	Ethiopia	Stratovolcano	13.975° N	40.408° E	287 m	?
Gedamsa	Ethiopia	Caldera	8.35° N	39.18° E	1984 m	?
Groppo	Ethiopia	Stratovolcano	11.73° N	40.25° E	930 m	?
Hayli Gubbi	Ethiopia	Shield volcano	13.50° N	40.72° E	521 m	?
Hertali	Ethiopia	Fissure vent	9.78° N	40.33° E	900? m	?
Hobicha Caldera	Ethiopia	Caldera	6.78° N	37.83° E	1800? m	?
Kone	Ethiopia	Calderas	8.80° N	39.692° E	1619 m	1820 ± 10 years
Korath Range	Ethiopia	Tuff cones	5.10° N	35.88° E	912 m	?
Kurub	Ethiopia	Shield volcano	11.88° N	41.208° E	625 m	?
Liado Hayk	Ethiopia	Maars	9.57° N	40.28° E	878 m	?
Ma Alalta	Ethiopia	Stratovolcano	13.02° N	40.20° E	1815 m	?
Manda Gargori	Ethiopia	Fissure vents	11.75° N	41.48° E	Unknown	?
Manda Hararo	Ethiopia	Shield volcanoes	12.17° N	40.82° E	600+ m	2009
Mat Ala	Ethiopia	Shield volcano	13.10° N	41.15° E	523 m	?
O'a Caldera	Ethiopia	Caldera	7.47° N	38.58° E	2075 m	?
Sodore	Ethiopia	Pyroclastic cones	8.43° N	39.35° E	1765 m	?
Tat Ali	Ethiopia	Shield volcano	13.28° N	41.07° E	700+ m	?
Tepi	Ethiopia	Shield volcano	7.42° N	35.43° E	2728 m	?
Tosa Sucha	Ethiopia	Cinder cones	5.93° N	37.57° E	1650 m	?
Tullu Moje	Ethiopia	Pumice cone	8.158° N	39.13° E	2349 m	1900 (?)
Unnamed	Ethiopia	Pyroclastic cones	8.70° N	39.63° E	1300 m	?
Unnamed	Ethiopia	Fissure vents	8.62° N	38.95° E	1800? m	?
Unnamed	Ethiopia	Fissure vents	8.07° N	39.07° E	1800? m	?
Unnamed	Ethiopia	Cinder cones	5.65° N	37.67° E	1200 m	?
Yangudi	Ethiopia	Complex volcano	10.58° N	41.042° E	1383 m	?
Gufa	Ethiopia/Djibouti	Volcanic field	12.55° N	42.53° E	600+ m	?
Manda-Inakir	Ethiopia/Djibouti	Fissure vents	12.38° N	42.20° E	600+ m	1928
Mousa Alli	Ethiopia/Djibouti	Stratovolcano	12.47° N	42.40° E	2028 m	?
Mallahle	Ethiopia/Eritrea	Stratovolcano	13.27° N	41.65° E	1875 m	?
Sork Ale	Ethiopia/Eritrea	Stratovolcano	13.18° N	41.725° E	1611 m	?
Mega Basalt Field	Ethiopia/Kenya	Pyroclastic cones	4.08° N	37.42° E	1067 m	?
Barrier, The	Kenya	Shield volcano	2.32° N	36.57° E	1032 m	1921
Central Island	Kenya	Tuff cones	3.50° N	36.042° E	550 m	?
Chyulu Hills	Kenya	Volcanic field	2.68° S	37.88° E	2188 m	1855 ± 5 years
Elmenteita Badlands	Kenya	Pyroclastic cones	0.52° S	36.27° E	2126 m	?
Emuruangogolak	Kenya	Shield volcano	1.50° N	36.33° E	1328 m	1910 ± 50 years
Homa Mountain	Kenya	Complex volcano	0.38° S	34.50° E	1751 m	?
Korosi	Kenya	Shield volcano	0.77° N	36.12° E	1446 m	?
Longonot	Kenya	Stratovolcano	0.914° S	36.446° E	2776 m	1863 ± 5 years
Marsabit	Kenya	Shield volcano	2.32° N	37.97° E	1707 m	?
Menengai	Kenya	Shield volcano	0.20° S	36.07° E	2278 m	6050 BC (?)
Namarunu	Kenya	Shield volcano	1.98° N	36.43° E	817 m	6550 BC ± 1000 years

Table 13.1. (*cont.*)

Name	Country	Type	Latitude	Longitude	Elevation above sea level (m)	Last known eruption
North Island	Kenya	Tuff cones	4.07° N	36.05° E	520 m	?
Nyambeni Hills	Kenya	Shield volcano	0.23° N	37.87° E	750+ m	?
Ol Doinyo Eburru	Kenya	Complex volcano	0.65° S	36.22° E	2856 m	?
Ol Kokwe	Kenya	Shield volcano	0.62° N	36.075° E	1130 m	?
Olkaria	Kenya	Pumice cones	0.904° S	36.292° E	2434 m	1770 ± 50 years
Paka	Kenya	Shield volcano	0.92° N	36.18° E	1697 m	7550 BC ± 1000 years
Segererua Plateau	Kenya	Pyroclastic cones	1.57° N	37.90° E	699 m	?
Silali	Kenya	Shield volcano	1.15° N	36.23° E	1528 m	5050 BC ± 1000 years
South Island	Kenya	Stratovolcano	2.63° N	36.60° E	800 m	1888
Suswa	Kenya	Shield volcano	1.175° S	36.35° E	2356 m	?
Haruj	Libya	Volcanic field	27.25° N	17.50° E	1200 m	?
Wau-en-Namus	Libya	Caldera	25.05° N	17.55° E	547 m	?
Ambre-Bobaomby	Madagascar	Volcanic field	12.60° S	49.15° E	1475 m	?
Ankaizina Field	Madagascar	Cinder cones	14.30° S	48.67° E	2878 m	?
Ankaratra Field	Madagascar	Cinder cones	19.40° S	47.20° E	2644 m	?
Itasy Volcanic Field	Madagascar	Scoria cones	19.00° S	46.77° E	1800 m	6050 BC (in or before)
Nosy-Be	Madagascar	Cinder cones	13.32° S	48.48° E	214 m	?
Tin Zaouatene Volc. Field	Mali	Volcanic field	19.83° N	2.83° E	Unknown	?
Todra Volc. Field	Niger	Cinder cones	17.68° N	8.50° E	1780 m	?
Biu Plateau	Nigeria	Volcanic field	10.75° N	12.00° E	Unknown	?
São Tomé	São Tomé and Príncipe	Shield volcano	0.20° N	6.58° E	2024 m	?
Marion Island	South Africa	Shield volcanoes	46.90° S	37.75° E	1230 m	2004
Prince Edward Island	South Africa	Shield volcano	46.63° S	37.95° E	672 m	?
Bayuda Volc. Field	Sudan	Cinder cones	18.33° N	32.75° E	670? m	850 AD ± 50 years
Jebel Marra	Sudan	Volcanic field	12.95° N	24.27° E	3042 m	2000 BC (?)
Jebel Umm Arafieb	Sudan	Shield volcano	18.17° N	33.83° E	Unknown	?
Kutum Volc. Field	Sudan	Scoria cones	14.57° N	25.85° E	Unknown	?
Meidob Volc. Field	Sudan	Scoria cones	15.32° N	26.47° E	2000? m	2950 BC ± 500 years
Igwisi Hills	Tanzania	Tuff cones	4.87° S	31.92° E	Unknown	?
Izumbwe-Mpoli	Tanzania	Pyroclastic cones	8.93° S	33.40° E	1568 m	?
Kilimanjaro	Tanzania	Stratovolcano	3.07° S	37.35° E	5895 m	?
Kyejo	Tanzania	Stratovolcano	9.23° S	33.78° E	2175 m	1800 (?)
Meru	Tanzania	Stratovolcano	3.25° S	36.75° E	4565 m	1910
Ngozi	Tanzania	Caldera	8.97° S	33.57° E	2622 m	1450 ± 40 years
Ol Doinyo Lengai	Tanzania	Stratovolcano	2.764° S	35.914° E	2962 m	2010
Rungwe	Tanzania	Stratovolcano	9.13° S	33.67° E	2956 m	1250 ± 40 years
SW Usangu Basin	Tanzania	Lava domes	8.75° S	33.80° E	2179 m	?
Unnamed	Tanzania	Pyroclastic cone	8.63° S	33.57° E	Unknown	?
Bufumbira	Uganda	Cinder cones	1.23° S	29.72° E	2440 m	?
Bunyaruguru	Uganda	Maars	0.20° S	30.08° E	1554 m	?
Fort Portal	Uganda	Tuff cones	0.70° N	30.25° E	1615 m	2120 BC ± 100 years
Katunga	Uganda	Tuff cone	0.471° S	30.191° E	1707 m	?
Katwe-Kikorongo	Uganda	Tuff cones	0.08° S	29.92° E	1067 m	?
Kyatwa	Uganda	Tuff cones	0.45° N	30.25° E	1430 m	?
Muhavura	Uganda/Rwanda	Stratovolcano	1.38° S	29.67° E	4127 m	?

Table 13.2. *The largest known Holocene eruptions in Africa (VEI of 4 and above), known or inferred (Siebert et al., 2010)*

Volcano Name	Location	Method	Date	VEI
Ngozi	Tanzania	Radiocarbon	8250 BC (?)	5?
Paka	Kenya	Ar-Ar	7550 BC ± 1000 years	C[a]
Meidob Volcanic Field	Sudan	Thermoluminescence	6050 BC ± 1000 years	P[b]
Menengai	Kenya	Radiocarbon	6050 BC (?)	6+
Meru	Tanzania	Radiocarbon	5850 BC (?)	P[b]
Meidob Volcanic Field	Sudan	Tephrochronology	4150 BC ± 1000 years	P[b]
Meidob Volcanic Field	Sudan	Tephrochronology	3050 BC (?)	P[b]
Meidob Volcanic Field	Sudan	Tephrochronology	3000 BC (?)	P[b]
Rungwe	Tanzania	Tephrochronology	2050 BC (?)	5
Jebel Marra	Sudan	Radiocarbon	2000 BC (?)	C[a]
Rungwe	Tanzania	Radiocarbon	50 BC	4
Dubbi	Eritrea	Historical	1861	6[c]

[a] Eruptions for which data are lacking to assign a specific VEI; eruptions associated with caldera collapse are normally large (probably VEI >4).

[b] Plinian eruptions for which there are no quantitive data.

[c] Based on magnitude, given by ($\log_{10}$[mass of deposit in kg] – 7) as opposed to VEI. Despite an intense explosive phase, this was predominantly an effusive eruption in terms of mass of erupted materials.

This chapter thus aims first and foremost to describe the most important areas of active volcanism in Africa, their most active or dangerous individual volcanoes, and the hazard types and societal implications related to them. Quite apart from the hazards associated with it, African volcanism is seriously under-represented in the academic literature. Much of the current understanding of volcanic dynamics, processes, and products is based on studies of a fraction of the terrestrial volcanoes around the globe, most of them in convergent tectonic settings. The volcanoes of Africa, situated in essentially rift tectonic settings, thus have much to contribute to an improved comprehension of fundamental volcanic processes at the global scale.

13.2 Zones of active volcanism in Africa

The African continent has three major zones of active volcanism, the EARS, the Cameroon Volcanic Line in West Africa, and the Indian Ocean Islands to the south and east of Africa (Fig. 13.1). There is no Holocene record for some other areas, including the Tibesti (e.g., Permenter and Oppenheimer, 2007), but this may simply reflect the paucity of studies that have been carried out on them.

13.2.1 The East African rift system (EARS)

The EARS is one of the major geological features in Africa (Fig. 13.2). It extends over 3000 km from the Afar Triple Junction in the Afar depression in the northeast of Africa southward to Mozambique, thereby crossing many countries including Djibouti, Eritrea, Ethiopia, Kenya, Uganda, Burundi, Rwanda, DR Congo, Tanzania, Zambia, Malawi, and Zimbabwe. The EARS overlies one of the most extensive seismic velocity anomalies in the Earth's mantle, extending from the core–mantle boundary beneath the South Atlantic into the upper mantle beneath East Africa (Grand *et al.*, 1997; Van der Hilst and Karason, 1999; Ritsema and van Heijst, 2000; Yirgu *et al.*, 2006). At the surface, the EARS is manifested as a series of several thousand-kilometre-long aligned successions of adjacent individual tectonic basins (rift valleys), separated from each other by relative shoals and generally bordered by uplifted shoulders. Each basin is controlled by faults and forms a subsiding graben or trough, nearly 100 kilometres long, a few tens of kilometres wide, and either empty or filled with sediments and/or volcanic rocks (Chorowicz, 2005).

The rift valleys form two main lines: the eastern and western branches of the EARS. A third, southeastern branch, running offshore of the coast of Mozambique along the Kerimba and Lacerda rifts, was suggested by Mougenot *et al.* (1986). The earliest volcanism in the EARS occurred in southwestern Ethiopia, southeastern Sudan, and northern Kenya approximately 40–45 My ago (Davidson and Rex, 1980; Ebinger *et al.*, 1993; George *et al.*, 1998). Between 31 and 22 My ago, volcanism was widespread throughout central Ethiopia, Eritrea, and Yemen, where flood basalts and associated silicic pyroclastic rocks were erupted (Zumbo *et al.*, 1995; Baker *et al.*, 1996; Hofmann *et al.*, 1997; Kieffer *et al.*, 2004). The

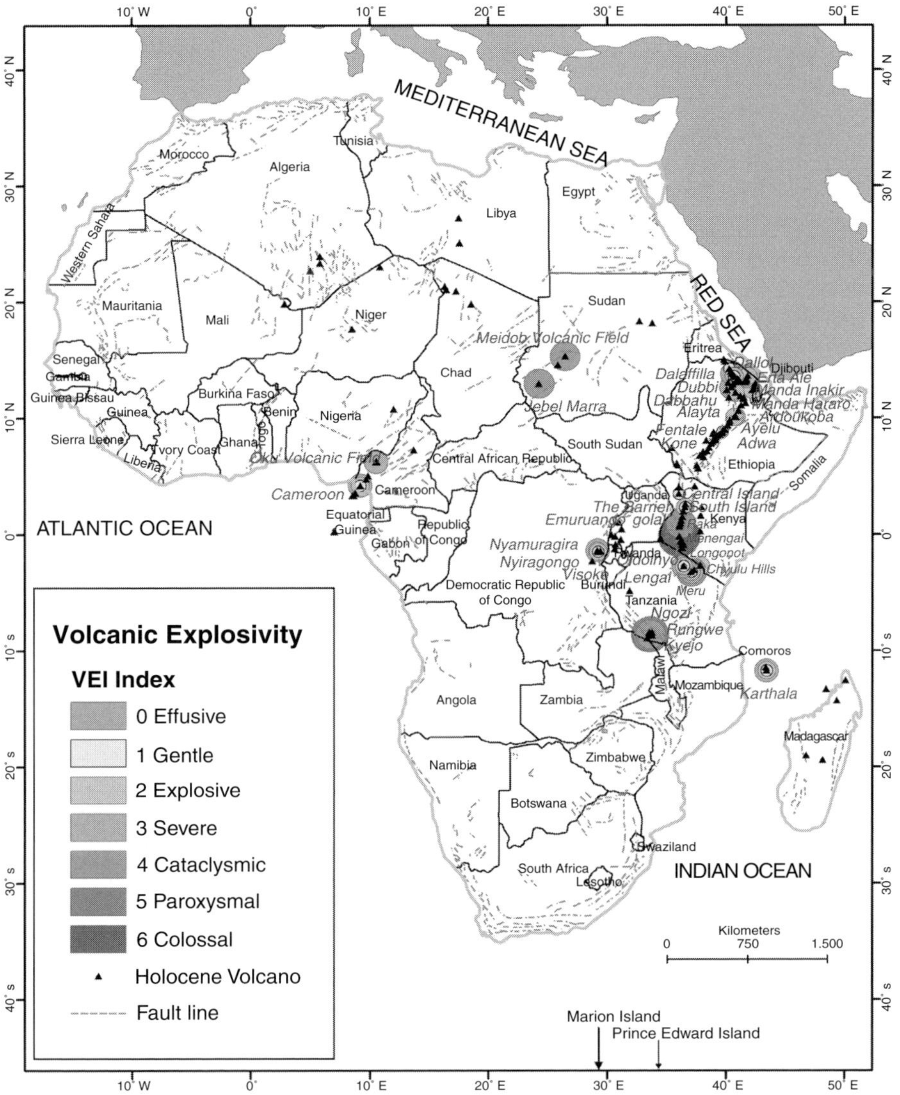

Figure 13.1. Density of volcanic eruptions based on the explosivity index for each eruption and the time period of the eruption. Modified from OCHA (2007) and Siebert *et al.* (2010). For colour version, see Plates section.

thickest sequences exposed along the conjugate margins of the southern Red Sea were erupted between 31 and 29 My ago, prior to or concurrent with the onset of rifting (e.g., Hofmann *et al.*, 1997; Ukstins *et al.*, 2002; Wolfenden *et al.*, 2005). Recent volcanism is mainly confined to magmatic segments within the Ethiopian rift, and incipient spreading zones in Afar (e.g., Vidal *et al.*, 1991; Schilling *et al.*, 1992; Boccaletti *et al.*, 1998; Ferguson *et al.*, 2010; see also Chapter 14). Thus, the Afar volcanic province has experienced a long, complex history of basaltic and explosive silicic volcanism for the past 45 Ma, with the volumetrically largest eruptions tied to the onset of rifting of Africa and Arabia at ca. 30 My ago (Yirgu *et al.*, 2006), and the eruption of the Afar Stratoid Basalts over the last ca. 4 My (Lahitte *et al.*, 2003).

13.2.1.1 *Eastern branch*

The eastern branch of the EARS runs over a distance of 2200 km, from the Afar triangle in the north, through the main Ethiopian rift, the Omo-Turkana lows, the Kenyan (Gregory) rifts, and ends in the basins of the North-Tanzanian divergence in the south (Chorowicz, 2005). The Afar triangle lies across the triple junction between the African, Arabian, and Somalian plates, and is characterised by Mio-Pliocene (24–5.4 Ma) red beds (Kazmin, 1976) and Quaternary tholeiitic volcanic rocks (Barberi *et al.*, 1972). Only the southern half of the Afar fully belongs to the EARS. It is an asymmetric rift comprising a series of east–southeast dipping tilted blocks forming plateaux, bounded by normal fault scarps, which are parallel to the rift axis. The northern corner is formed by the

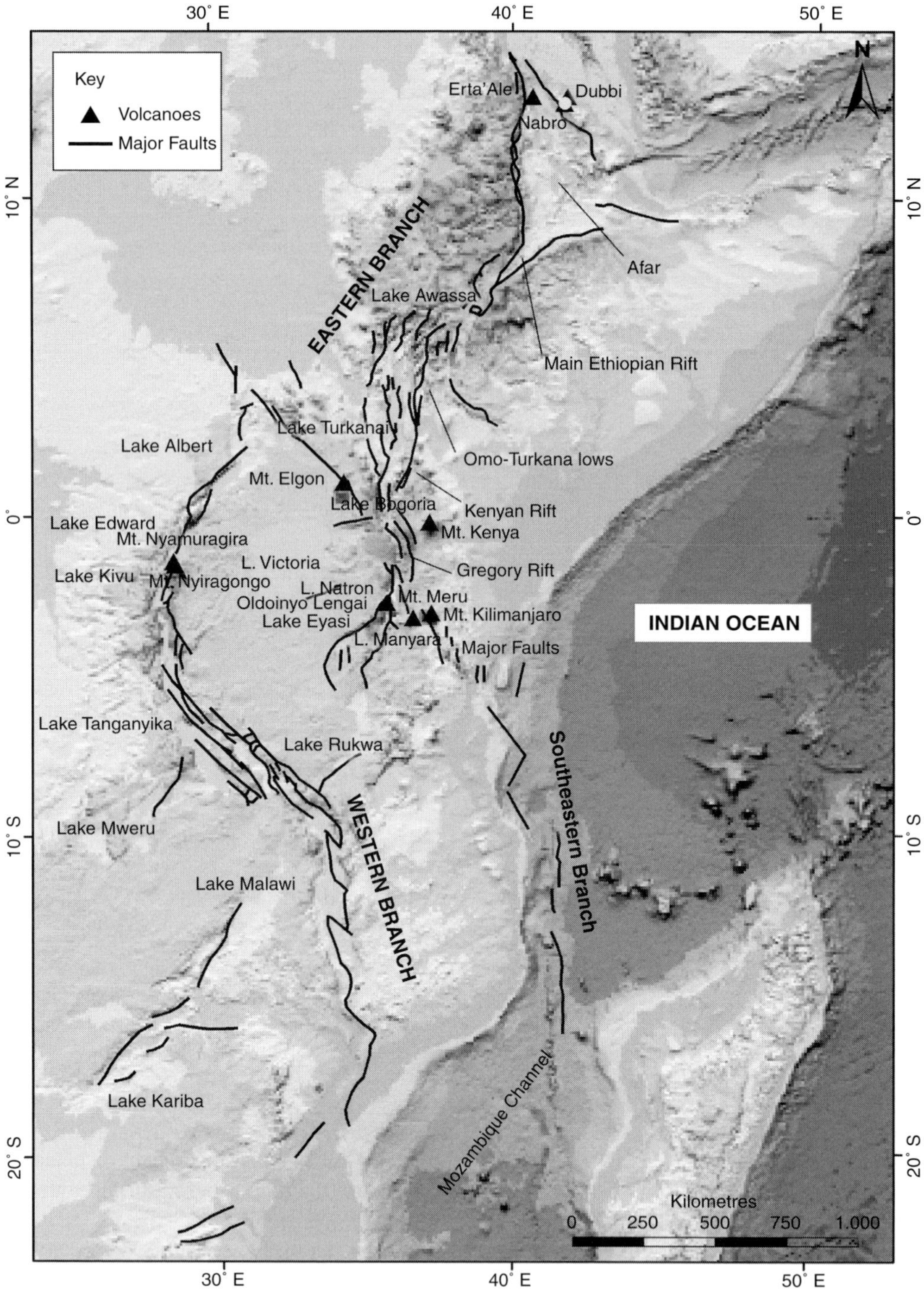

Figure 13.2. Digital elevation model of the EARS showing its eastern and western branches, major faults, and volcanos (modified after Chorowicz (2005)).

Erta 'Ale volcanic range, considered to represent an active proto-oceanic axial ridge constituting the northwestern end of the transitional Afar lithosphere (Tazieff, 1973; Barberi and Varet, 1975). The eastern corner of the triangle is an oceanic rift vanishing west of the Gulf of Tadjura, forming the western end of the oceanic ridge of the Gulf of Aden (Chorowicz, 2005).

Connected to the southern corner of the Afar is the main Ethiopian rift with a northeasterly strike direction, a length of about 330 km, and a rift floor about 50 km in width. The rift is segmented into graben, each one asymmetric, the graben fill dipping east, and the major fault lying along the eastern side. The graben are filled with lava flows, evaporites, and minor detrital sediments.

The elevations of the graben shoulders range from 2500 to 3000 m on the eastern crest line and from 1800 to 3500 m on the western side (Chorowicz, 2005).

The main Ethiopian rift contains at least 18 major caldera structures of probable Quaternary age (Newhall and Dzurisin, 1988; Woldegabriel *et al.*, 1990, 1992; Rampey *et al.*, 2010) and is considered to be the source of several Pliocene tephra layers that have been correlated between the Turkana and Afar basins, and DSDP cores in the Gulf of Aden and the Indian Ocean (Brown, 1982, 1983; Sarna-Wojcicki *et al.*, 1985; Hart *et al.*, 1992; Woldegabriel *et al.*, 1992). The main Ethiopian rift terminates near Lake Awassa.

The Omo-Turkana lows to the south of the main Ethiopian rift have no pronounced rift or shoulders, but several north- to northeast-striking half-graben (e.g., Omo valley, Lakes Turkana and Chew Bahir). Each half-graben is filled with late Cenozoic sediments, some up to 7 km thick. The first basins were initiated west of Lake Turkana during the late Oligocene–early Miocene, then faulting gradually shifted eastward with time (Morley *et al.*, 1992a,b; Morley, 1999).

Three major central volcanoes (Kenya 5200 m, Elgon 4321 m, Kilimanjaro 5964 m) are situated in the Kenyan (Gregory) rift region south of the Omo-Turkana lows but only Kilimanjaro shows some evidence for Holocene activity.

The northern Kenyan rift is composed of two parallel valleys at 1050 m elevation, both trending N10°E, separated by the Kamasia block at 2500 m elevation (Chorowicz, 2005). The eastern one (northern Kenyan rift) extends from the southern part of the Lake Turkana half-graben, and continues to the south, to reach the area of Lake Bogoria. The western Kerio graben ends abruptly west of Lake Bogoria. Both troughs are west-dipping roll-over half-graben, with major border faults running along the western shoulders. These major faults are sometimes accompanied by hydrothermal activity as is the case for Lake Bogoria (Renaut and Owen, 1988). The central Kenyan rift exhibits a sharp turn in rift direction at the southern end of the northern Kenyan rift valley, at the latitude of Lake Bogoria. Several authors have explained this bend as the intersection with a large northwest-striking basement structure called the Aswa lineament (e.g., Chorowicz and Vidal, 1987; Smith and Mosley, 1993).

The central and southern volcanoes of the Kenya rift have been particularly productive during the Quaternary, with many of the trachytic volcanoes undergoing cycles of collapse associated with both the generation of voluminous ignimbrite and fall deposits and debris avalanche deposits (e.g., Scott, 1980; Leat, 1984, 1991; Williams *et al.*, 1984; Macdonald, 1987). The plateaux, in particular to the west of the rift valley, are mantled with significant thicknesses of tephra fall deposits, yet, principally due to the large number of deposits and their closely similar chemical compositions (e.g., Leat, 1984), few individual deposits have been correlated, and no isopach maps have been published (Pyle, 1999). The time evolution of the Kenya rift comprises several periods (Hackman *et al.*, 1990; Smith and Mosley, 1993): (a) 23–11 Ma, first volcanism and formation of a low area in the middle of the future dome; (b) 11–5.3 Ma, individualisation of graben basins and uplift of shoulders; (c) 5.3–1.6 Ma, formation of the Kerio valley; (d) 1.6–0.01 Ma, concentration of activity in the eastern part of the northern Kenyan rift and the central Kenyan rift (Chorowicz, 2005).

Near Lake Natron, the southern Kenyan rift splays out southward to form the North-Tanzanian divergence, with three main branches: Lake Eyasi, Lake Manyara, and elongated plains south of Kilimanjaro (Chorowicz, 2005). In this sector of the rift, carbonatitic tephra from the Kerimasi and Oldoinyo Lengai volcanoes are considered to have been the sources for much of the surface limestone on the Crater Highlands and Serengeti plains, which extend up to 100 km west of the Gregory rift (e.g., Dawson, 1964; Hay, 1983, 1989). Oldoinyo Lengai is also one of the most frequently active volcanoes in Africa, erupting both explosively and effusively, with both carbonatite and silicate lavas (e.g., Dawson *et al.*, 1994; Oppenheimer, 1998a). Mount Meru, located east of the Tanzanian rift valley and south of Kilimanjaro, is one of the largest volcanoes in Africa. Meru has experienced major volcanic activity during the past 300 000 years (Wilkinson *et al.*, 1986), and a major episode of collapse at 7800 years BP (Hecky, 1971; cited in Wilkinson *et al.*, 1986), associated with extensive lahars to the east of the volcano, and a tephra blanket to the west (Pyle, 1999).

13.2.1.2 *Western branch*

The western branch of the EARS runs over a distance of 2100 km from Lake Albert in the north, to Lake Malawi (Nyasa) in the south. It comprises several segments: the northern segment includes Lake Albert, Lake Edward, and Lake Kivu basins, turning progressively in trend from north–northeast to north–south; the central segment trends northwest–southeast and includes the basins of Lakes Tanganyika and Rukwa; the southern segment mainly corresponds to Lake Malawi (Nyasa) and small basins more to the south. The southeastern branch comprises north-striking undersea basins located west of the Davie ridge (Chorowicz, 2005).

The northern rift segment of the western branch is an arched succession of half-graben, most of them hosting a lake. It is associated in particular with Nyamuragira (45 eruption episodes since 1550 (Siebert *et al.*, 2010)) and Nyiragongo (20 eruption episodes since 1884 (Siebert *et al.*, 2010)) volcanoes, well known for their persistent silica-undersaturated alkali-mafic magmas (e.g., Tedesco *et al.*, 2002; Chakrabarti *et al.*, 2009; Head *et al.*, 2011).

The Tanganyika rift of the central segment is more than 700 km long and up to 70 km wide, and almost entirely covered by the 650 km long Lake Tanganyika. The Rukwa rift basin is located in the relay zone between the Tanganyika and Malawi (Nyasa) rift valleys, which together form the northwest-trending Tanganyika–Rukwa–Malawi lineament (Delvaux, 2001). The Rukwa rift basin is bordered by strike-slip faults that are connected to those of Lake Tanganyika (Chorowicz *et al.*, 1983; Peirce and Lipkov, 1988; Tiercelin *et al.*, 1988; Wescott *et al.*, 1991; Morley *et al.*, 1992a,b; Kilembe and Rosendahl, 1992). The main basin is more than 200 km long and 60 km wide (Chorowicz, 2005).

The Malawi rift of the southern segment, 650 km long and 60 km wide, is mostly hidden by Lake Malawi (Nyasa). The rift is composed of four half-graben basins, separated by northwest-striking faults (Ebinger *et al.*, 1987; Flannery and Rosendahl, 1990; Chorowicz and Sorlien, 1992; Rosendahl *et al.*, 1992). The only Quaternary explosive volcanism, including Plinian-style eruptions, known from the western branch of the EARS occurred at the northern end of Lake Malawi (Nyasa), in the Rungwe Volcanic Province (e.g. Harkin, 1960; Bellon and Pouclet, 1980; Ebinger *et al.*, 1989; Kampunzu *et al.*, 1998; Fontijn *et al.*, 2010, 2011, 2012).

13.2.2 The Cameroon Volcanic Line (Western Africa)

The 1600-km-long Cameroon Volcanic Line (CVL) has been active for at least 64 Ma (Lee *et al.*, 1994a), and is a prominent, intraplate feature in central Africa that is essentially the result of passive rifting, i.e., mantle upwelling coupled with lithospheric extension (e.g., Meyers *et al.*, 1998; Montigny *et al.*, 2004; Pik *et al.*, 2006; Nkouathio *et al.*, 2008). The CVL shows a southwest–northeast alignment and is structurally subdivided by a network of faults associated with a system of alternating horsts and graben (Gèze, 1943; Déruelle *et al.*, 1991). It is exposed in both oceanic and continental settings extending from the islands of Annobón (also known as Pagalu Island, Equatorial Guinea), São Tomé, Príncipe (both Democratic Republic of São Tomé and Príncipe), and Bioko (Equatorial Guinea) to Lake Chad on the African continent (Fig. 13.3). The continental domain is characterised by (i) large stratovolcanoes and central type volcanic massifs, occasionally with calderas (Mounts Cameroon, Manengouba, Bambouto, Bamenda, Oku, Pico Basilé), (ii) plains which are filled with sediments and occupied by monogenetic volcanoes (Tombel, Noun, Kumba, Ngaoundéré) as well as flood basalts, (iii) volcanic necks and plugs (Kapsiki, Mandara) (Nkouathio *et al.*, 2008). Altogether, there are 12 major volcanic centres extending into the oceanic sector, ranging in age from 35 Ma to the present (e.g., Fitton *et al.*, 1983; Fitton, 1987). $^{40}Ar/^{39}Ar$ and K/Ar ages on basaltic (Fitton and Dunlop, 1985; Lee *et al.*, 1994b; Marzoli *et al.*, 2000) and silicic (Marzoli *et al.*, 1999) volcanics suggest a northeast–southwest younging trend from the continental to the oceanic sector; sporadic volcanism has continued into recent times in both sectors (Piper and Richardson, 1972; Grunau *et al.*, 1975; Dunlop and Fitton, 1979; Cornen and Maury, 1980; Fitton and Dunlop, 1985). The migration of volcanism suggests that the African plate is moving above a deep-seated mantle thermal anomaly. However, the age (11–7 Ma) and location of the Ngaoundéré volcanism does not conform to the northeast–southwest younging trend, implying that the continental sector of the CVL cannot be easily interpreted as the surface expression of a single hotspot system (Marzoli *et al.*, 1999, 2000).

The only currently active volcano is Mount Cameroon, which erupted last in 2000 (Suh *et al.*, 2003, 2008; Ateba *et al.*, 2009), situated on the boundary between the oceanic and the continental sectors of the volcanic chain. Pico Basilé may also have erupted in the historical period (1898?, 1903, 1923), though reports are ambiguous. Mafic lavas, ranging in composition from basanite to hypersthene-normative basalt, are present in all the volcanic centres except for Etinde (a large satellite peak of Mount Cameroon also known as Little Cameroon), at the southwestern flank of Mount Cameroon, which is composed entirely of nephelinite (Nkoumbou *et al.*, 1995). The evolved rocks in the continental volcanic centres comprise trachyte, trachyandesite, rhyolite, and rare phonolite (Déruelle *et al.*, 1991; Nono *et al.*, 1994; Marzoli *et al.*, 2000; Ngounouno *et al.*, 2000; Kagou Dongmo *et al.*, 2001; Nkouathio *et al.*, 2002).

Most of the CVL volcanoes tend to produce lavas, rather than large volume pyroclastic deposits (Pyle, 1999), which is relevant in considering the hazards posed to the surrounding populations. However, the catastrophic carbon dioxide emissions from Lake Monoun (so far undated but supposedly Holocene) in 1984 (Sigurdsson *et al.*, 1987) and Lake Nyos (3.7 ± 0.5 ka based on $^{230}Th/^{226}Ra$ age data of pyroclastic samples of the maar lake but supposedly older than 5 ka using the mean measured $^{230}Th/^{238}U$ activity ratio of the same samples; Aka *et al.*, 2008) in 1986 (Freeth and Kay, 1987; Kling *et al.*, 1987) of the Oku Volcanic Field showed that more attention has to be paid to the young volcanic fields of the CVL (Nkouathio *et al.*, 2002) and their secondary, geothermal activity such as carbon dioxide accumulation in deep lake waters long after a monogenetic eruption.

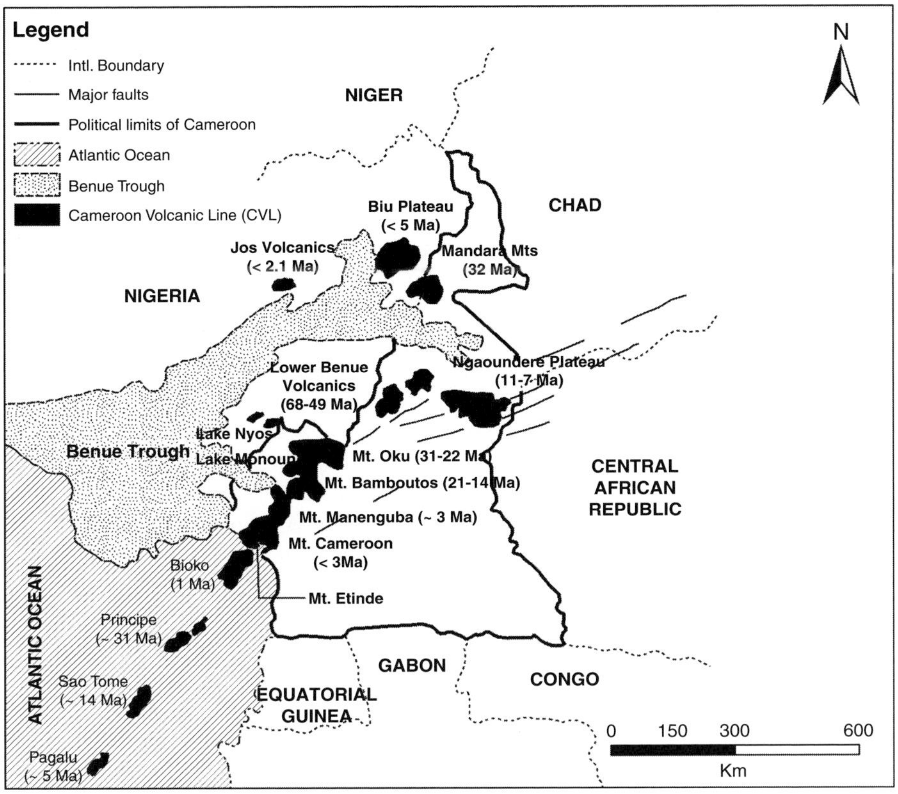

Figure 13.3. Geological map showing the CVL (modified from Suh *et al.*, 2011).

13.2.3 The Indian Ocean islands

The Indian Ocean islands discussed in this chapter encompass the Comoros Archipelago, Madagascar, and Marion and Prince Edward Islands, all of which are characterised by Holocene or ongoing volcanic activity.

The Comoros Archipelago, midway between Madagascar and northern Mozambique, consists of four islands: Grand Comoro, Moheli, Anjouan, and Mayotte, the last of which is a French dependent territory (Fig. 13.4). The islands form the summits of a northwest–southeast trending volcanic ridge that crosses the Mozambique Channel northwest of Madagascar (Frey, 2009). Grand Comoro, the westernmost and largest of the islands is formed by two shield volcanoes, La Grille and Karthala, the two most recently active centres of the Comoros Archipelago. Karthala is an active shield volcano erupting predominantly olivine-bearing alkali basalts and trachybasalts, whereas La Grille is composed mainly of a series of monogenetic cones of basanitic to alkali basaltic composition that erupted along fissures paralleling the north–south trending summit ridge and from radial fissures that reach as far as the coast (Strong, 1972; Cheminée, 1973; Bachèlery and Coudray, 1993; Späth *et al.*, 1996; Class and Goldstein, 1997). La Grille volcano is located in the northern part of the Grand Comoro Island and has not been active in the past 120 years. However, radiocarbon dating on charcoals found below ash-fall deposits (lapilli tuff) indicates eruptions within the past two millennia (740 $\pm$ 130 BP, 1300$\pm$65 BP) (Bachèlery and Coudray, 1993). The still active Karthala volcano (last eruption with VEI 2 in 2007 from the central vent and lava lake formation) forms the southern two-thirds of Grand Comoro Island. The structure of Karthala volcano is dominated by two large northwest–southeast oriented rift zones similar to those found on the Hawaiian shield volcanoes (Savin *et al.*, 2005), forming radial features at opposite sides of the large currently active summit caldera (which has a volume of 4 km^3) (De Saint-Ours, 1960; Esson *et al.*, 1970).

Several hypotheses for Comoros Island magmagenesis exist, ranging from reactivation of old lithospheric fractures at different periods (Nougier *et al.*, 1986) to hotspot volcanism (Duncan, 1981; Emerick and Duncan, 1982; Class and Goldstein, 1997; Class *et al.*, 1998, 2005; Claude-Ivanaj *et al.*, 1998).

Madagascar, the largest island in the Indian Ocean (Fig. 13.4(a)), is separated from Africa by the 250–400-km-wide Mozambique Channel. The early volcanism on Madagascar started 88 Ma ago during the separation from India and is inferred to be related to the Marion mantle plume (Storey *et al.*, 1995), which is currently creating Marion and Prince Edward Islands (Georgen and Lin, 2003; Le Roex *et al.*, 2012). The younger Pliocene–Holocene

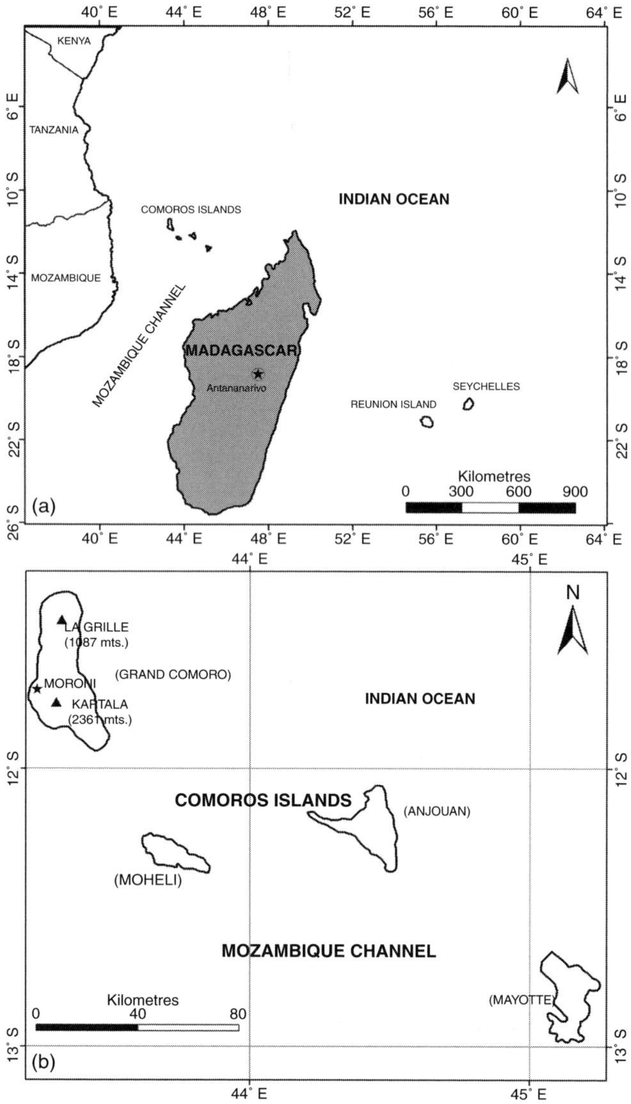

Figure 13.4. (a) Map showing the location of Madagascar and the Comoros Islands east of Mozambique in the Indian Ocean. (b) Map showing the four islands of the Comoros (Grand Comoro, Moheli, and Anjouan, and the island of Mayotte, belonging to France).

volcanic products (Besairie, 1964, 1972; Fourno and Roussel, 1994) of alkaline composition ranging from basanite to trachyte (Besairie, 1973; Melluso and Morra, 2000) are typically related to the Seychelles–Comores hotspot and exhibit a diverse assemblage of volcanic landforms, including scoria cones, lava domes, and maars (Karche, 1973; Nougier *et al.*, 1986; Rasamimanana, 1996; Melluso and Morra, 2000; Melluso *et al.*, 2007).

Marion and Prince Edward Islands are South African territories on the Antarctic plate, 160–200 km south of the actively spreading Southwest Indian ridge (Fig. 13.5). The pronounced north–south trending submarine Madagascar plateau, extending from the southern tip of Madagascar Island to the southwest Indian ridge, is inferred to be a hotspot track formed as the Indian plate migrated northward relative to the Marion hotspot (Frey, 2009).

Marion and Prince Edward Islands are formed of basaltic shield volcanoes with lava flows and fields of cinder cones indicating contemporary volcanism. Based on K/Ar data of whole rock basaltic lava flow samples, eight periods of effusive activity can be distinguished on Marion Island at approximately 450, 350, 240, 170, 110, 85, 50, and <10 ka (McDougall *et al.*, 2001). So far, no age data are available for Prince Edward Island.

13.3 Volcanic disasters in Africa

To illustrate the impacts of volcanic hazards on the African continent, several case studies are summarised in the following section. In each case it is clear that a particular set of factors influenced the impacts of the events on communities: tectonic setting, eruption style, preparedness and response, vulnerability, etc. The case studies are: (i) the Dubbi eruption of 1861 and the 2011 eruption of neighbouring Nabro in Eritrea, (ii) the Nyiragongo eruption of 2002, (iii) the eruptions of Mount Cameroon in 1999 and 2000 and subsequent lahar formation, and (iv) the Karthala eruption of 2005.

13.3.1 The 1861 Dubbi eruption and 2011 Nabro eruption, Eritrea

Dubbi volcano belongs to a large volcanic massif referred to as the Nabro Volcanic Range (Wiart and Oppenheimer, 2000) that trends south-southwest–north-northeast from the Red Sea coast of Eritrea and into Ethiopia. It reaches 1625 m above sea level and is located about 15 km from the southern margin of the crystalline basement rocks of the Danakil Alps. Some 20 small cinder cones are found near the summit, and basaltic lava fields to the north and northeast, known as the Edd lava fields, cover an area of 2700 km^2 reaching up to the Red Sea coast. The two most recent eruptive centres are fissure systems that extend northwest–southeast and north-northeast–south-southwest. The former produced lava flows that reached the Red Sea in 1400 AD. The second formed the small cones at the summit in 1861 (Siebert *et al.*, 2010). Dubbi's eruption in 1861 was the largest historical eruption recorded in Africa (Bell, 1861; Wiart and Oppenheimer, 2000). Captain R. L. Playfair, a British Resident at Aden (Yemen), described the eruption as follows (Wiart *et al.*, 2000):

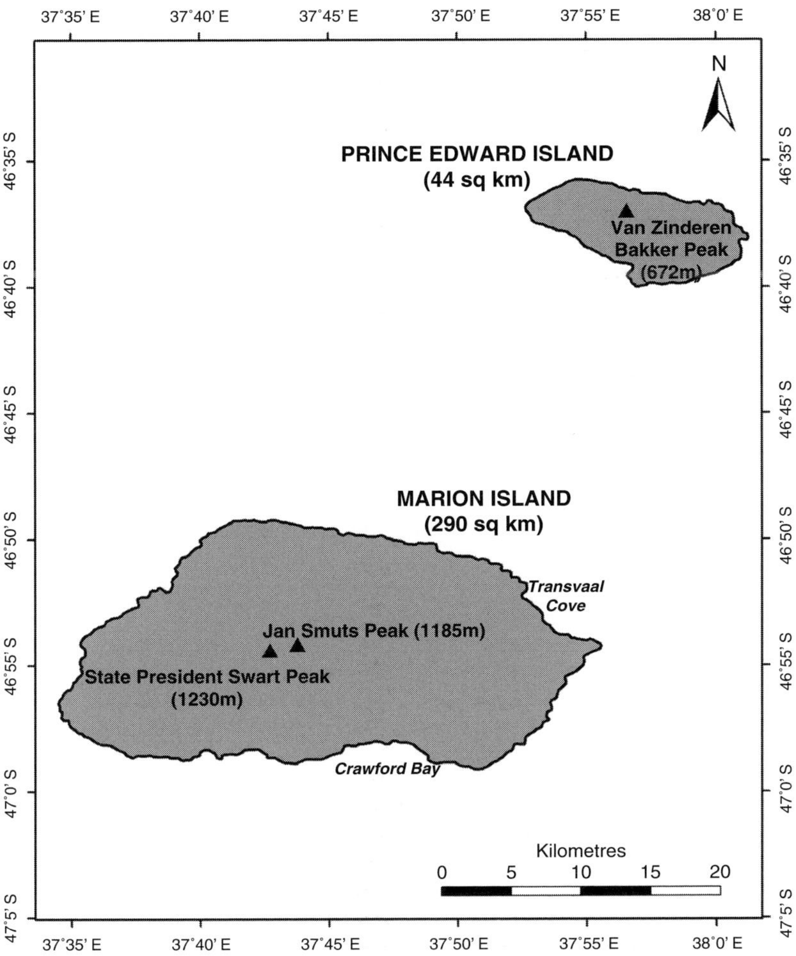

Figure. 13.5. Map showing the location of Marion and Prince Edward Islands in the southern Indian Ocean, ca. 160–200 km south of the actively spreading Southwest Indian ridge. Marion Island covers an area of 290 km^2, the neighbouring Prince Edward Island only 44 km^2.

> On the night of the 7th or the morning of the 8th of May, the people of Edd were awakened by the shock of an earthquake followed by others which continued with little intermission for about an hour. At sunrise, a quantity of fine white dust fell over the village . . . About noon, the character of this dust [. . .] resembled red earth. Shortly afterwards, it increased to such an extent that the air was perfectly darkened and we had to light lamps in our houses. It was darker than the darkest night, and the whole place was covered with dust, nearly knee-deep. At night [on May 9], we saw fire and dense smoke issuing from a mountain called Djebel Dubbeh.

The explosive eruption, unusual for the Afar region, showered maritime traffic in the Red Sea with pumice and plunged coastal settlements into darkness. Earthquakes associated with the opening phase of the eruption were felt 180 km distant in Hoddeida (Yemen), and explosions were heard as far as Massawa (Eritrea) at 330 km distance (Von Heuglin, 1862). Ash fell more than 300 km downwind from the volcano. Two villages were destroyed and 175 local inhabitants were reported killed, possibly as a result of pyroclastic flow emplacement (Beke, 1861; Perrey, 1864; Wiart and Oppenheimer, 2000). Herds of cattle were also killed. By October 1861, activity switched to basaltic fire-fountaining focused along a 4-km-long summit fissure that fed several lava flows that travelled as far as 22 km and reached the coast. Much of the tephra deposit from the explosive phase has been deflated through eolian erosion and no reliable figures have been reported for the tephra volume. However, the volume of the lava flows alone, 3.5 km^3, makes this the largest reported historical eruption in Africa (equivalent to a magnitude of ~6.0 (Wiart and Oppenheimer, 2000); for comparison, the 1783–4 eruption of Laki in Iceland has an estimated magnitude of ~6.6). An anomalously cold Northern Hemisphere summer in 1862, apparent in tree-ring records, could be the result of Dubbi's sulphate aerosol veil. Two further eruptions inferred to have taken place in 1863 and 1900 are uncertain and hard to verify.

While Dubbi shows limited signs of activity today (there are low temperature vents around the summit), the eruption of Nabro in 2011 provided a reminder that Eritrea has active volcanoes, and that they pose a direct threat to the surrounding communities and traffic routes. This was the first eruption of Nabro on record. Outside of the country, the first news circulating of the eruption suggested that Dubbi was the source. But as higher spatial resolution satellite data became available, it became clear that Nabro, 25 km to the southwest, was responsible.

Reaching 2218 m above sea level, Nabro is the highest volcano in the Afar. It has a summit caldera about 7.5 km across, with elevations around 1400–1500 m above sea level, and rises above a dissected plain of rhyolitic ignimbrite, which is presumably associated with the caldera-forming eruption(s) of both Nabro and the nearby Mallahle volcano (just across the border in Ethiopia). (Constraints on the ages of major episodes of volcanism of Nabro are anticipated at the time of writing, with a suite of samples under analysis by the argon–argon method.) Post-caldera rhyolitic obsidian domes and basaltic lava flows were erupted inside the caldera and along the volcano's flanks. Some of these lava flows, very recent in appearance, were erupted from north-northwest-trending fissures transverse to the trend of the Nabro (Siebert *et al.*, 2010).

The 2011 eruption began around 20:32 UTC (i.e., 23:32 local time) on 12 June following intense seismicity, including three $M_L > 5$ events. In hindsight, it appears that seismic unrest began as early as mid-February 2011, with one event on 31 March damaging buildings in and around the caldera. These signals were not read as precursors to an eruption (there was no seismic network operating in Eritrea at the time). Nevertheless a rapid response initiated by the local government led to timely evacuation of settlements on and around Nabro. Given that the eruption began around midnight local time, when most people might otherwise have been asleep, this likely saved many lives. Reports suggest that seven people were killed during the eruption on the Eritrean side of the border, and possibly more on the Ethiopian side (see Chapter 14). Several thousand people were displaced and cared for in temporary camps in the region; some have been resettled near the town of Asseb, more than 100 km away. One remarkable aspect of this is that while the region is sparsely populated, the caldera itself (and that of Mallahle, which received substantial tephra fall) was quite extensively settled owing to its high elevation and higher rainfall compared with the surrounding desert lowlands. The population of Sireru, a settlement at the southern end of Nabro caldera, was approximately 2700 at the time of the eruption.

The aviation hazard posed by Nabro's ash clouds was monitored by the Toulouse Volcanic Ash Advisory Centre (VAAC), and flight restrictions were imposed on regional aviation. Based on satellite observations, and reports from the ground, a second pulse of activity occurred on 17 June 2012. Meanwhile, several lava flows (trachybasalt to basaltic trachyandesite in composition) were extending from the vent area located in the centre of the Nabro caldera (Fig. 13.6). The volumetrically largest of these travelled more than 15 km, at first in a southwesterly direction, where it spilled over a low point on the western rim of the caldera from where it extended to the northwest, burying some previously settled areas.

Spaceborne and ground-based observations have indicated that the Nabro eruption produced the largest stratospheric aerosol perturbation since the 1991 Pinatubo eruption (Bourassa *et al.*, 2012; Sawamura *et al.*, 2012), reflecting a substantial sulphur yield to the atmosphere. (There is some controversy over whether the plume was initially emplaced into the upper troposphere or lower stratosphere (see Fromm *et al.*, 2013; Vernier *et al.*, 2013).) The petrological similarity between the 1861 lavas erupted at Dubbi and those erupted in 2011 from Nabro lends credence to the hypothesis that Dubbi's atmospheric sulphate veil was substantial and led to climate forcing.

13.3.1.1 *Volcanic hazards in the Afar*

Notwithstanding the sparse population of the Afar region, the volcanic hazards are significant (see also Chapter 14). In the last decade alone, there have been eruptions in the Manda Hararo rift (including the silicic eruption of Dabbahu in 2005 and subsequent fissure eruptions) and the Erta 'Ale range (the 2008 Dalla Filla eruption and overflows of the Erta 'Ale lava lake in 2010), and at Nabro (e.g., Ferguson *et al.*, 2010; Field *et al.*, 2012a,b; see also Chapter 14). In addition to these and other historically active volcanoes in Ethiopia and Eritrea, there are also numerous volcanoes in Djibouti, of which Ardoukôba was the last to erupt (in 1978). Volcanic hazards include inundation by lava flows, fumigation by grounding gas and aerosol plumes, pyroclastic density currents, and ash fall. These various effects have resulted in direct fatalities and injuries, damage to buildings and agricultural land, long-term displacement or resettlement of communities, and probable contamination of water supplies. In addition, volcanism in the region, along with the seismic risk, is an important consideration in the development of new infrastructure, mining activities, and geothermal exploration and exploitation. Longer-range hazards include the threat to aviation from explosive eruptions, and potentially to shipping in the Red Sea in the event of Plinian-style eruptions.

13.3.1.2 *Monitoring of the Afar volcanoes*

Satellite remote sensing has played a significant role in monitoring recent volcanic and tectonic events in the Afar region. While much of this work has involved scientists from the region, it has generally been carried out as a component of international research projects as opposed to being part of an operational monitoring programme. In

(a)

(b)

Figure 13.6. (a) Photograph of a part of the main lava flow erupted in June 2011 from Nabro, Eritrea. The location is outside the caldera and towards the distal end of the flow. Taken in October 2011, looking towards the caldera. (b) Photograph of the vent region of the 2011 Nabro eruption, Eritrea. Taken in October 2012. The ridge in the background is part of the caldera rim.

particular, a number of satellite geodetic studies have been carried out along with ground-based GPS observations. Operational satellite remote sensing products, such as the sulphur dioxide cloud detection and mapping based on the Ozone Monitoring Instrument, and thermal infrared alerts such as the MODVOLC system, have also aided the surveillance of volcanoes in the region.

In Djibouti, the Observatoire Géophysique d'Arta operates a seismic network, while in Ethiopia, seismic, magnetic, and geodetic observations are carried out by the Institute of Geophysics, Space Sciences and Astronomy, based in Addis Ababa. In Eritrea, plans are in development to establish a geophysical observatory at the Eritrea Institute of Technology, near Asmara. A significant quantity of seismic and other geophysical data has been acquired in Afar during field campaigns and using temporary networks commissioned under the auspices of assorted international collaborative projects (e.g., http://www.see.leeds.ac.uk/afar/). At the time of the 2011 eruption of Nabro in Eritrea, not one seismometer was operating in the country. Thanks to an 'urgency grant' awarded by the UK Natural Environment Research Council, basic regional and local networks were commissioned. These have acquired a substantial data set that is being studied at the time of writing.

13.3.2 The 2002 Nyiragongo eruption, DR Congo

Mount Nyiragongo (DR Congo) is notable among the volcanoes in the world for the risk associated with the propagation of its low viscosity nephelinite lava flows (Chirico *et al.*, 2009). The 3470-m-high active stratovolcano in the Virunga Mountains, 18 km north of the city of Goma (ca. 1 100 000 inhabitants) and Lake Kivu, is associated with the northern rift segment of the western branch of the EARS. Two older stratovolcanoes, Baruta and Shaheru, are partially overlapped by the Nyiragongo edifice on its northern and southern sides, respectively (Andersen *et al.*, 2012). About 100 parasitic cones are located primarily along radial fissures south of Shaheru, east of the summit, and along a northeast–southwest zone extending as far as Lake Kivu. Many cones are buried by voluminous lava flows that extend considerable distances down the flanks of the volcano. Nyiragongo has a ca. 1 km diameter summit crater, possibly the result of collapse above a shallow magma chamber (Platz *et al.*, 2004). The crater is the site of a persistent lava lake (Tazieff, 1984), currently the largest one on Earth (Fig. 13.7), that is known from the late nineteenth century, and that was drained catastrophically through its outer flanks in 1977. Since 1884 at least 20 eruptions have been

(a)

(b)

Figure 13.7. (a) The crater rim of Mt Nyiragongo, Democratic Republic of Congo, and (b) its active lava lake, which is currently the largest on Earth. The photos were taken in 2011. Photos courtesy of Morgan Trimble.

recorded according to the catalogue of the Global Volcanism Program of the Smithsonian Institute (Siebert *et al.*, 2010), most of them associated with effusion of unusually low-viscosity lava flows (melilite nephelinite). Isotopic and geochemical evidence indicates that these lavas were derived from a heterogeneous mantle plume (Chakrabarti *et al.*, 2009; Tedesco *et al.*, 2010).

After intermittent seismic tremors that began in March 2001, on 17 January 2002 Nyiragongo was the focus of global attention when several vents opened along a 10-km-long system of fractures, from which (14–34) $\times$ 10^6 m^3 of lava erupted (Tedesco *et al.*, 2007). The lava reached and destroyed a considerable part of the city of Goma (20–30% according to USAID (2002); ~13% according to the Global Volcanism Program (2002)), resulting in the most outstanding modern example of a lava flow inundating a large urban area. This eruption and the previous one at Nyiragongo in 1977 are also the only known cases in which people were directly killed by lava flows (Favalli *et al.*, 2009). During both eruptions the lava locally reached speeds of the order of 10–40 km h^{-1} (Tazieff, 1977; Kilburn, 2000; Komorowski *et al.*, 2002–2003), which suggests viscosities of the Nyiragongo lava were among the lowest in terrestrial magmas (with only carbonatites having lower viscosities; Dawson *et al.*, 1990; Giordano *et al.*, 2007). Fourteen villages and the city of Goma were inundated by two lava flows, causing substantial socio-economic disruption and a significant humanitarian crisis (Baxter *et al.*, 2002–2003; Komorowski *et al.*, 2002–2003; Tedesco *et al.*, 2007). One of these flows continued into Lake Kivu, forming a 100-m-wide delta where lava entered the lake.

CNN (2002) reported that lava flows up to a metre high moved through the city engulfing houses and shops. Many families fled spontaneously across the border into Gisenyi, Rwanda, and farther eastward to the Rwandan capital city, Kigali. According to the UN and USAID (2002), 147 people, many of them children from the villages on the volcano's flanks, were killed, and around 300 000 people were displaced by the volcanic eruption (IRIS, 2002). During the eruption, eye irritation and breathing difficulties were reported as a result of the ash and fumes in Goma. The eruption lasted only one day, after which 12 000–15 000 homes, water pumping stations, electrical facilities, the international airport, and the commercial centre had been destroyed (Baxter *et al.*, 2002–2003). The total number of people affected by the eruption was estimated to be 130 000 (USAID, 2002). According to OCHA and various news reports, refugees began to return to Goma just a few days after the eruption, despite the dangers that still existed in the area, such as building collapses due to around 100 tectonic events that occurred within the following five days with a magnitude of over 3.5 in the Goma area, and several small explosions due to methane combustion resulting from seepage of methane and carbon-dioxide-rich gases through tiny fissures.

During a visit to Nyiragongo's main crater on 28 January, the UN Volcano Surveillance team reported neither ongoing eruption nor any fumaroles at the bottom of the crater, although they could still smell sulphur dioxide gas. A few weak steam vents were visible on the inner crater wall and a small gas plume was seen above the crater rim drifting to the northeast. In July 2012, the Toulouse VAAC reported a series of ash plumes that reached up to an altitude of 5500 m above sea level, i.e. about 2000 m above the volcano's summit.

13.3.2.1 *Hazards at Mount Nyiragongo*

The two main volcanic hazards from Nyiragongo that could affect the local population and the surrounding environment are lava flows and the persistent halogen and sulphur-dioxide-rich gas and aerosol plume from the lava lake. Given the high fluidity of the Nyiragongo lavas (Giordano *et al.*, 2007) and the presence of a semi-persistent lava lake (Tazieff, 1995), this volcano presents a significant threat to the inhabitants of Goma. The low-viscosity lava from Nyiragongo promotes fast moving flows that are able to reach speeds of tens of kilometres per hour on the volcano's flanks. While Nyiragongo is mostly known for the presence of a semi-permanent lava lake inside its crater for nearly one century (Tazieff, 1977, and references therein), beginning in 1977, and with much more evidence in 2002, a transition from central crater to rifting activity with more frequent flank eruptions may be in process. The northern portion of a fracture system developed in 1977, at the time when the highly fluid lava flows were emitted (which terminated a few kilometres from Goma). In 1977, however, only degassed magma was erupted from the lava lake and from the upper conduit (Tazieff, 1977). In January 2002, the extent of fracturing with major syn-eruptive and post-eruptive subsidence became more prominent (Tedesco *et al.*, 2007).

Furthermore, as shown by the presence of several undated tuff cones along directions radial to the crater and extending mainly toward the south and southwest, eruptions from vents along the volcano flanks have occurred many times in the past (Tedesco *et al.*, 2007). If so, this would pose a significant escalation of volcanic hazards facing Goma and its environs, thus requiring the implementation of improved volcano-monitoring strategies to support volcanic risk management in the region (Tedesco *et al.*, 2007), bearing in mind that the neighbouring Nyamuragira volcano is also frequently in eruption.

The persistent plume emitted from the lava lake of Nyiragongo is responsible for a high flux of sulphur dioxide, hydrogen fluoride, and hydrogen chloride into the atmosphere (e.g., Sawyer *et al.*, 2008). It is associated with acidic rainfalls that affect vegetation, crops, human infrastructure, surface water quality, and human health (Cuoco *et al.*, 2013a,b). Another gas-related hazard is posed by the so-called mazuku ('evil winds' in the Kinyabwisha language), which are low-lying locations with dry carbon dioxide vents where carbon dioxide can, unnoticed, accumulate to lethal concentrations (Smets *et al.*, 2010). In addition, volcanic ash, phreatomagmatic eruptions induced by the presence of a high water table, earthquakes, high fluorine abundance in water supplies, and according to some assessments, the potential for release of carbon dioxide dissolved in the deeper waters of Lake Kivu (Lockwood and Casadevall, 1994) pose further hazards.

13.3.2.2 *Monitoring of Mount Nyiragongo*

Volcano monitoring and the related risk management in Goma is performed by three types of 'entities': (i) the Goma Volcano Observatory (OVG), a DR Congo state institute dedicated to the monitoring of the Virunga volcanoes, (ii) the scientific collaborators of the OVG, which are institutes from several countries around the world that work directly with the OVG, through different scientific projects, and (iii) the UN programs, which are dedicated to financially help the OVG (which is otherwise largely underfunded) develop risk management and create the interface between the scientific institutes, local authorities, and the humanitarian community (UN, non-governmental organisations (NGOs), etc.).

In 2004 a 'Disaster Management Unit' within the frame of the project 'Prevention and Analysis of Natural Disasters in DRC' was launched by the UN (United Nations Development Programme, UNDP). The project was funded until 2007 and was followed in the same year by a new initiative, the 'Volcano and Environmental Risk Management Unit' (UNOPS, 2013) to support OVG in the scientific observation of volcanic activity and estimate the risk levels in the area. The risk management unit is part of a project implemented by the United Nations Development Programme and executed by UNOPS (United Nations Office for Project Services) (funded by UK's Department for International Development, the Swiss Agency for Development and Cooperation, and the Ministry of Foreign Affairs of Luxembourg), and helps international experts to train local scientists and to install new monitoring equipment around the volcano. At the same time the project works in schools and public institutions (markets, churches, state enterprises, etc.) to inform and educate the local population. Local authorities have also been formed (through their Civil Defence branch) and a new 'Contingency Plan' was adopted in 2009 and revised on a quarterly basis.

Satellite remote sensing represents an important tool in monitoring the Virunga volcanoes, given the complexities of ground operations (e.g., Oppenheimer, 1998b; Wauthier *et al.*, 2012). In January 2007 the GORISK scientific project (GORISK, 2012) was initiated by scientists in Belgium and Luxembourg in order to combine the use of ground- and satellite-based remote sensing techniques as a tool for volcanic risk and health impact assessment. The project is a collaboration of the Royal Museum for Central Africa (RMCA) in Belgium, the National Museum of Natural History of Luxembourg (NMNH), the University of Luxembourg (UniLux), the Second University of Naples (UniNap), the OVG, the Volcanic Risk Mitigation Program (UNDP/UNOPS), and the Centre Scientifique et Médical de l'ULB pour ses Activités de Coopération (CEMUBAC). An important GORISK task, mostly undertaken by the RMCA and the NMNH, concerns the monitoring of the ground deformation. This is addressed by both spaceborne (satellite radar interferometry) and ground-based (tilt and

GPS) geodetic observations. Volcanic gas emissions are also a major concern as discussed above. Accordingly, measurements of water quality, Sulphur dioxide, carbon dioxide, and radon are regularly undertaken. Sulphur dioxide, carbon dioxide, and radon are continuously monitored through DOAS stations (sulphur dioxide) and a carbon dioxide–radon station installed in Goma. Water quality is monitored thanks to regular sampling campaigns (Benoît Smets, pers. comm., 2013).

The greatest problem facing those involved in assessing the hazards of the Virunga volcanoes, however, is the violence and insecurity that persists due in part to the legacy of the Rwandan genocide of 1994 and the struggle between various armed groups to control mining operations. Conflict and corruption have severely impacted efforts to monitor the volcanoes and to conduct the essential and everyday activities of the volcano observatory.

Since the end of the GORISK project in January 2010, the scientific collaboration initiated in this project has been maintained through a scientific network: the GORISK Scientific Network (more information on www.ecgs.lu/gorisk/). This scientific network aims to maintain the monitoring techniques developed during GORISK and still helps the OVG in its everyday work. While the GORISK network is focused on the scientific aspects, the EU/UNOPS programme aims to work on volcanic risk management. There have been several other projects concerned with monitoring the Virunga volcanoes, primarily utilising spaceborne sensors. However, until the situation on the ground improves and a more stable civil society resumes, the likely reality is that the threat of volcanic activity is not of paramount concern to those agencies still operating in eastern DR Congo.

13.3.3 The 1999 and 2000 Mount Cameroon eruptions, Cameroon

Mount Cameroon, on the shore of the Gulf of Guinea (Atlantic Ocean), in the Southwest Province of Cameroon, is a complex volcano with an elliptical plan view and a maximum height of 4095 m. It is the largest and most active volcano in West Africa, and a typical example of a hazardous volcano in a densely populated area, regularly inflicting damage by lava flows (Bonne *et al.*, 2008). Mount Cameroon has been described as a large horst south of the Tombel Graben (Déruelle *et al.*, 1987; Toteu *et al.*, 2004). The volcano displays a typical southwest–northeast alignment of vents, inherited from the CVL (Suh *et al.*, 2003). Approximately 400 parasitic cones are aligned along a rift zone that is narrow in the relatively flat summit area and widens out as an unfolding fan towards lower altitudes on both the southwestern and northeastern flanks (Bonne *et al.*, 2008). Inward facing topographic steps on the relatively flat summit plateau suggest that the volcano's summit has subsided (Mathieu *et al.*, 2011). A large satellite peak, Etinde, is located on the southern flank of the volcano near the coast.

Eruptions occur along southwest–northeast oriented fissures (more than 90 have been mapped (Thierry *et al.*, 2006)), implying that subvertical southwest–northeast-oriented dykes supply the magma (Suh *et al.*, 2003). The Mount Cameroon rocks are composed of (silica-undersaturated) alkali basalt and varieties of picrite, basanite, hawaiite, and mugearite, accumulating as large flows and cinder cones aligned along the gaping fissures that control the eruptive activity (Thierry *et al.*, 2008). The lavas rest on Miocene and Pliocene sedimentary successions making up the piedmonts (Dumort, 1968; Hedberg, 1968; Nni, 1984; Déruelle *et al.*, 1987; Zogning 1988; Nkoumbou *et al.*, 1995; Cruz Mermy *et al.*, 2006). The volcano has built up over the past 10 Ma through successive mafic, mostly effusive, volcanic eruptions (Fitton *et al.*, 1983; Suh *et al.*, 2003), and particularly intense eruptive activity with an increased frequency since 3 Ma ago (Marzoli *et al.*, 2000).

The edifice has mainly formed as a result of three types of volcanic activity: (i) effusive eruptions, responsible for the lava flows covering parts of the stratovolcano; (ii) relatively mild explosive activity which has built the cinder cones; (iii) hydromagmatic activity that formed the maars of Debundscha and Bomana in the sedimentary basement at the northwestern side of the volcano (Tsafack *et al.*, 2009; Ngwa *et al.*, 2010). Most eruptions involve moderate explosive and effusive activity from both summit and flank vents, mostly in the form of lava flows that sometimes reach the sea, as occurred in 1922 and 1999. Eruptions from higher-altitude vents (usually above 2000 m) are usually more explosive (violent Strombolian activity) and have mostly produced cinder cones (Njome *et al.*, 2010). Lower-altitude vents exhibit intermittent Strombolian to Hawaiian activity with the formation of hornitos (Suh *et al.*, 2003; Njome *et al.*, 2008; Ateba *et al.*, 2009). Voluminous lava flows are usually emitted from the base of spatter or scoria cones (Njome *et al.*, 2010).

Approximately 500 000 people live on the lower flanks or close to the volcano, in an area that represents one of Cameroon's main economic resources, and includes the cities of Buea (~90 000 citizens) and Limbe (~85 000 citizens) and numerous minor towns and villages scattered on the flanks of the volcano.

Mount Cameroon has erupted at least 18 times in the last 200 years (mostly with VEIs of 2), i.e. on average every 13 years, and most recently in 1999 and 2000 (Suh *et al.*, 2003, 2011; Njome *et al.*, 2008). According to Bonne *et al.* (2008), in a typical Mount Cameroon eruption, lava flows reach their maximum extent within 3–4 days and, after this, vents continue emitting lavas at gradually decreasing rates for several weeks or months (Suh *et al.*, 2003; Njome *et al.*, 2008). Mount Cameroon eruption volumes are typically of the order of 10^6–10^7 m^3 (Suh *et al.*, 2003).

The largest well-documented historical eruption of Mount Cameroon took place on 28 March–22 April 1999. From 1985 to 1998, local seismicity was characterised by discrete events and occasional swarms of about 15 events per month (Suh *et al.*, 2003). On 26 March 1999, over 15 events were recorded. Between 27 and 28 March 1999, the number of recorded events exceeded 200. A magnitude 4 event occurred beneath the summit around 20:00 local time on 28 March 1999, and coincided with the sighting of incandescent fumaroles at 2650 m altitude on the southwest flank of the mountain, marking the onset of the 1999 eruption. This earthquake (intensity VI–VII) was felt within a 100-km radius from the volcano, and destroyed buildings in Poto–poto quarters (Bokwango–Buea), leaving at least 10 families homeless. The eruption began on the southern flank at about 2650 m elevation and was marked by explosive magmatic activity emitting gases, lapilli, ash, and incandescent lava blocks. Scoriaceous bombs were ejected up to 500 m laterally from one vent and a'a flows emerged, reaching up to 3 km in length. These lava flows began to stagnate on 4 April when effusion ceased (Suh *et al.*, 2003).

Two days after the first eruption, on 30 March, a second vent opened at ~1400 m elevation, and sent a voluminous a'a flow, 5–20 m thick and 500–800 m wide, 14 km south-southwest through dense equatorial forest towards the coastal village of Bakingili. This flow travelled swiftly down the steep upper slopes, but slowed down to an estimated 10–25 m h^{-1} on the gentle flanks of the coastal plain (Global Volcanism Program, 1999). The flow eventually consumed hundreds of hectares of forest and destroyed several palm tree plantations as it moved towards the Atlantic Ocean, interrupting the coastal road that connected the villages on the eastern flanks of the volcano with those on the western flanks (Batoke and Bakingili; Déruelle *et al.*, 2000). On 17 April the lava flow stopped only 100 m from the ocean and appeared to have reached a stable position. The lava flow only narrowly missed the major, recently developed and thriving, Semme Beach tourist resort. This eruption necessitated, for the first time ever in the history of Mount Cameroon eruptions, the evacuation of over 600 inhabitants from the village of Bakingili along the southwestern flank of the volcano (Atanga *et al.*, 2010). The last glow from the 1400 m vent was seen on 14 April, and lava production, altogether ~8×10^7 m^3 with an average effusion rate of ~5.5 m^3s^{-1} and flow velocity of 5 m s^{-1}, probably ended about this time (Suh *et al.*, 2003).

13.3.3.1 *Hazards at Mount Cameroon*

The main hazards associated with the 1999 eruption were lava flows and volcanic ash, which was blown to the southwest into villages along the coast such as Bakingili, Batoke, Debundsha, and Idenau. Afane *et al.* (2001) observed eye and respiratory problems during medical consultations at the Batoke health centre due to the fine particle sizes of the ash, which was inhaled into the alveolar region of the lungs (Atanga *et al.*, 2010). Furthermore, this eruption and its associated events have had adverse effects on the environment as well as the local economy. Gas and ash emissions had polluted drinking water for about 2600 people living in the area. Forest ecosystems (ca. 1000 hectares were consumed), plantations and man-made structures were destroyed, in particular many private houses. Although many were immediately affected by the eruption and/or evacuated (more than 1000 people), there were no human casualties. The total economic loss was estimated at ~US$790 000.

A similar eruption in 2000 alarmed the population in the suburbs of Buea when a lava flow, originating from a fissure in the summit area above 4000 m above sea level, stopped just 4 km from the town (Favalli *et al.*, 2012). At Ekona village along the southeastern slopes of the volcano, Ateba *et al.* (2009) reported more than 30 fractured buildings due to associated earthquakes.

Although none of Mount Cameroon's historical eruptions have resulted in loss of life, the above events indicate that lava flows pose a significant threat to people and infrastructure around the volcano. On the other hand, landslides and mudflows (lahars) have resulted in fatalities on the flanks of the volcano. Ayonghe *et al.* (1999) found that since 1988 on average at least five people have been killed in Cameroon every year from landslide hazards with a substantial effect on infrastructure and the economy. The west coast of Limbe is one of the wettest places in the world with 10 000 mm of rainfall per year and is highly susceptible to (rainfall-triggered) landslides (Che *et al.*, 2012). Villages on the northeastern (leeward) side of the volcano like Mpundu and Muyuka are located in the rain shadow and have mean annual rainfall of 2000–2250 mm (Gaston, 2009). Pyroclastic material along the flanks of the volcano from explosive eruptions such as that of 1999 is easily eroded by torrential rainfall and transformed into lahars as occurred in 2001 when heavy downpours struck Limbe on 26–27 June. They led to a series of floods and landslides that killed ~23 people and destroyed homes, leaving over 1000 people homeless (Global Volcanism Program, 2001; Ayonghe *et al.*, 2004).

13.3.3.2 *Monitoring of Mount Cameroon*

The first seismograph to monitor the seismic activity along the central African shear zone system as well as the volcanic activity at Mount Cameroon was installed by the Geophysical and Volcanological Research Unit (ARGV) of the Institute for Geologic and Mining Research (IRGM) of Cameroon in Yaounde (330 km from the volcano) in 1982. The first permanent network of seismic stations was installed in 1984–1985, but due to difficulties in maintaining instruments, only one sensor was still operating during the 1999 and 2000 eruptions (Ateba *et al.*, 2009). Furthermore, at the time of eruption there was only limited information about the potential run-out and flow paths of Mount Cameroon lava flows; such information has only been published more recently (e.g., Bonne *et al.*, 2008; Thierry *et al.*, 2008; Favalli *et al.*, 2012; Wantim *et al.*, 2013).

The 1999 and 2000 eruptions revealed a lack of preparedness for such phenomena on the part of the communities and the Cameroonian administration, as a National Scientific Committee was only created following challenges from local scientists who began monitoring the event, motivated partly by scientific curiosity. Consequently, the evacuation camp was not prepared for emergency relief operations and there has remained a level of mistrust and miscommunication between the community, local scientists, and emergency managers (Atanga *et al.*, 2010; Njome *et al.*, 2010). Now, when an eruption begins, a crisis management committee is given responsibility for civil protection, for collection of relevant data, and for dissemination of information to the public. A permanent structure, however, is yet to be created.

Efforts are underway to develop a risk management strategy that incorporates aspects of the traditional socio-cultural beliefs of the population considered at risk. Two of these projects include the VLIR (Flemish Interuniversity Council) project on 'Capacity building in geohazards monitoring in volcanically active areas of southwest Cameroon' between the University of Buea (Cameroon) and the University of Ghent (Belgium) (running from 2006–2011) (some results were published in Njome *et al.* (2010)) and the Mitigate and Assess Risk from Volcanic Impact on Terrain and Human Activities; (MIAVITA; http://miavita.brgm.fr/pressroom/Pages/socio-economicvulnerabilitystudy.aspx) programme (financed by the European Commission from 2008 to 2012 under the 7th Framework Programme for Research and Technological Development), a socio-economic vulnerability study on the Mount Cameroon region that included the Italian and French Civil Defence Agencies, national geological and volcanological surveys, the Norwegian air quality survey, and five European universities and research centres from France, Germany, Portugal and the UK. Both programmes aim at developing tools and integrated cost-effective methodologies to mitigate risks from various hazards on active volcanoes (prevention, crisis management, and recovering), and in Cameroon they build upon a successful collaboration project between the University of Buea and the Department of Earth Sciences, University of Bristol, UK ('Understanding the Environment of Mount Cameroon') funded by the British Council (2002–2006).

The scientific monitoring programme around Mount Cameroon is based on studies of the local seismicity and deformation (Tabod and Bekoa, 2009). Mount Cameroon is permanently monitored with a network of six telemetred seismic broadband stations. Signals are interpreted in terms of magnitude and location of the earthquakes. Variations in the seismic activity in space and time are used for forecasting the eruptive activity of the volcano. All data are processed at the Geophysical and Volcanological Research Unit at Ekona, which is an arm of the Institute for Mining and Geological Research hosted by the Ministry of Scientific Research and Innovation.

13.3.4 The 2005 Karthala eruption, Comoros

Reaching 2361 m above sea level Mount Karthala is one of the largest active shield volcanoes in the world. Karthala's main structural features consist of two opposite rift zones diverging from a roughly elliptical summit caldera that is the result of successive collapses and the coalescence of at least seven pit craters formed by prehistoric eruptive activity (Lacroix, 1920; Pavlovsky and de Saint-Ours, 1953; Strong and Jacquot, 1971). The northern rift zone extends over 25 km to the north of the central caldera. An approximately 12-km-long southeast rift zone forms the Massif du Badjini, a peninsula at the southeastern tip of the island.

The centre of the 3.5 km × 2.8 km diameter caldera is occupied by the Choungou Chahale ('old crater'), a 1.4 km × 0.8 km and 290-m-deep crater (Morin *et al.*, 2009). The crater hosts an aqueous lake that is sometimes replaced by lava during effusive eruptions. Lava lake presence at Karthala was last reported in April and November 2005, May 2006, and January 2007 (Hirn *et al.*, 2008). A smaller cylindrical pit crater 250 m in diameter and 30 m deep, Changomeni, is found 1 km north of Chahale. The Hawaiian-style basaltic shield exhibits an asymmetrical profile that is steeper to the south.

Volcanic activity at Karthala has persisted for at least the past 130 000 years (Hajash and Armstrong, 1972; Emerick and Duncan, 1982, 1983). Since the beginning of the nineteenth century, at least 34 eruptions have been recorded at Mount Karthala from both summit and flank vents, occuring every 11 years on average (Siebert *et al.*, 2010). As is typical for a basaltic hotspot volcano, Mount Karthala's eruptions are mostly effusive with many lava flows reaching the sea on both sides of the island. The 1977 eruption generated the most destructive of Karthala's historic lava flows (Morin *et al.*, 2009), destroying approximately 280 houses, and leading to the evacuation of 4000 people. Phreatic explosions (involving rapidly heated groundwater) occurred in 1904 (uncertain), 1918, 1948, 1952, and 1991. Although this eruptive style is rare on basaltic volcanoes, phreatic eruptions are common on Grand Comoro Island, and represent a significant risk there (Savin *et al.*, 2005). The most spectacular manifestations of the active hydrothermal system are a solfatara, La Soufrière, located in the upper part of the northern rift zone, and the many fumaroles mainly located in the central part of the summit caldera. Also, water level changes of the 1991 Karthala crater lake that do not appear to be correlated with rainfall suggest the existence of an active hydrothermal system (Savin, 2001).

Two of the largest historic eruptions at Mount Karthala, probably of phreatomagmatic origin (involving rapidly heated groundwater and magma) (Morin *et al.*, 2009), occurred in 2005 when Karthala was disrupted by two strong explosive events. On 16 April at around 14:00 local time, inhabitants heard a rumbling noise coming from the volcano. A few minutes later they observed an ash column above the summit.

For the first time in living memory, the Comorian population experienced ash fall causing daily life disruption (the 1991 eruption's impacts were limited to the vicinity of the caldera) (Morin *et al.*, 2009). The first ash-fall deposits began to accumulate around 16:00, principally on the island's eastern side (Fig. 13.8). According to the first reports, ash deposition increased and continued through the night accompanied by a strong smell of sulphur. On the morning of the 17th a large plume together with liquid ejecta was observed over Chahale crater, developing into a buoyant column with an umbrella cloud and accompanied by lightning flashes. Ash started to fall on the island's western and northern sectors, the country's capital Moroni (~10 km northwest of the summit) and Hahaya airport (~20 km north of Moroni, ~25 km northwest of the summit). Approximately 10 000 people were evacuated by the authorities. During the evening, significant rainfall generated small mudflows along the island's southeast coast, and the rumbling noises became stronger. During the night of 17–18 April the plume and ash fall abated. In the meantime a lava lake had replaced the aqueous lake that had occupied the crater since 1991. On 19 April the lava lake had developed a chilled crust.

On 24–25 November, activity resumed with a phreatomagmatic eruption producing another ash cloud, ~11.6 km high, spreading southeast of the summit, with ash fall reaching Moroni. Ash first fell on the east coast of the island around 23:00 on 24 November. According to the local authorities, ~ 2000 people fled from their villages in the region of Bambao in the central part of the island, and

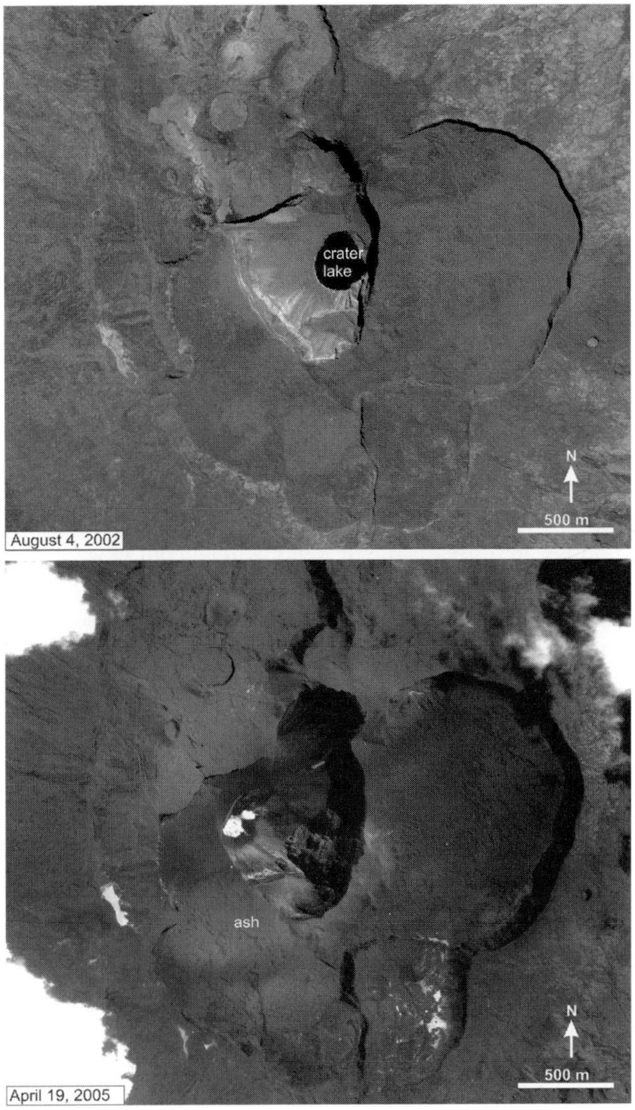

Figure 13.8. Mount Karthala before and after the 2005 eruption. The crater lake that formed after an eruption in 1991 (seen here in a photograph from 2002) dominated the caldera until the 2005 eruption when it disappeared completely. In its place rough, dark grey rocks, possibly cooling lava or rubble from the collapsed crater, were deposited. A grey field of ash surrounds the crater and the caldera itself seems larger and deeper. Photos courtesy of NASA Earth Observatory.

sought refuge in less exposed areas, such as Mitsamiouli, Mboudé, and Oichili. On the west part of the island, ash started to fall heavily on 25 November when the eruption probably entered a second, magmatic stage as suggested by the occurrence of a red cloud above the caldera and recharge of the lava lake. Ash deposits, as much as 10 cm thick in some places, covered three-quarters of the island. The international airport located in the northern part of the island remained free of ash deposits. In contrast to the crusted-over lava lake of April 2005, in November it was almost entirely liquid, with a very large sustained lava fountain in its centre. By 30 November, however, the lava lake had solidified across most of its surface. On 5 December it was almost entirely solid, with only two small spatter cones still active. The Karthala Volcano Observatory reported that the eruption ended, after 14 days, on 8 December 2005.

13.3.4.1 *Hazards at Mount Karthala*

The April 2005 eruption projected ash and volcanic debris predominantly across the eastern part of the island, affecting as many as 40 000 people in this region. In contrast, the November eruption was more explosive and of longer duration than the preceding eruption. The UN World Food Program (WFP) estimated that the November eruption affected approximately 245 000 people. Evacuation became very difficult, schools remained closed, and some people used masks to reduce inhalation of ash. Ash fall was so intense that the authorities asked people to remain in their homes. The UN Office for the Coordination of Humanitarian Affairs (OCHA) reported that, according to the local authorities, ~2000 people fled from their villages in the region of Bambao in the central part of the island, and sought refuge in safer areas, including Mitsamiouli, Mboudé, and Oichili. Several international and local flights were cancelled on 26–27 November.

According to OCHA, there were about 118 000 people living in 75 villages that were affected by contamination of water cisterns by ash (PIROI, 2005), raising concerns about the health of people and livestock, along with the effects on agriculture. This eruption placed Comoros sixth in country rankings of the total number of people affected by a volcano since 1900 (Center for Research on Epidemiology of Disasters, 2006). A further consequence of the 2005 eruptions was the sealing of the soil, reducing infiltration of rainfall and leading to lahars (Morin *et al.*, 2009). Lahars tend to occur after each period of rainfall, even after periods of moderate intensity. Lahars are more frequent on the western flank of Karthala, where the total precipitation is higher and the rain showers are stronger than on the eastern side (2695 mm per year in Moroni compared with 6000–9000 mm per year near the Karthala summit (Savin, 2001). Lahars now play an important role in the daily life and culture of Grand Comoro's inhabitants.

13.3.4.2 *Monitoring of Mount Karthala*

In 1988, the University of La Réunion, the Institut de Physique du Globe de Paris, and the Centre National de Documentation et de Recherche Scientifique des Comores (CNDRS) established a volcano observatory on Karthala volcano (OVK; http://volcano.ipgp.jussieu.fr/karthala/stationkar.html) (Savin *et al.*, 2005) with the aim of permanently monitoring the activity of the volcano and starting scientific research on the internal structures of the volcano, the geology, and the mechanisms of its activity. The observatory consists of a surface deformation network and a seismic network (five permanent seismic stations, including three summit stations (1-Hz vertical component L4C geophones, except for one summit station that is equipped with a three-component instrument); one clinometer; eleven 'dry-tilt' stations and two strain gauges). The 1991 eruption was the first volcanic event seismically recorded on Karthala volcano and is one of the few phreatic eruptions on a basaltic shield volcano ever geophysically recorded. When Karthala erupted on the evening of 28 May 2006, two of the three summit seismic stations were inoperative due to technical problems and lack of repair, so that – without being able to triangulate seismic activity or to corroborate data streams – it was almost impossible for OVK to build up a picture of precursors or of the volcanic activity, to issue advisories or warnings before the eruption, or to deliver any detailed information after it was clear that an eruption was in progress (Morin *et al.*, 2009).

13.4 Discussion

Population growth and expanding economies are increasing the exposure of people and assets to volcanic hazards across the world (see Chapter 3). The same is true on the African continent, but particular problems faced in characterising the levels of risk exposure are the generally limited knowledge of the volcanoes' past eruptive histories (either from historical sources or stratigraphic work), the deficiencies in or, in many cases, complete absence of operational volcano surveillance networks on the ground, and the poor overall resourcing of agencies tasked with geohazards management. High-resolution topographic maps, and geographic information system (GIS) databases of geological, volcanological, topographic, environmental, and human (i.e. civil constructions, economy) data are also generally lacking for most volcanoes in Africa (Bonne *et al.*, 2008).

While volcano monitoring capabilities may be underdeveloped, African geosciences thrive in other sectors, notably mineral exploration and extraction (in government ministries and companies), geological survey and mapping, and in academic institutions. There is a widespread recognition of the need to improve national and regional capacity for characterising geohazards and mitigating their impacts. This is especially the case where geotechnical issues need to be taken into account in major infrastructural projects, such as the construction of dams, but applies widely to the volcanically- and seismically-active regions discussed in this chapter. We suggest that these existing organisations within the geoscience sector are best placed to develop the geohazards agenda collectively, perhaps with the support of regional and pan-African entities such as the Organisation of African Geological Surveys and the Geological Society of Africa. In particular, we suggest there is an onus on, and strategic benefits for, the extractive and geothermal sector to encourage and facilitate geohazards work as an essential component of sustainable resource extraction, not least since these industries are often to be found exploiting resources associated with volcanic and tectonic activity. Major new infrastructure projects should go hand in hand with robust geotechnical surveys and characterisation of volcanic and seismic risks.

As we have discussed, important volcanic hazards include lava flows, earthquakes, landslides, and volcanic mudflows (lahars). In modern times, there has been relatively little experience within Africa of major pyroclastic eruptions (the explosive episodes of Nabro, Dubbi, and Karthala come closest perhaps). It would be easy, therefore, to dismiss hazards associated with Vulcanian and Plinian eruptions, or with hydrovolcanism, as such remote possibilities as to warrant their neglect in exploring future risk scenarios. However, the short historical records of volcanic activity in many of these volcanic regions should caution against such an approach. If the eruptive record around the world has taught us anything, it is that previously unheard of volcanoes can come back to life after centuries or millennia of repose (e.g., El Chichón, 1982, Mount Pinatubo, 1991, Chaitén, 2008, and Nabro, 2011). Considering the number of silicic volcanoes and caldera complexes, for instance associated with the Rift Valley, many of which sit above magma bodies (e.g., Biggs *et al.*, 2009) and host fumarole fields, it is surely prudent to regard these as potential sites of future, high-magnitude eruptions. Another possible threat is presented by the various volcanic lakes, notably those related to the CVL, a number of which may have the potential for catastrophic release of dissolved carbon dioxide (e.g., Lockwood and Rubin, 1989).

Effective strategies to prevent hazards from becoming disasters and to manage those disasters that do occur would make a lasting contribution to the quality of life and sustainable livelihoods of Africans (ICSU, 2005). Especially in countries where the geo-environmental database is still limited and rates of rural development are accelerated it is vitally important that authorities adopt cost-effective policies to lower risk and allocate appropriate resources for hazard and disaster mitigation. Accordingly, we conclude with three general recommendations for improving the state of geohazards management on the continent:

(i) Njome *et al.* (2010) among others have argued that the simplest and cheapest means for low-income countries to reduce the risk to and vulnerability of their populations to hazards is to increase educational levels. One approach involves developing an understanding of risk perception within the population. Many studies show that people's acceptance of vulnerability can increase warning effectiveness and improve preparedness (e.g., Mileti and Sorensen, 1990; Lindell and Perry, 1992; Lindell, 1994; Carlino *et al.*, 2008).

(ii) A clear institutional framework of specialised regional and global institutes with a clear division of responsibilities is needed for the collection of disaster information and to provide important links to national early warning systems. Here, many African countries might improve their use of NGO capacities. International bodies, regional and subregional organisations from academics to the civil protection corps should work together to maintain the vitally important timely exchange of and unrestricted access to observed data and information between countries, particularly when hazardous conditions affect neighbouring countries.

(iii) The monitoring networks in areas of active volcanism require development and maintenance, and systematic characterisation of geohazards and risk exposure needs to be undertaken. Furthermore, timely, accurate, and reliable warnings should be implemented in accordance with international standards, nomenclature, protocols, and reporting procedures. Established means of communication should be employed for the dissemination of any information to specific authorities designated in each country. At all levels of risk management, training and sharing of experience can help to achieve high competency. Furthermore, there should be a continuing education programme for authorities, institutional partners, and the population with the help of media. All this can only be achieved within the contexts of individuals, families, villages, and countries.

Acknowledgments. We thank Cheo E. Suh, Patrick Bachèlery, Stephen M. Njome, and Benoît Smets for sharing their information on the described volcanoes with us. Furthermore, we thank Patrick Eriksson and Amy Donovan for their helpful comments, and Sukanya Lenhardt for drafting the maps for this chapter. We also thank Karen Fontijn and an anonymous reviewer for their comments that led to improvements of the original manuscript. NL acknowledges a University of Pretoria Research Development Program (RDP) Grant. CO acknowledges support from the UK NERC for support with two projects: 'How does the Earth's crust grow at divergent plate boundaries? A unique opportunity in Afar, Ethiopia', and 'Mechanisms and consequences of the 2011 eruption of Nabro volcano, Eritrea'.

References

Afane, E. Z., Sende, N. C., Afane, E. A. *et al.* (2001). Irritation respiratoire Cameroun en Mars 1999. *Health Science and Disease; A Quarterly Publication of the Faculty of Medicine and Biomedical Sciences, University of Yaounde 1, Cameroon*, **3**, 62–63.

Aka, F. T., Yokoyama, T., Kusakabe, M. *et al.* (2008). U-series dating of Lake Nyos maar basalts, Cameroon (West Africa): Implications for potential hazards on the Lake Nyos dam. *Journal of Volcanology and Geothermal Research*, **176**, 212–224.

Alberico, I., Lirer, L., Petrosino, R. and Scandone, R. (2002). A methodology for the evaluation of long-term volcanic risk from pyroclastic flows in Campi Flegrei (Italy). *Journal of Volcanology and Geothermal Research*, **116**, 63–78.

Allard, P., Baxter, P., Hallbwachs, M. and Komorowski, J. C. (2002). *The January 2002 Eruptions of Nyiragongo Volcano (Dem. Repub. Congo) and Related Hazards: Observations and Recommendations*. Final report of the French-British scientific team, Paris, 8 March 2002.

Andersen, T., Elburg, M. and Erambert, M. (2012). Petrology of combeite- and götzenite-bearing nephelinite at Nyiragongo, Virunga Volcanic Province in the East African Rift. *Lithos*, **152**, 105–121.

Atanga, M. B. S., Merve, A. V., Njome, M. S., Kruger, W. and Suh, C. E. (2010). Health system preparedness for hazards associated with Mount Cameroon eruptions: a case study of Bakingili village. *International Journal of Mass Emergencies and Disasters*, **28**(3), 298–325.

Ateba, B., Dorbath, C., Dorbath, L. *et al.* (2009). Eruptive and earthquake activities related to the 2000 eruption of Mount Cameroon volcano (West Africa). *Journal of Volcanology and Geothermal Research*, **179**, 206–216.

Ayonghe, S. N., Mafany, G. T., Ntasin, E. and Samalang, P. (1999). Seismically activated swarm of landslides, tension cracks and rockfall after heavy rainfall in Bafaka, Cameroon. *Natural Hazards*, **19**, 13–27.

Ayonghe, S. N., Ntasin, E. B., Samalang, P. and Suh, C. E. (2004). The June 27, 2001 landslide on volcanic cones in Limbe, Mount Cameroon, West Africa. *Journal of African Earth Sciences*, **39**, 435–439.

Bachèlery, P. and Coudray, J. (1993). *Carte géologique des Comores: carte volcano-tectonique de la Grande Comore (Ngazidja)*, St. Denis de la Réunion, Laboratoire de cartographie et d'informations géographiques. CIRAD-Cultures.

Baker, J., Thirlwall, M. and Menzies, M. A. (1996). Sr–Nd–Pb isotopic and trace element evidence for crustal contamination of plume-derived flood basalts: Oligocene flood volcanism in western Yemen. *Geochimica et Cosmochimica Acta*, **60**, 2559–2581.

Barberi, F. and Varet, J. (1975). *Carte géologique de la chaîne volcanique de l'Erta'Ale (Dépression des Danakil-Ethiopie) au 1/100 000*. CNRS (France) and (CNR) Italy.

Barberi, F., Borsi, S., Ferrara, G. *et al.* (1972). Evolution of the Danakil depression (Afar, Ethiopia) in light of radiometric age determinations. *Journal of Geology*, **80**, 720–729.

Baxter, P., Allard, P., Halbwachs, M. *et al.* (2002–2003). Human health and vulnerability in the Nyiragongo volcano eruption and humanitarian crisis at Goma, Democratic Republic of Congo. *Acta Vulcanologica*, **14**(1–2)–**15**(1–2), 109–114.

Beke, C., 1861. Volcanic eruption of Djebel Dubbeh. London, *Times*, 16 October, p. 9.

Bell, C. (or Beke, C.), 1861. Eruption of a volcano on the East coast of Africa. London, *Times*, 20 June, p. 10.

Bellon, H. and Pouclet, A. (1980). Datations K–Ar de quelques laves du Rift-Ouest de Afrique Centrale; implications sur l'évolution magmatique et structurale. *Geologische Rundschau*, **69**(1), 49–62.

Besairie, H. (1964). *Geological Map of Madagascar.* Service Geologique de Madagascar, Tananarive.

Besairie, H. J. (1972). Géologie de Madagascar 1 – Les terraines sedimentaires. *Annales Géologiques de Madagascar*, **35**, 1–463.

Besairie, H. J. (1973). Précis de géologie Malgache. *Annales Géologiques de Madagascar*, **36**, 1–141.

Biggs, J., Anthony, E. Y. and Ebinger, C. J. (2009). Multiple inflation and deflation events at Kenyan volcanoes, East African Rift. *Geology*, **37**, 979–982.

Blomquist, J. (1979). *The Date and Origin of the Greek Version of Hanno's Periplus: with an Edition of the Text and a Translation*. Lund: CWK Gleerup.

Boccaletti, M., Bonini, M., Mazzuoli, R. *et al.* (1998). Quaternary oblique extensional tectonics in the Ethiopian Rift (Horn of Africa). *Tectonophysics*, **287**, 97–116.

Bonne, K., Kervyn, M., Cascone, L. *et al.* (2008). A new approach to assess long-term lava flow hazard and risk using GIS and low-cost remote sensing: the case of Mount Cameroon, West Africa. *International Journal of Remote Sensing*, **29**, 6539–6564.

Bourassa, A. E., Robock, A., Randel, W. J. *et al.* (2012). Large volcanic aerosol load in the stratosphere linked to Asian monsoon transport. *Science*, **337**, 78–81.

Brown, F. H. (1982). Tulu Bor Tuff at Koobi Fora correlated with the Sidi Hakoma Tuff at Hadar. *Nature*, **300**, 631–633.

Brown, F. H. (1983). Correlation of Tulu Bor Tuff at Koobi Fora with the Sidi Hakoma Tuff at Hadar: reply. *Nature*, **306**, 210.

Burton, R. F. (2011). *Abeokuta and the Cameroons Mountains: An Exploration*. Cambridge, UK: Cambridge University Press.

Carlino, S., Somma, R. and Mayberry, G. C. (2008). Volcanic risk perception of young people in the urban areas of Vesuvius: comparisons with other volcanic areas and implications for emergency management. *Journal of Volcanology and Geothermal Research*, **172**, 229–243.

Center for Research on Epidemiology of Disasters (2006). *EM-DAT: The OFDA/CRED International Disaster Database*, Université Catholique de Louvain, Brussels, online at: http://www.cred.be/emdat (accessed December 2012).

Chakrabarti, R., Basu, A. R., Santo, A. P., Tedesco, D. and Vaselli, O. (2009). Isotopic and geochemical study of the Nyiragongo and Nyamuragira volcanics in the western rift of the East African Rift System. *Chemical Geology*, **259**, 273–289.

Che, V. B., Kervyn, M., Suh, C. E. *et al.* (2012). Landslide susceptibility assessment in Limbe (SW Cameroon):

A field calibrated seed cell and information value method. *Catena*, **92**, 83–98.

Cheminée, J. L. (1973). Contribution à l'étude du potassium, de l'uranium et du thorium dans l'évolution des magmas. Doctural thesis, Paris VI 1.

Chirico, G. D., Favalli, M., Papale, P. *et al.* (2009). Lava flow hazard at Nyiragongo Volcano, DRC. 2. Hazard reduction in urban areas. *Bulletin of Volcanology*, **71**, 375–387.

Chorowicz, J. (2005). The East African rift system. *Journal of African Earth Sciences*, **43**, 379–410.

Chorowicz, J. and Sorlien, C. (1992). Oblique extensional tectonics in the Malawi Rift. *Geological Society of America Bulletin*, **104**, 1015–1023.

Chorowicz, J. and Vidal, G. (1987). Interprétation géologique d'une image SPOT XS du Kenya: les réseaux de failles du Gregory Rift. *Nouvelles de SPOT*, **9**, 16–17.

Chorowicz, J., Le Fournier, J., Le Mut, C. (1983). Observation par télédétection et au sol de mouvements décrochants NW–SE dextres dans le secteur transformant Tanganyika–Rukwa–Malawi du Rift Est-Africain. *Comptes Rendus Académie des Sciences Paris*, **296**(II), 997–1002.

Class, C. and Goldstein, S. L. (1997). Plume–lithosphere interactions in the ocean basins: constraints from the source mineralogy. *Earth and Planetary Science Letters*, **150**, 245–260.

Class, C., Goldstein, S. L., Altherr, R. and Bachèchlery, P. (1998). The process of plume–lithosphere interaction in the ocean basins – the case of Grande Comore. *Journal of Petrology*, **39**(5), 881–903.

Class, C., Goldstein, S. L., Stute, M., Kurz, M. D. and Schlosser, P. (2005). Grand Comore Island: A well-constrained "low $^{3}He/^{4}He$" mantle plume. *Earth and Planetary Science Letters*, **233**, 391–409.

Claude-Ivanaj, C., Bourdon, B. and Allègre, C. J. (1998). Ra–Th–Sr isotope systematics in Grande Comore Island: a case study of plume–lithosphere interaction. *Earth and Planetary Science Letters*, **164**, 99–117.

CNN (2002). More than 40 dead as volcano devastates Congolese city. Available at: http://europe.cnn.com/2002/WORLD/africa/01/19/congo.volcano/ (accessed August 2012).

Cornen, G. and Maury, R. C. (1980). Petrology of the volcanic island of Annobon, Gulf of Guinea. *Marine Geology*, **36**, 253–267.

Cruz Mermy, D., Lambert, A., Thierry, P. *et al.* (2006). *Projet GRINP-composante. 1 Notice des cartes geologiques du Mont Cameroun, Rapport final*. BRGM Report RC-54726-FR, Orléans, 216 pp.

Cuoco, E., Spagnuolo, A., Balagizi, C. *et al.* (2013a). Impact of volcanic emissions on rainwater chemistry: The case of Mt. Nyiragongo in the Virunga volcanic region (DRC). *Journal of Geochemical Exploration*, **125**, 69–79.

Cuoco, E., Tedesco, D., Poreda, R. J. *et al.* (2013b). Impact of volcanic plume emissions on rain water chemistry during the January 2010 Nyamulagira eruptive event: Implications for essential potable water resources. *Journal of Hazardous Material*, **244–245**, 570–581.

Davidson, A. and Rex, D. C. (1980). Age of volcanism and rifting in southwestern Ethiopia. *Nature*, **283**, 657–658.

Dawson, J. B. (1964). Carbonatitic ashes in northern Tanganyika. *Bulletin Volcanologique*, **27**, 81–91.

Dawson, J. B., Pinkerton, H., Norton, G. E. and Pyle, D. M. (1990). Physiochemical properties of alkali carbonatite lavas: data from the 1988 eruption of Oldoinyo Lengai, Tanzania. *Geology*, **18**, 260–263.

Dawson, J. B., Pinkerton, H., Pyle, D. M. and Nyamweru, C. (1994). June 1993 eruption of Oldoinyo Lengai, Tanzania: exceptionally viscous and large carbonatite lava flows and evidence for coexisting silicate and carbonate magmas. *Geology*, **22**, 799–802.

De Saint-Ours, J. (1960). *Etudes géologiques dans l'extrême nord de Madagascar et l'archipel des Comores*. These Sciences, II, Rapport Service Géologique Tananarive 221.

Delvaux, D. (2001). Tectonic and palaeostress evolution of the Tanganyika-Rukwa-Malawi rift segment, East African Rift System. In *Peri-Tethyan Rift/Wrench Basins and Passive Margins*, eds. Ziegler, P. A., Cavazza, W., Robertson, A. H. F. and Crasquin-Soleau, S. Peri-Tethys Memoir 6, Mémoires du Muséum National d'Histoire Naturelle, vol. 186 Paris: Publications scientifiques du Museum, pp. 545–567.

Demant, A., Lestrade, P., Lubala, R. T., Kampunzu, A. B. and Durieux, J. (1994). Volcanological and petrological evolution of Nyiragongo volcano, Virunga volcanic field, Zaire. *Bulletin of Volcanology*, **56**, 47–61.

Déruelle, B., Nni, J. and Kambou, R. (1987). Mount Cameroon: an active volcano of the Cameroon Line. *Journal of African Earth Sciences*, **6**, 197–214.

Déruelle, B., Moreau, C., Nkoumbou, C. *et al.* (1991). The Cameroon Line: a review. In *Magmatism in Extensional Structural Settings, the Phanerozoic African Plate*, eds. Kampunzu, A. B. and Lubala, R. T. Berlin: Springer-Verlag, pp. 274–327.

Déruelle, B., Bardintzeff, J. M., Cheminée, J. L. *et al.* (2000). Eruptions simultanées de basalte alcalin et de hawaiite au mont Cameroun (28 mars–17 avril 1999). *Comptes Rendus Académie Sciences Paris, Sciences de la terre et des planètes*, **331**, 525–531.

Dumort, J. C. (1968). Carte géologique de reconnaissance de la République Fédérale du Cameroun, Douala-Ouest, 1/500 000. Cameroon Mines and Geology Directorate, 1 map, 1 explanatory note, 69 pp.

Duncan, R. A. (1981). Hotspots in the southern oceans – an absolute frame of reference for motion of the Gondwana continents. *Tectonophysics*, **74**, 29–42.

Dunlop, H. M. and Fitton, J. G. (1979). A K-Ar and Sr-isotopic study of the volcanic rocks of the island of Principe, West Africa – evidence for mantle heterogeneity beneath the Gulf of Guinea. *Contributions to Mineralogy and Petrology*, **71**, 125–131.

Ebinger, C. J., Rosendahl, B. R. and Reynolds, D. J. (1987). Tectonic model of the Malawi rift, Africa. *Tectonophysics*, **141**, 215–235.

Ebinger, C. J., Deino, A. L., Drake, R. E. and Tesha, A. L. (1989). Chronology of volcanism and rift basin propagation: Rungwe volcanic Province, East Africa. *Journal of Geophysical Research*, **94**(B11), 15785–15803.

Ebinger, C. J., Yemane, T., WoldeGabriel, G., Aronson, J. L. and Walter, R. C. (1993). Late Eocene–Recent volcanism and rifting in the southern main Ethiopian rift. *Journal of the Geological Society, London*, **150**, 99–108.

Emerick, C. M. and Duncan, R. A. (1982). Age progressive volcanism in the Comores Archipelago, western Indian Ocean and implications for Somali plate tectonics. *Earth and Planetary Science Letters*, **60**, 412–428.

Emerick, C. M. and Duncan, R. A. (1983). Age progressive volcanism in the Comores Archipelago, western Indian Ocean and implications for Somali plate tectonics, Errata. *Earth and Planetary Science Letters*, **62**, 439.

Esson, J., Flower, M. F., Strong, D. F., Wadsworth, W. J. and Upon, B. G. (1970). Geology of the Comores archipelago, western Indian Ocean. *Geological Magazine*, **6**, 549–557.

Favalli, M., Chirico, G. D., Papale, P., Pareschi, M. T. and Boschi, E. (2009). Lava flow hazard at Nyiragongo volcano, D.R.C. 1. Model calibration and hazard mapping. *Bulletin of Volcanology*, **71**, 363–374.

Favalli, M., Tarquini, S., Papale, P., Fornaciai, A. and Boschi, E. (2012). Lava flow hazard and risk at Mt. Cameroon volcano. *Bulletin of Volcanology*, **74**, 423–439.

Ferguson, D. J., Barnie, T. D., Pyle, D. M. *et al.* (2010). Recent rift-related volcanism in Afar, Ethiopia. *Earth and Planetary Science Letters*, **292**, 409–418.

Field, L., Blundy, J., Brooker, R. A., Wright, T. and Yirgu, G. (2012a). Magma storage conditions beneath Dabbahu Volcano (Ethiopia) constrained by petrology, seismicity and satellite geodesy. *Bulletin of Volcanology*, **74**, 981–1004.

Field, L., Barnie, T., Blundy, J. *et al.* (2012b). Integrated field, satellite and petrological observations of the November 2010 eruption of Erta Ale. *Bulletin of Volcanology*, **74**, 2251–2271.

Fitton, J. G. (1987). The Cameroon Line, West Africa: a comparison between oceanic and continental alkaline volcanism. In *Alkaline Igneous Rocks*, eds. Fitton, J. G. and Upton, B. G. J. Geological Society of London Special Publication, vol. 30, London: Geological Society of London, pp. 273–291.

Fitton, J. G. and Dunlop, H. M. (1985). The Cameroon Line, West Africa, and its bearing on the origin of oceanic and continental alkali basalt. *Earth and Planetary Science Letters*, **72**, 23–38.

Fitton, J. G., Kilburn, C. R. J., Thirlwall, M. F. and Hughes, D. J. (1983). 1982 eruption of Mt. Cameroon, West Africa. *Nature*, **306**, 327–332.

Flannery, J. W. and Rosendahl, B. R. (1990). The seismic stratigraphy of Lake Malawi, Africa: implications for interpreting geological processes in lacustrine rifts. *Journal of African Earth Sciences*, **10**(3), 519–548.

Fontijn, K., Ernst, G. G. J., Elburg, M. A. *et al.* (2010). Holocene explosive eruptions in the Rungwe Volcanic Province, Tanzania. *Journal of Volcanology and Geothermal Research*, **196**, 91–110.

Fontijn, K., Ernst, G. G. J., Bonadonna, C. *et al.* (2011). The ~4-ka Rungwe Pumice (South-Western Tanzania): a wind-still Plinian eruption. *Bulletin of Volcanology*, **73**, 1353–1368.

Fontijn, K., Williamson, D., Mbede, E. and Ernst, G. G. J. (2012). The Rungwe Volcanic Province, Tanzania – a volcanological review. *Journal of African Earth Sciences*, **63**, 12–31.

Fournier d'Albe, E. M. (1979). Objectives of volcanic monitoring and prediction. *Journal of the Geological Society, London*, **136**, 321–326.

Fourno, J. P. and Roussel, J. (1994). Imaging of the Moho depth in Madagascar through the inversion of gravity data: geodynamic implications. *Terra Nova*, **6**, 512–519.

Freeth, S. J. and Kay, R. L. F. (1987). The Lake Nyos disaster. *Nature*, **325**, 104–105.

Frey, F. A. (2009). Indian Region. In *Encyclopedia of Islands*, eds. Gillespie, R. G. and Clague, D. A. Berkeley, CA: University of California Press, pp. 437–446.

Fromm, M., Nedoluha, G. and Charvát, Z. (2013). Comment on "Large volcanic aerosol load in the stratosphere linked to Asian monsoon transport". *Science*, **339**, 647.

Gaston, B. W. (2009). Geographic information systems based demarcation of risk zones: the case of the Limbe Sub-Division – Cameroon. *Journal of Disaster Risk Studies*, **2**, 54–70.

George, R., Rogers, N. and Kelley, S. (1998). Earliest magmatism in Ethiopia: evidence for two mantle plumes in one flood basalt province. *Geology*, **26**, 923–926.

Georgen, J. E. and Lin, J. (2003). Plume-transform interactions at ultra-slow spreading ridges: Implications for the Southwest Indian Ridge. *Geochemistry, Geophysics, Geosystems*, **4**, art. no. 9106.

Gèze, B. (1943). *Géographie physique et géologique du Cameroun Occidental*. Memoire du Muséum National d'Histoire, Paris, vol. XVII. Paris: Muséum National d'Histoire.

Giordano, D., Polacci, M. and Longo, A. (2007). Thermo-rheological magma control on the impact of highly fluid lava flows at Mt. Nyiragongo. *Geophysical Research Letters*, **34**, art. no. L06301.

Global Volcanism Program (1999). 03/1999 (BGVN 24:03) Eruptions and lava flows discharge from multiple S-flank vents. Available at: http://www.volcano.si.edu/world/volcano.cfm?vnum=0204-01=&volpage=var#bgvn_2404 (accessed January 2013).

Global Volcanism Program (2001). 11/2001 (BGVN 26:11) Late June non-volcanic floods and landslides take tens of lives; 1,000 homeless. Available at: http://www.volcano.si.edu/world/volcano.cfm?vnum=0204-01=&volpage=var#bgvn_2611 (accessed January 2013).

Global Volcanism Program (2002). 03/2002 (BGVN 27:03) Report of field work; MODIS imagery from January 2002 eruption. Available at: http://www.volcano.si.edu/world/volcano.cfm?vnum=0203-03=&volpage=var#bgvn_2703 (accessed January 2013).

GORISK (2012). GORISK scientific network website. Available at: http://www.ecgs.lu/gorisk/ (accessed January 2013).

Grand, S. P., Van der Hilst, R. D. and Widiyantoro, S. (1997). Global seismic tomography: a snapshot of convection in the Earth. *GSA Today*, **7**, 1–7.

Grunau, H. R., Lehner, P., Cleintuar, M. R., Allenbach, P. and Bakker, G. (1975). New radiometric ages and seismic data from Fuerteventura (Canary Islands), Maio (Cape Verde Islands), and São Tomé (Gulf of Guinea). In *Progress in Geodynamics*, eds. Borradaile, G. J., Ritsema, A. R., Rondeel, H. E. and Somin, O. J. New York, NY: North-Holland, pp. 90–118.

Hackman, B. D., Charsley, T. J., Key, R. M. and Wilkinson, A. F. (1990). The development of the East African Rift system in north-central Kenya. *Tectonophysics*, **184**, 189–211.

Hajash, A. and Armstrong, R. L. (1972). Paleomagnetic and radiometric evidence for the age of the Comores Islands, west central Indian Ocean. *Earth and Planetary Science Letters*, **16**, 231–236.

Harkin, P. A. (1960). The Rungwe volcanics at the northern end of Lake Nyasa. *Memoir Geological Survey of Tanganyika*, **2**, 172.

Hart, W. K., Walter, R. C. and WoldeGabriel, G. (1992). Tephra sources and correlations in Ethiopia: application of elemental and neodymium isotope data. *Quaternary International*, **13–14**, 77–86.

Hay, R. L. (1983). Natrocarbonatite tephra of Kerimasi volcano, Tanzania. *Geology*, **11**, 599–602.

Hay, R. L. (1989). Holocene carbonatite–nephelinite tephra deposits of Oldoinyo Lengai, Tanzania. *Journal of Volcanology and Geothermal Research*, **37**, 77–91.

Head, E. M., Shaw, A. M., Wallace, P. J., Sims, K. W. W. and Carn, S. A. (2011). Insight into volatile behavior at Nyamuragira volcano (D.R. Congo, Africa) through olivine-hosted melt inclusions. *Geochemistry, Geophysics, Geosystems*, **12** (10), art. no. Q0AB11.

Hecky, R. E. (1971). The palaeolimnology of the alkaline, saline lakes on the Mt. Meru lahar. Ph.D. Thesis, Duke University Durham, NC.

Hedberg, J. D. (1968). A geological analysis of the Cameroon trend. Ph.D. thesis, University of Princeton, Princeton, NJ.

Hirn, B., Di Bartola, C., Ferro, F. and Ferrucci, F. (2008). MODIS borne analysis of long series of radiant fluxes at active lava lakes. IGARSS 2008. *IEEE International*, **3**, 370–373.

Hofmann, C., Courtillot, V., Féraud, G. *et al.* (1997). Timing of the Ethiopian flood basalt event and implications for plume birth and global change. *Nature*, **389**, 838–841.

ICSU (International Council for Science) (2005). *ICSU Scoping Group on Natural and Human-induced Environmental Hazards, a Report to the ICSU 28th General Assembly*, Suzhou, China, October. 28 pp.

ICSU (International Council for Science) (2007). *ICSU Regional Office for Africa Science Plan – Natural and Human-induced Hazards and Disasters in Sub-Saharan Africa*, 29 pp.

IRIS (2002). *Nyiragongo Volcano. Special Report*. Available at: http://www.iris.washington.edu/news/events/congo/congo.htm (accessed January 2013).

Kagou Dongmo, A., Wandji, P., Pouclet, A. *et al.* (2001). Evolution volcanologique du mont Manengouba 'Ligne du Cameroun': nouvelles données pétrographiques, géochimiques et géochronologiques. *Comptes Rendus de l'Académie des Sciences Paris*, **333**, 155–162.

Kampunzu, A. B., Bonhomme, M. G. and Kanika, M. (1998). Geochronology of volcanic rocks and evolution of the Cenozoic Western Branch of the East African Rift System. *Journal of African Earth Sciences*, **26**, 441–461.

Karche, J. P. (1973). Le massif volcanique d'Ambre et les régions voisines de Madagascar. Etude volcanologique et pétrologique. *Annales Sciences Université Besançon*, **19**, 173 pp.

Kazmin, V. (1976). Geological map of Adigrat, 1:250,000. Ethiopia Geological Survey – Ministry of Mines, Addis Ababa, Ethiopia.

Kieffer, B., Arndt, N., Lapierre, H. *et al.* (2004). Flood and Shield basalts from Ethiopia: magmas from the African superswell. *Journal of Petrology*, **45**, 793–834.

Kilburn, C. R. J. (2000). Lava flows and flow fields. In *Encyclopedia of Volcanoes*, eds. Sigurdsson, H., Houghton, B., McNutt, S. R., Rymer, H. and Stix, J. San Diego, CA: Academic Press, pp. 291–305.

Kilembe, E. A. and Rosendahl, B. R. (1992). Structure and stratigraphy of the Rukwa rift. *Tectonophysics*, **209**, 143–158.

Kling, G. W., Clark, M. A., Compton, H. R. *et al.* (1987). The 1986 Lake Nyos gas disaster in Cameroon, West Africa. *Science*, **236**, 169–175.

Komorowski, J.-C., Tedesco, D., Kasereka, M. *et al.* (2002–2003). The January 2002 flank eruption of Nyiragongo volcano (Democratic Republic of Congo): chronology, evidence for a tectonic rift trigger, and impact of lava flows on the city of Goma. *Acta Vulcanologica*, **14**(1–2)–**15**(1–2), 27–62.

Lacroix, A. (1920). Une éruption du volcan Karthala, à la Grande Comore, en août 1918. *Comptes rendus hebdomadaires des séances de l'Académie des Sciences*, **171**, 5–10.

Lahitte, P., Gillot, P.-Y., Kidane, T., Courtillot, V. and Bekele, A. (2003). New age constraints on the timing of volcanism in central Afar, in the presence of propagating rifts. *Journal of Geophysical Research*, **108**, ECV 10–1 – 10–17.

Le Roex, A. P., Chevallier, L., Verwoerd, W. J. and Barends, R. (2012). Petrology and geochemistry of Marion and Prince Edward Islands, Southern Ocean: Magma chamber processes and source region characteristics. *Journal of Volcanology and Geothermal Research*, **223–224**, 11–28.

Leat, P. T. (1984). Geological evolution of the trachytic caldera volcano Menengai, Kenya rift valley. *Journal of the Geological Society, London*, **141**, 1057–1069.

Leat, P. T. (1991). Volcanological development of the Nakuru area of the Kenya rift valley. *Journal of African Earth Sciences*, **13**, 483–498.

Lee, D. C., Halliday, A. N., Fitton, J. G. and Poli, G. (1994a). Isotopic variations with distance and time in the volcanic islands of the Cameroon Line: Evidence of the mantle plume origin. *Earth and Planetary Science Letters*, **123**, 119–138.

Lee, D. C., Halliday, A. N., Hall, C. M. and Fitton, J. G. (1994b). Similarities and differences between continental and oceanic basalts in the Cameroon Line. Eighth International Conference on Geochronology, Cosmochronology and Isotope Geology, Berkeley. Denver, CO, USA. *US Geological Survey Circular*, **1107**, 188.

Lindell, M. K. (1994). Perceived characteristics of environmental hazards. *International Journal of Mass Emergencies and Disasters*, **12**, 303–326.

Lindell, M. K. and Perry, R. W. (1992). *Behavioral Foundations of Community Emergency Planning*. New York, NY: Hemisphere Publishing, 309 pp.

Lockwood, J. and Casadevall, T. (1994). Active volcanoes and associated hazards in the Goma area, eastern Zaire. *Hawaii Center for Volcanology Newsletter*, **2**, 1.

Lockwood, J. P. and Rubin, M. (1989). Origin and age of the Lake Nyos maar, Cameroon. *Journal of Volcanology and Geothermal Research*, **39**, 117–124.

Macdonald, R. (1987). Quaternary peralkaline silicic rocks and caldera volcanoes of Kenya. *Geological Society Special Publication*, **30**, 313–333.

Marzoli, A., Renne, P. R., Piccirillo, E. M. *et al.* (1999). Silicic magmas from the continental Cameroon Volcanic Line (Oku, Bambouto and Ngaoundere): ^{40}Ar–^{39}Ar dates, petrology, Sr–Nd–O isotopes and their petrogenetic significance. *Contributions to Mineralogy and Petrology*, **135**, 133–150.

Marzoli, A., Piccirillo, E. M., Renne, P. R. *et al.* (2000). The Cameroon volcanic line revisited: petrogenesis of continental basaltic magmas from lithospheric and asthenospheric mantle sources. *Journal of Petrology*, **41**, 87–109.

Mathieu, L., Kervyn, M. and Ernst, G. G. J. (2011). Field evidence for flank instability, basal spreading and volcano–tectonic interactions at Mt. Cameroon, West Africa. *Bulletin of Volcanology*, **73**, 851–867.

McDougall, I. A. N., Verwoerd, W. and Chevallier, L. U. C. (2001). K–Ar geochronology of Marion Island, Southern Ocean. *Geological Magazine*, **138**, 1–17.

Melluso, L. and Morra, V. (2000). Petrogenesis of Late Cenozoic mafic alkaline rocks of the Nosy Be archipelago (northern Madagascar): relationships with the Comorean magmatism. *Journal of Volcanology and Geothermal Research*, **96**, 129–142.

Melluso L., Morra V., Riziky H. *et al.* (2007). Petrogenesis of a basanite–tephrite–phonolite volcanic suite in the Bobaomby (Cap d'Ambre) Peninsula, northern Madagascar. *Journal of African Earth Sciences*, **49**, 29–42.

Meyers, J. B., Rosendahl, B. R., Harrison, C. G. A. and Ding, Z.-D. (1998). Deep-imaging seismic and gravity results from offshore Cameroon Volcanic Line and speculation of African hot-lines. *Tectonophysics*, **284**, 31–63.

Mileti, D. S. and Sorensen, J. H. (1990). *Communication of Emergency Public Warnings: A Social Science Perspective and State-of-the-Art Assessment. ORNL-6609*. Oak Ridge National Laboratory, Ork Ridge, TN.

Montigny, R., Ngounouno, I. and Déruelle, B. (2004). Ages K-Ar des roches magmatiques du fossé de Garoua (Cameroun): leur place dans le cadre de la "Ligne du Cameroun". *Comptes Rendus Géoscience*, **336**, 1463–1471.

Morin, J., Lavigne, F., Bachelery, F., Finizola, A. and Villeneuve, N. (2009). Institutional and social responses to hazards related to Karthala volcano, Comoros – Part I: Analysis of the May 2006 eruptive crisis. *Shima: The International Journal of Research into Island Cultures*, **3**, 33–53.

Morley, C. K. (1999). Basin evolution trends in East Africa. In: *Geosciences of Rift Systems-Evolution of east Africa.*, ed. Morley, C. K. The American Association of Petroleum Geologists Studies in Geology, vol. 44, pp. 131–150.

Morley, C. K., Cunningham, S. M., Harper, R. M., Wigger, S. T. and Karanja, F. M. (1992a). Geology and geophysics of the Rukwa rift, East Africa. *Tectonics*, **11**, 68–81.

Morley, C. K., Wescott, W. A., Stone, D. M. *et al.* (1992b). Tectonic evolution of the northern Kenyan Rift. *Journal of the Geological Society, London*, **149**, 333–348.

Mougenot, D., Recq, M., Virlogeux, P. and Lepvrier, C. (1986). Seaward extension of the East African Rift. *Nature*, **321**, 599–603.

Newhall, C. G. and Dzurisin, D. (1988). Historical unrest at large calderas of the world. *United States Geological Survey Bulletin*, **1855**, 1–1108.

Newhall, C. G. and Self, S. (1982). The volcanic explosivity index (VEI): an estimate of explosive magnitude for historical volcanism. *Journal of Geophysical Research*, **87**, 123–1238.

Ngounouno, I., Déruelle, B. and Demaiffe, D. (2000). Petrology of the bimodal Cainozoic volcanism of the Kapsiki plateau (northernmost Cameroon, Central Africa). *Journal of Volcanology and Geothermal Research*, **102**, 21–44.

Ngwa, C. N., Suh, C. E. and Devey, C. W. (2010). Phreatomagmatic deposits and stratigraphic reconstruction at Debunscha Maar (Mt Cameroon volcano). *Journal of Volcanology and Geothermal Research*, **192**, 201–211.

Njome, M. S., Suh, C. E., Sparks, R. S. J., Ayonghe, S. N. and Fitton, J. G. (2008). The Mount Cameroon 1959 compound lava flow field: morphology, petrography and geochemistry. *Swiss Journal of GeoSciences*, **101**, 85–98.

Njome, M. S., Suh, C. E., Chuyong, G. and De Wit, M. J. (2010). Volcanic risk perception in rural communities along the slopes of mount Cameroon, West-Central Africa. *Journal of African Earth Sciences*, **58**, 608–622.

Nkouathio, D. G., Ménard, J.-J., Wandji, P. and Bardintzeff, J. M. (2002). The Tombel graben (West Cameroon): a recent monogenetic volcanic field of the Cameroon Line. *Journal of African Earth Sciences*, **35**, 285–300.

Nkouathio, D. G., Kagou Dongmo, A., Bardintzeff, J. M. *et al.* (2008). Evolution of volcanism in graben and horst structures along the Cenozoic Cameroon Line (Africa): implications for tectonic evolution and mantle source composition. *Mineralogy and Petrology*, **94**, 287–303.

Nkoumbou, C., Déruelle, B. and Velde, D. (1995). Petrology of Mt Etinde nephelinites series. *Journal of Petrology*, **36**, 373–395.

Nni, J. (1984). Le volcan actif du Mont Cameroun (Ligne du Cameroun): géologie et pétrologie du volcan. Ph.D. thesis, University of Paris XI, Orsay.

Nono, A., Déruelle, B., Demaiffe, D. and Kambou, R. (1994). Tchabal Nganha volcano in Adamawa (Cameroon): petrology of a continental alkaline lava series. *Journal of Volcanology and Geothermal Research*, **60**, 147–178.

Nougier, J., Cantagrel, J. M. and Karche, J. P. (1986). The Comores archipelago in the western Indian Ocean: volcanology, geochronology and geodynamic setting. *Journal of African Earth Sciences*, **5**, 135–145.

OCHA (United Nations Office for the Coordination of Humanitarian Affairs) (2007). *Holocene Eruptions, Volcanoes and Faults in Africa*. Available at: http://ochaonline.un.org/rocea (accessed January 2013).

OFDA-CRED (2012). *The OFDA-CRED International Disaster Database 2002*. Centre for Research on the Epidemiology of Disasters (CRED). Available at: http://www.emdat.be/disaster-profiles (accessed January 2013).

Oppenheimer, C. (1998a). Satellite observation of active carbonatite volcanism at Ol Doinyo Lengai, Tanzania. *International Journal of Remote Sensing*, **19**, 55–64.

Oppenheimer, C. (1998b). Satellite observations of lava lake activity at Nyiragongo volcano, ex-Zaire, during the Rwandan refugee crisis. *Disasters*, **22**, 268–281.

Pavlovsky, R. and Saint-Ours, J. (1953). *Etude géologique de l'archipel des Comores*. Haut Commissariat de Madagascar et Dépendances, Direction des Mines et de la Géologie, Service Géologique Tananarive, Travaux du bureau géologique, 51.

Peirce, J. and Lipkov, L. (1988). Structural interpretation of the Rukwa rift, Tanzania. *Geophysics*, **53**, 824–836.

Permenter, J. L. and Oppenheimer, C. (2007). Volcanoes of the Tibesti massif (Chad, northern Africa). *Bulletin of Volcanology*, **69**, 609–626.

Perrey, A. (1864). Note sur les tremblements de terres de 1862, avec suppléments pour les années antérieures. *Mémoire de l'Académie Royale de Belgique*, **16**, 95–99.

Pik, R., Marty, B. and Hilton, D. R. (2006). How many mantle plumes in Africa? The geochemical point of view. *Chemical Geology*, **226**, 100–114.

Piper, J. D. A. and Richardson, A. (1972). The paleomagnetism of the Gulf of Guinea volcanic province, west Africa. *Royal Astronomical Society Geophysical Journal*, **29**, 147–171.

Platz, T., Foley, S. F. and André, L. (2004). Low-pressure fractionation of the Nyiragongo volcanic rocks, Virunga Province, D.R. Congo. *Journal of Volcanology and Geothermal Research*, **136**, 269–295.

Pyle, D. M. (1999). Widely dispersed Quaternary tephra in Africa. *Global and Planetary Change*, **21**, 95–112.

Rampey, M. L., Oppenheimer, C., Pyle, D. M. and Yirgu, G. (2010). Caldera-forming eruptions of the Quaternary Kone Volcanic Complex, Ethiopia. *Journal of African Earth Sciences*, **58**, 52–66.

Rasamimanana, G. (1996). Caractérisations géochimiques et géochronologiques de trois épisodes magmatiques (Crétacé, Miocène terminal et Quaternaire) à Madagascar, associés aux phénomènes de rifting. D.Sc. thesis. Université de Paris-Sud, Orsay.

Renaut, R. W. and Owen, R. B. (1988). Opaline cherts associated with sublacustrine hydrothermal springs at Lake Bogoria, Kenya Rift Valley. *Geology*, **16**, 699–702.

Ritsema, J. and van Heijst, H. J. (2000). New seismic model of the upper mantle beneath Africa. *Geology*, **28**, 63–66.

Rosendahl, B. R., Kilembe, E. and Kaczmarick, K. (1992). Comparison of the Tanganyika, Malawi, Rukwa and Turkana rift zones from analyses of seismic reflection data. *Tectonophysics*, **213**, 235–256.

Sarna-Wojcicki, A. M., Meyer, C. E., Roth, P. H. and Brown, F. H. (1985). Ages of tuff beds at East African early hominid sites and sediments in the Gulf of Aden. *Nature*, **313**, 306–308.

Savin, C. (2001). Circulation hydrothermale au sein du volcan Karthala (Ile de la Grande Comore – Océan Indien): apports de la géophysique. Ph.D. thesis. Université de La Réunion.

Savin, C., Grasso, J.-R. and Bachélery, P. (2005). Seismic signature of a phreatic explosion: hydrofracturing damage at Karthala volcano, Grande Comore Island, Indian Ocean. *Bulletin of Volcanology*, **67**, 717–731.

Sawamura, P., Vernier, J. P., Barnes, J. E. *et al.* (2012). Stratospheric AOD after the 2011 eruption of Nabro volcano measured by lidars over the Northern Hemisphere. *Environmental Research Letters*, **7**, 034013.

Sawyer, G. M., Carn, S. A., Tsanev, V. I., Oppenheimer, C. and Burton, M. (2008). Investigation into magma degassing at Nyiragongo volcano, Democratic Republic of Congo. *Geochemistry, Geophysics, Geosystems*, **9**, *art. no.* Q02017.

Schilling, J.-G., Kingsley, R., Hanan, B. and McCully, B. (1992). Nd–Sr–Pb isotopic variations along the Gulf of Aden: evidence for Afar mantle plume-continental lithosphere interaction. *Journal of Geophysical Research*, **97**, 10927–10966.

Scott, S. C. (1980). The geology of Longonot volcano, central Kenya: a question of volumes. *Philosophical Transactions of the Royal Society*, **A296**, 437–465.

Siebert, L., Simkin, T. and Kimberly, P. (2010). *Volcanoes of the World*, third edition. Berkeley, CA: University of California Press.

Sigurdsson, H., Devine, J. D., Tchoua, F. M. *et al.* (1987). Origin of the lethal gas burst from Lake Monoun, Cameroun. *Journal of Volcanology and Geothermal Research*, **31**, 1–16.

Smets, B., Tedesco, D., Kervyn, F. *et al.* (2010). Dry gas vents ("mazuku") in Goma region (North-Kivu, Democratic Republic of Congo): Formation and risk assessment. *Journal of African Earth Sciences*, **58**, 787–798.

Smith, M. and Mosley, P. (1993). Crustal heterogeneity and basement influence on the development of the Kenya Rift, East Africa. *Tectonics*, **12**(2), 591–606.

Späth, A., Le Roex, A. P. and Duncan, R. A. (1996). The geochemistry of lavas from the Comores archipelago, Western Indian ocean: petrogenesis and mantle source region characteristics. *Journal of Petrology*, **37**, 961–991.

Storey, M., Mahoney, J. J., Saunders, A. D. *et al.* (1995). Timing of hot spot-related volcanism and the breakup of Madagascar and India. *Science*, **267**, 852–855.

Strong, D. F. (1972). The petrology of the lavas of Grande Comore. *Journal of Petrology*, **13**, 181–217.

Strong, D. F. and Jacquot, C. (1971). The Karthala Caldera, Grande Comores. *Bulletin of Volcanology*, **34**, 663–680.

Suh, C. E., Sparks, R. S. J., Fitton, J. G. *et al.* (2003). The 1999 and 2000 eruptions of Mount Cameroon: eruption behaviour and petrochemistry of lava. *Bulletin of Volcanology*, **65**, 267–281.

Suh, C. E., Luhr, J. F. and Njome, M. S. (2008). Olivine-hosted glass inclusions from scoriae erupted in 1954–2000 at Mount Cameroon volcano, West Africa. *Journal of Volcanology and Geothermal Research*, **16**, 1–33.

Suh, C. E., Stansfield, S. A., Sparks, R. S. J. *et al.* (2011). Morphology and structure of the 1999 lava flows at Mount Cameroon Volcano (West Africa) and their bearing on the emplacement dynamics of volume-limited flows. *Geological Magazine*, **148**, 22–34.

Tabod, C. and Bekoa, A. (2009). *Monitoring Seismic and Volcanic Activity in Cameroon. Advanced Workshop on Evaluating, Monitoring and Communicating Volcanic and Seismic Hazards in East Africa*. ICTP Workshop, Trieste, Italy,

17–28 August 2009. Theste, Italy: The Abdus Salam International Centre of Theoretical Physics.

Tazieff, H. (1973). La signification tectonique de l'Afar. *Revue de Géographie Physique et de Géologie Dynamique*, **15**(2), 341–346.

Tazieff, H. (1977). An exceptional eruption: Mt. Nyiragongo, January 10th, 1977. *Bulletin of Volcanology*, **40**, 1–12.

Tazieff, H. (1984). Mt. Nyiragongo: renewed activity of the lava lake. *Journal of Volcanology and Geothermal Research*, **20**, 267–280.

Tazieff, H. (1995). Volcanic risk for Rwandan refugees. *Nature*, **376**, 394.

Tedesco, D., Papale, P., Vaselli, O. and Durieux, J. (2002). *The January 17th, 2002 Eruption of Nyiragongo, Democratic Republic of Congo, Final Report*. Geneva: UN Office for the Coordination of Humanitarian Affairs.

Tedesco, D., Vaselli, O., Papale, P. *et al.* (2007). January 2002 volcano-tectonic eruption of Nyiragongo volcano, Democratic Republic of Congo. *Journal of Geophysical Research B: Solid Earth*, **112**, B09202.

Tedesco, D., Tassi, F., Vaselli, O. *et al.* (2010). Gas isotopic signatures (He, C, and Ar) in the Lake Kivu region (western branch of the East African rift system): Geodynamic and volcanological implications. *Journal of Geophysical Research*, **115**, B01205.

Thierry, P., Stieltjes, L., Kouokam, E. *et al.* (2006). *Projet GRINP-composante 1 Réalisation d'une carte de zonage des risques du Mont Cameroun Rapport final*. BRGM Report RC-54727-FR, Orléans, 333 pp.

Thierry, P., Stieltjes, L., Kouokam, E., Nguéya, P. and Salley, P. M. (2008). Multi-hazard risk mapping and assessment on an active volcano: the GRINP project at Mount Cameroon. *Natural Hazards*, **45**, 429–456.

Tiercelin, J.-J., Chorowicz, J., Bellon, H. *et al.* (1988). East African Rift System: offset, age and tectonic significance of the Tanganyika–Rukwa–Malawi intracontinental transcurrent fault zone. *Tectonophysics*, **148**, 241–252.

Toteu, S. F., Penaye, J. and Poudjom Djomani, Y. H. (2004). Geodynamic evolution of the Pan-African belt in Central Africa with special reference to Cameroon. *Canadian Journal of Earth Sciences*, **41**, 73–85.

Tsafack, J.-P. F., Wandji, P., Bardintzeff, J.-M., Bellon, H. and Guillou, H. (2009). The Mount Cameroon stratovolcano (Cameroon Volcanic Line, Central Africa): Petrology, geochemistry, isotope and age data. *Geochemistry, Mineralogy and Petrology*, **47**, 65–78.

Ukstins, I., Renne, P., Wolfenden, E. *et al.* (2002). Matching conjugate volcanic rifted margins: $^{40}Ar/^{39}Ar$ chronostratigraphy of pre- and syn-rift bimodal flood volcanism in Ethiopia and Yemen. *Earth and Planetary Science Letters*, **198**, 289–306.

UNISDR, The United Nations Office for Disaster Risk Reduction (2009). 2009 UNISDR terminology on disaster risk reduction. Available at: http://www.unisdr.org/files/7817_UNISDRTerminologyEnglish.pdf (accessed May 2013).

United Nations (2011). *World Population Prospects, the 2010 Revision*. Department of Economic and Social Affairs, Population Division. CD-ROM Edition.

UNOPS (United Nations Office for Project Services) (2013). *Monitoring Volcanic Risk and Raising Awareness*. Available at: http://www.unops.org/english/whatwedo/UNOPSinaction/Pages/Monitoring-Volcanic-Risk.aspx (accessed May 2013).

US Population Reference Bureau (2012). *World Population Data sheet*. Available at: http://www.prb.org/pdf12/2012-population-data-sheet_eng.pdf (accessed May 2013).

USAID, US Agency for International Development (2002). *Democratic Republic of Congo – Volcano*. Fact Sheet #5, Fiscal Year (FY) 2002. Available at: http://www.usaid.gov/hum_response/ofda/situation.html (accessed January 2013).

Van der Hilst, R. D. and Karason, H. (1999). Composition heterogeneity in the bottom 1000 kilometres of Earth's mantle; toward a hybrid convection model. *Science*, **283**, 1885–1888.

Vernier, J.-P., Thomason, L. W., Fairlie, T. D. *et al.* (2013). Comment on "Large volcanic aerosol load in the stratosphere linked to Asian monsoon transport". *Science*, **339**, 647d.

Vidal, P., Deniel, C., Vellutini, P. J. *et al.* (1991). Changes of mantle source in the course of a rift evolution. *Geophysical Research Letters*, **18**, 1913–1916.

Von Heuglin, T. (1862). *Expedition nach Inner-Afrika: Justus Perthes*. Mittheilungen aus Justus Perthes' Geographischer Anstalt, Gotha, pp. 22–23.

Wantim, M. N., Kervyn, M., Ernst, G. G. J. *et al.* (2013). Numerical experiments on the dynamics of channelized lava flows at Mount Cameroon volcano with the FLOWGO thermo-rheological model. *Journal of Volcanology and Geothermal Research*, **253**, 35–53.

Wauthier, C., Cayol, V., Kervyn, F. and d'Oreye, N. (2012). Magma sources involved in the 2002 Nyiragongo eruption, as inferred from an InSAR analysis. *Journal of Geophysical Research B: Solid Earth*, **117**, art. no. B05411.

Wescott, W. A., Krebs, W. K., Engelhardt, D. W. and Cunningham, S. M. (1991). New biostratigraphic age dates from the Lake Rukwa rift basin in Western Tanzania. *American Association of Petroleum Geologists Bulletin*, **75**(7), 1255–1263.

Wiart, P. and Oppenheimer, C. (2000). Largest known historical eruption in Africa: Dubbi volcano, Eritrea, 1861. *Geology*, **28**(4), 291–294.

Wiart, P., Oppenheimer, C. and Francis, P. (2000). Eruptive history of Dubbi volcano, northeast Afar (Eritrea), revealed by optical and SAR image interpretation. *International Journal of Remote Sensing*, **21**, 911–936.

Wilkinson, P., Mitchell, J. G., Cattermole, P. and Downie, C. (1986). Volcanic chronology of the Meru–Kilimanjaro region, Northern Tanzania. *Journal of the Geological Society, London*, **143**, 601–605.

Williams, L. A. J., Macdonald, R. and Chapman, G. R. (1984). Late Quaternary volcanoes of the Kenya rift valley. *Journal of Geophysical Research*, **89**, 8553–8570.

Woldegabriel, G., Aronson, J. L. and Walter, R. C. (1990). Geology, geochronology and rift basin development in the central sector of the main Ethiopia Rift. *Geological Society of America Bulletin*, **102**, 439–458.

Woldegabriel, G., Walter, R. C., Aronson, J. L. and Hart, W. K. (1992). Geochronology and distribution of silicic volcanic rocks of plio-Pleistocene age from the central sector of the main Ethiopian rift. *Quaternary International*, **13–14**, 69–76.

Wolfenden, E., Ebinger, C., Yirgu, G., Renne, P. and Kelley, S. P. (2005). Evolution of the southern Red Sea rift: birth of a magmatic margin. *Geological Society of America Bulletin*, **117**, 846–864.

Yirgu, G., Ebinger, C. J. and Maguire, P. K. H. (eds.) (2006). *The Afar Volcanic Province within the East African Rift System.* Special Publications no. 25. London: Geological Society of London, pp. 1–6.

Zogning, A. (1988). Le Mont Cameroun, un volcan actif: contribution à l'étude de géographie physique appliquée. Ph.D. thesis, University of Yaoundé, Cameroon, 447 pp.

Zumbo, V., Feraud, G., Betrand, H. and Chazot, G. (1995). $^{40}Ar/^{39}Ar$ chronology of Tertiary magmatic activity in southern Yemen during the early Red Sea-Aden rifting. *Journal of Volcanology and Geothermal Research*, **65**(3–4), 265–279.

14

Recent volcanic eruptions in the Afar rift, northeastern Africa, and implications for volcanic risk management in the region

Gezahegn Yirgu, David J. Ferguson, Talfan D. Barnie, and Clive Oppenheimer

14.1 Introduction

The Afar region is a large, triangular low-lying (− 120–500 m elevation) area stretching from Eritrea in the north to Djibouti in the east and Ethiopia in the south. It is bordered by the Ethiopian plateau to the west (~2000–3000 m), the Somali plateau to the southeast, and the Danakil block to the northeast (Fig. 14.1). For the most part Afar is sparsely populated with a few large towns located along the major routes and at the principal ports on the Red Sea and the Gulf of Aden. Only small settlements are found close to some active volcanoes. Sparks *et al.* (2012) have identified 65 Holocene volcanoes in the Ethiopian and Afar rifts comprising Eritrea, Ethiopia, and Djibouti. Thirty-five of these are predominantly effusive in character while the rest show evidence for significant explosive eruptions. Many of these volcanoes are in the Afar region, but there are none that are operationally monitored.

In September 2005 a seismotectonic event began in west-central Afar accompanied by mafic dyking and a small silicic eruption. Since that time this sector of the rift has been the focus of major geoscientific studies using multi-disciplinary approaches including ground-based and satellite geodesy, potential-field gravity measurements, seismology, and petrology. Over this time period a series of rifting events has led to the emplacement of 14 basaltic dykes with a total width of 5–8 m (Hamling *et al.*, 2009, 2010). Between 2005 and 2012 there were four eruptions in this region: a rhyolitic ash and dome eruption following the 2005 dyking event at Da'Ure on the northern flanks of Dabbahu (Wright *et al.*, 2006), and three basaltic fissure eruptions in the Dabbahu-Manda-Hararo rift segment (Ferguson *et al.*, 2010). These volcanoes pose significant hazards to the population, infrastructure, and the economy. This study presents a brief overview of the eruptions in the region between 2005 and 2012 and their impacts, and presents the state of hazard awareness and future prospects.

14.2 Geological setting and characteristics of volcanism

Geologically, the Afar depression comprises the Afar rift triple junction where the Red Sea and Gulf of Aden spreading ridges emerge on land and meet the northern extension of the East African rift (see also Chapter 13). The triple junction formed as a result of rifting between Africa and Arabia over the past ~30 My (McKenzie *et al.*, 1970) (Fig. 14.1). The three arms of the triple junction formed within the Oligocene flood basalt province associated with the Afar mantle plume (Courtillot *et al.*, 1984; Schilling *et al.* 1992; Hofmann *et al.* 1997; Ebinger and Sleep 1998; Montelli *et al.*, 2004). Low P- and S-wave velocities in the upper mantle beneath the uplifted plateaux of Arabia and northeast Africa indicate that the thermal anomaly associated with the plume persists to the present (Benoit *et al.*, 2006; Bastow *et al.*, 2008).

The Arabian plate is currently separating from Africa at a rate of 16 mm y^{-1}, and across the northern sector of the Main Ethiopian Rift (MER) the rate of extension between the Somalia and African plates is 3–6 mm y^{-1} (Chu and Gordon, 1998, 1999). In the Gulf of Aden, the rifting has progressed to the stage of oceanic spreading (e.g., Manighetti *et al.*, 1997), and the Red Sea rift has opened southwards to 14° N (Prodehl *et al.*, 1997), where it is believed to have moved on land thus isolating the Danakil microplate. In Afar and the MER, these extensional tectonic processes are manifested by frequent and widespread seismic activity.

The floor of the depression is composed mostly of volcanic rocks with limited continental sediments. The major episodes of volcanism can be summarised as

Extreme Natural Hazards, Disaster Risks and Societal Implications, eds. Alik Ismail-Zadeh, Jaime Urrutia Fucugauchi, Andrzej Kijko, Kuniyoshi Takeuchi and Ilya Zaliapin. Published by Cambridge University Press.

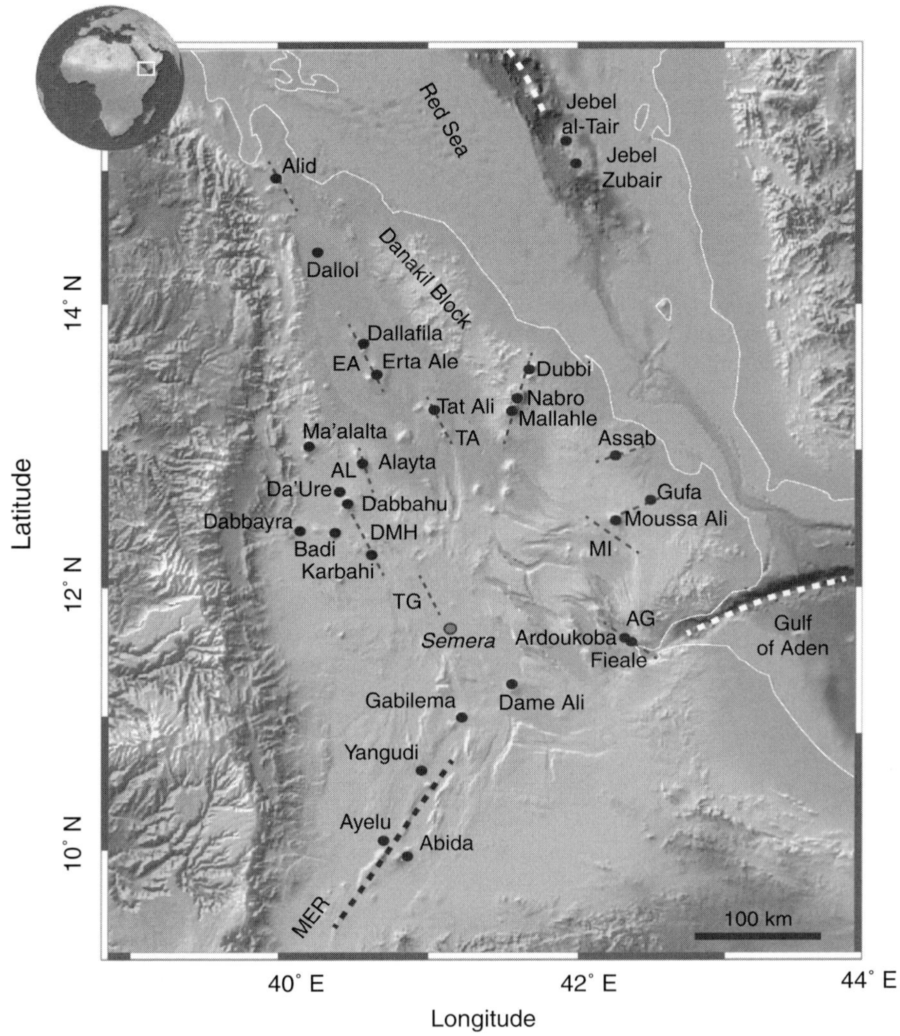

Figure 14.1. Topographic map of the Afar depression and surrounding areas showing the locations of significant or recently active volcanoes and volcanic centres. Current volcanism in Afar is concentrated along rift zones (e.g. Erta Ale, Dabbahu-Manda-Hararo, Asal-Ghoubbet, etc.) or at volcanic centres found around the margins of the depression (e.g. Nabro and Ma'alalta). Dashed lines show structural trends of rift zones or marginal volcanic lineations. At its southern end Afar meets structures associated with the Main Ethiopian Rift near the regional capital Semera. EA: Erta 'Ale, TA: Tat Ali, AL: Alayta, DMH: Dabbahu-Manda Hararo, TG: Tendaho Graben, AG: Asal-Ghoubbet, MI: Manda-Inakir, MER: Main Ethiopian Rift.

follows: (1) old flood basalts, older than ~25 Ma, formed by fissure flows overlying Mesozoic sedimentary rocks; (2) two phases of extensive rhyolitic volcanism that occurred between ~25 and ~19 Ma and ~14 and ~9.5 Ma (Mabla Series); (3) a thick basaltic succession emplaced between ~9 and ~1 Ma (Dalha and Afar Sratoid Series); and (4) recent fissure basalt flows (<1 Ma) that occur as elongated bodies within rift zones (Afar Axial Ranges). Nearly two-thirds of Afar is covered by a basalt-dominated sequence of lavas referred to as the Afar Stratoid Series, emplaced during Plio-Quaternary times over the last 3 Ma (e.g., Varet, 1978; Kidane *et al.*, 2003; Lahitte *et al.*, 2003). Individual basalt flows are between 1 and 6 m thick and the whole Series is up to 1500 m thick. In the east and west of the Afar, around the margins of the depression, are the Tranverse volcanic ranges and Marginal volcanic centres. These consist of east-northeast trending volcanic–tectonic structures that are intercalated with the top of the Afar Stratoid Series (Barberi *et al.*, 1974; Barberi and Varet, 1975; Varet, 1978).

Over the past ~1 Ma, faulting and volcanism in Afar have become localised along ~60-km-long, 20-km-wide axial volcanic ranges or magmatic segments of aligned magmatic centres, similar in size, morphology structure, and spacing to the second-order, non-transform offset segmentation of slow-spreading mid-oceanic ridges (Hayward and Ebinger, 1996; Manighetti *et al.*, 1998). The Quaternary magmatic segments are characterised by fissure eruptions, aligned basaltic scoria cones, and huge shield volcanoes with transitional alkali/tholeiitic composition and per-alkaline silicic rocks (Barberi *et al.*, 1972, 1975; Field *et al.*, 2012a) that occur along northwest–southeast trending narrow rift zones. These segments are underlain by heavily-intruded crust and constitute zones of shallow seismicity and positive gravity anomalies, collapsed graben, and domino-tilted blocks (Barberi and Varet, 1977;

Hayward and Ebinger, 1996; Acocella, 2010; Beutel *et al.*, 2010; Ebinger *et al.*, 2010). Extension across these segments is thought to occur chiefly via the intrusion of dykes into the upper crust (Wright *et al.*, 2006; Rowland *et al.*, 2007; Ebinger *et al.*, 2008). In northern and central Afar, the magmatic segments include Erta 'Ale, Tat 'Ali, Alayta, Dabbahu-Manda-Hararo, Manda Inakir and Asal-Ghoubbet (Fig. 14.1). The northernmost segments of Erta 'Ale and Tat 'Ali are characterised by axial volcanic ranges, south of these the rift axis moves westwards to the Alayta and Manda-Hararo magmatic segments and Tendaho graben (Barberi *et al.*, 1972). Unlike the more northerly magmatic segments, the Dabbahu-Manda-Hararo segment and Tendaho graben are characterised by negative topography controlled by normal faulting (Hayward and Ebinger, 1996; Rowland *et al.*, 2007). The marginal centres and transverse ranges are characterised by the occurrence of composite volcanoes with summit calderas, whose eruption products range from basalt to rhyolite via a range of intermediate magma types. (Barberi and Varet, 1975; Varet, 1978). Quaternary sedimentary rocks in the Afar depression are dominated by lacustrine deposits. Significant lacustrine sedimentary rocks were deposited in central Afar along the Manda Hararo–Gobaad rift zones between ~12 and 1 Ka (Rognon, 1975). Lacustrine sedimentary rocks of ~180–200 m thickness cover the Awsa plain in east-central Afar (Varet, 1978).

14.3 Chronology of eruptions in Afar between 2005 and 2012

Between 2005 and 2012, seven volcanic eruptions occurred at different centres in the Afar region. Due to the remoteness of many of these volcanoes much of the information concerning these events has come from satellite observations, which in some cases have been followed by field investigations and geochemical and geophysical studies. The initial identification of a new eruption has typically occurred via observations of volcanic sulphur dioxide emissions by the Ozone Monitoring Instrument (OMI), which is carried aboard the EOS-AURA satellite, or thermal anomalies detected by the MODIS instrument aboard NASA's Terra and Aqua satellites (e.g. Wright *et al.*, 2004). The thermal data from MODIS have allowed the locations of the eruption to be identified to within a few kilometres. As eruptions progressed, repeated observations by these space-borne instruments have allowed near real-time monitoring of volcanic phenomena such as gas, ash, and lava emissions. This method of detection has been of particular significance in the rapid recognition of new eruptive activity, permitting the scientific community to respond quickly to events, and allowing field expeditions to reach the eruption sites within a few days of the start of each event. Orbital remote sensing has also reduced the magnitude threshold for detecting eruptions in remote, sparsely populated regions. For instance, the May 2010 Manda Hararo eruption was of such small scale and duration that it may well have been missed by the scientific community were it not for a small number of hot pixels detected by MODIS. As a result, the historical survey of eruptions probably has a comparatively high magnitude threshold above which the catalogue can be considered 'complete'.

14.3.1 Historical eruptions pre-2005

14.3.1.1 *Dubbi volcano, May 1861*

Dubbi is a large stratovolcano located in the northeast part of the Afar triangle (Fig. 14.1). It has a maximum elevation of ~1625 m and in May 1861 produced a major explosive eruption from a 4-km-long fissure at the summit of the volcano. This is the earliest eruption in Afar for which historical and eyewitness accounts are known to exist (Wiart *et al.*, 2000; Wiart and Oppenheimer, 2000). This is partly due to the volcano's impact on maritime traffic in the Red Sea, which was disrupted by the 1861 activity. This eruption also has the highest number of reported fatalities of any known eruption in Afar, with reports of around 180 deaths in the surrounding area.

The initial explosive phase of the eruption began in early May 1861, with ash fall from the volcanic plume reported along the surrounding coastal region and across the Red Sea in Yemen, up to several hundred kilometres from the volcano. This initial phase also involved the emplacement of pyroclastic flows, which may have travelled up to ~25 km from the volcano's summit, and are likely to be responsible for most of the loss of life caused by this event (Wiart and Oppenheimer, 2000). It is not known precisely how long this initial explosive phase lasted; however, some reports suggest that further large explosions and ash fall occurred in mid-July, over a month after the start of the eruption. By mid-September the eruption had become more effusive and lava flows erupted from the summit fissure travelled up to ~20 km down the flanks of the volcano (Wiart *et al.*, 2000). On the volcano itself this more effusive phase was associated with lava fountain activity along the eruptive fissure. The eruptive products identified span trachytic pyroclastic rocks to basaltic trachyandesite lavas. The volume of lava flows alone has been estimated to be ~3.5 km^3 (Wiart and Oppenheimer, 2000), making it the largest documented historical eruption in Africa.

14.3.1.2 *Alayta, 1907, 1915*

There are limited reports of eruptions in the early twentieth century in the Alayta/Afdera region south of the Erta 'Ale axial volcanic range and north of the Manda Hararo rift (collated by the Smithsonian Institution's Global Volcanism Program). The eruptions were initially attributed to Afdera, but the apparent age of the most recently erupted material makes Alayta a more likely candidate. The Alayta 'rift segment' has been referred to as a single feature in the more recent literature, but was initially identified as two, distinct, sub-parallel volcano-tectonic features (e.g., Barberi *et al.*, 1974), a large shield volcano elongated north-northwest erupting basaltic through to intermediate products in the east and a more basic lava flow field in the west elongated north-northeast sourced from a separate, sub-parallel line of fissures. It is unclear which of these lineations was active (or both).

14.3.1.3 *Erta 'Ale lava lake, 1968–1974*

The Erta 'Ale magmatic rift segment is located in the northern part of the Afar region and is an 80-km-long, 35-km-wide continuous volcanic ridge made up of numerous distinct eruptive centres of picritic through transitional basalt to peralkaline rhyolite composition (Barberi and Varet, 1970; Bizourd *et al*, 1979) (Fig. 14.1). The rift segment forms part of the on-land section of the Red Sea rift system and marks the axis of extension between Africa and Arabia. The rift segment takes its name from the Erta 'Ale volcano itself, which lies in the centre of the segment and has historically hosted two lava lakes within pit craters sunk into the 1.5 km × 1 km summit caldera, of which only one is active at the time of writing. The first recorded observations of activity at the summit are thought to date back to 1907, but systematic observations by earth scientists began only in the late 1960s. During this period the lava levels in both pit craters were observed to have risen simultaneously by 160 m, resulting in overspills on to the caldera floor. In 1973 an overspill of the lava lakes produced a 1-km-long lava flow. The volcano is persistently active with sustained convection in the lava lake (Oppenheimer and Yirgu, 2002; Oppenheimer *et al.*, 2004) and fluctuations in magma levels at the southern pit crater, and fumarolic activity in the northern pit crater. This activity is believed to be due to the continuous degassing of the underlying magma reservoir. However, space-borne geodetic investigations have not revealed associated ground deformation (Amelung *et al.*, 2000).

14.3.1.4 *Ardoukoba fissure eruption, November 1978*

A small basaltic eruption took place in November 1978 at Ardoukoba (11.58° N, 42.47° E) (Fig. 14.1) in the Asal-Ghoubbet magmatic segment, Djibouti Republic. A description of this event can be found in Demange and Tazieff (1978). This eruption began on 7 November, following a series of earthquakes. These included a magnitude 5.3 event that occurred immediately prior to the onset of the eruptive activity. The eruption was localised along a 500-m-long fissure about 3 km southeast of Lake Asal and lasted less than a week. Two basaltic lava flows were reportedly emplaced: one flow travelled ~1 km to the southeast, the other ~0.5 km to the northwest. Lava effusion rates reached an estimated maximum of 1000 m^3 min^{-1}. Ash clouds reached ~300 m altitude while scoria and bombs were ejected to a height of ~70 m. The eruption formed a crater ~30 m in diameter and built a cone ~100 m high. The eruption ended after nine days of activity. No casualties or damage were reported. The total volume of lava emitted was estimated to be at least 12×10^6 m^3. Lava covered an area of slightly more than 3 km^2. Flow thicknesses reached 25 m but mostly ranged from 1 to 2 m. About 25 fissures, parallel to the northwest–southeast trend of the rift, opened during the activity. Most fissures were from 1–2 mm to about 1 m wide, but there was one short segment of 3 m width.

14.3.2 Eruptions since 2005

Since 2005 there has been a notable rise in volcanic activity in Afar, with six eruptions having occurred in different locations in the period 2005–2010. Following the onset of the rifting phase at the Dabbahu-Manda-Hararo (DMH) rift zone in central Afar in 2005 (e.g. Ebinger *et al.*, 2010) a small rhyolitic eruption (2005) and three dyke-fed basaltic fissure eruptions (2007, 2009, and 2010) have taken place along this magmatic rift segment. Further eruptions took place in 2008 at Alu Dallafila, part of the Erta 'Ale volcanic range in northern Afar, and also in 2011 at Nabro volcano in the Nabro Volcanic Range in eastern Afar. With the exception of the Nabro eruption, all of these have occurred along the on-land axis of the Red Sea rift system, which marks the axis of extension between Africa and Arabia. These events have been studied with a variety of field-based and remote-sensing techniques and have been well characterised with respect to magnitude, duration, and eruptive style.

14.3.2.1 *The September 2005 Da'Ure eruption*

In September 2005, a small explosive eruption took place at the northern end of the DMH rift segment (Fig. 14.1). The eruptive fissure was located on the southern flank of the Da'Ure silicic volcanic dome (12°39′ 01.60″ N, 40°31′ 09.05″ E, ~710 m elevation) around 5 km northeast from the summit of Dabbahu volcano (Yirgu *et al.*, 2006, Wright *et al.* 2006). This is believed to be the first rhyolitic eruption in this region since the 1861 eruption of Dubbi. The eruption was preceded by two strong earthquakes that

(a)

(b)

Figure 14.2. Images of volcanic activity at the DMH rift zone in central Afar. (**a**) Eruptive fissure from the 2005 Da'Ure rhyolitic eruption on the northern side of Dabbahu volcano (people for scale). (**b**) A section of a basaltic fissure eruption in the Karbahi region in June 2007. The 4-km-long fissure opened along the axis of the DMH rift zone. Both these eruptions were associated with dyke intrusions along the rift axis. Photos courtesy of Lorraine Field (a) and David Ferguson (b). For colour version, see Plates section.

shook the area on 26 September at about 12:30 p.m. local time (09:31 UTC, $m_b = 4.4$; 09:33 UTC, $m_b = 5.2$), causing surface displacements of up to 1 m on nearby fault scarps. During the eruption, local pastoralists reported witnessing a black column of smoke rising high into the atmosphere and spreading out to form an umbrella-shaped cloud that partly darkened the area for three days. Rhyolitic ash and pumice were deposited around the fissure and a small pumiceous lava dome was emplaced at the centre of the vent (Fig. 14.2).

The eruption was centred on a N10° W trending fissure around 400 m in length, 80-m-wide and up to 50-m-deep, which broke through pre-existing felsic lava flows and unwelded pyroclastic deposits (Fig. 14.2(a)). Proximal tephra deposits consisted largely of ash and minor pumice fragments and in the immediate vicinity of the vent (~500 m) these reached a maximum thickness of 15 cm. The Da'Ure ash cloud probably rose to a few kilometres above the eruptive fissure and thin distal deposits from the eruption consisted of fine white ash, which was reported by local pastoralists to have fallen around much of the surrounding region, including in the village of Teru some 40 km south-west of the vent. The main eruptive phase appears to have been relatively brief and had ceased by the time the first field observations were made on 5 October, nine days after the start of the eruption. Although no explosive or extrusive activity was observed during this visit, intense degassing accompanied by a strong smell of sulphur was recorded from the length of the fissure as well as a loud 'boiling sound' from beneath the newly emplaced lava dome. Both these phenomena had subsided by the time a subsequent visit was made on 15 October. The total erupted volume from the Da'Ure event was relatively low and has been estimated to be $<10^5$ m^3.

A notable feature of this eruption is that it accompanied a significant seismic-tectonic event, caused by the intrusion of a large dyke along the 60 km length of the DMH rift zone. This intrusion injected ~2 km^3 of (most likely basaltic) magma into the shallow crust and geodetic observations recorded several metres of surface extension perpendicular to the axis of intrusion (Wright *et al.*, 2006). Patterns of seismicity recorded prior to the eruption suggest that it was probably triggered by the thermal and/or mechanical perturbation of an existing shallow silicic magma body by this dyke, as it propagated northwards from the centre of the segment (Wright *et al.*, 2006).

14.3.2.2 *The August 2007 Karbahi eruption*

On 12 August 2007, a fissural basaltic eruption began in the southern part of the DMH rift zone. Details of this eruption (and a subsequent event here in 2009) can be found in Ferguson *et al.* (2010). The 2007 eruptive fissure was located along the rift axis, inside the axial graben at a location known locally as Karbahi. Outside of the local area, the first indications of the eruption were heat and gas emissions detected by satellite remote-sensing instruments. The first field visit occurred on 23 August, by which time the eruptive activity had ceased with the exception of low-level residual gas emissions. During this visit newly emplaced lava flows and scoria cones were observed along the fissure (Fig. 14.2(b)). The entire fissure system extended ~4 km in length and was divided into two parts, which fed different areas of new lava emplacement. In the northern section a small flow field was fed by an intermittently active fissure around 2 km in length, while in the southern part a ~1-km-long lava flow was fed from a 0.5-km-long fissure. Low (~10–30 m high) elongate scoria cones were constructed above these fissures during moderate lava fountaining, which was slightly more explosive in the southern section (as evidenced by higher scoria cones). Near the fissure, lavas flows were of pāhoehoe type. These became channelled downflow and made a transition to aā flows. The total duration of the eruption was ~48 h, with the bulk of material having erupted in the first 24 h.

The total area covered by the 2007 lavas was 2.2 km^2, while the bulk erupted volume is estimated, based on field observations and satellite imagery, to be $(4.4–8.8)\times10^6$ m^3. The erupted lava was basaltic in composition (~48 wt% silicon dioxide) and geodetic evidence suggests that it was fed by a dyke that propagated southwards from the centre of the segment (~10 km north of the fissure). This was the first eruption since the onset of the active rifting phase in 2005 at the Dabbahu rift zone to be directly fed by a dyke intruded along the rift axis.

14.3.2.3 *The November 2008 Alu-Dallafila eruption*

The Alu-Dallafilla eruption on the Erta 'Ale volcanic range began on 3 November 2008 and was first recognised by a large sulphur dioxide cloud over Afar (Pagli *et al.*, 2012). The Alu-Dallafilla volcanic sector lies 30 km north of the Erta 'Ale lava lake and caldera and includes three distinct eruptive centres: a basaltic volcanic cone (Alu), a 3-km-long basaltic volcanic range (Alu-South), and a silicic stratovolcano (Dallafilla). Fumarolic activity was observed at some faults in Alu and in the Dallafilla crater in 1968 (Barberi and Varet, 1977) but there has been no report of any volcanic eruption in the Erta 'Ale segment since the first observations in 1907, except for overflow events from the Erta 'Ale lava lake (Oppenheimer and Francis, 1998; Amelung *et al.* 2000).

Infrequent precursory seismicity was recorded on 30 October and again at 11:10 UTC on 3 November 2008 (Belachew *et al.*, 2011). By 12:22 UTC, this seismic activity had increased in frequency and magnitude and a nearly continuous swarm of local magnitude (M_L) < 4 earthquakes occurred between 12:22 and 12:55 UTC. These have been linked to surface faulting above a propagating dyke, which ultimately fed the eruption. Supporting evidence for this comes from the appearance of thermal anomalies in images acquired by the SEVIRI instrument (carried aboard the geostationary METEOSAT satellite) between 12:57 and 13:12 UTC on 3 November 2008. ASTER thermal infrared images acquired from NASA's Terra satellite showed that two en-echelon eruptive fissures opened southeast of Alu and lava flows from these flowed towards the east and north of Dallafila. OMI sulphur dioxide and MODIS thermal images show that a volcanic plume originated from the eruptive fissures on 3 November and drifted northeastwards. Direct observations of the eruption are lacking but based on a combined analysis of satellite images and seismicity, Pagli *et al.* (2012) deduced that the eruption started on 3 November and continued for at least three days. From thermal imagery, it is apparent that the Alu-Dallafila eruption emitted voluminous basaltic flows that travelled for more than 15 km. This has become Ethiopia's largest basaltic eruption on record. Because of the remoteness of the area and security concerns, it was not possible to organise field visits to the eruption site. There are no reports concerning the impacts of this eruption on life and property.

14.3.2.4 *The June 2009 Karbahi eruption*

In June 2009, a further basaltic fissure eruption occurred along the axis of the Dabbahu rift zone. Like the 2007 eruption, the eruptive fissure was located in the southern part of the rift segment and the 2007 lava flows were largely resurfaced during this eruption (Ferguson *et al.*,

2010). Both sulphur dioxide and thermal signals were detected with space-borne sensors on 28 June 2009 and the eruption is estimated to have begun around 17:00 UTC. The eruption was preceded 7 h earlier by a magnitude 4.4 earthquake, reported by the European-Mediterranean Seismological Centre. Although slightly longer, the eruptive fissure was located in the same location as the 2007 eruption, and the new eruption products almost entirely covered the August 2007 flows. The new lavas can be clearly seen by a comparison of satellite images acquired between the two eruptions (Ferguson *et al.*, 2010).

Near the fissure, lava was emplaced via pāhoehoe-type flows, which evolved into channelised aā-type flows with distance. Scoria cones, ~50 m in height, were also constructed along the fissure. The new lava flows and proximal scoria deposits (largely overlying the previous lava flows) cover an estimated area of 4.5 km^2 and represent a total bulk volume of $(13.5–18)\times10^6$ m^3. The eruption reactivated the entire length of the 2007 fissure and extended the system by ~500 m to the south and 1 km to the north. The active fissure length totalled 5.2 km. The sulphur dioxide plume from the eruption reached an altitude of ~3–5 km and during the most vigorous stage drifted ~380 km approximately southwest from the source. The mass of the sulphur dioxide emissions is estimated at around 15–36 Gg. The erupted lava was very similar in composition to that of 2007 and was fed by magma ascending from a reservoir beneath the centre of the Dabbahu rift zone. The mean effusion rates for both the ephemeral 2007 and 2009 Karbahi events (~55–70 m^3 s^{-1}) are similar to those previously estimated for fissure eruptions in Iceland (e.g., Harris *et al.*, 2000).

14.3.2.5 *The May 2010 Karbahi eruption*

A small volcanic eruption reportedly occurred at 12.33° N, 40.61° E in the Karbahi section of the DMH axial rift, approximately 2 km east of the central graben. This was preceded by intense seismic activity reported by the Institute of Geophysics, Space Science and Astronomy of Addis Ababa University. SEVIRI radiance time series show that lava eruption began on 21 May 2010 around 01:00 UTC, and finished by midday that day. Hotspots were detected by the Fire Information for Resource Management System (FIRMS, which is based on MODIS thermal observations) at the eruption site on 23 May. A small sulphur dioxide plume over the region was also observed by the OMI sensor. An EO-1 hyperion/ALI image (commissioned by Ashley Davies at the Jet Propulsion Laboratory of the California Institute of Technology, USA, acquired on the morning of 22 May) revealed a very small lava flow covering less than 0.25 km^2 in the said area. The geomorphology evident in the ALI image combined with the geological context suggests a basaltic composition. As far as the authors are aware the lava flow has not been sampled. Local residents reported having seen plumes of smoke and flowing lava. This eruption was the fourth to have occurred in the Dabbahu rift since the DMH rift episode began in September 2005.

14.3.2.6 *The November 2010 Erta 'Ale overspills*

Field *et al* (2012b) reported two days of field observations of the Erta 'Ale lava lake made in November 2010. These enabled the authors to relate variations in the thermal radiative output of the lava lake measured every 15 min by the SEVIRI instrument to overflows of the lake. The radiance time series indicates two periods associated with rising lake level, overflows, and falling lake level, one between 11 and 13 November, the other between 19 November and 15 December. The field observations were made between 21 and 23 November, when the pit the lava lake had formerly occupied had completely filled in and the lake had risen above the level of the caldera floor and was contained within an incipient scoria cone, flanked by overflows up to 90 m long. The hiatus between the two volcanic events coincided with a large seismic swarm in the Gulf of Aden. Coupled with the observation of increased lake activity from MODIS hotspots during the 2005 mega-dyke intrusion in the Manda Hararo rift (Ayele *et al.*, 2009), this suggests a link between lava lake activity and regional tectonic events (Barnie and Oppenheimer, pers. comm.).

14.3.2.7 *The June 2011 Nabro eruption*

Nabro caldera volcano (13° 22′ N, 041° 42′ E, 2218 m elevation, Eritrea) erupted explosively on 12 June 2011. This was the first eruption of the volcano in recorded history (Goitom *et al.*, 2013). Nabro is the highest of the stratovolcanoes in the Afar depression and belongs to a group of caldera volcanoes termed by Wiart and Oppenheimer (2005) the Nabro volcanic complex. The volcano is a predominantly trachytic edifice with abundant ignimbrites characterising the main periods of edifice growth, collapse, and post-collapse volcanism. Later erupted products include basaltic lava flows emitted from north-northwest–south-southeast vents or fissures. Nabro consists of two nested calderas breached on the southwestern side. The outer larger caldera is ~8 km across and opens to the south-southwest. Within this caldera, two much smaller craters are found, one inside the other. The inner wall of the larger caldera has steep sides, some as high as 400 m.

Wider reporting of the Nabro eruption followed recognition in satellite imagery of a large ash plume on 13 June 2011. The eruptive vent was identified at the centre of the main caldera. The explosions were heard in Ethiopia as far as Afdera village, 90 km away, while ash fallout was reported in the towns of Mekele, 240 km west, and Semera, 190 km south-southwest of Nabro. The eruption was preceded by an earthquake swarm, including two *M*5.7 earthquakes, in the vicinity of the volcano itself (as reported by the US Geological Survey (USGS)). Satellite observations showed an ash plume initially rising to altitudes of 9.1–13.7 km. The plume drifted over much of East Africa and the Middle East causing the cancellation and rerouting of hundreds of flights. The eruption released an estimated 1.3 Tg of sulphur dioxide, which resulted in a significant sulphate aerosol enhancement in the stratosphere (Bourassa *et al.*, 2012; Sawamura *et al.*, 2012). There is some controversy over whether the plume was initially emplaced into the upper troposphere (Bourassa *et al.*, 2012) or lower stratosphere (Fromm *et al.*, 2013; Vernier *et al.*, 2013). Nabro's sulphur dioxide emission is one of the largest since the 1991 eruption of Mt Pinatubo in the Philippines. Satellite images also showed a ca. 17-km-long lava flow that travelled first southwest from its source within the inner caldera and then northwest beyond the caldera. Ash eruption continued intermittently until 17 July.

Post-eruptive field observations confirmed tephra dispersal to the west and that there was essentially no trace of the eruption products even in the northeast sector of the main caldera floor. Up to a few metres of scoria accumulated in previously settled areas of the southern part of the caldera. The main lava flow extended within and beyond the caldera and its width exceeded 1 km in some parts. The flow margins were typically 5–10 m thick with thicknesses clearly increasing towards the flow interior. Within the caldera, three shorter flows were recognised. Lava compositions range from trachybasalt to basaltic trachyandesite.

The Nabro eruption resulted in considerable damage and losses within Eritrea and Ethiopia, and several thousand people were displaced as a consequence. By 20 June, eight villages in the Bidu district of Ethiopia were covered by volcanic ash, affecting at least 5000 people and polluting springs and streams. It was also reported that the Lake Afdera salt deposits were covered by volcanic ash and polluted with sulphuric acid. Large salt accumulations were subsequently discarded thus resulting in substantial economic losses. Ash had been deposited at least 60 km across the international border into Ethiopia along west-northwest and south-southwest directions. The main lava flow covered agricultural and settled areas within and beyond the caldera and also engulfed drainages.

The most affected area was arguably Sireru village, located inside the caldera and which had around 2700 residents. On 21 June, the Eritrean government reported seven people had been killed and three injured. In the Bidu administrative district, reports indicate 31 fatalities though we do not know the causes. On 6 July, unofficial reports suggest 48 000 people in the Bidu, Afdera, Erebti, Elidar, Teru, and Kori districts were in need of aid (with ash affecting food and water sources).

It is relevant to mention here that Chernet (2012), based on resident accounts that an unusual dark-coloured smoke was observed from Dallol volcanic crater in the first week of January 2011, considered that a small volcanic event possibly occurred there. Dallol volcano, located on the northern extension of the Erta 'Ale volcanic range, rises some 60 m above the Dallol salt plain (−120 m elevation). According to Chernet (2012), previous phreatic eruptions at Dallol and the surrounding salt plain have formed bubbling acid brine pools that probably lie above maar craters obscured by thick evaporite deposits and frequent marine invasions of the salt plain during the Holocene.

14.3.3 Red Sea eruptions

In addition to the volcanism on land, two island volcanoes in the Red Sea erupted for the first time in over a century between 2005 and 2011.

14.3.3.1 *The September 2007 Jebel Al-Tair eruption*

Jebel Al-Tair is a volcanic island in the southern part of the Red Sea (Fig. 14.1) and administratively part of the Republic of Yemen. Explosive eruptions from this island were recorded in the eighteenth and nineteenth centuries. After remaining dormant for 124 years, the stratovolcano erupted on 30 September 2007. According to unofficial reports, the eruption killed eight people. It released an estimated 80 Gg of sulphur dioxide into the atmosphere (Eckhardt *et al.*, 2008).

14.3.3.2 *The December 2011 Zubair eruption*

The Zubair Group consists of ten major volcanic islands, on top of an underlying shield volcano in the Red Sea (Fig. 14.1), reaching up to 191 m above sea level. The islands belong to the Republic of Yemen. The largest island is Zubair, which is one of the younger volcanic centres. The other young volcanic islands are Centre Peak, Saba, Haycock, and Saddle. A small eruption took place in 1824 on Saddle Island, with low-level explosive activity. An eruption is reported to have taken place on 14 August 1846, but this event is uncertain.

An undersea eruption began in mid-December 2011 located on the northern edge of the Zubair Islands, about 60 km off the coast of Yemen. Local fishermen reported lava fountaining near the island of Saba, while satellites captured a white plume rising from the sea as well as a pulse of sulphur dioxide. On 23 December, high-spatial-resolution satellite imagery revealed details of the eruption. Violent explosions, typical of shallow submarine (Surtseyan style) eruptions were observed. This submarine eruption has subsequently built a very small new island measuring about 530 m × 710 m between Rugged and Haycock Islands.

14.4 Future eruptive trends

Afar exhibits a range of volcano types and eruption styles. Basaltic lavas are most common, but the central volcanoes in particular also produce trachyte, dacite, and rhyolite, which erupt either effusively or explosively. Consequently, eruption styles have ranged from lava-flow-dominated eruptions of low explosivity through to Plinian and ignimbrite eruptions. The interaction of magma with surface and ground waters has also resulted in violent phreatomagmatic explosions from both basaltic and silicic magmas.

It has already been mentioned that 65 Holocene volcanoes have been identified in the Horn of Africa countries (Sparks *et al.*, 2012) with more than half of these in the Afar region. In Afar, the narrow magmatic segments and off-axis volcanic zones are clearly dominated by active volcanism. Some of these volcanic zones have erupted in historic times and many exhibit fumarolic and seismic activity as well as young surface deformation. Among the best-known volcanoes is Erta 'Ale renowned for its long-lived lava lake.

Investigations in Afar have added further evidence for the occurrence of shallow magma chambers beneath the active zones. Pham Van *et al.* (1981) identified a shallow (2–4 km) magma chamber beneath the axial zone of the Asal-Ghoubbet rift (Djibouti) with the roof of the chamber shallowing beneath Ardoukoba volcano, which was built during the November 1978 rifting episode. Peltzer *et al.* (2012) have further shown the presence of a shallow magma chamber beneath the Fieale caldera responsible for the observed sustained magmatic inflation there. Pagli *et al.* (2012) have demonstrated the presence of a shallow, 1-km-deep, elongated magma chamber (having two parts) beneath the Erta 'Ale volcanic ridge with the November 2008 Alu-Dallafila eruption triggered by a small influx of magma into the northern part of this magma chamber. These authors showed that the eruption was fed from both parts of the magma chamber and both parts have refilled rapidly since the eruption ended. Similarly, Field *et al.* (2012a) reported the occurrence of active fumaroles and earthquakes centred on Dabbahu volcano that indicate the volcano is still active. From results of integrated petrological, seismic, and InSAR studies, they have identified the presence of a magma chamber at 1–5 km depth beneath the volcano. According to Field *et al.* (2012a), magma lies in a series of stacked sills connected by a series of conduits or dykes. Furthermore, they suggested that these magmatic bodies are long-lived and have probably maintained a stable geometry over the past few thousand years.

In these and other mature rift segments, Biggs *et al.* (2009) expected the presence of large, shallow magma chambers and long, erupting dykes. Shallow magma chambers similar to those of Dabbahu and Erta 'Ale may also be present beneath other active zones in Afar such as Alayta, Tat 'Ali, Manda Inakir, Asal, Dama Ale, Gabilema, Yangudi, and Ayelu-Abida. The unexpected eruption of Nabro in May 2011 is strong evidence that there may be shallow magma chambers under the other caldera volcanoes lying within the Nabro volcanic range. These observations highlight two significant characteristic features of the Afar region noted by Pagli *et al.* (2012), namely the large amounts of magma production and the apparently frequent magma-chamber replenishment. The availability of large amounts of magma is a clear indication of magma storage at shallow depths beneath the active volcanic zones and thus that more frequent and probably more violent eruptions can be expected in the future.

The absence of a well-documented eruptive record and of detailed investigations or comprehensive monitoring of volcanoes in Afar limits the synthesis we can make of volcanism in the region. We can draw some useful analogies, however, with Iceland, which is much more systematically mapped and monitored. The similarity in rifting processes between Iceland and Afar has been recognised since the early 1970s (Barberi *et al.*, 1975 and references therein). More recently, Wright *et al.* (2012) pointed out that the Krafla (1975–1984, Iceland) and Dabbahu (2005–2011, Afar) rifting episodes share a number of similarities, including intermittent magma supply to the crust, multiple zones of magma storage feeding dykes that intrude the brittle upper crust. More specifically, Ferguson *et al.* (2010) recognised many similarities between the basaltic eruptions that occurred in August 2007, June 2009, and May 2010 on the DMH segment and those that occurred during the comparable stage of the Krafla rifting cycle (e.g., Bjornsson *et al.*, 1977; Tryggvasson, 1984; Einarsson, 1991). During nine years of activity at Krafla, 20 individual dykes were intruded along the rift, accompanied by nine basaltic eruptions. With 13 dyking and eruptive events between 2005 and 2011, Afar has clearly entered a new cycle of more frequent activity. To a first approximation, we consider Afar magmatic processes to be largely analogous to those taking place in Iceland, but that the contrasting meteorological and hydrological conditions can result in different eruptive styles. Although the flux of primary melts from the Icelandic mantle is likely to be greater than that of Afar (e.g. Maclennan *et al.*, 2001; Ferguson *et al.*, 2013), the trends and styles of eruptive activities observed in

Iceland provide a framework to assess the future locations and behaviour of volcanic eruptions in Afar.

As in Iceland, the following Afar axial magmatic segments and off-axis volcanic centres appear to be more likely sites to exhibit future eruptive phenomena: Erta 'Ale, Tat 'Ali, Alayta, Dabbahu-Manda-Harraro, Manda Inakir, Asal, the caldera volcanoes along the Nabro volcanic zone and Ma'alalta Volcanic Centre. Other potential eruptive sites in the Afar rift include the Alid and Dallol axial segments in northern Afar, the Dama Ale-Gabilema volcanic zone in central Afar and Yangudi and Ayelu-Abida volcanoes in southern Afar. The record in Iceland shows an average of 20–25 eruptions per century, which itself varies on a time scale of decades (Thordarson and Larsen, 2007). Taking the analogy further, we anticipate small eruptions in Afar (involving <0.1 km^3 dense rock equivalent (DRE) of lava) recurring roughly every 5–10 years, and larger flood-basalt eruptions (>10 km^3 DRE) with a return period of the order of 500–1000 years (Gudmundsson *et al.*, 2008). The most severe volcanic events to be expected in Afar are thus: (i) major basaltic eruptions similar to the Alu-Dallafila eruption of November 2008 and with less probability similar to that of the 1783 'Laki' eruption in Iceland, (ii) Plinian eruptions of large central volcanoes with associated pyroclastic flows and widespread tephra fallout, comparable to the May 1861 Dubbi eruption. Phreatic and phreatomagmatic eruptions in areas such as the Dallol are also possible.

The Krafla rifting episode involved a transition from dominantly intrusive (dyking) to a dominantly extrusive regime after ~4 years of activity (Harris *et al.*, 2000). Accordingly, Ferguson *et al.* (2010) suggested that an increase in volcanic eruptions and earthquakes on the DMH rift segment may occur over the next decade with a high potential to produce small–medium volume (i.e., $(10–30) \times 10^6$ m^3) basaltic eruptions with the possibility of larger magnitude events (i.e. an erupted volume of $>50\times10^6$ m^3). These authors further estimated that if eruptions do occur in the near term along the DMH rift, eruptions will be of increasingly greater magnitude, duration, and spatial extent. From the study of the 13 dyking and eruptive events that occurred on the DMH between 2005 and 2011, Ferguson *et al.* (2010) and Hamling *et al.* (2010) also expected that subsequent eruptions will more likely occur, if dyking activity continues, in the same area of the central rift, within 10–15 km of the rift centre, on either the northern or southern margins of the central Ado 'Ale volcanic complex.

Compared with basaltic volcanism associated with axial volcanic ranges such as the DMH and Erta 'Ale, there are few reliable data on which to base forecasts of likelihood and style of future eruptions of the central caldera volcanoes such as those located on the Nabro Volcanic Zone. Sparks *et al.* (2012) estimated the return periods of explosive eruptions in the Ethiopian region to be 7 years for eruptions of VEI 3, 29 years for those of VEI 4, and 100 years for those of VEI 5. It should be noted here that volcanism in Afar is, in part, pulsed and linked to major episodes of activity in particular segments, thus suggesting that simply using return periods for single eruptions might not capture the potential hazards when another segment ruptures.

14.5 Concluding remarks

We have provided an overview of historic volcanism in Afar, which demonstrates significant hazards in the region, notwithstanding the generally low population density. The main types of volcanic hazards expected include the direct effects of explosions, lava, pyroclastic flows, and volcanic gases and other far-field effects such as ash fallout. Their impacts can be widespread, even for eruptions of only moderate size. Both small and larger volumes of lava have been erupted from fissures more recently and in the past and are observed to flow for tens of kilometres away from the source. As the rate of advance of lava is relatively slow, except close to vents or in the case of much higher than typical effusion rates, the risk of fatalities in effusive eruptions is low. Pyroclastic density currents do not seem to be prominent in Afar due to the primarily basaltic nature of volcanism. However, major ignimbrites and minor pyroclastic flow deposits have formed throughout the geological history of Afar associated with silicic volcanoes. Several are found within the Nabro and Ma'alalta volcanic zones (Wiart and Oppenheimer, 2000, 2005). Although pyroclastic density currents often result in fatalities when they impact settled areas, the probability of such events reaching inhabited areas in Afar is low. The highest probability of this type of eruption occurring is from volcanoes on the Nabro and Ma'alalta volcanic zones, and Dabbahu and Ayelu-Abida volcanoes.

Explosive eruptions from Nabro (2011) and Da'Ure (2005) have signalled the potential hazard from tephra fallout in the region. Pumice deposits and volcanic bombs, together with the caldera structures on many volcanoes such as Dabbahu, indicate that some previous eruptions have been very explosive in nature, probably of Vulcanian type. With respect to potential hazards from tephra fallout, the prevailing wind field at the time of eruption is of great significance. Tephra fallout from Plinian and sub-Plinian eruption columns, which typically lasts an hour to several hours, is often confined to relatively narrow sectors but tephra thickness within these sectors can reach the decametre scale towards the vent. Other hazards associated with eruptions include earthquakes, faulting, and damming of rivers by lava flows or tephra. Seismic activity associated with the 2011 eruption of Nabro, for instance, resulted in significant structural damage to buildings. Faulting and fracturing may also damage roads and other man-made structures (e.g., dams).

Damage and loss from past volcanic eruptions in Afar have been limited mainly because of the sparse population in the region. More than 100 local inhabitants were reported killed in the 1861 Dubbi eruption (Wiart and Oppenheimer, 2005). The September 2005 Da'Ure explosive eruption and associated ground fissuring caused the death of hundreds of goats and cattle and the displacement of more than 10 000 people for several months (Yirgu *et al.*, 2006). Nothing is known about the effects incurred by the large lava eruption of Alu-Dallafila in 2008, which flowed in a particularly remote and inaccessible area. In June 2011, Nabro volcano erupted explosively and emitted lava, ash, and large amounts of sulphur dioxide gas resulting in a number of fatalities and the displacement of many thousands, in addition to much damage to pasture and cultivated lands, and losses of livestock.

According to Sparks *et al.* (2012), some 9.5 million people live within 30 km of an active volcano in the Ethiopian rift region (Ethiopia, Djibouti, and Eritrea). With the current, rather limited state of knowledge of these volcanoes, and the rudimentary or non-existent monitoring on the ground, it is impossible to make a robust appraisal of the hazards they pose, especially in a probabilistic sense. In the Afar region, natural hazards interact with widespread vulnerability in the context of rapid population growth and urbanisation, environmental degradation, food insecurity, poverty, and fragile economies, infrastructures, and institutions. The region is growing at a fast pace in economic terms, and the population is projected to pass 2 million. Further, seven major cities are within 30 km of the nearest active volcano (Sparks *et al.*, 2012). Because the population and urbanisation continue to grow at a fast rate (~3% and ~3.3% respectively), more and more construction is taking place on and near the active volcanic zones posing an increasing threat to lives and livelihoods. The ability of geoscientists in the countries constituting the Afar region to forecast the timing of an impending eruption and the impact on local communities is limited as demonstrated by events that have occurred with little or no identified warnings.

Part of the rural population in Afar depends on condensing fumarolic steam at *boinas* to secure water both for human and animal consumption. Thus far, few if any studies have been conducted to assess the potential health effects of associated trace metals or fluoride. Evaluation of water quality in the area should therefore be given high priority.

Despite these growing hazards, there are neither systematic studies nor concrete plans to implement monitoring and mitigation of active volcanic centres and zones in Afar. Awareness, planning, and preparedness remain inadequate. A single major volcano-seismic disaster is more likely than before to disrupt the life of the society and its economy, potentially for months or years. We consider that raising awareness of these issues with the local community would be very beneficial for future work. It is essential to build resilience in communities both as a means to mitigate the impacts of volcanism, as well as to protect social and economic development gains.

With the expansion of populations into hazardous volcanic areas, there is a growing need to develop appropriate scientific investigations and emergency response measures (see also Chapter 3). There is an urgent need for a volcanic hazard assessment and monitoring programme in the Afar rift to address volcanic risk in the region. Future studies of the long-term eruptive history of Afar volcanoes will lead to more accurate volcanic hazards assessments and enable refinement of eruption scenarios and planning for response. This requires international partnerships to strengthen the foundation for regional geohazard mitigation initiatives. We hope that this work will help to stimulate such initiatives, as well as serve as a reminder and guide for land-use planners and developers to make informed decisions for the future. Effective land-use planning is one of the most potent means of minimising risk in hazardous areas.

Acknowledgments. We acknowledge support from the UK NERC via the project 'How does the Earth's crust grow at divergent plate boundaries? a unique opportunity in Afar, Ethiopia'. CO additionally acknowledges support from the NERC through the project 'Mechanisms and consequences of the 2011 eruption of Nabro volcano, Eritrea'.

References

Acocella, V. (2010). Coupling volcanism and tectonics along divergent plate boundaries: Collapsed rifts from central Afar, Ethiopia. *Geological Society of America Bulletin*, **122**, 1717–1728.

Amelung, F., Oppenheimer, C., Segall, P. and Zebker, H. (2000). Ground deformation near Gada'Ale volcano, Afar, observed by radar interferometry. *Geophysical Research Letters*, **27**, 3093–3096.

Ayele, A, Keir, D., Ebinger, C. J. *et al.* (2009). The September 2005 mega-dike emplacement in the Manda-Hararo (Afar) nascent oceanic rift. *Geophysical Research Letters*, **36**, L20306.

Barberi, F. and Varet, J. (1970). The Erte 'Ale volcanic range (Danakil depression, northern Afar, Ethiopia). *Bulletin Volcanologique*, **34**, 848–917.

Barberi, F., and Varet, J. (1975). Recent volcanic units of Afar and their structural significance. In *Afar Depression of Ethiopia: Proceedings of an International Symposium on the Afar*

Region and Related Rift Problems, vol. 1. Stuttgart: Schweizerbart, pp. 174–178.

Barberi, F. and Varet, J. (1977). Volcanism of Afar: Small-scale plate tectonics implications. *Geological Society of America Bulletin*, **88**, 1251–1266.

Barberi, F., Tazieff, H. and Varet, J. (1972). Volcanism in the Afar depression: its tectonic and magmatic significance. *Tectonophysics*, **15**, 19–29.

Barberi, F., Bonatti, E., Marinelli, G. and Varet, J. (1974). Transverse tectonics during the split of a continent: data from the Afar rift. *Tectonophysics*, **23**, 17–29.

Barberi, F., Ferrara, G., Santacroce, R., Treuil, M., and Varet, J. (1975). A transitional basalt-pantellerite sequence of fractional crystallisation, the Boina Centre (Afar Rift, Ethiopia). *Journal of Petrology*, **16**, 22–56.

Bastow, I. D., Nyblade, A. A., Stuart, G. W., Rooney, T. O. and Benoit, M. H. (2008). Upper mantle seismic structure beneath the Ethiopian hot spot: rifting at the edge of the African low velocity anomaly. *Geochemistry, Geophysics, Geosystems*, **9**, Q12022.

Belachew, M., Ebinger, C., Coté, D. *et al.* (2011). Comparison of dike intrusions in an incipient seafloor-spreading segment in Afar, Ethiopia: Seismicity perspectives. *Journal of Geophysical Research*, **116**, B06405.

Benoit, M. H., Nyblade, A. A., and VanDecar, J. C. (2006). Upper mantle P wave speed variations beneath Ethiopia and the origin of the Afar hot spot. *Geology*, **34**, 329–332.

Beutel, E., van Wijk, J., Ebinger, C., Keir, D., and Agostini, A. (2010). Formation and stability of magmatic segments in the Main Ethiopian and Afar Rifts. *Earth and Planetary Science Letters*, **298**, 225–235.

Biggs, J., Amelung, F., Gourmelen, N., Dixon, T. H., and Kim, S.-W. (2009). InSAR observations of 2007 Tanzania rifting episode reveal mixed fault and dyke extension in an immature continental rift. *Geophysical Journal International*, **179**, 549–558.

Bizourd, H., Barberi, F., and Varet, J. (1979). Mineralogy and Petrology of Erta Ale and Boina Volcanic Series, Afar Rift, Ethiopia. *Journal of Petrology*, **21**, 401–436.

Bjornsson, A., Saemundsson, K., Einarsson, P., Tryggvason, E. and Gronvald, K. (1977). Current rifting episode in North Iceland. *Nature*, **266**, 318–323.

Bourassa, A. E., Robock, A., Randel, W. J. *et al.* (2012). Large volcanic aerosol load in the stratosphere linked to Asian monsoon transport. *Science*, **337**, 78–81.

Chernet, T. (2012). Dallol Volcano and Danakil Depression, Ethiopia. In *Proceedings of the Magmatic Rifting & Active Volcanism Conference*, Addis Ababa, Ethiopia, 11–13 January 2012. Addis Ababa: Afar Rift Consortium, p. 19.

Chu, D., and Gordon, R. G. (1998). Current plate motions across the Red Sea. *Geophysical Journal International*, **135**, 313–328.

Chu, D., and Gordon, R. (1999). Evidence for motion between Nubia and Somalia along the Southwest Indian ridge. *Nature*, **398**, 64–67.

Courtillot, V., Achache, J., Landre, F. *et al.* (1984). Episodic spreading and rift propagation: new paleomagnetic and geochronologic data from the Afar nascent passive margin. *Journal of Geophysical Research*, **89**, 3315–3333.

Demange, J. and Tazieff, H. (1978). L'éruption tectonique de l'Ardoukôba (Djibouti). *Comptes Rendus de l'Académie des Sciences Paris*, **287**, 1269–1272.

Ebinger, C. J. and Sleep, N. H. (1998). Cenozoic magmatism throughout East Africa resulting from impact of a single plume. *Nature*, **395**, 788–791.

Ebinger, C., Keir, D., Ayele, A. *et al.* (2008). Capturing magma intrusion and faulting processes during continental rupture: Seismicity of the Dabbahu Afar rift. *Geophysical Journal International*, **174**, 1138–1152.

Ebinger, C., Ayele, A., Keir, D. *et al.* (2010). Length and timescales of rift faulting and magma intrusion: the Afar Rifting Cycle from 2005 to present. *Annual Review of Earth and Planetary Sciences*, **38**, 439–466.

Eckhardt, S., Prata, A. J., Seibert, P., Stebel, K. and Stohl, A. (2008). Estimation of the vertical profile of sulfur dioxide injection into the atmosphere by a volcanic eruption using satellite column measurements and inverse transport modeling. *Atmospheric Chemistry and Physics*, **8**, 3881–3897.

Einarsson, P. (1991). The Krafla rifting episode 1975–1989. In *N'att'ura Myvatns, The Nature of Lake Myvatn*, eds. Gardarsson, A., and Einarsson, P. Reykjavik: Icelandic Nature Science Society, pp. 97–139.

Ferguson, D. J., Barnie, T. D., Pyle, D. M. *et al.* (2010). Recent rift-related volcanism in Afar, Ethiopia. *Earth and Planetary Science Letters*, **292**, 409–418.

Ferguson, D. J., Maclennan, J., Bastow, I. D. *et al.* (2013). Melting during late-stage rifting in Afar is hot and deep. *Nature*, **499**, 70–73.

Field, L., Blundy, J., Calvert, A. and Yirgu, G. (2012a). Magmatic history of Dabbahu, a composite volcano in the Afar Rift, Ethiopia. *Geological Society of America Bulletin*, **125**, 128–147.

Field, L., Barnie, T., Blundy, J. *et al.* (2012b). Integrated field, satellite and petrological observations of the November 2010 eruption of Erta Ale. *Bulletin of Volcanology*, **74**, 2251–2271.

Fromm, M., Nedoluha, G., and Charvát, Z. (2013). Comment on "Large volcanic aerosol load in the stratosphere linked to Asian monsoon transport". *Science*, **339**, 647.

Goitom, B., Ogubazghi, G., Blundy, J. *et al.* (2013). First recorded eruption of Nabro volcano, Eritrea, 2011. In *Abstracts of the IAVCEI Scientific Assembly, Kagoshima, Japan, 20–24 July 2013*. Kagoshima, Japan: International Association of Volcanology and Chemistry of the Earth's Interior.

Gudmundsson, M. T., Larsen, G., Höskuldsson, A. and Gylfason, G. (2008). Volcanic hazards in Iceland. *Jökull*, **58**, 251–268.

Hamling, I. J., Ayele, A., Bennati, L. *et al.* (2009). Geodetic observations of the ongoing Dabbahu rifting episode: new dyke intrusions in 2006 and 2007. *Geophysical Journal International*, **178**, 989–1003.

Hamling, I. J., Wright, T. J., Calais, E., Bennati, L. and Lewi, E. (2010). Stress transfer between thirteen successive dyke intrusions in Ethiopia. *Nature Geoscience*, **3**, 713–717.

Harris, A. J. L., Murray, J. B., Aries, S. E. *et al.* (2000). Effusion rate trends at Etna and Krafla and their implications for eruptive mechanisms. *Journal of Volcanology and Geothermal Research*, **102**, 237–270.

Hayward, N. J. and Ebinger, C. J. (1996). Variations in the along-axis segmentation of the Afar Rift System. *Tectonics*, **15**, 244–257.

Hofmann, C., Courtillot, V., Feraud, G. *et al.* (1997). Timing of the Ethiopian flood basalt event and implications for plume birth and global change. *Nature*, **389**, 838–841.

Kidane, T., Courtillot, V., Manighetti, I. *et al.* (2003). New paleomagnetic and geochronological results from Ethiopian Afar: block rotations linked to rift overlap and propagation and determination of a ~2 Ma reference pole for stable Africa. *Journal of Geophysical Research*, **108**, 2102. doi:10.1029/2001JB000645.

Lahitte, P., Gillot, P. Y. and Courtillot, V. (2003). Silicic central volcanoes as precursors to rift propagation: the Afar case. *Earth and Planetary Science Letters*, **207**, 103–116.

Maclennan, J., McKenzie, D. and Gronvold, K. (2001). Plume-driven upwelling under central Iceland. *Earth and Planetary Science Letters*, **194**, 67–82.

Manighetti, I., Tapponnier, P., Courtillot, V., Gruszow, S. and Gillot, P.-Y. (1997). Propagation of rifting along the Arabia-Somalia plate boundary: the Gulfs of Aden and Tadjoura. *Journal of Geophysical Research*, **102**, 2681–2710.

Manighetti, I., Tapponnier, P., Gillot, P. Y. *et al.* 1998. Propagation of rifting along the Arabia-Somalia Plate Boundary into Afar. *Journal of Geophysical Research*, **103**, 4947–4974.

McKenzie, D. P., Davies, D. and Molnar, P. (1970). Plate tectonics of the Red Sea and East Africa. *Nature*, **224**, 125–133.

Montelli, G., Nolet, G., Masters, F., Dahlen, A. and Hung, S -H. (2004). Global *P* and *PP* travel time tomography: rays versus waves. *Geophysical Journal International*, **158**, 637–654.

Oppenheimer, C. and Francis, P. (1997). Remote sensing of heat, lava and fumarole emissions from Erta 'Ale volcano, Ethiopia. *International Journal of Remote Sensing*, **18**, 1661–1692.

Oppenheimer, C. and Francis, P. (1998). Implications of the long-eval lava lakes for geomorphological and plutonic processes at Erta 'Ale volcano, Afar. *Journal of Volcanology and Geothermal Research*, **80** (1–2), 101–111.

Oppenheimer, C., and Yirgu, G. (2002). Thermal imaging of an active lava lake: Erta 'Ale volcano, Ethiopia. *International Journal of Remote Sensing*, **23**, 4777–4782.

Oppenheimer, C., McGonigle, A. J. S., Allard, P., Wooster, M. J. and Tsanev, V. I. (2004). Sulfur, heat and magma budget of Erta'Ale lava lake, Ethiopia. *Geology*, **32**, 509–512.

Pagli, C., Wright, T. J., Ebinger, C. J. *et al.* (2012). Shallow axial magma chamber at the slow-spreading Erta 'Ale Ridge. *Nature Geoscience*, **5**, 284–288.

Peltzer, G., Doubre, C., and Tomic, J. (2012). Sustained magmatic inflation and fault movement in the Asal-Goubbet Rift observed during 11 years of radar interferometry data. In *Proceedings of the Magmatic Rifting & Active Volcanism Conference, Addis Ababa, Ethiopia, 11–13 January 2012*. Addis Ababa: Afar Rift Consortium.

Pham Van, N., Boyer, D., Le Mouël, J -L., and Courtillot, V. (1981). Identification of a magma chamber in the Ghoubbet-Asal rift (Djibouti) from a magnetotelluric experiment. *Earth and Planetary Science Letters*, **52**, 372–380.

Prodehl, C., Fuchs, K., and Mechie, I. (1997). Seismic-refraction studies of the Afro-Arabian rift system – a brief review. *Tectonophysics*, **278**, 1–13.

Rognon, P. (1975). Tectonic deformation in central Afar basins in the upper Pleistocene and Holocene periods, from the study of lacustrine deposits. In *Afar Depression of Ethiopia*, eds. Pilger, A. and Rosler, A. *Proceedings of an International Symposium on the Afar Region and Rift Related Problems, Bad Bergzabren, Germany, 1974*, vol. 1. Stuttgart, Germany: E. Schweizerbartsche Verlagsbuchhandlung, pp. 198–200.

Rowland, J. V., Baker, E., Ebinger, C. J. *et al.* (2007). Fault growth at a nascent slow-spreading ridge: 2005 Dabbahu rifting episode, Afar. *Geophysical Journal International*, **171**, 1226–1246.

Sawamura, P., Vernier, J. P., Barnes, J. E. *et al.* (2012). Stratospheric AOD after the 2011 eruption of Nabro volcano measured by lidars over the Northern Hemisphere. *Environmental Research Letters*, **7**, 034013.

Schilling, J. G., Kingsley, R., Hanan, B. and McCully, B. (1992). Nd-Sr-Os isotropic variations along the Gulf of Aden: evidence for mantle-plume lithosphere interaction. *Journal of Geophysical Research*, **97**, 10927–10966.

Sparks, R. S. J., Aspinall, W. P., Auker, M. *et al.* (2012). Mapping and characterising volcanic risk. In *Proceedings of the Magmatic Rifting & Active Volcanism Conference, Addis Ababa, Ethiopia, 11–13 January 2012*. Addis Ababa: Afar Rift Consortium.

Thordarson, T. and Larsen, G. (2007). Volcanism in Iceland in historical time: Volcano types, eruption styles and eruptive history. *Journal of Geodynamics*, **43**, 118–152.

Tryggvason, E. (1984). Widening of the Krafla fissure swarm during the 1975– 1981 volcano tectonic episode. *Bulletin of Volcanology*, **471**, 47–69.

Varet, J. (1978). *Geology of Central and Southern Afar (Ethiopia and Djibouti Republic)*. Paris: CNRS.

Vernier, J.-P., Thomason, L. W., Fairlie, T. D. *et al.* (2013). Comment on "Large volcanic aerosol load in the stratosphere linked to Asian monsoon transport". *Science*, **339**, 647.

Wiart, P. A. M. and Oppenheimer, C. (2000). Largest known historic eruption in Africa: Dubbi volcano, Eritrea, 1861. *Geology*, **28**, 291–294.

Wiart, P., and Oppenheimer, C. (2005). Large magnitude silicic volcanism in north Afar: The Nabro Volcanic Range and Ma'alalta volcano. *Bulletin of Volcanology*, **67**, 99–115.

Wiart, P. A. M., Oppenheimer, C. and Francis, P. (2000). Eruptive history of Dubbi volcano, northeast Afar (Eritrea), revealed by optical and SAR image interpretation. *International Journal of Remote Sensing*, **21**, 911–936.

Wright, R., Flynn, L. P., Garbeil, H., Harris, A. J. L. and Pilger, E. (2004). MODVOLC: near-real-time thermal monitoring of global volcanism. *Journal of Volcanology and Geothermal Research*, **135**, 29–49.

Wright, T. J., Ebinger, C., Biggs, J. *et al.* (2006). Magma-maintained rift segmentation at continental rupture in the 2005 Afar dyking episode. *Nature*, **442**, 291–294.

Wright, T. J., Sigmundsson, F., Pagli, C. *et al.* (2012). Geophysical constraints on the dynamics of spreading centres from rifting episodes on land. *Nature Geoscience*, **5**, 242–250.

Yirgu, G., Ayele, A. and Ayalew, D. (2006). Recent seismovolcanic crisis in northern Afar, Ethiopia. *Eos, Transactions American Geophysical Union*, **87**, 325–326.

15

Large recorded earthquakes in sub-Saharan Africa

Vunganai Midzi and Brassnavy Manzunzu

15.1 Introduction

Earthquakes are the most terrifying and destructive among natural phenomena (e.g., floods, wind, or drought) in Africa. For many African countries, earthquake hazard constitutes a serious threat to human life and property, sometimes causing economic losses and disruption. The environmental concerns and an increased official and public awareness of earthquake hazards have led to a rapid rise of interest in seismicity, seismic hazard, and risk assessments in African countries. Sub-Saharan Africa is largely a stable intra-plate region (Hartnady, 1990, 2002; Calais *et al.*, 2006; Stamps *et al.*, 2008; Déprez *et al.*, 2013) characterised by relatively low levels of seismic activity, with earthquakes randomly distributed in space and time. The only parts of sub-Saharan Africa that do not display the characteristics of an intra-plate region are the East African Rift System (EARS) and the Cameroon Volcanic Line (Fig. 15.1), where earthquakes are associated with active plate-boundary fault zones and volcanic activity.

Earthquakes in Africa have been poorly understood mainly due to a lack of data and information. Krenkel (1923) produced one of the first seismicity maps of Africa, which clearly showed the concentration of major earthquakes along the East Africa Rift. Studies on various aspects of the African seismicity have been carried out (e.g., Fairhead and Henderson, 1977; Fairheard and Stuart, 1982; Yang and Chen, 2008, 2010). The rift trends largely north–south following the great lakes: Albert, Edward, Kivu, Tanganyika, and Nyasa (see Fig. 13.2), and is also assumed to extend to the southern part of Mozambique. An active branch on the eastern side of Lake Victoria follows the Rift Valley of Kenya, through Lake Rudolf into Ethiopia. In Ethiopia, seismic activities follow narrow zones associated with structures of the Afar depression and the main Ethiopian Rift (see Fig. 13.2). The Gregory Rift in Kenya and north Tanzania is almost devoid of strong seismic activity although micro-earthquake studies (e.g., Molnar and Aggarwal, 1971; Calais *et al.*, 2008; Ibs-von Seht *et al.*, 2008; Githiri *et al.*, 2012) show that the rift floor is seismically active. Although less well expressed, the Gregory Rift extends further south into Central Tanzania, where the region of Dodoma is more seismically active (Macheyeki *et al.*, 2008). The northern part of Tanzania is another highly active region on the eastern branch. The area has diffuse seismicity that starts close to Lake Victoria and stretches southeastwards to the Indian Ocean. The western branch is more seismically active than the eastern branch (Figs. 15.1 and 13.2). In the northern end, seismicity dies out abruptly in southern Sudan as the rift encounters the Aswa shear zone. West of Lakes Edward, Kivu, and Tanganyika in Democratic Republic of the Congo (DR Congo) and Zambia, the seismicity is diffuse. The southern extension of the western branch is well defined by a seismicity belt running through Malawi to southern Mozambique. Another active branch goes west just south of Lake Malawi into the Zambezi (Kariba) Rift (Figs. 15.1 and 13.2).

Large earthquakes in sub-Saharan Africa have occurred mostly along this major tectonic feature. The earthquakes recorded since 1900 with a magnitude greater than 6 that caused structural damage and/or deaths are discussed in this chapter.

15.2 The earthquakes

Some of the large earthquakes in sub-Saharan Africa are listed in Table 15.1 and their locations shown in Fig. 15.2. The largest of these earthquakes were the 13 December 1910 Rukwa (Tanzania) earthquake of magnitude $M7.3$,

Extreme Natural Hazards, Disaster Risks and Societal Implications, eds. Alik Ismail-Zadeh, Jaime Urrutia Fucugauchi, Andrzej Kijko, Kuniyoshi Takeuchi and Ilya Zaliapin. Published by Cambridge University Press.

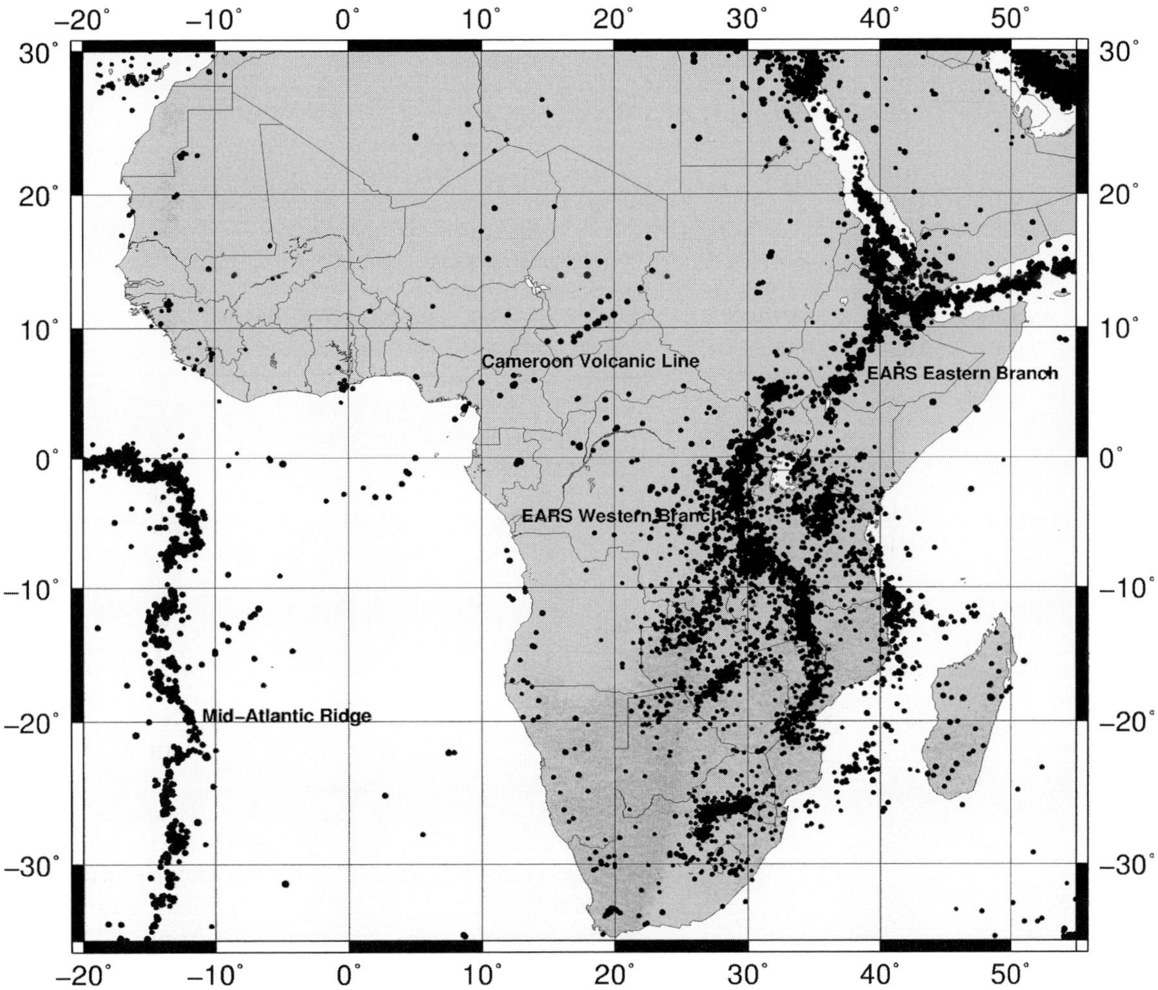

Figure 15.1. Earthquakes of magnitude greater than 3.0 on the African continent, which are dominated by events on the EARS.

the 20 May 1990 South Sudan *M*7.2 earthquake, and the 22 February 2006 Machaze (Mozambique) *M*7.0 earthquake. In this section we describe and discuss the individual large earthquakes that occurred, with exceptions of those events for which little information is available (e.g., 8 July 1919 and 9 October 1942 Tanzania earthquakes).

15.2.1 25 August 1906 Langano (Ethiopia) earthquake

The main event occurred on 25 August 1906 with a magnitude of *M*6.8 about 100 km south of Addis Ababa, Ethiopia. An aftershock, which occurred on 28 October 1906, caused great fear in the population of Addis Ababa. Though violent shaking was experienced during the main event, only slight damage was observed, mainly because the town was barely 10 years old and had not yet fully developed (Gouin, 1979). According to Ayele and Kulhánek (2000) the earthquake was the largest recorded in Ethiopia in the twentieth century and was widely felt in the country.

15.2.2 13 December 1910 Rukwa (Tanzania) earthquake

The 13 December 1910 Rukwa (Tanzania) *M*7.3 earthquake was the largest recorded earthquake in sub-Saharan Africa. The event was recorded by European seismic stations (Ayele and Kulhánek, 2000). It was felt throughout Tanzania and in all neighbouring countries (Fig. 15.3). The observed damage occurred in the epicentral area with many of the European type structures destroyed or severely damaged (Ambraseys, 1991a). The earthquake triggered landslides and set up water-waves in Lakes Tanganyika and Malawi. It caused springs of water in the Zimba and Mbosi areas to dry up (Ambraseys and Adams, 1991). According to Vittori *et al.* (1997) and Delvaux *et al.*

Table 15.1. *The large earthquakes in sub-Saharan Africa*

	Date	Epicentral location		Magnitude	Earthquake location and effects
		Latitude	Longitude		
1	25 Aug. 1906	7.020	40.090	6.8[a]	Langano, Ethiopia. Felt as far as Addis Ababa.
2	13 Dec. 1910	−8.000	31.000	7.3[a]	Rukwa, Lake Tanganyika. Caused damage in southern Tanzania.
3	09 Jul. 1912	3.000	33.000	6.8[a]	Kitgum, Aswa shear zone, northern Uganda. Partial destruction of buildings in northern Uganda.
4	08 May 1915	−26.000	38.000	6.8[a]	Mozambique Channel.
5	23 Sep. 1915	16.000	38.500	6.8[a]	Coast of Eritrea. Slight damage in Asmara.
6	08 Jul. 1919	−9.000	32.500	6.8[a]	Western Tanzania.
7	06 Jan. 1928	0.200	36.200	6.9[a]	Subukia, Kenya.
8	31 Dec. 1932	−28.500	32.800	6.7[a]	Coast of Natal, South Africa. Felt throughout KwaZulu-Natal and several other bordering provinces.
9	22 Jun. 1939	5.200	−0.100	6.5[a]	Ghana and 17 people killed.
10	09 Oct. 1942	−11.000	35.000	6.8[a]	Lake Nyasa
11	18 Mar. 1945	0.000	32.000	6.0[a]	Masaka, Uganda. 5 people killed, and buildings destroyed.
12	01 Jun. 1961	10.540	39.890	6.6[a]	Kara Kore, Ethiopia. 160 people killed. Towns of Majete and Kara Kore destroyed.
13	20 Mar. 1966	0.810	29.900	6.0[a]	Toro, Uganda. 150 people killed and over 1300 injuries; damage worth US$1.5million.
14	29 Mar. 1969	11.910	41.300	6.3[a]	Serdo, Djibouti. 40 people died, 160 injured. Village of Serdo destroyed.
15	29 Sep. 1969	−33.280	19.210	6.3[a]	Ceres, South Africa. 12 people killed; damage to structures worth $20 million.
16	10 Mar. 1989	−13.6965	34.4452	6.1[a]	Salima, Malawi. 9 people killed.
17	20 Aug. 1989	11.767	41.9601	6.5[a]	Dobi, Ethiopia. 6 people killed; several bridges damaged.
18	20 May 1990	5.0708	32.1588	7.2[b]	Mongola, southern Sudan. 31 people killed; some buildings damaged in the Juba area. Also some damage in Uganda and felt in Kenya.
19	24 May 1990	5.3589	31.866	7.1[b]	Mongola, Southern Sudan. Some buildings damaged in the Juba area. Felt in the Kapenguria area and at Nakuru, Kenya. Also felt in Uganda.
20	05 Feb. 1994	0.5918	30.0267	6.2[b]	Kismoro, Uganda. 9 people killed.
21	24 Oct. 2002	−1.951	29.050	6.3[b]	Kapili, Lake Kivu, Rwanda. 2 people killed and walls of buildings in nearby towns cracked and some destroyed.
22	05 Dec. 2005	−6.2415	29.6812	6.8[b]	Lake Tanganyika, Tanzania. 6 people killed.
23	22 Feb. 2006	−21.3113	33.5493	7.0[b]	Machaze, Mozambique. 4 people killed.

[a] Surface-wave magnitude M_s as published by the International Seismological Centre (ISC, 2011).
[b] Moment magnitude M_w as published by US Geological Survey. Epicentral locations were obtained from the ISC website apart from the 29 September 1969 Ceres earthquake, which was obtained from Green and Bloch (1971).

(2012), the earthquake is likely to have occurred along the 160-km-long Kanda fault system on the Ufipa plateau between Lakes Rukwa and Tanganyika.

15.2.3 9 July 1912 Kitgum (Uganda) earthquake

Ambraseys and Adams (1991) reported that the 9 July 1912 Kitgum earthquake was widely felt in DR Congo, Uganda, and Tanzania with the most severe shaking along the northwest coast of Lake Albert. Given the magnitude of the event (*M*6.8), there is surprisingly very little information published about this earthquake. This could be due to little data being collected after the event, though some stations in Europe recorded it enabling researchers to give an estimate of its location (Ambraseys and Adams, 1991).

15.2.4 8 May 1915 Mozambique Channel earthquake

In the case of the 8 May 1915 (Mozambique) event, Ambraseys and Adams (1991), using a standardised formula, recalculated the surface-wave magnitude from the

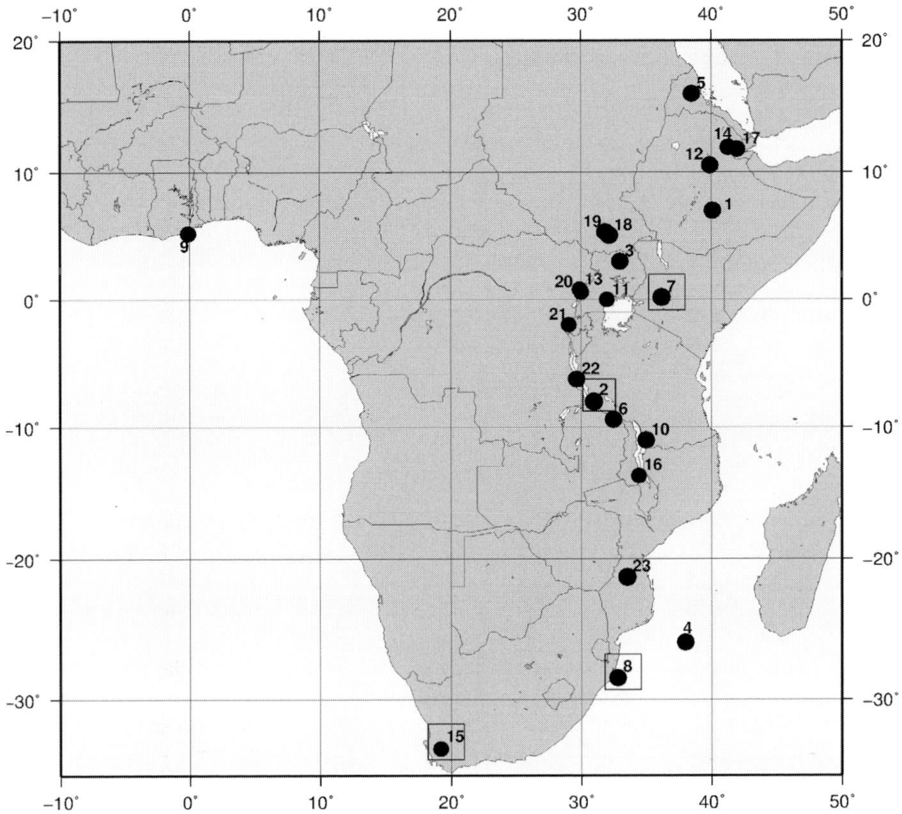

Figure 15.2. Location of the large earthquakes in sub-Saharan Africa. Earthquake locations are numbered according to the list in Table 15.1. The squares mark the locations of earthquakes with isoseismal maps presented in Figs. 15.3–15.6.

data of 12 stations and obtained a value of $M6.8$. No macro-seismic data were collected as there were no felt reports.

15.2.5 23 September 1915 Asmara earthquake

An earthquake of magnitude $M6.8$ occurred just outside Asmara, Eritrea, on 23 September 1915. Using macro-seismic information from 31 places (Gouin, 1979), the event was located in the Sabarguma graben on the margin of the Eritrean escarpment (e.g., Gutenberg and Richter, 1949). Shaking of intensity VI was felt in Asmara causing panic in the city and slight damage to some structures.

15.2.6 6 January 1928 Subukia (Kenya) earthquake

An earthquake of magnitude 6.9 occurred in Subukia valley in Kenya on 6 January 1928. This earthquake was the earliest and largest recorded earthquake in the Gregory Rift (Ambraseys, 1991b). No casualties were recorded since the epicentral area was very sparsely populated. Further away from the fault break, damage was widespread and mainly affected stone houses and adobe farmhouses. The shaking was widespread and felt as far as Lake Victoria (Fig. 15.4). The earthquake was associated with a normal fault surface rupture greater than 30 km long in the Kenya (Gregory) Rift (Ambraseys, 1991b; Ambraseys and Adams, 1991).

15.2.7 31 December 1932 coast of Natal earthquake

On New Year's Eve 1932, a powerful earthquake rocked the KwaZulu-Natal coast and was widely felt inland (Krige and Venter, 1933). Reports came in from as far as northern Johannesburg (Fig. 15.5). The epicentre was located off Cape Saint Lucia in South Africa. The magnitude was determined as $M_L6.3$ from local and $M_s6.8$ from teleseismic recordings. The earthquake led to the collapse of a homestead in Masunzeni near Newcastle, and less severe structural and non-structural damage in the epicentral area. Near Mtubutuba, a railway embankment failed, leading to the derailment of a cattle train. The earthquake also caused wide cracks to open along the banks of the Umfolozi river, probably as a result of liquefaction.

15.2.8 22 June 1939 Ghana earthquake

The largest and most destructive earthquake to be recorded in Ghana occurred on 22 June 1939 off the Ghana coast and had a magnitude of $M6.5$. Ambraseys and Adams

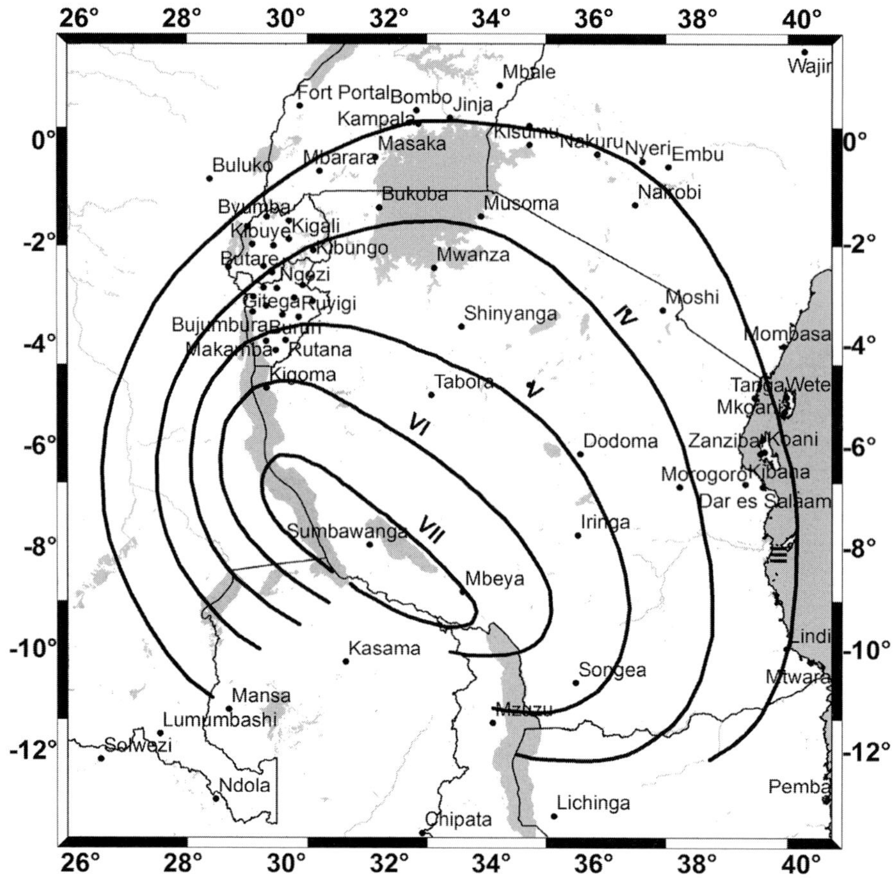

Figure 15.3. Isoseismals for the Rukwa earthquake of 13 December 1910 (adapted from Ambraseys (1991a)). Intensity values (marked by Roman numerals) were assigned using the Medvedev–Sponheur–Karnik (MSK) scale.

(1986) estimated the surface-wave and body-wave magnitudes to be 6.4 and 6.8 respectively. The earthquake was felt throughout the country but the highest intensity of VIII was felt in Accra. Seventeen people were killed, 133 injured, and an estimated one million pounds worth of damage done to buildings in Accra (Junner, 1941). Several buildings were destroyed including Osu Castle, the seat of government, which had its roof ripped off. According to Amponsah (2004), the most prominent geological effect of the 1939 earthquake was a line of fissures in the alluvium bordering the Akwapim scarp between Weija and Fete. Two minor shocks believed to be foreshocks of the main 1939 earthquake occurred in April of the same year. Aftershocks were also felt until July 1939.

15.2.9 18 March 1945 Masaka (Uganda) earthquake

The Masaka earthquake occurred on 18 March 1945. It had a magnitude of *M*6.0. Five people were killed during the earthquake and eight were injured. Some moderate damage to buildings was observed in Entebbe and the Entebbe seismograph station was put out of action.

15.2.10 1 June 1961 Kara Kore (Ethiopia) earthquake

A magnitude *M*6.6 earthquake occurred some 200 km south of Kara Kore on 1 June 1961. The main Kara Kore earthquake was preceded by many foreshocks from 19 May 1961, while aftershocks continued until September 1961. The shaking from the earthquake was felt all over central Ethiopia, resulting in the complete destruction of the nearby town of Majete. Most masonry houses in Kara Kore collapsed, yet local Ethiopian houses built of clay and reinforced with a eucalyptus frame withstood the shaking very well (Gouin, 1979). The earthquake shaking also caused rockslides and landslides in the Kara Kore area, which blocked roads and pavements. Sykes and Landisman (1964) and Fairhead and Girdler (1970) located the event and its aftershocks. Field observations of surface fractures also helped to constrain the locations accurately (Gouin, 1979).

15.2.11 20 March 1966 Toro (Uganda) earthquake

The 20 March 1966 Toro earthquake with a magnitude of *M*6.0 occurred in the Mt Ruwenzori region, which lies

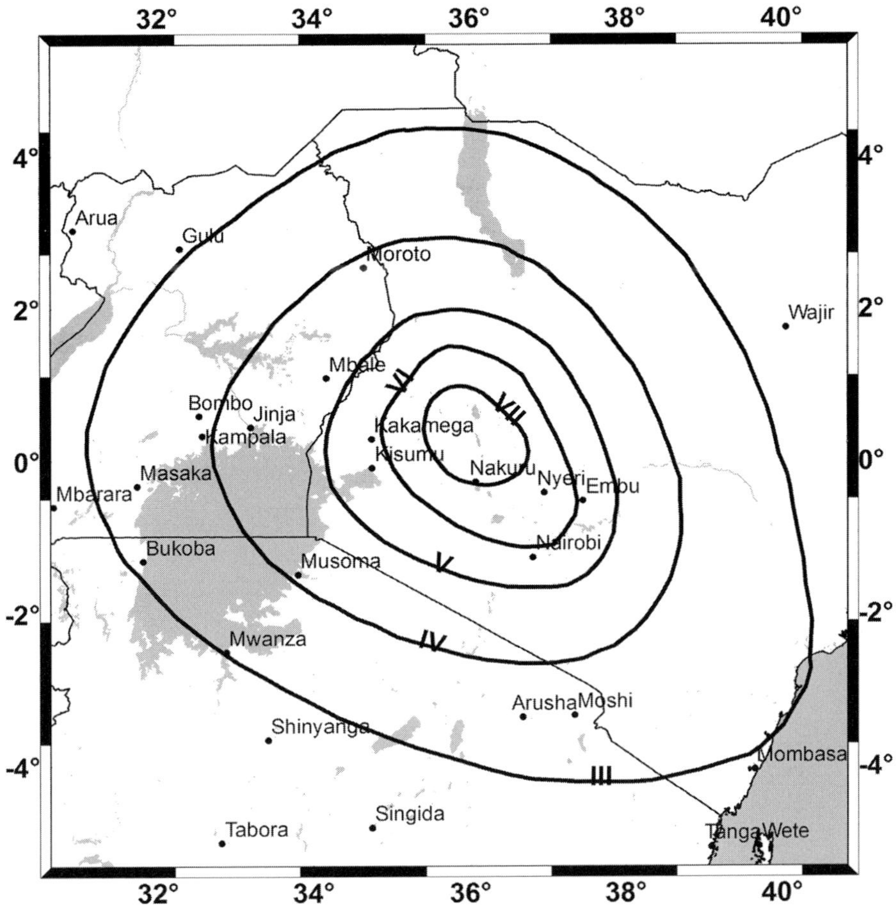

Figure 15.4. Isoseismal map of the 6 January 1928 Subukia earthquake. Intensity values are shown as Roman numerals on the map and were assigned using the MSK intensity scale (adapted from Ambraseys (1991b)).

within the western arm of the EARS, between Lake Edward and Lake Albert (see Fig. 13.2). The region is bounded by steep active normal faults along which the main event occurred, as shown in the calculated focal mechanism (Zana and Tanaka, 1981; Foster and Jackson, 1998). A maximum intensity value of VII (Modified Mercalli intensity MMI/scale) was observed in the Mt Ruwenzori area as well as in adjacent areas in the DR Congo. Extensive damage estimated at US$1.5 million was reported in both Uganda and DR Congo. The earthquake caused the deaths of 157 people, 104 in Uganda, 1 in Tanzania and 52 in DR Congo (Loupekine, 1966). The earthquake was also felt in Rwanda, Burundi, Tanzania, and Kenya. Cracks and fissures were observed on the ground and landslides were triggered. The earthquake was preceded by numerous foreshocks on 18 and 19 March 1966 and followed by many aftershocks during the next two months, nine of them with a magnitude greater than 5.0. One of the aftershocks (18 May 1966) caused the deaths of 90 people in DR Congo and destroyed many buildings in the country as well as in Uganda (Loupekine, 1966).

15.2.12 29 March 1969 Serdo (Djibouti) earthquake

The epicentre of this event was less than 100 km from the Djibouti–Ethiopia border in western Djibouti. It had a magnitude of $M6.3$. This event destroyed completely the Serdo village located close to the epicentre and resulted in the loss of 40 people and in injuries to 160 people (Gouin, 1979). Most of the masonry type structures were destroyed but the lighter structures made of wood actually survived. Cracks, fissures, landslides, and subsidence were observed on the ground in different sites. An aftershock of magnitude $M_s6.2$ occurred at the same place a few days later triggering rockslides over a large area. Over 250 more aftershocks were recorded at the WWSSN station in Addis Ababa with two reaching $M_b5.2$ (Gouin, 1979).

15.2.13 29 September 1969 Tulbagh–Ceres earthquake

A damaging earthquake occurred on 29 September 1969 in the Boland region of the Western Cape. Maximum MMI values reached IX in the epicentral area. The shaking was felt throughout the country, with felt reports from as far as

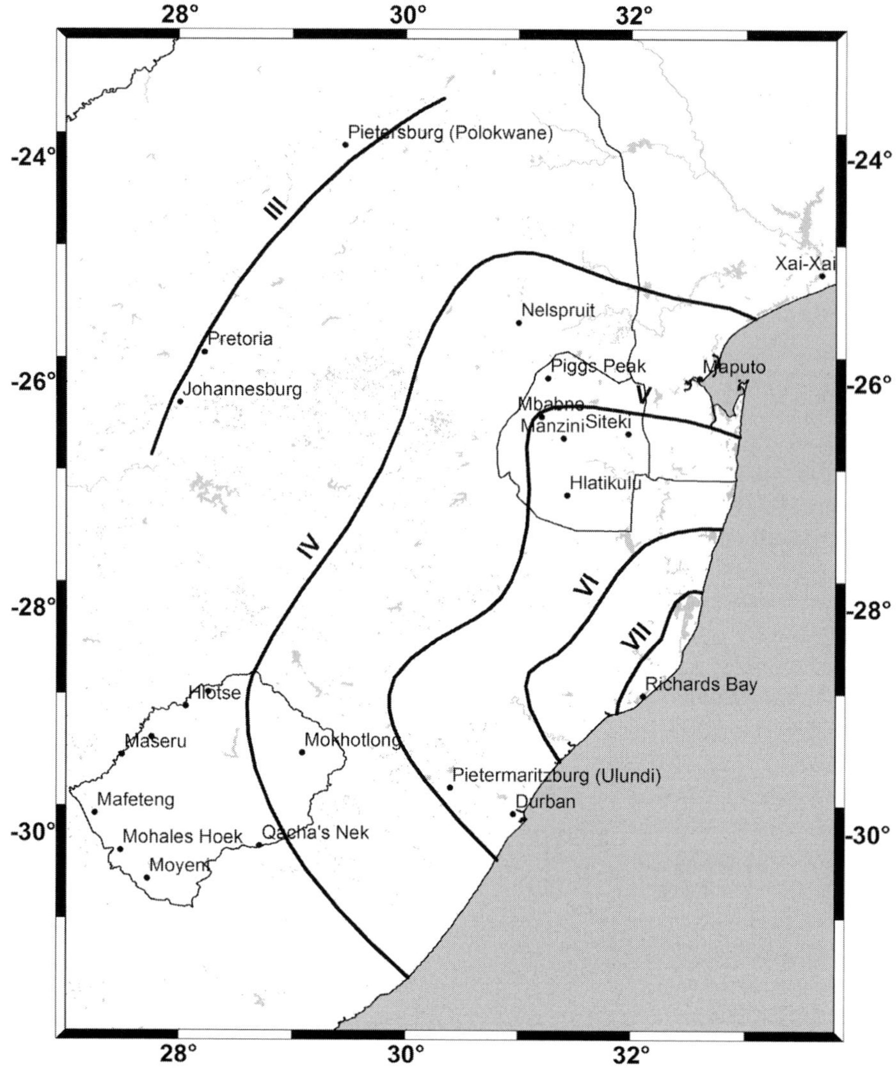

Figure 15.5. Isoseismal map (MMI-31 scale) of the Cape Saint Lucia event of 31 December 1932 (adapted from Krige and Venter (1933)).

Durban (Fig. 15.6). The local magnitude of the event was determined to be M_L6.3. This was one of the most destructive earthquakes in South Africa, resulting in the deaths of 12 people, with many more injured. Many buildings in the Ceres–Tulbagh area suffered serious damage, varying from the almost total destruction of old and poorly built buildings to large cracks appearing in those that were better constructed and in tarred roads (Keyser, 1974). The earthquake also caused numerous landslides and rock-falls in the mountains, some leading to bush fires. Damage worth about US$24 million was reported. The main earthquake was followed by a long sequence of aftershocks, the most severe of which occurred on 14 April 1970 and had a local magnitude of M_L5.7. A comprehensive report on the 1969 Boland earthquake was prepared by Van Wyk and Kent (1974) covering: the geology of the Ceres–Tulbagh area; the seismic history of the southwestern Cape; macroseismic observations; hydrological phenomena associated with the earthquake; the earthquake aftershocks; focal mechanism of the earthquake; geophysical implications of the whole earthquake sequence during 1969–1971; some elements of seismic risk assessment; and earthquake-resistant building recommendations.

15.2.14 10 March 1989 Malawi earthquake

The Malawi *M*6.1 earthquake occurred on 10 March 1989 in central Malawi. It was felt strongly in much of central Malawi, the Niassa and Tete provinces of Mozambique, and some parts of Zambia. At least 9 people were killed, 100 injured and 50 000 left homeless in Malawi. Damage was observed in the Salima, Dedza, and Mchinji areas. The event occurred at the southwestern end of Lake Malawi on a normal fault with nodal planes striking northwest (Foster and Jackson, 1998). Several studies have been carried out to analyse the source of this earthquake and its foreshocks

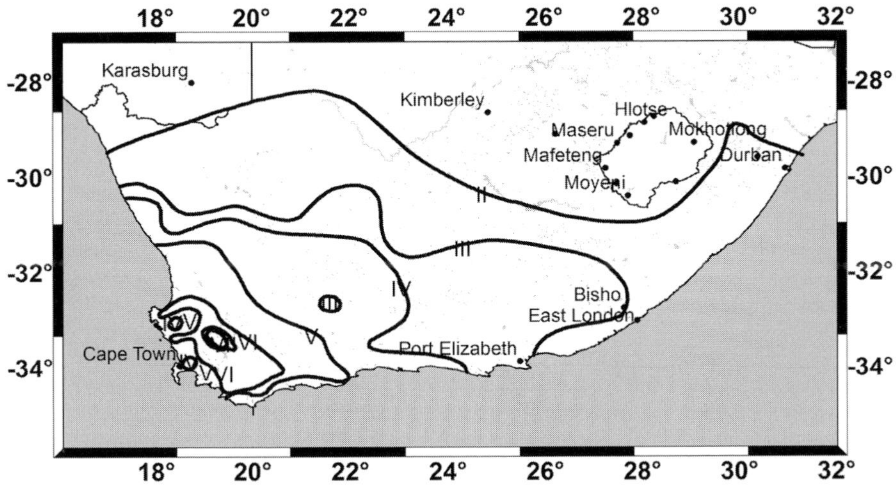

Figure 15.6. Isoseismal map of the 29 September 1969 earthquake (adapted from Keyser (1974)).

and aftershocks (e.g., Gupta, 1992; Jackson and Blenkinsop, 1993; Foster and Jackson, 1998). The hypocentre of the event was located at a depth of about 31 km and occurred on a major west-dipping fault that bounds the eastern side of the Malawi rift.

15.2.15 20 August 1989 Dobi (Ethiopia) earthquake

The 20 August 1989 Dobi earthquake occurred 20 years after the 1969 Serdo event. Both events were located on the same active tectonic zone. The 1989 earthquake was located in the Dobi Depression near the Djibouti–Ethiopia border and had a magnitude of *M*6.5. It caused widespread damage to buildings and bridges. A military camp located nearby was completely destroyed. According to Jacques *et al.* (2011), extensive damage was done to a nearby border post. The shaking at the Djiboutian border post, located about 5 km away from the epicentre of the largest shock of the sequence (*M*6.5), corresponds to an intensity of VIII–IX on the MMI scale. It was reported that two people were killed on the Djibouti side of the border (USGS, 1989) and four on the Ethiopian side (ISC, 1991). Landslides, subsidence, and fissures were observed in the area. This event was a part of the largest earthquake sequence in the Afar triangle since the March–April 1969 Serdo sequence, which also included several events with comparable magnitudes (McKenzie *et al.*, 1970; Gouin, 1979; Kebede *et al.*, 1989).

15.2.16 1990 South Sudan earthquakes

Several large earthquakes occurred in the south of the Sudan (now South Sudan) in May and July 1990. On 20 May 1990, a large earthquake of magnitude *M*7.2, shook the region located about 1120 km south of Khartoum and 100 km northeast of Juba, in the Mongola area. It was reported that 31 people lost their lives during this earthquake (http://earthquakes.findthedata.org/l/4608/Sudan-Juba-Maya-Uganda-Nakura). On 24 May 1990 another earthquake of magnitude *M*7.1 occurred in the same region, about 25 km south of the first one. This earthquake caused more substantial damage to private housing and government institutions than the first one but no loss of life was reported. Minor cracking in walls of some older buildings was observed in Juba. Most of the impact of the event fell on uninhabited areas.

The focal mechanism of the main shock (20 May 1990) shows left-lateral strike-slip faulting (Gaulon *et al.*, 1992, Girdler and McConnell, 1994). One of the nodal planes is parallel to the Aswa fault zone, a major Proterozoic tectonic feature. A line connecting the epicentres of the main events of May and July also strikes in the same direction. The rupture occurred in at least three steps, affecting a zone roughly 50 km long. However, the focal mechanism of the 20 May 1990 event was found to be different from the aftershocks, which all occurred on normal faults indicating a complicated tectonic regime in the epicentral area (Foster and Jackson, 1998).

15.2.17 5 February 1994 Kisomoro (Uganda) earthquake

The 5 February 1994 Uganda *M*6.2 earthquake caused extensive damage in the three western districts of Kabarole, Ibugyo, and Kasese in Uganda. While casualties were limited to nine deaths, heavy damage was inflicted on private, commercial, and public buildings (such as schools, churches, and hospitals). Damage was estimated at US$70 million with more than 50 000 people affected. The focal mechanism of the event was almost pure normal faulting with north–south striking nodal planes (Foster and Jackson, 1998).

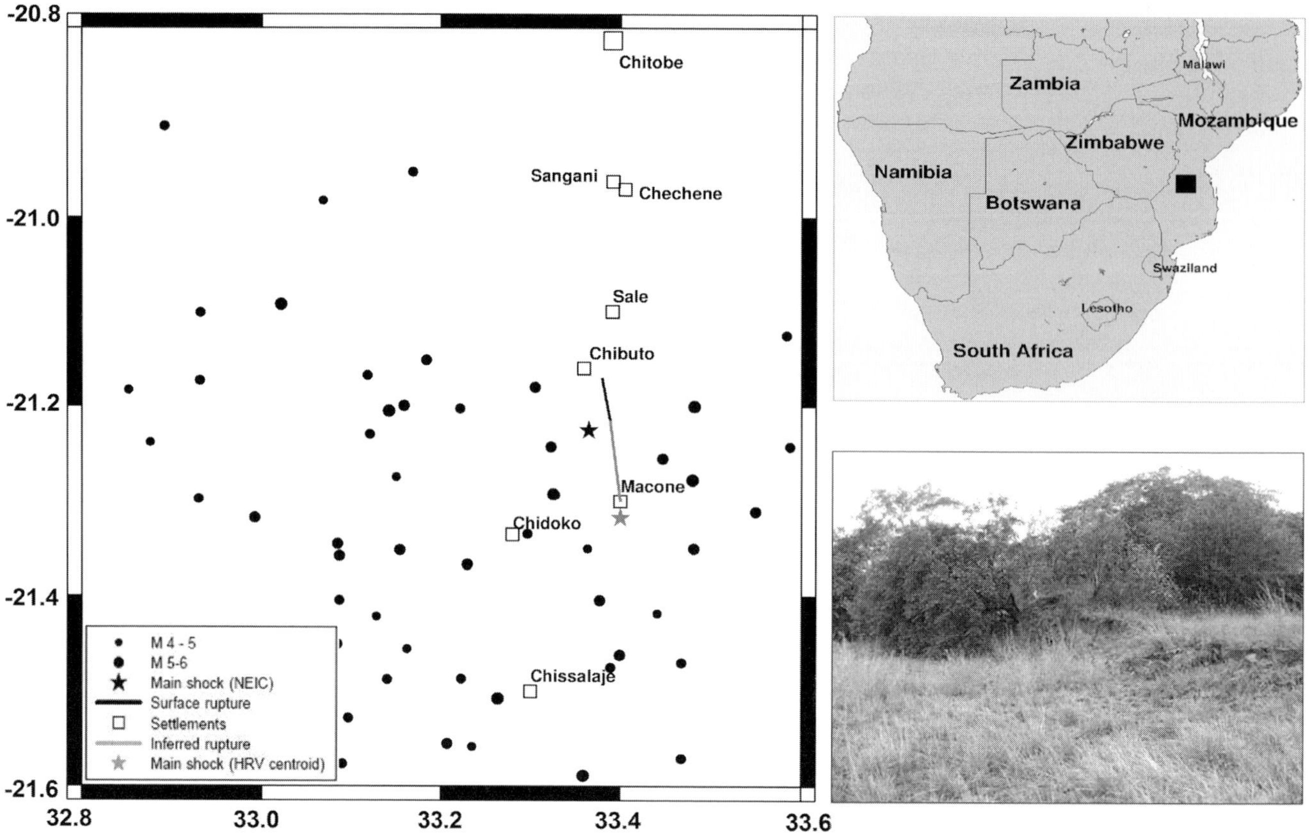

Figure 15.7. (left panel) Location map of the epicentral region, showing the observed surface rupture features, the main shock locations, and the aftershocks for a period of one month, as determined by the USGS's National Earthquake Information Centre (NEIC). (right lower panel) The main fault scarp at the northern extent of the mapped fault (after Fenton and Bommer, 2006). The location of the scarp marked by a blank square is shown in the right upper panel.

15.2.18 24 October 2002 Kalehe (Lake Kivu) earthquake

A magnitude *M*6.3 earthquake occurred at the western border of the central part of Lake Kivu in the EARS on 24 October 2002. According to the US Geological Survey (USGS), the earthquake had the shallow depth (10 km) typical of most local earthquakes. The effects of the event were felt in the DR Congo as well as in Rwanda, which resulted in the deaths of two people (Mavonga, 2007). Cracks on walls of buildings were observed in nearby towns, and some buildings were destroyed especially in Lwiro and Mugeri. This earthquake was followed by a *M*5.5 aftershock one hour later.

15.2.19 5 December 2005 Lake Tanganyika earthquake

An earthquake of local magnitude *M*6.8 occurred in the central part of Lake Tanganyika near the Mahale Mountain National Park in Tanzania on 5 December 2005. The USGS reported that 6 people were killed and 300 houses were destroyed in DR Congo including a church that collapsed at Kalamie. The shaking from the earthquake was felt in Tanzania and Rwanda with an intensity of IV and in Uganda and Kenya with an intensity of III. The earthquake was also felt as far as Luanda, Angola, sending people running in panic in search of shelter. However, no damage was reported there.

15.2.20 22 February 2006 Mozambique earthquake

A major earthquake of magnitude *M*7.0 occurred on 22 February 2006 in the western province of Manica in Mozambique, near the southern end of the EARS. The epicentre was located in Machaze, the southernmost district of Manica province, about 45 km due south of the district capital of Chitobe. The earthquake was felt in Beira, Inhambane, and Maputo with intensity V, and in Matola with intensity IV. It was also felt in Harare (intensity IV) and Mutare in Zimbabwe as well as at Louis

Trichardt and Phalaborwa (intensity IV), in Durban and Middelburg (intensity III), and in Johannesburg and Pretoria (intensity II) in South Africa. Although the earthquake was felt throughout southern Africa, it caused surprisingly little damage and a small number of casualties. One person was killed at Espungabera, another at Machaze, and two people in Beira. Twenty-seven people were injured and at least 160 buildings were damaged in the Espungabera, Beira, and Chimoio area. Surface rupture was mapped along the north-northwest-striking Machaze fault (Fig. 15.7) and liquefaction was observed throughout the epicentral region. The damage observed was much less than would be expected for such a large earthquake, because the epicentral area was sparsely populated with most people living in wattle-and-daub huts built using closely spaced timber verticals and a rather thick weave of material (Fenton and Bommer, 2006), which fared well during the earthquake.

15.3 Conclusion

Very little damage and few casualties have been observed during most of the large earthquakes in sub-Saharan Africa. However, with major population growth and increasing urbanisation in the region, vulnerability to earthquake hazards has greatly increased. The increased vulnerability is clearly observed, where smaller earthquakes have been resulting in fatalities and damage (e.g. 1989 Malawi, 1992 DR Congo, and 2004 Burundi earthquakes, the 2010 Karonga (Malawi) sequence, and other sequences on the continent). The Karonga earthquake sequence occurred between 6 and 19 December 2009 in Malawi, with its slip confined to the upper 6 km (Biggs *et al.*, 2010). Four of the sequence events had magnitudes of $M > 5.5$ with the largest having a magnitude of 6.0. According to the USGS archive of significant earthquakes, the largest event resulted in the deaths of three people, with at least 250 wounded, 1111 houses destroyed and 3565 damaged. It is clear from this description of the major earthquakes in sub-Saharan Africa that more research is required to understand the source and effects of such regional events in order to mitigate their effects.

References

Ambraseys, N. N. (1991a). The Rukwa earthquake of 13 Dec. 1910 in E. Africa. *Terra Nova*, **3**, 203–208.

Ambraseys, N. N. (1991b). Earthquake hazard in the Kenya Rift: the Subukia earthquake 1928. *Geophysical Journal International*, **105**, 253–269.

Ambraseys, N. N. and Adams, R. D. (1986). Seismicity of West Africa. *Annales Geophysicae*, **4**, 679–702.

Ambraseys, N. N. and Adams, R. D. (1991). Reappraisal of major African earthquakes, south of 20° N, 1900–1930. *Natural Hazards*, **4**, 389–419.

Amponsah, P. (2004). Seismic activity in Ghana: past, present and future. *Annals of Geophysics*, **47**, 539–543.

Ayele, A., and Kulhánek, O. (2000). Reassessment of source parameters for three major earthquakes in the East African rift system from historical seismograms and bulletins. *Annali di Geofisica*, **43**, 81–94.

Biggs, J., Nissen, E., Craig, T., Jackson, J. and Robinson, D. P. (2010). Breaking up the hanging wall of a rift-border fault: The 2009 Karonga earthquakes, Malawi. *Geophysical Research Letters*, **37**, L11305, doi:10.1029/2010GL043179.

Calais, E., Ebinger, C., Hartnady, C. J. H., and Nocquet, J. M. (2006). Kinematics of the East African Rift from GPS and earthquake vector data. In *The Afar Volcanic Province within the East African Rift System*, eds. Yirgu, G., Ebinger, C. J. and Maguire, P. K. H., Special Publication No. 259. London: Geological Society of London, pp. 9–22.

Calais, E., d'Oreye, N., Albaric, J. *et al.* (2008). Aseismic strain accommodation by dyking in a youthful continental rift, East Africa. *Nature*, **456**, 783–787.

Chorowicz, J., 2005. The East African rift system, *Journal of African Earth Sciences*, **43**, 379–410.

Delvaux, D., Kervyn, F., Macheyeki, A. S. and Temu, E. B. (2012). Geodynamic significance of the TRM segment in the East African Rift (W-Tanzania): active tectonics and paleostress in the Ufipa plateau and Rukwa basin. *Journal of Structural Geology*, **37**, 161–180.

Déprez, A., Doubre, C., Masson, F. and Ulrich, P. (2013). Seismic and aseismic deformation along the East African Rift System from a reanalysis of the GPS velocity field of Africa. *Geophysical Journal International*, advanced access, published 4 April, doi: 10.1093/gji/ggt085

Fairhead, J. D., and Girdler, R. W. (1970). How far does the rift system extend through Africa? *Nature*, **221**, 1018–1020.

Fairhead, J. D., and Henderson, N. B. (1977). The seismicity of southern Africa and incipient rifting. *Tectonophysics*, **41**, 19–26.

Fairhead, J. D. and Stuart, G. W. (1982). The seismicity of the East African Rift System and comparison with other continental rifts. In *Continental and Oceanic Rifts*, Geodyn. Ser., vol. 8. Washington, DC: AGU, pp. 41–61.

Fenton, C. H., and Bommer, J. J. (2006). The M_w7 Machaze, Mozambique, earthquake of 23 February 2006. *Seismological Research Letters*, **77**, 424–437.

Foster, A. N., and Jackson, J. A. (1998). Source parameters of large African earthquakes: implications for crustal rheology and regional kinematics. *Geophysical Journal International*, **134**, 425–438.

Gaulon, R., Chorowicz, J., Vidal, G., Romanowicz, B. and Roult, G. (1992). Regional geodynamic implication of the May-July 1990 earthquake sequence in southern Sudan. *Tectonophysics*, **209**, 87–103.

Girdler, R. W. and McConnell, D. A. (1994). The 1990 to 1991 Sudan earthquake sequence and the extent of the East African Rift System. *Science*, **264**, 67–70.

Githiri, J. G., Patel, J. P., Barongo, J. O., and Karanja, P. K. (2012). Trail analysis of ground magnetic data in Magadi area, southern Kenya Rift, *Tanzania Journal of Science*, **38**, 1–14.

Gouin, P. (1979). *Earthquake History of Ethiopia and the Horn of Africa*. Ottawa, Canada: International Development Research Centre.

Green, R. W. E. and Bloch, S. (1971). The Ceres, South Africa, earthquake of September 29, 1969. I. Report on some aftershocks. *Bulletin of the Seismological Society of America*, **61**, 851–859.

Gupta, H. K., 1992. The Malawi earthquake of March 10, 1989: a report of a microseismic survey. *Tectonophysics*, **209**, 165–166.

Gutenberg, B. and Richter, C. F. (1949). *Seismicity of the Earth and Associated Phenomena*: Princeton, NJ: Princeton University Press.

Hartnady, C. J. H. (1990). Seismicity and plate boundary evolution in southeastern Africa. *South African Journal of Geology*, **93**, 473–484.

Hartnady, C. J. H. (2002). Earthquake hazard in Africa: perspectives on the Nubia–Somalia boundary. *South African Journal of Geology*, **98**, 425–428.

Ibs-von Seht, M., Plenefisch, T. and Klinge, K. (2008). Earthquake swarms in continental rifts – a comparison of selected cases in America, Africa and Europe. *Tectonophysics*, **452**, 66–77.

ISC (1991). *Regional Catalogue of Earthquakes, 26 (July- December 1989)* Thatcham, UK: International Seismological Centre.

ISC (2011). *Online Bulletin*, http://www.isc.ac.uk, International Seismological Centre, Thatcham, UK.

Jackson, J. and Blenkinsop, T. (1993). The Malawi earthquake of March 10, 1989: deep faulting within the East African Rift System. *Tectonics*, **12**, 1131–1139.

Jacques, E., Kidane, T., Tapponnier, P. *et al.* (2011). Normal faulting during the August 1989 earthquakes in Central Afar: Sequential trigerring and propagation of rupture along the Dobi Graben. *Bulletin of the Seismological Society of America*, **101**, 994–1023.

Junner, N. R. (1941). The Accra earthquake of 22 June 1939. *Gold Coast Geological Survey Bulletin*, **13**, 3–41.

Kebede, F., Kim, W. and Kulhanek, O. (1989). Dynamic source parameters of the March–May 1969 Serdo earthquake sequence in Central Afar, Ethiopia, deduced from teleseismic body waves. *Journal of Geophysical Research*, **94**, 5603–5614.

Keyser, A. W. (1974). Some macroseismic observations of the Boland earthquake of 29 September 1969 in the southwestern Cape Province, South Africa. Seismological Series 4, Geological Survey of South Africa, 53 pp.

Krenkel, E. (1923). Die Seismizität Afrikas. *Zentralblatt für Mineralogie, Geologie und Paläontologie*, **6**, 173–183.

Krige, L. J., and Venter, F. A. (1933). The Zululand earthquake of the 31st December 1932. *Transactions of the Geological Society of South Africa*, **36**, 101–112.

Loupekine, I. S. (1966). *Earthquake Reconnaissance Mission: Uganda the Toro Earthquake of 20 March 1966*, UNESCO.

Macheyeki, A. S., Delvaux, D., De Batist, M. and Mruma, A. (2008). Fault kinematics and tectonic stress in the seismically active Manyara-Dodoma Rift segment in Central Tanzania – implications for the East African Rift. *Journal of African Earth Sciences*, **51**, 163–188.

Mavonga, T. (2007). Some characteristics of aftershock sequences of major earthquakes from 1994 to 2002 in the Kivu province, Western Rift Valley of Africa. *Tectonophysics*, **439**, 1–12.

McKenzie, D. P., Davies, D. and Molnar, P. (1970). Plate tectonics of the Red Sea and East Africa. *Nature*, **226**, 243–248.

Molnar, P. and Aggarwal, Y. (1971). A microearthquake survey in Kenya. *Bulletin of the Seismological Society of America*, **61**, 195–201.

Singh, M. and Hattingh, E. (2009). Short communication: Collection of isoseismal maps for South Africa. *Natural Hazards*, **50**, 403–408.

Stamps, D. S., Calais, E., Saria, E. *et al.* (2008). A kinematic model for the East African Rift. *Geophysical Research Letters*, **35**, L05304, doi:10.1029/2007GL032781.

Sykes, L. R., and Landisman, M. (1964). The seismicity of East Africa, the Gulf of Aden, the Arabian and the Red Seas. *Bulletin of the Seismological Society of America*, **54**, 1927–1940.

USGS (1989). US Geological Survey Earthquake Data Report, 8-89, part 2 of 2, 151–302.

Van Wyk, W. L. and Kent, L. E. (1974). *The Earthquake of 29 September 1969 in the Southwestern Cape Province, South Africa*. Seismological Series 4, Geological Survey of South Africa, 81pp.

Vittori, E., Delvaux, D., and Kervyn, F. (1997). Kanda fault: a major seismogenic element west of the Rukwa rift (East Africa, Tanzania). *Journal of Geodynamics*, **24**, 139–153.

Yang, Z. and Chen, W. P. (2008). Mozambique earthquake sequence of 2006: High-angle normal faulting in southern Africa. *Journal of Geophysical Research*, **113**, B12303, doi:10.1029/ 2007JB005419.

Yang, Z. and Chen, W. P. (2010). Earthquakes along the East African Rift System: A multiscale-system wide perspective. *Journal of Geophysical Research*, **115**, B12309, doi:10.1029/ 2009JB006779.

Zana, N. and Tanaka, K. (1981). Focal mechanism of major earthquakes in the Western Rift Valley of Africa. *Tohoku Geophysical Journal*, **28**, 126.

16

Tsunami impact on the African continent: historical cases and hazard evaluation

Viacheslav K. Gusiakov

16.1 Introduction

Africa is the world's second largest continent surrounded by the Mediterranean Sea to the north, the Indian Ocean to the east, and the Atlantic Ocean to the west. Its total area is about 30.2 million km^2 including adjacent islands, which is 20% of the total land area of the planet (Sayre, 1999). However, its overall coastline length is about 26 000 km, which is less than 13% of the total world coastline. The most prominent feature of the African coast is the absence of deep indentations of its shore; to a large extent the continent has one of the smoothest coastlines in the world. It is also the second most populous continent with nearly 1.0 billion people in 54 recognised sovereign states, which is about 15% of the world's human population.

Tectonically, the continent is located entirely within the African Plate that is bounded on the northeast by the Arabian Plate, on the east by the Indo-Australian Plate, on the south by the Antarctic Plate, on the west by the North and South American Plates, and on the north by the Eurasian Plate. The largest part of the African Plate boundary is a divergent type boundary that separates plates moving away from each other. This type of boundary has much lesser seismic potential than subduction boundaries, especially, in terms of the occurrence of large (magnitude 8 or higher) earthquakes (Fig. 16.1). The divergent boundaries are manifested in the oceanic lithosphere as rifts of the oceanic ridge system, and in the continental lithosphere as rift valleys like the East African Great Rift Valley.

The northern edge of the African Plate is a subduction-type boundary having much greater potential for large earthquakes. In addition, the northern plate boundary is located close to the coast. As a result, most of the known historical tsunamis affecting Africa occurred near to its northern part (Fig. 16.2). The collision of the African and Eurasian tectonic plates in the Mediterranean region results in a major belt of seismic and volcanic activity with strong earthquakes and volcanic eruptions repeatedly occurring during historical time. Numerous large destructive earthquakes and tsunamis occurred from antiquity up to the present in the eastern Mediterranean; however, only a few of them were documented as having impact on the African coast.

For the western world, the northern part of the continent has been known for thousands of years; however, its actual size and geographical boundaries remained unknown until the very end of the fifteenth century when the Portuguese first sailed around Africa. As a result, the written history for all parts of the continent, except its north, is very short.

At present, there are two global historical tsunami databases maintained separately by the NGDC/NOAA in Boulder, USA and by the NTL/ICMMG in Novosibirsk, Russia. The NGDC/NOAA database is maintained in the Oracle RDBMS from where the data can be accessed via web-based HTML forms and ArcIMS interactive maps as well as exported in several different formats (NGDC/NOAA Tsunami Database, 2013). The NTL/ICMMG tsunami database is maintained in MS Excel with a specialised graphic shell (PDM_TSU) providing screen forms for data editing, retrieval, and listing (NTL/ICMMG Tsunami Database, 2013). As a stand-alone application, the NTL/ICMMG database is distributed on a CD-ROM provided with a specially developed GIS-type graphic shell (WinITDB) allowing the fast and efficient manipulation of maps, models, and data (ITDB/WLD, 2012). The content of these two databases is fairly close in terms of the total number of historical events, temporal and spatial coverage, and basic source parameters. However, for many historical events they differ in the types of origin, the number of available run-up observations, resulting fatalities, and degree of validity for some older historical events. The present analysis of historical data on tsunamis

Extreme Natural Hazards, Disaster Risks and Societal Implications, eds. Alik Ismail-Zadeh, Jaime Urrutia Fucugauchi, Andrzej Kijko, Kuniyoshi Takeuchi and Ilya Zaliapin. Published by Cambridge University Press.

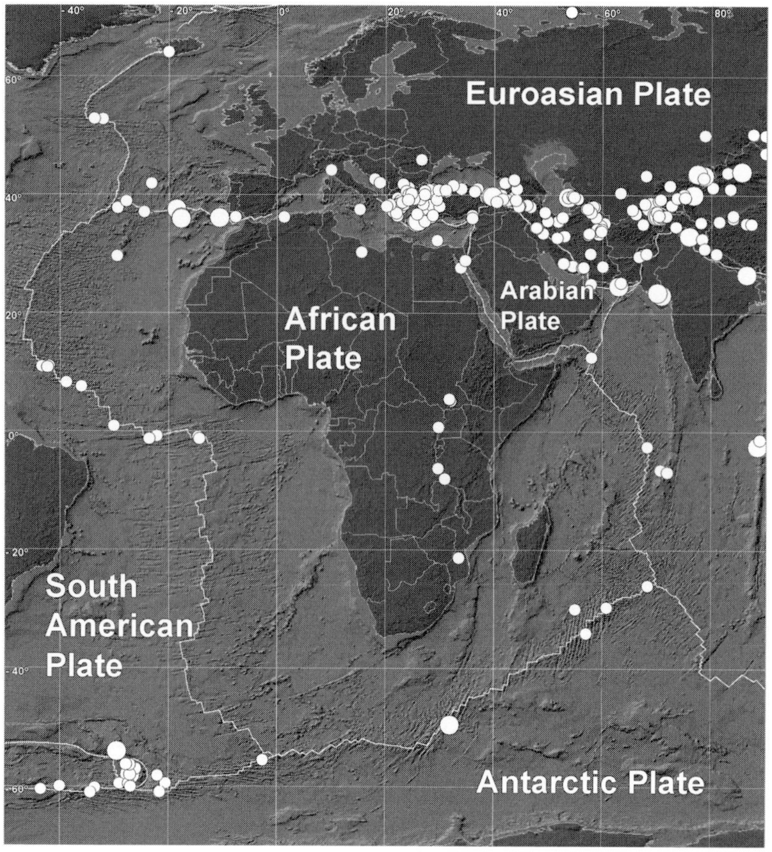

Figure 16.1. Tectonic settings and historical seismicity in and around the African continent. Earthquakes with magnitudes greater than 7.0 (small circles) and greater than 8.0 (large circles) that occurred from 1900 to 2013 are shown. The white solid lines show the main plate boundaries.

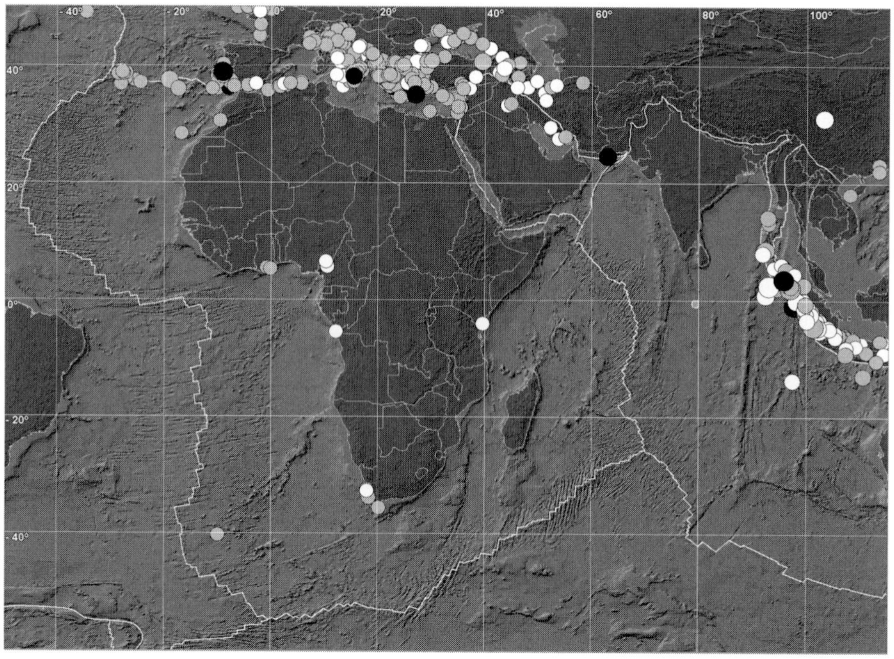

Figure 16.2. Map of historical tsunamigenic events that occurred in the geographical area within 60° S–50° N, 50° W–115° E from 2000 BC till present. 512 events are shown. The size of the circles is proportional to the event magnitude, the density of the grey tone shows the tsunami intensity I on the Soloviev–Imamura scale (Soloviev, 1972): black – destructive tsunamis with $I \geq 3$, grey – damaging tsunamis with $I \geq 1$, white – weak tsunamis with undetermined intensity. The white solid lines show the main plate boundaries.

affecting the African coast is based on the content of both databases, generally referred to as the Global Tsunami DataBase (GTDB) catalogue (Gusiakov, 2003).

As follows from analysis of the GTDB content, the most (up to 75%) of historical tsunamigenic events were generated by shallow earthquakes capable of transferring sufficient energy to the overlying water column (Gusiakov, 2009). Tsunamis of volcanic and landslide origins are essentially less frequent (4% and 10%, respectfully); however, they can produce the highest run-ups locally (in the immediate neighbourhood of a source area). In general, this ratio is true for the tsunamis affecting the African coast, but it varies depending on the part of the continent under consideration.

16.2 Historical events affecting the African coast

The earliest known tsunamigenic event that affected the African coast is probably the tsunami that resulted from the Etna flank failure that occurred ca. 6000 BC. This catastrophic collapse event with an estimated volume of displaced material of 35 km^3 generated a huge tsunami that devastated many areas in the eastern Mediterranean (Deeming *et al.*, 2010). There is a possibility that the Neolithic village of Atlit-Yam, Israel, now below the sea level, was abandoned around that time due to the impact of this tsunami (Pareschi *et al.*, 2006). If this was the case, the northern African coast, east of Benghazi, Lybia, which is directly exposed to the source area, would have been severely flooded by tsunami waves generated by the Etna volcano flank failure at about 6000 BC.

The next known large tsunamigenic event in the eastern Mediterranean was the volcanically generated tsunami that followed the famous 1628 BC Santorini mega eruption. The eruption with its 50–60 km^3 estimated volume of erupted material was one of the largest volcanic events in written history (Simkin and Siebert, 1994). Historians did not leave any eye witness reports on the actual run-ups or destructive effects of this tsunami, known as the Bronze Age or Minoan tsunami, but some geological evidence of flooding as high as 40–50 m above sea level has been found on the west side of the island of Anaphi, which is just 25 km east of the island of Santorini (Marinos and Melidonis, 1959). Other possible evidence of tsunami inundation was found as high as 160 m and 250 m on the northeastern side of the same island (Pararas-Carayannis, 1992). These high levels look incredible, but it is worth recalling that the maximum run-up for 1883 Krakatoa tsunami, generated by a much less voluminous eruption, reached 41 m. It is not known for sure how high the waves of the Bronze Age tsunami were in the immediate vicinity of the source area, but in the NGDC/NOAA Tsunami Database they are assumed to be as high as 90 m (NGDC/NOAA Tsunami Database, 2013).

At a greater distance away from Santorini, the evidence of the tsunami height has been found at 5 m above sea level north of Jaffa / Tel Aviv, Israel. There is also conclusive evidence that Minoan cities on the north and the east coasts of Crete were struck by huge tsunami waves (Marinatos, 1939). Undoubtedly, these waves destroyed most of the Minoan fleet, which played an important part in maintaining the prosperity and power of the Minoan Kingdom. This, together with the general environmental downturn that resulted from the giant eruption, could have led to the demise of the Minoan civilisation (Baillie, 1990). The part of the northern African coast nearest to Santorini (modern Cyrenaica) is protected from direct exposure to tsunami impact by the island of Crete but the areas adjacent to the Nile delta could have been seriously affected by the Minoan tsunami.

It is worth noting that the 27 August 1883 Krakatoa tsunami that resulted from the massive caldera collapse during the paroxysmal stage of the 1883 Krakatoa eruption which killed 36 416 people in Indonesia, did not cause significant waves near the African coast or in any other part of the Indian Ocean. Due to the fast decay of the amplitude of relatively short-period waves, the run-up heights outside the immediate vicinity of the source area, where its maximum height reached 41 m, turned out to be minor and did not produce any considerable damage.

The 365 AD earthquake, the most destructive earthquake in the eastern Mediterranean region, occurred in the early morning of 21 July and had its estimated epicentre near the western part of Crete. Based on the widespread destructive effect of the seismic shaking that extended from central Greece to northern Egypt and from Sicily to Syria, its macroseismic magnitude is estimated to have been 8 or greater (Stiros, 2001). In Crete, nearly all the towns were destroyed. The western part of Crete experienced a considerable coseismic uplift reaching 8 m. Ancient Falasarna, a city founded in the sixth century BC on the westernmost tip of the island and used as an intermediate port for vessels sailing from Egypt to Rome, was uplifted so that its harbour could no longer be used.

The earthquake generated a destructive tsunami that devastated almost the whole of the eastern Mediterranean, killing thousands and hurling ships as far as 3 km inland. The quake left a deep impression on the societies in the affected regions and many ancient historians refer to this event as a disaster of biblical proportions.

The combined impact of the earthquake and tsunami hitting Alexandria in the early hours of 21 July AD 365 was described by the Roman historian Ammianus Marcellinus in the following words:

> Slightly after daybreak, and heralded by a thick succession of fiercely shaken thunderbolts, the solidity of the whole earth was made to shake and shudder, and the sea was driven away, its waves were rolled back, and it disappeared, so that the abyss of the depths was uncovered and many-shaped varieties of sea-creatures were seen stuck in the slime; the great wastes of those valleys and mountains, which the very creation had dismissed beneath the vast whirlpools, at that moment, as it was given to be believed, looked up at the sun's rays. Many ships, then, were stranded as if on dry land, and people wandered at will about the paltry remains of the waters to collect fish and the like in their hands; then the roaring sea as if insulted by its repulse rises back in turn, and through the teeming shoals dashed itself violently on islands and extensive tracts of the mainland, and flattened innumerable buildings in towns or wherever they were found. Thus in the raging conflict of the elements, the face of the earth was changed to reveal wondrous sights. For the mass of waters returning when least expected killed many thousands by drowning, and with the tides whipped up to a height as they rushed back, some ships, after the anger of the watery element had grown old, were seen to have sunk, and the bodies of people killed in shipwrecks lay there, faces up or down. Other huge ships, thrust out by the mad blasts, perched on the roofs of houses, as happened at Alexandria, and others were hurled nearly two miles from the shore, like the Laconian vessel near the town of Methone which I saw when I passed by, yawning apart from long decay (cited by Kelly, 2004).

The tsunami in AD 365 was so devastating that the anniversary of the disaster was still commemorated annually at the end of the sixth century in Alexandria as a 'day of horror' (Stiros, 2001; Hetch, 2008).

The other great earthquake in the eastern Mediterranean occurred on 8 August 1303. The location of the earthquake source is largely uncertain, although it is generally agreed that it ruptured the eastern segment of the Hellenic arc somewhere between Crete and Rhodes (Guidoboni and Comastri, 1997). The earthquake caused damage over a wide area including Crete, the Peloponnese, Rhodes, Cairo, Acre, Damascus, Antioch, and Cyprus and was felt as far away as Constantinople and possibly Tunis. Based on this and the devastating damage to buildings and castles over the whole of Crete, its magnitude is estimated to have been 8 or greater. The earthquake caused severe damage in Cairo, Egypt, toppling minarets on many mosques. The city walls in Alexandria, Egypt, were mostly destroyed and the lighthouse was badly damaged (Papadopoulos *et al.*, 2007).

This earthquake generated a massive tsunami that impacted many coastal locations in the eastern Mediterranean. There was also severe flooding in Alexandria. Many ships were destroyed; some of them were carried up to 3 km inland. The sea level in the city rose, and it threw a French ship up onto the coast, together with many fish and sea animals and fragments of the city walls. The level of the Nile varied widely; ships sailing in the middle of the river and lying at anchor were thrown up onto the banks 15 m inland (Soloviev *et al.*, 2000). There are no quantitative data on tsunami run-up along the coast of Egypt, but based on this description and results of numerical modelling (Hamouda, 2006) it can be estimated to be as high as 10 m.

The other known destructive tsunami on the northern coast of Africa was generated by the 2 January 1365 earthquake in Algeria. The earthquake destroyed the town of Algiers almost completely; most of the buildings fell down, and many people were killed. The sea waves succeeded after the earthquake and flooded the coastal part of the town up to a 'significant height', and 'people were drowned' (Soloviev *et al.*, 2000). The magnitude of the earthquake is unknown because the data on its manifestation are available only for Algiers.

The modern history of the tsunamis affecting the African coast starts with the 1 November 1755 earthquake and tsunami that most severely struck Lisbon, a wealthy city of the time, legendary for its prosperity and sophistication. There are no instrumental data allowing the epicentre of this earthquake to be located, but it is believed to have occurred along the Azores–Gibraltar fracture zone (AGFZ) that marks the boundary between the African and the Eurasian plates. The earthquake magnitude has been estimated to be 8.6 or higher in modern seismic catalogues. However, the data of the far-field tsunami effect and the ground motion observed along the coast of Portugal suggest a moment magnitude closer to 9.0 (Pararas-Carayannis, 1997). The earthquake occurred in the morning of All Saints Day, a religious holiday. Most of Lisbon's population of 250 000 were praying in churches and cathedrals. Within minutes, the city was turned to debris by the two major seismic shocks and the waves of the subsequent catastrophic tsunami. A huge fire completed the destruction.

The height of tsunami waves in Lisbon is estimated to have been 20–30 m with a period of up to 60 min. The largest far-field wave height of 7 m was reported from the West Indies (the Saba Island). In Africa, the waves caused great damage and casualties on the western coast of Morocco, from Tangier, where the 16 m waves reached the walled fortifications of the town, to Agadir, where the waves passed over the city walls 'killing many'. So far, this 16 m run-up in Tangier remains the highest reported tsunami run-up along the African coast.

The earliest historical evidence of the tsunami impact on the African coast, other than its north coast, comes from the survey of the historical seismicity of West Africa made by Ambraseys and Adams (1986). According to their data, a strong earthquake that was felt over a large area of West Africa occurred around 1820 near the coast of Congo. 'The earthquake drove in upon the shore from the westward several tremendous waves, so as to cause some of the vessels when in the trough of them to strike the ground in five and a half fathoms water. Many of the natives, both here in Cabinda, and in Malembo, were drowned upon the beach, as consequence of it being very narrow, and the cliffs which bound it so extremely precipitous, that they could not escape from the overwhelming waves which came suddenly upon them. To this calamity, was added the destruction of many canoes, by this sudden inundation of the sea', according to Ambraseys and Adams (1986).

In the NDGC/NOAA Tsunami Database, this earthquake has a magnitude of only 6.2, which is too low to generate a destructive tsunami capable of killing 'many natives'. It is quite possible that the intensity of tsunamis was enhanced by a submarine landslide triggered by the earthquake. Its source area was located near the mouth of the Congo River, which brings a large amount of terrestrial sediment into the ocean. Normally, this sediment does not go far from the coast and is deposited near the edge of the continental shelf, where it is readily available to go down to the next level of sedimentation at the continent basement in repeated circles of so-called 'avalanche sedimentation' (Lisitsyn, 1988).

In the first half of the twentieth century, tsunamigenic events were observed in almost the same part of the Gulf of Guinea in 1911 and 1939 (Berninghausen, 1964). On 11 May 1911, a water wave submerged the coastal part of Lomé, the capital and main port of Togo, to a height of 5 m and destroyed the harbour wharf. The coast remained flooded for seven hours before the sea retired gradually to its former level (Ambraseys and Adams, 1986). There were no reports of a felt earthquake from the affected area. The second event occurred on 22 June 1939 just 50 km to the west on the coast of Ghana. The earthquake with an estimated magnitude of 6.5 devastated the coastal area near Accra. At Labadi and Tishi the sea 'was agitated' (Berninghausen, 1964).

There is a short report on some unusual wave action that was observed near Mombasa, Kenya, on 5 November 1930. Two tidal waves travelled towards Ras Serania as a front about 46 m long and possibly 6 m in amplitude. The waves rose to about half the height of the Mombasa cliff at the lighthouse (Loupekine, 1971).

A small tsunami was observed in several locations along the coast of Morocco after the $M7.8$ earthquake that occurred on 28 February 1969 off the coast of Portugal, near the AGFZ (Renou *et al.*, 2011). Its maximum amplitude (0.9 m) was recorded by a tide-gauge in the harbour at Casablanca (Kaabouben *et al.*, 2009).

Tsunamis on the African coast that were generated by nearby sources, both very small, were observed in 2003 on the Algerian coast and in 2008 in St Helena Bay, South Africa. The destructive Zemmouri earthquake of 21 May 2003 ($M_w6.8$) occurred 50 km east of Algiers, between the cities of Zemmouri and Boumerdès. The epicentre of this shock, the largest felt in this area since the $M_w7.3$ 1980 earthquake, was located very close to the coast. As a result of severe seismic shaking, more than 1243 buildings were completely or partially destroyed, 2266 people were killed, 10 261 injured, and 200 000 left homeless (Edwards, 2004). The earthquake resulted in 40–80 cm uplift of the coast between the cities of Reghaia and Zemmouri and generated a small tsunami that was noted along the coastal parts of Algiers and Boumerdès as a 100-m recession of the sea. Several fishing boats were seen resting on the sea bottom before the water returned (Sahal *et al.*, 2009).

It is interesting to note that much higher waves were observed along the southeastern coasts of Majorca and Minorca, and in Ibiza in the Spanish Balearic Islands located about 250 km north of the epicentre. In Majorca, they reached 3 m high and caused significant damage to hundreds of vessels in harbours and ports. Several underwater telephone cables were damaged at the time of the earthquake, cutting communications between Europe and Africa (ITIC, 2003). This indicates the possibility of bottom sediment slides triggered by the earthquake, which, in turn, could enhance tsunami heights in certain directions.

On 26 August 2008 sudden tidal changes caused damage to several factories in St Helena Bay, South Africa (*Daily Dispatch*, 2008). The magnitude of the associated seismic event was small and it was not capable of producing a measurable tsunami, but it is believed to have triggered a mudslide that generated water waves.

The South African coast can also be subject to the impact of meteo-tsunamis, which represent one of the major coastal hazards in some areas, such as the Balearic Islands and some parts of the Adriatic Sea coast (Rabinovich and Monserrat, 1996). These seiche-type oscillations are generated by some large-scale atmospheric disturbances like an atmospheric pressure front moving rapidly over a shallow sea at about the same speed as a tsunami allowing coupling of the air and water disturbance. Typically, they occur within straits and in particular bays and harbours; however, sometimes they can happen along a straight coast. Okal *et al.* (2013) reported on the unusual oscillations of water that occurred on 27 August 1969 in the village of Dwarskersbos, Western Cape, South Africa, with run-up reaching 2.9 m for a maximum inundation of 260 m from the sea shore. The remarkable aspect of this event is the extreme concentration of the flooding along a narrow stretch of coastline (less than 2 km).

The 26 December 2004 Indian Ocean tsunami generated by the $M_w9.0$ Aceh-Sumatra earthquake also affected the African coast. With its 228 000 estimated fatalities, this tsunami turned out to be the most deadly event in the 4000-year history of tsunamis. Maximum run-up heights at the nearest coast were up to 41–55 m. In seven hours the waves crossed the entire Indian Ocean and reached the eastern coast of Madagascar, where they produced 2–6 m run-up (Okal *et al.*, 2006). Despite the absence of any official warning, fortunately there were no fatalities in Madagascar. One fatality was reported in Kenya, where run-up was about 1.0–1.5 m and two in South Africa where the maximum wave height (2.7 m) was measured in Port Elizabeth. The largest waves (up to 9.6 m) of the Indian Ocean tsunami on the African coast were observed along the coast of Somalia, where they resulted in 289 fatalities (Fritz and Borrero, 2006).

16.3 Limnic eruptions

A specific, quite rare type of natural hazard presently known only in Africa is the so-called 'limnic eruption', also referred to as a 'lake overturn' or an 'exploding lake'. This disaster occurs when carbon dioxide or methane that is dissolved in the deep water layers is suddenly released and erupts to the surface, suffocating wildlife and humans. The specific gaseous composition of exploding lakes is unique to each lake, but carbon dioxide (CO_2) and methane (CH_4) are believed to be the main components of a released gas bubble. Due to the rarity of this type of event, the exact physical mechanism of generation of tsunami-type water waves during the limnic eruption is not known. It could be the dynamic impulse of the emerging voluminous gas bubble, displacing a large amount of water, a subsurface landslide resulting in disturbance of bottom sediments, or a combination of several mechanisms.

On 15 August 1984, Lake Monoun exploded in such a limnic eruption killing 37 people. Lake Monoun is located in the West Province of Cameroon within the Oku Volcanic Field. Its maximum length is about 1.5 km and it has a maximum depth of 60 m. Several people reported hearing a loud noise coming from the lake. A gas cloud reportedly emanated from a crater in the bottom of the eastern part of the lake. Survivors said that the whitish, smoke-like cloud, which smelled bitter and acidic, emerged from the water. Vegetation was flattened around the eastern part of the lake, probably by a tsunami-type wave that resulted from the large-scale disturbance of water produced by an emerging gas bubble.

The case was widely reported in the media. At first, the cause of the deaths was a mystery, and other reasons such as terrorism were suspected. Further investigation and a similar event that occurred two years later at Lake Nyos and killed about 1 700 people led to the currently accepted explanation (USGS Volcano Hazards Program, 2013).

With its reported 1700 fatalities, the 21 August 1986 limnic eruption of Lake Nyos is, at the time of writing, the most deadly case of this type of natural event. The lake is located on the flank of an inactive volcano in the Oku volcanic plain, northwest region of Cameroon, and has a maximum length of 2.0 km and a maximum depth of 207 m. Its estimated volume of water is 0.15 km^3. The event was triggered by a sudden release of about 100 000–300 000 m^3 carbon dioxide (Socolow, 2005). The carbon-dioxide-rich cloud was expelled rapidly from the floor at the southern end of Lake Nyos with an estimated speed of about 100 km per hour. Survivors from the valley recalled hearing a rumble that lasted about 20 seconds and seeing a fountain of water and foam that rose about 90 m above the lake's surface. A wave at least 24 m tall scoured the rocky lake shore (Brown, 2000). A large amount of gas escaped from the deep waters which resulted in the lake surface level dropping by 1 m. The cloud quickly enveloped houses within the crater that were 120 m from the shoreline of the lake. Due to the density of carbon dioxide being greater than that of air, the gaseous mass hugged the ground surface and descended down valleys along the north side of the crater. The deadly cloud was about 50 m thick and it advanced downslope at a rate of 20–50 km per hour. This deadly mist persisted in a concentrated form over a distance of 23 km, bringing sudden death to the villages of Nyos, Kam, Cha, and Subum. Today the lake also poses another type of threat, because the volcanic dyke along one side is weakening. If this natural dam breaks it will allow water to rush into downstream villages all the way into Nigeria and allowing much carbon dioxide to escape.

Lake Kivu located in the Democratic Republic of Congo is the largest and deepest of the three known exploding lakes in Africa that experience a violent lake overturn. It lies on the border between the Democratic Republic of Congo and Rwanda and is 1460 m above sea level. Its maximum length is 89 km, maximum depth is 480 m, and the estimated volume of water is 500 km^3. An analysis of Lake Kivu's geological history indicates sporadic massive biological extinctions on millennial timescales. The amount of methane in Lake Kivu is estimated to be 65 km^3. There is also an estimated 250 km^3 of carbon dioxide (Halbwachs *et al.*, 2002). A possible Lake Kivu overturn would be catastrophic, exceeding other known lake overturns at Lakes Nyos and Monoun, because about 2 million people presently live in the lake basin.

Another tsunami-related event occurred in Lake Kivu on 28 January 1938. A lava lake in the summit crater, which has been active since at least 1921, drained in 1938 during a major flank eruption. The lava flows from the eruption set hundreds of square kilometres of forest on fire destroying arable land. The flows were said to have reached Lake Kivu and the density contrast between the lava and the water generated a mini-tsunami (Venzke *et al.*, 2002).

16.4 Discussion

As shown in Fig. 16.1, the African continent lies entirely within a single African plate, which is separated from its neighbouring plates by boundaries of a divergent type, except for the northern border with the Eurasian Plate, which is a subduction type boundary. A subduction boundary has a much greater potential for tsunamigenic earthquakes. Accordingly, most of the tsunamigenic events known to have affected the African coast occurred near the northern coast of the continent, i.e. in the Mediterranean region. It is here that 20 out of the 46 known tsunamis affecting the African coast occurred. Of these 20 events, quantitative data on tsunami heights are available for only 9 events. For the other 11, we have only indirect estimates of tsunami intensity based on reports about property damage or human fatalities, or we can make assumptions based on the known physical mechanisms of excitation and propagation of tsunamis.

A list of tsunamis that produced highest run-up and/or fatalities on the African coast is shown in Table 16.1. The highest documented run-up of 16 m was observed in the area of Tangier, Morocco, after the Lisbon earthquake of 1755 (Blanc, 2009). Judging by the number of victims and the degree of destruction in Alexandria, one can conclude that the tsunami generated by the 365 AD Crete earthquake struck the coast of Egypt with a comparable height, i.e. 10–15 m. The tsunami generated by another strong earthquake in the eastern Mediterranean that occurred in 1303 could also have caused a devastating effect on the northern Egyptian coast (Guidoboni and Comastri, 1997).

These destructive tsunamis were generated by the strongest earthquakes that occurred during the last 2000 years in the Mediterranean. More frequent ordinarily tsunamigenic earthquakes with magnitudes from 7.0 to 7.8 that occurred in the Aegean Sea did not apparently produce a dangerous impact on the African coast. For example, the largest instrumentally recorded earthquake of 15 July 1956 in the Aegean Sea with a magnitude M_s7.5, but resulted in 30 m waves at the nearest islands, did not impact northern Africa.

Similarly, tsunamigenic earthquakes in the Caribbean region are not dangerous for the African continent. At least, for all the 400-year recorded history of the Caribbean tsunamis there have been no cases when a tsunami from this region reached the African continent. It should be noted, however, that the Caribbean region does not have a history of the occurrence of the *M*9 earthquakes that are capable of generating a trans-oceanic tsunami. Likewise, until December 2004 we did not have any examples of the impact of the Indonesian tsunamis on the coast of Africa. However, the destructive tsunami that resulted from the M_w9.0 mega-earthquake that occurred on 26 December 2004 reached the coast of Somalia, with heights of between 6 and 9 m and resulted in 289 deaths there (Fritz and Borrero, 2006). Based on the palaeo-tsunami data, the average return period of mega-quakes in the Andaman–Sumatra subduction zone is estimated to be 400–600 years (Jankaew *et al.*, 2008), so for the last 2000 years there could have been 3–5 cases of Indonesian tsunamis causing damage on the African continent.

When considering the Caribbean and the northwestern Atlantic earthquakes, one should also bear in mind the possibility of so-called tsunami earthquakes (Kanamori, 1972). Having a moderate magnitude, tsunami earthquakes are capable of generating a devastating tsunami, in both near-field and far-field areas. For example, the waves from

Table 16.1. *List of historical tsunamis that affected the African coast*

Date	Source location	Run-up, m	Fatalities
~6000 BC	Etna flank failure	unknown	unknown
1628 BC	Santorini eruption	unknown	unknown
21 Jul. 365	Crete	unknown	5000
08 Aug. 1303	Aegean Sea	unknown	unknown
2 Jan. 1365	Algeria	unknown	many
1 Nov. 1755	Lisbon, Portugal	16	60 000[a]
1820	Coast of Congo	unknown	several
11 May 1911	Lomé, coast of Togo	5	unknown
6 Nov. 1930	Mombasa, Kenya	6	none
22 Jun. 1939	Accra, coast of Ghana	unknown	none
15 Aug. 1984	Lake Monoun, Cameroon	5–6	37[a]
21 Aug. 1986	Lake Nyos, Cameroon	24	1700[a]
26 Dec. 2004	Sumatra, Indonesia	6–9	289

[a] Total number of fatalities resulting from the source event and tsunami waves.

the Aleutian earthquake of 1 April 1946, whose surface wave magnitude was only 7.4, were observed in Hawaii with heights up to 16 m and in the Marquesas Islands with heights up to 22 m (Okal *et al.*, 2002) and even produced some damage as far away as in Antarctica.

Another seismic zone with the potential for large submarine earthquakes which could possibly threaten the African coast is located in the southern Atlantic along the South Sandwich Trench. Eight submarine earthquakes with estimated magnitudes equal to or greater than 7.5 have been recorded there since the beginning of the instrumental era of seismological observations. Of them, one event (which occurred on 27 June 1929) had a magnitude of nearly 8.0. Magnitude 7.5 is adopted as the threshold value for issuing a regional warning in the Pacific and the Indian Ocean Tsunami Warning Systems. However, there are no historical records of tsunamis from the South Sandwich earthquakes being observed on the African coast or on any other coast. The reasons for this are two-fold: (1) the relatively flat bottom relief without steep slopes in this part of the southern Atlantic and (2) the absence of thick bottom sediments. The latter, as shown by Gusiakov (2001), can be an important factor in the generation of tsunamis. Another reason could be the earthquakes source mechanisms with a predominantly strike-slip component that lowers their tsunamigenic potential. A similar situation exists for the Macquarie seismic zone located south of New Zealand. The four submarine earthquakes with magnitude 8 or greater that occurred in 1989–2004 generated very minor tsunamis that were recorded by the New Zealand tide-gauges with amplitudes of no more than several tens of centimetres.

16.5 Conclusion

The Global Tsunami Database contains 46 historical events affecting the African continent of which only 20 have sources in the immediate proximity of the African coast. We can hence conclude that Africa is the only continent where the hazard from distant tsunamigenic sources may dominate that from local sources. Many parts of the African coast are vulnerable to the threat of landslide-generated tsunamis. Timely warnings of these tsunamis are especially difficult because they can occur with little or no seismic signal recorded in advance. A special type of hazard in Africa is limnic eruptions, which can result in tsunami-type water waves affecting the shoreline of lakes. A realistic assessment of the tsunami hazard for the coast of Africa is hampered by the lack of reliable historical data on tsunami manifestations. With the possible exception of its northern part, there has still been no systematic search for historical and geological data about tsunamis that impact the African coast. Such a search could greatly improve our understanding of the tsunami hazard on this continent.

Acknowledgments. The author thanks Efim Pelinovsky for careful reading of the manuscript and numerous suggestions for text improvement. The author also thanks T. Kalashnokova for assistance in graphics editing. This study was supported by the RFBR grants 12-07-00406, 12-05-00894, and 12-05-92697 and SD RAS Grants No. 37 and No. 117.

References

Ambraseys, N. N., and Adams, R. D. (1986). Seismicity of West Africa. *Annals of Geophysics*, **4**, 679–702.

Baillie, M. G. L. (1990). Irish tree-rings and an event in 1628 BC. In *Thera and the Aegean World III*, vol. 3, ed. Hardy, D. A. London: The Thera Foundation, pp. 160–166.

Berninghausen, W. H. (1964). Tsunamis and seismic seiches reported from the eastern Atlantic south of the Bay of Biscay. *Bulletin of the Seismological Society of America*, **54**, 439–442.

Blanc, P.-L. (2009). Earthquakes and tsunami in November 1755 in Morocco: a different reading of contemporaneous documentary sources. *Natural Hazards and Earth System Sciences*, **9**, 725–738.

Brown, D. (2000). Scientists hope to quiet Cameroon's killer lakes. *Washington Post*, 1 February 2000.

Daily Dispatch (2008). Mini-tsunami on Cape coast caused by quake. *Daily Dispatch Online*, downloaded 28 August 2008. http://www.dispatch.co.za/article.aspx?id=242455

Deeming, K. R., McGuire, B. and Harrop, P. (2010). Climate forcing of volcano lateral collapse: evidence from Mount Etna, Sicily. *Philosophical Transactions of the Royal Society*, **368**, doi:10.1098/rsta.2010.0054.

Edwards, C. L., ed. (2004) *Zemmouri, Algeria: Mw 6.8 Earthquake of May 21, 2003.* TCLEE Monographs, vol. 27 Reston, VA: ASCE.

Fritz, H. M. and Borrero, J. C. (2006). Somalia field survey after the December 2004 Indian Ocean tsunami. *Earthquake Spectra*, **22**, 219–233.

Guidoboni, E., and Comastri, A. (1997). The large earthquake of 8 August 1303 in Crete: seismic scenario and tsunami in the Mediterranean area. *Journal of Seismology*, **1**, 55–72. doi:10.1023/A:1009737632542.

Gusiakov, V. K. (2001). "Red", "Green" and "Blue" Pacific tsunamigenic earthquakes and their relation with conditions of oceanic sedimentation. In *Tsunamis at the End of a Critical Decade*, ed. Hebenstreit, G. Dordrecht: Kluwer Academic Publishers, pp. 17–32.

Gusiakov, V. K. (2003). NGDC/HTDB meeting on the historical tsunami database proposal. *Tsunami Newsletter*, **XXXV** (4), 9–10.

Gusiakov, V. K. (2009). Tsunami history – recorded. In *The Sea,* vol. 15, *Tsunamis*, eds. Robinson, A. and Bernard, E. Cambridge, MA: Harvard University Press, pp. 23–53.

Halbwachs, M., Tietze, K., Lorke, A. and Midaheranwa, C. (2002). *Investigations in Lake Kivu (East Central Africa) after the Nyiragongo Eruption of January 2002*. Final Report of the International Scientific Team, 9 March 2002, http://www.eawag.ch/forschung/surf/gruppen/kivu/publications/Halbwachs_2002_Investigations_in_Lake_Kivu.pdf

Hamouda, A. Z. (2006). Numerical computations of 1303 tsunamigenic propagation towards Alexandria, Egyptian Coast. *Journal of African Earth Sciences*, **44**(1), 37–44.

Hecht, J. (2008). Mediterranean's 'horror' tsunami may strike again. *New Scientist*, 10 March 2008

ITDB/WLD (2012). *Integrated Tsunami Database for the World Ocean,* Version 5.15 of December 31, 2012, CD-ROM, Tsunami Laboratory, ICMMG SD RAS, Novosibirsk, Russia.

ITIC, International Tsunami Information Center (2003). *Tsunami Newsletter*, **XXXV**(3), June.

Jankaew, K., Atwater, B., Sawai, Y. *et al.* (2008). Medieval forewarning of the 2004 Indian Ocean tsunami in Thailand. *Nature*, **455**, 1228–1231.

Kaabouben, F., Baptista, M. A., Iben Brahim, A., El Mouraouah, A. and Toto, A. (2009). On the Moroccan tsunami catalogue. *Natural Hazards and Earth System Sciences*, **9**, 1227–1236.

Kanamori, H. (2002). Mechanism of tsunami earthquakes. *Physics of Earth and Planetary Interiors*, **6**, 346–359.

Kelly, G. (2004). Ammianus and the Great Tsunami. *Journal of Roman Studies*, **94**, 141–167.

Lisitsyn, A. P. (1988). *Avalanche Sedimentation and Breaks in the Sediment Deposition in Seas and Oceans* (in Russian). Moscow: Nauka.

Loupekine, I. S. (1971). *A Catalogue of Felt Earthquakes in Kenya, 1892–1969*. Nairobi, Kenya: Geology Department, University of Nairobi, 128 pp.

Marinatos, S. (1939). The volcanic destruction of Minoan Crete. *Antiquity*, **13**, 425–439.

Marinos, G. and Melidonis, N. (1959). On the amplitude of the tsunami originating in the prehistoric eruption of Santorin. *Greek Geological Society*, **4**, 210–218.

NGDC/NOAA (2013). NGDC/NOAA Tsunami Database, 2013. http://www.ngdc.noaa.gov/hazard/tsu_db.shtml

NTL/ICMMG (2013). NTL/ICMMG Tsunami Database, 2013. http://tsun.sscc.ru/nh/tsunami.php

Okal, E. A., Synolakis, C. E., Fryer, G. J. *et al.* (2002). A field survey of the 1946 Aleutian tsunami in the far field. *Seismological Research Letters*, **73**, 490–503.

Okal, E. A., Fritz, H. M., Raveloson, R. *et al.* (2006). Madagascar field survey after the December 2004 Indian Ocean tsunami. *Earthquake Spectra*, **22**, 263–283.

Okal, E., de Beer, C., Visser, J. and Kalligeris, N. (2013). The rogue wave of 27 August 1969 at Dwarskersbos, South Africa: Field survey and simulation as a meteo-tsunami. *Geophysical Research Abstracts*, **15**, EGU2013–1571.

Papadopolous, G. A., Daskalaki, E., Fokaefs, A. and Giraleas, N. (2007). Tsunami hazards in the Eastern Mediterranean: strong earthquakes and tsunamis in the East Hellenic Arc and Trench system. *Natural Hazards and Earth Systems Sciences*, **7**, 57–64.

Pararas-Carayannis, G. (1992). The tsunami generated by the eruption of the volcano of Santotin in the Bronze Age (1650 B.C.+/-50 years). *Natural Hazards*, **5**, 115–123.

Pararas-Carayannis, G. (1997). The Great Lisbon earthquake and tsunami of 1 November, 1755. http://www.drgeorgepc.com/Tsunami1755Lisbon.html.

Pareschi, M., Bosch, E., and Favalli, M. (2006). Lost tsunami. *Geophysical Research Letters*, **33**, doi:10.1029/2006GL027790.

Rabinovich, A. B. and Monserrat, S. (1996). Meteorological tsunamis near the Balearic and Kuril Islands: Descriptive and statistical analysis. *Natural Hazards*, **13**, 55–90.

Renou, C., Lesne, O., Mangin, A. *et al.* (2011). Tsunami hazard assessment in the coastal area of Rabat and Sale, Morocco. *Natural Hazards and Earth System Sciences*, **11**, 2181–2191.

Sahal, A., Roger, J., Allgeyer, S. *et al.* (2009). The tsunami triggered by the 21 May 2003 Boumerdès-Zemmouri (Algeria) earthquake: field investigations on the French Mediterranean coast and tsunami modelling. *Natural Hazards and Earth System Sciences*, **9**, 1823–1834.

Sayre, A. P. (1999). *Africa*. Brookfield: Twenty-First Century Books.

Simkin, T., and Siebert, L. (1994). *Volcanoes of the World: A Regional Directory, Gazetteer, and Chronology of Volcanism During the Last 10,000 Years*. Tucson, AZ: Geoscience Press, Inc., in association with Smithsonian Institution, Global Volcanism Program, Washington, DC, second edn.

Socolow, R. H. (2005). Can we bury global warming? *Scientific American*, **293**, 49–55.

Soloviev, S. L. (1972). Recurrence of earthquakes and tsunamis in the Pacific ocean (in Russian). *Volny Tsunami, Trudy Sakhnii*, **29**, 7–47.

Soloviev, S. L., Solovieva, O. N., Chan, N., Go, Kh., Kim, S. and Shchetnikov, N. A. (2000). *Tsunamis in the Mediterranean Sea 2000 B.C.–2000 A.D.* Advances in Natural and Technological Hazards Research, vol. 13. Dordrecht: Kluwer Academic Publishers.

Stiros, S. C. (2001). The AD 365 Crete earthquake and possible seismic clustering during the fourth to sixth centuries AD in the Eastern Mediterranean: a review of historical and archaeological data. *Journal of Structural Geology*, **23**, 545–562.

USGS (2013). USGS Volcano Hazards Program web-site, 2013. Volcanic Lakes and Gas Releases, http://vulcan.wr.usgs.gov/Glossary/Lakes/description_volcanic_lakes_gas_release.html

Venzke, E., Wunderman, R. W., McClelland, L. *et al.*, eds. (2002). *Global Volcanism, 1968 to the Present*. Smithsonian Institution, Global Volcanism Program Digital Information Series, GVP-4, http://www.volcano.si.edu/reports/

17

Advancing disaster risk governance in Madagascar: the role of higher education institutions

Mahefasoa T. Randrianalijaona and Ailsa Holloway

17.1 Introduction

Although calamitous events such as the Great East Japan Earthquake or Chao Praya Floods have been described as 'natural disasters' (Okuzami *et al.*, 2013), there are strong calls to challenge the 'naturalness' of such occurrences (e.g., Abramovitz, 2001; Laska and Morrow, 2006; World Bank, 2010; Rahman, 2013). While phenomena such as hurricanes, earthquakes, and riverine floods clearly originate in nature, human action and inaction both contribute to exposure and vulnerability conditions that increase the likelihood of hardship and loss (IPCC, 2012; UNISDR, 2013a). Although the debates may remain open theoretically about the scientific formulation of disaster risks, there is wide consensus that they are a function of hazards and the associated vulnerabilities of assets (UNISDR, 2005; Amman *et al.*, 2006; Benson *et al.*, 2007)[1]. Such understanding has substantially advanced efforts in demystifying the causes of disasters, including the crucial role of human agency in reducing the likelihood and consequences of disasters.

The growing emphasis on the role of national and subnational authorities in reducing disaster risks (UNISDR, 2013b) has underlined important priorities for capacity building within the wider disaster risk domain. This is especially the case in Africa, where persistent gaps in skilled human resources prevail. These are in part attributed to decades of political instability in many countries, accompanied by protracted international humanitarian assistance programmes (Bhavnani, 2008; Holloway, 2009), which have actively discouraged the advancement and retention of skilled local capacity and disaster risk expertise. Moreover, while there have been recurrent attempts to strengthen national and subnational disaster risk management capability in Africa, these efforts have generally been confined to externally funded project time-scales and generated minimal traction and continuity (Holloway, 2009).

Historically, such capacity-building initiatives have also excluded engagement with Africa's higher education institutions (HEIs). This is despite arguments that underline that higher education collaborations constitute a recognised form of transnational partnership, especially when they aspire to advance 'sustained improvements in living conditions for vulnerable members of society in low-income and poverty-constrained contexts' (Koehn, 2013). In this context, African HEIs have key roles to play, not only in applied research that informs decision-making for averting disaster-related hardship, but also in formal and non-formal capacity-building to advance disaster-related resilience (UNISDR, 2013b).

This chapter specifically explores the role that HEIs might play in addressing the specialised human resources gap in the disaster risk domain and, as a result, enhancing disaster risk reduction efforts both nationally and transnationally. We consider the case of the University of Antananarivo with its introduction of a Multidisciplinary Disaster and Risk Management Masters Programme (DMGRC[2]). The chapter is divided into three main sections, and begins by describing the international disaster risk context, and its implications for the engagement of HEIs in the disaster risk domain. The second section provides a detailed description of the DMGRC, its rationale and focus, while the concluding component highlights illustrative outcomes of these efforts nationally in Madagascar.

17.2 The international context

17.2.1 An upward trend in economic losses caused by disasters

There are rising concerns at the upward trajectory in annual disaster losses. For instance, from 2001 to 2011, Munich

Extreme Natural Hazards, Disaster Risks and Societal Implications, eds. Alik Ismail-Zadeh, Jaime Urrutia Fucugauchi, Andrzej Kijko, Kuniyoshi Takeuchi and Ilya Zaliapin. Published by Cambridge University Press.

Reinsurance estimated disaster-related economic losses to be approximately US$1.68 trillion (Munich Re., 2012), while EM-DAT reported economic impacts of US$1.25 trillion (UNISDR, 2013a). Annual economic losses for 2011 and 2012 were US$400 billion and US$165 billion, respectively, with the ten-year average value for disaster-related deaths estimated at 106 000 (Munich Re., 2013). These figures contrast markedly with those compiled for the 1990s, which estimate the average number of lives lost annually to disasters triggered by natural hazards at 80 000 (Amman *et al.*, 2006).

While these figures reflect losses associated with major, 'intensive' disasters, they systematically exclude adverse effects of less accurately recorded, smaller extensive disasters that account for as much as 42% of total economic losses (UNISDR, 2013a). The report further argues that these effects are primarily associated with localised weather-related hazards with 'short return periods' and include surface water and flooding, landslides, fires, as well as agricultural and hydrological drought (UNISDR, 2013a).

In lower income countries, especially in Africa, these smaller, highly frequent, 'hidden' events create significant hardships and set-backs to local development. Examples include the high occurrence of market and informal settlement fires in Africa's rapidly growing towns and cities that spread quickly and lead to deaths and property loss. They include outbreaks of communicable disease, associated with weak public health systems and poor water and sanitation services. They are also reflected in flash flood events that are accompanied by little or no warning and that lead to loss of life, cause damage to crops and property, and result in livestock deaths (Periperi U, 2012). Such loss patterns are of particular concern, especially in countries likely to experience future climatic changes that include a greater frequency in heavy rainfall events and increased coastal high water levels (IPCC, 2012).

17.2.2 Advancing global coherence in disaster risk reduction efforts

Although a global initiative to reduce disaster losses emerged initially during the International Decade for Natural Disaster Reduction in the 1990s, international momentum to advance disaster risk reduction accelerated briskly in the mid-2000s. This was significantly enhanced by the World Conference on Disaster Reduction (WCDR) convened in Kobe, Hyogo, Japan, in January 2005, barely a month after the catastrophic Indian Ocean Earthquake and Tsunami of December 2004. The conference resulted in the adoption of the Hyogo Framework for Action (HFA), a global policy instrument to guide coherent global action for the ensuing decade. The HFA was developed with many partners, governments, international agencies, disaster experts, and others, who agreed on the need for implementing a shared system of coordination to reduce disaster risk both at national and global levels. Significantly, it represented the first international plan to describe, explain, and detail the efforts required across sectors and scales to reduce disaster losses (UNISDR, 2005).

By offering guiding principles and practical means for achieving disaster resilience, the HFA's goal was to substantially reduce disaster losses by 2015. It sought to achieve this by building the resilience of nations and communities to disasters, and by reducing loss of life as well as adverse social, economic, and environmental impacts when hazards strike (UNISDR, 2005). Although the HFA proposed five priorities for action, this chapter considers the three HFA components most relevant to capacity building in terms of education and training. These include Priority Action 2, which focuses on identifying, assessing, and monitoring disaster risks, as well as enhancing early warning. They also include Priority Actions 3 and 4 which foreground respectively the use of knowledge, innovation, and education to build a culture of safety and resilience, and efforts that reduce underlying risk factors (UNISDR, 2005).

The rationale for focusing on these particular HFA components is underpinned by their explicit emphasis on access to robust risk knowledge, innovation, and capacity for informed action. These concerns are centrally relevant to institutions of higher learning, especially those located in developing countries, where their core mission includes public engagement to advance national development goals (Varghese, 2013). Similarly, the identified HFA elements focus attention on the education sector's obligation to make disaster risk reduction knowledge accessible, relevant, and usable, recognising that a well-educated population constitutes a precondition for a 'productive, prosperous and resilient country' (Group of Eight, 2013).

These observations resonate profoundly in the case of Madagascar, especially due to its recurrent exposure to intense and damaging cyclones as well as other hydro-meteorological threats. Its risk profile is further exacerbated by a rapidly growing population characterised by persistent gaps in skilled human resources and a development context underpinned by protracted political uncertainty. They also underline the need for accelerated engagement by Madagscar's HEIs, recognising that strengthened educational capacity to reduce local vulnerability and risks is essential for social and economic development – especially in countries such as Madagascar that face recurrent threats.

17.3 Higher education capacity building: introducing a Disaster and Risk Management Masters Programme (DMGRC)

17.3.1 Madagascar: high vulnerability correlated with deep poverty

Madagascar joined at least 160 other states, when it adopted the HFA in January 2005, signalling willingness to advance efforts to protect its population and reduce disaster-related impacts. Yet, despite the recognised urgency to minimise disaster losses in Madagascar, risk reduction efforts have remained severely constrained. This is not only attributed to the recurrent character of the severe weather exposures, but also to the entrenched social and economic vulnerability that characterises the country's development profile. With a population of 21 million people in 2012 (UNDESA, 2013), Madagascar was ranked 151 in the Human Development Index, falling within the 'low human development' category of countries studied (UNDP, 2012). Moreover, its gross domestic product (GDP) per capita was estimated at only US$480 in 2012, equivalent to a mere US$1.32 per day, and constituting the second lowest value after Malawi in a study of 14 southern African countries (Holloway *et al.*, 2013).[3] It also exhibits steep population growth: between 2000 and 2012 the population rose by 38.7% from 15.4 million to 21.3 million people (UNDESA, 2013). This trend is expected to persist, with population estimates of 30.4 million and 43.2 million projected respectively for 2025 and 2040 (UNDESA, 2013). Across 14 southern African countries, Madagascar reports some of the lowest levels of participation in secondary as well as tertiary education. In 2012, this was reflected in secondary school enrolments of only 24% and 23% respectively for eligible girls and boys (Population Reference Bureau, 2012), while the country's 2010 gross enrolment ratio (GER) in higher education was reportedly a low 4% (Varghese, 2013). This compares with 25% in Mauritius, 7% for sub-Saharan Africa, and a global GER of 29%.

Similarly, nutrition indicators signal persisting social vulnerability, with sustained high levels of undernutrition in under five-year-old children at greater than 30% since 1988 (IFPRI, 2012). Consistent with trends elsewhere in Africa, Madagascar's historically rural population is expected to become more urban in forthcoming decades, rising from 4.1 million people (in 2000) to 10.8 million by 2025 (UN Habitat, 2010).

Such factors, combined with widespread rural poverty, constrained access to basic public services, poor infrastructure, and uncertain governance exacerbate Madagascar's risk profile. Significantly, they underline the role of developmental drivers in escalating national and subnational risk conditions, independent of recurrent cyclone and other exposures.

17.3.2 Disaster risk: vulnerable, exposed, and affected

Madagascar's recurrent exposure to hydro-meteorological extremes under conditions of profound socio-economic vulnerability is measurably reflected in significant disaster impacts. For instance, research on tropical cyclone impacts indicates that between 2000 and 2012, 31 loss-triggering cyclones were recorded for the 'island continent' of Madagascar. These reportedly resulted in 'cumulative impacts for 5.2 million people, affecting almost 170 000 on average per event' (Holloway *et al.*, 2013). Five droughts were also reported in the same period, affecting more than 2.1 million people. Altogether, 73 nationally significant disaster events were recorded between 2000 and 2012, with adverse effects reported for more than 7.6 million people (Holloway *et al.*, 2013).

At subnational scale, the recurrent character of severe weather shocks was measurably indicated in the south-western region of Atsimo-Andrefana and in the eastern region of Vatovavy Fitovinany, which, between 2000 and 2012, reported respectively 26 and 28 nationally significant disasters (Holloway *et al.*, 2013). These data highlight the short recurrence intervals for severe subnational shocks that may be concentrated in specific areas and which exact hardship and loss, undermining prospects for recovery and development.

17.3.3 The need for academic engagement: a contextually relevant master's degree

With growing awareness in Madagascar that costly disaster response to tropical cyclones and other threats has limited effects on social vulnerability, there has been an encouraging shift to more proactive reduction of disaster risks. Since 2008 specifically, there have been concerted national efforts to design sustainable strategies that incorporate relevant risk reduction and disaster management tools to improve decision-making. These have been accompanied by measures to reinforce the institutional frameworks to enable better disaster risk management.

Unfortunately, as is the case in many developing countries, Madagascar does not have enough suitably qualified people to support these efforts, in both the short- and the long-term. Moreover, the wide-ranging and cross-disciplinary character of the disaster risk domain, including its strategic planning dimensions, requires access to highly skilled people. This underlines an important emerging area of applied academic involvement for HEIs, as demonstrated in this instance by the University of Antananarivo.

Although efforts to establish a robust academic curriculum in disaster risk management gained momentum at the end of 2008, the need for a professional disaster and risk management (DRM) training programme had already been identified by the early 2000s. At that time, a UN

Development Program (UNDP) evaluation undertaken for the National Emergency Committee, now the Bureau National de Gestion des Risques et des Catastrophes (BNGRC, National Bureau of Disaster and Risk Management) proposed the establishment of a national training programme. Although this recommendation did not advance further then, at the end of 2008 the University of Antananarivo actively revisited the concept – with the support of the United Nations country office represented by UNDP, the United Nations Office for the Coordination of Humanitarian Affairs (UNOCHA), and the United Nations Children's Fund (UNICEF) – and additional enablement through the Periperi U consortium.[4]

From 2008 to 2010, extensive consultations took place with a wide range of academic, governmental, and non-governmental stakeholders to frame the emphasis and content of the new programme. A major focus of debate during this course conceptualisation phase concerned the qualification level to be offered, specifically whether to establish a shorter 'certificate' or longer 'degree' programme. The final decision in favour of a master's degree reflected recognition of the 'knowledge-intense' character of the disaster risk domain. This explicitly acknowledged the wide-ranging character of the natural and social science knowledge domains embraced by the field, as well as its application in the context of the complex risk in Madagascar. The decision to frame the academic programme at the master's level also reflected commitment to rapidly respond to gaps in skilled human capacity related to HFA Priorities 2, 3, and 4. This acknowledged the requisite role of robust scientific knowledge to advance efforts in risk identification, the coproduction of risk knowledge, and active reduction of underlying risk factors.

17.3.4 Multiple hazards and multidisciplinary learning

Madagascar is viewed as an 'island continent' and is the fifth largest island globally, with a surface area of approximately 587 041 km^2 (FAO, 2008). Located in the southwestern part of the Indian Ocean, Madagascar is separated from the southeastern coast of Africa by the Mozambique Channel (see Figs. 13.2 and 13.4). Located at the southernmost limit of the tropics, it is characterised by a coastline of 4828 km (Index Mundi, 2013b).

This vast surface and associated topography clearly result in a rich biodiversity of flora and fauna. However, they also pose major DRM challenges, due to the wide-ranging hydro-meteorological, geological, environmental, and human-induced conditions generated by such diverse landscapes. This varied risk profile was measurably reflected in a 12-year analysis of nationally significant disasters in Madagascar that included droughts, earthquakes, epidemics, fires, flooding, tropical cyclones, and locust infestations (Holloway *et al.*, 2013).

Such technical considerations, reinforced by the country's complex disaster history, vulnerable development profile, and climate change concerns, argued for a multidimensional and multisectoral approach to disaster risk management capacity building. They also underlined the importance of strong interdisciplinary thinking and applied problem-solving skills. As a result, the new postgraduate degree was designed to have an explicit multidisciplinary identity – *Master's Degree in Multidisciplinary Disaster and Risk Management* – and was the first of its kind for francophonic Africa.

Crafted collaboratively with guidance from multiple faculties and departments, including the Faculty of Science, the Medical College, and the Faculty of Law and Social Sciences, the new academic qualification has been situated institutionally within the Faculty of Law, Economy, Management, and Sociology. This has enabled a broader transdisciplinary application for the disaster risk management academic programme beyond naturally generated risks and explicitly extended access to the degree for students from diverse disciplines and professions.

These steps to institutionally embed and nurture robust disaster risk scholarship within the University of Antananarivo have been further reinforced through the establishment of a dedicated research centre to allow socially responsive engagement in the disaster risk domain. In this instance, the Economic Research and Studies Centre for Development (CERED in French) was officially recognised as the University of Antananarivo's designated research entity to include within its remit the socio-economic dimensions of risks and disasters, as well as their consequences.

17.4 Implementing the new education and training programmes

17.4.1 A multidisciplinary master's degree in Disaster and Risk Management (DMGRC)

After two years of feasibility studies and careful curriculum planning, the 18-month DMGRC was introduced in 2010. More than 220 applications were received from across Madagascar and from different sectors. Of these, 40 students were eventually selected to become the first cohort; they included both recent graduates and working professionals, with women constituting 35% of those enrolled. The first intake was highly diverse, representing non-governmental organisations, United Nations agencies, government institutions, and the private sector. The student profile also reflected a wide range of fields, including medicine, engineering, the armed services, geography, environmental management, law, and economics. Working professionals were drawn from two main groups, including those who were actively engaged in disaster risk management in Madagascar. Many had derived their professional knowledge primarily from 'working on the job', and now sought to deepen their understanding. A second group was drawn from the 'associated disaster risk management professions' who were interested in strengthening their knowledge of the disaster risk management field.

Student selection for the first intake was informed by the purposive and strategic intent of addressing prevailing disaster risk management capacity gaps as quickly as possible. This resulted in preferential recruitment of applicants who were already professionally qualified, with current work responsibilities. The decision not only addressed human capacity shortfalls in a relatively short period of time, but also provided scope for the exchange of cross-sectoral knowledge, which the feasibility study identified as essential for improving risk governance in Madagascar.

17.4.2 DRM certificate courses

While the DMGRC offers prospects of incrementally strengthening strategic DRM capacity over time, its annual intake is currently restricted to 40 students. With only limited places available for the DMGRC each year and high unmet demand for training in this field, it has been necessary to identify alternative and complementary capacity building options. Since 2010, this has resulted in the provision of two short certificate courses of 7–10 days duration annually, enabled through the collaborative efforts of the DMGRC, CERED, and Periperi U. As with the DMGRC, participants have been drawn from a wide range of fields and sectors (including civil society, United Nations agencies, and government), although to ensure the quality of training, the maximum number of participants per session has been capped at 25.

17.4.3 Progress so far?

With these programmes being implemented only from 2010, it is premature to be able to accurately assess their impact on national DRM capacity. It is also not possible to attribute national advancements in disaster risk reduction to these new initiatives. However, 88% of the first intake of DMGRC students completed their coursework and dissertations, and graduated as planned, and a further cohort of 40 students has been enrolled.

Qualitatively however, there have been identifiable changes in disaster-risk-related thinking and practice in Madagascar since the programme was introduced. These include reframing the scope of DRM by national and local officials that is observable in both governmental and non-governmental meetings focused on disasters and risks. While in the past, the concept of 'disaster risk management' was viewed as synonymous with 'emergency response', discussions now incorporate and apply concepts drawn from the disaster risk reduction and risk governance discourses. Similarly, the value of robust risk science is increasingly recognised by national and local officials as a prerequisite for informed decision-making. And, reportedly, since the introduction of the postgraduate DRM programme, international risk management specialists 'on mission' to Madagascar have commented on the growing number of national and local officials and others with whom they can engage constructively on DRM issues.

17.5 Advancement of disaster risk management through engaged scholarship

Consistent with efforts elsewhere to advance disaster risk scholarship (Holloway, 2009; Van der Walt, 2013), these encouraging signals from the University of Antananarivo underline the value of socially responsive academic programmes, particularly those that enable transdisciplinary scholarship. Typically, transdisciplinary learning approaches 'transcend' disciplinary boundaries (Horlick-Jones and Sime, 2004; Max-Neef, 2005), and are viewed as a 'robust means for understanding and addressing complex contemporary problems that are not amenable to disciplinary solutions alone' (Holloway, 2009).

In the risk context of Madagascar, the careful crafting of a multidisciplinary DRM programme that draws together practitioners and academics across diverse risk-related disciplines and sectors has provided a necessary platform for the joint production of risk knowledge. Not only has this advanced human capacity to address disaster risks but through academic requirements for postgraduate students to carry out disaster-risk-related research projects, the programme has generated the capacity to produce 35–40 dedicated studies a year focused on local risks. In resource-constrained disaster-prone countries, such as Madagascar, these represent a valuable source of risk knowledge, complementing often more coarsely generated consultancy reports commissioned by government or international agencies.

The purposive introduction of the DMGRC in Madagascar also coheres closely with the urgent call by the Southern African Regional Universities Association for greater 'knowledge diversity, interdisciplinary knowledge practices and southern scholarship' (Kotecha, 2012). This incisive review of trends in higher education across the Southern African Development Community (SADC) member states urged the region's institutions of higher learning to work more closely with a wider range of stakeholders. Similarly, it stressed the importance of stepping aside from a more traditional 'single disciplinary approach to scholarship, to applied interdisciplinary practices' (Kotecha, 2012).

The new postgraduate degree in Madagascar illustrates the value of such multistakeholder consultation. Specifically, the DRM qualification developed by the University of Antananarivo was not designed in isolation, but reflected several years of careful consultation with a wide range of stakeholders drawn from within and outside the university. This gradual and systematic approach in introducing the new programme also underlines the value of incremental capacity development (Brinkerhoff and Morgan, 2010). Such approaches are characterised by small and progressive adjustments that allow for flexibility and adaptiveness when embarking on new capacity building programmes, in contrast with more results-oriented, project-driven, and donor-supported initiatives.

The introduction of the new DRM capacity-building programme was also enabled through the University of Antananarivo's participation in Periperi U, a consortium comprising ten African universities committed to working collaboratively to advance disaster risk scholarship (Periperi U, 2012). Established in 2006, this stable and self-organising consortium spans the continent, and is committed to disaster risk-related capacity building matched to local needs. Increasingly, such transnational alliances are viewed as valuable vehicles for advancing innovation in contemporary scholarship, especially in low- and middle-income countries with limited access to resources (Koehn, 2013). In the case of Periperi U, the consortium's efforts at local and national levels, as well as across the continent, have helped to reframe Africa's self-view on its disaster reduction capacity, shifting from an entrenched history of aid dependency to an awareness of the emerging local agencies and capabilities.

17.6 Conclusion

With mounting concerns about Madagascar's rising disaster risk profile, particularly under conditions of climate variability and future change, there are pressing needs to strengthen the country's national and subnational disaster risk management capacity. The University of Antananarivo's successful introduction of a collaboratively crafted master's degree in disaster and risk management illustrates the value of university–society partnerships in generating relevant and sustainable capacity building, especially in countries that face profound development challenges. It underlines the important contribution that socially engaged universities can play in generating skilled human resources to address complex disaster risks and their potential to support risk reduction research and policy at all levels.

Endnotes

1 In UNISDR (2005) terminology, risk is defined as the combination of the probability of an event and its negative consequences. In general, the word 'risk' has two distinctive meanings. In daily usage the emphasis is usually placed on the concept of chance or possibility, such as in 'the risk of an accident'; whereas in technical settings the emphasis is rather placed on the consequences, in terms of 'potential losses'. However, people do not necessarily share the same perceptions of the significance and underlying causes of different risks.

2 DMGRC is the French abbreviation for Diplôme de Master Multidisciplinaire en Gestion des Risques et des Catastrophes

3 In terms of GDP per capita on purchasing power parity, Madagascar is ranked 218th out of 226 countries world-wide (Index Mundi, 2013a).

4 Partners Enhancing Resilience for People Exposed to Risks – with a focus on university engagement. For further information: www.riskreductionafrica.org (accessed on 6 August 2013).

References

Abramovitz, J. N. (2001). *Unnatural Disasters*. Worldwatch Paper 158. Washington, DC: Worldwatch Institute. Available at: http://www.worldwatch.org/system/files/EWP158.pdf. Accessed 11 July 2013.

Amman, W. J., Dannenmann, S. and Vulliet, L. (2006). *Coping with Risks Due to Natural Hazards in the 21st Century. Risk 21*. London, UK: Taylor & Francis.

Benson, C., Twigg, J. with Rosetto, T. (2007). *Tools for Mainstreaming Disaster Risk Reduction: Guidance Notes for Development Organisations*. Geneva, Switzerland: Prevention Consortium Secretariat. Available at: http://www.preventionweb.net/files/1066_toolsformainstreamingDRR.pdf. Acessed 11 July 2013.

Bhavnani, R., Vordzorgbe, S., Owor, M. and Bousquet, F. (2008). *Report on the Status of Disaster Risk Reduction in the Sub-Saharan Africa Region*. Nairobi: Commission of the African Union, World Bank, UNISDR.

Brinkerhoff, D. W. and Morgan, P. J. (2010). Capacity and capacity development: Coping with complexity. *Public Administration and Development*, **30**, 2–10.

FAO, Food and Agriculture Organization of the United Nations (2008). Fishery and Aquaculture Country Profiles. Madagscar (2008). Country Profile Fact Sheets. In *FAO Fisheries and Aquaculture Department*. Rome: FAO. Available at: http://www.fao.org/fishery/facp/MDG/en#CountrySector-GenGeoEconReport. Accessed on 6 August 2013.

Group of Eight (2013). *The Changing PhD, Discussion Paper*. Canberra: The Group of Eight House. Available at: http://www.go8.edu.au/__documents/go8-policy-analysis/2013/the-changing-phd_final.pdf. Accessed 11 July 2013.

Holloway, A. (2009). Crafting disaster risk science: Environmental and geographical science sans frontiers. *Gateways:*

International Journal of Community Research and Engagement, **2**, 98–118.

Holloway, A., Chasi, V., de Waal J. *et al.* (2013). *Humanitarian Trends in Southern Africa: Challenges and Opportunities*. Regional Interagency Standing Committee, Southern Africa. Rome: UN Food and Agriculture Organization (FAO), Rome.

Horlick-Jones, T. and Sime, J. (2004). Living on the border: Knowledge risk and transdisciplinarity. *Futures*, **36**, 441–456.

IFPRI, International Food Policy Research Institute (2012). Global Hunger Index (GHI) data. Available at http://www.ifpri.org/book-8018/ourwork/researcharea/global-hunger-index. Accessed on 6 August 2013.

Index Mundi (2013a). Country comparison. Economy: GDP – per capita. Available at: http://www.indexmundi.com/g/r.aspx?t=0&v=67. Accessed on 6 August 2013.

Index Mundi (2013b). Madagascar coastline. Available at: http://www.indexmundi.com/madagascar/coastline.html. Accessed on 6 August 2013.

IPCC, Intergovernmental Panel on Climate Change (2012). Summary for policy makers. In *Managing the Risks of Extreme Events and Disasters to Advance the Climate Change Adaptation*, eds. Field, C. B., Barros, V., Stocker, T. F., *et al.* A Special Report of Working Groups I and II of the Intergovernmental Panel on Climate Change. Cambridge, UK: Cambridge University Press, pp. 1–19. Available at: http://www.ipcc.ch/pdf/special-reports/srex/SREX_Full_Report.pdf. Accessed on 4 June 2013.

Koehn, P. (2013). Donor-supported transnational higher education initiatives for development and research: A framework for analysis and a call for increased transparency. *Higher Education Policy*, doi:10.1057/hep.2013.5.

Kotecha, P. (2012). Higher education in the Southern African region: current trends, challenges and recommendations – A contribution to the development of a regional strategic plan for revitalising and strengthening higher education in the SADC region. Paper presented to the extraordinary meeting of Ministers in Higher Education and Training, 2 June 2012, Johannesburg, South Africa.

Laska, S. and Morrow, B. H. (2006). Social vulnerabilities and Hurricane Katrina: An unnatural disaster in New Orleans. *Marine Technology Society Journal*, **40**, 16–26.

Max-Neef, M. A. (2005). Foundations of transdisciplinarity. *Ecological Economics*, **53**, 5–16.

Munich Re. (2012). *TOPICS GEO Natural Catastrophes 2011: Analyses, Assessments, Positions*. 2012 Issue. Munich, Germany: Munich Re.

Munich Re. (2013). Press Release: Natural catastrophe statistics for 2012 dominated by weather extremes in the USA. Munich Re, Munich, Germany. Available at: http://www.munichre.com/en/media_relations/press_releases/2013/2013_01_03_press_release.aspx. Accessed 11 July 2013.

Okazumi, T., Nakasu, T., Sugimoto, M. and Adikari, Y. (2013). Lessons learnt from two unprecedented disasters in 2011: Great East Japan Earthquake and Tsunami in Japan and Chao Phraya River Flood in Thailand. Background paper prepared for the *Global Assessment Report on Disaster Risk Reduction 2013*. Geneva, Switzerland: UNISDR. 17 pp. Available at: http://www.preventionweb.net/english/hyogo/gar/2013/en/bgdocs/Okazumi%20et.%20al.%202012.pdf. Accessed 11 July 2013.

Periperi U (2012). *An African Partnership: Building Capacity – Reducing Risk. Disaster Mitigation for Sustainable Livelihoods Programme*. Stellenbosch University. Available at: http://riskreductionafrica.org/en/rra-ddr-per/rra-whatisperiperi. Accessed on 11 July 2013.

Population Reference Bureau (2012). Population statistics. http://www.prb.org/Educators/TeachersGuides/HumanPopulation.aspx. Accessed on 6 August 2013.

Rahman, M. (2013). Indians question how far flash-flooding disaster was manmade. *The Guardian*, 28 June. Available at: http://www.guardian.co.uk/world/2013/jun/28/indians-flash-flooding-disaster-manmade. Accessed 11 July 2013.

UN Habitat, The United Nations Human Settlements Programme (2010). Statistical Annex, Table 1 for Angola, Botswana, Lesotho, Madagascar, Malawi, Mauritius, Mozambique, Namibia, Seychelles, South Africa, Swaziland, Zambia, Zimbabwe. In *The State of African Cities 2010: Governance, Inequality and Urban Land Markets*. UN-Habitat: Kenya: Human Settlements Programme. Available at: http://www.unhabitat.org/pmss/getElectronicVersion.aspx?nr=3034&alt=1. Accessed on 6 August 2013.

UNDESA (2013). United Nations Department of Economic and Social Affairs, 2013. Available at: http://unstats.un.org

UNDP (2012). Human Development Index. Available at: http://hdrstats.undp.org/en/indicators/103106.html

UNISDR (2005). Hyogo Framework for Action 2005–2015: Building the resilience of nations and communities to disasters, World Conference on Disaster Reduction, Kobe, Hyogo, Japan, 18–22 January 2005. Available at: http://www.unisdr.org/eng/hfa/hfa.htm. Accessed on 5 August 2013.

UNISDR (2013a). Global assessment report on disaster risk reduction. Geneva, Switzerland: The United Nations Office for Disaster Risk Reduction. Available at: http://www.preventionweb.net/english/hyogo/gar/2013/en/home/index.html. Accessed 11 July 2013.

UNISDR (2013b). Chair's Summary: Fourth Session of the Global Platform for Disaster Risk Reduction, Geneva 21–23 May 2013. Resilient People, Resilient Planet. Available at: http://www.preventionweb.net/globalplatform/2013/news/view/33306. Accessed 11 July 2013.

Van der Walt, G. (2013). Disaster risk management: Disciplinary status and prospects for a unifying theory. *Jàmbá: Journal of Disaster Risk Studies*, **5**, 1–11.

Varghese, V. N. (2013). Governance reforms in higher education: A study of selected countries in Africa. Paper presented at the Policy Forum on governance reforms in higher education in Africa, Nairobi Kenya. Paris: UNESCO, 30 p. Available at: http://www.iiep.unesco.org/fileadmin/user_upload/News_And_Events/pdf/2013/Governance_reforms_in_HE_paper_PF.pdf. Accessed on 6 August 2013.

World Bank (2010). *Natural Hazards and Unnatural Disasters. The Economics of Effective Prevention*. Washington, DC: United Nations and World Bank.

PART V

Case studies: the Middle East

18
Natural hazards in Saudi Arabia

Abdulaziz M. Al-Bassam, Faisal K. Zaidi, and Mohammad T. Hussein

18.1 Introduction

During the last few decades the frequency of the occurrence of natural hazards has increased mainly due to the inhabitation of zones prone to natural hazards. Population pressure has resulted in human settlements in hazardous environments such as on the flanks of volcanoes as in Yogyakarta, Indonesia (Chester *et al.*, 2000) or on flood plains as in the Ganges–Brahmaputra delta (Gracheva and Golyeva, 2010) or in low lying or subsiding coastal areas, which can be seen in parts of the Netherlands (Leroy, 2013). Natural hazards are threats to life and property only if they occur in populated or inhabited areas. Human intervention has increased the frequency and severity of disasters associated with natural hazards. Four factors which have the potential to change a natural hazard into a disaster include time, space, society, and the type of event (Leroy, 2013). The 13 March 2011 9.0 magnitude earthquake followed by tsunami waves and flooding in Japan, the 2004 Indian Ocean earthquake followed by tsunami waves and flooding in South-East Asia, the 2005 Hurricane Katrina in the USA, the 2005 Kashmir earthquake, and the 2008 Sichuan earthquake in China are just a few examples of the notable natural disasters in the recent past. According to the ISDR disaster statistics approximately 960 000 people were killed during the period 1991–2005 due to natural hazards. Earthquakes, tsunamis, floods, and hurricanes are the most commonly occurring natural hazards that cause wide spread destruction of lives and property on a frequent basis.

The Kingdom of Saudi Arabia is prone to natural hazards. The northwestern region of the Kingdom is prone to earthquakes and volcanic hazards, whereas the central and western regions of the Kingdom are exposed to floods especially during heavy rainfalls. Landslides are a common phenomenon in the inhabited mountainous regions of the southwest. Shifting sand dunes and dust storms are a serious natural hazard being faced by the cities mostly in central and eastern Saudi Arabia. Flash floods, occasional seismic events, and dust storms are the major natural hazards in Saudi Arabia. According to the EM-DAT (http://www.emdat.be/), 14 disasters due to natural hazards occurred in Saudi Arabia between 1980 and 2010 claiming a total of 484 lives and causing over US$1 billion of economic damage.

Arid and semi-arid regions are typically prone to flash floods during rainfall events mainly due to the lack of vegetation and poor infiltration capacity of the surface soils, which results in the generation of enormous volumes of surface runoff. Flash floods have become an issue of concern in the major urban centres around the world. This is mostly because of increasing population pressure, which requires more infrastructural development. This pressure has often resulted in the construction of buildings in topographically low-lying regions, which in turn has obstructed the natural rainwater drainage (Hussein and Zaidi, 2012). The increase of the constructed area has limited the natural land available for infiltration, thus resulting in the generation of more surface runoff, which at times exceeds the normal water bearing capacity of the urban sewer systems, thereby causing floods.

The seismic zones in Saudi Arabia are situated along the Red Sea and the Gulf of Aden in the west and south and the subduction zone associated with the Zagros suture in the north. In general, the damage and losses associated with earthquakes are low over Saudi Arabia; however, the regions along the Red Sea coast are vulnerable to earthquakes. The Dead Sea transform fault along the Gulf of Aqaba is the most seismologically active region of Saudi Arabia. The Gulf region has active sinistral transform faults with associated pull-apart basins and hence is the area where large damaging earthquakes occur.

Extreme Natural Hazards, Disaster Risks and Societal Implications, eds. Alik Ismail-Zadeh, Jaime Urrutia Fucugauchi, Andrzej Kijko, Kuniyoshi Takeuchi and Ilya Zaliapin. Published by Cambridge University Press.

Dust and sand storms are other major natural hazards, which are common in the arid and semi-arid regions. They occur when strong wind blows the dust and sand from dry surfaces. The Sahara Desert and the dry land around the Arabian Peninsula are the largest sources of airborne dust in the world (Junge, 1979; Morales, 1979; Ganor and Mamane, 1982). Though drought and winds are the major cause of dust storms, they are also caused by poor farming and grazing practices and infrastructure development projects.

18.2 Flood hazard

Flash floods are very common in arid regions and cause widespread destruction of life and property. The occurrence of flash floods is a complex phenomenon and depends on a number of factors such as regional geology, the morphometric characteristics of the drainage basin, and the flood plain area (Subyani and Al-Dakheel, 2009). Other important factors, which affect the occurrence and severity of flash floods, include the duration and the intensity of rainfall, surface runoff, evaporation, and infiltration rates (Nouh, 2006; Sen, 2008). Flash floods have become one of the significant threats to life and property in many parts of the world mainly due to the increase in population pressure and improper land use planning (Borga *et al.*, 2010).

Changes in the global precipitation pattern due to global warming (Groisman *et al.*, 2005) have resulted in a change of the planet's hydrological cycle (Huntington, 2006). This global change has resulted in the increase in frequency of flash floods in many parts of the world (Marchi *et al.*, 2010), including the arid regions. The magnitude of flash floods in such regions increases further if the soil is saturated due to previous rainfall events.

Saudi Arabia is one of the most arid regions of the world; however, this has not prevented the growth of big cities along its coasts, like Jeddah and Dammam, and along the ancient Wadi system, such as Riyadh, Madinah, and Makkah. Though the average annual rainfall in Saudi Arabia is only about 100 mm y^{-1} (Presidency of Meteorology and Environment, 2005), it is not free from hydrological hazards especially in the big cities like Jeddah and Riyadh mainly due to rapid urbanisation which has led to the development of housing in topographically low-lying regions and obstruction of the natural drainage systems. The severe flood in the city of Jeddah in November 2009 as a result of heavy rainfall and the blockage of natural drainage system is one example.

Because of the frequent occurrence of flash floods in the Kingdom of Saudi Arabia, a lot of studies have been focused on flood assessment and flood hazard estimations. Subyani and Al-Dakheel (2009) carried out a study on the hydrological behaviour and flood probability for a few selected basins in the Makkah region in western Saudi Arabia and established the rainfall frequency curves and flood probabilities based on the rainfall records. Al Saud (2010) used a Geographic Information System (GIS) based approach to assess the hazardous impact of flash floods that occurred in Jeddah in November 2009 and emphasised the relationship between urbanisation and the flooding pattern and pointed out the risk of allowing further development in areas at risk of flooding. Dawod *et al.* (2013) established a GIS based process to quantify the peak discharge and runoff volume in Makkah city based on curve number (CN) methodology. Al-Ghamdi *et al.* (2012) investigated the effects of urbanisation on the runoff volume using the CN flood modelling methodology and found a positive correlation between urbanisation, peak discharge, and flood volume.

The Jeddah flood, which occurred as a result of more than 90 mm of rainfall falling within a span of 4 hours on 25 November 2009, was described by the civil defence as the worst in the past 27 years. More than 100 people were reported dead and about 350 people missing. The business damages alone were reported to be about 1 billion Saudi Riyals (Al Saud, 2010). A similar heavy rainfall event of more than 110 mm within a short span flooded the city of Jeddah again on the 26 January 2011 (Fig. 18.1). Though there were only a few deaths, the loss to property was enormous as most of the city's low-lying areas were inundated.

Jeddah is situated on a narrow strip along the Red Sea coast with the width of the coastal zone within the city ranging from 5 to 10 km. The coastal plain gives way to the Asir mountain region, which presents an abrupt change in topography showing a very high relief. It was shown that the city of Jeddah serves as the mouth of as many as

Figure 18.1. Flooding in Jeddah, January 2011. (Photo: R. Awad; http://www.moveoneinc.com/blog/moving/jeddah-floods-cause-unknown-delays-to-shipping-times/)

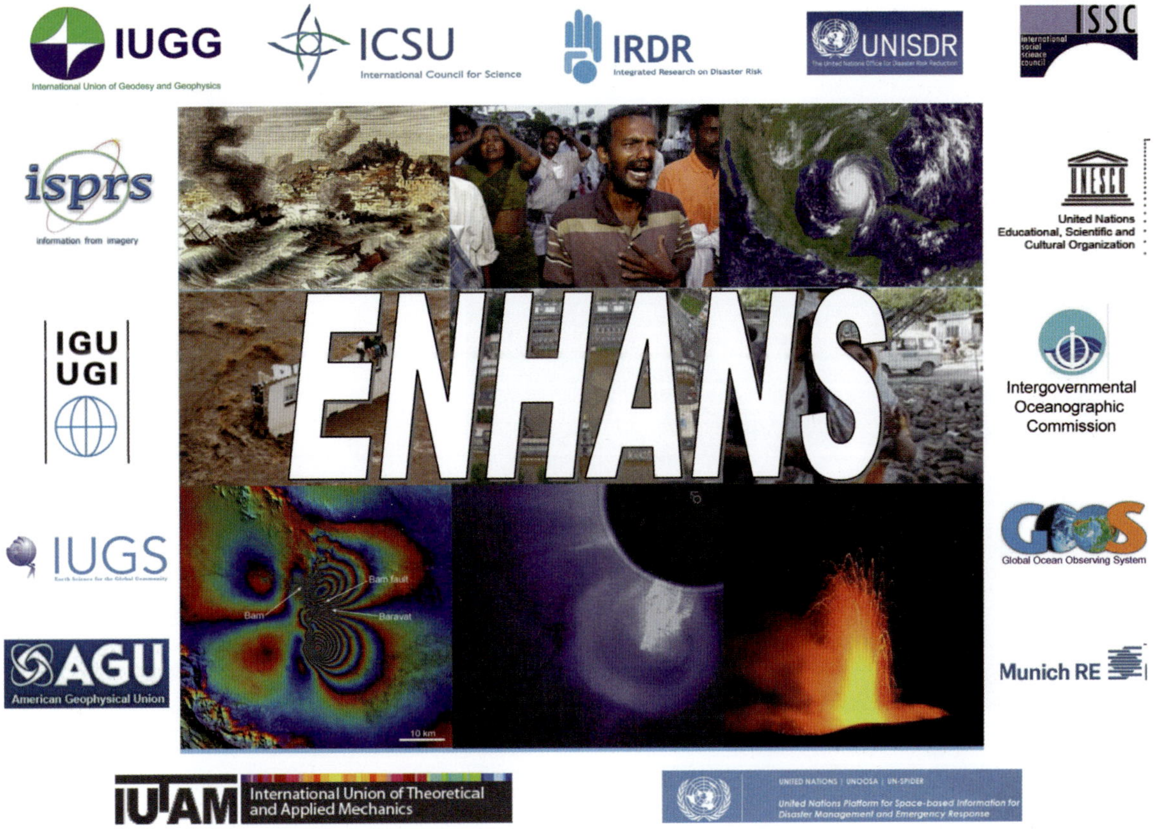

Figure P1. Graphical representation of the Supporting Partners of the ENHANS project.

Figure 1.1. Geographical illustration of the ENHANS project meetings.

Figure 1.2. Lake in the centre of Tswaing meteorite crater.

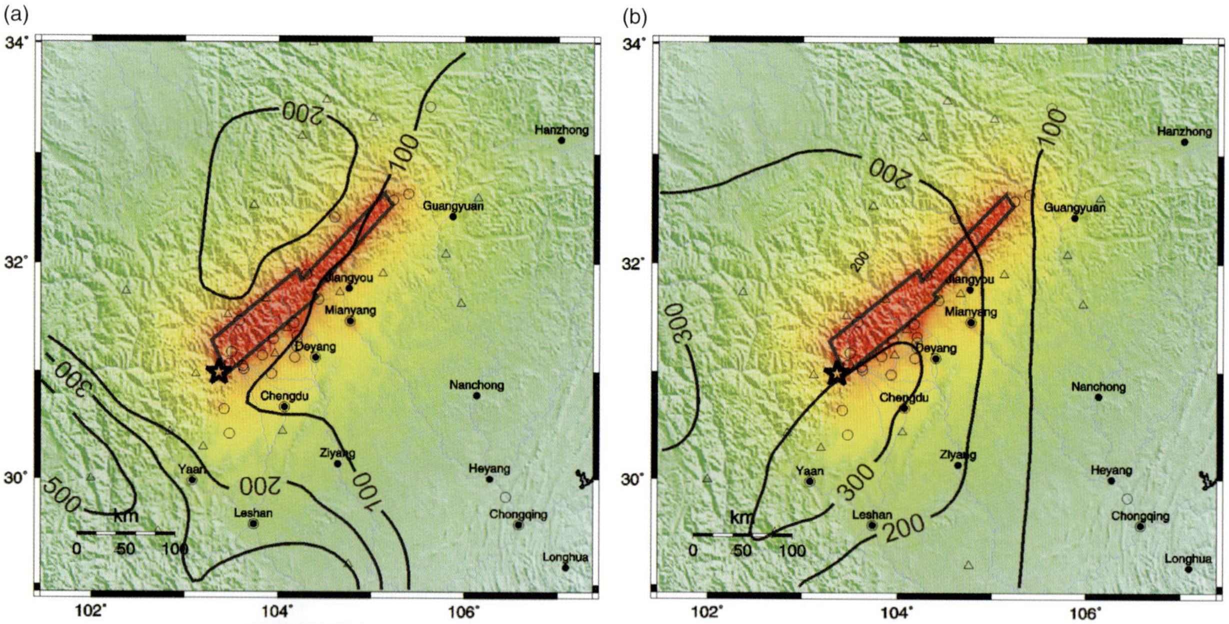

PERCETVED SHAKING	Not felt	Weak	Light	Moderate	Stong	Very strong	Serve	Violent	Extreme
POTENTIAL DAMAGE	none	none	none	Very light	Light	Moderate	Moderate/Heavy	Heavy	Very Heavy
PEAK ACC.(%g)	<.17	.17-1.4	1.4-3.9	3.9-9.2	9.2-18	18-34	34-65	65-124	>124
PEAK VEL.(cm/s)	<0.1	0.1-1.1	1.1-3.4	3.4-8.1	8.1-16	16-31	31-60	60-116	>116
INSTRUMENTAL INTENSITY	I	II-III	IV	V	VI	VII	VIII	IX	X+

Figure 4.2. Probabilistic SHA for the Eastern Sichuan region, China, for the return period of 475 y. Peak ground acceleration obtained using (a) the standard methodology (GSHAP map, http://www.seismo.ethz.ch/static/gshap/) and (b) the methodology accounting for modelled extreme seismic events. The background map is the United States Geological Survey United States Geological Survey (USGS) shake map for the 12 May 2008 Wenchuan $M_W = 7.9$ earthquake. Numbers at the contours are scaled in centimetres per second persecond.

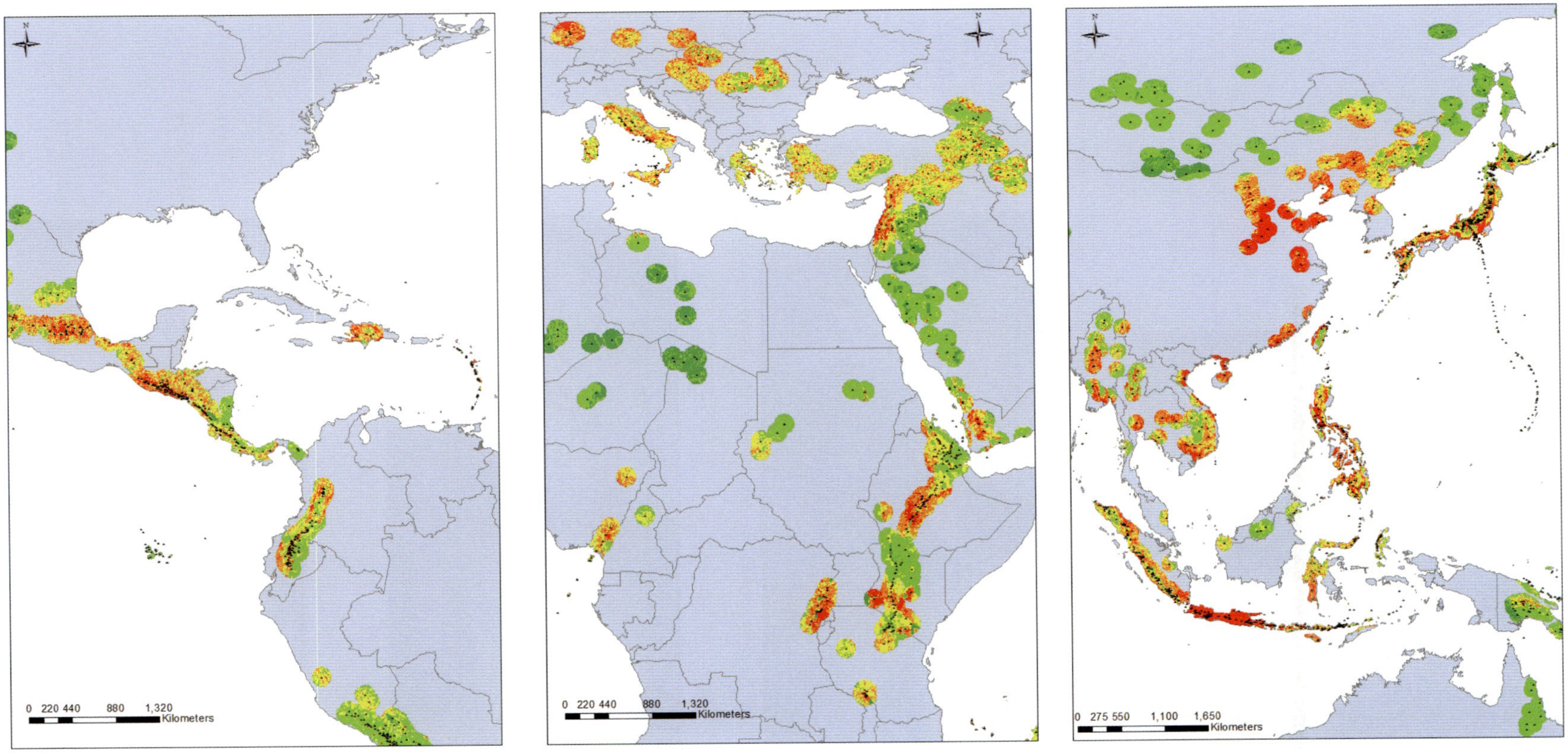

(a) (b) (c)

Figure 3.1. Population density within 100 km of volcanoes listed in the LaMEVE database (Crosweller *et al*., 2012): (a) Americas; (b) Europe, Near-East and Africa; and (c) East Asia and West Pacific. Population data obtained from the Center for International Earth Science Information Network / Columbia University, and Centro Internacional de Agricultura Tropical, 2005. Population density is shown in five classes: 0 people per km^2 (dark green), 1–9 per km^2 (light green), 10–49 per km^2 (yellow), 50–199 per km^2 (orange), and more than 200 per km^2 (red).

Figure 5.4. General view of the Vaiont landslide from Casso village, Italy. (Photo courtesy: A. P. Dykes, September 2011.)

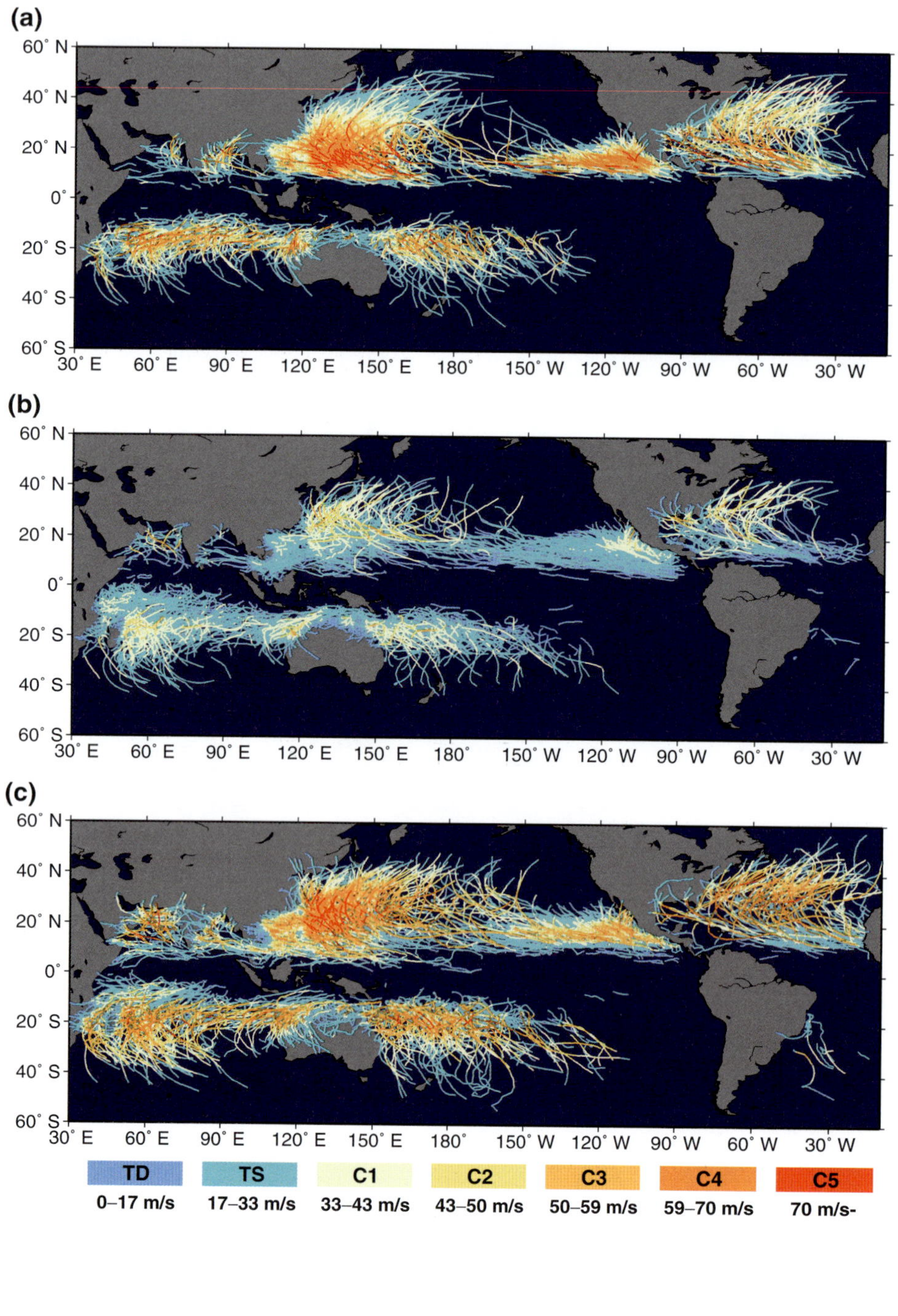

Figure 6.4. Global distributions of tropical cyclones for (a) observations, (b) the present day simulation using the MRI-AGCM3.1S, and (c) using the MRI-AGCM3.2S. Tropical cyclone tracks are coloured according to their intensity.

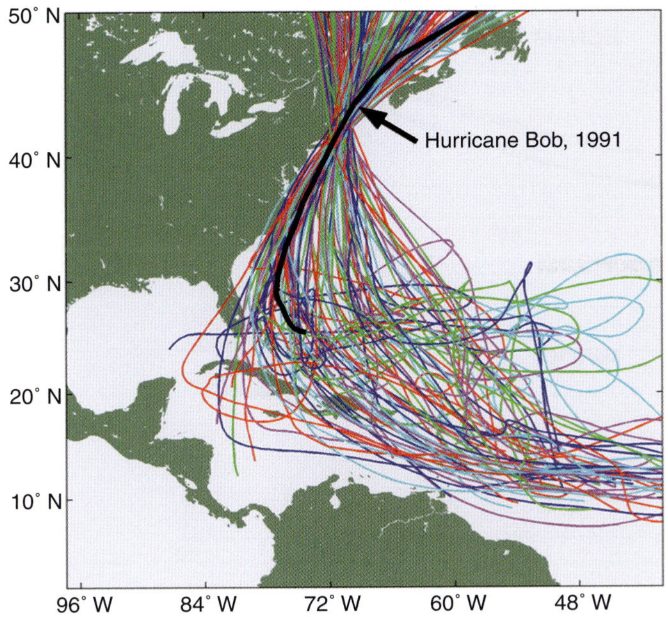

Figure 7.1. Tracks of the 100 most intense among 3000 synthetic storms affecting downtown Boston. Shown for comparison (in black) is the observed track of Hurricane Bob of 1991 (after Emanuel *et al.*, 2006).

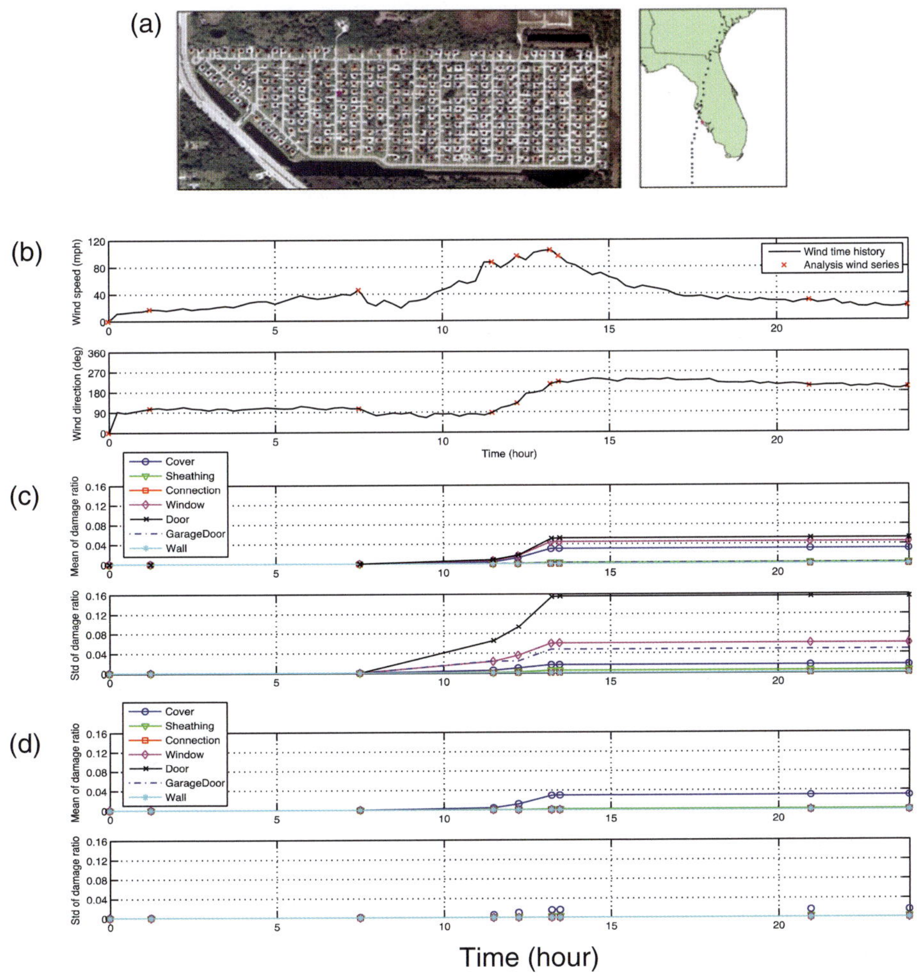

Figure 7.7. Case study of wind damage modelling for residential areas. (a) Study area of a residential development of 358 houses in Sarasota County, Florida (left; the dots represent the centres of the houses) and simulated storm track of Hurricane Charley of 2004 (right). The star (in both panels) represents the recording location of the wind time history. (b) Wind speed and direction time history during the passage of the simulated storm. (c) Estimated progressive damage in the study area, using the debris–pressure damage model. (d) Estimated progressive damage in the study area, using only the pressure damage model (after Lin *et al.*, 2010c).

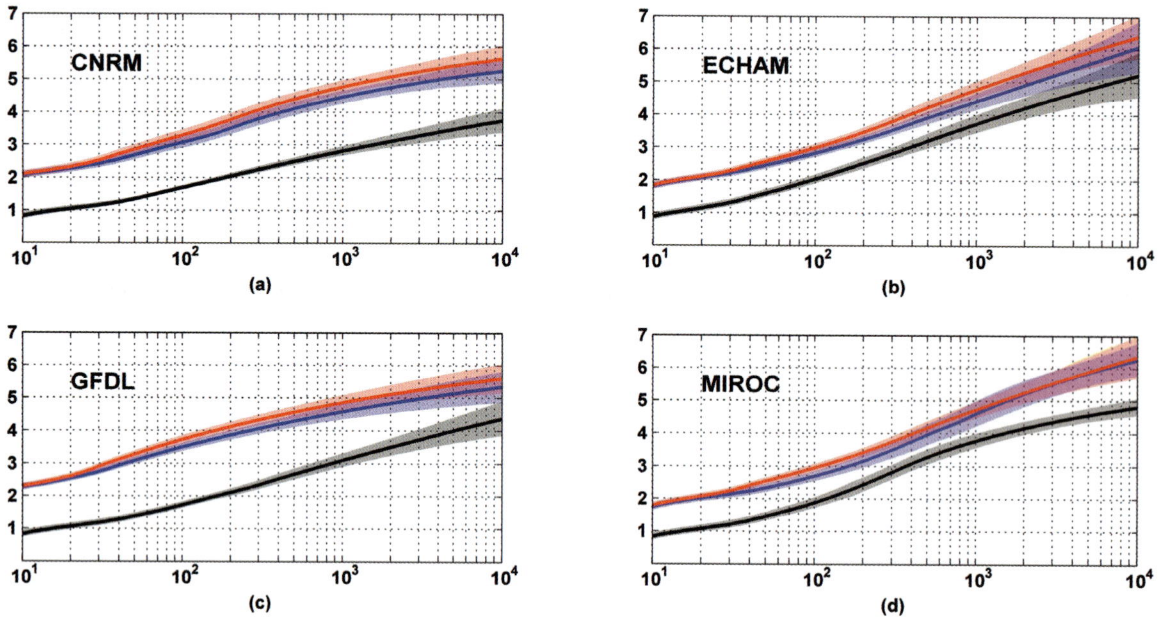

Figure 7.5. Estimated flood return levels for the Battery, NYC, predicted with each of the four climate models (CNRM-CM3, ECHAM5, GFDL-CM2.0 and MIROC3.2). The black is for the climate of 1981–2000 (current climate), the blue is for the IPCC A1B climate and the red is for the IPCC A1B climate with the storm outer radius increased by 10% and radius of maximum wind increased by 21% for 2081–2100 (future climate). The shading shows the 90% confidence interval. The estimation for each climate model was based on 5000 synthetic storm surge events for the current climate condition and 5000 synthetic storm surge events for the future climate condition. 1-m SLR was included for the future flood (after Lin *et al.*, 2012).

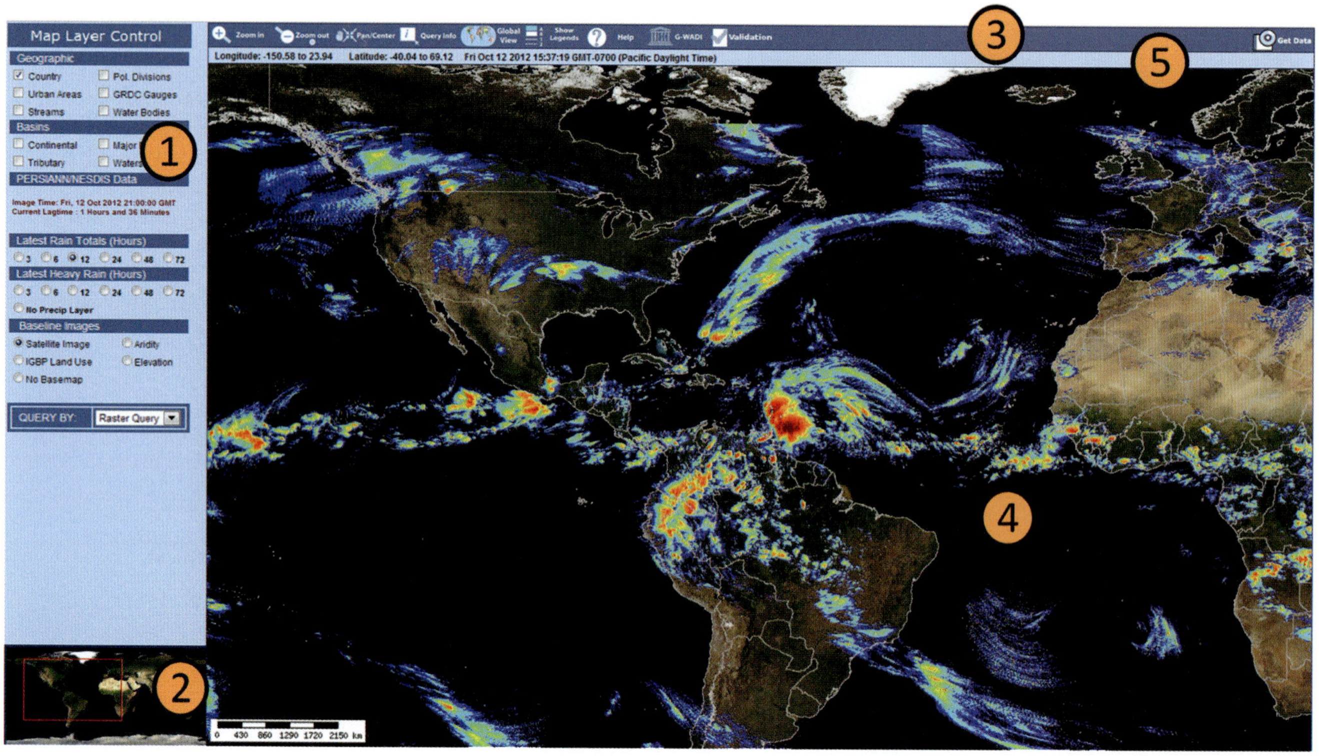

Figure 8.3. Components of the G-WADI GeoServer web-interface include: (1) map layer control menu, (2) interactive reference map, (3) map navigation and query menu, (4) map canvas, and (5) location bar.

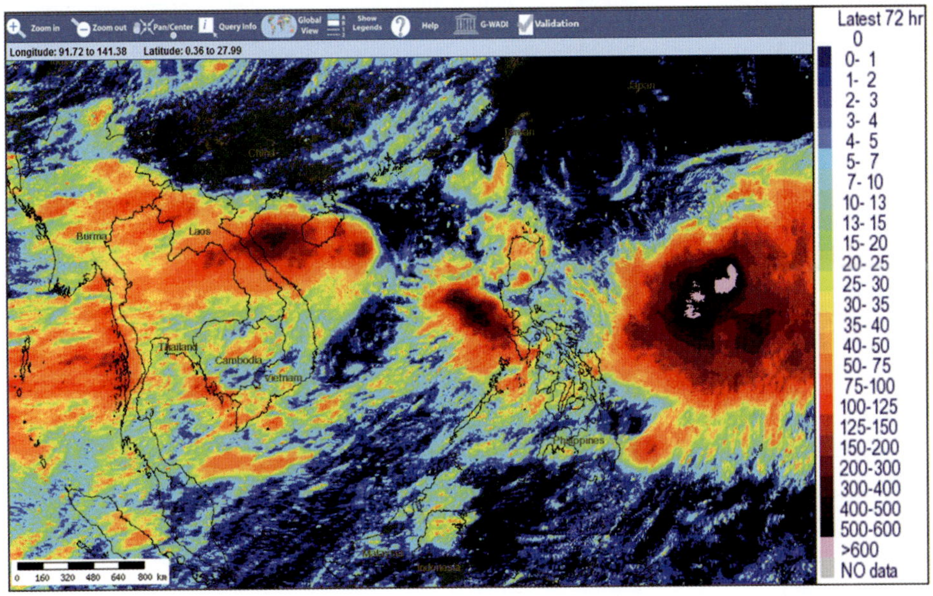

Figure 8.4. G-WADI GeoServer visualisation of 72-hour accumulated PERSIANN-CCS precipitation totals in millimeters for the downpour rainfall caused by the Nock-Ten depression in the Thailand 2011 flood from 00:00 UTC on 29 July 2011 up to 00:00 UTC on 1 August 2011 displaying the widespread heavy precipitation across Thailand.

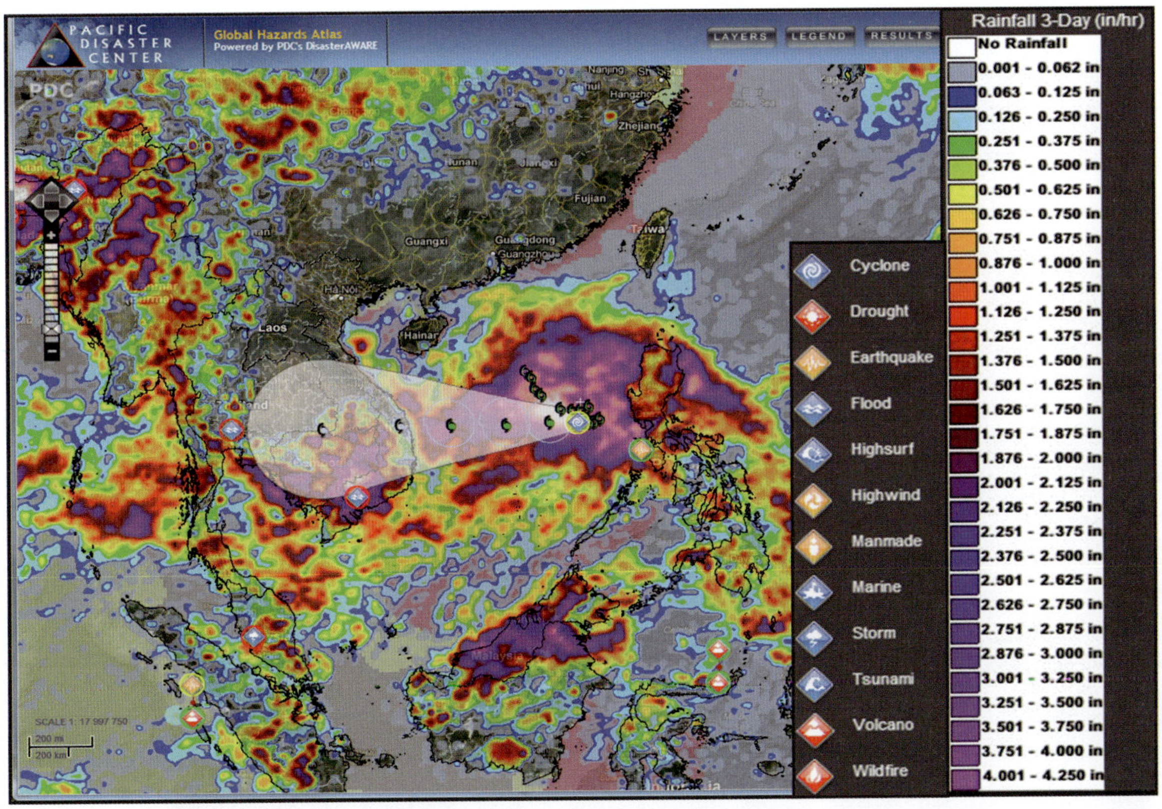

Figure 8.5. PDC Global Hazards Atlas showing rainfall (inches/hour) during 2–4 October 2012 (UTC) in the northwestern Pacific Ocean, with the Joint Typhoon Warning Center (JTWC) forecast positions and error cone for active storm Gaemi (source from http://www.pdc.org/atlas/).

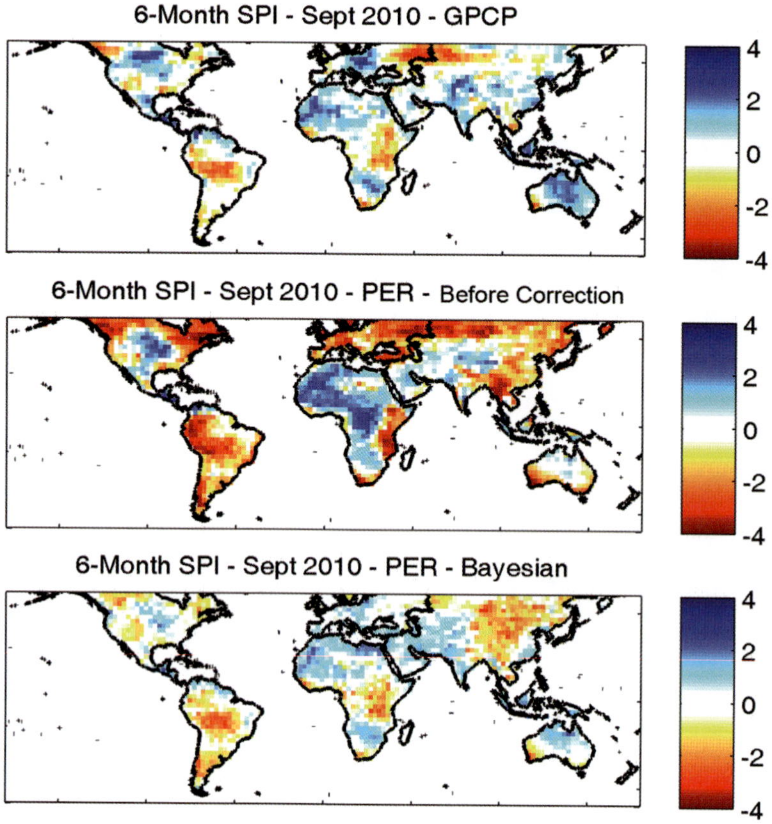

Figure 8.6. (a) Reference GPCP data, not used to derive the other two panels; (b) PERSIANN data before applying the Bayesian algorithm for making the climatology consistent; (c) combined GPCP (excluding 2010) and PERSIANN data, after the Bayesian correction for creating consistent climatology (AghaKouchak and Nakhjiri, 2012).

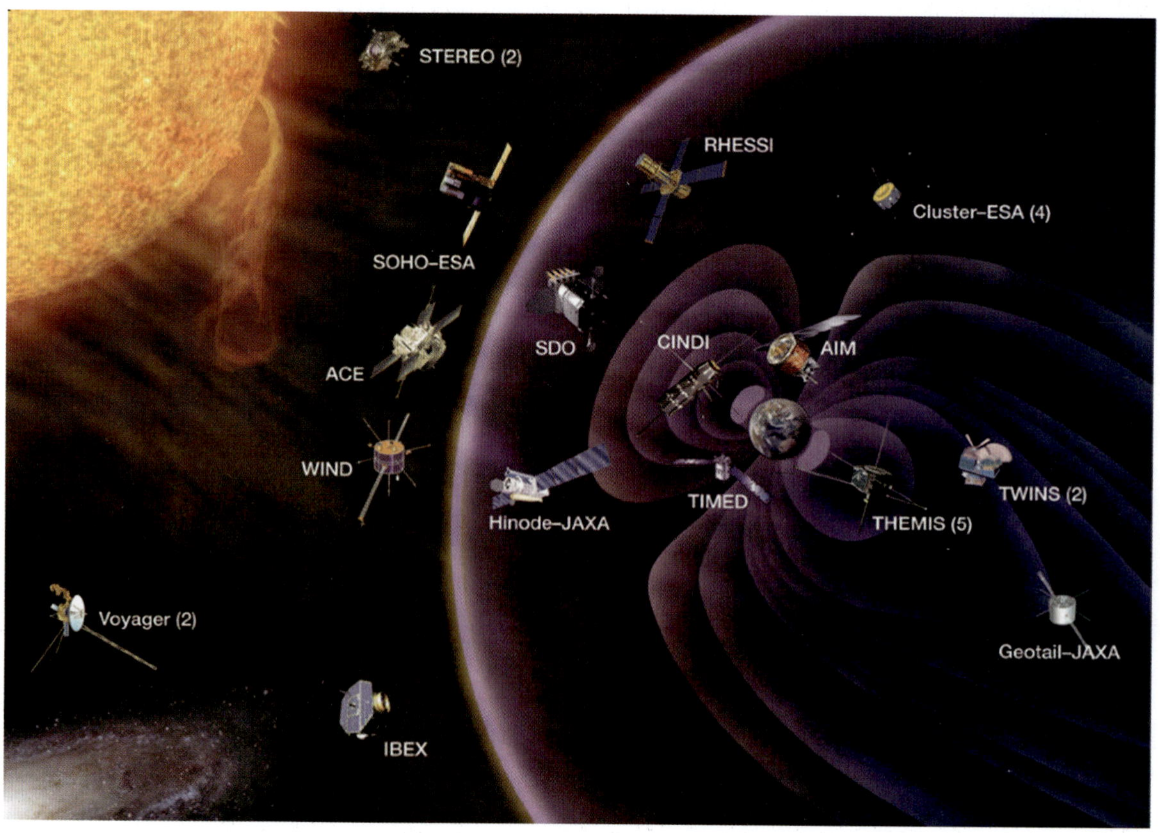

Figure 9.1. The Heliophysics System Observatory constellation of operating and planned NASA spacecraft. Together, these missions observe the Sun, the interplanetary medium, and the near-Earth space environment (courtesy of NASA Headquarters).

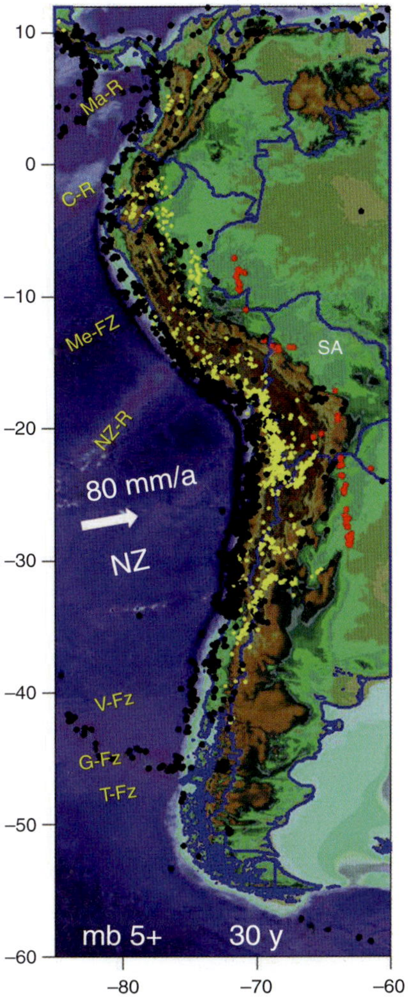

Figure 11.1. Epicentral locations (dots) of earthquakes with $m_b \geq 5$ in WSA during the period 1980–2010. Focal depths are indicated in black for $h < 70$ km, in yellow for 70 km $< h \leq 300$ km, and in red for $h > 300$ km. Tectonic plates: NZ Nazca, SA South American. Blue contours delineate the countries of the region. The depth distribution of these seismic events clearly defines a Benioff zone associated with the east-northeast subduction of the Nazca plate beneath South America that starts along the trench. Ridges (R): Ma, Malpelo; C, Carnegie; Nz, Nazca. Fracture zones (Fz): Me, Mendaña; V, Valdivia; G, Guafo; T, Taitao.

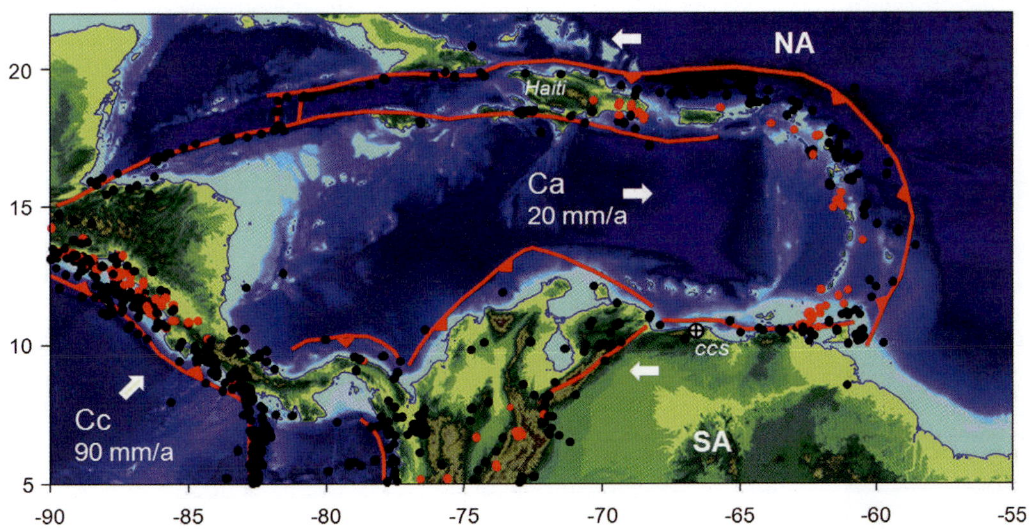

Figure 11.2. Epicentral locations (dots) of earthquakes with $m_b \geq 5$ in the Caribbean during the period 1980–2010. Focal depths are indicated in black for $h < 70$ km; and red for $h \geq 70$ km. Maximum focal depth reported for the region is 233 km. Tectonic plates: Ca, Caribbean; SA, South American; Cc, Cocos; NA, North American. CCS: City of Caracas. Red lines delineate the main plate boundaries of the Caribbean plate; triangles as the line show the areas where subduction takes place; they are strike-slip otherwise, left-lateral in the north and right-lateral in the south. The depth distributions of seismic events along the eastern and western Caribbean plate boundaries clearly define Benioff zones associated with the westerly and easterly subductions of the North Atlantic and Cocos plates, respectively, beneath the Caribbean plate.

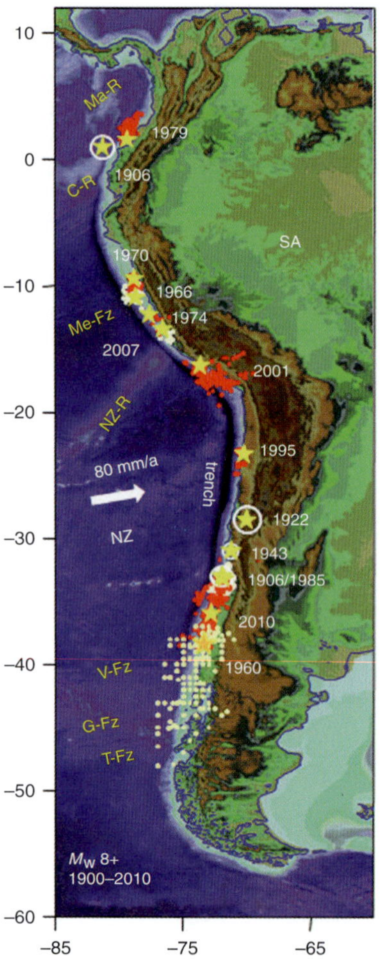

Figure 11.4. Main epicentres (stars) and aftershocks (dots, 5–8 days, $m_b \geq 5$) of great events ($M_w \geq 8$) that occurred in South America during the period 1900–2010. Other symbols as in Fig. 11.1.

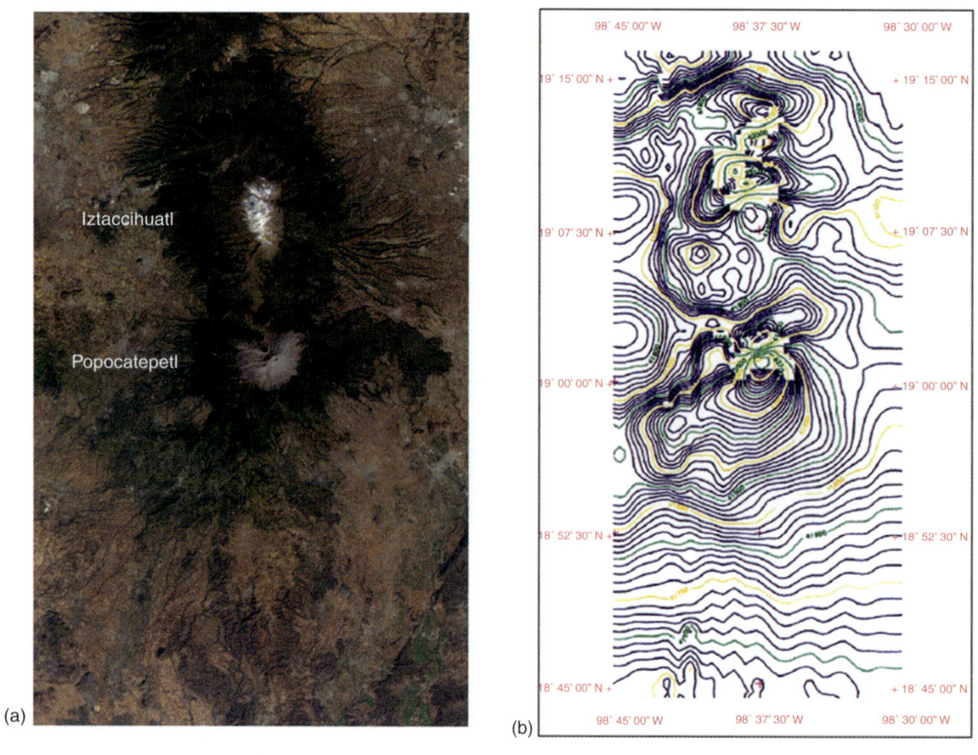

Figure 12.2. The Sierra Nevada volcanic complex, southern basin of Mexico. (a) Landsat 7-Enhanced Thematic Mapper Plus image of the basin of Mexico, showing the Iztaccihuatl and Popocatepetl volcanoes (courtesy NASA Earth Observatory, EROS Data Center; image 4, January 1999). (b) Aeromagnetic anomaly over the Iztaccihuatl and Popocatepetl volcanoes (Urrutia-Fucugauchi *et al.*, 2002a,b).

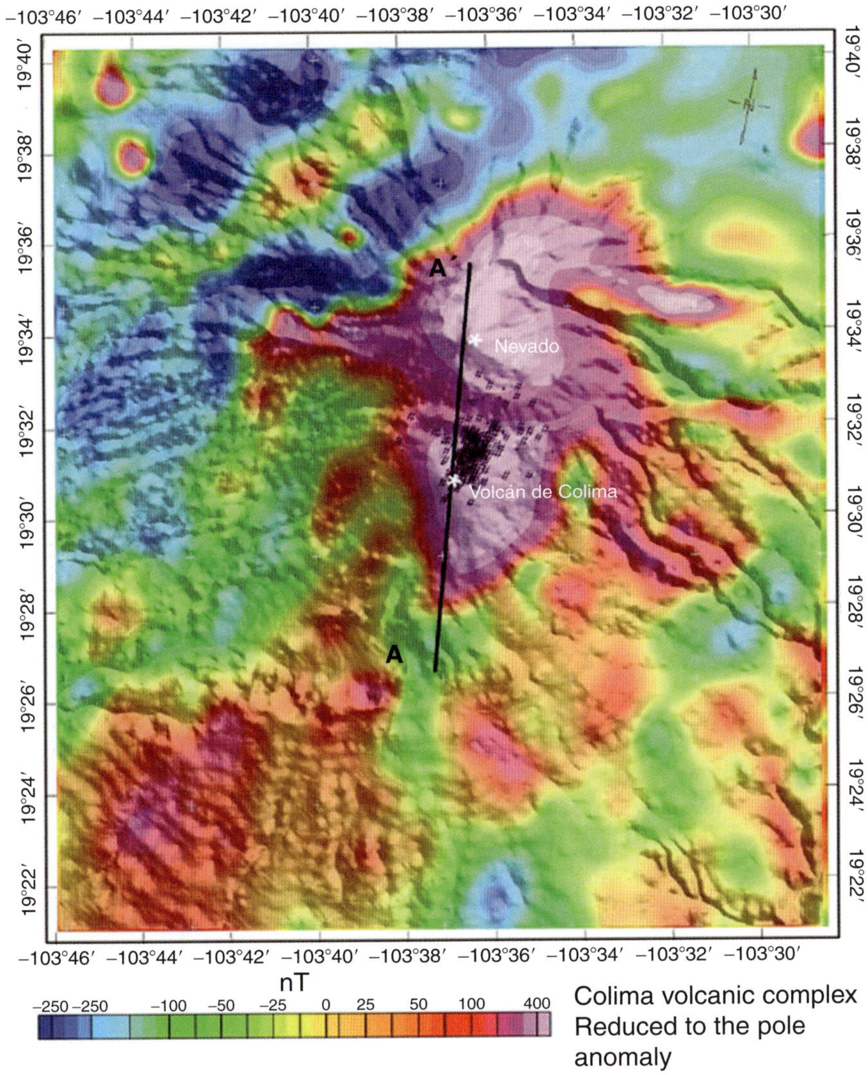

Figure 12.4. Reduced to the pole aeromagnetic anomaly over the Colima rift and volcanic complex. Observe the large amplitude anomaly over the Colima and Nevado de Colima volcanoes. The profile A–A′ shown in Fig. 12.5 is indicated by the line. Earthquake epicentres from the local RESCO 1997–1998 database of Colima volcano are indicated by crosses (Lopez-Loera *et al.*, 2011).

Figure 14.2. Images of volcanic activity at the DMH rift zone in central Afar. (**a**) Eruptive fissure from the 2005 Da'Ure rhyolitic eruption on the northern side of Dabbahu volcano (people for scale). (**b**) A section of a basaltic fissure eruption in the Karbahi region in June 2007. The 4-km-long fissure opened along the axis of the DMH rift zone. Both these eruptions were associated with dyke intrusions along the rift axis. Photos courtesy of Lorraine Field (a) and David Ferguson (b).

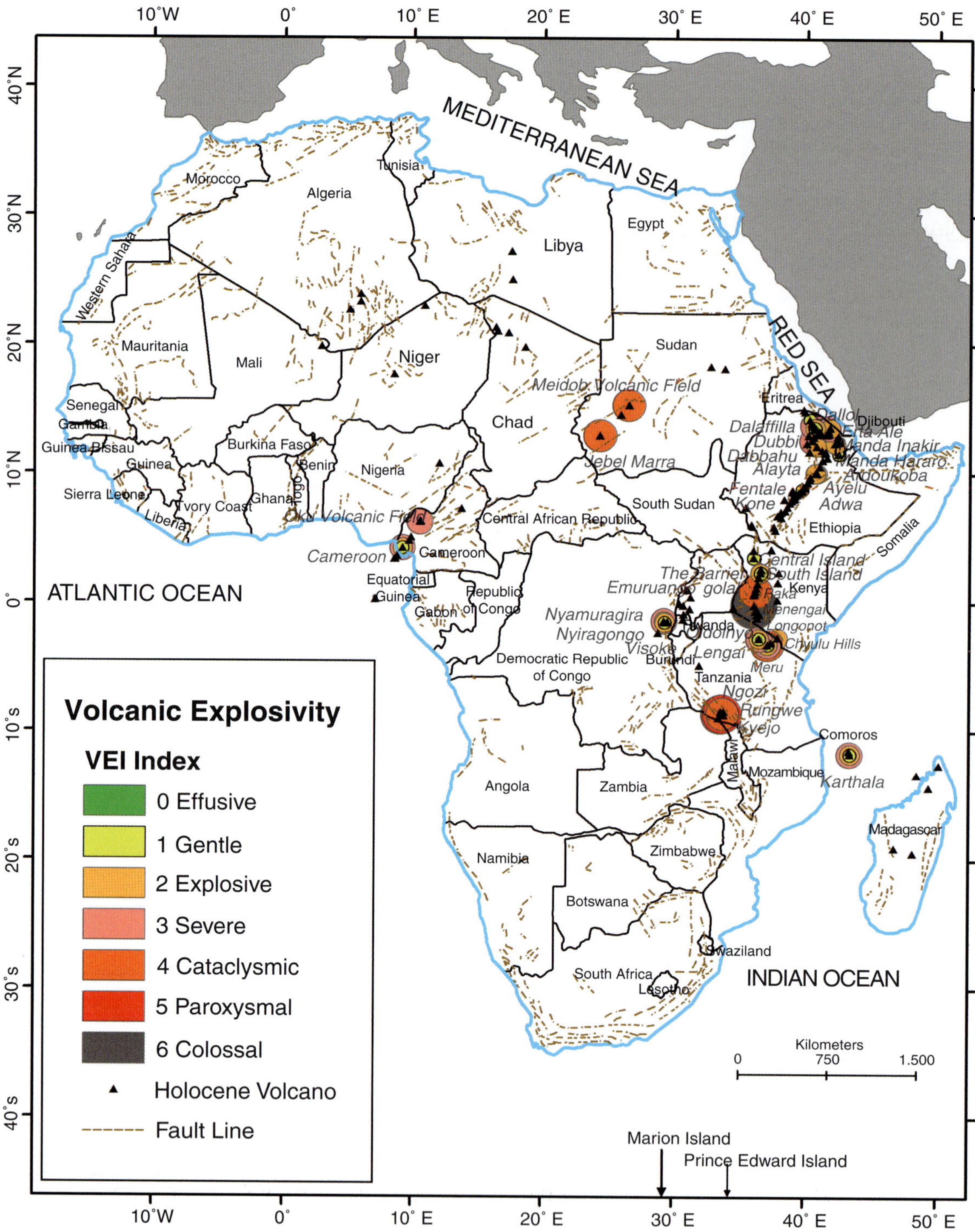

Figure 13.1. Density of volcanic eruptions based on the explosivity index for each eruption and the time period of the eruption. Modified from OCHA (2007) and Siebert *et al.* (2010).

Figure 18.5. A satellite image of the dust plumes (translucent blurs) over the Arabian Peninsula on 11 March 2009. (Photo: Jeff Schmaltz, MODIS Rapid Response Team, Goddard Space Flight Center, NASA.)

Figure 24.1 Tsunami inundation in the Sendai plain (copyright: *Mainichi Newspaper*).

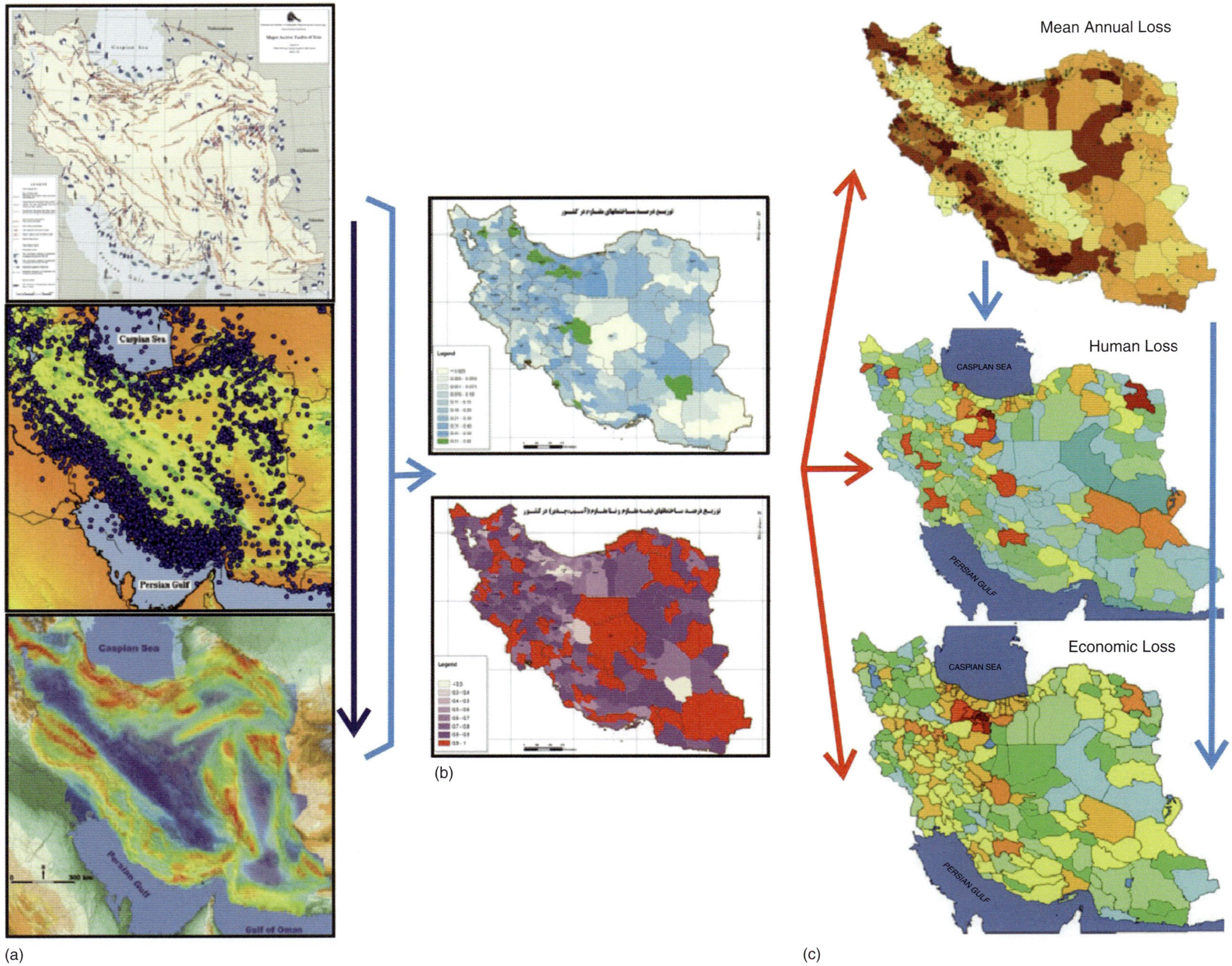

Figure 20.2. An overview of earthquake risk in Iran. (a) Hazard: active fault (top panel), seismicity (middle), and seismic hazard (bottom) maps of Iran. (b) Exposure: safe (top) and vulnerable (bottom) building exposure. (c) Distribution of the expected mean annual losses: buildings (top), humans (middle), and economic (bottom) due to the expected risk of failure of buildings.

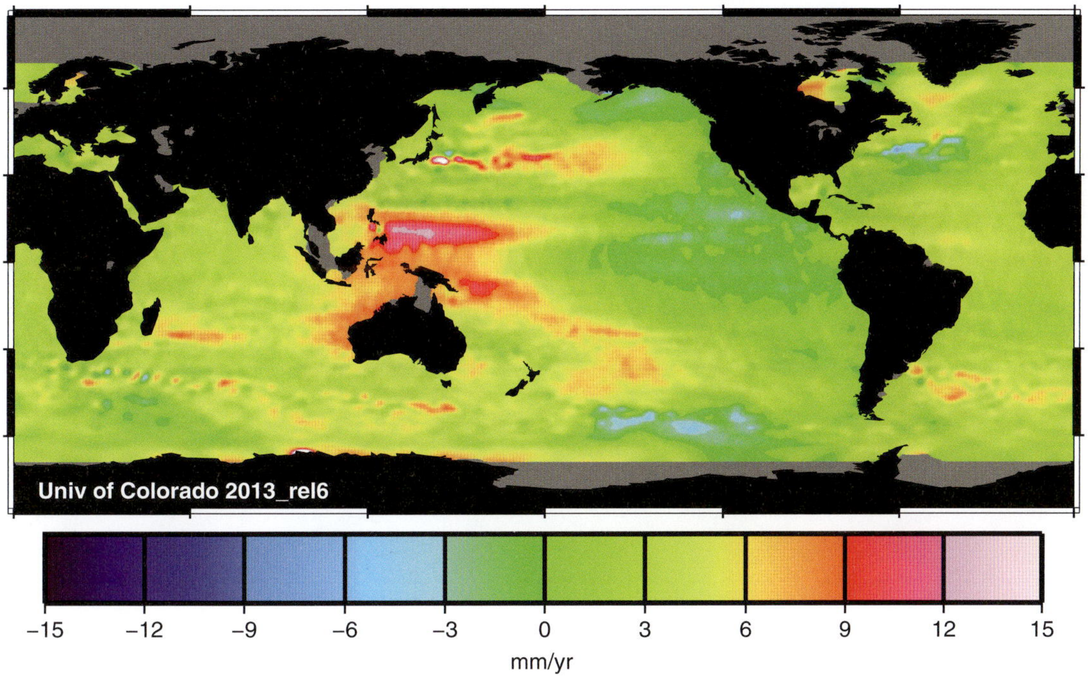

Figure 28.2. Global pattern of sea level rise trends over the past 20 years as measured by satellite altimetry (Nerem *et al.*, 2010). Source: http://sealevel.colorado.edu.

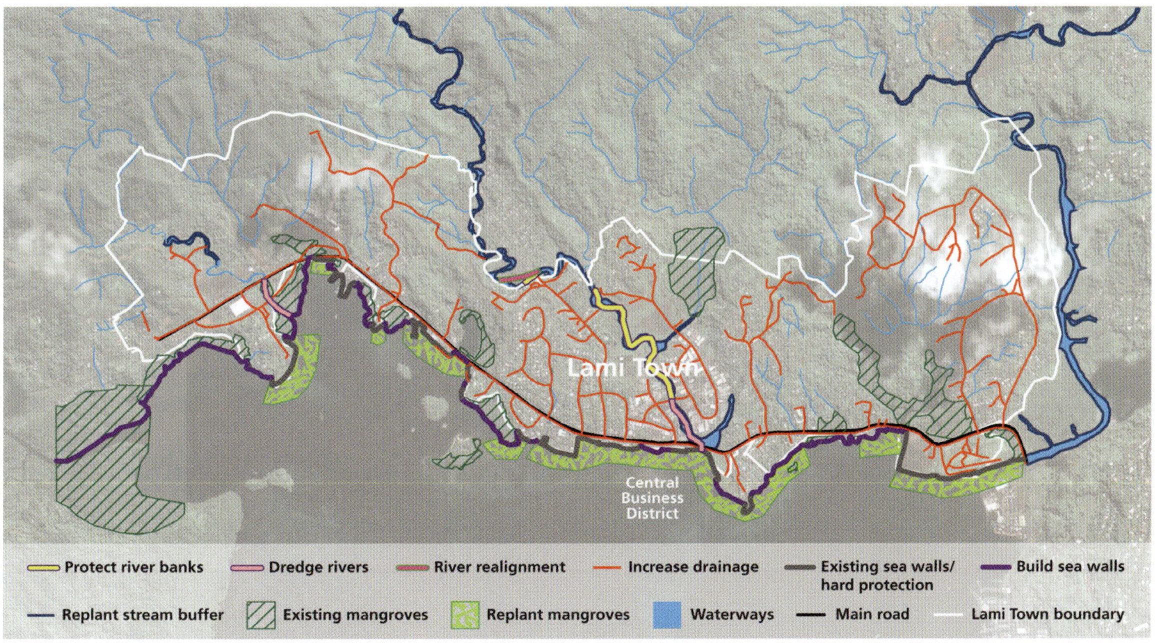

Figure 28.3. Proposed integrated ecosystem-based and engineering approach to reducing vulnerability to sea level hazards in the face of climate change induced sea level rise in Lami Town, Fiji. Source: UNEP and SPREP (Rao *et al.*, 2012)..

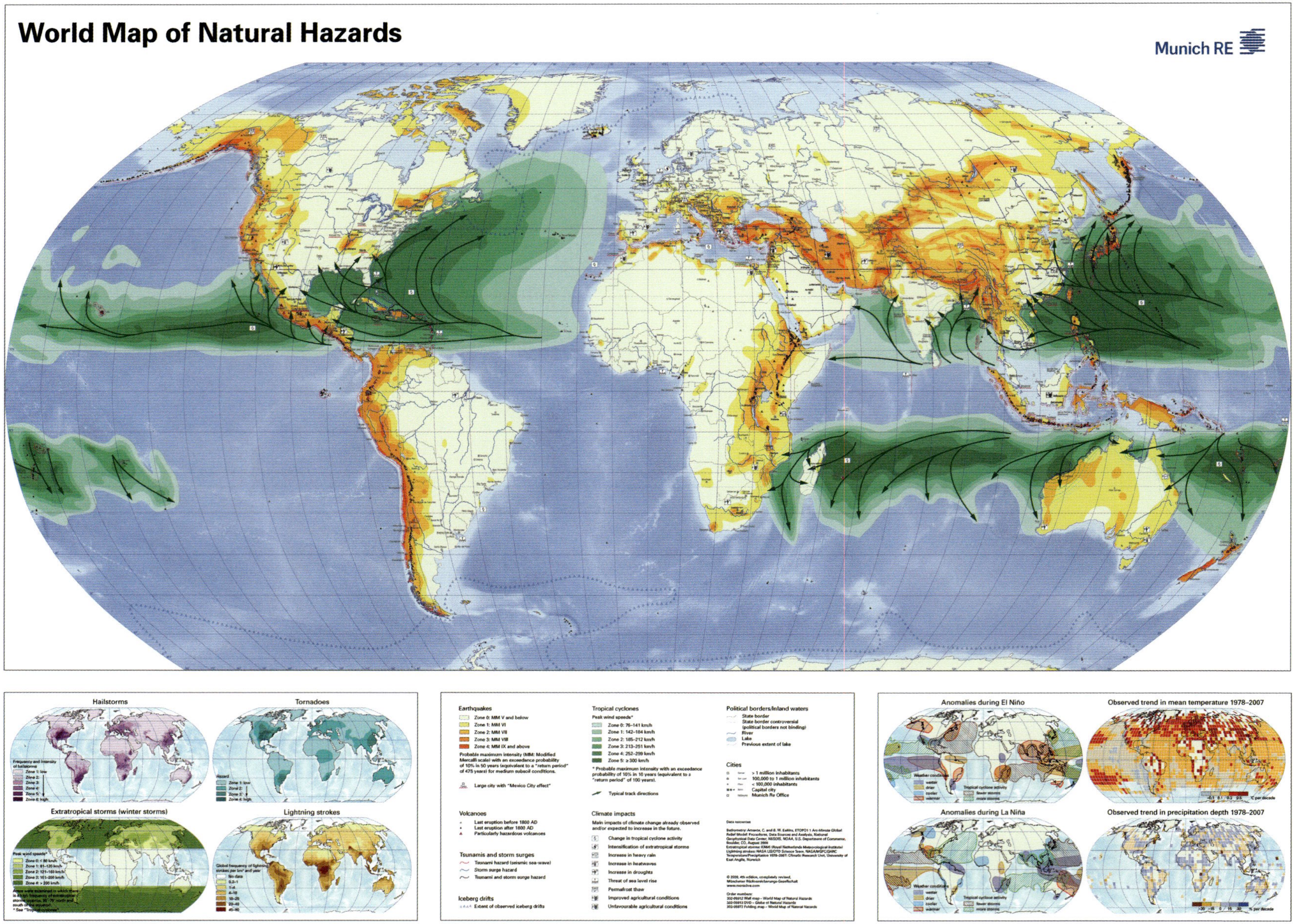

Figure 29.1. World Map of Natural Hazards. Earthquake hazard is shown in yellow–brownish colours and has been classified into five grades according to the minimum intensity (modified Mercalli scale) to be reached or exceeded once in 475 years. Darkest colour signifies the highest hazard. The hazard of tropical windstorms is presented in shades of green; again, the darkest colour corresponds to the highest hazard. Classification is according to the five-degree Saffir–Simpson scale. Green arrows represent the main cyclone tracks. Other hazards shown are extratropical storms (grey shading) and active volcanoes (small black symbols).

16 watersheds (Al Saud, 2010). The dense drainage networks coupled with thick alluvium deposits in the valley courses were some of the natural causes that attenuated the flooding process. However, the main concern has been the flooding of the city due to anthropogenic causes which resulted from unplanned urbanisation. Construction activities in the topographically low-lying areas have obstructed the natural flow paths of the streams causing inundation, as was the case in November 2009.

The southwestern region of Saudi Arabia generates about 60% of the total wadi flow in the region between the Red Sea coast and the adjacent mountains (Nouh, 2006). Interestingly, this volume of flow is generated in only 10% of the total surface area of the country. The enormous runoff generated during the occasional heavy rainstorms in the region, coupled with the unplanned urban growth, are the main causes of the occurrence of flash floods in western and south-western Saudi Arabia. The volumes of the wadi flows in the Arabian Platform, east of the Arabian Shield, are much lower with the maximum volume of runoff generated in Wadi Al-Dwasir and Wadi Najran.

18.3 Earthquake and volcanic hazards

The Arabian Peninsula is composed of the Arabian Shield (in the west) and the Arabian Platform (in the east) (Rodgers *et al.*, 1999). The Proterozoic basement is exposed in the Arabian Shield part, whereas the Platform is covered by Phanerozoic sediments showing a gradual thickening from west to east (Şeber and Mitchell, 1992).

The Arabian Peninsula forms a single tectonic plate (Fig. 18.2) surrounded by active boundaries, where earthquakes occur (Adams and Barazangi, 1984). Following the rifting of the Red Sea Basin some 30 million years ago (Camp and Roobol, 1992), the Red Sea region became a broad zone of active deformation between Africa and Arabia. The Late Cenozoic evolution of Saudi Arabia was mainly controlled by: (i) extensional processes in the Red Sea Basin (Bellahsen *et al.*, 2003), (ii) continental collision between Arabia and Eurasia to the east since the Middle Miocene (Ring and Layer, 2003), and (iii) the left lateral strike-slip boundary to the northwest represented by the Dead Sea Transform Fault System (Garfunkel, 1981).

Seismicity in the region (Fig. 18.2(b)) is dominated by the collision of the Arabian plate with the Eurasian plate along the Zagros and Bitlis thrust system (Stocklin, 1968), rifting and sea floor spreading in the Red Sea and Gulf of Aden (Stoeser and Camp, 1985), and strike-slip faults along the Gulf of Aqaba and the Dead Sea Transform Fault System (Garfunkel, 1981; Girdler, 1991). The Dead Sea Transform Fault System connects the active spreading centres in the Red Sea to the convergent zone between the Arabian and Eurasian plates in southern Turkey. The seismicity in its vicinity is characterised by swarm and mainshock–aftershock types of earthquake activities (Al Amri and Rodgers, 2013).

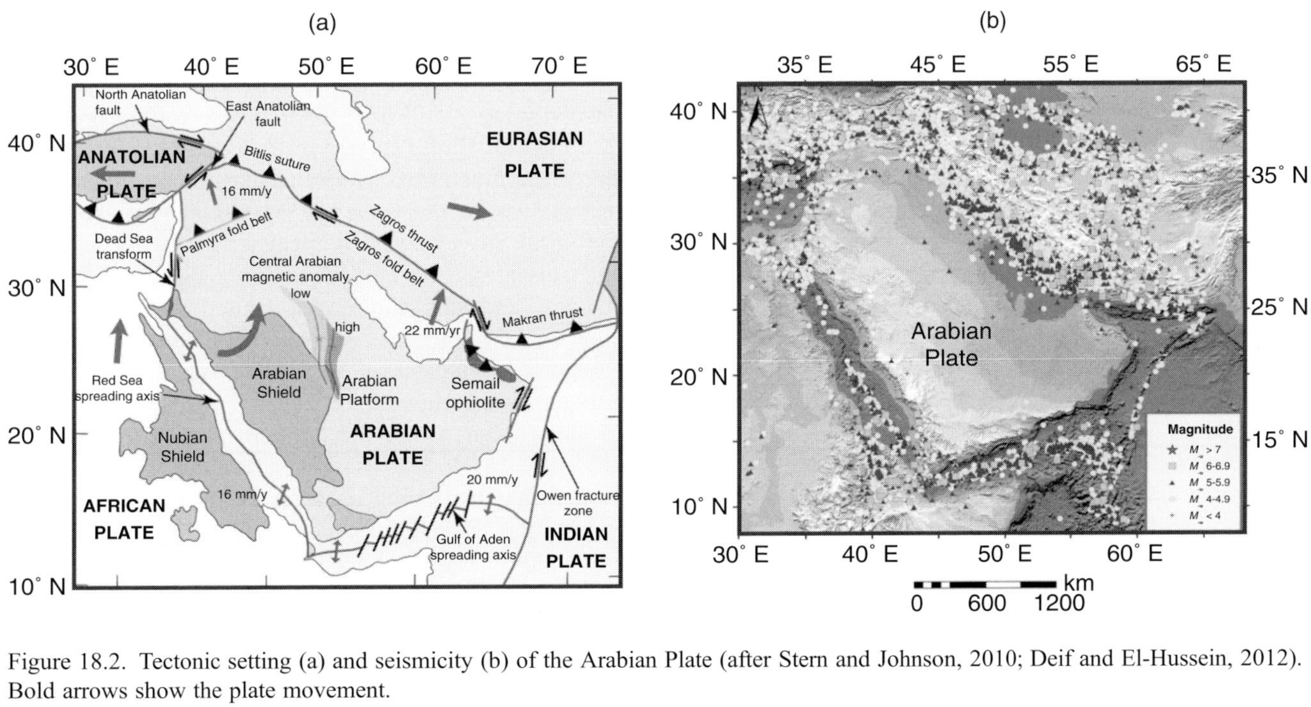

Figure 18.2. Tectonic setting (a) and seismicity (b) of the Arabian Plate (after Stern and Johnson, 2010; Deif and El-Hussein, 2012). Bold arrows show the plate movement.

The main zone of active tectonics in the Red Sea region lies along a belt that extends from the central Red Sea to the south of Afar and then eastward through the Gulf of Aden. There is little seismic activity in the northern Red Sea (Al Amri, 1995). The main threat to Saudi Arabia in terms of seismic hazards is from the significant number of earthquakes occurring in the Gulf of Aqaba including the large earthquakes and swarms in 1983, 1990, 1993, 1995, and 2004 (Al Amri, 1995; Klinger *et al.*, 1999; Ambraseys *et al.*, 2005; Al Damegh *et al.*, 2009).

The lack of seismicity in the interior of the Arabian Peninsula suggests that internal deformation of the Arabian plate is insignificant. However, the major cities in the Gulf region are situated along the eastern margin of the Arabian plate and are within 300 km of the collision zone between the Arabian and Eurasian plates. Strong earthquakes of magnitude greater than 5 are common along the Zagros thrust belt (a collision zone between the Arabian and Eurasian plates). Though the epicentres of these events are more than 200 km from the cities situated along the Arabian coast, seismic waves travelling through the sedimentary structures of the Arabian platform are capable of long duration ground motions (between 3 and 10 seconds) and are a potential threat to the engineering structures in the region (Al Amri *et al.*, 2008). Al Amri and Rodgers (2013) delineated the seismic zones in the Arabian Platform based on seismological and geological parameters. The seismological parameters included the spatio-temporal distribution of the seismic events indicating the seismogenic provinces and seismoactive faults and the occurrence of large earthquakes. The geological parameters included the location of the major faults, lineaments, joints, and rift systems that are associated with seismic activities.

The southwestern Arabian Shield and the southern Red Sea region gained importance in terms of seismic hazard assessment because of the extensive urbanisation in the region in the past years including strategic development projects and new urban communities (Al Amri and Rodgers, 2013). Historical and instrumental records of the earthquakes in the region show that most of the seismic events occur along the axial trough of the Red Sea (Al Amri, 1994). However, four strong earthquakes (*M*6.2 in 1941, *M*5.7 in 1955, *M*4.7 in 1962, and *M*6 in 1982) occurred away from this axial trough (Langer *et al.*, 1987). Based on seismicity, the region is divided into four seismic source zones: the Sanaa-Dhamar, the southern Red Sea, northern Yemen, and the middle of the Red Sea zone (Al Amri and Rodgers, 2013).

The tectonic evolution of the Arabian plate resulted in large volumes of Cenozoic volcanism, mainly along the western margin of the Arabian plate (Segev, 2005; Trifonov *et al.*, 2011; Chang and Van der Lee, 2011; Moufti *et al.*, 2012). The volcanic fields are situated in the rifted neo-Proterozoic crust and include parts of Syria, Lebanon, Jordan, Saudi Arabia, and Yemen (Fig. 18.3) forming one of the world's largest intraplate volcanic provinces (Coleman *et al.*, 1983). These lava flows are mostly dominated by alkali olivine basalts and hawaiite (Moufti *et al*, 2013).

The presence of recent volcanism in the Arabian Shield makes it different from the other shield areas of the world (Rodgers *et al.*, 1999). The opening of the Red Sea and the corresponding upwelling of the asthenosphere beneath western Saudi Arabia is likely to be the most probable cause for the volcanic activity (Camp and Roobol, 1992). The volcanic sequences are represented by a thick sequence of basaltic lava flows including pyroclastic cones, shield volcanoes, and hydro-magmatic craters (Al Kwatli *et al.*, 2012) ranging in age from 30 Ma to recent (Shaw *et al.*, 2003; Ibrahim *et al.*, 2003; Weinsten *et al.*, 2006).

Based on ages and the structural pattern of the vents, the Arabian Shield volcanism is divided into two periods (Coleman *et al.*, 1983; Camp and Roobol, 1992). The first period of volcanism (30–15 Ma) was associated with the doming and rifting of the Proterozoic basement of the present Arabian Nubian Shield along the north-northwest trending rift system leading eventually to the opening of the Red Sea Basin. The second period of volcanism ($<$ 12 Ma) is characterised by north–south trending vent system associated with the onset of a new north–south trending 900 km long crustal rift system passing through the 600 km long Makkah-Madinah volcanic line (Camp and Roobol, 1992). Individual volcanic fields can be very large, such as Harrat Ash Shaam and Harrat Rahat. Harrat Al Madinah volcanic province (a part of Harrat Rahat) is an active volcanic field characterised by two historical eruptions, one in AD 641 and another in AD 1256 (Moufti *et al.*, 2013).

Northwestern Saudi Arabia experienced notable earthquakes during April–June 2009. These earthquakes took place beneath Harrat Lunayyir (Palister *et al.*, 2010). The peak seismic activity (19 earthquakes of magnitude 4 or greater) was recorded on 19 May 2009 by the Saudi Geological Survey's (SGS) telemeteric network of broadband seismometers. The maximum magnitude recorded was 5.4, and this earthquake caused minor structural damage in the town of Al-Eis about 40 km from the city of Madinah (Pallister *et al.*, 2010). As a result of this earthquake, a northwest trending 8-km-long surface rupture propagated across the northern part of the volcanic field (Fig. 18.4). The shallow crustal earthquakes were grouped into two depth clusters: 5–10 km and 13–20 km suggesting a dyke intrusion could be responsible for this earthquake swarm (Mukhopadhyay *et al.*, 2013).

The origin of the 2009 earthquake swarm is still unknown. The key question is whether the earthquakes were of a tectonic origin or were triggered by a magma movement. Crustal expansion was responsible for

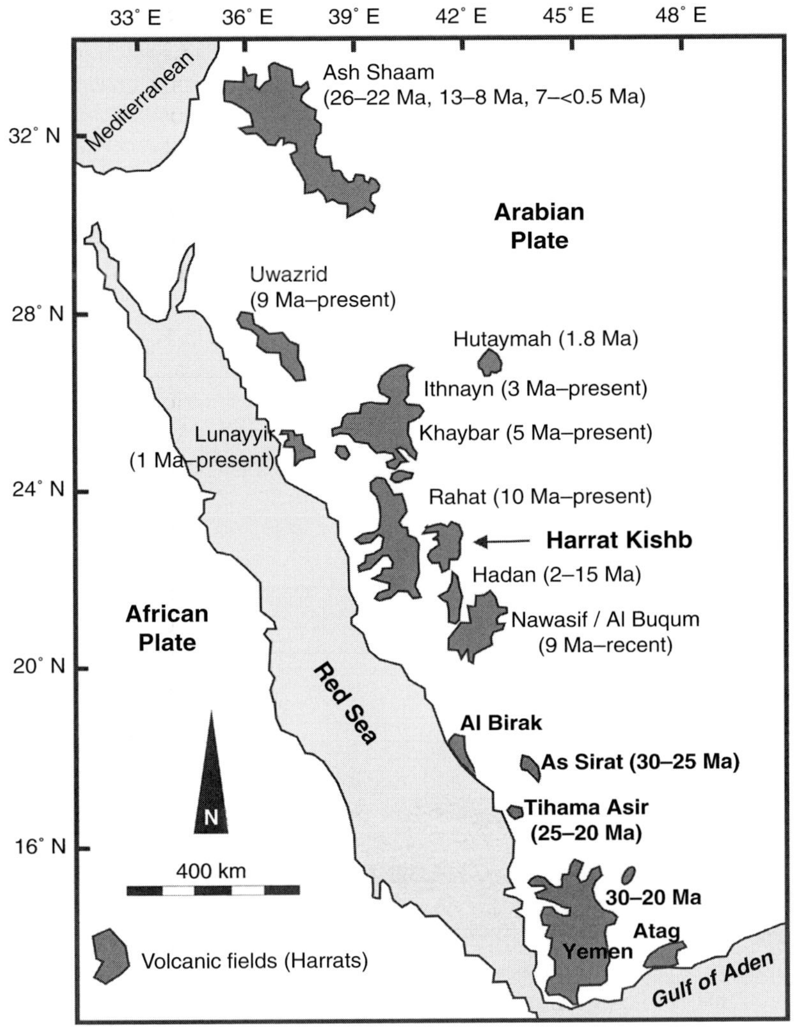

Figure 18.3. Cenozoic volcanic fields of the Arabian Peninsula (after Abdel Wahab *et al.* (2013)).

Cenozoic volcanism in central western and northwestern Saudi Arabia. Historically the area has witnessed volcanic eruptions, with the last recorded eruption taking place about 1400 years ago (Grainger, 2007). If a magma movement triggered the 2009 earthquakes in the region, then some magmatic processes are still active in the region.

18.4 Dust storms

Shifting sand dunes and dust storms are a serious natural hazard being faced by cities mostly in central and eastern Saudi Arabia. The phenomenon has become more common in recent years due to the expansion of cities, road, and infrastructure development. The Middle East region is the world's second (after Africa) most dust-affected area (Kutiel and Furman, 2003). The quantity of dust present in the air at any given time is directly related to the wind velocity. Therefore, understanding the atmospheric air circulation in the region provides a clue to the probable times of dust storms. Dust storms affect daily life for from a few hours to a few days depending on the storm severity (Maghrabi *et al.*, 2011). The most important issue associated with dust storms is the reduction of visibility that increases the incidence of traffic accidents and may increase the occurrence of vertigo in aircraft pilots (Dayan *et al.*, 1991; Kutiel and Furman, 2003). The suspended air particles have a significant impact on human health (Bennett *et al.*, 2006; Bennion *et al.*, 2007) especially for patients suffering from lung disease. Other environmental impacts may include damage to telecommunication and mechanical systems, reduction in soil fertility, and damage to crops.

The frequency of sandstorms increases between March and May. The dust originates mainly from the arid areas across the Arabian Peninsula and is transported by the southwestern winds towards the east (Ackerman and Cox, 1989). Since dust storms in Saudi Arabia are a

Figure 18.4. Rupture created by the Al-Eis earthquake in May 2009. (Photo: Omar Lafouza, KACST, Riyadh, Saudi Arabia.)

common phenomenon, their impact on the environment has been intensively studied (e.g., Smirnov *et al.*, 2002; Al Harbi and Moeid, 2005; Badarinath *et al.*, 2010; Maghrabi *et al.*, 2011).

The very severe dust storm of 10 March 2009 in the Riyadh region was visible from space (Fig. 18.5). The dust storm disrupted air traffic in Saudi Arabia for several hours. Magrabhi *et al.* (2011) carried out a detailed analysis of this storm event in terms of meteorological parameters: the aerosol optical depth, the ångström exponent, the infrared sky temperature, and the emissivity. Their findings showed that the effects of the storm were associated with an increase in both atmospheric pressure and humidity and a reduction in temperature and visibility, two days after the storm as compared to the conditions before it. The infrared sky temperatures and the emissivity remained at higher levels after the storm as compared to the pre-storm conditions.

18.5 Disaster risk reduction (DRR) in the Kingdom of Saudi Arabia

DRR strategies involving governmental and non-governmental/private organisations, communities, and individuals have resulted in reducing the risks associated with human and economic losses, thus reducing the overall impacts of natural hazards (Innocenti and Albrito, 2011). There are a number of government agencies in Saudi Arabia continuously engaged in research on natural hazards, prevention, and mitigation, and the development of early warning systems in case of a natural calamity.

The Directorate General of Civil Defence is the premier national agency responsible for prevention and mitigation of disasters due to natural hazards in the country. The Directorate issues early warnings in the case of dust storms, heavy rainfall, and flash floods, and adopts and hosts a number of initiatives to protect lives and property in the Kingdom from all types of natural hazards. Several awareness programmes are organised by the Directorate to try to educate the general public and especially the children on how to respond before and after a natural disaster.

The SGS is another government agency mainly engaged in research and gathering primary information on natural events such as earthquakes and volcanic activity in the Kingdom. Due to the increase in population and the expansion of the urban areas, more and more areas are being exposed to the risks of natural hazards. The SGS is continuously working towards development of methodologies and strategies to reduce or mitigate the effects of the natural hazards. The National Centre for Earthquakes and Volcanoes at the SGS is engaged in close monitoring of earthquake and volcanic activity within the Kingdom and cooperation with the international Earthquake Data Centre for exchange of data. The centre is also involved in raising public awareness related to earthquake hazards in the Kingdom and the risks related to flash floods within the major cities. The SGS Research Chair on Natural Hazards housed within the Department of Geology and Geophysics at King

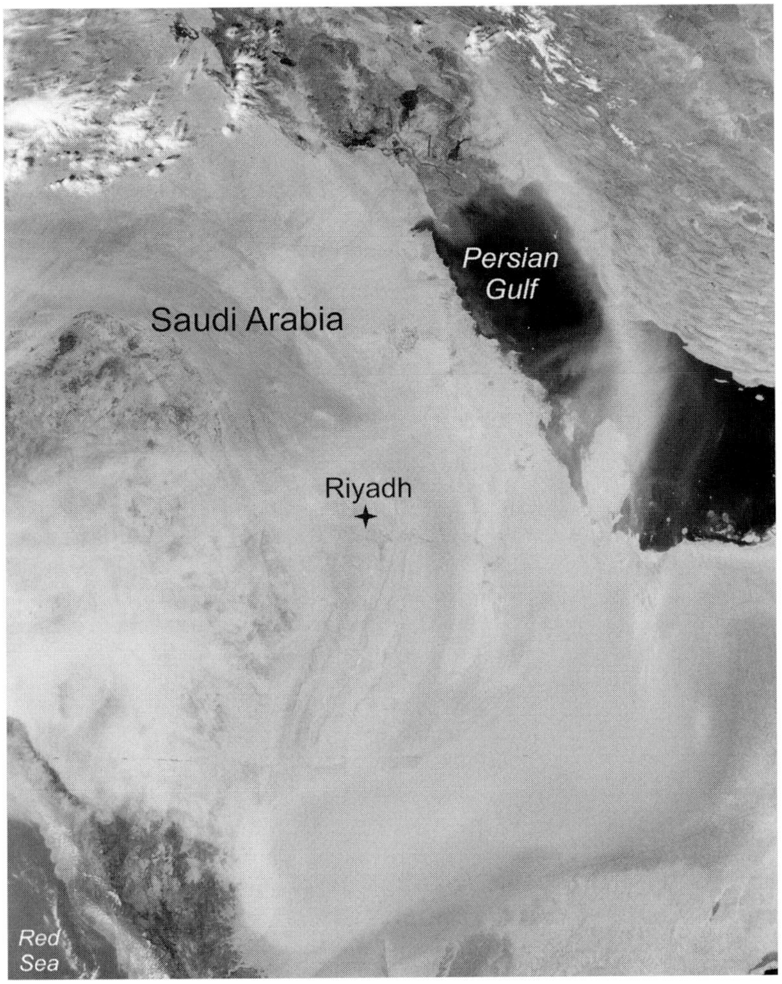

Figure 18.5. A satellite image of the dust plumes (translucent blurs) over the Arabian Peninsula on 11 March 2009. (Photo: Jeff Schmaltz, MODIS Rapid Response Team, Goddard Space Flight Center, NASA.) For colour version, see Plates section.

Saud University is involved in innovative research and education in the field of natural hazards and disaster risk reduction and mitigation in Saudi Arabia, addressing the essential issues and problems related to the prediction of natural hazards, early warning systems, and hazards risk assessment.

Acknowledgment. The authors thank the SGS Research Chair for Natural Hazards and the King Saud University for providing technical and financial support.

References

Abdel Wahab, A., Abul Maaty, M. A., Stuart, F. M., Awad, H., and Kafafy, A. (2013). The geology and geochronology of Al Wahbah maar crater, Harrat Kishb, Saudi Arabia. *Quaternary Geochronology*, in press, http://dx.doi.org/10.1016/j.quageo.2013.01.008.

Ackerman, S. A. and Cox, S. K. (1989). Surface weather observations of atmospheric dust over the southwest summer monsoon region. *Meteorology and Atmospheric Physics*, **41**, 19–34.

Adams, R. D. and Barazangi, M. (1984). Seismotectonics and seismology in the Arab region: a brief summary and future plans. *Bulletin of the Seismological Society of America*, **74**(3), 1011–1030.

Al Amri, A. (1994). Seismicity of the south-western region of the Arabian shield and southern Red Sea. *Journal of African Earth Sciences*, **19**(1), 17–25.

Al Amri, A. (1995). Recent seismic activity in the northern Red Sea. *Journal of Geodynamics*, **20**(3), 243–253.

Al Amri, A. M. and Rodgers, A. J. (2013). Improvement of seismicity parameters in the Arabian Shield and Platform using earthquake location and magnitude calibration. In *Lithosphere Dynamics and Sedimentary Basins: The Arabian Plate and Analogues*, eds. Al Hosani, K., Roure, F., Ellison, R. and Lokier, S. Berlin: Springer, pp. 281–293.

Al Amri, A. M., Rodgers, A. J. and Al-Khalifah, T. A. (2008). Improving the level of seismic hazard parameters in Saudi Arabia using earthquake location. *Arabian Journal of Geosciences*, **1**(1), 1–15.

Al Damegh, K. S., AbouElenean, K. M., Hussein, H. M. and Rodgers, A. J. (2009). Source mechanisms of the June 2004 Tabuk earthquake sequence, Eastern Red Sea margin, Kingdom of Saudi Arabia. *Journal of Seismology*, **13**(4), 561–576.

Al-Ghamdi, K. A., Elzahrany, R. A., Mirza, M. N. and Dawod, G. M. (2012). Impacts of urban growth on flood hazards in Makkah City, Saudi Arabia. *International Journal of Water Resources and Environmental Engineering*, **4**(2), 23–34.

Al Harbi, B. H. and Moied, K. (2005). Riyadh Air Quality Report (1999–2004), King Abdulaziz City for Science and Technology. Report No. 279-25-ER.

Al Kwatli, M. A., Gillot, P. Y., Zeyen, H., Hildenbrand, A. and Al Gharib, I. (2012). Volcano-tectonic evolution of the northern part of the Arabian plate in the light of new K–Ar ages and remote sensing: Harrat Ash Shaam volcanic province (Syria). *Tectonophysics*, **580**, 192–207.

Al Saud, M. (2010). Assessment of flood hazard of Jeddah area 2009, *Saudi Arabia. Journal of Water Resource and Protection*, **2**, 839–847.

Ambraseys, N. N., Melville, C. P. and Adams, R. D. (2005). *The Seismicity of Egypt, Arabia and the Red Sea: A Historical Review*. Cambridge: Cambridge University Press.

Badarinath, K. V. S., Kharol, S. K., Kaskaoutis, D. G. *et al.* (2010). Long-range transport of dust aerosols over the Arabian Sea and Indian region–A case study using satellite data and ground-based measurements. *Global and Planetary Change*, **72**(3), 164–181.

Bellahsen, N., Faccenna, C., Funiciello, F., Daniel, J. M. and Jolivet, L. (2003). Why did Arabia separate from Africa? Insights from 3-D laboratory experiments. *Earth and Planetary Science Letters*, **216**(3), 365–381.

Bennett, C. M., McKendry, I. G., Kelly, S., Denike, K. and Koch, T. (2006). Impact of the 1998 Gobi dust event on hospital admissions in the Lower Fraser Valley, British Columbia. *Science of the Total Environment*, **366**(2–3), 918–925.

Bennion, P., Hubbard, R., O'Hara, S. *et al.* (2007). The impact of airborne dust on respiratory health in children living in the Aral Sea region. *International Journal of Epidemiology*, **36**(5), 1103–1110.

Borga, M., Anagnostou, E. N., Blöschl, G. and Creutin, J. D. (2010). Flash floods: observations and analysis of hydrometeorological controls. *Journal of Hydrology*, **394**, 1–3.

Camp, V. E. and Roobol, M. J. (1992). Upwelling asthenosphere beneath western Arabia and its regional implications. *Journal of Geophysical Research*, **97**(15), 255–271.

Chang, S. J. and Van der Lee, S. (2011). Mantle plumes and associated flow beneath Arabia and East Africa. *Earth and Planetary Science Letters*, **302**(3), 448–454.

Chester, D. K., Degg, M., Duncan, A. M. and Guest, J. E. (2000). The increasing exposure of cities to the effects of volcanic eruptions: a global survey. *Global Environmental Change, Part B: Environmental Hazards*, **2**(3), 89–103.

Coleman, R. G., Gregory, R. T. and Brown, G. F. (1983). Cenozoic volcanic rocks of Saudi Arabia. USGS Open-File Report: 83–788. 86 pp.

Dawod, G. M., Mirza, M. N. and Al-Ghamdi, K. A. (2013). Assessment of several flood estimation methodologies in Makkah metropolitan area, Saudi Arabia. *Arabian Journal of Geosciences*, **6**(4), 985–993.

Dayan, U., Heffter, J., Miller, J. and Gutman, G. (1991). Dust intrusion events into the Mediterranean Basin. *Journal of Applied Meteorology*, **30**, 1185–1199.

Deif, A. and El-Hussain, I. (2012). Seismic moment rate and earthquake mean recurrence interval in the major tectonic boundaries around Oman. *Journal of Geophysics and Engineering*, **9**(6), 773–783.

Ganor, E. and Mamane, Y. (1982). Transport of Saharan dust across the eastern Mediterranean. *Atmospheric Environment*, **16**, 581–587.

Garfunkel, Z. (1981). Internal structure of the Dead Sea leaky transform (rift) in relation to plate kinematics. *Tectonophysics*, **80**(1), 81–108.

Girdler, R. W. (1991). The Afro-Arabian rift system – an overview. *Tectonophysics*, **197**(2), 139–153.

Gracheva, R. and Golyeva, A. (2010). Landslides in mountain regions: hazards, resources and information. In *Geophysical Hazards: Minimising Risk, Maximizing Awareness*, ed. Beer, T. Dordrecht: Springer, pp. 248–260.

Grainger, D. (2007). *The Geologic Evolution of Saudi Arabia*. Jeddah: Saudi Geological Survey.

Groisman, P. Y., Knight, R. W., Easterling, D. R. *et al.* (2005). Trends in intense precipitation in the climate record. *Journal of Climate*, **18**(9), 1326–1350.

Huntington, T. G. (2006). Evidence for intensification of the global water cycle: review and synthesis. *Journal of Hydrology*, **319**(1–4), 83–95.

Hussein, M. T. and Zaidi, F. K. (2012). Assessing hydrological elements as key issue for urban development in arid regions. In *Urban Development*, ed. Polyzos, S. Rijeka, Croatia: InTech, pp. 129–158.

Ibrahim, K. M., Tarawneh, K. and Rabba, I. (2003). Phases of activity and geochemistry of basaltic dike systems in northeast Jordan parallel to the Red Sea. *Journal of Asian Earth Sciences*, **21**(5), 467–472.

Innocenti, D. and Albrito, P. (2011). Reducing the risks posed by natural hazards and climate change: the need for a participatory dialogue between the scientific community and policy makers. *Environmental Science & Policy*, **14**(7), 730–733.

Junge, C. (1979). The importance of mineral dust as an atmospheric constituent. In *Saharan Dust, Scope 14*, ed. Morales, C. New York: John Wiley and Sons, pp. 49–60.

Klinger, Y., Rivera, L., Haessler, H. and Maurin, J. C. (1999). Active faulting in the Gulf of Aqaba: New knowledge from the MW 7.3 earthquake of 22 November 1995. *Bulletin of the Seismological Society of America*, **89**(4), 1025–1036.

Kutiel, H. and Furman, H. (2003). Dust storms in the Middle East: sources of origin and their temporal characteristics. *Indoor and Built Environment*, **12**(6), 419–426.

Langer, C. J., Bollinger, G. A. and Merghelani, H. M. (1987). Aftershocks of the 13 December 1982 North Yemen earthquake: Conjugate normal faulting in an extensional setting.

Bulletin of the Seismological Society of America, **77**(6), 2038–2055.

Leroy, S. A. G. (2013). Natural hazards, landscapes, and civilizations. In *Treatise on Geomorphology*, vol. 13, ed. Shroder, J. F. San Diego: Academic Press, pp. 190–203.

Maghrabi, A., Al Harbi, B. and Tapper, N. (2011). Impact of the March 2009 dust event in Saudi Arabia on aerosol optical properties, meteorological parameters, sky temperature and emissivity. *Atmospheric Environment*, **45**(13), 2164–2173.

Marchi, L., Borga, M., Preciso, E. and Gaume, E. (2010). Characterisation of selected extreme flash floods in Europe and implications for flood risk management. *Journal of Hydrology*, **394**(1), 118–133.

Morales, C., ed. (1979). *Saharan Dust, Scope 14*. New York: John Wiley and Sons.

Moufti, M. R., Moghazi, A. M. and Ali, K. A. (2012). Geochemistry and Sr–Nd–Pb isotopic composition of the Harrat Al-Madinah Volcanic Field, Saudi Arabia. *Gondwana Research*, **21**(2), 670–689.

Moufti, M. R., Moghazi, A. M. and Ali, K. A. (2013). $^{40}Ar/^{39}Ar$ geochronology of the Neogene–Quaternary Harrat Al-Madinah Intercontinental Volcanic Field, Saudi Arabia: implications for duration and migration of volcanic activity. *Journal of Asian Earth Sciences*, **62**, 253–268.

Mukhopadhyay, B., Mogren, S., Mukhopadhyay, M. and Dasgupta, S. (2013). Incipient status of dyke intrusion in top crust – evidence from the Al-Ays 2009 earthquake swarm, Harrat Lunayyir, SW Saudi Arabia. *Geomatics, Natural Hazards and Risk*, **4**(1), 30–48.

Nouh, M. (2006). Wadi flow in the Arabian Gulf states. *Hydrological Processes*, **20**, 2393–2413.

Pallister, J. S., McCausland, W. A. and Jónsson, S. (2010). Broad accommodation of rift-related extension recorded by dyke intrusion in Saudi Arabia. *Nature Geoscience*, **3**(10), 705–712.

Presidency of Meteorology and Environment (2005). First National Communication, Kingdom of Saudi Arabia, submitted to the United Nations Framework Convention on Climate Change, PME, Jeddah.

Ring, U. and Layer, P. W. (2003). High-pressure metamorphism in the Aegean, eastern Mediterranean: Underplating and exhumation from the Late Cretaceous until the Miocene to Recent above the retreating Hellenic subduction zone. *Tectonics*, **22**(3), 1022, doi:10.1029/2001TC001350.

Rodgers, A., Walter, W., Mellors, R., Al-Amri, A. and Zhang Y. (1999). Lithospheric structure of the Arabian shield and platform from complete regional waveform modeling and surface wave group velocities. *Geophysical Journal International*, **138**(3), 871–878.

Şeber, D. and Mitchell, B. J. (1992). Attenuation of surface waves across the Arabian Peninsula. *Tectonophysics*, **204**(1), 137–150.

Segev, A. (2005). Magmatic rocks. In *Geological Framework of the Levant*, vol. II, Part 4, eds. Krasheninnikov, V. A., Hall, J. K., Hirsch, F., Benjamini, C. and Flexer A. Jerusalem: Historical Productions-Hall, pp. 553–576.

Sen, Z. (2008). *Wadi Hydrology*. Boca Raton, FL: CRC Press.

Shaw, J. E., Baker, J. A., Menzies, M. A., Thirlwall, M. F. and Ibrahim, K. M. (2003). Petrogenesis of the largest intraplate volcanic field on the Arabian Plate (Jordan): a mixed lithosphere–asthenosphere source activated by lithospheric extension. *Journal of Petrology*, **44**, 1657–1679.

Smirnov, A., Holben, B. N., Dubovik, O. *et al.* (2002). Atmospheric aerosol optical properties in the Persian Gulf. *Journal of the Atmospheric Sciences*, **59**(3), 620–634.

Stern, R. J. and Johnson, P. (2010). Continental lithosphere of the Arabian Plate: a geologic, petrologic, and geophysical synthesis. *Earth-Science Reviews*, **101**(1), 29–67.

Stocklin, J. (1968). Structural history and tectonics of Iran: a review. *AAPG Bulletin*, **52**(7), 1229–1258.

Stoeser, D. B. and Camp, V. E. (1985). Pan-African microplate accretion of the Arabian Shield. *Geological Society of America Bulletin*, **96**(7), 817–826.

Subyani, A. M. and Al-Dakheel, A. M. (2009). Multivariate geostatistical methods of mean annual and seasonal rainfall in southwest Saudi Arabia. *Arabian Journal of Geosciences*, **2**(1), 19–27.

Trifonov, V. G., Dodonov, A. E., Sharkov, E. V. *et al.* (2011). New data on the Late Cenozoic basaltic volcanism in Syria, applied to its origin. *Journal of Volcanology and Geothermal Research*, **199**, 177–192.

Weinstein, Y., Navon, O., Altherr, R. and Stein, M. (2006). The role of lithospheric mantle heterogeneity in the generation of Plio-Pleistocene alkali basaltic suites from NW Harrat Ash Shaam (Israel). *Journal of Petrology*, **47**(5), 1017–1050.

19

Large earthquakes and tsunamis in the Mediterranean region and its connected seas

Gerassimos A. Papadopoulos and Antonia Papageorgiou

19.1 Introduction

In the Mediterranean region subduction of oceanic crust and/or plate collision takes place along a series of orogenic belts. From west to east they are the Gibraltar Arc, the Calabrian Arc, the Hellenic Arc, and the Cyprean Arc. This is due to the active convergence between the African and Eurasian plates, which is about 4–5 mm y^{-1} in the western Mediterranean and 30–50 mm y^{-1} in the eastern Mediterranean (e.g. De Mets *et al.*, 2010). The active Mediterranean geodynamics is also controlled by ongoing movements along transform boundaries of adjacent plates, such as the Arabian plate and the Anatolian 'micro-plate'.

Historical, geological, and instrumental data show that tsunamis in the Mediterranean region and in its connected seas are mainly generated by earthquakes, but also by landslides (e.g. Lo Iacono *et al.*, 2012) and volcanic processes (e.g. Stromboli volcano; Tinti *et al.*, 2005). Regardless of their generation mechanisms the tsunami sources are concentrated in 22 distinct zones as illustrated in Fig. 19.1. In this chapter, we focus our attention on the largest earthquakes and their tsunamigenic potential, because large magnitude earthquakes produce the most frequent tsunamis. From an overview of the earthquake and tsunami history we determined the largest tsunamigenic earthquakes (LTEs) in the Mediterranean region and examined their focal mechanisms, source areas, and reoccurrence times.

19.2 A historical overview

Tsunamis in the Mediterranean and its connected seas were well documented in the form of descriptive and/or parametric catalogues including those of Ambraseys (1962, 2002, 2009), Antonopoulos (1980), Papadopoulos and Chalkis (1984), Soloviev (1990), Guidoboni *et al.* (1994), Tinti and Maramai (1996), Guidoboni and Comastri (1997, 2005), Altinok and Ersoy (2000), Soloviev *et al.* (2000), Papadopoulos (2001, 2003, 2009, 2011), Tinti *et al.* (2001, 2004), Yalciner *et al.* (2002), Papazachos and Papazachou (2003), Papadopoulos and Fokaefs (2005), Fokaefs and Papadopoulos (2007), Papadopoulos *et al.* (2007, 2010), Salamon *et al.* (2007), Baptista and Miranda (2009), and Altinok *et al.* (2011). In an extensive review, Papadopoulos *et al.* (2013) presented the updated tsunami source zonation map (shown in Fig. 19.1).

In this section we examine the history of tsunamis generated by large earthquakes in the entire Mediterranean region. Since no standard methods for measuring tsunami magnitude are in place, tsunami intensity has been used as a proxy for the tsunami size. The new 12-point tsunami intensity scale introduced by Papadopoulos and Imamura (2001) was applied (see also full reproduction of the scale in Yalciner *et al.* (2002)). Event dates are given in New Style (Gregorian calendar) unless it is specified that they are in Old Style (Julian calendar).

19.2.1 Western Mediterranean

There is a zone of LTEs in the offshore southwest Iberia region (Baptista and Miranda, 2009). The most famous was the disastrous 1755 earthquake ($M_w 8.5 \pm 0.3$) and tsunami event which affected all the countries around the Gulf of Cadiz as well as a large part of the northeast Atlantic Ocean coasts (e.g. Johnston, 1996; Baptista *et al.*, 1998; Buforn *et al.*, 2004; Stich *et al.*, 2007). Wave heights up to 15 m were reported in Cape St Vincent and along the Gulf of Cadiz. In Lisbon, the number of casualties due exclusively to the tsunami is estimated as close as to 900 and the run-up in Lisbon downtown is estimated to have been as great as 250 m (Baptista et al., 1998).

Extreme Natural Hazards, Disaster Risks and Societal Implications, eds. Alik Ismail-Zadeh, Jaime Urrutia Fucugauchi, Andrzej Kijko, Kuniyoshi Takeuchi and Ilya Zaliapin. Published by Cambridge University Press.

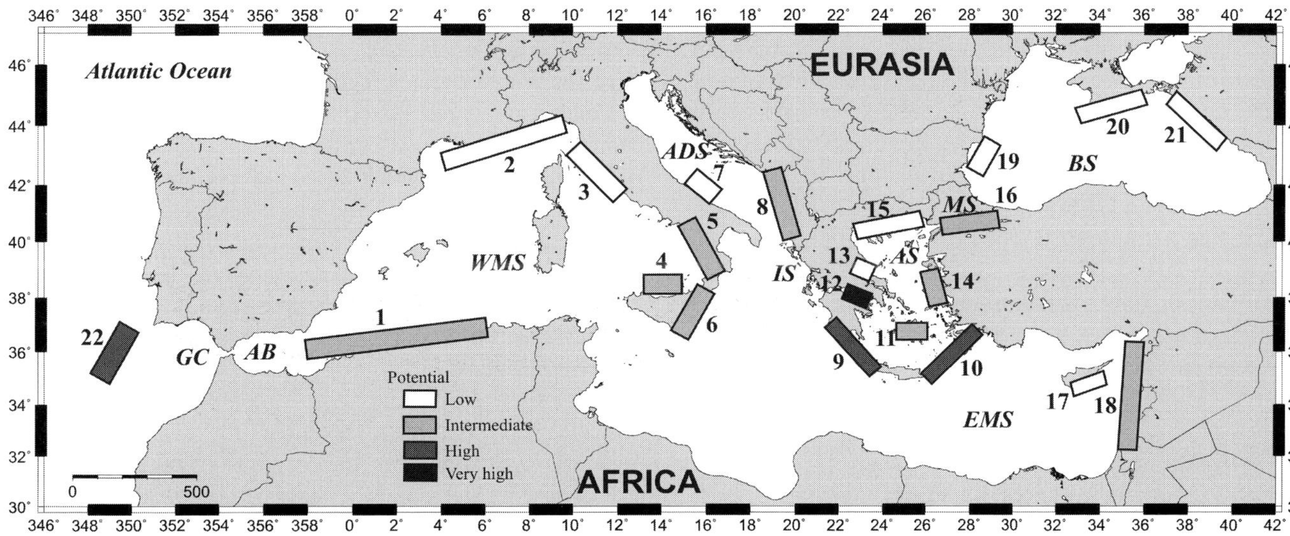

Figure 19.1. Tsunamigenic zones defined from documentary sources and their relative tsunami potential classification: WMS = Western Mediterranean Sea, GC = Gulf of Cadiz, AB = Alboran Basin, EMS = Eastern Mediterranean Sea, AS = Aegean Sea, ADS = Adriatic Sea, MS = Sea of Marmara, BS = Black Sea. Zonation key: 1 = Alboran Sea, 2 = Liguria and Côte d'Azur, 3 = Tuscany, 4 = Aeolian islands, 5 = Tyrrhenian Calabria, 6 = Eastern Sicily and Messina Straits, 7 = Gargano, 8 = East Adriatic Sea, 9 = West Hellenic Arc, 10 = East Hellenic Arc, 11 = Cyclades, 12 = Corinth Gulf, 13 = Maliakos Bay, 14 = East Aegean Sea, 15 = North Aegean Sea, 16 = Sea of Marmara, 17 = Cyprus, 18 = Levantine Sea, 19 = Bulgaria, 20 = Crimea, 21 = East Black Sea, 22 = SW Iberia.

In the western Mediterranean basin, large earthquakes and tsunamis are mainly associated with the North Algeria margin. Before 1811 some sporadic earthquake events are known, but more abundant data are available for the period after 1812. Algiers was destroyed by a very strong earthquake that occurred on 2 January 1365. Many people were killed and a tsunami flooded part of the city (Ambraseys and Vogt, 1988; Guidoboni and Comastri, 2005). During an earthquake that occurred on 6 May 1773, the sea rose 6 feet at Algiers and 30 feet (about 9 m) at Tangiers and then fell so low as to leave the boats aground (Anonymous, 1774). This action continued with decreasing amplitude. A remarkable earthquake sequence occurred on 21 and 22 August 1856 near the town Jijel (Djidjelli) situated about 300 km to the east of Algiers (Ambraseys, 1982; Soloviev *et al.*, 2000; Harbi *et al.*, 2003; Yelles-Chaouche *et al.*, 2009). Both earthquakes caused the collapse of many buildings, which buried several inhabitants. After the first shock, the sea receded a certain distance from the coast in Jijel and suddenly returned and flooded low-lying areas of the shore. Again in Jijel, a tsunami 2–3 m high was generated and flooded the shore several times after the second shock. Sea flooding was also reported from Philippeville and Bougie. According to Ambraseys (1982), the tsunamis associated with the two earthquakes were not due to fault dislocation but to associated turbulent flows. However, a seismic source comprising a set of three submarine *en échelon* fault segments, with dimensions and geometry in agreement with a $M_w7.2$ earthquake, was introduced in tsunami numerical modelling and reproduced the tsunami wave heights observed in Balearic Islands well (Roger and Hébert, 2008).

On 15 January 1891 the central-north region of Algeria, particularly the coastal area of Gouraya, suffered two very strong earthquakes, which were 10 minutes apart (Ambraseys and Vogt, 1988). Houses collapsed and about 40 people were killed. Several types of ground failures were observed. At Villebourg the coast uplifted by 30 cm. Possibly this was related to an uplift in Gouraya, just to the east of Villebourg, where after the second shock the sea retreated for about 30 m and then, on its return, flooded the shore but without causing damage. Two important seismic events occurred in El Asnam to the west of Algiers in the central part of the Chelif valley, the first on 9 September 1954 ($M_s6.6$) and the second on 10 October 1980 ($M_s7.3$) (Ambraseys, 1981). The 1954 and 1980 earthquakes killed about 1200 and 2500 people, respectively. The principal mechanism for both quakes consisted of the activation of a northwest–southeast trending reverse fault accompanied by a left-lateral motion. Soloviev *et al.* (2000) reported that both the 1954 and 1980 earthquakes generated tsunami waves recorded with maximum amplitudes, h, of 27 cm and 48 cm, respectively, at the tide-gauge station of Alicante (Spain). In addition, the 1980 tsunami was also recorded in Cartagena (Spain) with $h = 70$ cm. The generation of both tsunamis was attributed to seismically-triggered submarine mass movements. There was a tsunamigenic earthquake (M_w6.8) in Boumerdès Province, Algeria, about 50 km to the east of Algiers on 21 May 2003. The earthquake caused about 2500 casualties. From combined local and teleseismic seismic records and joint inversion of ASAR, RADARSAT, and GPS data it seems that the main shock rupture was a shallow (8–10 km) thrust, 50–60 km long located 15–20 km offshore causing an average 0.50 m coastal uplift (e.g. Braunmiller and Bernardi, 2005; Belabbès *et al.*, 2009). Tsunami waves up to 3 m in height with a mean period of 10–12 min were reported on Majorca, Minorca, and Eivissa in the Balearic Islands (Spain) at an epicentral distance of about 250 km (Anonymous, 2003). The waves caused significant damage to hundreds of vessels in harbours and ports. According to eyewitness accounts, in coastal sections of Algiers and Boumerdès a 100 m recession of the sea occurred, while in the port of Zemmouri-El Bahri the reported drop in the water level was less than 1 m, with boats ending up resting on the sea bottom before the water returned. Meghraoui *et al.* (2004) showed that non-uniform slip occurred in the seismic fault and that the earthquake magnitude was larger than the M_w6.8 assigned by the Harvard centroid-moment-tensor (CMT) solution. However, the modelling performed by Alasset *et al.* (2006) concluded that the best fit between synthetic and real data is obtained for a thrust rupture of M_w6.9, which is comparable to the earthquake fault inferred from seismotectonic studies and located within 15 km offshore.

The Mediterranean coasts of Morocco, Spain, and France, as well as Tunis, Libya, and Egypt, are in general considered to be of low to intermediate seismicity and no LTEs have been reported to generate from local sources. However, in the margins of the Alboran Sea (the westernmost Mediterranean), there is evidence of instrumental and historical seismic activity with associated tsunamis. For instance, the 1994 (M_w6.0) and 2004 ($M_w > 6.4$) large earthquakes occurred in the northern Moroccan margin near Al-Hoceima (e.g. Akoğlu *et al.*, 2006). In the south Spanish margin, the 22 September 1522 Almeria earthquake (with macroseismic intensity I_{EMS98} IX) destroyed the city of Almeria (with more that 2500 deaths) and an associated tsunami was reported in anonymous German chronicles from that period (e.g. Reicherter and Hübscher, 2007). The offshore Carboneras Fault has been suggested as a potential source of this event (Gràcia *et al.*, 2006; Reicherter and Hübscher, 2007), and tsunami deposits have also been found in coastal lagoons onshore (Reicherter and Becker-Heidmann, 2009). Nearby, the small Spanish town of Adra was destroyed in 1910 by a M_w6.1 earthquake with an associated tsunami (e.g. Stich *et al.*, 2003). On the basis of acoustic and seismic imaging offshore, the tectonic source of this event, the normal-dextral Adra Fault, has been found and characterized (Gràcia *et al.*, 2012).

Finally, the largest historical earthquake in Spain, the 21 March 1829 Torrevieja earthquake and tsunami (intensity MSK I_0 X, M_w6.6), which killed 800 people and destroyed 3000 homes, very possibly was associated with a large blind fault of thrust type running from onshore to offshore (Perea *et al.*, 2012). In the Ligurian Sea, the 23 February 1887 strong, damaging earthquake (Eva and Rabinovich, 1997) triggered a tsunami that inundated the coast from Genoa to Cannes.

In the Tyrrhenian Sea, the historical record of tsunamis indicates no significant potential for seismic tsunami generation. However, an important tsunamigenic zone is that of eastern Sicily and the Strait of Messina. An extreme event occurred in Tyrrhenian Calabria on 6 February 1783 (Tinti and Guidoboni, 1988) when a large earthquake induced rockfall and triggered an intensity VI tsunami at the Scilla beach. Inundation heights of 6–9 m were observed and more than 1500 lives were lost. In northeast Sicily, the 11 January 1693 earthquake that claimed about 70 000 victims caused an intensity VII tsunami. Sea level oscillations destroyed many boats and ships, while flooding was reported from Catania, Augusta, and Messina.

The 28 December 1908 earthquake (M_w7.2) was one of the largest and most lethal ever reported in Italy. Major towns in southern Italy, including Messina and Reggio Calabria, were completely destroyed with a death toll of ~60 000. A violent tsunami was generated in the Strait of Messina. At least three large waves were observed that caused many deaths and severe damage to ships, buildings, and property. Tsunami wave heights up to 13 m were observed (see the review in Tinti and Maramai (1996)), while the maximum tsunami intensity has been estimated to be X (Papadopoulos, 2009). In the Adriatic Sea, although several small-to-moderate tsunamis were reported in the north and east (Soloviev *et al.*, 2000; Maramai *et al.*, 2007; Paulatto *et al.*, 2007), the only significant seismic tsunami source is controlled by the seismicity of the Gargano promontory in the Italian coast (Tinti *et al.*, 1995). The strong, destructive earthquake of 30 July 1627, which may have been associated with the Apricena normal fault on land (Patacca and Scandone, 2004), caused a local but still powerful tsunami (Guidoboni and Tinti, 1988) with an estimated intensity of VI (Papadopoulos, 2009).

19.2.2 Eastern Mediterranean

The Hellenic Arc and Trench (HA-T) system is a major tectonic structure that produces large earthquakes, volcanic eruptions, and tsunamis. Geological and archaeological evidence collected in the Hellenistic/Roman harbour of Phalasarna, northwest Crete (Hadjidaki, 1988; Pirazzoli *et al.*, 1992; Dominey-Howes *et al.*, 1998) as well as several documentary sources (Sieberg, 1932; Guidoboni *et al.*, 1994; Ambraseys, 2009) implies that around AD 66 a tsunami flooded the southwestern part of Crete after a strong earthquake (Papadopoulos, 2011). Notable tsunamis caused by strong earthquakes in the western segment of the Hellenic Arc were also observed between the Peloponnese and Crete on 9 March 1630, 6 February 1866, and 20 September 1867 (Papadopoulos *et al.*, 2010). The large ($M_w 8.3 \pm 0.3$), subduction-related tsunamigenic earthquake of 21 July 365, ruptured the western segment of HA-T possibly on a northeast-dipping fault within the overriding plate (Shaw *et al.*, 2008). Documentary sources leave no doubt that a large area was affected by that basin-wide tsunami, which caused damage in the southwestern Peloponnese. In Alexandria many people were reportedly drowned and the tsunami reached as far as east Sicily and Dalmatia in the Adriatic Sea (Guidoboni *et al.*, 1994; Ambraseys *et al.*, 1994; Shaw *et al.*, 2008; Papadopoulos, 2011). The AD 365 tsunami event was probably the largest experienced historically in the entire Mediterranean.

In the eastern HA-T segment, Byzantine, Venetian, and Arabic sources indicate that the 8 August 1303 tsunamigenic earthquake, which ruptured between Crete and Rhodes, was also one of the largest reported historical events in the Mediterranean (e.g. Guidoboni and Comastri, 1997; Ambraseys, 2009; Papadopoulos, 2011). Similarly to the AD 365 event, the AD 1303 tsunami propagated through an extensive part of the eastern Mediterranean Sea. The sea swept violently into the capital city of Heraklion, north Crete, with such force that it destroyed buildings and killed inhabitants. In Acre, Israel, people were swept away and drowned by the huge wave. In Alexandria, the sea destroyed port facilities.

In the easternmost Hellenic Arc strong earthquakes producing tsunamis occurred in AD 148 in Rhodes and in AD 556 in Cos. LTEs were reported in Rhodes on 3 May 1481, April 1609, and 31 January 1741 (Ambraseys and Finkel, 1995; Papadopoulos *et al.*, 2007). The 1481 tsunami very probably reached the coast of Israel. Palaeotsunami coastal trenches in late Holocene sediments at Dalaman, southwest Turkey, reveal three sand layers which have been attributed to the 1303, 1481, and 1741 tsunamis (Papadopoulos *et al.*, 2012a), which makes that site one of the few places where correlations between onshore tsunami sediment deposition and historical record of tsunamigenic earthquakes can be tested.

Large tsunamis were produced by Santorini (Thera) volcano during its eruptive episodes in the seventeenth century BC and by Columbo volcano on 29 September 1650. However, purely tectonic earthquakes take place there as well. The most recent large Mediterranean tsunami occurred on 9 July 1956 after a $M_w 7.5$ crustal earthquake (Galanopoulos, 1957; Ambraseys, 1960; Papazachos *et al.*, 1985) associated with normal faulting (Papadopoulos and Pavlides, 1992; Perissoratis and Papadopoulos, 1999; Okal *et al.*, 2009; and references therein). The tsunamigenic source was about 100 km in length, located within a northeast–southwest trending basin (Papadopoulos and Pavlides, 1992). Initial estimates of the near-source wave height varied between 15 and 30 m in Amorgos and Astypalaea (Galanopoulos, 1957; Ambraseys, 1960), which are probably overestimates. A more realistic maximum height is about 12 m, as concluded from a survey by Papadopoulos *et al.*, (2005). Extensive destruction of port facilities, small and large vessels, cultivated land and property was noted, while four people were drowned. The tsunami was recorded by near-field tide-gauges in Laki, the island of Leros, and in Souda, Crete (Galanopoulos, 1957), and as far as Yafo (Jaffa), Israel (Goldshmith and Gilboa, 1985) with wave amplitudes of ~4 m, 1.5 m and 28 cm, respectively.

In the northern and eastern Aegean Sea, a few earthquake sources have historically produced powerful but very local tsunamis (see Papadopoulos, 2009, and references therein). In contrast, the rift in the Corinth Gulf, central Greece, is characterised by the highest rate of tsunami production in the European-Mediterranean region. This can be explained by the concurrence of several predisposition factors, such as high seismicity, susceptibility to coastal/submarine landsliding, and steep slopes (Papadopoulos, 2003). The tsunami rate is higher on the west side of the Gulf, which is consistent with its higher seismicity rate in comparison with the eastern side. However, the Corinth Gulf tsunamis are not the largest in the Mediterranean and are incapable of propagating outside the gulf. Large earthquakes producing local but powerful tsunamis on the western side of the Gulf were reported in 373 BC as well as on 25 May 1748, 23 August 1817, and 26 December 1861 (Schmidt, 1879, Galanopoulos, 1960; Ambraseys, 1962, 2009; Papadopoulos and Chalkis, 1984; Guidoboni *et al.*, 1994; Papadopoulos, 2003, 2009; Guidoboni and Comastri, 2005). However, the largest tsunami was generated in June 1402 at the centre of the eastern side, and caused significant damage on both eastern and western coasts (Evagelatou-Notara, 1986). The causative earthquake was of an estimated magnitude $M_s 6.8$ (Papazachos and Papazachou, 2003).

About 90 km northeast of Corinth Gulf, the northwest–southeast trending tectonic rift of the Evoikos Gulf, which terminates to the northwest at Maliakos Bay, hosts the sources of large earthquakes, such as the ones of the

20 (M_s6.4) and 27 April 1894 (M_s6.9) associated with the Atalanti normal fault (Ambraseys and Jackson, 1990). A likely predecessor was the LTE that occurred in 426 BC according to the Greek historian Thucydides (460–403? BC) and geographer Strabo (64 BC–AD 19). The strong tsunami violently inundated the coastal localities of the Maliakos Bay causing serious damage.

The tsunami frequency is relatively low in Cyprus and in the Levantine Sea region. In Cyprus two remarkable tsunamis were reported in AD 1202 and 1222 (Ambraseys, 1965, 2009). From Arabic sources, one may conclude that the very large (M_w7.6) destructive earthquake of 20 May 1202 that ruptured along the Dead Sea Transform Fault (Guidoboni and Comastri, 2005) caused a strong tsunami that violently hit the southern and eastern coasts of Cyprus with an estimated intensity VII (Fokaefs and Papadopoulos, 2007). The tsunami was also reported from the Syrian coast. However, the triggering of a tsunami by an earthquake that according to all the evidence was generated inland, is still puzzling. In the Levantine Sea the 1202 event is not unique and the generation of tsunamis from inland earthquakes has been a matter of discussion for a long time. A possible explanation is that inland earthquakes produce strong ground shaking, which triggers submarine landslides eventually generating tsunamis (e.g. Salamon *et al.*, 2009).

On 11 May 1222 a very strong earthquake struck the southwestern coast of Cyprus causing extensive damage, casualties, and a tsunami: '. . . and the sea withdrew from the shore, leaving a large part of the sea bed quite dry; and Paphos and Limassol were completely submerged by this sea-wave' (Guidoboni and Comastri, 2005). The fact that 'the sea withdrew from the shore' before Paphos and Limassol were temporarily submerged due to tsunami flooding could be interpreted as due to a dip-slip coseismic fault movement occurring offshore Paphos. If the port dried up permanently, as Rosser (1985) suggested, then ground uplift along the coast may have accompanied the earthquake. The imminent result was the triggering of a local tsunami, which shortly afterwards inundated both Paphos and Limassol.

In the left-lateral strike-slip Levantine rift, tsunami-generating earthquakes have been revealed by documentary sources (Guidoboni *et al.*, 1994; Darawcheh *et al.*, 2000; Elias *et al.*, 2007; Salamon *et al.*, 2007; Ambraseys, 2009). On 9 July 551 the sea retreated for a mile and then the tsunami flooded the coasts of Beirut and Tripoli, Lebanon, where many ships were destroyed. The tsunami was also reported from Syria and Palestine. After an earthquake on 18 January 746 or 749, which affected Palestine, Jordan, and Syria, the waves 'rose up to the sky' and destroyed cities and villages (e.g. Guidoboni *et al.*, 1994; Salamon *et al.*, 2007). That event, however, is rather puzzling with respect to the localities hit by the tsunami. This is one of the historical events that certainly deserve further examination for better understanding the tsunami source.

In the Sea of Marmara basin, the most tsunami-prone area is its eastern part due to its higher seismicity compared with its western side. Historical earthquakes triggered local tsunamis in Constantinople (Istanbul), in the Bosporus Straits, and in the western Sea of Marmara (Ambraseys, 1962, 2002; Papadopoulos and Chalkis, 1984; Guidoboni *et al.*, 1994; Altinok and Ersoy, 2000; Yalciner *et al.*, 2002; Altinok *et al.*, 2003, 2011). The large M_w7.4 Izmit earthquake of 17 August 1999, caused by right-lateral strike-slip faulting with a significant normal component, generated damaging waves up to 2.5 m high in Değirmendere, the south coast, and elsewhere in Izmit Bay (Altinok *et al.*, 1999; Yalciner *et al.*, 2002; Tinti *et al.*, 2006) with intensity of up to 4 (Papadopoulos, 2001). However, it has been found that the LTE in the Sea of Marmara was the one of 18 October 1343. In Constantinople, the tsunami grew to the height of a man, penetrated inland up to eight stadia (1.8 km), and set several boats adrift, crushing them against the land (Evagelatou-Notara, 1993; Ambraseys, 2002; Yalciner *et al.*, 2002).

An updated data compilation and critical evaluation of tsunamis that occurred from antiquity up to the present indicated that the tsunami hazard is very low but still not negligible in the Black Sea (see Papadopoulos *et al.* (2011), and references therein). Byzantine sources say that a strong sea wave, very likely a tsunami, that hit coastal settlements of Bulgaria, was caused by a very strong earthquake that occurred in AD 544 or 545 offshore from modern Varna.

19.3 Mechanisms of tsunamigenic earthquakes

Based on the historical overview we have reached the conclusion that large tsunamis caused by earthquakes can be found in 13 out of 22 tsunami zones illustrated in Fig. 19.1. From west to east the 13 zones are the Gulf of Cadiz, the margin of north Algeria, the Strait of Messina and northeastern Sicily, the western and eastern segments of the Hellenic Arc, the Cyclades in the south Aegean Sea, the southwestern side of the Cyprian Arc, offshore from the Mount Lebanon thrust including the Corinth Gulf and the northern Evoikos Gulf in central Greece, the east side of the Sea of Marmara, and offshore from Bulgaria in the Black Sea. For each of the 13 zones we selected the LTEs historically reported (Table 19.1). The selection was based on the maximum tsunami intensity, which, however,

Table 19.1. List of the 13 large tsunamigenic earthquakes selected as explained in the text. M_w = *earthquake moment-magnitude,* K = *tsunami intensity in the 12-grade scale of Papadopoulos and Imamura (2001),* h = *tsunami wave height. For data references see in the text. Earthquake magnitude was selected as follows: 1, assumed equal to the magnitude of the 27 April 1894 event (see text); 2, 6, 13 (Papadopoulos, 2011), 3 (Nikonov, 1997); 4 (Elias* et al., *2007); 5, calculated in this chapter from empirical relationships between seismic intensity and magnitude developed by Papadopoulos (2011); 7, 9 (Papazachos and Papazachou, 2003); 8, assumed equal to the magnitude of the El Asnam 1980 earthquake (see text); 10 (Papadopoulos* et al., *2012a, b); 11 (Baptista* et al., *1998); 12 (Amoruso* et al., *2002). Fault length,* L, *and width,* W, *were calculated from empirical relationships* $\log L = -1.49 + 0.47Mw$ *and* $\log W = -1.07 + 0.34Mw$ *found by Konstantinou et al. (2005) for all fault types in the Mediterranean, where* L = *length of subsurface seismic fault in kilometres,* W = *width of subsurface seismic fault in kilometres.*

n	Year	Mo	Day	*Mw*	Region	Area, *K*	*h*(cm)
1	−426			6.9±0.3	Central Greece	Maliac Bay, 8	
2	365	Jul.	21	8.3±0.3	Crete	Alexandria, 10	
3	544			7.5±0.5	Bulgarian Black Sea	Odessus, 8–9	
4	551	Jul.	09	7.5±0.3	Lebanon	Beirut, 8	
5	1222	May	11	6.5±0.3	Southwest Cyprus	Paphos, 6	
6	1303	Aug.	08	8.0±0.3	Crete	Heraklion, 10	
7	1343	Oct.	18	7.2±0.3	Sea of Marmara	Constantinople, 8	200
8	1365	Jan.	02	7.3±0.3	North Algeria	Algiers, 8	
9	1402	Jun.		6.8±0.3	Corinth Gulf	Xylokastro, 8	
10	1481	May	03	7.2±0.3	Rhodes	Rhodes, 7	
11	1755	Nov.	01	8.3±0.3	Southwest Iberia	Lisbon, 10	1500
12	1908	Dec.	28	7.2±0.1	Straits of Messina, Italy	Strait of Messina, 10	1300
13	1956	Jul.	09	7.4±0.2	Cyclades	Astypalaea, 9	1500

partly depends on the tsunami size, but also on the extent of the area of tsunami propagation as well as the magnitude of the causative earthquake. If there was more than one LTE of the same magnitude in the same area, then the most recent one was chosen. This is the case for the 1693 and 1908 events in eastern Sicily and the Strait of Messina. The 1908 event was selected as the representative LTE of the area because more abundant data were available for that event. The seismic sources of the largest tsunamis were mapped with dimensions estimated from the magnitude of the respective causative earthquake (Figs. 19.2–19.4).

From the tsunami history, the Hellenic Arc has the highest tsunami potential. This can be explained by the fact that the Hellenic subduction zone is the most active in the entire Mediterranean. An additional indicator of the tsunami potential is the reoccurrence time, *t*, of LTEs. Parameter *t* can be estimated if occurrence times of LTE predecessors (i.e. past similar tsunamigenic earthquakes) are known. Unfortunately, for all the seismic areas mapped in Figs. 19.2–19.4 no historical evidence for earthquake predecessors is available. In a few areas a little historical information exists and it is quite puzzling. For example, in the Strait of Messina a tsunami probably similar to that of 1908 was reported to hit the area in 1169 but very little is known about it. In Maliac Bay, central Greece, a tsunami quite similar to that of 426 BC was reported in AD 552 by the contemporaneous Byzantine historian Procopius. However, a critical review showed that the narration of Procopius is very probably just an imitation of the classic Greek authors who reported on the 426 BC event, thus producing a fake report of a tsunami in AD 552. The lack of LTE predecessors is a clear indication that reoccurrence times of LTEs are quite long. We assume that *t* is different in different areas but the minimum in all areas is equal to several hundred years.

Reoccurrence times of LTEs in the Mediterranean have not been well studied using geological observations and or earthquake and tsunami statistics. Numerous and accurate historical, archaeological, and geoscientific results leave very little doubt that the AD 365 earthquake and tsunami event was one of the largest ever observed in the Mediterranean Sea. Of great significance to future research is the estimation of the reoccurrence time of AD 365-type events in the western Hellenic Arc. This is very important for the assessment of seismic and tsunami hazards that a future repeat of such an event may imply for a large part of the eastern and central Mediterranean Sea basin. Papadopoulos and Kijko (1991), based on applying a maximum-likelihood method to earthquake magnitude–frequency statistics, calculated that in the HA-T the maximum

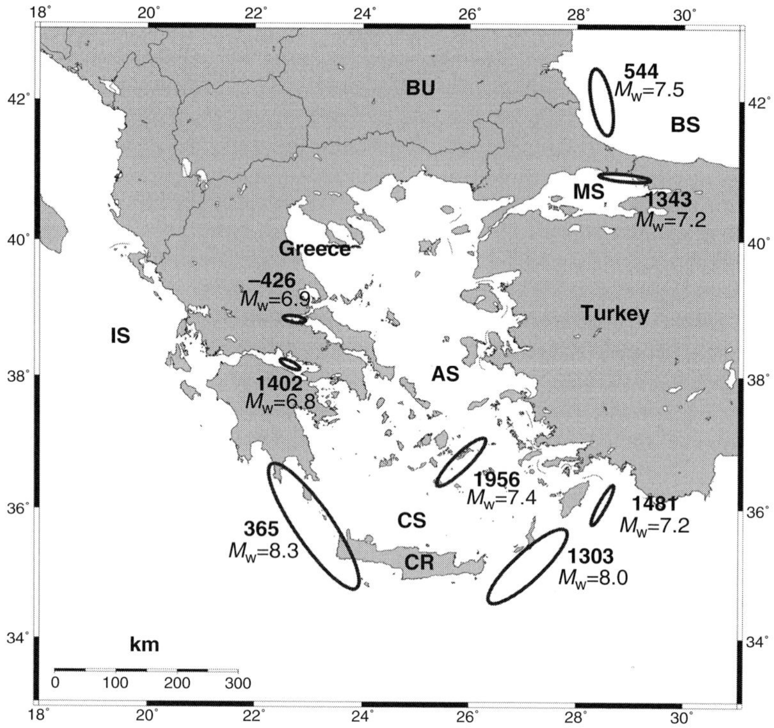

Figure 19.2. Source areas of the largest tsunamigenic earthquakes historically known in Greece, Turkey, and the surrounding regions. For explanations of the calculations see the text and Table 19.1. Key for geography: AS = Aegean Sea, BS = Black Sea, BU = Bulgaria, CR = Crete, CS = Cretan Sea, IS = Ionian Sea, MS = Sea of Marmara. Symbol key: figure near source area = year of earthquake occurrence (see Table 19.1), – means BC date, M_w = earthquake moment-magnitude.

magnitude expected is about 7.5, with a reoccurrence time of about 1000 years. More recently, Papadopoulos *et al.* (2010) using an intensity of 10 in the 12-point scale of Papadopoulos and Imamura (2001) for the AD 365 tsunami and from tsunami intensity–frequency statistics calculated that in the HA-T the reoccurrence time of such large tsunamis is about 1300 years. Shaw *et al.* (2008) suggested that measurement of the present rate of crustal shortening near Crete yields an estimate of ~5000 years for the reoccurrence time of tsunamigenic events on the single fault that generated the AD 365 event, but if the same process takes place along the entire Hellenic subduction zone, such events may occur approximately once every 800 years.

In some cases, geological data may provide estimates of the reoccurrence times of large earthquakes. For example, Wegmann (2008) presented a marine terrace chronostratigraphy extending back 125 ka for western Crete. The stratigraphy is anchored by ^{14}C dating. Terraces older than the limit of radiocarbon dating are correlated to known eustatic changes. The derived long-term coastal uplift of Crete is between 1 and 2 mm/y, the terrace decreases in elevation 26 m in a length of 55 km northwards, while an erosive notch uplifted and tilted by the AD 365 earthquake decreases 5.7 m along the same distance. Based on this evidence, Wegmann (2008) concluded that about five earthquakes of similar magnitude have occurred on this causative fault in the last 75 ka, i.e. 1 event per 15 000 years.

In the Strait of Messina the pattern of deformation in the last 125 ka has been examined from elevated coastal features, such as stratigraphic levels of marine terraces, marine sedimentary units, and landscape features, with the aim of estimating the reoccurrence time of the 1908 earthquake (Valensise and Pantosti, 1992). It was found that the 1908 earthquake was characteristic and that the reoccurrence time of a 1908-type earthquake is estimated to be from 700 to 1500 years. Thus, we may assume that the 1169 earthquake was perhaps a predecessor of the 1908 one.

Since only very few large tsunamigenic earthquakes occurred in the last century or so, our knowledge regarding the earthquake focal mechanisms is quite limited. Based on tectonic observations and fault-plane solutions, a consensus has been reached that the causative fault of the 1956 south Aegean earthquake was a northeast–southwest trending normal fault dipping either to the southeast or to the northwest (Papadopoulos and Pavlides, 1992; Perissoratis and Papadopoulos, 1999; Okal *et al.*, 2009; and references therein). As regards the 1908 Strait of Messina earthquake, no general agreement has been reached but most models favour a possibly east-dipping normal fault striking either north-northeast or north-northwest (e.g. Capuano *et al.*, 1988; Amoruso *et al.*, 2002; Neri *et al.*, 2004).

Different rupture scenarios have been proposed as the source of the great Lisbon earthquake of 1755 (e.g. Fukao,

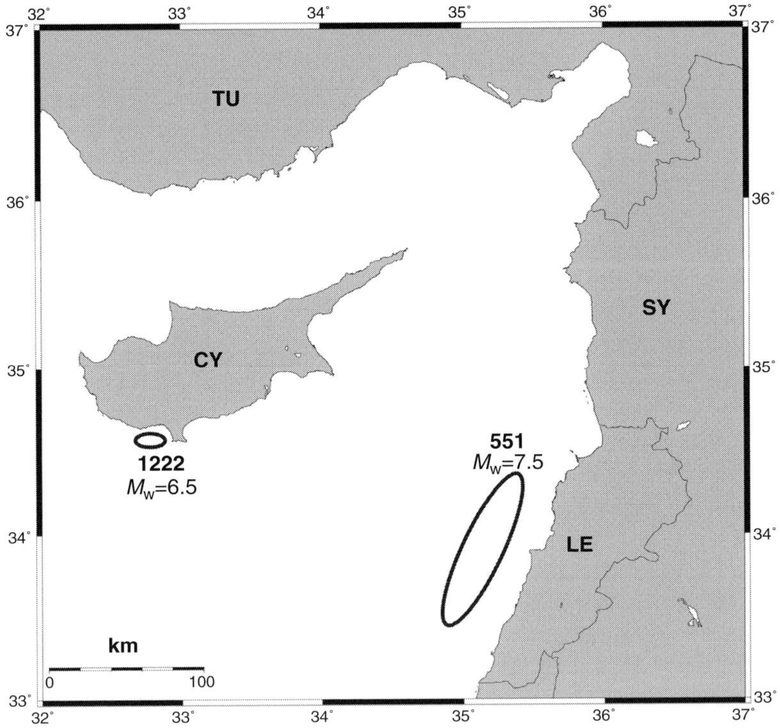

Figure 19.3. As in Fig. 19.2 for the Levantine Sea and Cyprus. Key for geography: CY = Cyprus, LE = Lebanon, SY = Syria, TU = Turkey.

1973; Johnston, 1996; Baptista *et al.*, 1998, 2003; Gutscher *et al.*, 2002; Gràcia *et al.*, 2003a,b; Terrinha *et al.*, 2003; Zitellini *et al.*, 2004, 2009; Stich *et al.*, 2007), although none of them satisfactorily accounts for the estimated magnitude of the earthquake and tsunami arrival times onshore. All the models include a thrust fault although the fault position and geometry varies between the models. However, wide-angle regional seismic profiles suggesting the presence of Early Cretaceous serpentinised mantle under the Horseshoe Abyssal Plain (Sallarès *et al.*, 2013; Martínez-Loriente *et al.*, 2013), and data from a local ocean bottom system (OBS) network installed in the area that demonstrates that the earthquake nucleated at depths of between 45 and 55 km (e.g. Geissler *et al.*, 2010; Bartolome *et al.*, 2012), shed light on the source properties and mechanism of the 1755 Lisbon earthquake. Regarding reoccurrence times, on the basis of seismically triggered turbidites found in deep basins, the regional reoccurrence interval of great earthquakes ($M_w \geq 8.0$), such as the 1755 event, is approximately 1800 years for the Holocene period (Gràcia *et al.*, 2010).

No evidence is available about the mechanism of the 1481 earthquake off Rhodes. Papadopoulos *et al.* (2012a) speculated that this earthquake probably occurred in the Rhodes Abyssal Plain where a thrust component prevails in modern seismic activity. As for the 1402 Corinth Gulf earthquake, we suggest that it may have been associated with the east-southeast–west-northwest trending Xylokastro fault on the southeast coast, given that this is the area that suffered most from the earthquake. No evidence is available about the mechanism of the 1365 Algiers and 1343 Constantinople earthquakes. The large 1303 earthquake caused severe damage in eastern Crete as well as in Rhodes (Guidoboni *et al.*, 1997). Therefore it is reasonable to consider that its source trended northeast–southwest along the HA-T between the two islands (Fig. 19.2, see review in Papadopoulos (2011)). A thrust-type mechanism, associated with the HA-T subduction zone provides efficient tsunami energy propagation towards both northeast Crete and Alexandria (e.g. Tinti *et al.*, 2005) where indeed a strong tsunami was reported in the documentary sources.

Little can said about the mechanism of the 1222 tsunamigenic earthquake in Paphos, southwest Cyprus. As noted earlier, if as according to historical information port of Paphos dried up permanently due to the earthquake, then there may have been coseismic ground uplift along the seashore, which is consistent with the thrust seismotectonics of the area (e.g. Salamon *et al.*, 2003). A submarine geophysical survey performed off Lebanon revealed an active, east-dipping, ~100–150-km-long thrust, which has been suggested as the possible source of the large 551 tsunamigenic earthquake (Elias *et al.*, 2007). However, nothing is known about the mechanism of the

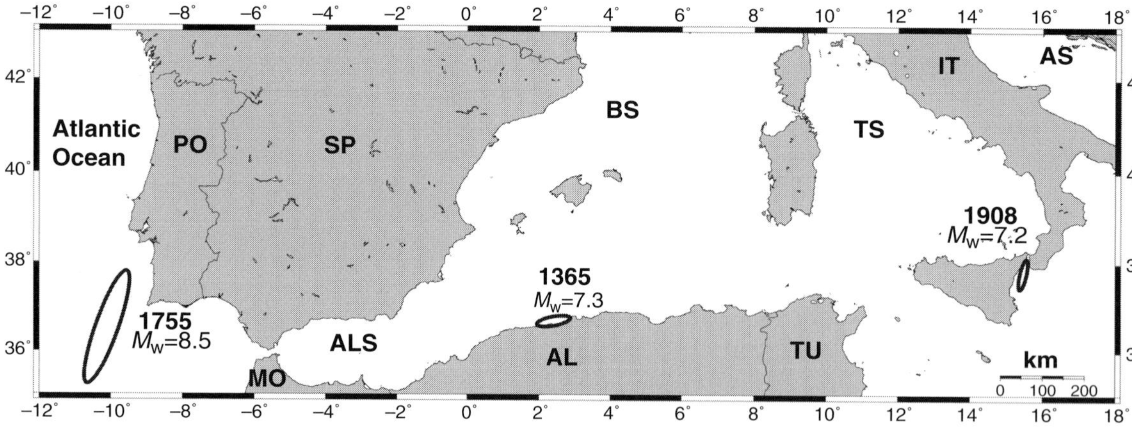

Figure 19.4. As in Fig. 19.2 for the Western Mediterranean Sea and Gulf of Cadiz. Key for geography: AL = Algeria, ALS = Alboran Sea, AS = Adriatic Sea, BS = Balearic Sea, IT = Italy, MO = Morocco, PO = Portugal, SP = Spain, TS = Tyrrhenian Sea, TU = Tunisia.

544 earthquake that supposedly occurred off modern Varna in the Bulgarian Black Sea.

The very large earthquake of 365 was an exceptional case. Coastal tectonic elevations of up to 9 m in western Crete and 3 m in Antikythira (e.g. Pirazzoli *et al.*, 1992) were carbon-dated to the calibrated calendar ranges of AD 341–439 and AD 265–491, respectively. Therefore, the 365 event possibly corresponds to the tremendous uplift which raised the Hellenistic/Roman harbour of Phalasarna in northwest Crete by ~6.6 m. However, Shaw *et al.* (2008) argued that the 365 earthquake originated not on the subduction interface beneath Crete, but on a fault with a dip of about 30° within the overriding plate. The 426 BC earthquake in the north of the Evoikos Gulf, central Greece, was very likely associated with a north-northwest–south-southeast trending normal fault, dipping towards the northeast, if the event was a predecessor of the 1894 Atalanti fault earthquake.

Our conclusion is that from all the evidence most of the largest historical tsunamigenic earthquakes in the Mediterranean region were associated with dip-slip faulting, either reverse or normal. For a few other events, however, no evidence about their focal mechanism is available. It is worth noting that in highly seismogenic areas that are dominated by strike-slip faulting, such as the Cephalonia–Lefkada fault system in the Ionian Sea and the North Aegean Sea trough, tsunami activity is historically low (Papadopoulos, 2009). In fact, large well-documented modern earthquakes, such as those of 12 August 1953 (M_s7.2) and 17 January 1983 (M_w7.0) in the Ionian Sea as well as of 18 January 1982 (M_w7.0) and 6 August 1983 (M_w 6.8) in the North Aegean Sea trough did not produce tsunamis.

19.4 Discussion

Most of the LTEs that occur in the Mediterranean are probably associated with dip-slip tectonics does not necessarily imply that the resulting tsunamis were due to coseismic dislocation. Tsunamigenic landslides could be also triggered by strong ground shaking, while a combined mechanism involving seismic dislocation and landsliding components should not be ruled out. Large magnitude earthquakes may trigger associated slope failures, such as the North Gorringe Avalanche (Lo Iacono *et al.*, 2012), increasing the overall tsunami risk. A good historical example of a tsunami triggered by a coseismic landslide was the one of 6 February 1783 in Calabria, southern Italy. Historical examples in the western Mediterranean were the submarine tsunamigenic landslides triggered in the margin of north Algeria by the large earthquakes of 1954 and 1980 (Soloviev *et al.*, 2000).

The seismic or landsliding mechanism of the instrumentally recorded tsunamis produced by the large earthquakes of 1908 and 1956 in the Strait of Messina and the south Aegean Sea, respectively, are still open to debate although little doubt exists that both earthquakes were associated with normal faulting. For both events the collected seismological, tectonic, geodetic, and tide-gauge data as well as results from numerical tsunami simulations have so far proved insufficient to distinguish between seismic, landslide, or combined source tsunami mechanisms.

In fact, Gerardi *et al.* (2008) concluded that the 1693 and 1908 tsunamis were caused by seismic dislocation sources, while the 1783 one was due to earthquake-triggered submarine landslide. In contrast, Billi *et al.* (2010) found that the three tsunami events were produced by submarine landslides. The real scenario becomes more complex if the seafloor displacement and gravitational failure are considered to be concurrent components of the tsunami generating mechanism. As regards the case of the 1956 earthquake, both a tectonic and a landslide origin of the tsunami have been proposed on the basis of the seismological and marine geophysics of the area and tested through numerical tsunami simulations (Galanopoulos, 1957; Ambraseys, 1960; Papadopoulos and Pavlides, 1992; Perissoratis and Papadopoulos, 1999; Okal *et al.*, 2009). The 'slowness' factor, θ, which is a seismological criterion characterising the seismic slip in the earthquake causative fault for source discrimination (Newman and Okal, 1998) is determined by the energy radiated by an earthquake over the low-frequency seismic moment. Using this approach for the 1908 and 1956 earthquakes, values of $\theta = -5.39$ and $\theta = -4.68$ respectively were obtained. This result implies that neither earthquake exhibits the exceptionally slow source behaviour ($\theta < -6.0$) expected for 'tsunami earthquakes' in the terminology introduced by Kanamori (1972) to characterise the 1992 Nicaragua earthquake source and other earthquakes whose tsunamis are disproportionately large with respect to their seismic moment.

Numerical modelling of tsunamis is an additional powerful tool in the effort to characterise the source and to discriminate seismic from aseismic sources. Extensive numerical simulation studies, involving both tectonic and landslide components for the tsunami generation, have been carried out with the aim of understanding the sources of the 1908 tsunami (e.g. Tinti *et al.*, 1999; Billi *et al.*, 2008, 2010; Gerardi *et al.*, 2008; Argnani *et al.*, 2009; Favali *et al.*, 2009) and the 1956 tsunami (e.g. Yalciner *et al.*, 1993; Beisel *et al.*, 2009; Okal *et al.*, 2009; Papadopoulos *et al.*, 2012b). In both cases, however, the results obtained do not allow any definitive conclusion to be reached about the source mechanism, which clearly indicates the complexity of the source and the need for further research.

Assessment of the reoccurrence of LTEs is of crucial importance for seismic and tsunami hazard assessment. We have shown that our knowledge on this issue in the Mediterranean and its connected seas is extremely limited probably because the long reoccurrence times in most of the areas considered have meant that no historical and geological traces of LTE predecessors have been left. However, strong earthquake clustering in space and time should not be ignored when considering tsunamigenic earthquake reoccurrence. For example, in the HA-T easternmost segment, it is historically documented that large earthquakes that occurred in 1481, 1609, and 1741 between Rhodes and the opposite mainland of Asia Minor caused strong tsunamis (Ambraseys, 2009). The 1481 and 1741 waves were also documented by tsunami sediment layers identified on the Dalaman coast, southwest Turkey (Papadopoulos *et al.*, 2012a). There have been no tsunamigenic earthquakes in the area since 1741. The large earthquake of 1851 triggered only a small tsunami-like sea disturbance. Prior to 1481 the only documented LTE occurred in AD 148 in the Rhodes. Consequently, the concentration of those three large events in a short narrow time interval of only 260 years in exactly the same area, if not on the same fault, provides evidence of strong earthquake and tsunami clustering which complicates the calculation of reoccurrence times and the assessment of tsunami hazard.

Of particular interest is the great similarity between the 1748 and 1817 earthquakes and the respective tsunamis in western Corinth Gulf (Papadopoulos, 2003). Having in mind the similarity between the 1481, 1609, and 1741 tsunamigenic earthquakes in Rhodes, one might think that earthquake similarity perhaps is evidence for the validity of the characteristic earthquake model, which thus introduces for further examination the idea of a characteristic tsunami model. The case of the 1908-type earthquakes and tsunamis in the Strait of Messina was examined earlier.

19.5 Conclusion

Based on an overview of the tsunamigenic earthquakes documented historically and/or geologically in the Mediterranean Sea and its connected seas in the last 2.5 millenia or so, we were able to identify the 13 out of the 22 tsunami zones illustrated in Fig. 19.1 that have generated the largest tsunamigenic earthquakes. Estimated earthquake magnitudes range from $M_w 6.5$ to about $M_w 8.5$. The geographical distribution of the earthquake source areas indicates that the highest potential for tsunami generation is concentrated in the Hellenic Arc and Trench. From the evidence, the earthquakes considered were as a rule shallow and in some instances, such as the 365 and 1303 ones in the HA-T, possibly interplate. No tsunamis generated by intermediate-depth earthquakes (focal depth $> \sim 70$ km) were reported.

An alternative indicator of the tsunami potential is the LTE repeatability. However, in the 13 areas no historical or geological evidence is available about possible predecessors of the large tsunamigenic earthquakes identified in this study. This implies that reoccurrence times of LTEs in the Mediterranean region are long, being at least equal to several hundred years. The lack of earthquake predecessors makes it difficult to check whether the characteristic earthquake model is valid or not in the tsunamigenic areas considered. A possible exception is the 1908-type earthquake in the Strait of Messina. As a consequence, we are unable to say whether the tsunamigenic earthquake magnitudes identified are the maximum ones that could be expected in each one of the areas examined.

From geotectonic data, historical descriptions, and the few instrumental records available it appears that most of the LTEs in the Mediterranean Sea and its connected seas are very probably associated with dip-slip seismic faulting, either reverse or normal. For a few areas, however, no evidence is available about the earthquake focal mechanism due to the lack of relevant observations. On the other hand, highly seismogenic areas, producing earthquakes with magnitudes exceeding 7, and dominated by strike-slip faulting, such as the Cephalonia-Lefkada transform in the Ionian Sea and the north Aegean Sea trough, are characterised by very low tsunami potential. We may assume, therefore, that dip-slip faulting is critical for the process of tsunamigenesis.

However, the association of tsunamis with dip-slip seismic faulting does not necessarily imply that the resulting tsunamis were generated by coseismic dislocation. In fact, tsunamigenic landslides could be also triggered by strong ground shaking, while a combined mechanism of seismic dislocation and landslide should not be ruled out. This is the lesson learned from the best studied cases of 1908 in the Strait of Messina and of 1956 in the south Aegean Sea. Tsunami source identification based on reliable discrimination criteria is one of the most important fields for future research in the Mediterranean Sea and its connected seas.

From the tsunami hazard point of view it is of great importance that all the 13 LTE source areas are situated very close to coastal zones, which implies that the resulting tsunamis are of the very near-field type having travel times of the first wave arrivals on the order of a few minutes. Therefore, the development of local tsunami warning systems is an important component in the tsunami risk mitigation effort.

Acknowledgments. This is a contribution to the project NEARTOWARN supported by the DG-ECHO, European Commission, contract No. 230301/2011/614039/SUB/A5 (http://www.neartowarntsunami.com).

References

Akoğlu, A., Çakır, Z., Meghraoui, M. *et al.* (2006). The 1994–2004 Al Hoceima (Morocco) earthquake sequence: Conjugate fault ruptures deduced from InSAR. *Earth and Planetary Science Letters*, **252**, 467–480.

Alasset, P. J., Hébert, H., Maouche, S., Calbini, V. and Meghraoui, M. (2006). The tsunami induced by the 2003 Zemmouri earthquake ($M_W = 6.9$, Algeria): modelling and results. *Geophysical Journal International*, **166**, 213–226.

Algerian earthquake (2003). *Tsunami Newsletter*, **35**(3), 2.

Altinok, Y. and Ersoy, Ş. (2000). Tsunamis observed on and near the Turkish coast. *Natural Hazards*, **21**, 185–205.

Altinok, Y., Alpar, B., Ersoy, Ş. and Yalçiner, A. C. (1999). Tsunami generation of the Kocaeli Earthquake (August 17th 1999) in the Izmit Bay; coastal observations, bathymetry and seismic data. *Turkish Journal of Marine Science*, **5**(3), 131–148.

Altinok, Y., Alpar, B., Özer, N. and Yaltirak, C. (2003). Şarköy-Mürefte 1912 earthquake's tsunami, extension of the associated faulting in the Marmara Sea, Turkey. *Journal of Seismology*, **7**, 329–346.

Altinok, Y., Alpar, B., Özer, N. and Aykurt, H. (2011). Revision of the tsunami catalogue affecting Turkish coasts and surrounding regions. *Natural Hazards and Earth System Science*, **11**, 273–291.

Ambraseys, N. N. (1960). The seismic sea-wave of July 1956 in the Greek Archipelago. *Journal of Geophysical Research*, **65**(4), 1257–1265.

Ambraseys, N. N. (1962). Data for the investigation of the seismic sea-waves in the Eastern Mediterranean. *Bulletin of the Seismological Society of America*, **52**, 895–913.

Ambraseys, N. N. (1965). The seismic history of Cyprus. *Revue pour l'étude des Calamites*, **40**, 25–48.

Ambraseys, N. N. (1981). The El Asnam (Algeria) earthquake of 10 October 1980; conclusions drawn from a field study. *Quarterly Journal of Engineering Geology London*, **14**, 143–148.

Ambraseys, N. N. (1982). The seismicity of North Africa: the earthquake of 1856 at Jijeli. *Bolletino Di Geofiscia Teorica ed Applicata*, **24**(93), 31–37.

Ambraseys, N. N. (2002). The seismic activity of the Marmara Sea region over the last 2000 years. *Bulletin of the Seismological Society of America*, **92**, 1–18.

Ambraseys, N. N. (2009). *Earthquakes in the Mediterranean and Middle East, A Multidisciplinary Study of Seismicity up to 1900*. Cambridge: Cambridge University Press.

Ambraseys, N. N. and Finkel, C. F. (1995). *The Seismicity of Turkey and Adjacent Areas, a Historical Review, 1500–1800*. Istanbul: EREN Ltd.

Ambraseys, N. N. and Jackson, J. A. (1990). Seismicity and associated strain of central Greece between 1890 and 1988. *Geophysical Journal International*, **101**, 663–708.

Ambraseys, N. and Vogt., J. (1988). Material for the investigation of the seismicity of the region of Algiers. *European Earthquake Engineering*, **3**, 16–29.

Ambraseys N. N., Melville, C. P. and Adams, R. D. (1994). *The Seismicity of Egypt, Arabia and the Red Sea: A Historical Review*. Cambridge: Cambridge University Press.

Amoruso, A., Crescentini, L. and Scarpa, R. (2002). Source parameters of the 1908 Messina Straits, Italy, earthquake from geodetic and seismic data. *Journal of Geophysical Research*, **107**, B4, 2080, doi:10.1029/2001JB000434.

Anonymous (1774). *The Annual Register*, **16**, 105.

Anonymous (2003). Algerian Earthquake. *Tsunami Newsletter*, **XXXV**(3), 2.

Antonopoulos, J. (1980). Data for investigation on seismic sea-waves events in the eastern Mediterranean from the Birth of

Christ to 1980 AD (six parts). *Annali di Geofisica*, **33**, 141–248.

Argnani, A. F., Chiocci, L., Tinti, S. *et al.* (2009). Comment on "On the cause of the 1908 Messina tsunami, southern Italy" by Andrea Billi et al. *Geophysical Research Letters*, **36**, L13307, doi:10.1029/2009GL037332.

Baptista, M. A. and Miranda, J. M. (2009). Revision of the Portuguese catalog of tsunamis. *Natural Hazards and Earth System Sciences*, **9**, 25–42.

Baptista, M. A., Heitor, S., Miranda, J. M., Miranda, P. and MendesVictor, L. (1998). The 1755 Lisbon tsunami: evaluation of the tsunami parameters. *Journal of Geodynamics*, **25**, 143–157.

Baptista, M. A., Miranda, J. M., Chierici, F. and Zitellini, N. (2003). New study of the 1755 earthquake source based on multichannel seismic survey data and tsunami modeling. *Natural Hazards and Earth System Sciences*, **3**, 333–340.

Bartolome, R., Gràcia, E. and Stich, D. (2012). Evidence for active strike-slip faulting along the Eurasia-Africa convergence zone: Implications for seismic hazard in the SW Iberian Margin. *Geology*, **40**, 495–498.

Beisel, S., Chubarov, L., Didenkulova, I. *et al.* (2009). The 1956 Greek tsunami recorded at Yafo, Israel, and its numerical modeling. *Journal of Geophysical Research*, **114**, C09002, doi:10.1029/ 2008JC005262.

Belabbès, S., Wicks, C., Cakir, Z. and Meghaoui, M. (2009). Rupture parameters of the 2003 Zemmouri (Mw 6.8), Algeria, earthquake from joint inversion of interferometric synthetic aperture radar, coastal uplift, and GPS. *Journal of Geophysical Research*, **114**, B03406, doi:10.1029/ 2008jb005912.

Billi, A., Funiciello, R., Minelli, L. *et al.* (2008). On the cause of the 1908 Messina tsunami, southern Italy. *Geophysical Research Letters*, **35**, doi:10.1029/2008GL033251.

Billi, A., Minelli, L., Orecchio, B. and Presti, D. (2010). Constraints to the cause of three historical tsunamis (1908, 1783, and 1693) in the Messina Straits Region, Sicily, Southern Italy. *Seismological Research Letters*, **81**, 907–915.

Braunmiller, J. and Bernardi, F. (2005). The 2003 Boumerdes, Algeria earthquake: Regional moment tensor analysis. *Geophysical Research Letters*, **32**, L06305, doi:10.1029/2004 GL022038.

Buforn, E., Bezzeghoud, M., Udias, A. and Pro, C. (2004). Seismic sources on the Iberia–African plate boundary and their tectonic implications. *Pure and Applied Geophysics*, **161**, 623–646.

Capuano, P., De Natale, G., Gasparini, P., Pingue, F. and Scarpa, R. (1988). A model for the 1908 Messina Straits (Italy) earthquake by inversion of levelling data. *Bulletin of the Seismological Society of America*, **78**, 1930–1947.

Darawcheh, R., Sbeinati, M. R., Margottini, C. and Paolini, S. (2000). The 9 July 551AD Beirut earthquake, Eastern Mediterranean region. *Journal of Earthquake Engineering*, **4**, 403–414.

DeMets, C., Gordon, R. G. and Argus, D. F. (2010). Geologically current plate motions. *Geophysical Journal International*, **181**, 1–80.

Dominey-Howes, D., Dawson, A. and Smith, D. (1998). Late Holocene coastal tectonics at Falasarna, W. Crete: a sedimentary study. In *Coastal Tectonics*, ed. Vita-Finzi, C., Special Publication No. 146. London: Geological Society of London, pp. 343–352.

Elias, A. Tapponnier, P., Singh, S. C. *et al.* (2007). Active thrusting offshore Mount Lebanon: source of the tsunamigenic AD 551 Beirut-Tripoli earthquake. *Geology*, **35**, 755–758.

Eva, C. and Rabinovich, A. B. (1997). The February 23, 1887 tsunami recorded on the Ligurian coast, western Mediterranean. *Geophysical Research Letters*, **24**, 2211–2214.

Evagelatou-Notara, F. (1986). The earthquake of 1402 in Achaia and in other regions (in Greek). In *Proceedings of Local Meeting Achaian Studies, Kalavryta 25–27 June 1983*. Athens: Peloponnisiaka, pp. 241–251.

Evagelatou-Notara, F. (1993). Earthquakes in Byzantium from the 13th to the 15th century: A historical examination (in Greek). *Parousia*, **24**, 1–179.

Favalli, M., Boschi, E., Mazzarini, F. and Pareschi, M. T. (2009). Seismic and landslide source of the 1908 Straits of Messina tsunami (Sicily, Italy). *Geophysical Research Letters*, **36**, L16304, doi:10.1029/2009GL039135.

Fokaefs, A. and Papadopoulos, G. A. (2007). Tsunami hazard in the Eastern Mediterranean: strong earthquakes and tsunamis in Cyprus and the Levantine Sea. *Natural Hazards*, **40**, 503–526.

Fukao, Y. (1973). Thrust faulting at a lithospheric plate boundary: The Portugal earthquake of 1969. *Earth and Planetary Science Letters*, **18**, 205–216.

Galanopoulos, A. G. (1957). The seismic sea-wave of 9 July 1956 (in Greek). *Praktika Academy Athens*, **32**, 90–101.

Galanopoulos, A. G. (1960). Tsunamis observed on the coasts of Greece from Antiquity to present time. *Annali di Geofisica*, **13**, 369–386.

Geissler, W. H., Matias, L., Stich, D. *et al.* (2010), Focal mechanisms for sub-crustal earthquakes in the Gulf of Cadiz from dense OBS deployment. *Geophysical Research Letters*, **37**, L18309.

Gerardi, F., Barbano, M. S., De Martini, P. M. and Pantosti, D. (2008). Discrimination of tsunami sources (earthquake versus landslide) on the basis of historical data in eastern Sicily and southern Calabria. *Bulletin of the Seismological Society of America*, **98**, 2795–2805.

Goldsmith, V. and Gilboa, M. (1985). Development of an Israeli tidal atlas and comparison with other Mediterranean tidal data. *Israeli National Oceanography Institute Report* H8/85, 28.

Gràcia, E., Dañobeitia, J. J., Vergés, J. and PARSIFALTeam (2003a). Mapping active faults offshore Portugal (36°N–38°N): Implications for seismic hazard assessment along the southwest Iberian margin. *Geology*, **31**, 83–86.

Gràcia, E., Danobeitia, J. J., Verges, J. and Bartolome, R. (2003b). Crustal architecture and tectonic evolution of the Gulf of Cadiz (SW Iberian margin) at the convergence of the Eurasian and African plates. *Tectonics*, **22**(4), 1033, doi:10.1029/2001TC901045.

Gràcia, E., Pallàs, R., Soto, J. I. *et al.* (2006). Active faulting offshore SE Spain (Alboran Sea): Implications for earthquake hazard assessment in the Southern Iberian Margin. *Earth and Planetary Science Letters*, **241**, 734–749.

Gràcia, E., Vizcaino, A., Escutia, C. (2010). Holocene earthquake record offshore Portugal (SW Iberia): Testing turbidite paleoseismology in a slow-convergence margin. *Quaternary Science Reviews*, **29**, 1156–1172.

Gràcia, E., Bartolome, R., Lo Iacono, C. *et al.* (2012). Acoustic and seismic imaging of the Adra Fault (NE Alboran Sea): in search of the source of the 1910 Adra earthquake. *Natural Hazards Land Earth and System Sciences*, **12**, 3255–3267.

Guidoboni, E. and Comastri, A. (1997). The large earthquake of 8 August 1303 in Crete: seismic scenario and tsunami in the Mediterranean area. *Journal of Seismology*, **1**, 55–72.

Guidoboni, E. and Comastri, A. (2005). *Catalogue of Earthquakes and Tsunamis in the Mediterranean Area from the 11th to the 15th century.* Rome-Bologna: INGV-SGA Italy, 1037 pp.

Guidoboni, E. and Tinti, S. (1988). *A Review of the Historical 1627 Tsunami in the Southern Adriatic. Vol. 6 of Science of Tsunami Hazards, 1.* Bologna, Italy: Universita di Bologna.

Guidoboni, E., Comastri, A. and Traina, G. (1994). *Catalogue of Ancient Earthquakes in the Mediterranean Area up to the 10th Century.* Rome-Bologna: ING-SGA.

Gutscher, M. A., Malod, J., Rehault, J.-P. *et al.* (2002). Evidence for active subduction beneath Gibraltar. *Geology*, **30**, 1071–1074.

Hadjidaki, E. (1988). Preliminary report of excavations at the harbor of Phalasarna in West Crete. *American Journal of Archaeology*, **92**, 463–479.

Harbi, A., Benouar, D. and Benhallou, H. (2003). Re-appraisal of seismicity and seismotectonics in the North-eastern Algeria. Part I: Review of historical seismicity. *Journal of Seismology*, **7**, 115–136.

Johnston, A. C. (1996). Seismic moment assessment of earthquakes in stable continental regions: III. New Madrid 1811–1812, Charleston 1886 and Lisbon 1755. *Geophysics Journal International*, **126**, 314–344.

Kanamori, H. (1972). Mechanism of tsunami earthquakes. *Physics of the Earth and Planetary Interiors*, **6**, 346–359.

Konstantinou, K. I., Papadopoulos, G. A., Fokaefs, A. and Orphanogiannaki, K. (2005). Empirical relationships between aftershock area dimensions and magnitude for earthquakes in the Mediterranean Sea region. *Tectonophysics*, **403**, 95–115.

Lo Iacono, C., Gràcia, E., Zaniboni, F. *et al.* (2012). Large, deep water slope failures: implications for landslide-generated tsunamis. *Geology*, **40**, 931–934.

Maramai, A., Graziani, L. and Tinti, S. (2007). Investigation on tsunami effects in the central Adriatic Sea during the last century- a contribution. *National Hazards and Earth System Sciences*, **7**, 15–19.

Martínez-Loriente, S., Gràcia, E., Bartolome, R. *et al.* (2013). Active deformation in old oceanic lithosphere and significance for earthquake hazard: Seismic imaging of the Coral Patch Ridge area and neighboring abyssal plains (SW Iberian Margin). G3, in press, DOI:10.1002/ggge.20173.

Meghraoui, M., Maouche, S., Chemaa, B. *et al.* (2004). Coastal uplift and thrust faulting associated with the $M_w = 6.8$ Zemmouri (Algeria) earthquake of 21 May, 2003. *Geophysical Research Letters*, **31**, L19605, doi:10.1029/2004GL020466.

Neri, G., Barberi, G., Oliva, G. and Orecchio, B. (2004). Tectonic stress and seismogenic faulting in the area of the 1908 Messina earthquake, south Italy. *Geophysical Research Letters*, **31**, L10602, doi:10.1029/2004gl019742, 2004.

Newman, A. V. and Okal, E. A. (1998). Teleseismic estimates of radiated seismic energy: The E/Mo discriminant for tsunami earthquakes. *Journal of Geophysical Research*, **103**, B11, 26885–26898.

Nikonov, A. A. (1997). Tsunamis occurrence on the coasts of the Black Sea and the Sea of Azov. *Izvestiya. Physics of the Solid Earth*, **33**, 77–87.

Okal, E. A., Synolakis, C. E., Uslu, B., Kalligeris, N. and Voukouvalas, E. (2009). The 1956 earthquake and tsunami in Amorgos, Greece. *Geophysical Journal International*, **178**, 1533–1554.

Papadopoulos, G. A. (2001). Tsunamis in the east Mediterranean: a catalogue for the area of Greece and adjacent seas. In *Proceedings of Joint IOC-IUGG International Workshop on Tsunami Risk Assessment Beyond 2000. Theory, Practice and Plans, Moscow, June 14–16, 2000.* Moscow: Russian Academy of Sciences, pp. 34–42.

Papadopoulos, G. A. (2003). Tsunami hazard in the eastern Mediterranean: Strong earthquakes and tsunamis in the Corinth Gulf, central Greece. *Natural Hazards*, **29**, 437–464.

Papadopoulos, G. A. (2009). Tsunamis, Chapter 17. In *Physical Geography of the Mediterranean*, ed. Woodward, J. Oxford: Oxford University Press, pp. 493–512.

Papadopoulos, G. A. (2011). *A Seismic History of Crete: Earthquakes and Tsunamis, 2000 B.C. – A.D. 2010.* Athens: Ocelotos Publ.

Papadopoulos, G. A. and Chalkis, B. (1984). Tsunamis observed in Greece and the surrounding area from antiquity up to the present times. *Marine Geology*, **56**, 309–317.

Papadopoulos, G. A. and Fokaefs, A. (2005). Strong tsunamis in the Mediterranean Sea: a re-evaluation. *ISET Journal of Earthquake Technology*, **42**, 159–170.

Papadopoulos, G. A. and Imamura, F. (2001). A proposal for a new tsunami intensity scale. In *Proceedings of the International Tsunami Symposium 2001, Seattle, 7–10 August 2001.* Seattle: NOAA, pp. 569–577.

Papadopoulos, G. A. and Kijko, A. (1991). Maximum likelihood estimation of earthquake hazard parameters in the Aegean area from mixed data. *Tectonophysics*, **185**, 277–294.

Papadopoulos, G. A. and Pavlides, S. (1992). The 1956 large shock in the South Aegean: Macroseismic field, faulting and neotectonics of the Amorgos Island. *Earth and Planetary Science Letters*, **113**, 383–396.

Papadopoulos, G. A., Imamura, F., Minoura, K. *et al.* (2005). The 9 July 1956 large tsunami in the South Aegean Sea: compilation of a data basis and re-evaluation. In *Proceedings of the 22nd International IUGG Tsunami Symposium, June 27–29, 2005, Chania, Crete*, eds. Papadopoulos, G. A. and Satake, K. Chania, Greece: IUGG Tsunami Commission, pp. 173–189.

Papadopoulos G. A., Daskalaki, E., Fokaefs, A. and Giraleas, N. (2007). Tsunami hazard in the eastern Mediterranean: Strong earthquakes and tsunamis in the East Hellenic Arc and Trench System. *Natural Hazards and Earth System Sciences*, **7**, 57–64.

Papadopoulos, G. A., Daskalaki, E., Fokaefs, A. and Giraleas, N. (2010). Tsunami hazard in the Eastern Mediterranean sea: strong earthquakes and tsunamis in the west Hellenic arc and trench system. *Journal of Earthquakes and Tsunamis*, **4**, 145–179.

Papadopoulos, G. A., Diakogianni, G., Fokaefs, A. and Ranguelov, B. (2011). Tsunami hazard in the Black Sea and the Azov Sea: a new tsunami catalogue. *Natural Hazards and Earth System Sciences*, **11**, 1–19, doi:10.5194/nhess-11-1-2011.

Papadopoulos, G. A. *et al.* (2013) Historic and pre-historic tsunamis in the Mediterranean and its connected seas: documentation, geological signatures, generation mechanisms and coastal impacts. *Marine Geology*, under review.

Papadopoulos, G. A., Minoura, K., Imamura, F. *et al.* (2012a). Strong earthquakes and tsunamis in the East Hellenic Arc. *Research in Geophysics*, **2**:e12, 90–99, doi:10.4081/rg.2012.e12.

Papadopoulos, G. A., Novikova, T. and Diakogianni, G. (2012b). The large tsunami of 1956 in the South Aegean Sea, Greece, revisited. Paper presented at The Deep-Sea & Sub-Seafloor Frontier Conference, Barcelona, March 11–14, 2012. Sitges: University of Barcelona.

Papazachos, B. C. and Papazachou, C. (2003). *The Earthquakes of Greece* (in Greek). Thessaloniki: Ziti Publ.

Papazachos, B. C., Koutitas, C., Hatzidimitriou, P., Karacostas, B. and Papaioannou, Ch. (1985). Source and short-distance propagation of the July 9, 1956 southern Aegean tsunami. *Marine Geology*, **65**, 343–351.

Patacca, E. and Scandone, P. (2004). The 1627 Gargano earthquake (Southern Italy): Identification and characterization of the causative fault. *Journal of Seismology*, **8**, 259–273.

Paulatto, M., Pinat, T. and Romanelli F. (2007). Tsunami hazard scenarios in the Adriatic Sea domain. *Natural Hazards and Earth System Sciences*, **7**, 309–325.

Perea, H., Gràcia, E., Alfaro, P. *et al.* (2012). Quaternary active tectonic structures in the offshore Bajo Segura basin (SE Iberian Peninsula – Mediterranean Sea). *Natural Hazards and Earth System Sciences*, **12**, 3151–3168.

Perissoratis, C. and Papadopoulos, G. A. (1999). Sediment instability and slumping in the southern Aegean Sea and the case history of the 1956 tsunami. *Marine Geology*, **161**, 287–305.

Pirazzoli, P. A., Ausseil-Badie, J., Giresse, P., Hadjidaki, E. and Arnold, M. (1992). Historical environmental changes at Phalasarna harbour, West Crete. *Geoarchaeology*, **7**, 371–392.

Reicherter, K. and Becker-Heidmann, P (2009). Tsunami deposits in the western Mediterranean: remains of the1522 Almería earthquake? *Geological Society, London, Special Publications*, **316**, 217–235.

Reicherter, K. and Hübscher, C. (2007). Evidence for a seafloor rupture of the Carboneras Fault Zone (southern Spain): Relation to the 1522 Almería earthquake? *Journal of Seismology*, **11**, 15–26.

Roger, J. and Hébert, H. (2008). The 1856 Djijelli (Algeria) earthquake and tsunami: source parameters and implications for tsunami hazard in the Balearic Islands. *Natural Hazards and Earth System Sciences*, **8**, 721–731.

Rosser, J. (1985). Excavations at Saranda Kolones, Paphos, Cyprus, 1981–1983. *Dumbarton Oaks Papers* **39**, 81–97, doi:10.2307/1291516.

Salamon, A., Hofstetter, A., Garfunkel, Z. and Ron, H. (2003). Seismotectonics of the Sinai subplate- the eastern Mediterranean region. *Geophysical Journal International*, **155**, 149–173.

Salamon, A., Rockwell, T., Ward, S. N., Guidoboni, E. and Comastri, A. (2007). Tsunami hazard evaluation of the eastern Mediterranean: historical analysis and selected modeling. *Bulletin of the Seismological Society of America*, **97**, 705–724.

Salamon, A., Rockwell, T., Guidoboni, E. and Comastri, A. (2009). A critical evaluation of tsunami records reported for the Levant coast from the second millennium BCE to the present, Israel. *Israel Journal of Earth Science*, **58**, 327–354.

Sallarès, V., Martínez-Loriente, S., Prada, M. *et al.* (2013). Seismic evidence of exhumed mantle rock basement at the Gorringe Bank and the adjacent Horseshoe and Tagus abyssal plains (SW Iberia). *Earth and Planetary Sciences Letters*, **365**, 120–131, doi:10.1016/j.epsl.2013.01.021.

Schmidt, J. F. (1879). *Studien über Erdbeben*. Leipzig: Carl Scholtze.

Shaw, B., Ambraseys, N. N., England, P. C. *et al.* (2008). Eastern Mediterranean tectonics and tsunami hazard inferred from the AD 365 earthquake. *Nature Geoscience*, **1**, 268–276.

Sieberg, A. (1932). Untersuchungen über Erdbeben und Bruchschollenbau im östlichen Miettelmeersgebiet. *Denkschr. Mediz.-Naturwiss. Gesell. Jena*, **2**, 184–224.

Soloviev, S. L. (1990). Tsunamigenic zones in the Mediterranean Sea. *Natural Hazards*, **3**, 183–202.

Soloviev, S. L., Solovieva, O. N., Go, C. N., Kim, K. S. and Shchetnikov, N. A. (2000). *Tsunamis in the Mediterranean Sea 2000 BC–2000 AD*. Advanced Natural and Technological Hazards Research, vol. 13. London: Kluwer Academic Publishers.

Stich, D., Batlló, J., Morales, J., Macià, R., and Dineva, S. (2003). Source parameters of the Mw = 6.1 1910 Adra earthquake (southern Spain). *Geophysical Journal International*, **155**, 539–546.

Stich, D., Mancilla, F., Pondrelli, S. and Morales, J. (2007). Source analysis of the February 12th 2007, Mw 6.0. Horseshoe earthquake: Implications for the 1755 Lisbon earthquake. *Geophysical Research Letters*, **34**, L11208, doi:10.1029/2007GL030012.

Terrinha, P. *et al.* (2003). Tsunamigenic seismogenic structures, neotectonics, sedimentary processes and slope instability on the southwest Portuguese Margin. *Marine Geology*, **195**, 1–4, 55–73.

Tinti, S. and Guidoboni, E. (1988). Revision of the tsunamis that occurred in 1783 in Calabria and Sicily (Italy). *Sci. Tsunami Hazards*, **6**, 17–22.

Tinti, S. and Maramai, A. (1996). Catalogue of tsunamis generated in Italy and in Cote d' Azur, France: a step towards a unified catalogue of tsunamis in Europe. *Annali di Geofisica*, **39**, 1253–1299.

Tinti, S., Maramai, A. and Favali, P. (1995). The Gargano promontory: An important Italian seismogenic-tsunamigenic area. *Marine Geology*, **122**, 227–241.

Tinti, S., Armigliato, A., Bortolucci, E. and Piatanesi, A. (1999). Identification of the source fault of the 1908 Messina earthquake through tsunami modelling. Is it a possible task? *Physics and Chemistry of the Earth (B)*, **24**, 417–421.

Tinti, S., Armigliato, A. and Bortolucci, E. (2001). Contribution of tsunami data analysis to constrain the seismic source: the case of the 1693 eastern Sicily earthquake. *Journal of Seismology*, **5**, 41–61.

Tinti, S., Maramai, A. and Graziani, L. (2004). The new catalogue of Italian tsunamis. *Natural Hazards*, **33**, 439–465.

Tinti, S., Pagnoni, G. and Zaniboni, F. (2005). The landslides and tsunamis of the 30th of December 2002 in Stromboli analysed through numerical simulations. *Bulletin of Vulcanology*, **68**, 462–479.

Tinti, S., Armigliato, A., Manucci, A., *et al*. (2006). The generating mechanisms of the August 17, 1999 Izmit bay (Turkey) tsunami: Regional (tectonic) and local (mass instabilities) causes. *Marine Geology*, **225**, 311–330.

Valensise, G. and Pantosti, D. (1992). A 125-kyr long geological record of seismic source repeatability: the Messina Straits (southern Italy) and the 1908 earthquake (Ms 7½). *Terra Nova*, **4**, 472–483.

Wegmann, K. W. (2008). Tectonic geomorphology above Mediterranean subduction zones: Northeastern Apennines of Italy and Crete, Greece. Ph.D. thesis, Earth and Environmental Science, Lehigh University, PA, pp 169.

Yalciner, A. C., Kuran, U., Akyarli, A. and Imamura, F. (1993). An investigation on the propagation of tsunamis in the Aegean Sea by mathematical modeling. In *Proceedings of the IUGG/IOC International Tsunami Symposium Wakayama, Japan: 23–27 August 1993*. Watayama, Japan: IUGG Symposium Commission, pp. 65–75.

Yalciner, A. C., Alpar, B., Altinok, Y., Özbay, I. and Imamura, F. (2002). Tsunamis in the Sea of Marmara, Historical documents for the past, models for the future. *Marine Geology*, **190**, 445–463.

Yelles-Chaouche, A. K., Roger, J., Déverchère, J. *et al*. (2009). The 1856 tsunami of Djidjelli (Eastern Algeria): Seismotectonics, modelling and hazard implications for the Algerian coast. *Pure and Applied Geophysics*, **166**, 283–300.

Zitellini, N. M., Rovere, P., Terrinha, F. *et al*. (2004). Neogene through Quaternary tectonic reactivation of SW Iberian Passive Margin, *Pure and Applied Geophysics*, **161**, 565–587.

Zitellini, N. M., Gràcia, E., Matias, L. *et al*. (2009). The quest for the Africa-Eurasia plate boundary west of the Strait of Gibraltar. *Earth and Planetary Science Letters*, **280**, 13–50.

20 Earthquake risk and risk reduction capacity building in Iran

Mohsen Ghafory-Ashtiany

20.1 Introduction

History shows that earthquakes are among the most destructive natural hazards and have caused extensive human and economic losses, especially in developing countries. Despite the major progress made over the past 30 years in earthquake engineering research, the world is still vulnerable to earthquakes. Iran, located in a region of active faults and hence high seismicity (over more than 90% of its land), is one of the highest-seismic-hazard areas in the world. For the past several decades the country has experienced devastating earthquakes, which have caused high human, social, and property losses stemming from the vulnerability of the built environment, rapid population growth, and seismically-incompatible urban development. Low-quality building construction and the vulnerability of existing structures in the country increase the risk of casualties and extensive economic losses due to the occurrence of an earthquake. The consequences of this vulnerability are shown by the 1978 Tabas, the 1990 Manjil-Rudbar, and the 2003 Bam earthquakes.

During the past two decades, especially after the destructive Manjil-Rudbar earthquake, a new initiative on the reduction of earthquake risk was started in Iran to develop and expand earthquake engineering know-how. Specialised institutions, such as the International Institute of Earthquake Engineering and Seismology (IIEES), were established. Universities and research centres expanded their earthquake engineering research and educational programmes and made a noticeable contribution to increasing the knowledge required for the country's earthquake safety. As a result, the number of earthquake engineering specialists, technical reports, scientific papers, and specialist consulting services has grown. This capacity building initiative provided an excellent platform for the enhancement of the demand for safety by developing the disaster reduction programme and providing the required know-how and expertise for hazard and risk mapping, vulnerability reduction, and public awareness and preparedness. However, despite this success in resource and knowledge capacity building to deal with earthquake risk reduction and significant scientific and technical achievements since the 1990s, there is still a wide gap between practice and knowledge. This is the biggest challenge in the earthquake engineering in Iran and elsewhere.

In this chapter after a brief overview of exposure to and risk from natural hazards in Iran with special emphasis on earthquakes, I shall discuss Iran's achievements in risk reduction during three main 'eras': (1) before the 1990 Manjil earthquake, (2) after the Manjil earthquake and before the 2003 Bam earthquake, and (3) after the Bam earthquake.

20.2 Natural hazards and risks in Iran

The Iranian plain along the Alpine–Himalayan mountain range (with an arid and vibrant four-season climate), surrounded by mountainous rims of the Alborz in the north and the Zagros in the west, forests and the Caspian Sea in the north, and dry lands and the Persian Gulf in the south, is exposed to natural hazards such as earthquakes, floods, droughts, dust and sand storms. Since 1900, almost all parts of Iran have suffered high physical, human, and economic losses due to natural disasters (as shown in Fig. 20.1). On average, a large earthquake occurs every five years, and there are two floods per year and a drought every four years. Among the natural hazards, earthquakes have caused most of the physical and social losses (Ghafory-Ashtiany, 2010a).

Being located in an active tectonic and seismic region, Iran has experienced strong earthquakes in the recent past,

Extreme Natural Hazards, Disaster Risks and Societal Implications, eds. Alik Ismail-Zadeh, Jaime Urrutia Fucugauchi, Andrzej Kijko, Kuniyoshi Takeuchi and Ilya Zaliapin. Published by Cambridge University Press.

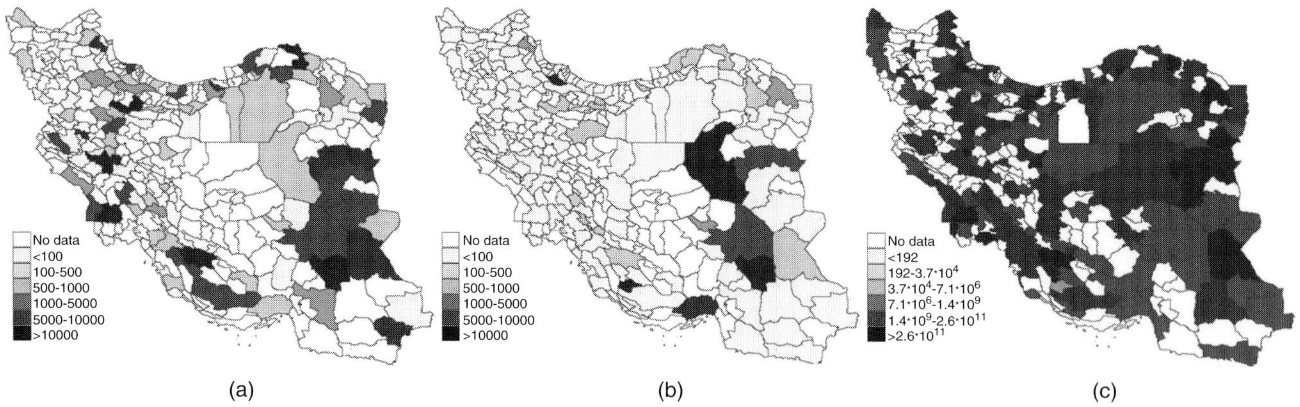

Figure 20.1. Distribution of (a) collapsed building, (b) human losses, and (c) economic losses due to natural disasters in Iran from 1900 to 2010 (source: DesInventar, http://www.desinventar.net).

namely: the Manjil-Rudbar (June 1990, magnitude $M_b = 7.2$), Darab (November 1990, $M_b = 6.6$), Bojnoord (February 1997, $M_b = 6.1$), Ardebil (February 1997, $M_b = 5.5$), Ardekul (May 1997, $M_b = 6.6$), Golbaf (March 1998, $M_b = 6.0$), Kazeroon (May 1999, $M_b = 6.0$), Bam (December 2003, $M_b = 6.6$), Zarand (February 2004, $M_s = 6.5$), Baladeh (May 2004, $M_b = 6.1$), Silakhor (March 2006, $M_L = 6.1$), and Varzeghan (August 2012, $M_w = 6.2$). These earthquakes have claimed around 80 000 lives, destroyed many towns and thousands of villages, and caused extensive economic damage. Figure 20.2(a) presents the active fault map, seismicity map, and seismic hazard map of Iran indicating that the big cities of Iran are located in a high hazard zone. The building exposure in terms of its safety and vulnerability is shown in Fig. 20.2(b), and the expected building losses and their consequences in terms of human and economic losses in Fig. 20.2(c). In all the earthquakes that have occurred in the past, human and economic losses were due to the vulnerability of the built environment and incompatible expansion of the urban and rural areas.

20.3 The Iran's disaster risk reduction (DRR) programme and its achievements

The development and implementation of DRR programmes is of great importance for natural hazard prone regions. In Iran, earthquake disasters triggered this development. Figure 20.3 shows the process and history of risk reduction in Iran, which can be divided into three main periods: before the 1990 Manjil earthquake, after the Manjil earthquake and before the 2003 Bam earthquake, and after the 2003 Bam earthquake (Ghafory-Ashtiany, 2005).

20.3.1 Before the 1990 Manjil earthquake

Before the 1990 Manjil earthquake, in the absence of a national programme to deal with earthquake disasters, most of the activities were limited to scientific research and studies, and individual initiatives like the establishment of the Iran strong ground motion network and the earthquake engineering programmes at Sharif University. In general, the view of the state to earthquake hazard and seismic safety before 1990 can be summarised as follows (Ghafory-Ashtiany, 2009):

- Very limited investment in research, education, and institutional activities. Few earthquake-related experts without an integrated programme or research agenda. No specific educational programme in universities (except the geophysics programme at Tehran University).
- Poor seismic monitoring with few analogue seismic stations, although a good coverage of strong ground motion stations.
- Limited knowledge of seismic hazard in Iran and aseismic design of structures.
- Publication of the first Building Code and enacting the first engineering law after the 1978 Tabas earthquake. Poor engineering practice and code enforcement, and almost no implementation.
- Rapid development of the country, especially in the rural area with the main objective of reconstructing of the country after the revolution (1979) and the Iran–Iraqi War (1980–1988) to answer the public demand.
- Lack of awareness about and preparedness for seismic risk. Very low political will for disaster prevention, although the disaster response (due to the eight-year war experience), i.e. the rescue and reconstruction processes, was accepted.

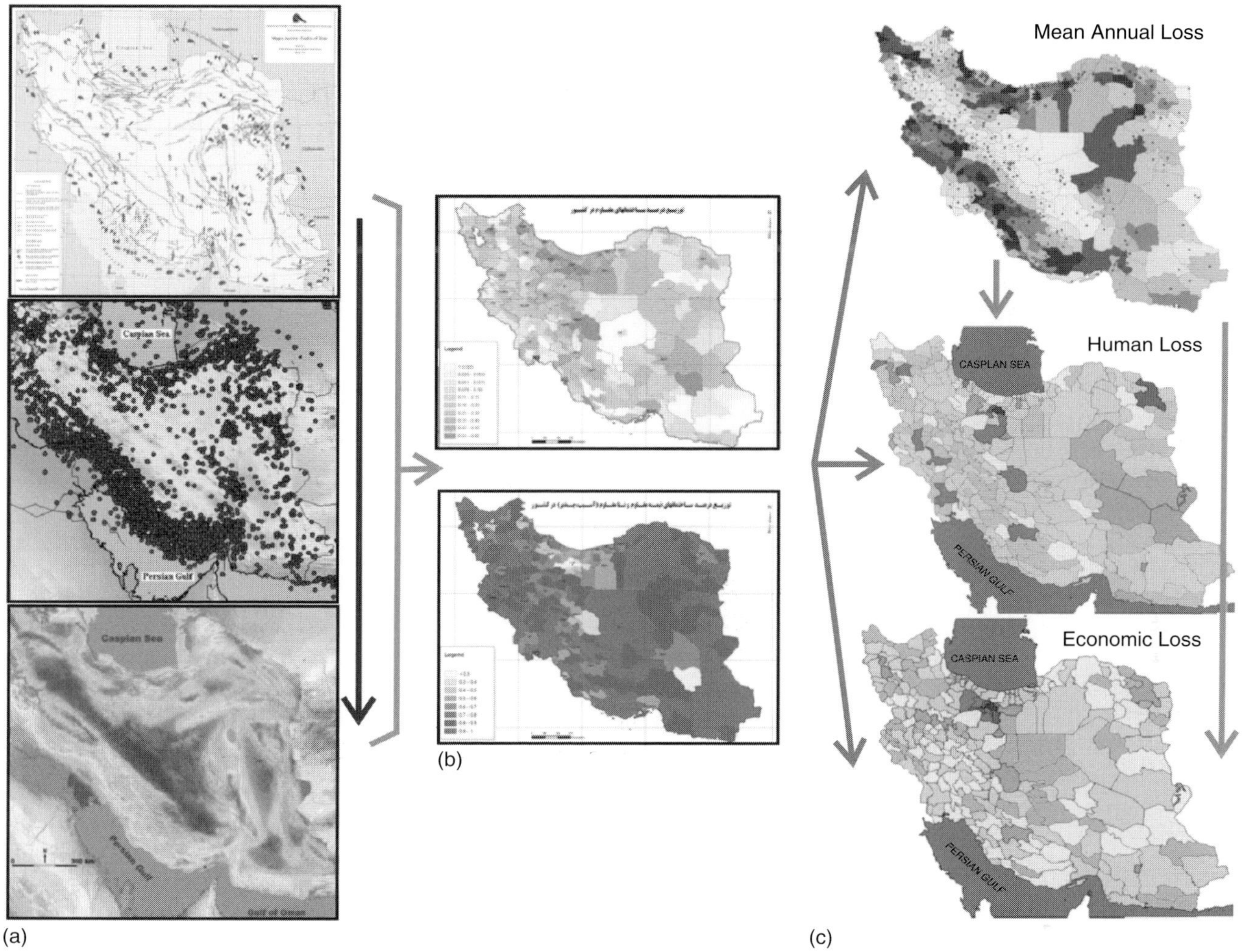

Figure 20.2. An overview of earthquake risk in Iran. (a) Hazard: active fault (top panel), seismicity (middle), and seismic hazard (bottom) maps of Iran. (b) Exposure: safe (top) and vulnerable (bottom) building exposure. (c) Distribution of the expected mean annual losses: buildings (top), humans (middle), and economic (bottom) due to the expected risk of failure of buildings. For larger colour version, see the Plates section.

20.3.2 After the Manjil earthquake (1990–2003)

The Manjil earthquake (30 June 1990, magnitude 7.2, peak ground acceleration (PGA) of 0.69 g) in northern Iran, which resulted in the loss of about 14 000 people and heavy damage to infrastructure, significantly impacted socio-economic life and became a turning point in earthquake hazard mitigation activities in the country (Ghafory-Ashtiany, 2009). After this event and at the beginning of the United Nations International Decade for Natural Disaster Reduction (IDNDR), the government decided to implement a multidisciplinary research and mitigation plan entitled 'Iran Earthquake Risk Mitigation Program (IERMP)' with the objectives of (i) increasing the scientific knowledge regarding earthquake hazard mitigation; (ii) reducing the risk of failure in different types of constructions and building safer structures; (iii) increasing public awareness of seismic hazards and promoting a collective prevention culture; and (iv) developing plans for post-earthquake actions.

The comparison of the several risk indices before and after the implementation of the programme indicates the significant (though not yet sufficient) steps that have been taken towards risk reduction, especially in the areas of research, education, building codes, construction quality, and public awareness. Notable IERMP achievements are:

- Recognition of hazard, vulnerability, and risk in Iran by decision-makers, engineers, and the public.
- Movement towards planning and long-term actions in all levels of decision-makers.

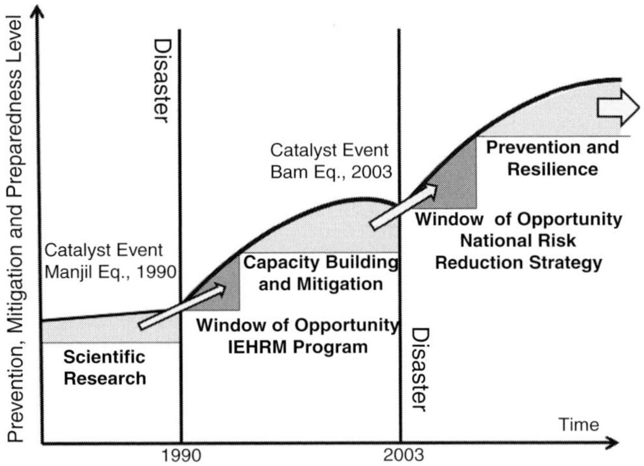

Figure 20.3. Risk reduction planning and activities in Iran (courtesy of D. Alexander).

- Improvement of technical knowledge, capabilities, research, education, engineering practice, etc.
- Improvement of preparedness and help towards the implementation of mitigation actions.
- Construction quality: due to the gradual code and law enforcement and construction controls as well as the training programme and increased awareness, the quality of construction in urban areas has improved with the trend being towards aseismic construction of public and private buildings. More incentive and enforcement is required.
- Disaster management: the attitude to disaster management had started to change towards more long-term planning and prevention.
- Strengthening of public-funded buildings and infrastructures. A more detail description of this programme is given below.

Because of the high vulnerability of public facilities, the government allocated a special fund to strengthen important public buildings. For this purpose, the first guidelines for reinforcing existing buildings were developed by IIEES based on the latest advances in earthquake engineering. The performance-based design philosophy adopted in the guidelines provides a rational framework to address different performance or damage levels under multiple seismic events. The results of this programme can be outlined as:

- Increased technical knowledge of aseismic design and construction, vulnerability assessments, and building structure reinforcement.
- Capacity building: training qualified engineers, employing consulting firms, the introduction of reinforcement projects (like reinforcement of school buildings).
- Performing pilot projects for various types of buildings (such as hospitals, fire stations, schools) demonstrating the applicability of the vulnerability analysis and reinforcement.
- Development of the guidelines for the protection of historical buildings.
- Vulnerability analysis applied to pipelines, bridges, dams, offshore structures, power plants, oil and chemical facilities, and development of the required technical capacity.
- Development of seismic vulnerability functions for some of the typical buildings in Iran.
- Development of an awareness programme for oil and chemical industry managers on earthquake risk and its reduction.
- Helping and advising authorities to develop a work plan for vulnerability analysis of important facilities.
- Improvement of the quality of the buildings and reinforcement of public-funded buildings and infrastructures.

20.3.3 After the 2003 Bam earthquake

The Bam earthquake disaster of 26 December 2003 ($M_b = 6.5$, PGA = 0.7g), which caused catastrophic destruction of the buildings and infrastructure of the ancient city of Bam, became the second turning point in the history of DRR in Iran (Ghafory-Ashtiany and Hosseini, 2008). Even though DRR efforts after the Manjil earthquake were quite successful, these efforts turned out not to be effective enough. The government needed to make more optimal use of existing know-how on earthquake risk reduction and its integration into the country's development programmes to ensure the sustainable development of the country. This event also compelled the scientific and engineering community to provide more doable, usable, and socio-economic and culturally compatible solutions to national needs. Moreover, it showed that the public at large should have been more concerned about seismic hazards and have increased its own preparedness level. Based on 13 years of the risk reduction experience, the lessons learned from the Bam earthquake, and the integrated risk management concept (shown in Fig. 20.4), the action-oriented strategy entitled 'Natural Disaster Risk Reduction and Prevention Strategy of Iran: IRAN-DRRPS' was developed, and approved by Iran's Council of Expediency and ratified by the leadership of Iran in 2005. This strategy covers main natural hazards, namely earthquakes, floods, and droughts with a special emphasis on earthquakes (Kishor *et al.*, 2004; Ghafory-Ashtiany, 2010a).

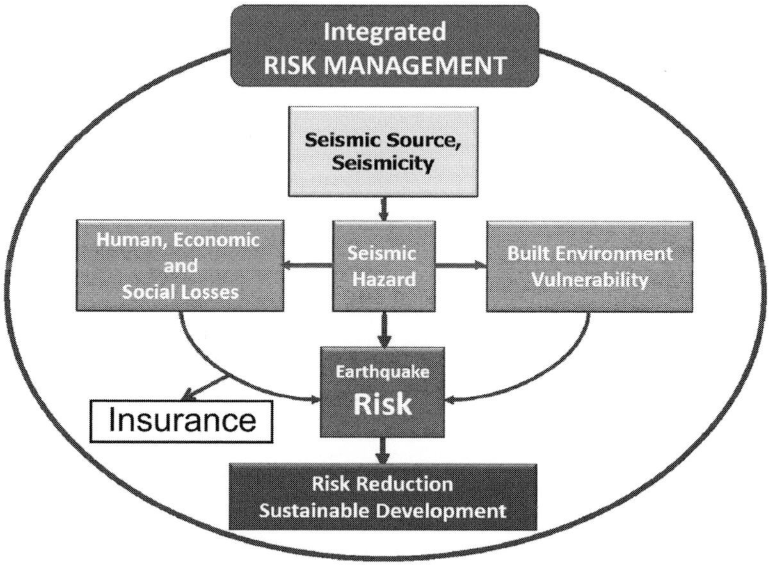

Figure 20.4. Integrated earthquake risk management concepts.

The main concepts of this strategy can be outlined as follows:

1. Increasing public awareness and promoting a collective prevention and safety culture at all levels of society through an extensive educational programme.
2. Increasing the scientific and technical know-how required for risk mitigation through increasing support and expansion of existing research institutions, specially related to earthquakes.
3. Development of a united command and management system to be well prepared for an effective disaster response (rescue and relief operations) using all types of civil and military resources.
4. Development of a comprehensive disaster information database to facilitate research and an early warning system.
5. Development of a comprehensive and scientifically sound programme for the management and implementation of effective plans for post-earthquake recovery (mental, social, and physical) as well as for rehabilitation and reconstruction of the damaged area. This answers the need for a more effective and timely reconstruction process with a special emphasis on a rehabilitation process.
6. Development of an effective system for financial recovery and compensation through insurance, financial incentives, and supporting funds.
7. Reduction of the risk for future construction and developments of urban and rural areas and an increase in the safety level through adequate urban and rural planning and management; ensuring safer buildings by full implementation of building codes with 'zero tolerance' for any type of violation.
8. Substantial risk reduction for existing structures with the main aim of saving human lives, especially schools, hospitals, public buildings, pipelines, and infrastructure, as well as providing special loans and incentives for the strengthening of the private buildings.
9. Improvement in the quality of city planning and management; and the establishment of a legal framework such that violations of the building code are treated as criminal acts and not simple civil violations.
10. Improvement of the understanding of climate change and its environmental effects, and development of an effective programme for reducing the effects of flood and drought disasters.

It should be noted that the above-mentioned strategy was implemented in executable programmes by respective organisations in Iran. The scientific communities took a multidisciplinary approach in striving to translate the current technical know-how into a doable format with simplified solutions that are socio-economically and culturally acceptable to the general public. The following examples of the implemented cases are discussed in Sections 20.4–20.7: School Safety Act; Tehran and megacities reurbanisation; Iran earthquake insurance; and technical capacity building programmes.

20.4 School Safety Act

The loss of and injury to children, a binding factor in each family, can create a long-term psychosocial disorder in society; this has been observed and experienced in Iran, China, India, and Pakistan. Thus the safety of children,

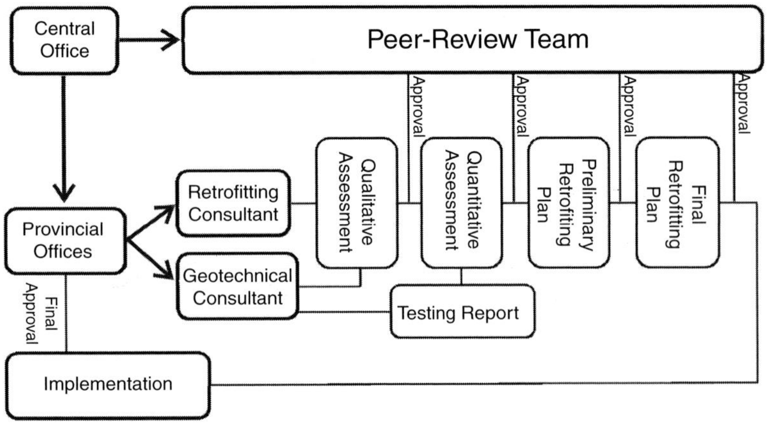

Figure 20.5. OSRDM-ME peer-review process for the school-strengthening programme.

which should have the highest priority in any risk reduction programme, has been well addressed in all of the Iran risk reduction programmes, especially through safe school buildings. Building safe schools, which began in 1992 with the enforcement of building codes, is key to having a successful earthquake education programme in schools (Parsizadeh *et al.*, 2007; Parsizadeh and Ghafory-Ashtiany, 2010). In 2000, the State Organisation of School Renovation, Development, and Mobilisation of the Ministry of Education (OSRDM-ME) implemented a more rigorous inspection and peer-review process in order to make sure schools were built according to the codes. However, this programme was not fully implemented, especially in rural areas and small towns.

With this background and experience, one of the first and most important outcomes of the new strategy was the School Safety Act passed by Iran's Parliament in 2006 with a budget of US$4 billion (actually following the IIEES campaign for school safety since 1993). This law calls for the reconstruction and strengthening of 258 000 vulnerable classrooms from (64% of the total 394 000 public classrooms in Iran), which 132 000 classrooms (35% of total) needed to be demolished and reconstructed, and 126 000 classrooms (29% of total) needed to be retrofitted. Based on Article 3 of the law, IIEES, as the technical body, and the Ministry of Urban Development and Transportation became responsible for ensuring the quality control and safety of the reconstructed schools. Also an equal budget was allocated from the regular development fund for the construction of new, safe school buildings (Mahdizadeh, 2011). This shows the awareness and concern of policy makers regarding earthquake safety in Iran.

School buildings with a low vulnerability were identified as needing to be retrofitted and upgraded to become resistant; unsafe and highly vulnerable school buildings were to be demolished and rebuilt. The highest priority was given to elementary schools and to those in the provinces with more vulnerable school buildings (mainly in the border provinces). From the beginning of this process a technical committee was formed to supervise the reinforcement process and to provide the required guidelines and technical specifications (as shown in Fig. 20.5 (Mahdizadeh, 2011)). More than 20 guidelines were developed, and many national and regional seminars and workshops were organised to address the related issues. This is a good example of building capacity within the school system.

The school buildings taxonomy can be summarised as:

- Vulnerable schools: masonry (89%), steel (8%), concrete (2%), and others (1%).
- Location with respect to seismic hazard zone: very high (15%), high (67%), moderate (16%), and low (1%).
- Year on construction: 68% of buildings was built before 1990 (the year of building code enforcement).
- Number of storeys: single storey (86%); two storeys (12%); more than two storeys (1%).

Since the start of the project, 59 376 classrooms have been reconstructed, 12 193 are under reconstruction, and 24 013 classrooms (of 30% vulnerability) are undergoing the retrofitting process, with a budget of around US$3.1 billion throughout the 30 provinces of Iran. Of the 59 376 reconstructed schools, 23 321 were elementary schools, 11 007 middle schools, and 1752 high schools. As of November 2012, the implementation process was behind schedule due to the complexity of the issues, the difficulty and underestimation of the cost for the reinforcement of the vulnerable schools, inflation, etc. Also after a detailed study, it was found that more schools require full reconstruction than reinforcement. With an average cost of US$150 per m^2 for retrofitting and US$400 per m^2 for building new classrooms and average of US$100 per m^2 for each classroom (room + auxiliary space), at least US$7.2 billion is needed to fully implement the School Safety Act (Mahdizadeh, 2011).

The implementation of this important programme along with the ongoing Earthquake Safety Education programme, which has been in operation in Iranian schools since 1994, will hopefully ensure the safety of the future generations of Iran and consequently the sustainable development of the country. Figure 20.6 shows a reconstructed school and the old one that it replaced. Some examples of the strengthening process of vulnerable school buildings using different techniques are presented in Fig. 20.7. A new strengthening technique (Fig. 20.8) for single storey masonry schools was adopted for some schools in rural areas.

20.5 Tehran and megacities reurbanisation

Tehran, the capital of Iran and one of the world's megacities with a population of more than 9 million (15% of Iran's population and 25% of Iran's urban population) and covering an area of 62 100 hectares is exposed to high seismic risk. The main reasons for its high risk are listed below (Ghafory-Ashtiany, 2010a):

1. High seismic hazard due to the existence of 15 active faults around the city; there have been more than 10 earthquakes of magnitude greater than 7 and it has the

Figure 20.6. An example of a new school that has been reconstructed through the School Safety Act and the old unsafe one that it replaced. (image source: OSRDM-ME).

Figure 20.7. Some examples of the strengthening process of vulnerable schools, using shear wall strengthening techniques (image source: OSRDM-ME).

Figure 20.8. A new strengthening process for vulnerable schools that has been adopted for some schools in rural areas (image source: OSRDM-ME).

potential for an earthquake occurrence of magnitude greater than 7.5.

2. Rapid growth and expansion of the city from a village in the twelfth century to a megacity by the end of the twentieth century; the city crosses active faults.
3. Inadequate urban planning and development.
4. Rapid growth of social, economic, cultural, and security infrastructure.
5. Rapid growth of the pipeline systems and the incompatibility of their design and construction with the level of seismic hazard causing high vulnerability, especially in the southern part of Tehran with old buildings and narrow streets.
6. Insufficient preparedness and insufficient effective disaster response capability.
7. Lack of an appropriate and integrated risk management system.

Thus the only option is to reduce the risk to an acceptable and affordable level, which can be supplemented by hazard-compatible urban planning and development and the implementation of engineering know-how.

Because one of the main reasons for the high risk is inappropriate urban growth, the reurbanisation of major cities in Iran has been emphasised in the IRAN-DRRPS as an effective and long-term approach for risk reduction. Here as a case study, the reurbanisation process for Tehran is discussed (Master Plan of Tehran, 2007). Tehran's first master plan (1970) divided the city into ten regional centres, each serving 500 000 people separated by landscaped open spaces. Each community would have had its own commercial, employment, and industrial complexes. This plan was never fully implemented effectively, and the ten regional centres became twenty municipality districts. The second master plan (1991), which took no account of natural hazards, due to rapid economic growth of the city and high land price, was subject to many changes through time. To gain control of the existing unacceptable situation, in 2007, the new comprehensive master plan for Tehran was developed by Tehran Municipality with emphasis on the city's ability to cope with natural hazards including earthquakes and floods. Based on the IIEES land use planning criteria (Table 20.1) and recommendations, the following principal safety guidelines were used in the new urban plan (Kamalian and Jafari, 2007):

1. No constructions are permitted within a minimum distance of 15 m from an active fault and a special design and permit is required for construction within 300 m of a fault in order to reduce the surface faulting effect.
2. No important structures are permitted in an area with high geotechnical hazard (landslide, liquefaction, amplification, soil subsidence, etc.), and other types of structure should be constructed with a special foundation design or the land is to be used for parks or for low human occupancy buildings such as storage.
3. Urban setting, building height, type of the use (medical, school, commercial, housing) are to be based on seismic hazard and geotechnical microzonation.
4. Important and high occupancy structures are to be built or located in the vicinity of wide streets with easy access.
5. As far as possible, pipelines are not to cross active faults or areas with high geotechnical instability.
6. Critical and service facilities are to be distributed within the city as uniformly as possible and within 2 km of each other in order to provide the easiest and open access in the time of disasters.
7. Hazardous facilities, such as petro-chemical plants, oil and gas stores, are to be surrounded by green and/or open areas; etc.

The major emphasis of the new master plan is on the sustainable development of the city, physical vulnerability reduction, and disaster management. The new master plan was put into the force in 2011. It will be a long time before Tehran changes and fits into the new plan. Similar master plans are being developed for other major cities such as Mashhad, Isfahan, Tabriz, Shiraz,

Table 20.1. *General geotechnical criteria for urban land use planning and type of buildings (courtesy of IIEES, M. Kamalian)*

Site zonation		Building type									
		Less than 6 floors (short period)				6–13 floors (mid period)			More than 13 floors (long period)		
		Intermediate importance									
		Single construction	Multiple construction	High importance	Very high importance	Intermediate importance	High importance	Very high importance	Intermediate importance	High importance	Very high importance
Surface faulting hazard zonation		Prefer. not allowed	Not allowed	Not allowed	Not allowed	Not allowed	Not allowed	Not allowed	Not allowed	Not allowed	Not allowed
Near fault hazard zonation		Conditional allowance	Conditional allowance	Conditional allowance	Prefer. not allowed	Conditional allowance	Prefer. not allowed	Prefer. not allowed	Conditional allowance	Prefer. not allowed	Prefer. not allowed
High-intensity ground motion zonation		Conditional allowance	Conditional allowance	Conditional allowance	Prefer. not allowed	Conditional allowance	Prefer. not allowed	Prefer. not allowed	Conditional allowance	Prefer. not allowed	Prefer. not allowed
Amplification hazard zonation	Short period	Allowed	Allowed	Conditional allowance	Conditional allowance	Allowed	Allowed	Allowed	Allowed	Allowed	Allowed
	Mid period	Allowed	Allowed	Allowed	Allowed	Conditional allowance	Conditional allowance	Prefer. not allowed	Allowed	Allowed	Allowed
	Long period	Allowed	Allowed	Allowed	Allowed	Allowed	Allowed	Allowed	Conditional allowance	Prefer. not allowed	Prefer. not allowed

Qom, and Bam. In the short term, the government has set a priority for reconstruction and reurbanisation of the old fabrics of the big cities by providing special funds and financial incentives.

20.6 Earthquake insurance in Iran

Despite the fact that a voluntary earthquake insurance has been offered by insurance companies in Iran since 1992, in practice the government has acted as a free insurer for the private and public owners providing reconstruction funds to those who lost their properties (Ghafory-Ashtiany, 2010b). To change this trend and use insurance as an effective risk transfer mechanism, Article 5 of the IRAN-DRRPS states the need to: 'Develop an effective system for financial recovery and compensation through insurance, financial incentives, and supporting funds, as well as promoting and regulating financial incentives for the general public and integrating earthquake risk insurance into the construction process'.

As the first step, the Iran Central Insurance Authority requested to revise the building loss index to be used as a basis for earthquake insurance premium calculation. For this purpose the mean annual loss zonation for different building typologies in Iran was evaluated and used. To do this:

1. The earthquake hazard model of Iran has been modified based on latest information on seismic sources and seismicity to obtain earthquake hazard curves for 328 city provinces of Iran.
2. An updated building inventory and its classifications has been obtained.
3. Twenty-five building types have been identified according to the construction material, year of construction, and number of storeys.
4. A fragility function has been assigned according to the available information.
5. The mean annual loss of each building type has been evaluated for each city province.

Based on the results obtained, the cities have been categorised in five hazard zones and the mean annual loss for each building type in each zone has been estimated (Ghafory-Ashtiany and Naser-Asadi, 2011). Table 20.2 shows the final results of the earthquake insurance risk index.

Table 20.2. *Earthquake insurance risk index for different building typologies and hazard zones*

				Hazard zone				
Type	Construction material	Year of construction	Number of storeys	1	2	3	4	5
1	Steel structures	Before 1991	1–3	15	17	20	22	24
2			4–7	17	20	23	25	27
3			>8	19	21	25	27	29
4		Between 1991 and 2003	1–3	7	9	10	11	15
5			4–7	6	8	9	10	13
6			>8	7	9	10.5	11.5	14
7		After 2003	1–3	3	4	5	6	7
8			4–7	4	4.5	5.5	7	8
9			>8	5	6	7	8	9
10	Concrete structures	Before 1991	1–3	14	16	19	21	23
11			4–7	15	18	21	23	25
12			>8	17	20	23	25	27
13		Between 1991 and 2003	1–3	8.5	11	12	13	16
14			4–7	6.5	7.5	8.5	9.5	12
15			>8	7	8.5	9.5	10.5	13
16		After 2003	1–3	3.5	4.5	5	6	7
17			4–7	4	5	5.5	6.5	7.5
18			>8	5.5	6	7	8	9.5
19	Semi steel structures	All times	1–2	16	18	21	23	25
20			3–5	18	20	24	26	29
21	Confined masonry	Before 1991	All buildings	12	14	17	19	21
22		1991–2003		8	10	12	14	17
23		After 2003		5	7	9	10	12
24	Adobe URM	All times	1 and 2	18	22	25	32	35
25	Wood	All times	All	2	3	4	5	6

20.7 Technical capacity building programme

In the risk reduction initiatives after the Manjil and Bam earthquakes, a special emphasis was placed on 'increasing the scientific and technical know-how required for risk mitigation' and 'technical capacity and institutional building' with the consideration of the fact that without national technical capability no DRR programme can be implemented and be sustainable (Ghafory-Ashtiany, 2005). All countries should help, share, and learn from each other's experiences, but each nation must develop its own specially tailored risk reduction programme, and this cannot be duplicated. With this fact in mind, the scientific communities started serious efforts to develop and expand earthquake engineering know-how with the objective of reducing the earthquake risk. IIEES and other universities and research centres have either initiated or expanded their earthquake engineering research and educational programme and made a major and noticeable contribution towards the development of the knowledge required for earthquake safety in Iran. As a result, the number of earthquake engineering experts, research reports, scientific papers, guidelines, codes, and specialist consulting services has increased.

To evaluate and quantify progress and development in earthquake engineering, with special emphasis on technical capacity building in Iran, a quantitative and qualitative analysis of various indicators such as the number of Institute for Scientific Information (ISI) papers, journals, students, faculties, and researchers was performed by the Iranian Earthquake Engineering Association (IEEA) in 2011 (Ghafory-Ashtiany *et al.*, 2012). The overall results of this independent study show that Iran has been successful in acquiring the required resources and knowledge. It has built the technical capacity, and has provided an excellent platform for answering the increasing demand for safety, the development of the required disaster reduction programme and has provided the required know-how and expertise for hazard and risk mapping, vulnerability reduction, and public awareness and preparedness. Scientific communities have responded very well to the demand for seismic safety in Iran, by directing civil engineering towards aseismic design and construction and by training a new generation of engineers to meet the needs of the country. The result of this effort is the existence of an earthquake engineering and seismology curriculum in many universities, and a number of earthquake related publications. For example, less than 10 papers by Iranian experts were presented at the 7th World Congress of Earthquake Engineering in 1984 compared with about 350 papers presented at the 15th Congress in 2012.

However, despite the great scientific achievements since the 1990s that can be easily adopted in engineering practice, there is a wide gap between the practice and implementation of know-how and theoretical knowledge. This is the biggest challenge for the earthquake engineering community in Iran and elsewhere.

20.8 Concluding remark

The establishment of IIEES, the timely response of scientific communities to the safety demand, the decision by Iran's government after the Manjil earthquake to implement earthquake risk reduction activities, good cooperation among the scientific communities, and the excellent effort on internal capacity building have all contributed to the move towards a seismically safe Iran. I strongly believe that the full implementation of the new strategy together with the existence of excellent human resources at all levels would ensure the country's seismic safety. Iran's strategy has been a success and can be applied to the Middle East and North Africa region and other developing countries.

References

Ghafory-Ashtiany, M. (2005). Information, knowledge and experience: International Institute of Earthquake Engineering and Seismology, an Iranian Experience. In *Know Risk*, ed. UNISDR. Leicester, UK: Tudor Rose, pp. 125–126.

Ghafory-Ashtiany, M. (2009). Iran earthquake risk reduction strategy and the International Institute of Earthquake Engineering and Seismology. In *Proceedings of the First Iran-US Workshop, Tehran, Iran*. Tehran, Iran: International Institute of Earthquake Engineering and Seismology

Ghafory-Ashtiany, M. (2010a). *Iran National Disaster Report* (in Farsi). Tehran, Iran: UNDP and Management and Planning Organization of Iran.

Ghafory-Ashtiany, M. (2010b). Earthquake insurance in Iran. In *Proceedings of the Iran National Insurance Day Conference, Tehran, Iran*. Tehran, Iran: Bimeh Markazi

Ghafory-Ashtiany, M., Aziminejad, A., Parsizadeh, F. *et al.* (2012). *Status of Earthquake Engineering in Iran*. The Iranian Earthquake Engineering Association Report, Tehran, Iran.

Ghafory-Ashtiany, M. and Hosseini, M. (2008). Post Bam recovery and reconstruction. *Natural Hazard*, **44**, 229–241.

Ghafory-Ashtiany, M. and Naser-Asadi, K. (2011). *Earthquake Insurance Index for the Buildings in Iran*. Tehran: Insurance Research Centre.

Kamalian M. and Jafari, M. K. (2007). General principal guidelines for seismically safe urban planning (in Farsi). *Sofheh Journal*, **45**, 140–164.

Kishore, K., Jha, S. K., Bagha, Z. *et al.* (2004). A United Nations strategy for support to the Government of the Islamic Republic of Iran following the Bam Earthquake of 26 December 2003. *Journal of Seismology and Earthquake Engineering*, **5**, 217–227.

Mahdizadeh, A. (2011). *Report on Retrofit Procedure of School Buildings in Islamic Republic of Iran; Reconstruction of School Offices*. Tehran, Iran: State Organisation of School Renovation, Iranian Ministry of Education.

Master Plan of Tehran (2007). *Strategic and Structural Plan for Development and Expansion of Tehran City*. Tehran, Iran: Municipality of Tehran.

Parsizadeh, F. and Ghafory-Ashtiany, M. (2010). Iran public education and awareness program and its achievements. *Disaster Prevention and Management*, **19**, 32–47.

Parsizadeh, F., Ghafory-Ashtiany, M., Heshmati, V. and Seif, A.-E. (2007). Iran's school earthquake safety initiative. *Journal of Regional Development Dialogue*, **28**, 35–46.

PART VI

Case studies: Asia and the Pacific Region

21
The Chao Phraya floods 2011

Sucharit Koontanakulvong

21.1 Introduction

The Central Plain of Thailand suffered from serious flooding during July–November 2011 (Koontanakulvong and Sriariyawat, 2011). It started in the northern area in July and reached the midstream area in September and Bangkok in October. It was the first time that flood waters had reached the inner city of Bangkok and caused flash flooding in many industrial estates in the suburbs of Bangkok. The damage from these floods was estimated to be ten times that of any previous major flooding, especially to the industrial sector. The World Bank estimated there was 1425 billion baht (US$45.7 billion) worth of economic damage and losses due to flooding, and it was recorded as the world's fourth most damaging disaster as of 2011, surpassed only by the 2011 earthquake and tsunami in Japan, the 1995 Kobe earthquake, and Hurricane Katrina in 2005 (World Bank, 2011).

The total catchment area of the Chao Phraya watershed is about 158 000 km^2. There are two main dams, i.e., Bhumibol and Sirikit dams, which control upstream runoff before it is released into the Chao Phraya Plain downstream of the main dams. The Chao Phraya Plain can be divided into an upper area with the slope of 1:2000 and a lower area with the slope of 1:8000 to 1:25 000 as shown in Fig. 21.1. I focus here on the hydrological aspects of the flood and the flood effects in the Chao Phraya Plain.

To understand clearly the 2011 flood phenomenon, the time sequence of monsoon, rainfall, runoff, and flood area from satellite images was collected and interpreted, together with dam operation data. The locations of where irrigation control structures broke down in the main streams were also collected and presented to explain the inundation phenomena in the flood plains. The collected data helped to explain what happened during the 2011 floods and gave information useful for future flood planning.

21.2 Monsoon trough and storms

During the rainy season, Thailand has annual monsoon troughs in the northern and central plains from June till October (normally two monsoons per year on average). A number of depressions or typhoons from the South China Sea occur annually during August and September and blow through the upper central plain and northeast area, although sometimes the monsoon may not pass through the country but instead the tropical storms that originate in the Pacific Ocean and the South China Sea turn into low-pressure cells and monsoon troughs. These low-pressure cells and monsoon troughs also affect the climate and rainfall pattern in Thailand.

In 2011 these storms occured before and during the floods as shown in Fig. 21.2. In May, the southwest monsoon from the Andaman Sea moved into Thailand and the Gulf of Siam and induced continuous rainfall, which was the starting point of the rainy season for the year (May 6). In June, tropical storm Haima 1104 (Fig. 21.2(a)), which originated in the upper part of the South China Sea, moved onshore in the north of Vietnam. The storm decayed to a depression by 24 June and moved to Laos as a low-pressure cell on 26 June and diminished over Nan Province. In July, Thailand was affected by a strong southwest monsoon. At the end of July, the tropical storm Nock-Ten 1108 (Fig. 21.2(b)), which originated in the middle of the South China Sea, moved onshore in the north of Vietnam, passing through Laos and decayed to a depression over Nan Province (on the eastern side of the northern part of Thailand) on 31 July. Between the beginning and middle of August, the trough passed over the northern and upper northeastern area and over the following month weakened in the central area of the country and in the Andaman Sea. In September, tropical storm Haitang 1118 (Fig. 21.2(c)) moved onshore at Wae, Vietnam, on

Extreme Natural Hazards, Disaster Risks and Societal Implications, eds. Alik Ismail-Zadeh, Jaime Urrutia Fucugauchi, Andrzej Kijko, Kuniyoshi Takeuchi and Ilya Zaliapin. Published by Cambridge University Press.

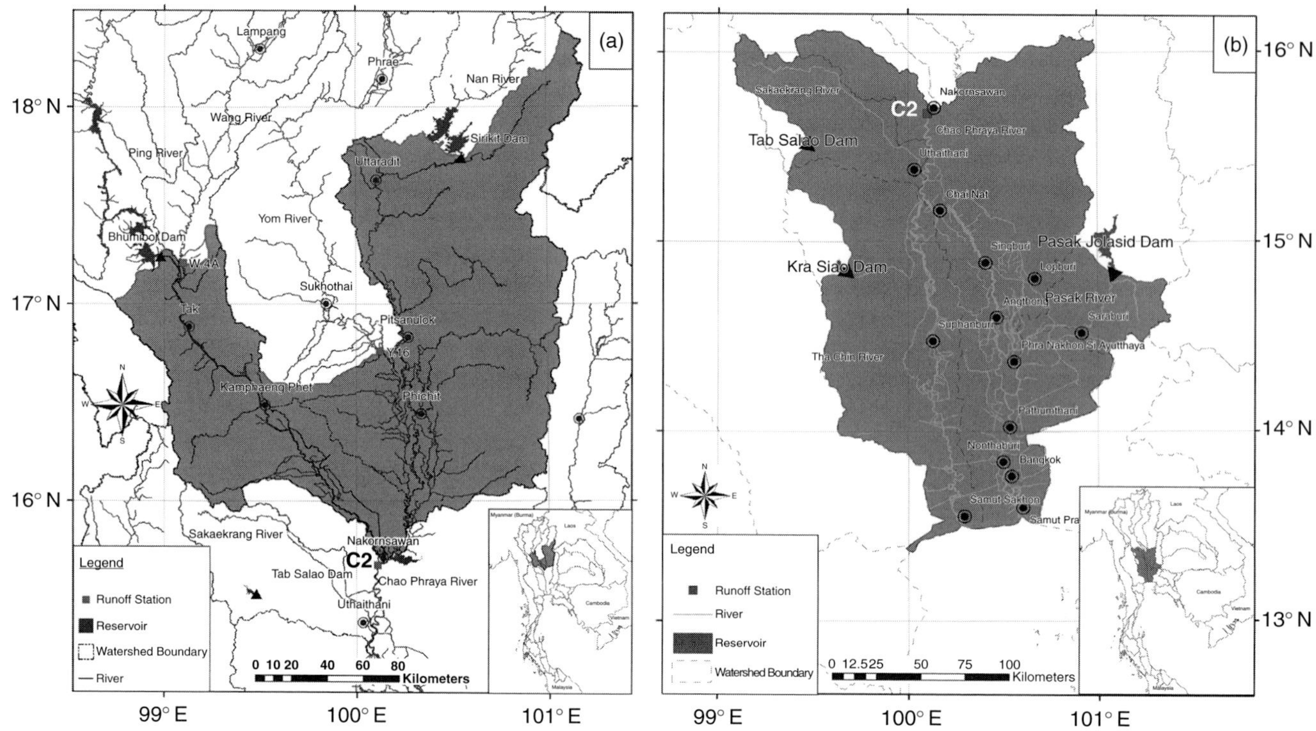

Figure 21.1. The Chao Phraya Plain ((a): upper, and (b): lower) extent and the main dams.

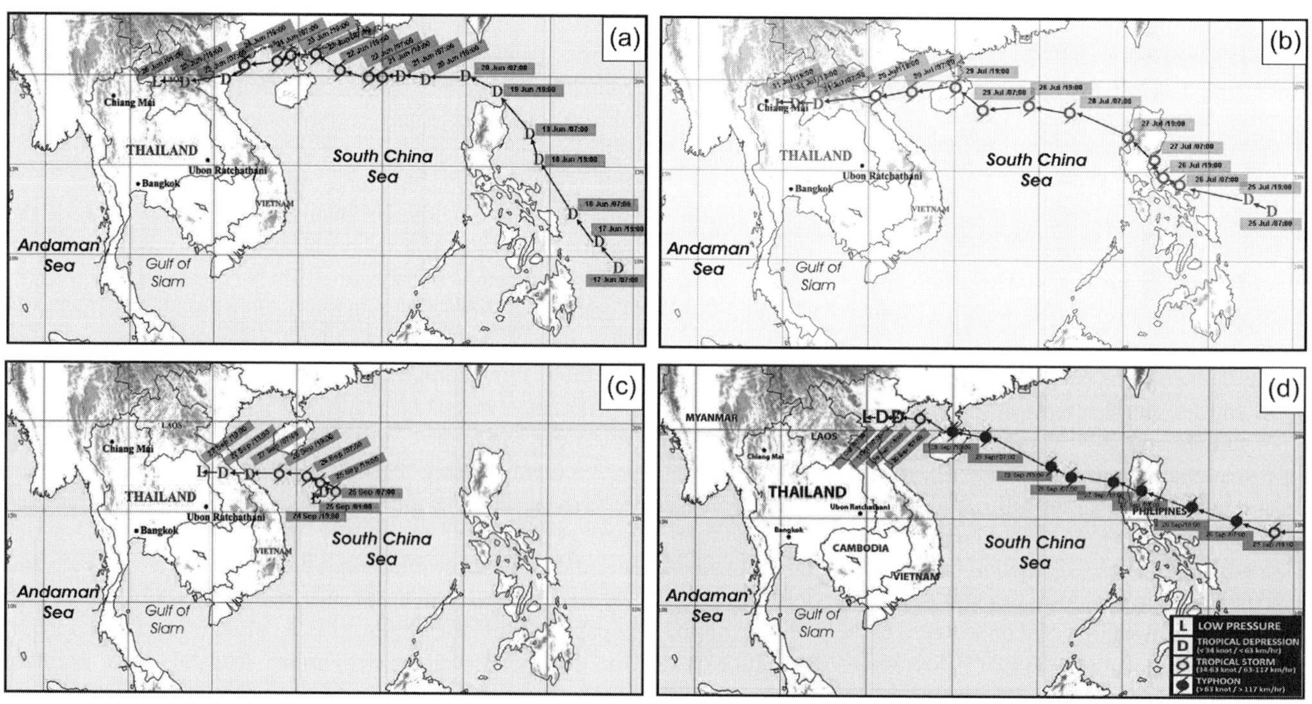

Figure 21.2. Storms movement during rainy season in 2011. (a) Haima 1104 storm track during 17–26 June 2011. (b) Nock-Ten 1108 storm track during 25–31 July 2011. (c) Haitang 1118 storm track during 25–27 September 2011. (d) Nesat 1117 storm track during 25 September–1 October 2011.

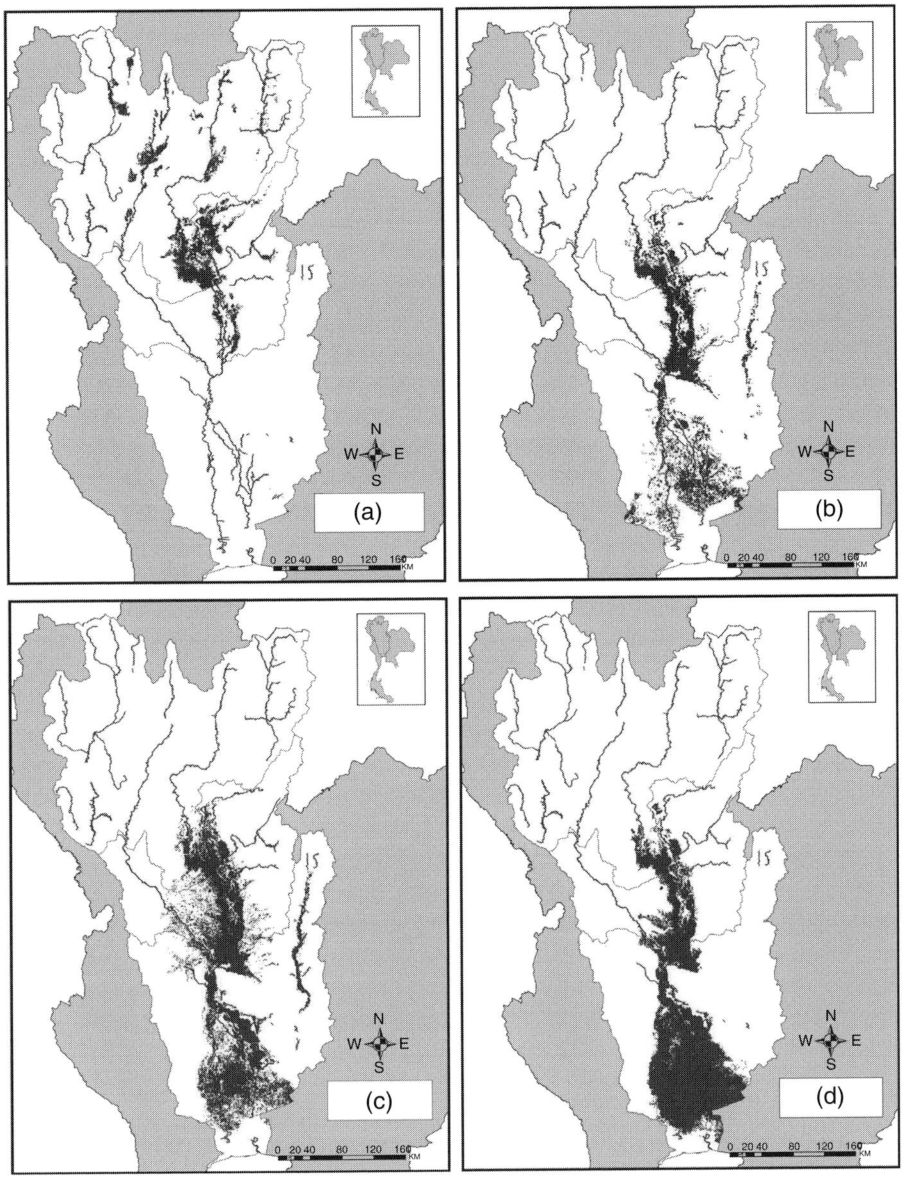

Figure 21.3. The flooded area (shown in black) in July–November 2011: (a) 2001 km^2 on 6 August; (b) 4895 km^2 on 11 September; (c) 8402 km^2 on 28 September; and (d) 14241 km^2 on 9 November.

25 September and decayed to be a depression in Laos on 27 September. The depression moved and covered northeast and north Thailand on 28 September as low-pressure cell. Typhoon Nesat 1117 (Fig. 21.2(d)) made landfall over northern Vietnam on 30 September, before decaying to a tropical storm, depression, and low-pressure cell on 1 October. The tropical storm Nakae was generated on 5 October but it decayed to a depression and low-pressure cell over Vietnam (and thus is not shown in Fig. 21.2).

During the floods 2011, Thailand was affected by a low-pressure cell over the Mataban Bay, Burma, in the early rainy season and this was followed by four tropical storms: Haima in June, Nock-Ten in July, Haitang in September, and Nesat in October. These continuous tropical storms resulted in high rainfall over a wide area of the Chao Phraya Plain.

21.3 Rainfall distribution

The guidelines for water management and operation in Thailand are basically set from the 30-year rainfall average value, thus the actual amount and distribution of rainfall in 2011 should be compared with the 30-year average to understand the flood phenomena. Based on the rainfall data from 65 stations in the central plain, the distribution of accumulated rainfall during the rainy season (from May to October 2011) was compared with the 30-year average distribution. The accumulated rainfall during the rainy season in the north of Chao Phraya Basin (the upper Yom, Nan, the main part of Ping and Wang, and the upper part of Sakaekrang Basin, see Fig. 21.1 for the location of the river basins) was higher than the 30-year rainfall average. The rainfall distribution in the Yom and Nan basins was about 70% above average from June till October especially downstream of the Sirikit dam. The widespread and high rainfall during the 2011 rainy season in the Chao Phraya Plain was a result of the continuous storms (as discussed in Section 21.2).

21.4 Flooded area

Satellite images reveal that the flooding in the Chao Phraya Plain had already started in July 2011 in northern areas like the Nan and Yom basins. Floods occurred in the north area (in the Ping River and in the confluence of the Yom and the Nan) in August. In September, the flood area moved down to Nakornsawan (station C2, see Fig. 21.1) and the central plain. In October the flood inundated the whole of the central plain, especially the low terrain area, and it lasted until November 2011. Satellite images were used to estimate the monthly flood area as shown in Fig. 21.3. The flood area in July is estimated to be 2001 km^2, with the affected area mainly in the lower Yom Basin. The flood area (4895 km^2) extended to the lower Chao Phraya in August, and the total area of flooding was about 14 241 km^2 in November.

21.5 Dam operations

Dam operation is a factor in flood enhancement or mitigation. There are five main dams in the central plain (as shown in Fig. 21.1), which affected the flood management in 2011. The dam operation data, i.e., effective storage, dam inflow, release, and actual storage, at the end of each month during the 2011 rainy season were gathered and are shown in Table 21.1. The effect of dam release on the downstream flood was analysed from inflow and release time analysis. The operation data of the Bhumibol and Sirikit dams are shown in Fig. 21.4.

The inflow to the Bhumibol dam (Fig. 21.4(a)) was higher than the flows recorded in the past, especially in the late rainy season (September and October). The peak inflows to the Bhumibol dam reservoir matched well with the storm events. As a result, the stored volume of water stored in the dam increased rapidly, and the reservoir water level exceeded the upper rule curve level, which prompted a rapid release of water in October to ensure the safety of the dam.

Table 21.1. *Dam operation data during May–October 2011 (in units of 10^6 m^3)*

Dam	Effective storage	Inflow	Release	Actual storage
May 2011				
Bhumibol	9,662	1010.08	112.51	6459.54
Sirikit	6,660	511.8	212.21	4877.29
Pasak Cholasid	785	123.87	39.03	316.61
Kra-sioa	200	82.1	45.06	174.45
Tap Salao	143	12.74	0.08	53.61
June 2011				
Bhumibol	9,662	908.42	98.1	7327.77
Sirikit	6,660	1078.95	187.64	5206.27
Pasak Cholasid	785	386.57	276.87	480.77
Kra-sioa	200	66.05	59.77	214.47
Tap Salao	143	16.42	6.84	70.93
July 2011				
Bhumibol	9,662	991.65	176	8090.94
Sirikit	6,660	2044.8	473.64	6730.48
Pasak Cholasid	785	320.66	496.33	389.61
Kra-sioa	200	24.04	55.262	184.55
Tap Salao	143	3.87	6.3	62.97
August 2011				
Bhumibol	9,662	2610.57	625.97	9583.81
Sirikit	6,660	3194.78	1559.31	8355.55
Pasak Cholasid	785	782.01	554.62	355.71
Kra-sioa	200	31.47	45.73	157.00
Tap Salao	143	9.03	0.4	67.71
September 2011				
Bhumibol	9,662	2842.51	776.03	11 487.17
Sirikit	6,660	2340.88	1830.5	9118.27
Pasak Cholasid	785	1749.23	1093.3	737.43
Kra-sioa	200	92.07	38.62	172.83
Tap Salao	143	51.28	0	96.00
October 2011				
Bhumibol	9,662	2895.05	1987.81	13 286.87
Sirikit	6,660	1062.78	945.77	9433.09
Pasak Cholasid	785	1530.35	1356.03	1050.32
Kra-sioa	200	113.97	36.67	228.77
Tap Salao	143	77.1	32.31	150.97

In the case of the Sirikit dam (Fig. 21.4(b)), the inflow to the dam reservoir and the storage level increased continuously from June until October. The level exceeded the upper rule curve level in the middle of July. The inflow peak for the dam matches well the storms that hit the region. The dam released a lot of water during August and September, which affected the flood situation downstream, i.e., the lower Yom area, even though the operators tried not to release water during this period.

Other small dams in the plain also contributed to the flooding although only in a small way. In the case of the

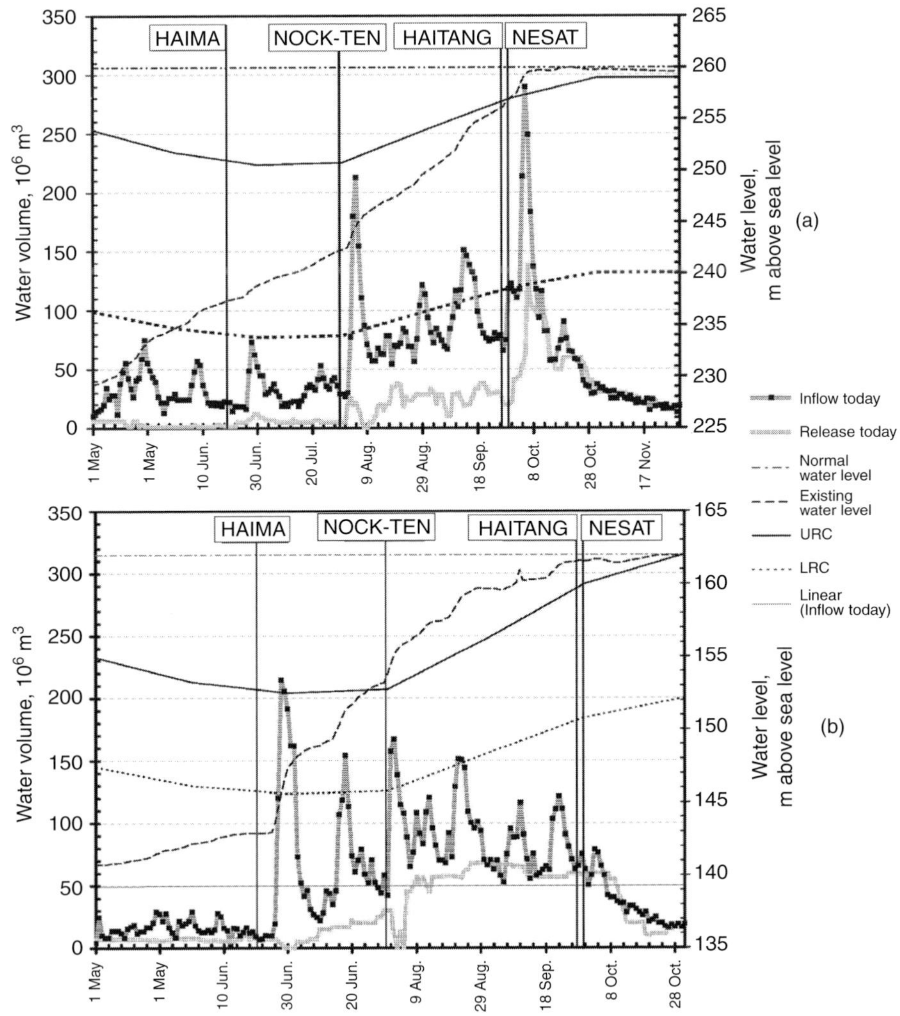

Figure 21.4. Inflow and water level of (a) the Bhumibol dam and (b) the Sirikit dam during the 2011 starts. URC: Upper Rule Curve; LRC: Lower Rule Curve.

Pasak dam, the inflow started in May and lasted till the end of October with the peak in October. The storage level was above the upper rule curve level for May–October, and the dam released a lot of water in September and October with the peak release in October. In the case of the Kra-sioa dam, the inflow was high in September and October with its peak in October. The water storage was still within the rule curve range with the high release from the Pasak dam in June. In the case of the Tap Salao dam, the inflow was high in September and October with the peak in October. The dam released water in October.

In general, the peak inflow for the dams matched the monsoon events and the dams released a lot of water in the late rainy season, which coincided with the peak runoff from rainfall on the plain and caused more serious floods in the downstream area. The continuous intense rainfall from August made dam operation during September–October quite difficult with regard to making decisions about water release.

21.6 Flood hydrographs

The runoff from the upper Chao Phraya watershed comprises flows from four tributaries (Ping, Wang, Yom, Nan) which converge in Nakornsawan to form the Chao Phraya River (station C2). The flood hydrographs at C2 for four flood years including 2011 are shown in Fig. 21.5. The floods in the lower central plain (downstream of C2) are controlled by the river flow and bypass canals (there are two bypass canals, i.e., Tachin canal to the west and Chainart-Pasak canal to the east) at the Chainart Barrage and a flood will occur downstream of the Chainart Barrage if the rate of flow at C2 is above 3600 m^3 s^{-1}. It can be seen in Fig. 21.5 that the rates of flow in the past exceeded the level of 3600 m^3 s^{-1} for about three–four weeks. The flood hydrograph peak in 2011 was 4600 m^3 s^{-1}, which is not high compared to that of the year 2006, when floods occurred in the irrigation area to the north of Ayuthaya, but not in the Bangkok area. The flow remained higher than 3600 m^3 s^{-1}

continuously for about 2 months in 2011, which is the longest duration of high flow recorded. This has been attributed to the early rainfall and continuous typhoon hits during the rainy season as explained in Section 21.2.

21.7 Irrigation control structures damaged

Another main factor for flooding is the breakdown of irrigation infrastructure along the river. The high water level in the Chao Phraya River, due to the runoff from upstream (recorded at station C2 as described in Section 21.6), caused overflow and embankment damage at many gates along the river as shown in Fig. 21.6. The chronology of the damage is presented in Fig. 21.6. The runoff record at C2 on 14 September was about 4000 $m^3\ s^{-1}$ which was higher than the control peak of 3600 $m^3\ s^{-1}$. In early October, another 11 gates on the west side of Patumthani Province, downstream of Chao Phraya River, were also damaged due to the high water level in the river. In total, more than 30 water gates were destroyed during the flood. This caused inundation in the low terrain area (mainly paddy fields) due to the overflow water from the river. Since all monitoring stations were located along the river, these overflow flood water volumes could not be accurately monitored and estimated. The overflow water inundated the flood plain over the low terrain and contributed to river flooding in many municipal areas and industrial estates.

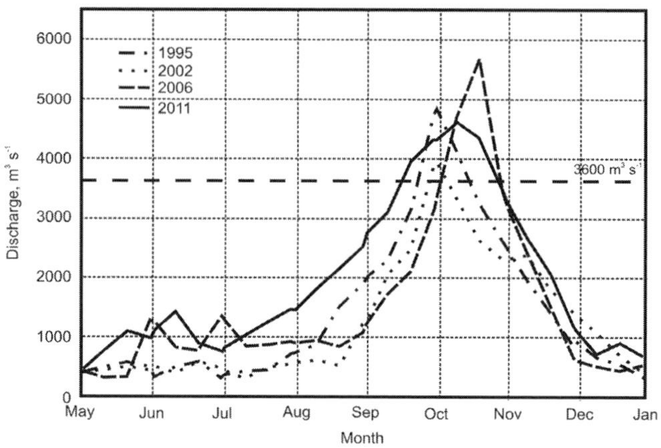

Figure 21.5. Comparison of flood hydrographs at Nakornsawan (station C2) in 1995, 2002, 2006, and 2011.

21.8 Conclusions

From the hydrological investigations, the floods of 2011 were caused by a number of consecutive monsoon hits which showed a different pattern of rainfall compared with those in the past, i.e., the severity of the typhoons exceeded expectation and the typhoons were earlier than usual, arriving in June. Dam release also affected the flood situation because it occurred during September and October. With classical water management rules (dam and diversions) based on the 30-year average rainfall and the existing physical limitations (i.e., drainage capacity, land use change), water control could not reduce the peak flow

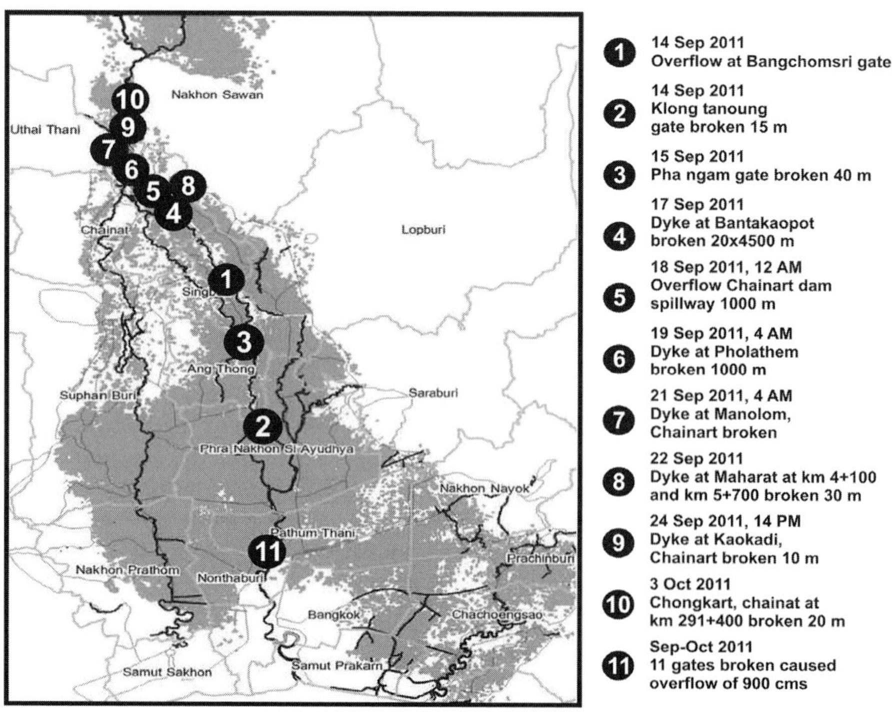

Figure 21.6. Damaged structures along the Chao Phraya River during the floods.

and there were a number of structural (dykes, levees, diversion works) breakdowns. This caused the water to overflow into the flood plain and inundate the low terrain in the flood plain. The floods lasted for more than two months in many areas and caused damage to the socio-economics of the country.

There are other issues which have not been mentioned here but which affected the flooding and flood rescue, i.e., an inadequate flood prediction system, an inadequate real time flood warning system, an unsystematic flood defence system, poor facility maintenance, social and political involvement etc., which need further considerations for future flood planning. In addition to the development of main structures (dams, dyke, floodway etc.) by the government of Thailand, the introduction of modern technologies for weather prediction, a flood-warning system (RTG-SFCWRM, 2012,a,b; Koontanakulvong, 2012), the impact of climate change (which may induce even more extreme weather conditions), and community (who are normally affected directly) involvement should be considered in long-term mitigation planning of future flood management. Timely preventive measures against flood disasters would reduce flood losses considerably.

Acknowledgment. The data used in the study were collected from various agencies, i.e., rainfall data from Thai Meteorological Department and Royal Irrigation Department, runoff data from Royal Irrigation Department, dam operation data from Electricity Generating Authority. The rainfall runoff estimation procedure and coefficients used in this study were based on the assessment study report on floods 2006 (Chulalongkorn University, 2007). The author thanks the mentioned agencies for providing data and comments.

References

Chulalongkorn University (2007). Flood 2006 Assessment Report, submitted to the Royal Irrigation Department, September 2007 (in Thai).

Koontanakulvong, S. and Sriariyawat A. (2011). Thailand flood 2011–causes and future scenario. *Proceedings of USMCA 2011*, Chiengmai, Thailand, October 2011.

Koontanakulvong, S. (2012). Thailand floods 2011 – causes and future management system. *Proceedings SSMS2012*, Taiwan, May 2012.

Royal Thai Government – SFCWRM (2012a). Immediate Term Flood Mitigation Report, January 2012 (in Thai).

Royal Thai Government – SFCWRM (2012b). Long Term Strategy for Flood Prevention and Mitigation in Chao Phraya Basin, 14 January, 2012.

World Bank (2011). The World Bank Supports Thailand's Post-Floods Recovery Effect, 13 December 2011 (retrieved on 25 January 2012).

22

Environmental risk management in Australia: natural hazards

Tom Beer

22.1 Introduction

There are two aspects to environmental risk management. The first is the protection of the environment from human-induced activity. The second is the protection of humans from the damage that the environment may cause – especially as a result of the manifestation of extreme natural hazards.

This chapter, though dealing cursorily with protection of the natural environment, will focus on the second aspect that is specifically manifest in the question: how can we protect people from extreme natural hazards? This expression of the question emphasises that the problem is one that links science and society. Knowledge of the extreme natural hazard is an outcome of focussed scientific knowledge. Protection is an outcome of social organisation, community structures, communication, and preparedness.

The two are, of course, related. Social organisation develops as a result of the occurrence of past events and disasters. Scientific knowledge, especially when in the form of forecasts, provides the information needed to mobilise societal response.

22.1.1 State of the Environment reporting

The success of the Australian society in protecting the environment from human-induced activity is reported every five years in a document known as the State of the Environment Report. The latest State of the Environment Report was issued in 2011 and is abbreviated as SoE2011.[1] SoE2011 has volumes that deal with the following aspects of the Australian environment: atmosphere, inland water, land, marine environment, Antarctic environment, biodiversity, heritage, built environment, and coasts.

The framework on which SoE reporting is based is the pressure–state–response framework originally promulgated by the Organisation for Economic Co-operation and Development (OECD). This has been modified to a process whereby the development of the SoE Report involves a range of data, information, and conceptual tasks, including:

- defining the key environmental issues to be addressed, which are:
  - descriptive of problems or challenges related to the environment,
  - significant at a national scale, and
  - important for current and/or future generations;
- developing relevant indicators, which describe:
  - driving forces and pressures giving rise to an issue,
  - environmental conditions or states associated with an issue,
  - impacts of the issue on natural and human systems, and
  - responses by human society to the issue;
- identifying and acquiring appropriate data, information, and research to underpin the indicators, seeking a balance between relevance, practicality, and credibility; and
- developing supplementary materials on thematic, integrative, and emerging issues and developing and managing an independent review of the report.

22.1.2 Risk

Operational experience has found that the pressure–state–response framework, though adequate to deal with protection of the environment from the damage that humans may cause, is not well suited to dealing with the protection of humans from the damage that the environment may cause. Risk-based frameworks (Beer, 2003) have proved more useful and are now regularly used.

Extreme Natural Hazards, Disaster Risks and Societal Implications, eds. Alik Ismail-Zadeh, Jaime Urrutia Fucugauchi, Andrzej Kijko, Kuniyoshi Takeuchi and Ilya Zaliapin. Published by Cambridge University Press.

Following a NATO Advanced Study Meeting held in Budapest, Hungary, the International Union of Geodesy and Geophysics and Euroscience issued the Budapest Manifesto, which provided guidance on undertaking studies of risks of the uncertainty of nature (Beer and Ismail-Zadeh, 2003). The general concepts of the Budapest Manifesto were incorporated into the Australia/New Zealand Standard on Risk, known at AS/NZS 4360, which was issued in 2004 and eventually incorporated into the International Standard ISO 31000.

The key concept is that any evaluation of a risk needs to examine two variables – the likelihood of the occurrence of an event, and the consequences of the event should it occur. The standard uses the term likelihood, rather than the probability, as a means of emphasising that the likelihood can be a qualitative estimate as well as a quantitative measure of the probability.

Beer (2003) has given some of the history of the development of AS/NZS 4360 and pointed out that different communities use different risk frameworks depending on whether the community is dealing with episodic or chronic events, and whether the community is concerned with the effects on individuals or on populations.

A hazard is an event or occurrence that may generate or pose a risk if it interacts with vulnerable entities. The natural hazards community examines episodic events in which populations are affected. Though the natural hazards community agrees on the importance of likelihood and consequences, in practice the community has tended to deal with multi-variable measures of risk that combine concepts of probability of occurrence, the severity of the hazard measured in physical units, the exposure to the hazard that is a time-based measure of the event, and the vulnerability of the population.

22.1.3 Likelihood and probability

The use of the term likelihood by the risk community rather than the term probability is, as indicated above, generally done because of the expectation that a probability must be a numerical value that lies between 0 and 1. If the expression of the probability is derived by subjective means rather than objective means or if the probability is given in terms other than a numerical value then, for example, rather than stating that the probability is high the risk community would state that the likelihood is high.

There is also a more subtle reason for the risk community to prefer the term likelihood. The likelihood function is well defined in theoretical statistics (Cox and Hinkley, 1974) as the function that defines how the parameters of a particular random variable vary for a fixed value of the variable. This use of likelihood permits an analyst to draw subjective inference from observational data by using Bayes' theorem (Vose, 2000), and to do so in a quantitative manner. This approach (the Bayesian approach) is rejected by those statisticians who espouse what is called the frequentist approach – namely that science should be objective and that observations are a basis for possibly rejecting a hypothesis, but in no case for supporting a hypothesis thus making inductive inference impossible. Thus in quantitative risk analysis likelihood is used as a synonym for a generalised probability. Vose (2000, p. 171) claims that 'since likelihood functions are calculating probabilities (or probability densities) they can be combined in the same way as we usually do in probability calculus'. This claim is misleading as shown by Barnard and Sprott (1983, p. 640) who give an example in which the addition of likelihoods is shown to be a weighted mean rather than a simple sum.

22.2 Australian natural hazards

Australia is dominated by hydro-meteorological hazards. Volcanic eruptions have not occurred during the history of European settlement. The last volcanic eruption on the Australian mainland was that of Mount Gambier about 4500 years ago, which indicates that the volcano is dormant rather than extinct. The Australian continent is completely within the Australian plate, so that even though earthquakes occur, they are intra-plate earthquakes whose magnitude is far less than those of earthquakes resulting from tectonic plate movements. Nevertheless, the insurance damage from the magnitude 5.6 Newcastle earthquake[2] of 28 December 1989 with 13 fatalities was, at US$670 million (Munich Re, 2000), the largest insurance payout at the time – because the soil under the town of Newcastle was honeycombed with disused coal mines thus reducing the soil stability.

Because Australia is not a particularly mountainous country – despite the presence of the Great Dividing Range on the east coast – landslides, though they occur, are rare and usually not associated with fatalities. Despite this, the landslide at the ski resort of Thredbo on 30 July 1997 killed 18 people, the 27 September 1996 cliff-face collapse at Gracetown, Western Australia, killed 9, and the March 1997 landslide resulting from Tropical Cyclone Justin also killed one person.[3]

The Australian government has published an eclectic list of Australian natural disasters,[4] which groups natural disasters into those associated with drought, fire, heat waves, floods, and cyclones; again emphasising that hydro-meteorological disasters are the natural disasters of greatest concern. Thus, in the rest of this chapter the hydro-meteorological hazards, which are the ones that dominate Australian emergency management services, will be examined.

22.2.1 Drought

Beer (1991) pointed out that there are numerous definitions of drought, none completely satisfactory. The reason is that the term drought is used with at least three different meanings. A meteorological drought (normally measured in terms of rainfall deciles) occurs when the rainfall, over a period of time, is substantially below normal levels. An agricultural drought (measured, for example, by the moisture deficit in the top 200 mm of soil) occurs when the soil moisture, over a period of time, is substantially below normal levels. A hydrological drought (measured, for example, by water availability) occurs when drinking water impoundments, over a period of time, are substantially below normal levels.

Droughts in Australia are associated with El Niño, with prolonged El Niño events leading to prolonged droughts. The two most extreme drought periods in Australia are the Federation drought[5] of 1896–1902, and the recent Millennium drought[6] that lasted from 1995 to 2009. There is some controversy over the start and end dates of these droughts with some authors extending the Federation drought from 1895 to 1903 and some starting the Millennium drought as early as 1991 with others claiming a start date of 1997. In the case of the Millennium drought there is agreement that it ended with the two back-to-back La Niña years of 2010–2011 and 2011–2012 (CSIRO, 2012).

22.2.1.1 *Science*

Climatologists have long known that the Hadley circulation, which consists of warm air rising in the tropical regions and descending in the mid-latitudes, plays a dominant role in the global climate because the world's major deserts all lie in the descending region of the Hadley circulation. The circulation can be quantified by measuring the location and strength of the subtropical high-pressure system (nowadays known as the subtropical ridge) and Pittock (1973) confirmed its importance for Australian climate.

During the 1970s and 1980s the work of oceanographers (e.g. Wyrtki *et al.*, 1976) was influential in establishing the El Niño as an oceanic phenomenon with profound, large-scale climatic consequences. Climatologists (Troup, 1965; McBride and Nicholls, 1983) had found that the El Niño could be quantified using the southern oscillation index, and that it had a profound effect on southeastern Australian rainfall. Consequently, the phenomenon is frequently referred to as ENSO, which stands for El-Nino Southern Oscillation. Basically when ENSO is in an El Niño phase the eastern equatorial Pacific is warmer than usual, pressures are higher across much of Australia, and rainfall is reduced across much of southeastern Australia. When ENSO is in a La Niña phase the equatorial Pacific is colder than usual and rainfall is increased across much of southeastern Australia.

More recent research found that sea surface temperatures in the Indian Ocean, as measured by the Indian Ocean dipole (IOD), and winds in the Southern Ocean, as measured by the southern annular mode (SAM), also play a role in determining the climate and rainfall of southeastern Australia (Gallant *et al.*, 2012). The IOD refers to an out of phase variation in sea-surface temperatures between the western and eastern tropical Indian Ocean. It is quantified using the dipole mode index (DMI) that is based on the measured difference between sea-surface temperature in the western (50° E–70° E and 10° S–10° N) and eastern (90° E–110° E and 10° S–0° S) equatorial Indian Ocean. The IOD typically develops in early winter, peaks in October–November, and disappears by January. A negative IOD is associated with wetter conditions across southeastern Australia, while a positive IOD is associated with drier conditions. The IOD is the most dominant large-scale influence on climate in southeastern Australia in winter and spring, when it explains up to 40 per cent of the rainfall variance in parts of southeastern Australia.

22.2.1.2 *Management*

Because the dominant effect of prolonged drought is the reduction of water supplies (i.e. hydrological drought), the management of drought in Australia is primarily a role for water authorities, which are entities that are controlled by state governments, rather than the Federal Government. In this respect the Western Australian water authorities have accepted that even though Western Australia has not had the high-profile droughts of eastern Australia, there has been a steady decline in rainfall (and available water resources) over the past 40 years (Fig. 22.1). As a result the state government has built a desalination plant to protect the community against such water shortages and the Victorian and New South Wales Governments have also done so. Australia has a system whereby a state government or the Federal Government can formally declare a region to be drought affected[7] and thus under 'Exceptional Circumstances' so as to thereby enable eligible farmers and their families to receive financial assistance.

There is awareness that current approaches to drought and Exceptional Circumstances are no longer the most appropriate in the context of a changing climate. Drought policy needs to be improved to create an environment of self-reliance and preparedness, and to encourage the adoption of appropriate climate change management practices. It is necessary to review and improve drought policy in parallel with the development of climate change policies and programmes.

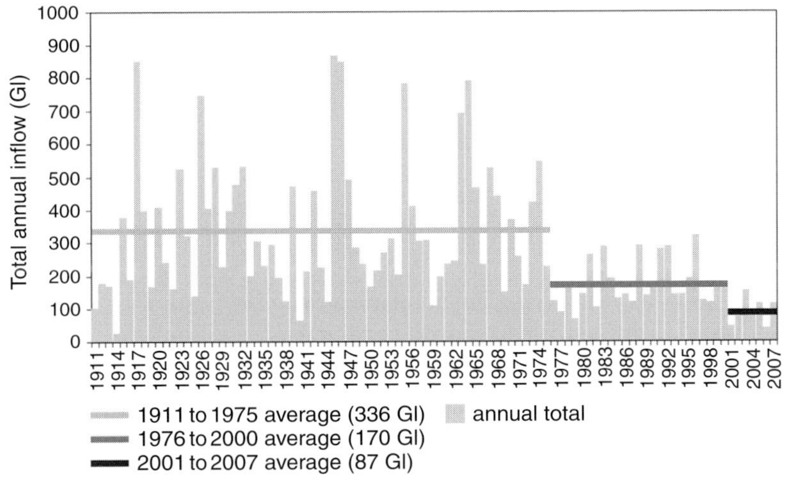

Figure 22.1. Annual stream flows into Perth's dams in Western Australia (excluding the Stirling and Samson dams) (adapted from Fig. 5.2 in Garnaut (2008)).

The Australian Government conducted a comprehensive national review of drought policy through three separate assessments. The review includes:

- an *economic assessment*[8] of drought support measures by the Productivity Commission;
- an assessment by an expert panel of the *social impacts*[9] of drought on farm families and rural communities;
- a *climatic assessment*[10] by the Bureau of Meteorology and the Commonwealth Scientific and Industrial Research Organisation (CSIRO) of the likely future climate patterns and the current Exceptional Circumstances standard of a one-in-20-to-25-year-event.

22.2.2 Fire

In poorer parts of the world, one of the tragic consequences of long-term drought is that of famine. Fortunately this does not occur in Australia where the most common consequence of a long-term drought is severe fire. In Australia the term 'bushfire' is used as a generic term for any wildfire, with forest fires and grass fires being the two major categories.

22.2.2.1 *Science*

Bushfires are a manifestation of extreme drought. Vegetation will ignite if it is sufficiently dry and is raised to a sufficiently high temperature. Both of these necessary conditions are directly or indirectly related to meteorological variables. Beer *et al.* (1988) showed that atmospheric humidity was the most important variable in determining the risk of fire danger under changing climatic regimes. The reason for this is that during days of extreme temperature the humidity will be very low. However, in addition, low humidity acts to dry the surface vegetation so that the combination of high temperatures and dry vegetation can produce a dangerous combustible mix.

Following the catastrophic Ash Wednesday bushfires of Victoria and South Australia that occurred in February 1983, the CSIRO set up a programme of research in bushfire meteorology. The objective of this research was to understand better the mechanism of fire spread so as to provide improved predictions of fire propagation. This objective was achieved by developing tools with which to study fire spread, and undertaking laboratory experiments to improve the tools.

Fire agencies in Australia at the time of the Ash Wednesday fires used the McArthur fire danger meters (Luke and McArthur, 1978), which have now partly been replaced by the CSIRO Fire Danger and Fire Spread Calculator.[11] Such calculators use meteorological information on wind speed, temperature, humidity, and, in the case of forests, rainfall to determine fire-front rate of spread. This velocity is considered a fundamental variable of fire behaviour. At the time of the Ash Wednesday fires, these meters existed only in the form of a circular slide rule. Beer (1990, 1991) converted the McArthur meters and the Rothermel model used in the USA into computer-based tables. The impetus originally came from a need to convert between the 10 m wind speed at which meteorological measurements are made, and the mid-flame wind speed experienced by a fire. The interaction is complex (Beer, 1991) and depends as much on wind speed as it does on the distribution of the fuel. Laboratory experiments indicated that simple empirical formulae, such as those used by McArthur and Rothermel did not perform well (Beer, 1993). By extending the simple model of Weber (1990) that dealt with fire propagation in vertical arrays to deal with wind, the reason for the poor performance of empirical formulae became apparent (Beer, 1995). Determination of fire spread involves a differential equation whose solutions are not adequately represented by simple analytical forms.

Such developments in the understanding of fire spread and fuel properties were incorporated into a two-dimensional computer-based model of fire spread known as Sirofire[12] that is available for training and simulation.

22.2.2.2 *Management*

Following the devastating Black Friday bushfires of 13 January 1939, the State Government of Victoria set up a Royal Commission,[13] known as the Stretton Commission because it was chaired by Justice Leonard Stretton, to determine 'The Measures to be Taken to Prevent Bush Fires in Victoria and to Protect Life and Property in the Event of Future Bush Fires'. The *Australian Dictionary of Biography*[14] notes that the 'report was acknowledged for its literary qualities and anthologised in a collection of Australian nature writing. Stretton's recommendations officially sanctioned the common bush practice of controlled burning for fuel reduction, and led the government to broaden the responsibilities of the Forests Commission of Victoria and to form (1945) the Country Fire Authority.'

The idea that bushfire management cannot be tackled by a local authority but needs to be dealt with at a state government level was adopted by all the Australian states which emulated the Victorian Country Fire Authority and established state-based bushfire agencies. Nevertheless there have been many serious bushfires in Australia (Table 22.1) and two occasions in Victoria when the magnitude and scale of bushfires have been such as to overwhelm the resources and capabilities of the Country Fire Authority: the Ash Wednesday fires of 1983, and the Black Saturday fires of 2009.

Eburn and Dovers (2012) note that the response of the fire authorities, and of volunteers, is coming under closer judicial scrutiny and express concern over the lengthy post-event inquiry process that has now become standard procedure after a significant bushfire.

As a result of the 2003 Canberra fires, the Australian Government established the Bushfire Cooperative Research Centre (CRC) to study bushfires which brings together researchers and the fire agencies. This research body annually issues a seasonal bushfire outlook by using the long-term seasonal forecasts from the Bureau of Meteorology and combining them with estimates of forest and grassland growth conditions.

Figure 22.2 depicts the Southern Australian Bushfire Assessment that was issued at the end of August 2012 – indicating that during the southern Australian bushfire season (from January to March 2013) there were expected to be large areas with an above average fire potential, particularly in relation to grass fires. In the light of subsequent events during January 2013 when there were significant fires in Tasmania that destroyed many homes, severe bushfires in much of Victoria, and over 150 fires in various parts of New South Wales, it would appear that the assessment was correct for New South Wales but underestimated the fire danger in Victoria and Tasmania.

22.2.3 Severe storms and tropical cyclones

The 18 March 1990 Sydney hailstorm[16] is noteworthy because the insurance payout for the storm exceeded US$1100 million dollars (Munich Re, 2005), making it the most costly insurance payout in Australian history – and a relatively unusual situation in which the estimated economic losses approximately equalled the insured losses. The more normal situation is that insurance losses are an order of magnitude less than the estimated economic losses.

Hailstorms are one consequence of severe storms. Local flash flooding, which will be discussed in Section 22.2.4,

Table 22.1. *List of some significant or noteworthy Australian bushfires*

	Location	Date	Lives lost	Homes destroyed	Ref.
Black Friday	Victoria	13 January 1939	71	700	1
Hobart Fires	Tasmania	7 February 1967	62	1300	2
Ash Wednesday	Victoria and South Australia	16 February 1983	75	2463	3
Canberra Fires	Canberra	18 January 2003	4	500	4
Black Saturday	Victoria	7 February 2009	173	2133	5

1. http://www.voltscommissar.net/docs/Leonard_Stretton-1939_Bush_Fires_Royal_Commission_Report.pdf
2. Chambers, D. M. and Brettingham-Moore, C.G. (1967) *The Bush Fire Disaster of 7 February 1967: Report and Summary of Evidence* (excluding appendices) (Solicitor-General (Mr. D. M. Chambers) and the Master and Registrar of the Supreme Court (Mr. C. G. Brettingham-Moore) Office of the Solicitor-General. Hobart, Tasmania.
3. http://www.dse.vic.gov.au/fire-and-other-emergencies/major-bushfires-in-victoria/ash-wednesday-1983
4. http://www.cmd.act.gov.au/functions/publications/archived/mcleod_inquiry
5. http://www.royalcommission.vic.gov.au/Commission-Reports/Final-Report/Summary/Interactive-Version

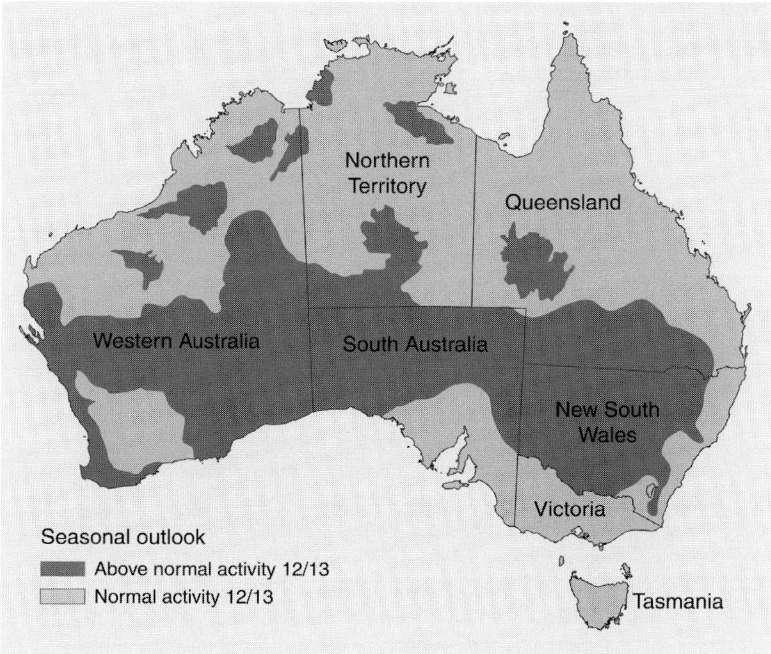

Figure 22.2. Southern Australian Bushfire Assessment that was issued at the end of August 2012.[15] Reproduced with permission of the Bushfire CRC.

can be another consequence of severe storms. The most severe storms to impact the Australian coastline and inland regions are tropical cyclones, the local name for a meteorological phenomenon that is called a typhoon in the North Pacific, and a hurricane in the Atlantic.

On 25 December 1974, Tropical Cyclone Tracy[17] destroyed virtually all of the Northern Australian city of Darwin causing the deaths of 71 people (49 on land and 22 at sea) and the evacuation of 75% of the city's residents. This event shocked the Australian public and encouraged the serious scientific study of Australian tropical cyclones. This was not the first time Darwin had been severely damaged by a tropical cyclone: in both January 1897 and March 1937 the city was badly damaged, but only after Tracy was more attention given to building codes and other social aspects of disaster planning. Darwin was rebuilt and is now a thriving city of 128 100 people as at June 2011.

The Bureau of Meteorology[18] provides a database of past tropical cyclones, histories of tropical cyclones, and a library of individual cyclone reports. These reports, my own research, plus the list of Australian tropical cyclones given on the web[19] were used to compile the list of significant tropical cyclones of Table 22.2.

22.2.3.1 *Science*

Australian tropical cyclones[20] are classified according to the modified Saffir–Simpson scale shown in Table 22.3, with Category 1 tropical cyclones having winds below 42 m/s but above 33 m/s, which is the minimum wind speed needed for a tropical storm to be classified as a tropical cyclone. Category 5 tropical cyclones are the most intense and will cause catastrophic damage to structures.

There are, on average, approximately 12 tropical cyclones per year that are identified as occurring within the Australian region. Of these about 40% (~5) make landfall over the Australian continent. Tropical cyclones and tropical storms provide a large proportion of rainfall in tropical Australia; it ranges from 40% in tropical Queensland to 60% in tropical Western Australia (Lavender and Abbs, 2013). Tropical cyclone climatologies for Australia have been used to determine the tropical cyclone hazard.

Numerical weather prediction models have not, as yet, reached sufficiently fine resolution that they can predict the formation and subsequent strengthening and motion of a tropical cyclone. They are, however, able to identify tropical lows so that a sufficiently skilled forecaster is able to use such numerical weather prediction models, along with satellite photographs of tropical cyclone clouds that position the tropical cyclone, to assist with forecasts of tropical cyclone tracks. Once the tropical cyclone has made landfall, there are four particular impacts that need to be considered: strong winds; extreme rainfall; the flooding associated with the cumulative rainfall; and the short-term rise of sea level (known as storm surge).

22.2.3.2 *Management*

With the reduction of maritime traffic around the coast of Australia, most of the risks associated with tropical

Table 22.2. *List of some significant or noteworthy Australian tropical cyclones*

	Location	Date	Comments	Ref.
Louisiade Archipelago	Papua New Guinea	16–22 April 1952	Papua New Guinea's most severe tropical cyclone with a central pressure of 940 hPa and a 24-hour rainfall of 914 mm – both of which constituted a Papua New Guinea record, and because Papua New Guinea at that time was an Australian territory the rainfall was also probably an Australian record at the time, albeit an unrecognised record.	1
Althea	Queensland – Townsville	19–24 December 1971	Severe Tropical Cyclone Althea was at the time considered to be one of the strongest cyclones to affect the Queensland coast. There were three deaths in Townsville and damage costs in the region reached AU$50 million (1971 value). Many houses were damaged or destroyed, including 200 Housing Commission homes. On Magnetic Island 90% of the houses were damaged or destroyed.	2
Joan	Western Australia – Port Hedland	30 Nov–10 December 1975	Joan was a severe tropical cyclone with maximum measured wind gusts of 208 km/h. It damaged 85% of the houses in Port Hedland, 600 km southwest of Broome, WA, and other settlements along its path. Subsequent flooding damaged roads and sections of the iron ore railways, particularly that of Hamersley Iron Pty Ltd. Sheep losses were heavy but there was no loss of human life or serious injury. The estimated damage to private property and public facilities is believed to have exceeded AU$25 million.	3
Tracy	Northern Territory	21–25 December 1974	Tracy, a Category 4, struck Darwin in the Northern Territory on Christmas Eve 1974. It remains Australia's most destructive with winds of around 250 km/h. Seventy-one people were killed, and many thousands injured. Of a population of 43 000, more than 25 000 were left homeless, according to the Australian government.	4
Alby	Western Australia – Perth	27 March–4 April 1978	Tropical Cyclone Alby passed close to the southwest corner of Western Australia on 4 April 1978, killing five people and causing widespread but mostly minor damage to the southwest. The damage bill was estimated to be AU$39 million (2003 dollars). Storm surge and large waves caused coastal inundation and erosion from Perth to Busselton, while fires fanned by the very strong winds burned an estimated 114 000 hectares of forest and farming land.	5
Yasi	Queensland	3 February 2011	Cyclone Yasi crossed the Queensland coast at Mission Beach and Tully, south of Cairns. It is the most severe cyclone to hit Australia since 1918, with 290 km/h winds but there was only one known fatality. It destroyed much of Australia's banana crop.	6

1 Beer (1986).

2 Hannan, F. T. and Shields, A. J. (1972). Report by Director of Meteorology on Cyclone Althea, Australian Government Publishing Service, Canberra. Available at: http://www.bom.gov.au/cyclone/history/pdf/althea.pdf; http://en.wikipedia.org/wiki/1970%E2%80%9375_Southern_Hemisphere_tropical_cyclone_seasons

3 Beer (1981); http://en.wikipedia.org/wiki/Cyclone_Joan

4 http://www.bom.gov.au/cyclone/history/pdf/tracy.pdf; http://en.wikipedia.org/wiki/Cyclone_Tracy

5 http://www.bom.gov.au/cyclone/history/pdf/alby.pdf; http://en.wikipedia.org/wiki/Cyclone_Alby

6 http://www.bom.gov.au/cyclone/history/yasi.shtml; http://www.bom.gov.au/cyclone/history/yasi.shtml

Table 22.3. *Modified Saffir–Simpson tropical cyclone scale used by the Australian Bureau of Meteorology*

Category	Wind description	Wind speed (km/h)	Typical effects
1	Gales	118–125	Minimal house damage. Damage to some crops, trees, and caravans. Boats may drag moorings
2	Destructive winds	125–164	Minor house damage. Significant damage to signs, trees, and caravans. Heavy damage to some crops. Risk of power failure. Small boats may break moorings
3	Very destructive winds	164–224	Some roof and structural damage. Some caravans destroyed. Power failure likely.
4	Very destructive winds	224–279	Significant roofing and structural damage. Many caravans destroyed and blown away. Dangerous airborne debris. Widespread power failures.
5	Extremely destructive winds	>280	Extremely dangerous with widespread destruction.

cyclones occur when they make landfall. In this respect three regions have been identified as being at risk: Northern Australia, Western Australia, and northern Queensland (i.e. eastern Australia).

Watch and warnings

Figure 22.3 shows the tropical cyclone forecast issued for Cyclone Yasi at 0400 on 2 February 2011, by which stage it had intensified to a Category 5 tropical cyclone. The dark grey area of land on the map is the region designated by the Bureau of Meteorology as a cyclone warning area[21] whereas the grey area marks a cyclone watch area. During an active tropical cyclone the Bureau of Meteorology co-ordinates with the police and emergency services to provide information to the public in relation to the areas liable to be affected, and the need for evacuation.

Tropical cities in Australia have Tropical Cyclone Disaster Plans that consist of evacuation of the population to cyclone-proof structures and these plans are made known to the public through education campaigns in schools, and through publicity in the media during periods of active tropical cyclone warnings. The experience of Tropical Cyclone Tracy and overseas experience such as Hurricane Katrina mean that such warnings are taken seriously. During the cyclone warning period for Tropical Cyclone Yasi (Fig. 22.3) the patients at Cairns Hospital started to be evacuated because of the possibility that Yasi would pass directly over Cairns and flood the hospital. Later, when it became clearer that Yasi would pass over Innisfail rather than Cairns, the evacuation was stopped.

Concatenated hazards and Tropical Cyclone Yasi

Concatenated hazards refers to situations where one extreme event precipitates one or more other extreme events. The exemplar is the tsunami and the resulting nuclear accident that occurred in Japan following the 11 March 2011 magnitude 9 earthquake. In the case of Tropical Cyclone Yasi, a major portion of the Australian banana crop was wiped out causing extreme spikes in the banana price (Fig. 22.4) – repeating the situation of 2006 when Tropical Cyclone Larry made landfall on 20 March 2006 and also destroyed 80–90% of Australia's banana crop. Australia is relatively free of banana pests and diseases, and therefore does not allow bananas to be imported. Bananas were in short supply throughout Australia for the remainder of both 2011 and 2006, which increased prices across the country by 400–500%.

22.2.4 Floods

The term 'flood' covers a large variety of different hydrological events that can range from local flooding due to a blocked drain or culvert, short-term flash flooding of creeks as a result of a heavy, but localised storm to the long-term flooding that occurs as rivers overtop their banks and inundate the floodplain that surrounds them. The dramatic effects of localised severe storms and flash flooding were graphically illustrated in January 2011, when the continuation of heavy rainfall that had started over much of Queensland in December 2010 caused severe flash flooding in the centre of the city of Toowoomba.

The Flood Commission of Inquiry in its interim report[22] states that:

> On 10 January, the Gowrie Creek catchment experienced intense rainfall between 1.00 pm and 2.30 pm. In the city of Toowoomba itself, heavy rain began falling at about 12.45 pm, and peaked between 1.45 pm and 2.15 pm. The most severe rain fell in a northeast–southwest band that covered the middle and lower parts of East and West Creeks, where they crossed Toowoomba's central business district. This concentration of rain in the East Creek and West Creek catchments continued for approximately 60 to 90 minutes. It had largely ceased between 2.15 pm and 2.45 pm.
>
> The intense rainfall over the catchment of the three creeks caused a severe flash flood in the city between 1.30 pm and 2.45 pm.

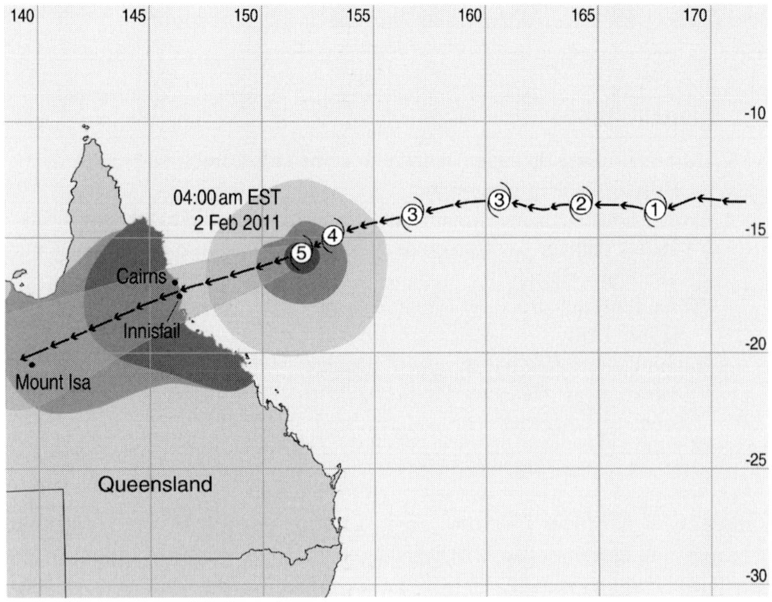

Figure 22.3. Forecast track of Tropical Cyclone Yasi as at 4 am on 2 February 2011. An animated version of the tropical cyclone is available at: http://www.bom.gov.au/cyclone/history/yasi.shtml#loops. Reproduced with permission from the Bureau of Meteorology, Melbourne, Australia.

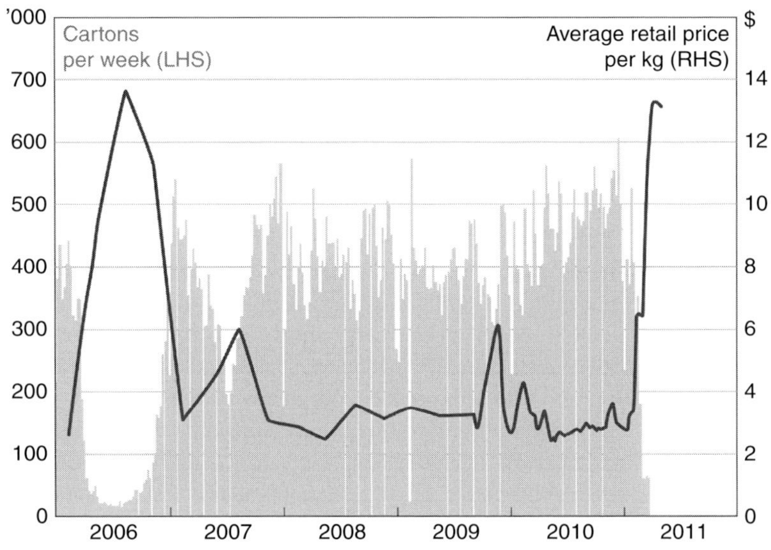

Figure 22.4. Graph of North Queensland banana price and banana supply indicating the effects of Cyclone Larry in 2006 and Cyclone Yasi in 2011. (Reserve Bank of Australia, 2011). Reproduced with permission of the Reserve Bank of Australia with acknowledgement to the Australian Bureau of Statistics, Australian Banana Growers' Council Inc., and the Reserve Bank of Australia for provision of data.

Closed circuit television footage provided by Toowoomba Regional Council shows water rising at extraordinary speed and flowing over the roadways. It also demonstrates the speed with which the water rose. It is clear that this was not a situation in which any agency could have effectively warned residents of what was to come.

Water covered all the roadway crossings of East, West and Gowrie Creeks, making them impassable to pedestrians and vehicles. The rapidity of the flooding caught people by surprise: in city streets they found themselves surrounded by water, or were trapped in their vehicles. A woman and her teenaged son lost their lives when their car was caught in the flooding in a city intersection. A number of buildings in and around the city were extensively damaged, and numerous parked cars were swept away or inundated by the flooding.

On the same day the Lockyer Creek, in the Lockyer Valley to the east of Toowoomba, flooded and residents suggest that flooding in the town of Grantham occurred between approximately 3.20 pm and 4.00 pm. The flood appeared as a wave, sweeping from Lockyer Creek across the paddocks and through the town. Thirty-seven people died in the 2010/2011 floods with an estimated reconstruction cost of AU$5 billion.

22.2.4.1 *Science*

Large-scale floods in Australia (Table 22.4) tend to be associated with the higher rainfalls that occur as a result

Table 22.4. *List of some significant or noteworthy Australian floods*[28]

	Location	Date	Lives lost	Homes flooded	Ref
Three separate tropical cyclones	Brisbane	February 1893	35	AU$4 million (1893 values)	1
Hunter Valley floods	Hunter Valley (especially Maitland)	February 1955	25	7,000 homes (1.3 billion in 2005 AU$)	2
Flash flooding A well-known example of overland flows occurred in Elizabeth Street on 17 February 1972, when 78.5 mm of rain fell in one hour over the Central Business District. This flooding caused significant disruption to the public and damage to businesses.	Melbourne	17 February 1972			3
Brisbane floods of 1974	Brisbane	January 1974	14	6,700 homes AU$200 M damage	4
The La Niña floods that broke the millennium drought.	Queensland and Victoria	December–January 2010/11 (Queensland) 12–14 January 2011 (Victoria)	35 (Qld) 2 (Victoria)	AU$1.51 billion (Qld) 1730 homes (Vic) AU$2 billion	5

1. http://en.wikipedia.org/wiki/1893_Brisbane_flood
2. http://www.hunterweather.com/events.php?id=41 and http://en.wikipedia.org/wiki/Maitland_Flood
3. http://www.melbournewater.com.au/content/library/publications/reports/flood_management_and_drainage_strategy/flood_management_and_drainage_strategy_part_2.pdf
4. http://en.wikipedia.org/wiki/1974_BrisbaneBrisbane_flood
5. http://en.wikipedia.org/wiki/2011_Victorian_floods and http://en.wikipedia.org/wiki/2010%E2%80%932011_Queensland_floods

of La Niña – the opposite meteorological situation to an El Niño. La Niña occurs when the temperatures of the western part of the tropical Pacific are higher than usual. The record rainfalls and flooding experienced across much of the region, and throughout Australia, in the 2010/11 and 2011/12 summers highlighted that the tropical Pacific ocean temperatures are not the only sea-surface temperatures of importance but one needs to consider the status of all three oceans surrounding Australia – Pacific, Indian, and Southern – in influencing seasonal and inter-annual rainfall variability. The spring/summer of 2010/11 saw one of the strongest La Niña events on record combined with a negative IOD event and a positive SAM, i.e. all three key influences were in their wet phases in terms of expected rainfall impacts on southeastern Australia.

SAM describes the north–south movement of the westerly wind belt that circles Antarctica, dominating the middle to higher latitudes of the Southern Hemisphere. The changing position of this westerly wind belt influences the strength and position of cold fronts and mid-latitude storm systems, and is an important driver of rainfall variability in southern Australia. In a positive SAM event, the belt of strong westerly winds contracts towards Antarctica. This results in weaker than normal westerly winds and higher pressures over southern Australia, restricting the penetration of cold fronts inland. The positive phase of SAM is typically associated with wetter and cooler conditions over much of Australia during the spring and summer but with drier and cooler conditions over the southwest and southeast coasts of the continent during winter (Hendon *et al.*, 2007).

These conditions, coupled with the warmest sea-surface temperatures on record to the north of the Australian continent, contributed to making 2010–2011 Australia's wettest two-year period on record. While the extensive flooding of 2010/2011 was, by and large, consistent with natural variability, it is possible that ongoing global warming contributed to the magnitude of the event through its impact on ocean temperatures. The spring and summer of 2011/2012 saw the re-emergence of another (weaker) La Niña event, combined with a positive SAM during early summer, but this time the Indian Ocean played a lesser role.

22.2.4.2 *Management*

Short-term flood assistance is under the control of the State Emergency Service. This means that floods are a state responsibility rather than a federal responsibility. The web site for the Victorian State Emergency Service[23] states that:

> The Victoria State Emergency Service (SES) is the control agency for flood, storm, tsunami and earthquake in Victoria, managing responses to these emergencies and helping communities to prepare through planning and education. As well as operating the largest road crash rescue network in Australia, we also assist the Victoria Police in search and rescue operations and play an important support role during major bushfire responses.

The Australian insurance industry has a narrow view of floods and their narrow view has caused widespread community concern. Unlike most other countries, water damage arising from a storm (i.e. if the water comes from above rather than below) is not considered to be a flood for insurance purposes. A technical committee of the Insurance Council of Australia (ICA) produced a definition of flood that in 1984 became widely adopted in the insurance industry as an exclusion.

The definition was: 'the inundation of normally dry land by water escaping from the normal confines of any natural watercourse or lake whether or not altered or modified, or any reservoir, canal or dam'. This exclusion became the base from which changes were made by companies to grant partial or fuller 'flood' cover, or which they further refined to make the exclusion more watertight. As more events occurred and greater interpretations were applied to what was included as storm and excluded as flood the ICA issued further advisory wordings, including a 'plain language' version prepared by Professor Robert Eagleson in 1999. However, in the absence of an authorisation under the Trade Practices Act (1974) these were not generally adopted and each insurer developed its own definition. The difference between the term inundation and the term flood was discussed by Flick *et al.* (2012) who proposed that the term inundation be used for a dry area being permanently drowned or submerged whereas the term flooding be used when dry areas become wet temporarily.

Heavy storms and rain in the Hunter and Central Coast regions during early June 2007 caused extensive flooding over a 36-hour period. Ten people were killed, 105 000 homes were without power and insured losses of AU$1170 million were recorded.[24] Many insurance claims were rejected on the basis that the cause of inundation was riverine flooding or storm surge. The Australian Competition and Consumer Commission (ACCC) acknowledged at the time that there were unsatisfactory levels of consumer confusion on insurance cover for flood and granted the ICA conditional authorisation for a proposed common definition. The controversy resulting from the Hunter and Central Coast floods in 2007 is reported to have led the insurance industry to again propose a standard definition for flood in 2008.[25]

The ICA proposed a definition in the following form:

> inland flood is the covering of land that is not normally underwater by:
>
> - water that overflows or escapes from a naturally occurring or man made inland watercourse (such as a river, creek, canal or storm water channel) or water pool (such as a lake, pond or dam), whether it is in its original state or it has been modified; or
> - water released from a dam whether it be accidentally released or intentionally released to control, mitigate, regulate, or otherwise respond to excess water; or
> - water that cannot drain or run off as a result of water that is overflowing or escaping from an inland watercourse or water pool preventing the escape of water.

The ACCC later rejected the proposed definition and stated that the form proposed by the ICA would be unlikely to lessen consumer confusion. In its press release, the ACCC said the proposed definition had raised 'too many uncertainties' and that it was particularly concerned 'that the ICA definition of flood introduced a range of new concepts the legal implications of which are not clearly understood'. Consumer groups argued that the definition would increase, rather than decrease, consumer confusion. The ICA then stated it would undertake no further work on the standard definition. The greatest problem faced by consumers is understanding the boundary between flood and storm. Even where insurers include flood there is a new boundary between flood and actions of the sea, storm surge, tsunami, and similar excluded perils.[26]

22.3 Discussion and conclusion

It has been shown that science and natural hazards management, or emergency management as it is generally known in Australia, need to work together. Science on its own is sterile. Management that is not informed by science is directionless. At the most elementary level this is done by the interchange of information. Science agencies provide forecasts of the natural hazards. Emergency management agencies provide observations that are then fed into the next round of forecasts.

More sophisticated collaboration involves close collaboration at both the institutional level and the individual level. Thus even though the Bureau of Meteorology and the Victorian Country Fire Authority have a long history of working together, for a long time it was in the form of a one-way communication. The Bureau of Meteorology provided special fire-weather forecasts to the Country Fire Authority, which then based operational decisions on the basis of the forecasts. The level of collaboration improved markedly when dedicated fire-weather meteorologists from the Bureau of Meteorology were seconded to the Country Fire Authority during the fire season so that they could understand the way in which the forecasts were used, and could then determine how better to tailor the forecasts so as to have greater utility for the fire-fighters.

Such collaboration cannot take place without suitable funding mechanisms. Kundzewicz and Takeuchi (1999) have noted the situation, which they call the hydro-illogical cycle, that funding for flood prevention and mitigation is provided *after* a catastrophic flood rather than *before* the flood, which is the time when it is most needed. Though as scientists we may rail against the lack of logic, as managers we should recognise the window of opportunity and use it to accomplish benefits for society as a whole.

In the final reckoning, the success of emergency management depends on the avoidance of a catastrophe. This is a goal to which all of society needs to work. Thus Australian society can be proud that major Australian droughts have not led to famine within the nation. The Australian record of bushfire response is good, but the Black Saturday bushfires of 2009 indicate that both scientists and emergency managers need to improve their skills – and their ability to envisage and prepare for natural hazards of unprecedented magnitude. In part improved risk forecasts can play a role. The UK Meteorological Office presently has a system in which risk matrices for high impact weather are determined based solely on physical variables (Swinbank *et al.*, 2012)[27]. The likelihood is based on the probability of occurrence, and the consequence is based on the size of the wind. Though this does indeed follow a risk assessment methodology, the utility of such a product can be greatly improved when it is extended to incorporate vulnerable infrastructure and vulnerable people.

Beer (2004) notes that the terms likelihood and consequences (as well as time) have both scientific and social aspects. When dealing solely with past events – as in the construction of a hazard map – it is possible to use probability as an exact measure of likelihood. But when attempting to safeguard against possible future events it is necessary to construct scenarios of the future, which is primarily a social science activity. In a similar vein, determining the overall consequences of a geophysical hazard requires consideration of the magnitude of the event, and the severity of the event. When one considers only the hazard, then it is legitimate to treat the severity of the hazard as being identical to the magnitude of the hazard. However, when the focus of interest shifts to environmental risk, then the consequences can still be considered as a combination of magnitude and severity, but in this case the severity depends on the vulnerability of the elements that are at risk from the hazard.

Endnotes

1 http://www.environment.gov.au/soe/2011/index.html
2 http://en.wikipedia.org/wiki/1989_Newcastle_earthquake
3 http://en.wikipedia.org/wiki/List_of_disasters_in_Australia_by_death_toll
4 http://australia.gov.au/about-australia/australian-story/natural-disasters
5 http://en.wikipedia.org/wiki/Federation_Drought
6 http://en.wikipedia.org/wiki/2000s_Australian_drought
7 http://www.daff.gov.au/agriculture-food/drought
8 http://www.daff.gov.au/agriculture-food/drought/national_review_of_drought_policy/economic_assessment
9 http://www.daff.gov.au/agriculture-food/drought/national_review_of_drought_policy/social_assessment
10 http://www.daff.gov.au/agriculture-food/drought/national_review_of_drought_policy/climatic_assessment
11 http://www.csiro.au/en/Outcomes/Safeguarding-Australia/Fire-Danger-And-Spread-Calculator.aspx
12 http://www.csiro.au/en/Outcomes/Safeguarding-Australia/SiroFire-Overview.aspx
13 http://www.voltscommissar.net/docs/Leonard_Stretton-1939_Bush_Fires_Royal_Commission_Report.pdf
14 http://adb.anu.edu.au/biography/stretton-leonard-edward-bishop-11793
15 http://www.bushfirecrc.com/managed/resource/southern_outlook_f95_final.pdf
16 http://www.bom.gov.au/nsw/sevwx/18mar90.shtml
17 http://www.bom.gov.au/cyclone/history/pdf/tracy.pdf
18 http://www.bom.gov.au/cyclone/history/index.shtml
19 http://www.australiangeographic.com.au/journal/timeline-australias-worst-cyclones.htm
20 http://www.bom.gov.au/cyclone/about/intensity.shtml
21 http://www.bom.gov.au/cyclone/about/warnings/advice.shtml
22 http://www.floodcommission.qld.gov.au/publications/interim-report
23 http://www.ses.vic.gov.au/about/what-we-do
24 http://www.swissre.com/rethinking/natcat/Flooding_in_Australia_The_reality_of_a_secondary_peril.html
25 http://afr.com/p/national/easy_definition_of_flood_is_yet_H07VEJWagzfoGrE8WNc5dL
26 http://www.ndir.gov.au/content/commissionedpapers/2011/Flood_Insurance_Aus.doc
27 http://icdm2012.csp.escience.cn/dct/page/65537
28 http://en.wikipedia.org/wiki/Floods_in_Australia

References

Barnard, G. A. and Sprott, D. A. (1983). Likelihood. In *Encyclopedia of Statistical Sciences*, vol. 4, eds. Kotzand, S. and Jonson, M. L. New York, NY: Wiley, pp. 639–644.

Beer, T. (1981). Satellite microwave estimation of tropical cyclone rainfall: A case study of Cyclone Joan. *Australian Meteorology Magazine*, **28**, 155–166.

Beer, T. (1986). The Louisiade Archipelago Tropical Cyclone of April 1952. *Meteorology Australia*, **4**(2), 6–14.

Beer, T. (1990). *The Applied Environmetrics Meteorological Tables*. Melbourne: Applied Environmetrics.

Beer, T. (1991). *The Applied Environmetrics Hydrological Tables*. Melbourne: Applied Environmetrics.

Beer, T. (1993). The speed of a front and its dependence on wind speed. *International Journal of Wildland Fire*, **3**, 193–202.

Beer, T. (1995). Fire propagation in vertical stick arrays: the effects of wind. *International Journal of Wildland Fire*, **5**, 43–49.

Beer, T. (2003). Environmental risk and sustainability. In *Risk Science and Sustainability: Science for Reduction of Risk and Sustainable Development of Society*, eds. Beer, T. and Ismail-Zadeh, A. Dordrecht: Kluwer Academic, pp. 39–61.

Beer, T. (2004). Geophysical risk, vulnerability, and sustainability. In *The State of the Planet: Frontiers and Challenges in Geophysics*, eds. Sparks, R. S. J. and Hawkesworth, C. J. Washington, DC: American Geophysical Union, pp. 375–385.

Beer, T. and Ismail-Zadeh, A., eds. (2003). *Risk Science and Sustainability*. Dordrecht: Kluwer Academic Publishers.

Beer, T., Gill, A. M. and Moore, P. H. R. (1988). Australian bushfire danger under changing climate regimes. In *Greenhouse*, ed. Pearman, G. Melbourne: CSIRO, pp. 421–427.

Cox, D. R. and Hinkley, D. V. (1974). *Theoretical Statistics*. London: Chapman and Hall.

CSIRO (2012). *Climate and Water Availability in South-eastern Australia: A Synthesis of Findings from Phase 2 of the South Eastern Australian Climate Initiative (SEACI)*, CSIRO, Australia, September 2012, 41 pp.

Department of Water (2010). *The Effects of Climate Change on Streamflow in South-west Western Australia: Projections for 2050,* Surface Water Hydrology Series Report HY34, Department of Water, Perth, 55 pp.

Eburn, M. and Dovers, S. (2012). Australian wildfire litigation. *International Journal of Wildland Fire*, **21**, 488–497.

Flick, R. E., Chadwick, D. B., Briscoe, J. and Harper, K. C. (2012) "Flooding" versus "Inundation". *EOS Transactions of the American Geophysical Union*, **93**, 365–366.

Gallant, A. J. E., Kiem, A. S., Verdon-Kidd, D. C., Stone, R. C. and Karoly, D. J. (2012). Understanding hydroclimate processes in the Murray-Darling Basin for natural resources management. *Hydrology and Earth Systems Sciences*, **16**, 2049–2068.

Garnaut, R. (2008). *The Garnaut Climate Change Review*. Cambridge: Cambridge University Press.

Hendon, H., Thompson, D. W. J. and Wheeler, M. C. (2007). Australian rainfall and surface temperature variations associated with the Southern Hemisphere annular mode. *Journal of Climate*, **20**, 2452–2467.

Houghton, J. T., Ding, Y., Griggs, D. J. *et al.* (2001). *Climate Change 2001: The Scientific Basis*. Cambridge: Cambridge University Press.

Kundzewicz, Z. W. and Takeuchi, K. (1999). Flood protection and management: quo vadimus? *Journal of Hydrological Sciences*, **44**(3), 417–432.

Lavender, S. L. and Abbs, D. J. (2013). Trends in Australian rainfall: contribution of tropical cyclones and closed lows. *Climate Dynamics*, **40**, 317–326.

Luke, R. H. and McArthur, A. G. (1978). *Bushfires in Australia*. Canberra: Australian Government Publishing Service.

McBride, J. L. and Nicholls, N. (1983). Seasonal relationships between Australian rainfall and the Southern Oscillation. *Monthly Weather Review*, **111**, I998–2004.

McCarthy, J. J., Canziani, O. F., Leary, N. A., Dokken, D. J. and White, K. S. (2001). *Climate Change 2001: Impacts, Adaption, and Vulnerability*. Cambridge: Cambridge University Press.

Munich Re (2000). *Topics, Natural Catastrophes – the Current Position*. Munich: Münchener Rückversicherungs-Gesellschaft.

Munich Re (2005). *Weather Catastrophes and Climate Change: Is There Still Hope for Us?* Munich: Münchener Rückversicherungs-Gesellschaft.

Pittock, A. B. (1973). Global Meridional interactions in stratosphere and troposphere. *Quarterly Journal of the Royal Meteorological Society*, **99**, 424–437.

Reserve Bank of Australia (2011). *Statement on Monetary Policy – Box B: An Update on the Impact of the Natural Disasters in Queensland*, Reserve Bank of Australia, Sydney, May 2011.

Swinbank, R., Mylne, K., Tennant, W. *et al.* (2012), *Probabilistic Forecasting of High-impact Weather*. Proceeding of the ICDM Workshop on Dynamics and Predictability of High Impact Weather and Climate Events, Kunming, August 2012. Available at: http://icdm2012.csp.escience.cn/dct/page/65537

Troup, A. J. (1965). The 'southern oscillation'. *Quarterly Journal of the Royal Meteorological Society*, **91**, 490–506.

Vose, D. (2000). *Risk Analysis: a Quantitative Guide*. Chichester: John Wiley and Sons Ltd.

Weber, R. O. (1990). A model for fire propagation in arrays. *Mathematical and Computer Modelling*, **13**, 95–102.

Wyrtki, K., Stroup, E. Patzert, W. Williams, R. and Quinn, W. (1976). Predicting and observing El Niño. *Science*, **191**, 343–346.

23
The 2008 Wenchuan, China, earthquake

Zhongliang Wu and Tengfei Ma

23.1 Introduction

The 12 May 2008 Wenchuan $M_S8.0$/ $M_W7.9$ earthquake in southwest China was one of the great earthquakes, and was the most devastating earthquake to occur in China since the beginning of the twenty-first century. According to the results of field investigation and seismological/geological studies (Department for Earthquake Monitoring and Prediction of the China Earthquake Administration, 2009), the earthquake occurred at 14:28:04 pm (local time), with an intensity of up to XI on the modified Mercalli intensity (MMI) scale and a peak ground acceleration (PGA) close to 1 g at the strong-motion recording station closest to the epicentre (~23 km). The narrow-ellipsoid-shaped MMI-IX isoseismal is almost collocated with the rupture zone, being about 300 km long striking northeastward from Dujiangyan to Guangyuan, Sichuan Province. The earthquake occurred along the Longmenshan fault zone, rupturing two parallel faults with lengths of about 240 km and 72 km. According to official data on 25 August 2008, the earthquake disaster toll was 69 226 dead, 374 643 injured, and 17 923 missing, with a significant portion of the loss of life being due to earthquake-induced landslides and/or rock falls – a typical case of concatenated disasters. The direct economic loss was estimated as RMB Yuan 845.1 billion. The anatomy of the disasters has been summarised in detail by Chen and Booth (2011).

23.2 Before the earthquake

23.2.1 Seismic zonation and disaster preparedness

Seismic zonation of the middle-to-north Longmenshan fault area was assigned as Chinese intensity (similar to MMI) VI–VII, much lower than the actual intensity, reflecting the blind spots of seismology and the geology of earthquakes. Since the 1950s, four generations of seismic zonation maps have been published in China (in 1957, 1977, 1990, and 2001). Codes for the seismic design of industrial and civilian buildings, as well as their revisions, came into effect in 1974, 1978, 1984, 1989, 1990, and 2001, based on the zonation results (Chen and Booth, 2011). Before the Wenchuan earthquake, a nationwide project for the reinforcement of rural residences had been underway. The Wenchuan earthquake occurred just at the time when this reinforcement was being put into practice in Sichuan Province.

China has suffered from destructive earthquakes for a long time. Reduction of earthquake disasters plays an important role in its sustainable development. *The Law of the People's Republic of China on Protecting Against and Mitigating Earthquake Disasters*, adopted at the 29th Meeting of the Standing Committee of the Eighth National People's Congress of the People's Republic of China on 29 December 1997, came into force on 1 March 1998. From over half a century of experience of dealing with earthquake disasters, the China Earthquake Administration (CEA), the successor of the State Seismological Bureau (SSB), has gradually established the 'three-plus-one system', that is, a system for earthquake monitoring and forecasting, a system for earthquake disaster preparedness, a system for earthquake emergency response, rescue, and relief, and a system for innovation in earthquake science and technology. Since 2005, CEA has set as its dual missions, increasing its capacity to serve society, and increasing the capacity of the whole of society to withstand earthquake disasters; its aim is to reduce losses of life and property due to earthquakes to the largest possible extent.

Research and development (R&D) plays an essential role in the reduction of earthquake disasters. In August 2007, co-sponsored by CEA, the Ministry of Science and

Extreme Natural Hazards, Disaster Risks and Societal Implications, eds. Alik Ismail-Zadeh, Jaime Urrutia Fucugauchi, Andrzej Kijko, Kuniyoshi Takeuchi and Ilya Zaliapin. Published by Cambridge University Press.

Technology of China (MOST), the National Natural Science Foundation of China (NSFC), and the Chinese Academy of Sciences (CAS), the National Conference on Earthquake Science and Technology was held in Beijing, with the official document 'The Strategic Plan of National Earthquake Science and Technology Development (2007–2020)' published, outlining the scientific challenges and suggested actions in science and technology for the reduction of earthquake disasters in China. Areas identified as priorities for the reduction of earthquake disasters include: (1) theory and techniques for earthquake monitoring; (2) continental tectonics and continental earthquakes; (3) oceanic seismology; (4) earthquake forecasting and prediction; (5) preparedness for earthquake disasters; (6) earthquake emergency response and rescue techniques; and (7) seismological information service to scientific communities and the public. Remarkably, since the end of 2006, China has established a special mechanism of supporting science and technology for social sustainability, such as meteorological and seismological works.

23.2.2 Time-dependent seismic hazard: long to short term

China can be divided into several 'tectonic blocks', with most of the major and all of the great earthquakes distributed along the block boundary zones (Zhang *et al.*, 2003). From a large-scale geodynamic perspective, the occurrence of the Wenchuan earthquake is closely related to the interaction between the Indian plate and the Eurasian plate (Burchfiel *et al.*, 2008). However, the likelihood that the Longmenshan fault system would be able to produce a great earthquake, like the Wenchuan earthquake, was significantly underestimated.

From 1997 to the time of writing, the Bayan Har block in western China has been seismically active. A series of major to great earthquakes have occurred along the boundary zone of this tectonic block, which is the sole contributor to the major to great earthquakes in continental China in this period. These earthquakes include the 1997 Mani/Manyi (Tibet) $M_S7.5$ earthquake, the 2001 Kunlunshan/Kokoxili (Qinghai) $M_S8.1$ earthquake, the 2008 Yutian (Xinjiang) $M_S7.3$ earthquake, the 2008 Wenchuan (Sichuan) $M_S8.0$ earthquake, and the 2010 Yushu (Qinghai) $M_S7.1$ earthquake, as shown in Fig. 23.1.

At the time scale of 5–15 years, a national research project developed by CEA (Research Group, 2007) using geological, seismological, and geophysical observations identified several 'key regions subject to special monitoring', which are estimated to be the regions with the potential for large earthquakes of magnitude greater than M_S7 in the western part of China (west of 107° E) and earthquakes over M_S6 in the eastern part of China (economically more developed) for the period 2006–2020, as shown in Fig. 23.2. Among the 17 key regions identified in western China, the south-to-mid Longmenshan fault zone was identified as probable for earthquakes over M_S7.

The CEA has been organising an annual consultation on the likelihood of earthquakes since the early 1970s (see Appendix in Zhao *et al.*, (2010)). At the beginning of 2008, the annual consultation failed to identify the middle-to-north Longmenshan fault area as a potential

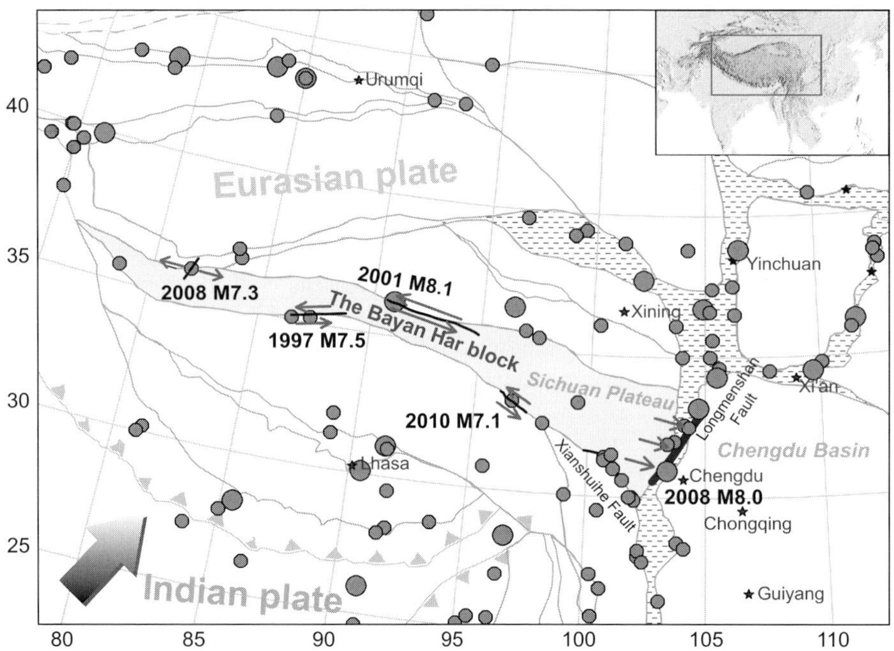

Figure 23.1. Distribution of large earthquakes between 1997 and 2013 (marked by grey dots with year and magnitude and showing the event ruptures (black lines) with grey arrows indicating their focal mechanism) around the Bayan Har block and historical big earthquakes (grey dots). The 2008 Wenchuan $M_s8.0$ earthquake was located to the east along the Longmenshan fault zone between the Sichuan plateau and the Chengdu basin. The insert in the upper right corner is the position map. Grey solid lines indicate the boundaries of the tectonic blocks (Zhang *et al.*, 2003). Note that some of the boundary zones are quite wide (marked by horizontal grey dashed lines). The shaded arrow at the bottom left shows the relative motion of the Indian plate with respect to the Eurasian plate.

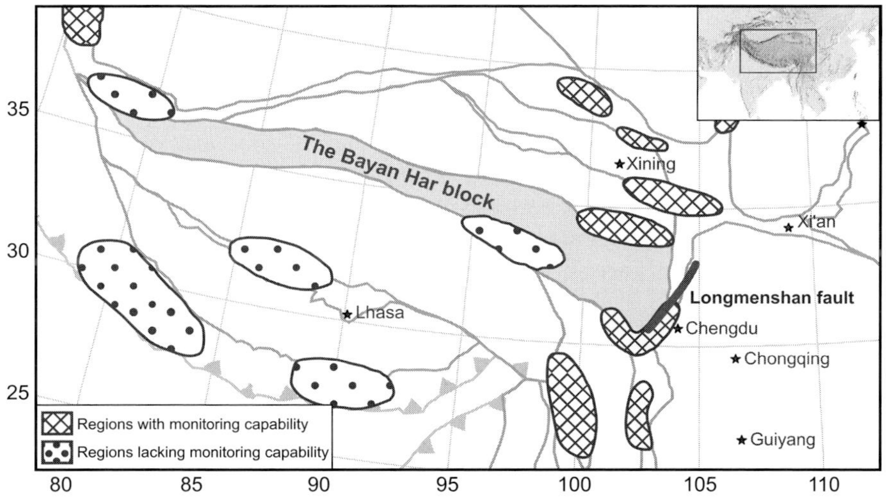

Figure 23.2. Identified regions (marked by crosses and dots to represent the monitoring capacity) with the potential for large earthquakes of magnitude greater than M_S7 in western China (west of 107° E) for the period 2006–2020 (adapted from Research Group (2007)). Grey lines indicate the boundaries of the tectonic blocks. Note that the south-to-mid Longmenshan fault zone (near Chengdu) was identified as one of the key regions.

region for earthquakes over M_S7, and underestimated the potential maximum magnitude in 2008 to be ~M_S7. In retrospect, several precursor-like anomalies were observed (see Ma and Wu, (2012) for a review of the anomalies), but none of these anomalies was so conclusive as to lead to an intermediate-term or short-term warning of the earthquake.

23.2.3 Observational infrastructure upgrading underway

Infrastructure for geodetic, seismological, and other geophysical observations and monitoring has been developed and modernised constantly since the 1990s. Supported by the central and local government, CEA began to establish the 'China Digital Seismological Observation System' in 1996, which was completed and began operation at the end of 2000. This includes the National Digital Seismograph Network, 20 regional digital seismograph networks, and a mobile digital seismograph network comprising 100 portable digital seismographs. Between 1999 and 2001, the Beijing Capital-Circle Digital Seismograph Network (covering the Beijing Municipality, the Tianjin Municipality, and Hebei Province) with real-time data transmission was established and comprises 107 seismic stations. Having been approved in July 2001, launched in June 2004, and passed the acceptance inspection in April 2008 (just one month before the Wenchuan earthquake), the 'China Digital Earthquake Observation Networks Project' marked a new stage in earthquake science and technology of China. The system consists of six components (three networks and three systems): a geophysical/geochemical-anomaly monitoring network, a digital seismograph network, and a digital strong motion network; an active fault mapping system, an earthquake emergency commanding system, and an earthquake information service system. The project has significantly enhanced the density of seismic stations, geophysical/geochemical monitoring stations, and strong-motion stations in China, with four geophysical/geochemical-anomaly monitoring stations, more than eight seismic stations, and twelve strong-motion stations for every 0.1 million km^2. The monitoring equipment is 95% digital, and the time of quick-epicentre-determination (QED) has been reduced from 30 min in the 1990s to 10 min in 2008. The completeness of the earthquake catalogue has improved from $M_L4.5$ to $M_L2.5$ nationwide, and to $M_L1.5$ in the regions near megacities.

At about the same time, infrastructure for global positioning system (GPS) monitoring was constructed jointly by CEA and other geodetic and/or geoscience agencies (the Crustal Movement Observation Network of China, CMONOC, Phase I, 1997–2001; campaigns 1999, 2001, 2004, 2007; Phase II, 2007–2011). Within 800 km of the epicentre of the Wenchuan earthquake, there were 300 regional GPS stations by early 2008. This infrastructure was severely tested during the Wenchuan earthquake.

Starting in 2005, and supported by MOST, a national project (2005–2010) was conducted to study geodynamics and seismicity in the western Sichuan region. From a seismic hazard perspective, this basic research project paid most attention to the Xianshuihe fault zone (see Figs. 23.1 and 23.2), which is near to the Longmenshan fault zone, and seemed the more seismically active of the two – in fact, the Xianshuihe fault does have a rapid slip rate (close to 1 cm y^{-1}), and can thus be expected to have earthquakes more frequently than the Longmenshan fault. The project deployed a seismic array with 200 mobile broadband seismic stations, and was the first large-scale dense seismic array experiment in China. The array captured the mainshock and the aftershocks of the Wenchuan earthquake.

23.3 Response to the earthquake

23.3.1 Governmental response and rescue actions

Prime Minister Wen Jiabao flew to the earthquake area just 90 minutes after the earthquake. The National Earthquake Relief Headquarters was established on 12 May. The emergency contingency plan rose to Level I from Level II at 22:15 on 12 May. A national earthquake emergency rescue and relief team of 184 people (consisting of 12 from CEA, 150 from the Beijing Military Area Command, and 22 from the Armed Police General Hospital) arrived in the earthquake area 10 hours after the mainshock. The Chengdu Military Area Command dispatched 50 000 troops and armed police to facilitate the relief.

Saving lives was the first priority in the emergency response. The national rescue team (working at 48 sites for 216 hours) together with 19 provincial rescue teams (about 4000 people) saved 322 people from the ruins (note that the apparently small number of survivors rescued by these teams was due to the fact that they were deployed in the most difficult areas). In total, there were 84 017 people rescued from the ruins, 4.3 million injured were provided with medical care, and about 10 000 seriously injured were moved to hospitals all over the country (data from CEA, 2009).

Foreign rescue teams from Japan, Korea, Russia, and Singapore played an active role in the rescue and relief process. Countries including Cuba, France, Italy, and the UK and organisations such as the Red Cross sent medical teams to the field. The USA shared satellite images of the disaster areas with the Chinese authorities. This marked the first time that China accepted international aid in the rescue and relief effort, and was also the first time that China allowed foreign media into the earthquake area during the emergency stage of the rescue operation.

China's first nationally organised event mourning earthquake victims was on 19 May 2008. The Wenchuan earthquake also marked the first time that volunteers and non-government organisations (NGOs) joined in the rescue widely, and was the first time that the Internet played an important role in the earthquake emergency response. All these firsts indicate the great impacts that this severe earthquake disaster had on Chinese society.

23.3.2 Aftershocks, quake lakes, and beyond: avoiding concatenated disasters

Table 23.1 lists the aftershocks over M_S5.5 in 2008. The whole aftershock zone is basically consistent with the rupture zone of the mainshock, which can be divided into northern and southern parts. Estimates of the duration, maximum magnitude, rate, and most-likely locations of the strong aftershocks, based on the presently available knowledge of seismicity, such as the Omori's law and Gutenberg–Richter's law (e.g. Reasenberg and Jones, 1989, 1994), were shown to be consistent with the observed aftershocks. Scientists of the Union Commission on Geophysical Risk and Sustainability of the International Union of Geodesy and Geophysics (IUGG-GRC) and the US Geological Survey (USGS) were invited to join the predictive study of the aftershock tendency. These estimates greatly helped in the planning for rescue and recovery. The estimates were especially important because of the landslides and landslide lakes, which threatened relief and reconstruction efforts. Also the estimates

Table 23.1. *The Wenchuan earthquake sequence in 2008 (data from the China Earthquake Networks Center – CENC)*

No.	Date	Origin time (local)	Lat.	Lon.	M_S	Note
1	2008–5–12	14:28:04	31.0	103.4	8.0	(mainshock)
2	2008–5–12	14:36:39	31.3	103.6	5.8	
3	2008–5–12	14:43:14	31.3	103.7	6.0	
4	2008–5–12	19:11:00	31.3	103.4	6.0	
5	2008–5–13	04:08:49	31.4	103.8	5.6	
6	2008–5–13	15:07:07	31.0	103.2	6.1	
7	2008–5–14	10:54:36	31.3	103.4	5.6	
8	2008–5–16	13:25:46	31.4	103.2	5.9	
9	2008–5–18	01:08:25	32.3	104.9	6.0	
10	2008–5–25	16:21:49	32.6	105.3	6.4	(the largest aftershock)
11	2008–5–27	16:37:51	32.8	105.6	5.7	
12	2008–7–24	03:54:44	32.8	105.5	5.6	
13	2008–7–24	15:09:29	32.8	105.5	6.0	
14	2008–8–01	16:32:42	32.1	104.7	6.1	
15	2008–8–05	17:49:16	32.8	105.5	6.1	
16	2008–9–12	01:38:59	33.0	105.6	5.7	

indicate that proper use of earthquake predictability, even though the scientific capability for prediction is limited, may contribute to the reduction of earthquake disasters. However, on 19 May, a false alarm of a 'magnitude 7 aftershock' broadcast on TV caused widespread social disorder in Chengdu, Xi'an, and other cities near the Wenchuan earthquake, reflecting the public understanding of earthquake hazard information, especially when such information is expressed as a probability, or more often, as the '(time of the) increase of the probability'.

The Wenchuan earthquake received widespread attention from the public. As an example, Figure 23.3(a) shows the access to the Earthquake Data Centre's website after the Wenchuan earthquake. This was the first time that the emergency response team and the scientific community has been exposed to the media in the emergency response stage (note that the first CEA Training Course of Media Spokesman was held on 7–10 May 2008), and neither the performance of scientists nor that of the Chinese news media was as good as expected. The widespread misunderstanding about the predictability of earthquakes led to seismologists being blamed for not predicting the earthquake, and to claims that the CEA had not taken any notice of predictions by amateurs. The vitriolic coverage in the media led to a significant setback in the public understanding of earthquake forecasting and distracted attention from the emergency response, rescue, and relief efforts. It took a long time to recover from this setback, which was mainly initiated and promoted by news media. About one month after the Wenchuan earthquake, news media reported that Japanese seismologists could 'predict' an earthquake several seconds before its occurrence, and this caused a lot of discussion in China. This was actually a misunderstanding of the earthquake early warning system (EEWS) used during the Iwate-Miyagi Nairiku earthquake on 14 June 2008.

One of the causes for such rampant irresponsible media coverage was that the real experts had been working on countermeasures to prevent potential concatenated disasters. As a result of the mainshock and the strong aftershocks, many rivers became blocked by large landslides. This resulted in several quake lakes, which threatened places downstream. As of 27 May 2008, 34 lakes had formed due to earthquake debris blocking and damming rivers, 28 of which were estimated as of potential danger. Downstream, people were evacuated in anticipation of these dams bursting. The largest of these quake lakes was located in Tangjiashan (31.84° N, 104.44° E) and was accessible only on foot or by air. The machinery and the fuel used to facilitate the safe discharge of the bottlenecked water had to be air lifted to the site. Fortunately, because they were engineered with the local seismicity in mind, the 1996 reservoirs and 495 dams in the disaster region withstood the Wenchuan earthquake, and there were no collapses at all during the earthquake. Engineering measures to prevent the collapse of these reservoirs and dams were another of the major concerns during the emergency response. Despite the ragged rivers, complex topography, and widespread landslides in the disaster region, cascading disasters were successfully avoided.

In order to avoid concatenation of disasters, the government organised prompt and comprehensive responses to

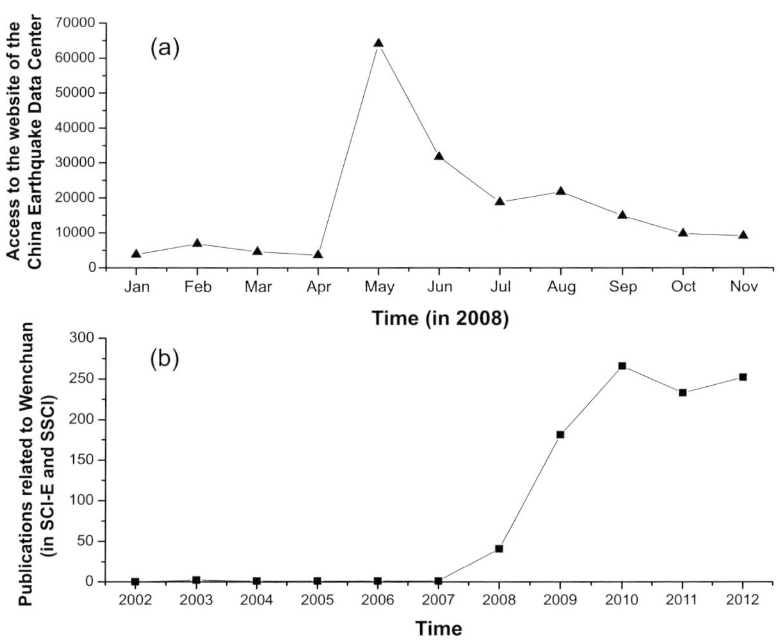

Figure 23.3. (a) Statistics of the access to the website of the China Earthquake Data Centre (http://data.earthquake.cn) in 2008, and (b) the scientific publications related to the Wenchuan earthquake. Data related to the website access are from the China Earthquake Networks Centre (courtesy of Huang Zhibin), and the data related to scientific publications are from the Science Citation Index Expanded (SCI-E) and Social Sciences Citation Index (SSCI) databases.

prevent epidemics. Communication networks were repaired and restored shortly after the earthquake. The earthquake had damaged 28 000 km of road and 656 bridges. To respond to this situation, large-scale engineering works were organised immediately to restore transportation links, despite the harsh environment with on-going landslides and rock-falls, and with the risk amplified by the on-going aftershocks and, occasionally, heavy rainfall. Temporary settlements in the disaster regions also had to be assessed for seismic and environmental safety, and Chinese engineering seismologists and geologists contributed to this assessment.

23.3.3 Technological aids to the emergency response and scientific investigations

Scientific investigations on the regional geology and on the earthquake destruction were organised almost immediately after the earthquake by CEA, the Ministry of Land and Resources, and the CAS institutions. Surface faulting, intensity, engineering destruction, and secondary hazards such as landslides, were mapped by field studies. Aftershocks and after-slip were recorded by mobile seismic and GPS stations. The NSFC and MOST supported major research projects related to the Wenchuan earthquake including a cross-Taiwan-Strait joint project on the Wenchuan earthquake studies. Several special issues of international scientific journals were published providing rich scientific results on the earthquake (e.g. Klinger *et al.*, 2010; Yin *et al.*, 2010). The statistics on the publications related to the Wenchuan case studies shows an increase in the number of articles by a factor of 5 within three years (Fig. 23.3(b)). Special sessions were run at several international symposia reflecting the on-going studies of the Wenchuan earthquake.

On 21 May, the National Expert Committee for the Wenchuan Earthquake was established by the State Council to coordinate the emergency response and reconstruction after the earthquake and to provide advice to the Council. Under the umbrella of the Committee, six working groups were set up looking at: (1) the strong aftershock tendency; (2) the seismotectonics; (3) the earthquake mechanism; (4) the evaluation of losses; (5) secondary hazards; and (6) reconstruction and recovery.

The upgraded observational infrastructure played an important role in aiding the earthquake emergency response, although it had evident shortcomings with regard to its scientific products and data service due to its short time of operation. The Wenchuan earthquake provides a typical case of the measurable complexity of the seismic source. The epicentre location or the surface projection of the rupture initiation point of this great earthquake was determined to be at 30° 57′ N×103° 24′ E. The centroid of the rupture located about 200 km northeastward clearly indicates the rupture propagation process. The focal mechanism of this earthquake was shown to be thrust-predominant with the dip angle of the earthquake fault about 60°. With noticeable uncertainties, the initiation point as indicated by the hypocentre was determined to be about 15 km by near-source seismic recordings, while the rupture spanned from the surface to the lower crust about 30 km deep. The moment magnitude of the earthquake was measured as $M_W = 7.9$. Disruption of the 'magnitude saturation' for this earthquake (M_W is smaller than $M_S = 8.0$), similar to the 2001 Kunlunshan (Kokoxili) $M_S8.1/M_W7.8$ earthquake, indicates the complexity of the source rupture process, in which the whole rupture process can be divided into several stages corresponding to a series of subevents with two main asperities (located in Wenchuan and Beichuan) broken during the earthquake. The rupture duration was determined to be about 105 s with about 50% of the whole energy released by seismic waves at between 12 and 45 s.

The surface wave magnitude was initially estimated to be $M_S7.8$ by the quick-and-rough report and was revised to $M_S8.0$ based on more data, checked recordings, and expert advice. This was the first time that the Chinese seismological agency officially revised the magnitude of an earthquake. Before it, quick estimates of earthquake magnitudes were never revised (even though they were issued with a significant bias/uncertainty) in order not to cause public misunderstanding.

23.4 After the earthquake

23.4.1 Reconstruction and recovery

The Wenchuan earthquake affected 417 county-level units (county, county-level-city, or district, according to the terminology of the Chinese governmental system) of 10 provinces, of which 51 county-level units were hard hit and/or extremely hard hit, mainly in Sichuan, Gansu, and Shaanxi Provinces, with a population about 20 million. On 19 September 2008 the directive on *The Overall Planning for Post-Wenchuan Earthquake Restoration and Reconstruction* was ratified by the State Council with the aim that basic living conditions and economic development would reach or surpass the pre-disaster level. The reconstruction used the characteristic Chinese model that 19 provinces/cities were assigned to assist their counterparts in the reconstruction (for example, Zhejiang Province was partnered with Qingchuan County), and had the modern vision that living conditions and economic development were to be considered at the same time. In mid-2011, after three years of hard work, the State Council announced that the reconstruction has been successfully completed. Remarkably, this may be the first time that China included psychological and social recovery work in post-earthquake recovery.

Taking the lessons of the Wenchuan earthquake that 7444 schools and about 100 million m^2 rural residences were damaged, a national project of reinforcement of schools was launched after the Wenchuan earthquake, and the pre-earthquake-launched project for the safety of rural residences was accelerated and strengthened. Two important lessons from the Wenchuan earthquake were that (i) active faults were the major cause of destruction of buildings with a hanging-wall effect related to the thrust rupture, and (ii) quake-induced landslides caused tremendous disasters, indicating that 'secondary' hazards are by no means of secondary importance. These lessons were taken into full account in the reinforcement.

On 27 December 2008 *The Law of the People's Republic of China on Protecting Against and Mitigating Earthquake Disasters* which was amended and adopted at the 6th Meeting of the Standing Committee of the Eleventh National People's Congress of the People's Republic of China, was promulgated, and came into effect on 1 May 2009. On 2 March 2009 the State Council ratified that, 12 May was to be the National Day for Protecting Against and Mitigating Natural Disasters, the day to enhance the public understanding of preparedness for, mitigation against, and management of natural disasters.

23.4.2 Scientific response to and new understandings of the Wenchuan earthquake

According to field investigations and geological studies, the tectonic context that the western Sichuan plateau is being thrust over the Chengdu basin at the Longmenshan fault zone (Burchfiel *et al.*, 2008) determined the behaviour of observed precursory anomalies. Earthquake-forecast-oriented Chinese literature (for example, Department for Earthquake Monitoring and Prediction of the China Earthquake Administration, 2009) described the preparation process of the Wenchuan earthquake by the 'three-units model': the deformation unit (the western Sichuan plateau), the locking unit (the Longmenshan fault zone), and the resisting unit (the Chengdu basin). 'Monitoring and modelling for prediction' became an operational goal in designing an earthquake forecasting experiment site. It has also been argued that the Zipingpu reservoir may have triggered the Wenchuan earthquake, but the evidence is not persuasive, and the study is far from conclusive.

The surface shortening versus lower crustal flow debate for the Longmenshan fault zone and its surrounding region has lasted years. Evidence of many different types (from Cenozoic tectonics to GPS measurements) suggests a low present-day convergence rate along the Longmenshan fault zone (Gan *et al.*, 2007). The convergence rate across the Longmenshan is quite low, and great earthquakes are very rare. Assessing the seismic hazard along a fault, one has to take both moment release and moment accumulation into account at the same time, that is, to consider the moment deficit (Wang *et al.*, 2010). The moment deficit indicates that very slowly slipping faults can still result in great earthquakes although with a very long recurrence time. Palaeoseismic studies that have been done since the earthquake showed clear evidence for several previous great earthquakes of similar magnitude as that of the 2008 Wenchuan earthquake, but they only occur once every 2000–3000 years. The important lesson to be learned from this event whose degree of hazard was greatly underestimated is that the earthquake potential associated with this type of fault should never be overlooked.

On 12 July 2008, two months after the Wenchuan earthquake, CAS Academician Xu Zhiqin sent a project proposal on the Wenchuan earthquake Fault Scientific Drilling (WFSD) directly to Prime Minister Wen Jiabao who ratified the proposal on 13 July. It is worth noting that in *The Strategic Plan of National Earthquake Science and Technology Development (2007–2020)*, scientific drilling of an active fault after an earthquake was listed as one of the priorities for the period 2007–2020. The geological background of the Prime Minister of China, who worked in geology in his early career, might have played an important role in his quick response. MOST approved the finance for the project in October 2008. The WFSD project was organised jointly by the Ministry of Land and Resources, CEA, and MOST. On 6 November 2008, 185 days after the Wenchuan mainshock, the WFSD-1 drilling was started at Hongkou, the southwest part of the Wenchuan earthquake fault. This was the fastest drilling response to a great earthquake, the first scientific drilling of a continental thrust earthquake fault, and the first drilling of an earthquake fault on the Chinese mainland. On 13 July 2009, the WFSD-1 completed its drilling to a depth of 1201.15 m. The earthquake rupture was identified to be at a depth of 589m. The drill was destroyed by a giant flood and landslide on 17 July, just four days after the completion of the project. Aftershocks and landslides plus floods continued throughout the drilling process. Due to engineering difficulties, the WFSD-2 drilling started on 6 July 2009 at Hongkou near the WFSD-1 drilling site and although designed to reach 3000 m, it was adjusted to reach 2200 m. The WFSD-2 drilling was completed on 5 April 2012, at a depth of 2283.56 m. On 15 December 2009, the WFSD-3 drilling (designed to reach 1400 m) was started at Jiulong, Mianzhu, the middle part of the Wenchuan earthquake fault, and was completed on 21 February 2012 at a depth of 1502.30 m. On 6 August

2012, the WFSD-4 drilling (designed to reach 3500 m) was started at Pingwu, the northeast part of the Wenhuan earthquake fault.

23.4.3 Scientific review and reflection

In July 2008, CEA organised the *Scientific Review and Reflection of the Wenchuan Earthquake*, a campaign joined by hundreds of experts and tens of the institutions and/or provincial earthquake administrations within the framework of CEA. The reports were completed in mid-2009, and four reports were issued: (1) overall summary report; (2) a summary report of the work of CEA; (3) a summary report to the central government; and (4) a summary outline to the society for public understanding. The reports included six topical issues: (1) social management and public service; (2) seismological monitoring; (3) earthquake forecasting; (4) preparedness and engineering; (5) earthquake emergency response; and (6) science and technology. Guided by a Working Committee composed of 105 experts, and working with a Consultation Committee of 66 senior experts, the reports proposed several operational suggestions, among which some of the general comments seem useful not only for China but also for other countries:

- Earthquake safety is a special type of national security, which needs the attention of the whole nation.
- Great earthquakes with extreme disaster risk are one of the threats to the sustainability of society and have to be taken into special consideration, even though they generally have long recurrence times.
- Efforts to reduce earthquake disaster risks have to be conducted comprehensively and systematically.
- National seismological agencies play an important role in coordinating and assisting the capacity building of the whole society for seismic safety.

Looking back at the forecasting of time-dependent seismic hazard at different time scales, it can be seen that the decade-scale forecast has a relatively sound scientific/empirical basis and is potentially useful for the preparedness. After the Wenchuan earthquake, CEA organised a working group on the long- to intermediate-term forecasting of major to great earthquakes (Working Group of M7, 2012). Aiming at deepening the understanding of seismotectonics, CEA launched a long-term national project which will last until 2020, which contains three components: (1) nation-wide active fault mapping; (2) geophysical background field mapping; and (3) the ChinArray (similar to the USArray) for the geophysical exploration of lithosphere structure. Phase I of the ChinArray was conducted in Yunnan-Sichuan region, the northern part of 'the central north–south seismic zone', covering the Wenchuan earthquake fault.

The seismological information service provided to the government and the public was far from satisfactory during the Wenchuan earthquake, partly because the upgraded seismological monitoring system had begun being tested less than 2 months before the Wenchuan earthquake occurred. A quick loss estimation (e.g. fatalities) gave a greatly underestimated result because (at least in part) of the use of a 'point source' model. The seismic recordings of the Wenchuan mainshock in the Beijing Capital-Circle Region mis-triggered an event report of 'an M_L3.9 earthquake in Beijing'. The fast focal mechanism solution (automatic regional CMT solution) of the mainshock provided by the China Earthquake Networks Center (CENC) was shown to be incorrect, at least partly due to the lack of an analyst-review procedure. After the Wenchuan earthquake, taking these lessons into account, CEA led a campaign to improve the seismological information service based on up-to-date science and technology, learning from the experiences of the international advanced seismological agencies such as NEIC. CEA distributed the tasks of seismological information products between its institutions and established round-the-clock monitoring on a routine basis. This provides rapid reports on the earthquake rupture process, a refined location of aftershock sequence, focal mechanisms of the mainshock and strong aftershocks, ground motion characterisation and loss estimation, and tectonic background information, among other things. A Chinese version of the 'earthquake poster' has been produced and published online since 2009. CEA has also established a mechanism to cooperate with news media for the seismological information service. Micro-blogs and mobile phone messages are used to meet the needs of fast earthquake information. This endeavour was shown to be successful during the emergency response to the 2010 Yushu M_S7.1 earthquake. Aiding the earthquake information service not only for earthquake emergencies but also for earthquake disaster preparedness, CEA launched Social Earthquake Safety Engineering Service, based on the nodes of regional and local earthquake administrations.

23.5 Concluding remarks

Disaster risk can be characterised by the convolution of natural hazards, exposure, and vulnerability/capacity: *Disaster risk = Hazard * Exposure * (Vulnerability – Capacity)*. To reduce disaster risk, an increase of capacity and a decrease of vulnerability are the key issues. The lessons learned from the 2008 Wenchuan earthquake tell us that all four terms on the right hand side of the equation are time-dependent, and the total disaster risk depends on the competition between these four time-dependent terms. Reduction of earthquake disasters should be based on scientific and technological knowledge and the social preparedness for earthquake risk. The 2008 Wenchuan earthquake was a major test of the progress made in China regarding efforts to ameliorate the effects of earthquakes: nine months before the earthquake there was the National Conference on Earthquake Science and Technology, the new digital earthquake observation system had been in operation just one month at the time of the earthquake, and the national project for the seismic safety of rural residences had just begun when the earthquake struck – as a result, those areas in which seismic reinforcement had been successfully completed withstood the shaking, while those areas where reinforcement was not complete suffered a tremendous disaster. Prepared and well-trained rescue teams played an important role in the relief, although the size of the team and its equipment were far from sufficient for a magnitude 8 earthquake. National mobilisation seemed to be carried out efficiently, but the scientific and technological support for such a mobilisation was weak. The 2008 Wenchuan earthquake provided many lessons and experiences on earthquake disaster reduction and mitigation. Based on the review (note that this chapter is far from a comprehensive and balanced review) and our analysis, we conclude that the most important lesson learned is the necessity to strengthen the efforts to cope with earthquake disasters and to be prepared for the next big earthquake.

Acknowledgments. Data collection and analysis were supported by the CEA Field Investigation of the 2013 Lushan Earthquake. ZLW thanks the CEA for being assigned to be a member of the National Expert Committee for the Wenchuan Earthquake and the group leader of the Summary Outline to the Society for Public Understanding of the *Scientific Review and Reflection of the Wenchuan Earthquake*, and the authorization for presentation at international symposia.

References

Burchfiel, B. C., Royden, L. H., van der Hilst, R. D. *et al.* (2008). A geological and geophysical context for the Wenchuan earthquake of 12 May 2008, Sichuan, People's Republic of China. *GSA Today*, **18**, 4–11.

CEA (2009). *Open File on the Scientific Review and Reflection of the Wenchuan Earthquake* (in Chinese). Beijing: China Earthquake Administration (CEA).

Chen, Y. and Booth, D. C. (2011). *The Wenchuan Earthquake of 2008: Anatomy of a Disaster*. Beijing: Science Press (in cooperation with Springer).

Department for Earthquake Monitoring and Prediction of the China Earthquake Administration (2009), *Science Report on the Wenchuan Magnitude 8.0 Earthquake* (in Chinese). Beijing: Seismological Press.

Gan, W., Zhang, P., Shen, Z.-K. *et al.* (2007). Present-day crustal motion within the Tibetan Plateau inferred from GPS measurements. *Journal of Geophysical Research*, **112**, B08416, doi:10.1029/2005JB004120.

Klinger, Y., Ji, C., Shen, Z.-K. and Bakun, W. H., eds. (2010). *Special Issue on the 2008 Wenchuan, China, Earthquake. Bulletin of the Seismological Society of America* **100**(5B), 2353–2872.

Ma, T. F. and Wu, Z. L. (2012). Precursor-like anomalies prior to the 2008 Wenchuan earthquake: a critical-but-constructive review. *International Journal of Geophysics*, **2012**, article ID 583097, doi:10.1155/2012/583097.

Reasenberg, P. A. and Jones, L. M. (1989). Earthquake hazard after a mainshock in California. *Science*, **243**, 1173–1176.

Reasenberg, P. A. and Jones, L. M. (1994). Earthquake aftershocks: update. *Science*, **265**, 1251–1252.

Research Group (2007). *Researches on Earthquake Risk Regions and Losses Prediction of China Continent from 2006 to 2020* (in Chinese). Beijing: Seismological Press.

Wang, H., Liu, M., Shen, X. H. and Liu, J. (2010). Balance of seismic moment in the Songpan-Ganze region, eastern Tibet: Implications for the 2008 great Wenchuan earthquake. *Tectonophysics*, **491**, 154–164.

Working Group of M7 (2012). *Study on the Mid- to Long-term Potential of Large Earthquakes on the Chinese Continent* (in Chinese). Beijing: Seismological Press.

Yin, A., Freymueller, J. T., Keller, G. R., Ni, S. D. and Song, X. D., eds. (2010). *Special Issue on the Great 12 May 2008 Wenchuan Earthquake (M_W7.9), China. Tectonophysics*, **491** (1–4), 1–276.

Zhang, P., Deng, Q., Zhang, G. *et al.* (2003). Active tectonic blocks and strong earthquakes in the continent of China. *Science in China. Series D: Earth Sciences*, **46**, 13–24.

Zhao, Y. Z., Wu, Z. L., Jiang, C. S. and Zhu, C. Z. (2010). Reverse tracing of precursors applied to the annual earthquake forecast: Retrospective test of the Annual Consultation in the Sichuan-Yunnan Region of southwest China. *Pure and Applied Geophysics*, **167**, 783–800.

24
The 2011 Tohoku, Japan, earthquake and tsunami

Kenji Satake

The 2011 Tohoku earthquake (magnitude 9.0) was the largest earthquake in Japan's history. It caused nearly 20 000 casualties, mostly from devastating tsunamis. The earthquake and tsunami also seriously damaged the Fukushima Dai-ichi nuclear power station (NPS), causing meltdown of the reactor, a hydrogen explosion, and the release of radioactive materials. The Japanese long-term forecast of large earthquakes failed to foresee the gigantic size of this earthquake, because it was based on a characteristic earthquake model and past historical data of a few centuries. Recently installed extensive land and marine geophysical observation networks have revealed huge seafloor displacement and details of earthquake fault motion. While the promptly issued tsunami warning saved many lives, several issues have been identified to further improve the tsunami warning system. Similar tsunami disasters have occurred in the past, but that experience was not fully utilised to reduce the 2011 disaster. The lessons learned include the importance of geophysical observations, improvement of long-term forecasting of earthquakes, effective tsunami warning, assessment for future tsunami hazard, and preparedness and education for infrequent hazards.

24.1 Introduction

A giant earthquake occurred off northern Honshu, Japan, on 11 March 2011. This earthquake, with magnitude *M*9.0, was the largest in Japan's history, and caused a devastating tsunami disaster (Fig. 24.1), as well as serious damage to the nearby Fukushima Dai-ichi NPS. The earthquake was officially named the 'off the Pacific coast of Tohoku earthquake' by the Japan Meteorological Agency (JMA), but abbreviated as the Tohoku (-oki) earthquake in scientific literature including this report. The Japanese Cabinet Office called the earthquake and tsunami disaster, including the NPS accident, 'the Great East Japan Earthquake disaster'.

The earthquake and tsunami caused approximately 15 900 deaths, with a further 3100 missing, and 6000 injured. It resulted in the total destruction of 129 000 houses, with 255 000 being partially destroyed and a further 697 000 sustaining some damaged. Among these, some 7600 houses were destroyed by ground shaking, 19 000 were damaged by liquefaction, and the rest (nearly 98%) were affected by the tsunami. Approximately 500 000 people lost their houses and were displaced. The total economic loss is estimated as 16 900 billion yen (about US$200 billion), which is about 20% of the annual national budget of Japan, according to the Central Disaster Management Council.

The Great East Japan Earthquake disaster was the worst natural disaster in Japan since the 1923 Kanto earthquake, which caused 105 000 casualties (Moroi and Takemura, 2004) and destroyed the Tokyo metropolitan area. Japan also experienced the 1995 Kobe earthquake, which claimed about 6500 human lives. The main causes of casualties were different for each of these earthquake disasters. During the 1923 Kanto earthquake, approximately 90% of casualties were due to fires caused after the ground shaking, while approximately 85% of victims from the 1995 Kobe earthquake were due to collapse of houses. For the 2011 Tohoku earthquake, 92% of casualties were due to drowning by the tsunami.

24.2 The earthquake

24.2.1 Mainshock

The earthquake source parameters provided by JMA are: the epicentre at 38° 06.2′ N, 142° 51.6′ E, depth 24 km, the origin time 14h 46m 18.1s Japan Standard Time (UT + 9 h).

Extreme Natural Hazards, Disaster Risks and Societal Implications, eds. Alik Ismail-Zadeh, Jaime Urrutia Fucugauchi, Andrzej Kijko, Kuniyoshi Takeuchi and Ilya Zaliapin. Published by Cambridge University Press.

Figure 24.1 Tsunami inundation in the Sendai plain (copyright: *Mainichi Newspaper*). For a colour version, see the Plates section.

The earthquake was felt in most of the Japanese islands, with the largest seismic intensity of 7 (the highest) on JMA's scale, which corresponds to 11 or 12 on the modified Mercalli scale or nearly 3000 Gals (30 m s^{-2}) of peak ground acceleration.

The 2011 Tohoku earthquake occurred on the boundary between the Pacific plate and the overlying plate. The Pacific plate subducts beneath northern Honshu along the Japan Trench at a rate of about 8 cm per year. This subduction causes westward movement and east–west compression of northern Honshu, as observed before 2011 by the land-based global positioning system (GPS) network. The Geospatial Information Authority of Japan (GSI) has deployed a nationwide GPS network consisting of approximately 1200 stations. The focal mechanism solution of the 2011 Tohoku earthquake shows a thrust-type fault movement on a shallow dipping plane, indicating that it was a typical interplate earthquake. Along the Japan Trench, many interplate earthquakes have occurred to release strain stored by the subducting plate (see Section 24.2.3).

24.2.2 Large offshore slip

At the time of the 2011 Tohoku earthquake, the Pacific coast of the Tohoku region moved eastward by up to several metres. In addition, the coast subsided by 1 m, which has resulted in flooding during high tide. The nationwide GPS network recorded large movements in directions opposite to the previous observations, with a maximum of 5.3 m eastward and 1.2 m downward motions (Fig. 24.2; (Ozawa *et al.*, 2011)).

In addition to the land-based network, marine geophysical measurements undertaken before the 2011 Tohoku earthquake detected huge seafloor displacement. The repeat marine geodetic measurements, using onboard GPS and acoustic signals (GPS/A), showed that the seafloor near the epicentre moved by 31 m horizontally and by 3 m vertically (Fig. 24.2 (Kido *et al.*, 2011; Sato *et al.*, 2011)). The off-line bottom pressure gauges also recorded a huge horizontal displacement of 60 m and vertical displacement of 5 m (Ito *et al.*, 2011). Repeat multi-beam sonar soundings indicated that the seafloor moved nearly 50 m horizontally near the trench axis (Fujiwara *et al.*, 2011). Such huge seafloor movements have never been recorded globally.

The 2011 Tohoku earthquake source has been modelled with various geophysical data, including seismic waves (e.g., Ide *et al.*, 2011), land-based GPS data (e.g., Ozawa *et al.*, 2011), tsunami data (e.g., Fujii *et al.*, 2011), and a combination of these (e.g., Simons *et al.*, 2011). These studies show that the seismic moment is approximately 4×10^{22} N m. The corresponding moment magnitude

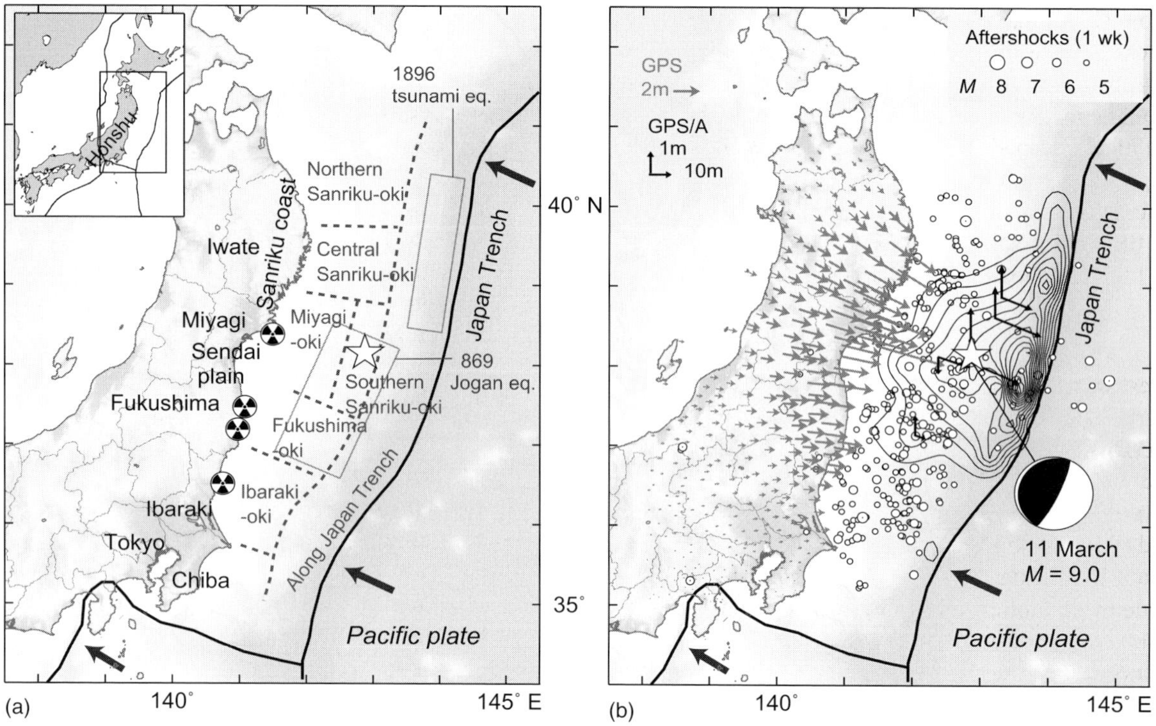

Figure 24.2. The source region of the 2011 Tohoku earthquake (the epicentre is shown by the white star). The plate boundaries and the relative motions are shown by thick curves and arrows. (a) The regions for long-term forecast (grey dashed lines; Earthquake Research Committee (2009) and the proposed fault models (grey rectangles) of the 1896 Sanriku earthquake (Tanioka and Satake, 1996) and the 869 Jogan earthquake (Satake *et al.*, 2008) are shown. The locations of four nuclear power stations are also shown by the radioactive symbols. (b) The focal mechanism solution of the mainshock (beach ball) and the locations of earthquakes with $M \geq 5.0$ (according to JMA) that occurred within a week (white circles) are shown. The slip distribution estimated from the tsunami waveforms is shown by contours at 4 m intervals (Satake *et al.*, 2013). Coseismic movements of land-based GPS stations (grey arrows, (Ozawa *et al.*, 2011)) and marine GPS/acoustic measurements (black arrows (Sato *et al.*, 2011)) are also shown.

M_w is 9.0. A common feature of the source models is that a huge (approximately 50 m) slip occurred on a shallow plate interface near the trench axis. This was a surprising result to many geophysicists, because the shallow plate interface had been considered to be weakly coupled and unable to accumulate strain.

Large interplate earthquakes occur at seismogenic zones, typically at a depth range of 10–50 km, where the plate interface is strongly coupled. The patches where large slips occur during great earthquakes are known as asperities. At shallow plate interfaces, unusual ‘tsunami earthquakes’ that generate weak seismic waves but large tsunamis, occur (Kanamori, 1972; Satake and Tanioka, 1999).

During the Tohoku earthquake, high-frequency seismic waves, which cause strong ground shaking, were generated from four small patches, called strong-motion generation areas, at the deeper plate interface. These patches correspond to asperities of previous $M{\sim}7$ earthquakes in the 1930s. The size of the Tohoku earthquake inferred from high-frequency (0.1–10 Hz) strong-ground motion records is only $M8$ (e.g., Asano and Iwate, 2012), much smaller than the moment magnitude M_w9 estimated from long-period seismic waves, GPS, or tsunami data. This indicates that the giant earthquake size and huge seafloor displacement could not be detected without modern geophysical observations. It further implies that historically similar gigantic earthquakes could have been missed, as the assessment of the size of the past earthquakes has been based on macroseismic data, namely ground shaking and tsunamis.

24.2.3 Foreshock

The 2011 Tohoku earthquake was preceded by foreshocks that occurred near the epicentre. A swarm activity including four earthquakes with $M5.2$–5.3 started on 13 February 2011 and continued until the end of February 2011. On 9 March, two days before the mainshock, an $M7.3$ earthquake occurred at about 45 km northeast of the mainshock epicentre. This earthquake is now recognised as a foreshock, because it occurred before the larger mainshock. However, between 9 and 11 March 2011, many seismologists considered that the 9 March earthquake was the anticipated earthquake in southern Sanriku-oki region (see Section 24.2.5), and did not expect that a larger earthquake would follow within a few days. The foreshock had a mechanism similar to the mainshock, with east–west compression, and produced a minor tsunami with a maximum height of 0.6 m at Ofunato. Many aftershocks of the foreshock, including an *M*6.6 event, continued until the mainshock on 11 March.

24.2.4 Activated seismicity in Japan

The Tohoku earthquake produced numerous aftershocks. Three large aftershocks with $M > 7$ occurred 22, 29, and 39 min after the mainshock. The first one (*M*7.4) occurred near the northern end of the source region, the second one (*M*7.7) occurred near the southern end. These aftershocks had focal mechanisms similar to the mainshock, and probably represent delayed rupture along the plate boundary. The third one (*M*7.5), in contrast, occurred on the east side of the Japan Trench with a normal fault mechanism. Within two months (until 10 May), five (including the above mentioned three that occurred within an hour) $M \geq 7$ earthquakes, 76 $M \geq 6$ earthquakes, and 449 $M \geq 5.0$ earthquakes occurred in the 500 km long and 200 km wide source region. The large 7 April aftershock (*M*7.1) was deeper than the plate interface and is considered to have occurred within the subducting Pacific plate.

The Tohoku earthquake also triggered earthquakes in central to eastern Japan. Three inland $M > 6$ earthquakes occurred at large distances (more than a few hundred kilometres away) within 7 days of the mainshock, the focal mechanism being different from typical east–west compression. These have been interpreted as events triggered by an increase of east–west extensive stress. In Fukushima prefecture, an *M*7.0 normal-fault earthquake occurred on 11 April. It was followed by many aftershocks, also caused by the east–west extensive stress, which was the opposite of the stress condition before 11 March. Within one year of the 11 March event, 116 large ($M \geq 6$) earthquakes occurred in and around Japan, including the Tohoku aftershocks. This number is much larger than the annual average of 14 such earthquakes.

24.2.5 Long-term forecast

Was the gigantic earthquake foreseen in Japan? The Japanese government made long-term forecasts of large earthquakes in and around Japan based on past earthquake records (Table 24.1), and published national seismic hazard maps (Earthquake Research Committee, 2006). The future probabilities of earthquake occurrence, for example, in the next 30 years, are calculated from the data of past large earthquakes by fitting inter-earthquake times by a probabilistic density function. If similar sized earthquakes recur more or less regularly (called characteristic earthquakes), and the average recurrence interval and the dates of the most recent events are known, a renewal process model called the Brownian passage time model (e.g., Matthews *et al.*, 2002) is used to calculate the

Table 24.1. *Long-term forecast of large earthquakes along Japan Trench*

Region	Type	As of 1 January 2011		As of 1 January 2012	
		M	Probability	M	Probability
Northern Sanriku	Interplate	8.0	0.5–10%	8.0 (M_t8.2)	0.7–10%
	Others	7.1–7.6	90%	7.1–7.6	90%
Central Sanriku		-	-	-	-
Miyagi-oki	Interplate	7.5	99%	7.4	Unknown
	Others			7.0–7.3	60%
Southern Sanriku	Interplate	7.7	80–90%	7.9	0%
	Others			7.2–7.6	50%
	2011 Tohoku-type	Not estimated		M_w 8.4–9.0	0%
Fukushima-oki	Interplate	7.4 (swarm)	< 7%	7.4 (swarm)	10%
Ibraki-oki	Interplate	6.7–7.2	> 90%	6.7–7.2	>90%
	Others			6.9–7.6	70%
Japan Trench	'Tsunami eq.'	M_t8.2	20%	M_t 8.6–9.0	30%
Japan Trench	Normal-fault	8.2	4–7%	8.2 (M_t 8.3)	4–7%

Interplate: recurrent characteristic earthquakes with interplate type. M_t and M_w are tsunami magnitude and moment magnitude, respectively.
Data source: Earthquake Research Committee (2009, 2011).

time-dependent probabilities. If past data indicate that the temporal distribution is random, or if there are not enough available data to indicate regular occurrence, the Poisson process is assumed to compute the time-independent probabilities.

Off Miyagi prefecture (Miyagi-oki region), to the west of the 2011 Tohoku earthquake epicentre, the probability of the occurrence of a great ($M{\sim}7.5$) earthquake between 2010 and 2040 was estimated as 99% (Earthquake Research Committee, 2009). This estimate was based on the facts that large ($M{\sim}7.4$) earthquakes have repeatedly occurred there with an average interval of 37 years since 1793, and that 34 years had passed since the last earthquake occurred in 1978. Just to the east, in the southern Sanriku-oki region, where the 2011 epicentre is located, the 30-year probability of an earthquake with $M{\sim}7.7$ was estimated as 80–90%. The forecast also stated that if an earthquake ruptures these two regions simultaneously, the earthquake size would be $M{\sim}8.0$. To the north, in the central Sanriku-oki region, no large earthquake occurrence was known hence the future probability could not be calculated. To the south, off Fukushima prefecture (Fukushima-oki), a swarm activity in which the largest earthquake was $M7.5$ was recorded in 1938 (Abe, 1977). Assuming that these were the only large earthquakes since the 1600s, when the historical records became available for the region, the 30-year probability of an earthquake of $M7.4$ was estimated as <7%.

Along the Japan Trench, long-term forecasts were made for two types of earthquakes, 'tsunami earthquakes' and normal-fault earthquakes. A 'tsunami earthquake' refers to an earthquake with relatively minor ground shaking but which causes a very large tsunami (Kanamori, 1972). As will be described in Section 24.3.2, the 1896 Sanriku earthquake was a typical example of a 'tsunami earthquake'. The 30-year probability of a tsunami earthquake with tsunami magnitude (a magnitude indicating the tsunami size) $M_t8.2$ was estimated as 20%, based on the past history of three 'tsunami earthquakes' recorded in 1611, 1677, and 1896 somewhere along the Japan Trench. A normal-fault earthquake occurs outside the trench axis, often called the outer-rise region. The 1933 Sanriku earthquake ($M8.1$) was the only known great normal-fault earthquake (Kanamori, 1971), and the 30-year probability of similar earthquake along the Japan Trench was estimated as 4–7%.

The 2011 Tohoku earthquake was much larger than the forecast, both in magnitude and the source area. The rupture started in the southern Sanriku region and propagated into the neighbouring regions. The source area was about 500 km long and 200 km wide, including the central Sanriku, Miyagi-oki, southern Sanriku, Fukushima-oki regions as well as parts of Ibaraki-oki and along the Japan Trench regions (Fig. 24.2). The long-term forecast thus failed to predict the size (magnitude) and the source region of the 2011 Tohoku earthquake. The long-term forecast was based on past earthquake records but there was no record of such a giant ($M{\sim}9$) earthquake occurring in Japan, except for the 869 Jogan earthquake, which will be discussed in Section 24.3.4. After the Tohoku earthquake, the Earthquake Research Committee (2011) tentatively revised the long-term forecast, including the Tohoku earthquake (Table 24.1).

24.3 The tsunami

24.3.1 Tsunami disaster and measurements

The tsunami arrived at the Sanriku coast about 30 min after the earthquake, with a maximum height of nearly 40 m. The tsunami reached the Sendai plain about 1 hour after the earthquake. Despite the considerable time delay between the earthquake and tsunami arrival, nearly 20 000 people lost their lives. More than two thirds of the casualties were elderly (60 years or older), who experienced difficulties in promptly evacuating. The total area of the 2011 tsunami inundation was estimated as 561 km^2 by the Geospatial Information Authority of Japan, and the population in the inundated area was about 600 000. Had the earthquake and the resultant tsunami occurred at night time, the casualties would have been even higher.

The 2011 tsunami heights on the Japanese coasts were measured by the Joint Survey Group, which consisted of ~300 researchers from more than 60 organisations (Fig. 24.3). The total number of measurement points exceeded 5900. The density of measurements far exceeds any previous tsunamis and provides invaluable data for future research. The surveys show that the tsunami heights varied locally, particularly on the saw-tooth-shaped Sanriku coast. The tsunami inundation on land extended up to 5 km. For comparison, the dataset collected by international research teams following the 2004 Indian Ocean tsunami, which affected more than 14 countries around the Indian Ocean, had approximately 1000 observation points.

24.3.2 Recorded tsunamis

The 2011 tsunami was first recorded at offshore gauges (Fig. 24.4). The ocean bottom pressure gauge TM1, located about 76 km off Sanriku coast at 1600 m water depth and connected with seafloor cable, recorded ~2 m water rise starting immediately after the earthquake, followed by an impulsive wave with an additional 3 m rise within 2 min (Fujii *et al.*, 2011). At TM2, located about 47 km from the coast at 1000 m water depth, very similar two-stage tsunami waveforms were recorded starting 7 min

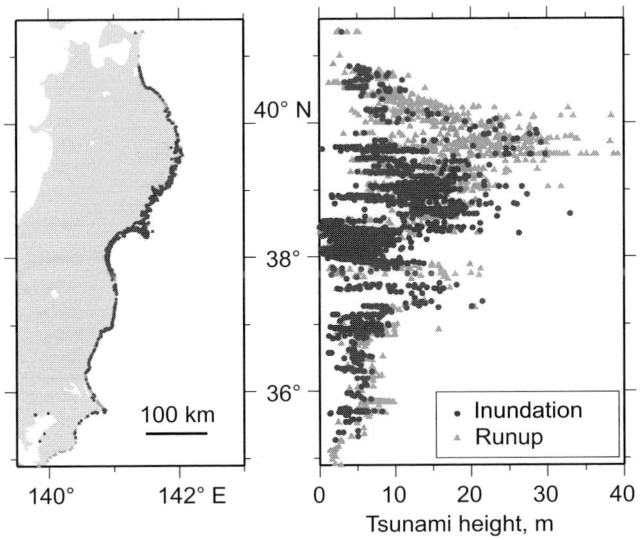

Figure 24.3. Distribution of measured tsunami heights along the Pacific coast of northern Japan by the Joint Survey Group. The original data are available at http://www.coastal.jp/tsunami2011.

after the earthquake. On a near-shore GPS wave gauge, located about 10 km offshore at a water depth of 200 m, similar tsunami waveforms with larger amplitudes were recorded starting 12 min after the earthquake. The arrival of the tsunami on the coast was recorded on the tide gauge at the port of Kamaishi starting about 30 min after the earthquake, but the large tsunami destroyed the system and hence the maximum amplitude was not recorded. The offshore tsunami generation and propagation towards the Sanriku coast were thus recorded on the offshore bottom pressure and GPS wave gauges, and the data were sent to JMA in real time to update the tsunami warning, as described in the next section.

The tsunami was also recorded on deep ocean bottom tsunami gauges (the deep-ocean assessment and recording of tsunami, or DART system operated by NOAA and other agencies) and coastal gauges around the Pacific Ocean. The maximum trough-to-crest tsunami heights recorded were 3–4 m on the west coast of the USA and the southern Chilean coast. The tsunami caused two fatalities outside Japan: one in the USA and the other in Indonesia. Vivid images of the tsunami destruction in Japan were rapidly

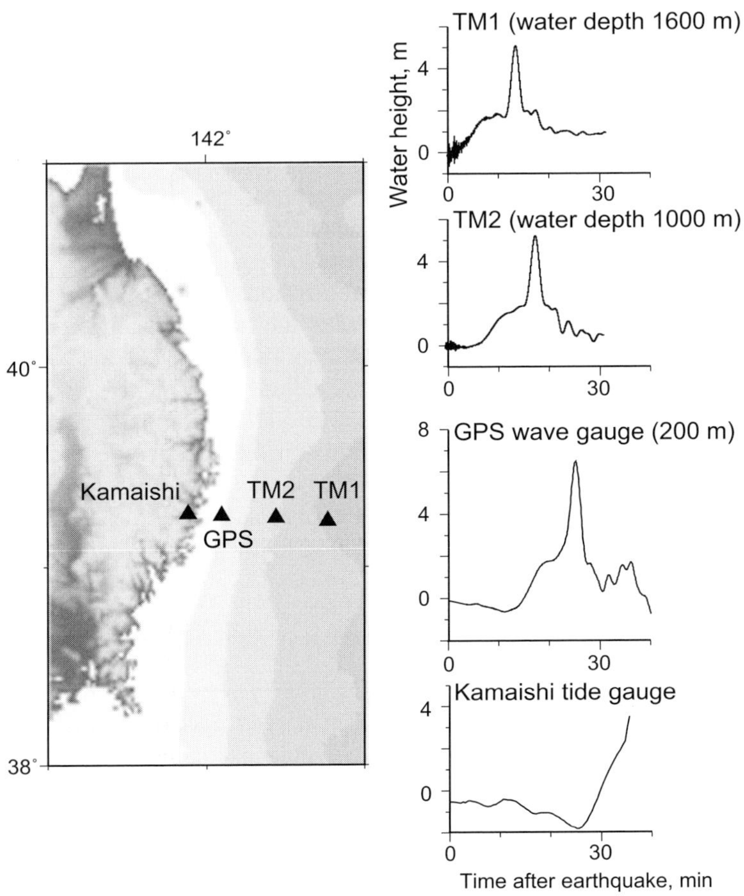

Figure 24.4. Tsunami waveforms recorded on offshore bottom pressure gauges (TM1, TM2), GPS wave gauge, and a coastal tide gauge at Kamaishi (based on Fujii *et al.*, (2011)).

Table 24.2. *Tsunami warning messages issued from the JMA*

Date	Time	After earthquake	M	Tsunami warning
11 March	14:46	0		Earthquake
	14:49	3 min	7.9	6 m Miyagi, 3 m Iwate and Fukushima
	15:14	28 min	7.9	> 10 m Miyagi, 6 m Iwate and Fukushima
	15:30	44 min	7.9	> 10 m Iwate, Fukushima, Ibaraki, and Chiba
	16:08	1.3 h	8.4	> 10 m Aomori
	18:47	∼ 4 h	8.8	Advisory region expanded
	21:35	∼ 7 h	8.8	Warning/advisory region expanded
	22:53	∼ 8 h	8.8	Warning/advisory region expanded
12 March	3:20	∼ 13 h	8.8	Warning or advisory for entire coast of Japan
	13:50	∼ 23 h	9.0	Advisory partially cleared
	20:20	1.2 days	9.0	Warning/advisory partially cleared
13 March	7:30	1.7 days	9.0	Warning all cleared
	17:58	2.1 days	9.0	Advisory all cleared

broadcast around the world, alerting the public and emergency management officials to the severity of the event. There may have been more tsunami victims around the Pacific if such images had not been immediately broadcast globally.

24.3.3 Tsunami warning

The JMA issued a tsunami warning at 14 h 49 m, or 3 min after the earthquake (Table 24.2). The JMA's tsunami warning system consists of three categories: major tsunami warning (estimated tsunami heights > 3 m), tsunami warning (1 m < tsunami height < 3 m), and tsunami advisory (0.2 m < height < 1 m). The estimated tsunami heights for the Tohoku tsunami were 6 m on the Miyagi coast, 3 m on the Fukushima and Iwate coasts, 2 m on the Ibaraki coast and the Pacific coast of Chiba, and 1 m on the Pacific coasts of Hokkaido, Aomori, and Izu islands (Ozaki, 2011). These were based on the initial estimate of magnitude (*M*7.9) and tsunami numerical simulation results stored in a database. Because the initial magnitude estimate was much smaller than the actual earthquake size (*M*9.0), the levels of tsunami warning and the estimated tsunami heights were underestimated. Nevertheless, very strong ground shaking and the rapid tsunami warning urged many coastal residents to evacuate to high ground and thus saved their lives.

In order to promptly issue tsunami-warning messages, most near-field tsunami warning systems including that of JMA rely on seismic observations to estimate the location, depth, and size of the earthquake that will potentially generate a tsunami. The tsunami-warning messages are then updated based on more comprehensive seismic observation and confirmation of a tsunami by sea-level measurements. At JMA, the epicentre and magnitude of earthquakes are first estimated within a few minutes of occurrence using a seismographic network consisting of several hundred stations. For large earthquakes, the size, or moment magnitude M_w, is usually reestimated using broadband seismograph data within 15 min. However, because of the gigantic size of the Tohoku earthquake, most domestic broadband seismograms went off the scale and JMA could not estimate M_w from Japanese data. It took them almost 1 hour to estimate the gigantic size (M_w∼8.8) using foreign seismographic observations.

The JMA observed the first tsunami arrival at Ofunato with 0.2 m amplitude, and announced this at 14 h 59 m. While this was precursor to the following large tsunami, some people erroneously interpreted the initial announcement to mean that the actual tsunami was small and slowed down their evacuation.

After detecting the large offshore tsunami on the GPS wave gauges, JMA upgraded the tsunami warning messages at 15 h 14 m (28 min after the earthquake but before the tsunami arrived at the coast) to a higher level of estimated tsunami heights. The tsunami warning messages were further updated at 15 h 30 m, and again several times later (Table 24.2). However, the updated tsunami-warning message did not reach all the coastal communities, because a power failure occurred and many people had already started evacuation.

24.3.4 Past tsunamis

The Sanriku coast had been devastated by previous tsunamis. The 1896 Sanriku earthquake generated a large tsunami, with a maximum height of 38 m, similar to the 2011 Tohoku earthquake tsunami. The 1896 tsunami caused about 22 000 casualties, somewhat more than the 2011 tsunami. One notable difference was that the ground shaking was much weaker for the 1896 Sanriku earthquake. It was a typical 'tsunami earthquake' (Kanamori, 1972), and its seismic magnitude was only *M*7.2 while the tsunami magnitude, which indicates the tsunami size, was estimated as M_t8.6–9.0. Study of tsunami waveforms indicates that the 1896 earthquake was generated from a fault motion near the trench axis (Tanioka and Satake, 1996). The 1933 Sanriku earthquake (*M*8.2) generated a tsunami up to 24 m high on the Sanriku coast, and claimed about 3000 human lives. This earthquake had a normal-fault mechanism and occurred on the eastern, or outer-rise, side of the Japan Trench. The large tsunami, damage, and casualties from the 1896 and 1933 Sanriku earthquakes were limited along the Sanriku coast.

On the Sendai plain, south of the Sanriku coast, the above Sanriku earthquakes produced only a few metres of tsunami and did not flood the plain. The 2011 tsunami inundation stretched up to 5 km from the coast in the Sendai plain (Fig. 24.1). While no similar tsunami had been recorded in the Sendai plain during the last few centuries, a similarly large earthquake and tsunami occurred there more than 1000 years ago. A national history book *Nihon Sandai Jitsuroku* depicts strong shakings, ground fissures, collapse of houses, and kilometres of tsunami flooding with 1000 people drowned in the Sendai plain in AD 869 during the Jogan era. In addition, sand brought by this tsunami, called tsunami deposits, had been found in geological layers in coastal lowland, more than 4 km from the current coast in the Sendai plain (Minoura and Nakaya, 1991; Sawai *et al.*, 2008, 2012). Based on the distribution of tsunami deposits, the 869 Jogan earthquake was modelled as an interplate earthquake with *M*8.4 (Satake *et al.*, 2008). Younger and older tsunami deposits have also been found, indicating a recurrence interval of 500–800 years if they are from the same type of earthquakes.

24.3.5 Tsunami hazard maps

Tsunami hazard maps had been made and distributed to the coastal residents, before 11 March 2011, in most of the affected coastal communities. Tsunami hazard maps show anticipated flood, or inundation, areas as well as other information such as tsunami arrival time or evacuation facilities. The tsunami inundation areas are usually estimated from past tsunami data or computed from a scenario earthquake based on long-term forecasts.

On the Sanriku coast, most hazard maps showed the inundation areas from the past tsunamis, typically the 1896 Sanriku tsunami. Because the 1896 and 2011 tsunami heights and inundation areas were roughly similar for most of the Sanriku coast, the 2011 tsunami inundation areas were also similar to those shown in tsunami hazard maps.

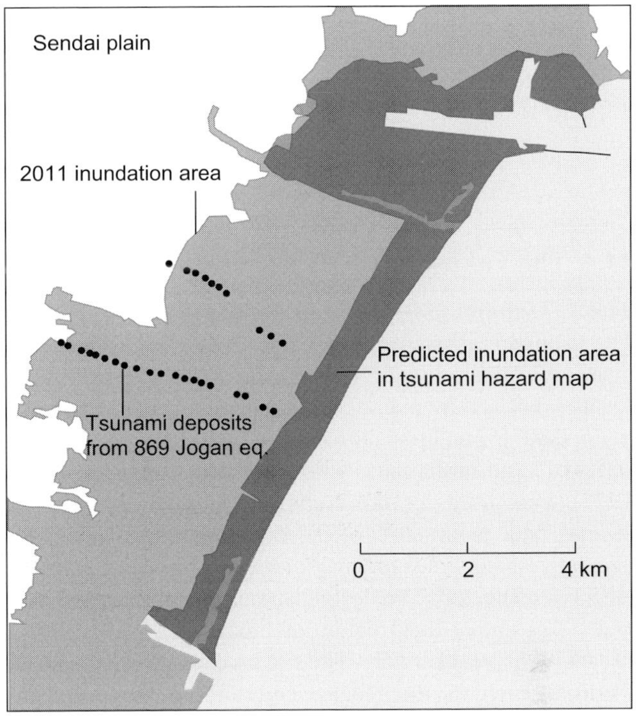

Figure 24.5. Tsunami hazard map for the Sendai plain. The expected tsunami inundation zone assuming a Miyagi-oki earthquake (*M*8.0) is shown in dark grey. The actual tsunami inundation area of the 2011 Tohoku earthquake (based on GSI data) is shown in lighter grey. The distribution of 869 tsunami deposits (Sawai *et al.*, 2012) is also shown. Modified from Central Disaster Management Council (2011).

In the Sendai plain, including Sendai city, tsunami hazard maps and other countermeasures had been prepared for a Miyagi-oki earthquake of $M{\sim}8$ which was forecast to occur with 99% probability in the next 30 years. The predicted inundation area for the $M{\sim}8$ earthquake was mostly within less than 1 km from the coast, and much smaller than the actual tsunami inundation area from the 2011 $M{\sim}9$ earthquake (Fig. 24.5). On the other hand, a distribution of the 869 tsunami deposits was similar to the inundation area of the 2011 tsunami.

24.3.6 The 2011 tsunami source

The tsunami modelling indicates that the 2011 earthquake was a combination of a 1896-type 'tsunami earthquake' and a 'Jogan-type' deeper interplate earthquake (Fujii *et al.*, 2011; Satake *et al.*, 2013). A huge slip near the trench axis (Fig. 24.2), similar to the 1896 Sanriku tsunami earthquake, caused the impulsive tsunami waves recorded on the bottom pressure and GPS gauges (Fig. 24.4), and was responsible for the high tsunami along the Sanriku coast (Fig. 24.3). The fault motion and large slip along the deeper plate interface, similar to the previous model of the 869 Jogan earthquake (Fig. 24.2), produced a long-wavelength seafloor deformation and caused the first gradual rise at the gauges (Fig. 24.4), as well as the large tsunami inundation in the Sendai plain (Fig. 24.5).

24.4 The Fukushima NPS accident

24.4.1 Tsunami assessment at Japanese NPSs

The tsunami assessments for the Japanese NPSs were made by a deterministic method, based on the past large earthquakes and by making parameter studies (Yanagisawa *et al.*, 2007). For Fukushima Dai-ichi and Dai-ni NPSs, a series of large earthquakes with magnitudes up to 7.5 (Abe, 1977) were used for the parameter studies. By varying several fault parameters of these earthquakes, the design tsunami, which yields the maximum tsunami height at each NPS, was selected. The design tsunami heights were 6.1 m and 5.2 m above local datum at Fukushima Dai-ichi and Dai-ni NPSs, respectively. The site levels, on which critical facilities such as the reactor and turbine buildings are located, were 10–13 m and 12 m, respectively, at these sites (Table 24.3).

At Onagawa NPS, the northernmost and the nearest NPS to the epicentre, the design height was 13.6 m above local datum, based on the 1896 Sanriku tsunami, and the site level was 14.8 m. At Tokai NPS, the southernmost NPS around the source area, the design tsunami height was 5.8 m above local datum, and the site level was 8.9 m.

24.4.2 Accident sequence at Fukushima Dai-ichi NPS

Fukushima Dai-ichi NPS had six units, but only three units were in operation on 11 March 2011. The strong ground shaking at 14 h 46 m automatically shut down the reactors. The strong ground shaking damaged the electricity transmission facilities on site, resulting in a total loss of external electric power supply. The emergency diesel power generators then started to pump up seawater and cool down the reactors. The tsunami, with a height of up to 15.5 m, flooded and damaged the emergency diesel generators, the seawater cooling pumps, and electric wiring system, resulting in the loss of all power and station black out. Without power, the reactors were not cooled, leading to a meltdown of the reactors followed by hydrogen explosions in three reactors and the release of radioactive materials. The area within 20 km of the Dai-ichi NPS became off limits due to high radioactivity, and the residents were forced to evacuate from the area.

24.4.3 Other NPSs

At Fukushima Dai-ni NPS, located 10 km south of Fukushima Dai-ichi NPS, strong ground shaking also shut down the system, and a similar tsunami, with a maximum height of 14.5 m, hit the site and damaged the emergency diesel generators. However, one of the external power supply systems survived, and it was used to cool down reactors of all the units.

At Onagawa NPS, strong ground shaking shut down the system and damaged four (out of five) external power supply systems. The emergency diesel generator started automatically. The maximum tsunami height was 13 m, but the critical facilities were located at 14.8 m altitude while coseismic crustal subsidence lowered the site by about 1 m. The main tsunami did not reach the facilities, although the tsunami inundated the basement of the reactor building through a seawater pump conduit. The emergency diesel generators continued to cool the reactors. When the tsunami receded, the intake of the seawater pump was exposed for several min, but water left in the piping system was used to continuously cool the reactors.

At Tokai Dai-ni NPS, the system was also shut down by the ground shaking. This site also lost all the external power supplies, and the emergency diesel generator was

Table 24.3. *Comparison of tsunami effects at four NPSs*

	Fukushima-1	Fukushima-2	Onagawa	Tokai-2
# of units	6	4	3	1
Design tsunami height	6.1 m	5.2 m	13.6 m	5.8 m
Site level	10–13 m	12 m	14.8 m	8.9 m
Tsunami inundation height	15.5 m	14.5 m	13 m	6.3 m
External power	None (of 6)	1 (of 4)	1 (of 5)	None (of 3)
Diesel generators (DG)	Damaged	Partly damaged	No damage	No damage
Power supply	None	External and DG	DG	DG
Reactor cooling	Failed	Succeeded	Succeeded	Succeeded

Data source: Nuclear and Industrial Safety Agency, Japan, and International Atomic Energy Agency (IAEA) (2011)

continuously used to cool the reactor. The tsunami height was 6.3 m and the tsunami partially flooded the pump area, but did not reach the critical facilities located at 8.9 m above the sea level.

In summary, the other NPSs than Fukushima Dai-ichi were also shut down by strong ground shaking, and were hit by similar tsunamis. Nevertheless, the reactors at these NPSs were cooled by using external power or diesel generators, and fatal accidents were avoided (Table 24.3).

24.5 Lessons learned

24.5.1 Geophysical monitoring

The 2011 Tohoku earthquake is probably the best-monitored subduction-zone earthquake. The land-based and densely-installed seismic and GPS networks recorded the propagation of seismic waves and coseismic movements of the gigantic ($M \sim 9$) earthquake. In addition, marine geophysical measurements, such as those at GPS/A measurement stations, ocean bottom pressure gauges, and multi-beam sonar sounding carried out before the earthquake, provided invaluable data to compare with the post-earthquake measurements, and coseismic changes have been detected. These measurements confirmed the huge seafloor displacement near the trench axis, and forced geophysicists to revise their views on the shallowest part of subduction zones.

The offshore gauges, i.e., the cabled ocean bottom pressure gauges and the GPS wave gauges, monitored the tsunami generation and propagation before its arrival at the coasts. Such offshore tsunami gauges provided important data for updating the tsunami warning messages. The DART system, deployed over the entire Pacific Ocean as well as in the Indian Ocean after the 2004 Indian Ocean tsunami, also monitored the tsunami propagation across the Pacific Ocean. For future tsunami warnings, multiple cabled networks of 150 bottom pressure gauges with a total length of several thousand kilometres are now being installed along the Japan Trench. This system will help the delivery of rapid and accurate tsunami warnings for Japan.

24.5.2 Long-term forecast of earthquakes

While the 2011 Tohoku earthquake occurred in a region where the estimated 30-year probability was highest (99%), the long-term forecast failed to estimate the size of this earthquake. The forecast was based on the characteristic earthquake model and past earthquake data for the last 400 years. The long-term forecast should have incorporated the 869 Jogan earthquake data, both its historical description and the palaeoseismological data such as tsunami deposits. While the simplified assumption that similar earthquakes repeat at the same location is required for the calculation of probability, this basic assumption should be also re-evaluated. Variability of recurrent large earthquakes has been pointed out for many subduction zones in the world (e.g., Satake and Atwater, 2007).

24.5.3 Tsunami warning

The tsunami warning was issued 3 min after the earthquake and saved many lives. The 2011 tsunami casualties numbered nearly 20 000, but the population in the tsunami inundation zone was 600 000. If the tsunami warning had not been issued, the number of casualties would have been much larger. One of the reasons for the large number of tsunami victims was that people did not evacuate even though a tsunami warning had been issued. This was also true for the 2010 Chilean tsunami, for which JMA issued a tsunami warning with the maximum estimated heights of up to 3 m. The 2010 tsunami height was less than 2 m, which gave the public the impression that JMA's tsunami warning was overestimated. The accuracy and reliability of a tsunami warning are very important. Besides having an accurate and reliable tsunami warning system, it is also important to educate coastal residents to escape to higher ground whenever they feel strong ground motion near coastal lowlands.

The JMA, based on lessons from the 2011 Tohoku earthquake, has made efforts to improve the tsunami warning system. This includes technical developments to estimate the earthquake size in a few minutes by using a suite of information, with some being redundant, and deploying and utilising the offshore tsunami observation systems. On the dissemination aspect, if there is a danger that the earthquake size will be potentially underestimated, the warning issued will be based on the worst-case scenario. Expected tsunami heights will be given more simply, especially with regard to the uncertainties involved.

24.5.4 Tsunami hazard assessment

After the 2011 Tohoku earthquake and tsunami, the Japanese government established a general policy for future tsunami hazards (Central Disaster Management Council, 2011). Possible future tsunamis are classified into two levels: L1 and L2. L2 tsunamis are the largest possible tsunamis which have a low frequency of occurrence, but cause devastating disasters when they do occur. For such events, saving people's lives is the first priority and adequate measures such as tsunami hazard maps, evacuation facilities, and disaster education are needed. L1 tsunamis are expected to occur more frequently, typically once in a few decades, and for these relevant countermeasures, such as breakwaters, must be constructed to protect lives and properties of residents as well as economic and industrial activities.

24.5.5 Tsunami assessment at NPSs

The estimated maximum tsunami, or design tsunami height, at Fukushima Dai-ichi NPS was 6.1 m, based on the *M*7.5 earthquake in the 1930s. The long-term forecast of Earthquake Research Committee (2009) indicated that a 'tsunami earthquake' may occur anywhere along the Japan Trench. Palaeoseismological studies proposed a fault model for the 869 Jogan earthquake. Tokyo Electric Power Company estimated that the tsunami heights at Fukushima Dai-ichi NPS from these earthquakes would be 15.7 m and 8.9 m, respectively, but did not take preventive action because they considered that such earthquakes were hypothetical (Tokyo Electric Power Company, 2012). For critical facilities such as NPSs, the seismological progress should be closely monitored and reflected in safety preparedness.

While similar tsunamis also hit other NPS, the reactors were successfully cooled using external or backup power, and fatal accidents were avoided. For such a beyond-design incident, 'defence-in-depth' countermeasures are needed to avoid a severe accident.

24.5.6 Infrequent hazard and education

Gigantic earthquakes and tsunamis are rare phenomena. Only a few $M \sim 9$ earthquakes have occurred in the world in the last century. While the Sanriku coast has experienced tsunami disasters repeatedly, the average interval between each tsunami is a few decades, i.e. once or twice in an individual's life. For such infrequent hazards, education is most important and effective to reduce disasters. For example, coastal residents should have been informed of the fact that old tsunami deposits were distributed far beyond the estimated tsunami inundation limit in the hazard map in the Sendai plain. The experience from the past tsunamis needs to be taught to next generations and to be utilised to reduce tsunami disasters.

References

Abe, K. (1977). Tectonic implications of the large Shioya-oki earthquakes of 1938. *Tectonophysics*, **41**, 269–289.

Asano, K. and Iwata, T. (2012). Source model for strong ground motion generation in the frequency range 0.1–10 Hz during the 2011 Tohoku earthquake. *Earth Planets and Space*, **64**, 1111–1123.

Central Disaster Management Council (2011). *Report of the Committee for Technical Investigation on Countermeasures for Earthquakes and Tsunamis Based on the Lessons Learned from the "2011 off the Pacific coast of Tohoku Earthquake"*, 46 pp. http://www.bousai.go.jp/kaigirep/chousakai/tohoku-kyokun/ pdf/Report.pdf.

Earthquake Research Committee (2006). *Report: National Seismic Hazard Maps for Japan (2005)*. Headquarters for Earthquake Research Promotion, 162 pp.

Earthquake Research Committee (2009). *Long-term forecast of earthquakes from Sanriku-oki to Boso-oki (revised)* (in Japanese). Headquarters for Earthquake Research Promotion, 63 pp.

Earthquake Research Committee (2011). *Long-term forecast of earthquakes from Sanriku-oki to Boso-oki* (2nd version) (in Japanese). Headquarters for Earthquake Research Promotion, 70 pp.

Fujii, Y., Satake, K., Sakai, S., Shinohara, M. and Kanazawa, T. (2011). Tsunami source of the 2011 off the Pacific coast of Tohoku, Japan, earthquake. *Earth Planets and Space*, **63**, 815–820.

Fujiwara, T., Kodaira, S., No, T. *et al.* (2011). The 2011 Tohoku-Oki earthquake: displacement reaching the trench axis. *Science*, **334**, 1240.

Ide, S., Baltay, A. and Beroza, G. C. (2011). Shallow dynamic overshoot and energetic deep rupture in the 2011 Mw 9.0 Tohoku-oki earthquake. *Science*, **332**, 1426–1429.

International Atomic Energy Agency (IAEA) (2011). *Report of Japanese Government to IAEA Ministerial Conference on Nuclear Safety – Accident at TEPCO's Fukushima Nuclear Power Station*, http://www.iaea.org/newscenter/focus/fukushima/japan-report/

Ito, Y., Tsuji, T., Osada, Y. *et al.* (2011). Frontal wedge deformation near the source region of the 2011 Tohoku-oki earthquake. *Geophysical Research Letters*, **38**, L00G05, doi:10.1029/2011GL048355.

Kanamori, H. (1971). Seismological evidence for a lithospheric normal faulting – The Sanriku earthquake of 1933. *Physics of Earth and Planetary Interiors*, **4**, 289–300.

Kanamori, H. (1972). Mechanism of tsunami earthquakes, *Physics of Earth and Planetary Interiors*, **6**, 246–259.

Kido, M., Osada, Y., Fujimoto, H., Hino, R. and Ito, Y. (2011). Trench-normal variation in observed seafloor displacements associated with the 2011 Tohoku-Oki earthquake. *Geophysical Research Letters*, **38**, L24303, doi:10.1029/2011GL050057.

Matthews, M. V., Ellsworth, W. L. and Reasenberg, P. A. (2002). A Browning model for recurrent earthquakes, *Bulletin of the Seismological Society of America*, **92**(6), 2233–2250.

Minoura, K. and Nakaya, S. (1991). Traces of tsunami preserved in intertidal lacustrine and marsh deposits. *Journal of Geology*, **99**, 265–287.

Mori, N., Takahashi, T., Yasuda, T., and Yanagisawa, H. (2011). Survey of 2011 Tohoku earthquake tsunami inundation and run-up. *Geophysical Research Letters*, **38**, L00G14, doi:10.1029/2011GL04921011-12.

Moroi, T. and Takemura, M. (2004). Mortality estimation by causes of death due to the 1923 Kanto earthquake (in Japanese with English abstract), *Journal of the Japan Association of Earthquake Engineers*, **4**(4), 21–45.

Ozaki, T. (2011). Outline of the 2011 off the Pacific coast of Tohoku earthquake (Mw 9.0) -Tsunami warnings/advisories and observations-. *Earth Planets and Space*, **63**, 827–830.

Ozawa, S., Nishimura, T., Suito, H. *et al.* (2011). Coseismic and postseismic slip of the 2011 magnitude-9 Tohoku-Oki earthquake. *Nature*, **474**, 373–376.

Satake, K. and Atwater, B. F. (2007). Long-term perspectives on giant earthquakes and tsunamis at subduction zones. *Annual Review of Earth and Planetary Sciences*, **35**, 349–274.

Satake, K. and Tanioka, Y. (1999). Sources of tsunami and tsunamigenic earthquakes in subduction zones, *Pure and Applied Geophysics*, **154**, 467–483.

Satake, K., Namegaya, Y. and Yamaki, S. (2008). Numerical simulation of the AD 869 Jogan tsunami in Ishinomki and Sendai plains (in Japanese with English abstract), *Annual Report of Active Fault Paleoearthquake Research*, **8**, 71–89.

Satake, K., Fujii, Y., Harada, T. and Namegaya, Y. (2013). Time and space distribution of coseismic slip of the 2011 Tohoku earthquake as inferred from tsunami waveform data. *Bulletin of the Seismological Society America*, **103**, 1473–1492.

Sato, M., Ishikawa, T., Ujihara, N. *et al.* (2011). Displacement above the hypocenter of the 2011 Tohoku-oki earthquake. *Science*, **332**, 1395.

Sawai, Y., Fujii, Y., Fujiwara, O. *et al.* (2008). Marine incursions of the past 1500 years and evidence of tsunamis at Suijin-numa, a coastal lake facing the Japan Trench. *Holocene*, **18**, 517–528.

Sawai, Y., Namegaya, Y., Okamura, Y., Satake, K. and Shishikura, M. (2012). Challenges of anticipating the 2011 Tohoku earthquake and tsunami using coastal geology. *Geophysical Research Letters*, **39**, L21309, doi:10.1029/2012GL053692.

Simons, M., Minson, S. E., Sladen, A. *et al.* (2011). The 2011 magnitude 9.0 Tohoku-Oki earthquake: Mosaicing the megathrust from seconds to centuries. *Science*, **332**, 1421–1425.

Tanioka, Y. and Satake, K. (1996). Fault parameters of the 1896 Sanriku tsunami earthquake estimated from tsunami numerical modeling. *Geophysical Research Letters*, **123**, 1549–1552.

Tokyo Electric Power Company (2012). *Fukushima Nuclear Accident Analysis Report*. http://www.tepco.co.jp/en/press/corp-com/release/2012/1205638_1870.html.

Yanagisawa, K., Imamura, F., Sakakiyama, T. *et al.* (2007). Tsunami assessment for risk management at nuclear power facilities in Japan. *Pure and Applied Geophysics*, **164**, 565–576.

25

India's tsunami warning system

T. Srinivasa Kumar, Shailesh Nayak, and Harsh K. Gupta

25.1 Introduction

The 26 December 2004 Sumatra earthquake of moment magnitude (M_w) 9.2 is the second largest earthquake ever recorded. The tsunami created by this earthquake was the deadliest known, having claimed over 250 000 lives and causing immense financial losses in south and southeast Asian countries (Gupta, 2008). The tsunami magnitude (M_t) for the tsunami is estimated as 9.5. It claimed about 15 000 lives in India. Amplitudes of 4.5 and 3.2 m were estimated at Port Blair and Chennai. A tsunami of this magnitude had been never recorded before on the Indian coasts. Table 25.1 provides information on known tsunamis in the north Indian Ocean.

As far as the Indian Ocean rim countries are concerned, there was no tsunami warning network available for the region before the occurrence of the 26 December 2004 earthquake. The Pacific Tsunami Warning Center (PTWC), West Coast and Alaska Tsunami Warning Center (WC/ATWC), and North West Pacific Tsunami Advisory Center (NWPTAC) of the Japan Meteorological Agency (JMA) were operational in the Pacific Ocean to provide tsunami warning in their respective areas of responsibility based on earthquake parameters and water level observations. After the Indian Ocean Tsunami, PTWC and JMA started providing interim advisories to the Indian Ocean region.

25.2 Sources for tsunamigenic earthquakes for the Indian region

Global seismic activity is depicted in Fig. 25.1. The circum-Pacific belt accounts for about 75% of the global seismic energy release, the Alpine-Himalayan belt for 20%, and the mid-oceanic ridges and the stable continental regions for the remaining 5%. A tsunamigenic earthquake could occur almost anywhere along the circum-Pacific seismic belt. There were more than 100 large and small tsunamis in the Pacific as against only three in the Indian Ocean during the twentieth century. Our initial assessment showed that there are only two areas capable of generating tsunamis that could cause damage to the Indian Ocean rim countries. These are an area from Sumatra to the Andaman Islands (about 4000 km in length) and a patch with a length of some 1000 km off the Makran coast in the Arabian Ocean, as shown in Fig. 25.1 (Gupta, 2005). This finding was presented at the Inter-governmental Commission on Oceanography (IOC) meeting held at Paris soon after the 26 December 2004 earthquake, and finally formed a part of the communiqué at the IOC meeting at Mauritius on 14–16 April 2005 (http://ioc3.unesco.org). It reads as follows:

> Recognize the unique tectonic plate structure of the Indian Ocean, and that there are primarily two tsunamigenic sources that could affect the coastlines of the Indian Ocean, namely the Indonesian seismic zone and its extension, about 4000 km in length, and the Makran source.

25.3 Initiation of the Indian programme

Unlike for the Pacific Ocean rim countries, which could be affected by tsunamis generated from earthquakes anywhere on the Pacific seismic belt, for the Indian Ocean rim countries, including India, the parts of the seismic belts that are tsunamigenic are limited. The Indian programme was therefore envisaged to focus on these two seismic areas, namely Sumatra to the Andaman Islands and the area off the Makran coast. One specific aim was the deployment of ocean bottom pressure recorders to cover both the tsunamigenic source areas. Early in 2005 the Government of India approved a programme to set up an end-to-end system to mitigate tsunamis with a direct cost of US$30 million and within a time frame of 30 months.

Extreme Natural Hazards, Disaster Risks and Societal Implications, eds. Alik Ismail-Zadeh, Jaime Urrutia Fucugauchi, Andrzej Kijko, Kuniyoshi Takeuchi and Ilya Zaliapin. Published by Cambridge University Press.

Table 25.1. *Tsunamis in the Indian Ocean (after Murty et al., 1999, Ramanamurthy et al., 2005, Shailesh Nayak and Srinivasa Kumar, 2008, Borrero et al., 2011, Srinivasa Kumar et al., 2012a,b)*

Date	Remarks
326 BC	Army of Alexander the Great suffered
Between 1 April and 9 May 1008	Tsunami on the Iranian coast from a local earthquake
27 August 1883	Krakatoa: 1.5 m tsunamis at Madras, 0.6 m at Nagapattinam, 0.2 m at Arden
1884	Earthquake in the western part of the Bay of Bengal, tsunamis at Port Blair, Dublet (mouth of Hooghly River)
26 June 1941	8.1 earthquake in the Andaman Sea at 12.9° N, 92.5° E, tsunamis on the east coast of India with amplitudes from 0.75 to 1.25 m
27 November 1945	8.25 earthquake 70 km south of Karachi at 24.5° N, 63.0° E, tsunami amplitude at Kutch was 11.0–11.5 m
19 August 1977	8.0 earthquake in Sunda Islands of Indonesia at 10.5° S, 118.8° E, tsunami heights were 5–8 m at Lunyuk on Sumbawa Island, and 5 m at Leterua on the south coast of Sumba Island
26 December 2004	9.2 earthquake off coast of northern Sumatra, Indonesia, at 3.3° N, 95.8° E. Amplitudes of 4.5 and 3.2 m were estimated at Port Blair and Chennai
28 March 2005	8.6 earthquake off coast of northern Sumatra, Indonesia, at 2.1° N, 97.0° E, tsunami run-up about 2–4 m observed on southern Nias and Simeulue Islands, Indonesia
12 September 2007	8.5 earthquake off coast of Southern Sumatra, Indonesia, at 4.5° S, 101.3° E, tsunami amplitude about 0.60 m at Padang, Indonesia; 0.08 m at Port Blair and 0.18 m at Chennai, India
11 April 2012	8.6 earthquake off west coast of northern Sumatra, Indonesia, at 2.3° N, 93.1° E, tsunami amplitude about 1.06 m at Meulaboh, Indonesia, 0.30 m at Campbell Bay and 0.18 m at Chennai, India

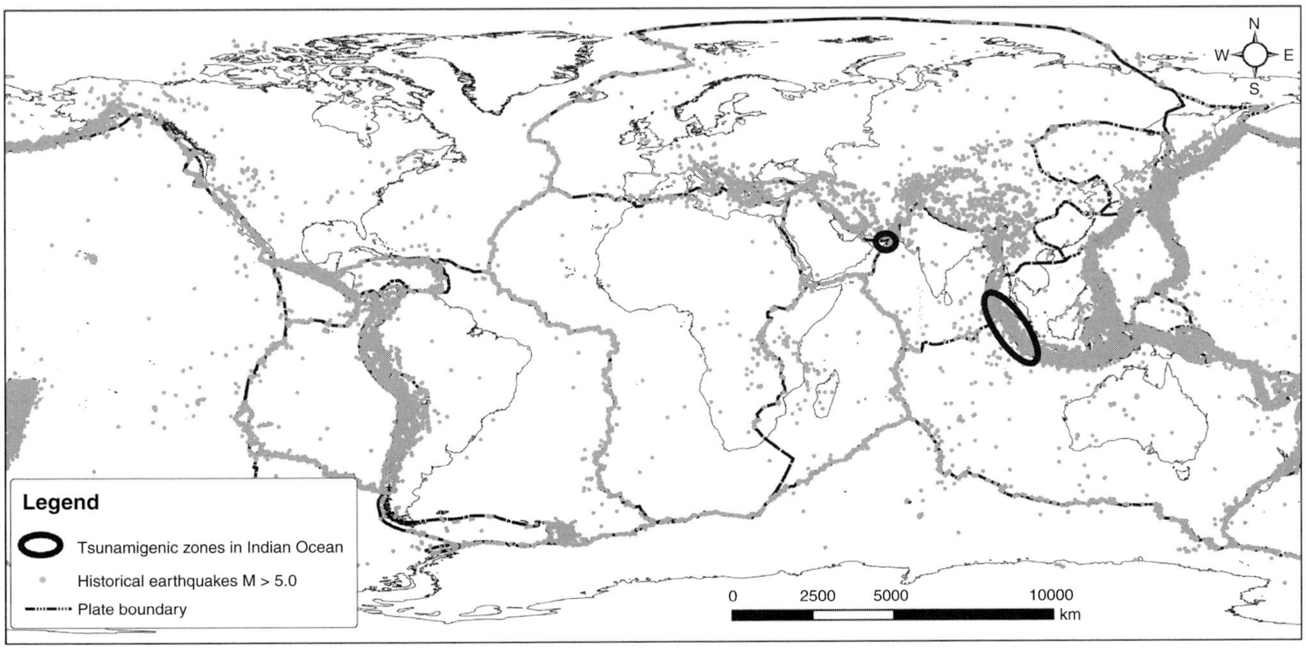

Figure 25.1. Global seismic activity map.

25.4 Indian tsunami early warning system (ITEWS)

The mission of the Indian tsunami early warning system (ITEWS) is to provide early warnings of potentially destructive tsunamis. It provides this information to emergency officials, and as appropriate, directly to the public. In order to carry out its mission, the ITEWS consists of a real-time network of local and global seismic stations that monitor seismicity to locate and size potentially tsunamigenic earthquakes, a sea-level network of bottom pressure recorders

(BPR), and tide gauges to verify the generation and evaluate the severity of a tsunami. It operates as a 24 hours a day, 7 days a week operational tsunami warning centre to analyse and provide timely advisories to the vulnerable community based on the back-end support of the scenario database, vulnerability modelling, and decision support system by means of various available communication methods (Gupta 2005). The performance of the system has been tested during all major earthquake events since its inception and it has been found that ITEWS earthquake parameters as well as its tsunami advisories are comparable to or better than those estimated by international agencies, and fall within the targets set by ICG/IOTWS (IOC-ICG/IOTWS-V/13, 2008), as described in Section 25.5. The components of the ITEWS are described in the following sections.

25.4.1 Seismic network

A real-time network of seismic stations is one of the prime requirements for the warning centre to detect earthquakes and estimate focal parameters in the two known tsunamigenic zones in the Indian Ocean. As part of the Indian tsunami early warning system, a real-time seismic monitoring network of 17 broadband seismic stations has been established. In addition to these stations, data from about 300 international seismic stations are also received at the warning centre in real-time, which enables autolocation of earthquakes within 5–10 minutes of any event.

25.4.2 Sea-level network

In order to confirm whether the earthquake has actually triggered a tsunami, it is essential to measure the changes in water level with high accuracy. A network of BPRs installed close to the tsunamigenic source regions is used to detect the propagation of tsunami waves in the open ocean and consequent sea-level changes. These BPRs can detect changes of 1 cm at water depths up to 6 km. Currently four out of the seven planned BPRs are operational. Two BPRs, TB09 and STB01, are deployed in the Bay of Bengal at average depths of 2297 m and 3793 m respectively. Another two BPRs, TB12 and STB02, are deployed at average depths of 3094 m and 2614 m, respectively, in the Arabian Sea. These BPRs send their data to the Indian Tsunami Early Warning Centre (ITEWC) via satellite on a 6-hour transmission schedule. They also have an internal tsunami-mode triggering mechanism that activates more frequent transmissions and a higher sampling rate if a tsunami is detected. The data from the BPRs are supplemented by a network of 36 tidal gauges which monitor the progress of tsunami waves, and which are used to validate the models. Currently, 26 tide gauges are operational. In addition to the instruments installed by India, near-real-time data are also received from stations operated by several other countries. The software necessary for real-time reception, display, and archiving of sea-level data has been developed at the ITEWC. The network of seismic and sea-level sensors is shown in Fig. 25.2.

25.4.3 Quantitative tsunami forecast system using numerical models

To overcome the problems associated with extremely small timelines and low accuracies associated with qualitative tsunami warning, the concept of a scenario database using tsunami pre-run simulations was implemented at ITEWC using the TUNAMI-N2 (Tohoku University's numerical analysis model for investigation of near-field tsunamis, version 2) model, customised for the Indian Ocean. The implementation approach required complete scientific understanding of the problem – from tsunami generation to propagation to inundation. The model uses earthquake source parameters such as earthquake location, fault length, fault width, slip, strike, dip, rake, and depth to compute the initial water level changes (Mansinha and Smylie, 1971). For tsunami warning purposes, a strike angle along the trench or parallel to the nearest coast, a dip angle of 45 degrees, and a slip angle of 90 degrees constitute the worst case representing a pure dip slip reverse fault mechanism. Fault length, width, and slip are calculated from the magnitude using global relations (Papazachos *et al.*, 2004). The model simulates wave propagation based on this worst case and estimates possible tsunami travel times and run-up heights at about 1800 coastal forecast points covering all Indian Ocean rim countries. The ITEWC maintains a large database of pre-run scenarios of all possible earthquakes in the two subduction zones for timely generation and dissemination of quantitative tsunami advisories. Following a large under sea earthquake, the closest scenario is picked from the pre-run scenario database to provide tsunami threat information at spatial units called 'coastal forecast zones' (CFZs), thus enabling administrators to take appropriate action to mitigate the impact of tsunamis. CFZs are high-resolution spatial areas of 100 km × 50 km, extrapolated to district-level. In the case of India these CFZs cover a total length of 7517 km of Indian coastline.

As an improvement to the concept of a static scenario database, ITEWC now uses 1000 unit sources covering both tsunamigenic regions in the Indian Ocean. Each unit source is 100 km long and 50 km wide, which represents a rupture caused by an *M*7.5 earthquake with a slip of 1 m. During any earthquake event, the basic unit source open-ocean propagation scenarios are selected from the scenario database which are dynamically combined and scaled up or down with a slip parameter (Gica *et al.*, 2008) depending on the earthquake's location and magnitude to resemble the

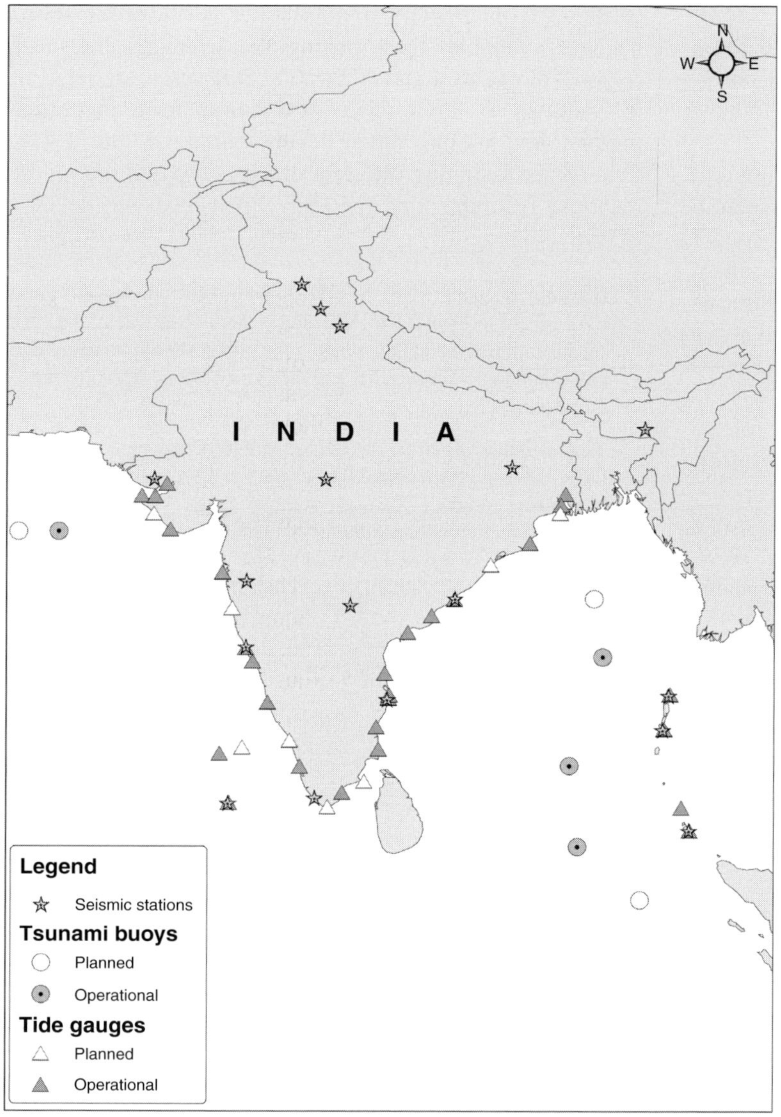

Figure 25.2. Network of Indian seismic and sea-level sensors.

actual situation. The goal is to generate only one final scenario with a finite number of scenarios for infinite combinations of earthquake locations and magnitudes. The main advantage with this kind of approach is that it eliminates the need for accessing a huge static scenario database for scenario matching. In addition, for higher magnitudes, such as 9.0, the rupture can be represented more realistically with multiple rupture zones of *M*7.5, instead of a single uniform rupture zone. Real-time observations of sea level can then be used to 'invert' for the slip parameter and update the forecast.

25.4.4 Standard operating procedure (SOP) and decision support system (DSS)

From a tsunami warning perspective, different Indian coastal areas can be categorised as being near-source or far-source based on their proximity to the tsunamigenic earthquake source regions. For example, for earthquakes happening in the Sunda subduction zones, coastal areas in the Andaman and Nicobar Islands with tsunami travel times of less than 60 minutes are categorised as near-source and the areas on the mainland are categorised as far-source. The ITEWC has a very uniquely designed standard operating procedure (SOP) and decision support system (DSS) for the generation of timely and accurate tsunami bulletins to handle both near-source and far-source coastal regions appropriately. The SOP of ITEWC begins with the detection of an earthquake of magnitude $\geq$ 6.5 within the Indian Ocean and magnitude $\geq$ 8.0 outside the Indian Ocean recorded by the seismic network. The first bulletin issued immediately after an earthquake is called an earthquake information bulletin. This is immediately followed by a second bulletin containing quantitative tsunami threat information from model scenarios. Subsequent bulletins are issued at regular intervals as and when updated earthquake information and/or water level observations become available until the threat is deemed to have passed. The SOP flowchart is given in Fig. 25.3.

As soon as the earthquake is detected, the matching unit source scenarios are picked from the pre-run scenario database and scaled dynamically to generate a resultant scenario that matches the earthquake event. This resultant scenario is used to generate a quantitative tsunami forecast with an estimated time of arrival (ETA), estimated wave amplitude (EWA), and a threat level (Warning / Alert / Watch / No threat) for all coastal forecast zones.

The accuracy of tsunami warnings is directly related to the available public response time. Since near-source coastal regions have very much less response time, they have to be warned based solely on the initial earthquake parameters and scaled model scenarios. However, far-source regions can be warned after updated earthquake parameters and/or water level observations become available, thus making the warnings for those regions more accurate. ITEWC's SOP uses an ETA of 60 minutes to differentiate near-source from far-source regions and assign threat levels based on the EWA. Criteria for assigning different threat levels and the public response recommended for each threat level are detailed below:

Warning is issued when the ETA is within 60 minutes and the EWA is greater than 2.0 m. Warning is the highest threat level that calls for immediate action wherein the public should be advised to move inland towards higher grounds and vessels should move into deep ocean.

Alert is issued when (i) the ETA is within 60 minutes and the EWA is between 0.5 and 2.0 m or (ii) the ETA is greater than 60 minutes and the EWA is greater than 2.0 m. Alert is the second highest threat level that calls for immediate action wherein the public should be advised to avoid beaches/ low-lying coastal areas and vessels should move into deep ocean.

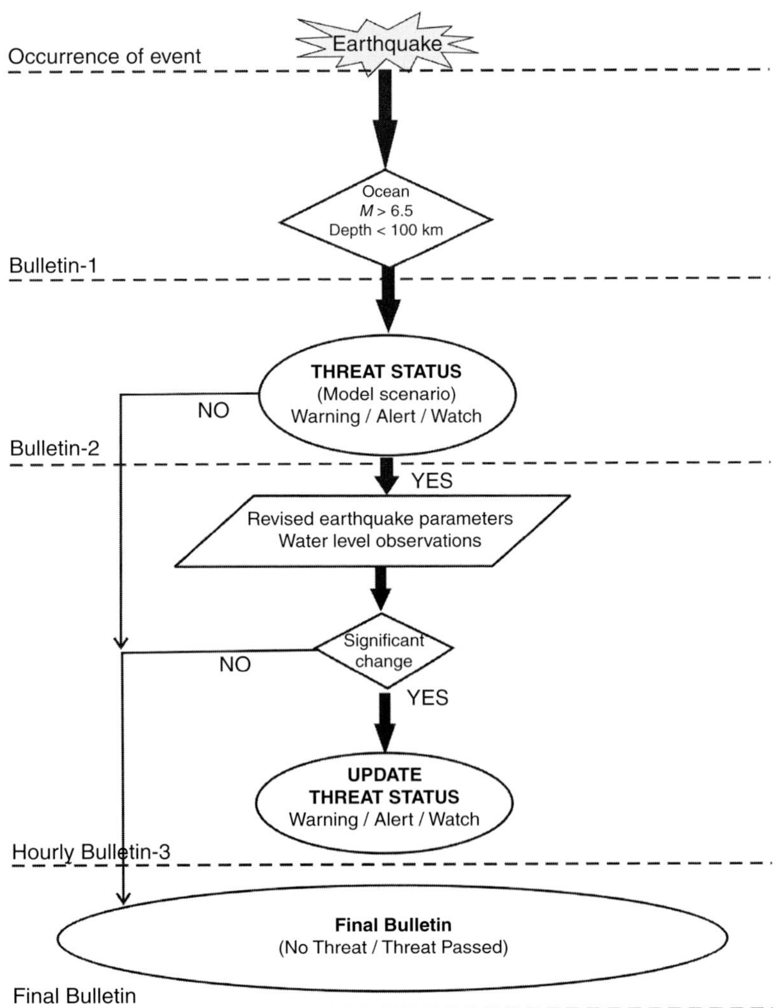

Figure 25.3. SOP of the ITEWC.

Watch is issued when (i) the ETA is within 60 minutes, or it is greater than 60 minutes and the EWA is between 0.2 and 0.5 m or (ii) the ETA greater than 60 minutes and the EWA is between 0.5 and 2.0 m. Watch does not call for any immediate public action and is only meant as to alert emergency management officials. Public action may be required only if the Watch is upgraded to an Alert/Warning.

No threat is issued when the EWA is less than 0.2 m and no action is required.

The advantage of this approach is that while the possibility of false alarms is higher for areas close to the earthquake source, for the far-source regions since the advisories are given only after confirmation of water-level data, false alarms are minimal. This concept has proven to be highly accurate compared to the traditional tsunami warning method that generates bulletins based solely on earthquake magnitude.

25.4.5 Warning centre infrastructure

The ITEWC established at the Indian National Centre for Ocean Information Services (INCOIS) serves as the 24×7 operational centre responsible for issuing tsunami bulletins to India as well as to other Indian Ocean rim countries. The state-of-the-art early warning centre has all the necessary computational and communication infrastructure that enables real-time data reception, display, analysis, modelling as well as generation and dissemination of tsunami advisories. Seismic and sea-level data are continuously monitored in the early warning centre using a custom-built software application that generates alarms/ alerts in the warning centre whenever a pre-set threshold is crossed. A display wall facilitates the visualisation of different data streams, pre-run model outputs, and vulnerability maps. Tsunami advisories like warnings/ alerts/watches are generated on the basis of pre-set decision support rules and disseminated to concerned authorities for action, following SOP.

25.4.6 Dissemination system

To ensure timely and effective dissemination of tsunami advisories, communication methods capable of rapidly reaching all stakeholders are essential. A real-time dissemination system has been established for disseminating the bulletins to the Ministry of Home Affairs (MHA), Ministry of Earth Sciences (MoES), National Disaster Management Authority (NDMA) as well as State and District Administrations through the global telecommunication system (GTS), fax, phone, SMS messages, and emails. The earthquake information, tsunami bulletins, and real-time sea-level observations are also made available on a dedicated website (http://www.incois.gov.in/Incois/tsunami/) for officials, the public, and the media. Users can also register on the website to receive earthquake alerts and tsunami bulletins through emails and SMS messages.

25.4.7 Capacity building, training, and education

While a tsunami cannot be prevented, its impact can be mitigated through community and emergency preparedness, timely warnings, an effective response, and public awareness. An effective tsunami early warning system requires continuous capacity building and training to support its day-to-day progress. The training should enable the public to understand tsunami risk and know ways and means by which they can mitigate the hazard. ITEWC conducts training and capacity building workshops regularly and also whenever new methods, technologies, and practices are identified. These programmes help in facilitating an effective and efficient response and coordination mechanism. Easily understandable publicity material on earthquakes, tsunamis, and storm surges in vernacular languages is being created for distribution to all stakeholders. A dedicated multilingual website is also being developed to provide information on tsunamis.

As a part of its regional operations, the ITEWC actively participates in communications tests (COMMs Tests), SOP workshops, tabletop exercises and mock tsunami drills (Indian Ocean wave exercises) conducted by the International Coordination Group for the Indian Ocean Tsunami Warning and Mitigation System (ICG/IOTWS) of UNESCO, which are carried out to validate the dissemination and reception of tsunami notification messages, access to the website containing bulletins, threat maps, and tsunami information tables with wave arrival and amplitude details. The mock tsunami drills not only emphasise the testing of communications from the warning centre to its stakeholders, but also provide an opportunity for testing national/state/local chains of command and decision-making, including the alerting and evacuation of people from selected coastal communities. They also provide the prospect of promoting emergency response plans and tsunami preparedness. The latest (at the time of writing) Indian Ocean wide mock tsunami drill was conducted on 12 October 2011.

25.5 Performance of the system

One of the most critical aspects of a tsunami warning system is to estimate earthquake parameters with reasonable accuracy in the shortest possible time (i.e., within 10 minutes for local events). The earthquake location and its magnitude are two other critical parameters to be estimated so that the right scenario can be chosen.

An acceptable error of ±0.2 in magnitude and 30 km in location during initial estimates is considered good for tsunami warning applications and it is also extremely important to remain within this error consistently. From its inception in October 2007 untill April 2012, the ITEWC successfully monitored 306 earthquakes of $M > 6.5$, out of which 56 were in the Indian Ocean region (both on land and under-sea). For all these events, the average elapse time of ITEWC for earthquake detection was 6 min (target 10 minutes). The average difference in magnitude of the ITEWC estimate with those of international agencies such as the United States Geological Survey (USGS) was 0.20 (target 0.3), the average focal depth difference was 22.5 km (target 25 km), and the average earthquake location difference was 30 km (target 30 km). The above statistics indicate that the earthquake parameters estimated by ITEWC are comparable with or better than those estimated by international agencies and fall within the target set by ICG/IOTWS.

During the above period, tsunami bulletins were issued by ITEWC for 30 tsunamigenic events in the Indian Ocean. The capabilities of this warning system have been well proven during all the major tsunamigenic earthquake events in the Indian Ocean. It should be noted that a tsunami Warning/Watch/Alert has been issued by ITEWC on only five occasions (Table 25.2). A warning generated solely on the basis of earthquake parameters, as is the case for many traditional warning systems, would have issued an Indian-Ocean-wide tsunami warning for several more events in the selected near-source areas in the Andaman and Nicobar Islands and the east coast of India. Thus, the timely advisories generated based on ETAs and EWAs and revised bulletins based on water level observations avoided false alarms and unnecessary public evacuations for all major events in the Indian Ocean.

Case studies for two major tsunamigenic earthquakes are described below.

25.5.1 Case study 1: 12 September 2007 earthquake

On 12 September 2007 at 11:10:26 UTC an earthquake of magnitude M_w8.4 occurred off the west coast of Sumatra, Indonesia, as the result of thrust faulting on the boundary between the Australia and Sunda plates. The displaced seafloor generated a tsunami of about 60 cm near the source. On the same day at 23:49:04 UTC a second event of M_w 7.9 also occurred 225 km northwest of the first event.

As per ITEWC's SOP the first event resulted in the issue of an Alert only for the Andaman and Nicobar Islands based on the ETA and EWA from the pre-run model scenario. All other regions were issued a No threat thus avoiding unnecessary evacuations. ITEWC monitored the real-time sea-level gauges and observed 1–2 cm wave amplitudes at three Indian tsunami buoys. Additionally, wave amplitudes of 60 cm at Padang, 50 cm at Cocos Island, 30 cm at Sabang were observed by the international tide gauge network. After confirmation that a significant tsunami had not been generated, ITEWC issued a Threat passed bulletin.

The estimated tsunami wave arrivals at Indian BPRs compared well with real-time observations. The observed water levels at coastal tide gauge locations also matched well (within 15–20% error) with those predicted using the model (Shailesh Nayak and Srinivasa Kumar, 2008).

25.5.2 Case study 2: 11 April 2012 earthquake

On 11 April 2012 at 08:38 UTC a great earthquake of magnitude M_w 8.5 occurred off the west coast of northern Sumatra, Indonesia, as a result of strike-slip

Table 25.2. *Major events where ITEWC issued tsunami Warning/Watch/Alert (Srinivasa Kumar et al., 2012a,b)*

Number of the issued alert	Magnitude and origin time	Region Name	Remarks
1	M8.5 on 12 September 2007, 11:10:26	Southern Sumatra, Indonesia	Tsunami Alert for a few regions in the Andaman and Nicobar Islands
2	M7.8 on 12 September 2007, 23:49:05	Kepulauan Mentawai region, Indonesia	Tsunami Watch for a few regions in the Andaman and Nicobar Islands
3	M6.9 on 30 March 2010, 16:54:50	Andaman and Nicobar Islands, India	Tsunami Watch for a few regions in the Andaman and Nicobar Islands
4	M7.5 on 12 June 2010, 19:26:47	Nicobar Island, India	Tsunami Watch for a few regions in the Andaman and Nicobar Islands
5	M_w8.5 11 April 2012, 08:38:38	Off west coast of northern Sumatra, Indonesia	Tsunami Warning for a few regions in the Andaman and Nicobar Islands, Alert for the remaining parts of the Andaman and Nicobar Islands and Watch for Indian mainland

faulting. This earthquake generated a tsunami measuring 106 cm near the source. The earthquake was followed by another great shallow strike-slip earthquake (aftershock) of magnitude $M_w8.2$ at 10:43 UTC, southwest of the main shock.

As per ITEWC's SOP the first event resulted in the issue of a Warning only for the Nicobar Islands, an Alert for a few areas in Andaman, Andhra Pradesh, Tamil Nadu, and a Watch for rest of the mainland, based on ETA and EWA from the pre-run model scenario. On the regional front, Threat (EWA > 0.5 m) was issued for the coast of the Sumatra region, Oman, Somalia, Kenya, the east coast of Madagascar, and the west coast of Australia.

ITEWC monitored the real-time sea-level gauges and observed a 6-cm wave amplitude at one of the Indian tsunami buoys, 30 cm at the tide gauge in Campbell Bay, 35 cm at Sabang, 18 cm at Telukdalam, 106 cm at Meulaboh, and 12 cm at Nancowry. ITEWC issued regular bulletins for this event revising the alert levels based on updated earthquake parameters and water level observations. After confirmation that a significant tsunami had not been generated, ITEWC issued a Threat passed bulletin.

All the systems i.e. the automatic location of the earthquake, the estimation of the tsunami arrival time and amplitude, the dissemination of messages through SMS, email, fax, GTS and the website, as well as tsunami buoys and tide gauges performed as envisaged. All six bulletins generated by ITEWC for this event were disseminated to both the national and regional contacts. ITEWC disseminated regional tsunami advisories to 21 countries in the Indian Ocean region as a part of its Regional Tsunami Service Provider (RTSP) operations (Srinivasa Kumar *et al.*, 2012b). It may be noted that the second earthquake occurred while the tsunami monitoring for the first event was in progress. This event was also monitored simultaneously and bulletins issued as per the SOP.

25.6 India's role in the international framework

The UNESCO's IOC formed the International Coordination Group for the Tsunami Warning System in the Indian Ocean (ICG/IOTWS) in July 2005, as a coordinated international response to the 2004 tsunami event. The ICG/IOTWS is serving as the regional body to plan and coordinate the design and implementation of an effective and durable tsunami warning and mitigation system. The system is fully owned by the 28 member states of the Indian Ocean Tsunami Warning and Mitigation System (IOTWS), and is based on international and multilateral cooperation with governance provided by IOC.

The ICG/IOTWS works as a 'system of systems' with operational robustness achieved through a number of RTSPs that communicate with the national tsunami warning centres (NTWCs) of the Indian Ocean countries. RTSPs do not issue warnings; they only provide information and data on tsunami threats to NTWCs. It remains the responsibility of NTWCs to issue tsunami warnings for their countries.

The major capability requirements for any centre to act as an RTSP include that it should have access to real-time seismic and sea-level data sources, and pre-calculated numerical model simulations, and it should provide timely and effective tsunami advisories to NTWCs. Accordingly, the tsunami warning centres of Australia, India, and Indonesia have developed the capabilities that are required to act as RTSPs. Being one of the RTSPs in the ICG/IOTWS framework, the RTSP services of India for an earthquake event commence whenever earthquakes are recorded with magnitudes ≥ 6.5 within the Indian Ocean and magnitudes ≥ 8.0 outside the Indian Ocean. The information is disseminated to 28 Indian Ocean rim countries in various ways as mentioned in section 25.4.6. The types of products issued by ITEWC as an RTSP include public and exchange tsunami bulletins. Public products contain earthquake information, tsunamigenic potential based on earthquake magnitude, and a summary of NTWC warnings. Exchange products contain CFZs tagged with results of forecast models, such as EWAs, ETAs and threat level. As the exchange products need expert interpretation they are provided only to NTWCs and other RTSPs, through secure means. The RTSP operations of ITEWC (along with RTSP-Australia and RTSP-Indonesia) were operationalised on 12 October 2011 during the IOWave11 exercise, with the Director General of IOC/UNESCO congratulating the RTSPs for achieving this capability.

25.7 Future directions

While the ITEWC has been performing as per its design standards, the Japan tsunami of March 2011 has brought to the fore several important issues that have to be addressed to improve the accuracy of the tsunami warning systems. Water-level inversion, real-time inundation modelling, real-time estimation of the focal mechanism of earthquake to show the style of faulting, and incorporation of GPS data into the warning chain are a few key issues that ITEWC has taken up as a priority. Communication tests and mock tsunami drill have also brought to the fore several issues related to the communication of warnings, as well as the awareness and response mechanisms. The ITEWC is working with all stakeholders involved to improve these aspects.

References

Borrero, J., McAdoo, B., Jaffe, B. *et al.* (2011). Field survey of the March 28, 2005 Nias-Simeulue earthquake and tsunami. *Pure and Applied Geophysics*, **168**(6–7), 1075–1088.

Gica, E., Spillane, M. C., Titov, V. V, Chamberlin, C. D. and Newman, J. C. (2008). *Development of the forecast propagation database for NOAAs short-term inundation forecast for tsunamis (SIFT)*. NOAA technical memorandum oar mel-139, Pacific Marine Environmental Laboratory, Seattle, WA.

Gupta, H. K. (2005). Mega-tsunami of 26th December, 2004: Indian initiative for early warning system and mitigation of oceanogenic hazards. *Episodes*, **28**(1), 2–5.

Gupta, H. K. (2008). India's initiative in mitigating tsunami and storm surge hazard. *Journal of Earthquake and Tsunami*, **2**(4), 287–295.

ICG/IOTWS-V (2008). *Implementation Plan for Regional Tsunami Watch Providers (RTWP)*. Fifth Session of the Intergovernmental Coordination Group for the Indian Ocean Tsunami Warning and Mitigation System. IOC-ICG/IOTWS-V/13, Malaysia.

Mansinha, L. and Smylie, D. E. (1971). Displacement fields of inclined faults. *Bulletin of the Seismological Society of America*, **61**, 1433–1440.

Murty, T. S. (1999). Tsunamis on the coastlines of India. *Science of Tsunami Hazards*, **17**(3), 167–172.

Papazachos, B. C., Scordilis, E. M., Panagiotopoulos, D. G., Papazachos, C. B., and Karakaisis, G. F. (2004). Global relations between seismic fault parameters and moment magnitude of earthquakes. *10th Congress of the Hellenic Geological Society, 14–17 April 2004*, Thessaloniki, Greece: Hellenic Geological Society, pp. 539–540.

Ramanamurthy, M. V., Sundaramoorthy, S., Pari, Y. *et al.* (2005). Inundation of sea water in Andaman and Nicobar Island, and part of Tamil Nadu coast during 2004 Sumatra tsunami. *Current Science*, **88**(11), 1736–1740.

Shailesh Nayak, and Srinivasa Kumar, T. (2008). Addressing the risk of tsunami in the Indian Ocean. *Journal of South Asia Disaster Studies*, **1**(1), 45–57.

Srinivasa Kumar, T., Shailesh Nayak, Patanjali Kumar, Ch. *et al.* (2012a). Performance of the tsunami forecast system for the Indian Ocean. *Current Science*, **102**(1), 110–114.

Srinivasa Kumar, T., Shailesh Nayak, Patanjali Kumar, Ch. *et al.* (2012b). Successful monitoring of April 11, 2012 off coast of Sumatra tsunami by Indian Tsunami Early Warning Centre. *Current Science*,**102**(11), 1519–1526.

PART VII

Disaster risks and societal implications

26

The International Strategy for Disaster Reduction and the Hyogo Framework for Action (2005–2015): essential tools for meeting the challenges of extreme events

Sálvano Briceño

26.1 Introduction

Following the ten-year review of the 1994 Yokohama Strategy[1] (United Nations, 1994), and at the conclusion of two years of negotiations, governments adopted the *Hyogo Framework for Action (2005–2015): Building the Resilience of Nations and Communities to Disasters* (United Nations, 2005) at the second World Conference on Disaster Reduction (WCDR) in Kobe, Hyogo, 18–22 January 2005.[2] This agreement represents an historic engagement of governments and the international community to respond forcefully to the increasing vulnerability to natural events or phenomena around the world. In doing so, governments agreed on a series of specific policies and measures to be taken to substantially reduce disaster losses by 2015 (in terms of loss of life and social, economic, and environmental assets of communities and countries).

The collective decision of governments to engage at this level was made just a few weeks after the giant Indian Ocean tsunami on 26 December 2004. While governments' adoption of this set of specific recommendations has not yet translated into bold changes in policies and investments around the world, much has been advanced and the process towards implementing such changes is now irreversible.

A few pioneering countries began immediately to review their policies and investments in order to make risk reduction an objective and risk management a tool in the implementation of the Hyogo Framework. In particular, members of the Association of South East Asian Nations (ASEAN) were the first to commit an implementation of the Hyogo Framework in a legal agreement that was subsequently ratified by their respective parliaments.[3]

At present, almost all regional and subregional intergovernmental organisations have included disaster risk reduction (DRR) as a priority in their programmes, with the Hyogo Framework as their policy guide. Some are facilitating its implementation more effectively than others.[4] As a consequence, a large majority of countries are implementing at least one of the Hyogo Framework's five priorities for action and, in some cases, even several of its recommendations. However, very few countries have undertaken to implement the Hyogo Framework in its entirety, which is essential to the achievement of this important tool's objectives.[5]

The key message that emerged from the Kobe-Hyogo Conference is that DRR needs to be mainstreamed into development agendas, considering 'DRR is a cross-cutting issue in the context of sustainable development and therefore an important element for the achievement of internationally agreed development goals, including those contained in the Millennium Declaration (2000). In addition, every effort should be made to use humanitarian assistance in such a way that risks and future vulnerabilities will be lessened as much as possible' (United Nations, 2005; page 5, A. General Considerations (k)).

At the 2012 United Nations Conference on Sustainable Development (UNCSD, Rio+20), DRR was recognised as a requirement for sustainable development in the outcome document *The Future We Want*, and will thus be considered for inclusion in the set of goals to follow the Millennium Development Goals (MDGs), to be adopted in 2015, with the new title of Sustainable Development Goals (SDGs) (United Nations, 2012; paragraphs 186–189).

Although at the international level the direct relationship between risk reduction and development has been clearly recognised, at the national level, many governments still perceive it only as part of the narrower disaster management field and mainly as a humanitarian issue.

Extreme Natural Hazards, Disaster Risks and Societal Implications, eds. Alik Ismail-Zadeh, Jaime Urrutia Fucugauchi, Andrzej Kijko, Kuniyoshi Takeuchi and Ilya Zaliapin. Published by Cambridge University Press.

Transforming risk reduction into a development objective and risk management into a development tool remains a challenge requiring the full attention of policy-makers and other stakeholders. The challenging risk element here is not the natural events or phenomena but rather the increasing social and human vulnerability, mainly in urban and coastal areas, that is defying current technical and policy capacities worldwide.

Taking into account natural phenomena, all of them, earthquakes, volcanic eruptions, landslides, floods, droughts, tropical cyclones, typhoons, hurricanes, wild land fires, heat and cold waves, tornados, dust storms, etc., in all development planning and projects is as important as considering ecosystems, watersheds, and other natural resources, before deciding where and how to develop any social or economic activity[6] (Sudmeier-Rieux, 2013).

This chapter is aimed primarily at identifying key policy actions still needed to ensure greater effectiveness in reducing risk and vulnerability to natural hazards. Such policy actions should be considered in the process leading to the third WCDR, planned for 2015 in Japan. They should provide the basis for the new agreement to follow the 2005 *Hyogo Framework for Action*. Although governments will be the main negotiators of such an agreement, other actors, civil society, scientific and academic institutions, the private sector, parliamentarians, local authorities, media, among others, will also play important roles in advancing risk reduction. They are all expected to be a part of, and contribute to, new forms of risk governance mechanisms required for more effective risk reduction at all levels.

Following the IDNDR (1990–1999), carried out by the UN at the request of governments in the UN General Assembly, the UN launched in 2000 the International Strategy for Disaster Reduction (ISDR), moving into a new era of international engagement at the global level to support and facilitate more intensive work. By then, it had become evident that many lives and livelihoods can be saved if risk reduction is addressed as a specific policy objective and risk management knowledge (e.g. risk assessments, early warning systems, building and infrastructure safety, preparedness and recovery plans, risk transfer mechanisms such as insurance) is more systematically integrated in key development policy processes, such as land-use planning, environmental management, fiscal management as well as in health, education, agriculture, etc.

The 2009, 2011, and 2013 UNISDR Global Assessments Reports on Disaster Risk Reduction,[7] prepared jointly by key experts and institutions, provide a full picture of the situation globally and nationally regarding technical and institutional progress in this field. Other relevant reports include those by the non-governmental organisation (NGO) community,[8] the World Bank's Global Facility for Disaster Reduction and Recovery,[9] and the International Federation of Red Cross and Red Crescent Societies (IFRC).[10] The United Nations Development Programme (UNDP) also produces relevant reports at the country level on the specific national or local experiences on DRR.

Overall, massive investments have been made since 2000 in promoting and facilitating disaster and hazard risk reduction by many governments and other institutions in various sectors and levels and by various stakeholders.

26.2 Background

26.2.1 Looking back

The experience and knowledge accumulated on DRR since the 1970s has been staggering. A great deal of work has emerged since authors like Gilbert F. White, Ian Burton, Ben Wisner, Ken Hewitt, and many others recognised social vulnerability as a main cause in disasters and developed the now widely accepted and understood concept of DDR (see Wisner *et al.* (2012) for a comprehensive bibliography on the topic).

Developed initially by scientific and academic experts, risk reduction knowledge has evolved and been transformed into specialised institutional mechanisms (policies, legislation, and organisational arrangements) taking different forms in many local settings, national, sub-regional, regional, and international levels. Some are focused on how to prepare better to respond to the impact of natural events or phenomena, others, more advanced, are focused on how to reduce various sources of hazards and vulnerability: human, social, economic, ecological, or physical.

26.2.2 Looking at the present

At the same time, and despite all the progress made, we are witnessing a rapid increase in disasters and negative impacts of natural events or phenomena, mainly in loss of lives and livelihoods. Although these impacts are growing mainly in developing countries, developed countries are also suffering from such increases. These impacts are becoming a greater obstacle to sustainable development as they also divert valuable resources that otherwise could be used for development activities.

The knowledge available on DRR is, however, sufficient to address and amend the trend of rapid expansion of impacts. Policy and technical advice is available and has been applied in many places with positive results. For example, mortality has been notably reduced in situations related to tropical storms, floods, droughts, and other hydro-meteorological events due to improvements in early warning systems in many developing countries (IPCC-SPEX, 2012). The good practices of Bangladesh, Cuba, Chile, Jamaica, Mozambique, some regions in

China, India, USA, Japan, and Europe are regularly disseminated worldwide.[11] Bangladesh and Cuba have developed community-based early warning and preparedness systems and organise recurrent awareness raising campaigns and educational activities. These examples remain, however, still too few. At present, it is also possible to estimate the cost of disasters as well as the corrective and prospective risk reduction measures (mitigation, prevention, and preparedness) based on probabilistic risk assessment methodologies that are available, although still not widely utilised.[12]

A more substantive reduction in the loss of lives, livelihoods, and property will, however, require much greater and systematic efforts, with a long-term perspective of sustainable development, integrating risk management in all development sectors and, at the same time, promoting it as a specific policy objective. For the latter to happen strong leadership by high-level authorities will be required, aimed at counteracting the current increasingly negative trends in urbanisation, especially in coastal and delta areas as well as in mountainous regions, where declining snow coverage is creating greater hazards for populations up and downstream.

26.2.3 Looking forward

In 2015, the opportunity will arise for more aggressive action to reduce the risk of disasters. The Hyogo Framework reaches its deadline and a new agreement will emerge from the third WCDR in early 2015. The MDGs also reach their end and are expected to be followed by new SDGs, to be adopted some time during the UN General Assembly session in the last quarter of 2015. The climate change negotiations, in the late part of 2015, will focus on deciding how to replace or complement the Kyoto Protocol, including climate change adaptation in addition to mitigation. All three processes will take place at different times of the year, each will have its own negotiating process, facilitated by different UN offices or programmes, and each will involve different actors and delegates from each government, as well as other intergovernmental organizations and NGOs.[13] These major international negotiations, like the previous ones (Hyogo, Kyoto, and MDGs), will certainly have an impact at national and local levels. Governments, private sector and civil society organisations will utilise their decisions as guidance in developing further institutional and technical capacities to address the sustainable development needs in their respective jurisdictions and activities.

Strong leadership by public authorities and the private sector is still needed to explain and persuade citizens to engage in risk management related to natural events, similarly to the ways we have learned to manage risk in areas such as health and traffic, going beyond the short-term collective and selfish interests. Understanding that resilience to natural events can be developed and risk reduced by our actions and decisions represents a paradigm shift in the way we relate and integrate with nature (IPCC-SREX, 2012). It is about understanding fully that although we have been able to modify nature in many, sometimes positive, ways, there are still natural events we cannot, and should not try to, control or ignore, but rather adapt by reducing vulnerability and building resilient behaviour.

Such leadership also requires effective team-oriented coordination at the international level, in particular within the UN, which can influence actors at regional and national levels facilitating the development of common positions at all three international processes with regard to new risk governance mechanisms. Building such a team effort among international organisations will be a major challenge. It is only with a united voice that a substantive impact on awareness and policy-making can be expected.

As a consequence, integrated approaches are more than ever required in addressing risk reduction and resilience in sustainable development policies and programmes.[14] However, the hurdles remain huge and elusive as priority is usually given to short-term economic goals in policy-making. Addressing the particular challenge of DRR could provide an exercise and test for addressing larger and more complex needs thereafter. 'In many ways, resilience emerges as a topic that unites different groups with the goals of creating a common dialogue, reducing losses, and decreasing vulnerability to hazards and disasters' (National Research Council, 2012).

Disasters are not only a great obstacle to development but also an opportunity for improving public governance in general, with the introduction of risk management as an essential policy tool applicable to many other potential threats to development. Given the critical need for addressing risk in the context of sustainable development and climate change and within each development sector, it would be helpful if a high-level leadership initiative,[15] of an intergovernmental nature, were to undertake to assess and provide evidence-based and policy-oriented scientific results to address risk and vulnerability related to extreme natural events. Such a high-level assessment would contribute to enhancing: knowledge of disaster risk at global, regional, and local levels; awareness of the people living with risk; and identification of the appropriate risk governance mechanisms to guide institutional development in this field, including required policies, legislation, and organisational arrangements for risk management.

Such an initiative should bring together expertise in physical science and engineering, geographical science, social and behavioural science, economics, political and administrative sciences, and public health with professional experience from research, public policy, emergency and disaster management, NGOs, the private sector, and government service. It should also take into account the experience of similar initiatives such as the Intergovernmental Panel on Climate Change (IPCC) and the Intergovernmental Panel on Biodiversity and Ecosystem Services (IPBES). The decision to develop such an initiative remains however, in the hands of governments in the context of the UN. The preparations for the post-2015 negotiations provide a unique opportunity to consider such an initiative. The Integrated Research on Disaster Risk (IRDR) programme, cosponsored by the International Council for Science (ICSU), the International Social Science Council (ISSC), and the UNISDR, jointly with other relevant institutions and programmes, has started to address these needs and will be presenting its recommendations to the post-2015 processes.

26.3 The International Strategy for Disaster Reduction (ISDR) and the Hyogo Framework

The expected outcome of the Hyogo Framework up to 2015 is 'the substantial reduction of disaster losses, in lives and in the social, economic and environmental assets of communities and countries', recognising that the achievement of this outcome will require the full commitment and involvement of all actors concerned, including governments, regional and international organisations, civil society including volunteers, the private sector, and the scientific community (United Nations, 2005; section II B).

The *Hyogo Framework for Action's* strategic goals are:

(a) the more effective integration of disaster risk considerations into sustainable development policies, planning, and programming at all levels, with special emphasis on disaster prevention, mitigation, preparedness, and vulnerability reduction;
(b) the development and strengthening of institutions, mechanisms, and capacities at all levels, in particular at the community level, that can systematically contribute to building resilience to hazards; and
(c) the systematic incorporation of risk reduction approaches into the design and implementation of emergency preparedness, response, and recovery programmes as part of the reconstruction of affected communities.

The five priorities for action are to:

1. ensure that DRR is a national and a local priority with a strong institutional basis for implementation;
2. identify, assess, and monitor disaster risks and enhance early warning systems;
3. use knowledge, innovation, and education to build a culture of safety and resilience at all levels;
4. reduce the underlying risk factors;
5. strengthen disaster preparedness for effective response at all levels.

In 2010–2011, a mid-term review of the Hyogo Framework was carried out with ample consultations.[16] By the first anniversary of the 2011 Great East Japan earthquake and tsunami, the Government of Japan announced their interest in hosting the third WCDR[17] to be held in early 2015. In preparation for the third WCDR, a number of consultations are being organised by, among others, the ISDR Secretariat and the Global Network of Civil Society Organisations for Disaster Reduction (GNDR).

The 2010–2011 mid-term review of the Hyogo Framework identified a number of areas requiring either an update in terms of approach or targets or new treatment that have emerged as new priorities. In the next section I address each of the priority areas identified in the Hyogo Framework proposing modifications or corrections.

26.4 Reviewing the Hyogo Framework priorities

26.4.1 Ensuring the right policy priority for reducing and managing risk at all levels, from local to global, with the appropriate governance and institutional mechanisms

The development of institutional capacities focusing on risk management and vulnerability reduction is needed for more effective results in reducing disasters. Just as public health policies that focus on prevention have been able to reduce society's risk of diseases and illnesses, similar policies focusing on risk reduction would more effectively reduce mortality and other negative impacts arising from extreme natural events.

Paragraph 16 of the Hyogo Framework recognises that 'countries that develop policy, legislative and institutional frameworks for DRR and that are able to develop and track progress through specific and measurable indicators have greater capacity to manage risks and to achieve widespread consensus for, engagement in and compliance with DRR measures across all sectors of society' (United Nations, 2005).

This priority has been misconstrued, however, and, in some cases, reduced to merely setting up new institutional mechanisms for disaster reduction without providing the right policy and political priorities. It is clear that without the personal involvement and engagement of high-level authorities at national and community levels, from both public and private sectors, it will not be possible to change the attitudes and behaviour that this new focus requires.

Also greater responsibilities and resources for risk management should be decentralised to local levels as the intervention of local authorities, in both risk reduction and response and recovery, is usually of greater impact and effectiveness.

The focus on reducing risk and vulnerability to natural events requires a paradigm shift or fundamental change in the understanding and perception of risk. The traditional belief that disasters are 'acts of God or Nature', hence providing a fatalistic perception, can only be overcome with strong awareness and educational efforts involving high-level leadership from both the public and private sectors. People usually assume that there is little they can do to avoid natural hazards turning into disasters. Natural events such as earthquakes or storms are part of nature; they are not preventable but their potential impacts are manageable by citizens, communities, and governments. It is important to understand that it is vulnerability that turns natural events into disasters, becoming, therefore, the main cause of disasters. Vulnerability, be it human, social, economic, institutional, ecological, or physical, can be surmounted by the joint efforts of governments, community organisations, and citizens by (i) including risk management in land-use planning, environmental management, and all other development sectors, (ii) choosing the location and materials to build houses, schools, and other critical infrastructure, and (iii) avoiding expanding urban centres into potential risk prone areas.

The engagement of the highest levels of government in awareness-raising was emphasised by the mid-term review of the Hyogo Framework in 2010–2011. It must be followed by appropriate awareness-raising campaigns that target political leaders. Such high-level awareness is beneficial and can contribute greatly to other priority processes such as adaptation to climate change, achievement of the MDGs, and ensuring sustainable development. The role of the UNISDR with its awareness-raising campaigns has been instrumental in motivating such dedication by parliamentarians, city mayors, heads of governments, and other political and business leaders.[18]

The fact that financial savings can be realised by investing in DRR, thereby avoiding or reducing the destruction of development gains, should be more than sufficient to convince top-level decision-makers, in particular ministers of finance, international financial institutions, and other multilateral and bilateral development investors. The World Bank's Global Facility for Disaster Reduction and Recovery (GFDRR) has played a key role in this regard, gradually motivating ministries of finance to focus on this policy priority and providing tools to assess and estimate such costs and savings.[19]

At the global level the challenge will be to maintain the focus on DRR while at the same time integrating it in other sectors. Discussions are on-going in various consultations, forums, and networks[20] about the need to identify DRR as a specific policy objective as well as integrating it as an objective in each development sector. The role of the Global Network for Disaster Risk (GNDR)[21] in facilitating discussions among NGOs, in particular, to provide an alternative view to that of governments is to be commended. The GNDR includes a wide range of actors at all levels and in many countries around the world.

In the end it will be the community of aware leaders in all sectors representing relevant stakeholders and working together that will make a decisive impact an changing people's attitudes and behaviour towards reducing vulnerability and managing risk in their daily lives and, in consequence, reducing the negative impacts of extreme natural events. Hence, multi-stakeholder institutional mechanisms, such as the national and local platforms for DRR, are relevant first steps in advancing appropriate risk governance institutions (policies, legislation, and organisational arrangements). Basically, 'disaster resilience is everyone's business and is a shared responsibility among citizens, the private sector, and government' (National Research Council, 2012).

At the same time, governance for risk reduction or resilience building should be developed with a rights-based approach. Not only is reducing risk related to natural hazard events a basic human right, it is also essential that public authorities deploy transparency and accountability when managing it, especially with regard to housing and critical infrastructure building policies (more in Section 26.4.4). Such ethical approaches should also include addressing the needs of those with disabilities or in greater need for special attention and priority. Understanding that disasters only ensure that the poor remain poor is essential for effective poverty elimination plans. These elements should be top priorities in the upcoming 2015 agreements.

Last but certainly not least, the next generation of risk reduction agreements to be reflected in the post-2015 regime (post-Hyogo, post-Kyoto, and post-MDGs) will need to have a stronger legally binding nature. Not only should facilitating risk reduction or resilience building with regard to natural events become an obligation for governments and no longer a voluntary task, it should also impose enforcement mechanisms that can be monitored and assessed regularly by society, at national as well as local levels.

26.4.2 Identifying, assessing, and monitoring natural events, vulnerability, and risk including developing early warning mechanisms for vulnerability development

Paragraph 17 of the Hyogo Framework recognises that 'the starting point for reducing disaster risk and for promoting a culture of disaster resilience lies in the knowledge of the hazards and the physical, social, economic and environmental vulnerabilities to disasters that most societies face, and of the ways in which hazards and vulnerabilities are changing in the short and long term, followed by action taken on the basis of that knowledge' (United Nations, 2005).

The notion that risk assessment is primarily a technical exercise to understand natural events or phenomena needs to be complemented by a wider and more intense understanding of social vulnerability. Assessing risk for a specific location or community involves identifying the potential hazards, the exposure of the population, their degree of vulnerability, the institutional and technical capacities available for risk reduction (prevention, mitigation, and preparedness) and estimating the potential impacts (see National Research Council (2012) for more detail, pp. 27–35).

Understanding and assessing the natural events and their hazardous potential is, of course, essential. Although knowledge is widely available on most natural hazards (hydro-meteorological as well geophysical), understanding their nature and magnitude as well as monitoring and assessing their potential impacts remain important scientific and technical challenges. However, besides understanding them better, there is little we can do to reduce them or their potentially hazardous nature.

On the other hand, understanding and assessing vulnerability, which is composed of human, social, institutional, economic, physical, and ecological elements, is a much less developed exercise. Vulnerability, being a human creation, is the risk component to which changes can be made. In order to be addressed, it requires the wider involvement of community and stakeholder representatives in addition to technical experts. In other words, there is no better contribution to understanding the vulnerability of a given city, town, or neighbourhood than the participation of those who are exposed and vulnerable, and similarly for identifying and implementing solutions. A well-informed and aware community is the first requirement to building their resilience. Although new social media can be extremely helpful in disseminating information on time, it is essential that understanding risk and vulnerability is well engrained in the community in advance.

Assessing risk can and should therefore become an exercise for every person and every family as citizens are often the first responders in times of disaster and they are the ones who have a vested interest in understanding and managing their own vulnerability. Therefore, risk assessment should be promoted as an exercise in which every person, family, and community must participate regularly with regard to their homes, offices, livelihoods, schools, etc. This risk assessment exercise has been referred to as 'self-assessment' or 'citizen-based vulnerability assessment' (Wisner and Walker, 2005).

The mid-term review of the Hyogo Framework highlighted the relevance of understanding risk assessment by the public, requiring broader media and educational campaigns at all levels in response to the Hyogo Framework's third priority area (see next item). In addition to the work being done by local governments and intergovernmental institutions, NGOs and other civil society organisations have advanced a huge amount of work at the community and local levels.[22]

Further, as an essential component of sustainable development, DRR provides an opportunity to address other related topics. 'In education and public awareness, DRR can be a window through which discussion of less dramatic but important development issues gain visibility – concerns such as urban sprawl, management of water resources, teacher and health provider pay and conditions of work, even very broad issues such as fair and free trade' (Wisner, 2006; p. 6).

In this priority area, two items require further research and policy implementation: (a) the development of a global early warning system, utilising existing international mechanisms such as World Meteorological Organization (WMO), United Nations Environment Programme (UNEP), and others; with participation by and contributions from all countries; and (b) the development of early warning capacities to provide warnings in respect of vulnerability development in communities and nations as part of the ongoing land-use and other development plans. Probabilistic risk assessment methodologies would enable such capacity in countries and communities.

On the former item, UNISDR, in concert with other UN and non-UN partners, conducted a global survey of early warning systems in 2006, and issued a report assessing its viability based on existing organisational arrangements at international and regional levels, and providing specific recommendations for its development. However, it still needs to be fully implemented.[23]

Further thinking is required on the latter item, which speaks to the provision of warnings in a rather systematic manner and not only at the time of the event's occurrence. It entails providing warnings much in advance of natural events and focusing on identifying and raising awareness on vulnerability creation (e.g., wrongly located urban developments, poor construction practices, ecosystem depletion, etc.). The 2009 Global Assessment Report on Disaster Risk Reduction (United Nations, 2009; Chapter 7) provided preliminary analysis and advice in this regard.

26.4.3 Using knowledge and education to explain, in particular, risk, vulnerability, and resilience

Paragraph 18 of the Hyogo Framework for Action states that 'disasters can be substantially reduced if people are well informed and motivated towards a culture of disaster prevention and resilience, which in turn requires the collection, compilation and dissemination of relevant knowledge and information on hazards, vulnerabilities and capacities'. Reflections in the previous item are also relevant to the third priority of the Hyogo Framework since understanding risk and vulnerability is mainly a research and educational task.

Developing new knowledge through integrated research and communicating it through educational and awareness-raising mechanisms, in particular schools and new social media, must be high priority tasks in all societies. This requires substantive investments, both financial and in terms of transmitting the right concepts and approaches. Academic institutions, of course, play a central role in knowledge generation, while team efforts among stakeholders (government, business, and civil society organisations) are required for communicating and raising risk awareness.

The specific role and capacities of the private sector, at all levels, are of particular relevance, as reducing risk and vulnerability to natural events or phenomena is particularly relevant for business continuity and job security. All these specific roles need to be highlighted in the post-2015 mechanism to follow the Hyogo Framework.

At the government and public policy levels, investing in education is perhaps the most relevant action for reducing risk. Investment in education must address two different but complementary objectives. First, introducing DRR into school curricula (formal education) coupled with ensuring the safety of schools and other educational facilities. Secondly, undertaking awareness-raising campaigns (in formal and non-formal education) by authorities and leaders at all levels, promoting a change of attitudes and behaviour with regard to vulnerability and resilience, in the same manner that prevention has been advocated for health, traffic accidents, and other hazardous situations.

The Hyogo Framework, building on numerous studies and reports, articulates a worldwide consensus that DRR is an integral part of sustainable human development, not a side issue of limited merely specialist/technical interest or concern. The UNISDR released a short video in 2006, on the occasion of their annual campaign, based on the concept that DRR is 'Everybody's Business'. If this is so, and DRR is an essential part of sustainable development, then education for sustainable development should logically include risk reduction in the school curriculum (see Wisner (2006) for more detail).

On school safety, the role and responsibilities of ministries of education and international organisations such as the United Nations Educational, Scientific and Cultural Organization (UNESCO), United Nations Children's Fund (UNICEF), and others, must be identified more clearly in education policies. Despite facilitating many discussions on DRR among educational authorities at national, regional, and global levels, in particular at the 48th International Conference on Education (Geneva, 25–28 November 2008), this topic still does not feature prominently in the programmes of organisations involved in education policy, and when it does, it is mainly for preparedness and emergency management purposes. By extension, ministries of education do not always feel responsible for ensuring the safety of schools, which is perceived by many in the education sector as the responsibility of other branches of government (ministries of construction, public works, emergency management, or local authorities). The next priority area also addresses this issue.

26.4.4 Reducing the underlying risk factors in social and economic development sectors and in recovery processes

Paragraph 19 of the Hyogo Framework recommends that 'disaster risks related to changing social, economic, environmental conditions and land use, and the impact of hazards associated with geological events, weather, water, climate variability and climate change, are addressed in sector development planning and programmes as well as in post-disaster situations' (United Nations, 2005).

This priority area has been identified as the one currently requiring more development, primarily because it groups together many different issues, whereas each of these issues would benefit from more focused treatment. This priority area includes everything from introducing DRR into various development policies and management and recovery efforts (environment, health, agriculture, energy, transport, infrastructure, tourism, and other sectors) to ensuring building code enforcement and enhancing insurance and other financial risk sharing and transfer mechanisms.

With regard to environmental policy, a specific recommendation that was not developed well enough in the Hyogo Framework, and has gradually become urgent, is the concept that DRR is an essential service provided by ecosystems. All ecosystems perform a buffer function to natural phenomena, and this function needs to be recognised through environmental policies and legislation in order to protect ecosystems, thereby reducing risk and vulnerability to natural events or phenomena. In this capacity, DRR would contribute, therefore, to reducing ecological vulnerability, which is a significant key factor in the intensity and frequency of disasters. Given the rapid increase in ecosystem degradation around the world (as a consequence of rapid urban and population expansion as well as sea-level rise and subsidence) the role of ecosystems in DRR needs to be recognised urgently and this needs to be accompanied by appropriate policies and measures to reverse the trend and ensure sound ecosystem management.[24]

Recognising DRR as an ecosystem service would have the additional benefit of providing ministries of environment and other governmental organisations and NGOs working on environmental matters with guidance and tools to become DRR actors and promoters. This would have the effect of raising people's awareness and collective consciousness, which is urgently required to ensure a paradigm shift in this field.

Another key issue in this priority area is the need to enhance building codes and other safety regulations. The introduction of a multi-hazard approach to developing and implementing building codes requires greater recognition as a building code is frequently linked to specific fields of expertise, for example, to seismic codes, fire codes, or wind codes. Greater interaction between the fields of earthquake engineering, wind engineering, and fire management, among others, and among policy-makers and the media would be beneficial. A consolidated and integrated approach to building codes and safety still needs further development and wider promotion. It needs to become a central component in awareness-raising activities, at the same time, with government officials responsible for housing and infrastructure building and with the public through the media.

Furthermore, in the same way that people are increasingly aware of factors that affect their health, it is important to develop awareness-raising programmes for people to learn about the factors that affect the 'health' of their houses, their offices, and their children's schools and to develop systematic approaches to examining vulnerability trends with regard to the location of these structures, the manner in which they are built, and the materials used to construct them.

Specific recommendations must be developed in the post-2015 institutional mechanisms for each development sector, in particular the integration of risk management in all development sector plans. A good example of such sectorial advice is provided by the Climate Development Knowledge Network (CDKN), which has published recommendations on climate resilience for agriculture, ecosystems, health, and water[25] based on the findings of the IPCC's *Special Report on Managing the Risk of Extreme Events and Disasters to Advance Climate Change Adaptation* (IPCC-SREX, 2012).

A key occasion for advancing resilience building or risk reduction is the recovery process following a disaster. During these difficult moments, communities are more sensitised and alert, and thus amenable to supporting special measures and investments towards more resilient development. It is also during these occasions that citizens become more generous and large donations are obtained that could be utilised to enhance resilience capacities. However, these opportunities are seldom utilised to enhance awareness or institutional mechanisms for vulnerability reduction or 'building back better'. Unfortunately, authorities tend to either simply recover to the same situation as before the disaster or worse, often just maintain affected populations in shoddier situations of displacement for longer periods.

In order to manage recovery more effectively it is indispensable to develop pre-disaster recovery planning[26] capacities in exposed and vulnerable communities, which would increase their resilience. This is a new concept that needs to be widely disseminated, facilitated, and included in the post-2015 regime for DRR.

26.4.5 Strengthening preparedness capacities at all levels, from families to local and national levels for effective response to natural events

Paragraph 20 of the Hyogo Framework recognises that 'at times of disaster, impacts and losses can be substantially reduced if authorities, individuals and communities in hazard-prone areas are well prepared and ready to act and are equipped with the knowledge and capacities for effective disaster management' (United Nations, 2005).

The 2009 *Global Assessment Report on Disaster Risk Reduction* (United Nations, 2009) describes the worldwide progress made in disaster preparedness, and details the contributions of many emergency management and humanitarian organisations including the Red Cross and Red Crescent Societies and their International Federation, who play a pioneering role in promoting worldwide preparedness for response to disasters. A direct consequence has been the significant reduction of the mortality rate associated with some types of hazards and a reduction in mortality rate for many communities, with the cases of Bangladesh and Cuba serving as noteworthy examples. However, many other communities around the world still need to enhance their disaster preparedness capacity. The anticipated impacts of climate change enable us to see the challenges ahead, thus requiring a more comprehensive preparedness approach, going beyond traditional preparedness that focuses on preparing to respond, and rather aimed at preparing to be resilient. Such resilience preparedness would require the full implementation of the Hyogo Framework recommendations.

In this field, in addition to enhancing preparedness at the community level, the boldest action required is to expand disaster preparedness programmes to include more thorough risk and vulnerability assessments, early warning systems, and awareness-raising activities engaging community leaders, including development of pre-disaster recovery plans. Similar to early warning systems, many of the preparedness programmes still focus too heavily on hazard assessments without paying sufficient attention to identifying vulnerability and its possible reduction.[27]

Thus, vulnerability reduction must become a key component of disaster preparedness programmes at all levels with a certain degree of universally-accepted guidance. On this latter element, the role of scientific and academic expert institutions is crucial. In order to develop a 'preparedness for resilience' approach it is necessary to understand the root causes of disasters, for which the Forensic Investigations of Disasters (FORIN) methodology[28] produced by the Integrated Research on Disaster Risk (IRDR) programme provides relevant guidance.

26.5 Review of progress in implementation of the Hyogo Framework and the current status of risk

Implementing the five priorities for action of the Hyogo Framework enables any country or community to significantly reduce its risk related to natural hazard events, hence reducing both mortality caused by disasters and social and economic impacts that ensue from disasters, which are increasingly affecting great numbers of people and hampering development efforts.

While most governments have undertaken some type of action over the last eight years in response to the recommendations arising from the Hyogo Framework, it is more difficult to assess whether and how many of these have addressed the entire set of recommendations, and whether they have done so in a consistent and effective manner. Many governments have developed policies, enacted legislation, and set up organisational arrangements to undertake DRR. Their initial steps have been encouraging.[29] Countries that have experienced large disasters have moved faster in this regard, while others still lag behind, with some notable exceptions in both cases.

A preliminary global review of progress was carried out in 2007 in preparation for the first session of the Global Platform for Disaster Risk Reduction held in Geneva (5–7 June 2007). At that time, it was too early to identify any significant trend in the implementation of the Hyogo Framework and this first Global Platform session served mainly to assess how governments were preparing to address the issue as well as to provide an initial guidance in setting up national mechanisms to guide and support the process. In 2008, a second set of reports was requested for the production of the first *Global Assessment Report on Disaster Risk Reduction* (United Nations, 2009). The assessment served as the basis for discussions at the second session of the Global Platform for Disaster Risk Reduction, held in Geneva (16–19 June 2009) and was instrumental in devising the key recommendations flowing from that session.[30] A new review of implementation progress took place during 2010 and is contained in the second version of the *Global Assessment Report on Disaster Risk Reduction* (GAR 2011) produced for the third session of the Global Platform for Disaster Risk Reduction (Geneva, 8–13 May 2011). Governments reported directly through the HFA Monitoring facility;[31] additionally, international organisations, NGOs, and other institutions reported on their progress in implementing the Hyogo Framework.

As part of GAR 2011, an independent assessment was carried out in concert with expert institutions around the world to identify the evolution and trends of vulnerability, hazards, and overall risk. Results were compared with contributions made to the mid-term review of the Hyogo Framework, when it reached its fifth year. The mid-term review process benefitted from various consultations among expert institutions, inter-agency and intergovernmental actors at the global, regional, and national levels.[32]

At the local level, many community-based organisations and local governments are taking bold steps to reduce risk and vulnerability, although their efforts remain largely scattered and are of varying degrees of effectiveness, particularly with regard to the increasing vulnerability of communities to natural events. At the international level, several important initiatives have been developed. In addition to the ISDR launched in 2000, and its Hyogo Framework adopted in 2005, three major initiatives have been developed at the global level focusing on DRR, namely: (a) the World Bank's GFDRR, focusing on advising ministries of finance and donor agencies;[33] (b) the GNDR facilitating and supporting the work of NGOs and community-based organizations;[34] and (c) the ICSU/ISSC/UNISDR IRDR programme focusing on promoting and facilitating integrated approaches in policy-oriented research.[35] Many other programmes have been developed by international agencies at global, regional, and sub-regional levels, as well as other academic, scientific, private sector, and stakeholder organisations, most of them as part of their larger mandates. As a general feature, in most of these latter programmes, investments in risk reduction remain less important than the investments these organisations make on disaster preparedness and response efforts.

Eight years later, the world continues to witness ever increasing impacts and losses due to social vulnerability to natural events. While the number of geological and hydro-meteorological hazards has not increased on average, the number of victims, people and communities, affected by these hazards continues to grow enormously due primarily to three factors:

(a) the increase in urban density, mainly in coastal areas, with increasing numbers of migrants moving to urban areas, amplified by a lack of adequate policies that integrate risk reduction coupled with increasing poverty and economic vulnerability in many developing countries;[36]
(b) the increasing deterioration of natural ecosystems,[37] all of which provide an essential service as buffers (directly to climate or hydro-meteorological events and indirectly to geological events) as well as supporting the livelihoods of communities and nations; and
(c) the increasing impact of climate change, which is changing weather patterns, increasing the number and intensity of some hazards and forcing communities to move inland because of rising sea levels and sea temperature along many coastal areas around the world, thus increasing the already worrying urban migration trend.[38]

The combination of increasing urban social vulnerability with the changes in climate and ecosystems is expected to result in more severe and frequent disasters in the coming years, a fact that governments will have to face in the upcoming post-2015 negotiations on DRR, climate change, and sustainable development. Hence, the focus on developing risk governance mechanisms, which is becoming urgent. A strong recommendation in this regard in the above policy areas could signify a major step forward in reducing risk while contributing significantly to climate change and sustainable development goals.

26.6 Relevance of the multi-hazard approach in climate change negotiations for risk related to geological events and hazards

One of the most frequently asked questions in this field is whether any agreement on climate change that includes DRR can also help to reduce risk from earthquakes and other geological or non-climate related events and hazards. If these hazards are not related to climate change, it is assumed that agreements on climate change should not apply. However, since an essential principle of the Hyogo Framework is that risk reduction policies need to have a multi-hazard approach, and with the Hyogo Framework being promoted as a guidance for climate change adaptation policies and measures, it is expected that such policies and measures will also contribute to reducing risk from other non-climate related hazards.

For example, if risk assessments are introduced as an essential tool in land-use planning, environmental planning, and planning in all sectors (health, agriculture, tourism, etc.), as guided by the Hyogo Framework, hence following a multi-hazard approach, it would become routine to address all hazards or natural events, not only the climate-related ones, through such assessments. This would then inevitably lead to assessing and preparing for geological risks as well as climate ones in an integrated and more effective manner.

Furthermore, educational, awareness-raising, and community-led preparedness programmes should be developed with a multi-hazard approach. If new risk governance mechanisms following the Hyogo Framework advice include multi-sectorial, multi-disciplinary, multi-stakeholder, and multi-hazard platforms for DRR at national and local levels, it can be expected that such institutional mechanisms would address both climatic and non-climatic extreme events.[39]

A multi-hazard approach is, of course, needed for general DRR policies and measures when addressing specific areas such as awareness-raising, primary basic education, institutional development, general risk assessments, building codes and safety regulations, most insurance and other risk transfer and sharing mechanisms, etc. On the other hand, some particular measures will need to be hazard-specific as they address specific challenges posed by each type of hazard, for example, forecasting and early warning systems for natural events must be adapted to each type of hazard, scientific research and preparedness programmes must also focus on specific hazards, etc.

The mid-term review process of the Hyogo Framework identified climate change as a major area for the post-2015 successor arrangement. Climate change was clearly not fully addressed at the time of the adoption of the Hyogo Framework. To this end, it will require a comprehensive review of decisions made during the climate change negotiations, such as the Cancun Adaptation Framework, to ensure coherence, compatibility, and complementarity between the two processes and to provide more effective guidance to governments and to organisations working on climate change and on DRR issues.

The UNISDR has advised the 1992 UN Framework Convention on Climate Change (UNFCCC) process on the need for making vulnerability reduction and building resilience to extreme events a priority in the immediate and short terms for adaptation to climate change. Assigning these priorities would help to avoid humanitarian and economic losses in the short term, as well as secure development gains and provide a more sustainable basis for other adaptation actions over the long term. It would also capitalise on the knowledge and capacities that are currently available, especially in the fields of DRR and risk management.[40]

In addition to the recommendation related to highlighting DRR as the first line of defence in climate change adaptation, it would be important to explain the value of such programmes in addressing the risks related to non-climatic hazards, such as earthquakes and other geological events. Haiti would be a good example, for this, but is not the only one.

Although struck by a strong earthquake on 12 January 2010, Haiti continues to suffer from the impacts of hurricanes and heavy rains that occur each year with greater frequency and intensity. The vulnerabilities that made the January 2010 earthquake so devastating are the same human, social, institutional, physical, economic, and ecological vulnerabilities that transform hurricanes and heavy rains into large-scale disasters. The same hurricanes that result in either no victims or very small losses of life in neighbouring countries (Dominican Republic, Cuba, Jamaica, Bahamas, Cayman Islands, and the USA) can cause large numbers of victims and great destruction in Haiti. It would be unimaginable for the new set of building codes and construction practices to be developed for Haiti not take into account all types of natural (climatic or non-climatic) hazards.

The Hyogo Framework will require updating to reflect advances made in climate change negotiations. At the 13th Session of the Conference of the Parties (COP) to the UNFCCC in Bali, December 2007, DRR was recognised by the negotiators as an essential component for climate change adaptation, ultimately becoming part of the Bali Action Plan.

The Bali Action Plan provided the roadmap towards the new international climate change agreement supposed initially to be concluded by 2009, then by 2012, now expected for 2015, leading to a post-Kyoto Protocol international agreement on climate change. In paragraph 1c, the Bali Action Plan highlights the significance of DRR as a part of the enhanced action on climate change adaptation. Among other things, it calls for the consideration of: 'Risk management and risk reduction strategies, including risk sharing and transfer mechanisms such as insurance; and disaster reduction strategies and means to address loss and damage associated with climate change impacts in developing countries that are particularly vulnerable to the adverse effects of climate change.'[41]

Despite the lack of a final agreement being reached by the Parties at the 15th Session of the COP to the UNFCCC in Copenhagen in December 2009, delegates took note of the recommendations of the Ad-hoc Working Group on Long-term Cooperative Action on the Bali Action Plan. In the text agreed upon at Copenhagen, the elements addressing DRR were expanded to include a reference to the Hyogo Framework, which makes this Framework all the more meaningful. Moreover, such references were not placed in brackets, which allowed agreement to be reached on this subject under the Cancun Adaptation Framework at the 16th session of the COP to the UNFCCC in Cancun, Mexico, in 2010.

At the Durban Climate Change Conference (COP 17th session in 2011), the Parties advanced the implementation of the Cancun Adaptation Framework by agreeing on: (a) modalities, procedures, and composition of the Adaptation Committee; (b) activities to be undertaken under the work programme on loss and damage; and (c) modalities and guidelines for the national adaptation plans.[42]

At the Doha Climate Change Conference (COP 18th session in 2012), the Parties advanced the negotiations, agreeing to continue their work to enhance action on adaptation as provided in the Cancun Adaptation Framework, and relevant decisions of the COPs adding specific actions to be undertaken. Also at Doha, a new text was adopted to address loss and damage associated with climate change impacts, with references to the IPCC's *Special Report on Managing the Risks of Extreme Events and Disasters to Advance Climate Change Adaptation* (SREX), the UN *Global Assessment Report on Disaster Risk Reduction*, the Hyogo Framework, and the WMO Global Framework for Climate Services, among other international instruments.[43]

When a final agreement is adopted in late 2015, it is expected to include DRR and therefore reducing risk from natural events would become a required element for adaptation to climate change, in turn making it a legal obligation for governments as well as a strong incentive for other relevant organisations in various sectors at all levels around the world. Further, if the Hyogo Framework and its successor arrangement are used to guide these efforts, risk reduction policies and measures will address all natural events and hazards (climatic and non-climatic), since the multi-hazard approach is an essential component of this guiding document.

Governments that are Parties to the UNFCCC and to the new post-Kyoto agreement will have to develop more intensive DRR policies and measures. As a result, those countries that are prone to geological hazards (earthquakes, tsunamis, and volcanic eruptions with their associated landslides) will also benefit from a legally binding commitment as these policies will need to be multi-hazard oriented following the guidance provided in the Hyogo Framework.

26.7 Challenges ahead

A cursory reflection on the first eight years of the Hyogo Framework reveals that larger and more intensive efforts are needed if a paradigm shift (that is, a fundamental change in mentalities, attitudes, and behaviour) is to be achieved that will reduce risk from and vulnerability to extreme natural events. Knowledge is available to understand risk, hazards, and vulnerability, but the wider public perception is still tainted by the term 'natural' disaster, and, therefore, efforts by governments, communities, and individuals need to go beyond simple awareness-raising to more deep consciousness-raising.[44]

Public authorities expend greater efforts and resources in preparing for the disaster than they do on the long-term need of reducing vulnerability and increasing building resilience to natural events. Coupled with the short life span of governments, the traditional fatalistic belief that disasters are somehow inevitable makes it easier to end up blaming them on God or Nature, rather than addressing the root causes, which requires investing in vulnerability reduction and developing a culture of prevention, a longer-term and not so visible investment.

For all of the above, it is urgent to invest in raising consciousness of the fact that it is not the earthquake (or the cyclone, etc.) that kills people but rather the poorly constructed buildings. In other words, although there is a natural element involved, it is the social behaviour that creates the vulnerability leading to the disaster, and this is an element on which we can act. It is essential to move beyond the term 'natural' disasters, which tends to promote the perception that disasters are somehow inevitable. Social and behavioural sciences addressing cultural and ethical values and their translation into attitudes and behaviours have a vast research area in this regard.

The thousands of experts and representatives of governments and organisations from around the world who discussed and negotiated the Hyogo Framework and who managed to avoid the use of the term 'natural' disasters serve as a good example that must be encouraged and replicated (see endnote 2).

Facing the upcoming negotiations for the post-2015 regime (post-Hyogo, post-Kyoto, and post-MDGs), a broader and more aggressive campaign needs to ensure that reducing risk and vulnerability to extreme natural events receives a higher priority from government authorities at all levels and greater consciousness from citizens. More specific recommendations for stronger policies and measures will need to be adopted at the third WCDR in Japan in 2015. Moreover, specific targets, benchmarks, and indicators need to be identified to ensure that routine monitoring and assessments lead to effectively reducing risk and to building resilience, hence obtaining a substantive reduction in losses of lives and livelihoods and facilitating a more sustainable development.

First and foremost, risk governance must include specific mechanisms – policies, legislation, and organisational arrangements – focusing on risk reduction in close partnership with climate change adaptation. Substantive educational and awareness- or consciousness-raising programmes aiming at reducing vulnerability and at building resilience to natural events should represent a major investment in time, not necessarily in money, by political and private sector leaders.

In this regard, campaigns on prevention and risk reduction can benefit from the experiences gained by the health sector in raising awareness on the importance of preventive measures for reducing the risk of diseases. Natural events could be compared to nature's viruses for which individuals and society at large must learn to develop risk reduction approaches and resilient attitudes and behaviour.[45]

It is not enough to manage disaster risk by simply putting aside funds in order to respond when hazards strike; yet this is what is still being done in many places. Much more attention and resources are still dedicated to preparing for disaster response than to reducing risk and vulnerability. In other words, it is like 'bringing ambulances to the bottom of the cliff rather than building fences at the top'.

As vulnerabilities in urban areas continue to increase through population migration due to economic, environmental, conflict, and other reasons and as new vulnerabilities prompted by climate change also increase, further research coupled with a greater application of existing knowledge is required. The scope of such a large task inevitably extends beyond the time frame established for the Hyogo Framework. Hence laying the foundation for a new and wider framework is expected to be implemented more intensively in the period post-2015 as it links policies and measures more closely to climate change adaptation and the new SDGs (that are expected to replace the MDGs in 2015).

Like the 1994 Yokohama Strategy and Plan of Action, the expected outcome, goals, and priority areas for action of the Hyogo Framework will remain valid over the long term, if not for ever. Governments will need to work more closely with the scientific and academic communities to identify and address the challenges in risk reduction and resilience building, gradually involving other stakeholders whose experience and knowledge are essential for devising solutions to managing risk. Risk will simply move into more demanding stages as populations, particularly in urban and coastal areas, increase creating greater and more complex risks to assess, monitor and manage.

Reducing risk and increasing resilience to extreme natural events will require bold decisions and actions with a longer-term sustainable development perspective that will probably be in opposition to short-term gains and interests. Maintaining the status quo cannot be an option as disasters have become too expensive to be allowed and trends show they will only increase; addressing specific local challenges with specific solutions will not stop disasters from happening in different unprepared locations.

If we want to address the root causes, we need to embark on a new path, recognising and rewarding the value of resilience to citizens, families, communities, nations, and at the global level. It requires a new vision that includes shared responsibility for managing risk, reducing vulnerability, building resilience, and integrating these objectives in many of the public policies that have both direct and indirect effects on risk creation or reduction and enhancing resilience.

'Disaster resilience may not be on the forefront of a political or institutional agenda until a disaster strikes one's own community. Political will and strong leadership are therefore essential to building resilience at any level. Enhancing the nation's resilience will not be easy, nor will it be cheap. But the urgency is there and we need to begin the process now in order to build an ethos that will make nations and communities safer, stronger, more secure, and more sustainable for our children and grandchildren' (National Research Council, 2012).

Endnotes

1 The Yokohama Strategy was adopted at the first World Conference on Natural Disaster Reduction (Yokohama 23–27 May 1994), building on the mid-term review of the International Decade for Natural Disaster Reduction (IDNDR, 1990–1999).

2 At the time of the first World Conference, the concept of 'natural' disasters still prevailed among experts and thus was included in the title of the conference. By the time the second conference was held in 2005, the adjective 'natural' had been dropped in order to emphasise that disasters are caused by the intersection of social vulnerability and natural events and not only by natural events or phenomena. The scope of the Hyogo Framework was defined in its preamble as encompassing disasters caused by hazards of natural origin and related environmental and technological hazards and risks. It thus reflects a holistic and multi-hazard approach to disaster risk management and the relationship between these hazards, which can have a significant impact on social, economic, cultural, and environmental systems, as stressed in the Yokohama Strategy (section I, part B, letter I, p. 8).

3 http://www.aseansec.org (accessed on 19 August 2013).

4 Regional and subregional intergovernmental organisations with a DRR programme include: African Union/NEPAD, League of Arab States, ECOWAS, ECCAS, IGAD, SADC, Indian Ocean Commission, SAARC, ASEAN, ECO, South Pacific Forum/SOPAC/SPREP, Organization of American States, SICA/CEPREDENAC, CARICOM/CDEMA, CAN/CAPRADE, MERCOSUR, Council of Europe/EUR-OPA, DPPI-SEE, CEUDIP. Additionally, most development banks also have developed a DRR activity, including the World Bank, the Inter-American Development Bank, the Caribbean Development Bank, BCIE, CAF, the Islamic Development Bank, the African Development Bank, the Asian Development Bank, the Council of Europe Development Bank, EBRD, IFAD.

5 For details on the status of implementation by each country, see http://www.preventionweb.net/english/countries (accessed on 19 August 2013).

6 In the early 1980s, Venezuela, through its brand new Ministry of Environment and Natural Renewable Resources (MARNR) created in 1977, following the recommendations of the 1972 Stockholm Environment Conference, pioneered in this regard by developing the concept of natural homogenous zones ('áreas naturales homogéneas') as a pre-condition in land-use planning, itself a pre-condition to all social and economic development planning (UNDP Project VEN/79/001: Sistemas Ambientales Venezolanos).

7 http://www.preventionweb.net/english/hyogo/gar/ (accessed on 19 August 2013).

8 http://www.globalnetwork-dr.org/views-from-the-frontline.html (accessed on 19 August 2013).

9 http://www.gfdrr.org/gfdrr/sites/gfdrr.org/files/GFDRR_2011_Annual_Report.pdf (accessed on 19 August 2013).

10 http://www.ifrc.org/publications-and-reports/world-disasters-report (accessed on 19 August 2013).

11 More detailed information on these experiences and good practices can be found on the websites mentioned in endnotes 7–10 and throughout the article.

12 See CAPRA methodology developed by the World Bank and UNISDR (http://www.ecapra.org) and the Global Earthquake Model Program (GEM) (http://www.globalquakemodel.org). These are open-source systems, additional to those available through private risk assessment companies.

13 A fourth process is relevant although it does not involve intergovernmental negotiations, the International Decade on Education for Sustainable Development (2005–2015) will also come to an end and some follow-up is expected under the coordination of United Nations Educational, Scientific and Cultural Organization (UNESCO). Such a follow-up should also include DRR as a key component.

14 The academic and scientific community working on DRR have agreed to promote and develop integrated approaches in research and education on the topic. Since 2010, the International Council for Science (ICSU), together with the International Social Science Council (ISSC) and the ISDR Secretariat (UNISDR), put in place the Integrated Research on Disaster Risk (IRDR) programme to promote this task, working with academic and research institutions around the world: http://www.irdrinternational.org.

15 Various experts have promoted this idea in the past. More recently, the International Union of Geodesy and Geophysics (IUGG; http://www.iugg.org) has made specific recommendations in the context of the International Council for Science (ICSU, http://www.icsu.org), where it is currently being discussed. See also the Declaration on Disaster Risk Reduction through Research and Assessments adopted by the Partners of the Project 'Extreme Natural Hazards and Societal Implications – ENHANS' and calling for 'a negotiation on setting up a process of assessing and synthesizing the

policy-relevant results of peer-reviewed published research . . . by a high-level intergovernmental body' (http://www.enhans.org/about/Declaration.pdf).

16 http://www.unisdr.org/we/inform/publications/18197 (accessed on 19 August 2013).

17 The first World Conference was held in Yokohama, May 1994, less than a year before the Great Hanshin-Awaji (Kobe) earthquake, and adopted the Yokohama Strategy and Plan of Action that served as the first policy guide on the topic. Ten years later, and to commemorate the tenth anniversary of the Kobe earthquake, the second World Conference was held in Kobe, Hyogo, January 2005, and adopted the *Hyogo Framework for Action*. The third World Conference is planned in 2015 for the twentieth anniversary of the Great Hanshin-Awaji (Kobe) Earthquake and fourth anniversary of the Great East Japan Earthquake and Tsunami.

18 See http://www.preventionweb.net and http://www.unisdr.org for examples of specific engagements in this regard (the websites were accessed on 19 August 2013).

19 See http://www.oecd.org/gov/risk/G20disasterriskmanagement.pdf for specific actions and examples in this regard. Also more information at http://www.gfdrr.org (the websites are accessed on 19 August 2013).

20 Relevant consultations, forums, and networks at the global level include those facilitated by: the UNISDR (http://www.preventionweb.net); UN/DESA (http://sustainabledevelopment.un.org/rio20.html); the World Bank's GFDRR (http://www.understandrisk.org); GNDR (http://www.globalnetwork-dr.org); IRDR (http://www.irdrinternational.org); Radical Interpretations of Disasters and Radical Solutions (RADIX, http://www.radixonline.org); emBRACE (http://www.embrace-eu.org), among others (the websites were accessed on 19 August 2013).

21 www.globalnetwork-dr.org (accessed on 19 August 2013).

22 The GNDR, through its 'Views from the Frontline', has collected valuable information on NGOs and other local organisations working on DRR (http://www.globalnetwork-dr.org; accessed on 19 August 2013).

23 http://www.unisdr.org/ppew/info-resources/ewc3/Global-Survey-of-Early-Warning-Systems.pdf. The WMO has developed the Global Framework for Climate Services (GFCS), which contains elements required for developing such a global early warning system, http://www.gfcs-climate.org, at least for climate-related events (the websites were accessed on 19 August 2013).

24 http://www.iucn.org/about/work/programmes/ecosystem_management/disaster/ecosystems/ (accessed on 19 August 2013). Additional source: Sudmeier-Rieux (2013).

25 http://www.cdkn.org (accessed on 19 August 2013).

26 For more information, see: http://www.fairfaxcounty.gov/oem/pdrp/; http://www.recoveryplatform.org/; http://www.preventionweb.net/files/globalplatform/entry_presentation~15h00%5B1%5D.pdf; http://www.oas.org/pgdm/document/preplan.htm; http://www.huduser.org/portal/publications/pre_disasterplanning.html (all websites were accessed on 19 August 2013).

27 A large number of current preparedness programmes around the world can be found at http://en.wikipedia.org/wiki/Category:Disaster_preparedness_by_country; and a specific example at http://www.ready.gov/about-us (the websites were accessed on 19 August 2013).

28 http://www.irdrinternational.org/wp-content/uploads/2012/06/FORIN-REPORT_web.pdf (accessed on 19 August 2013).

29 More detailed information on each country's experience may be found at: http://www.preventionweb.net/english/hyogo/progress/?pid:3&pil:1 (accessed on 19 August 2013).

30 The full report of the second session of the Global Platform for Disaster Risk Reduction can be found at http://www.preventionweb.net/english/professional/publications/v.php?id=11963 (accessed on 19 August 2013).

31 http://www.preventionweb.net/english/hyogo/hfa-monitoring (accessed on 19 August 2013).

32 http://www.preventionweb.net/files/18197_midterm.pdf (accessed on 19 August 2013).

33 http://www.gfdrr.org (accessed on 19 August 2013).

34 http://www.globalnetwork-dr.org (accessed on 19 August 2013).

35 http://www.irdrinternational.org (accessed on 19 August 2013).

36 Regarding poverty expansion China, India, and Brazil are notable exceptions, which, by the sheer size of their populations, distort the average figures for the world, while numbers are increasing in many other countries.

37 Although apparently redundant, the term 'natural' is utilised here to differentiate these ecosystems from those that have been created or enhanced by human intervention and which tend to be in a better state.

38 In addition to the IPCC SREX, the following sources provide specific examples:http://www.epa.gov/climatechange/impacts-adaptation/coasts.html#impactssea; https://darchive.mblwhoilibrary.org/bitstream/handle/1912/2273/SEALEV~1.pdf?sequence=1; http://www.fao.org/sd/EIdirect/EIre0047.htm; http://www.academia.edu/1224958/CLIMATE_CHANGE_SEA_LEVEL_RISE_AND_COASTAL_VULNERABILITIES_OF_BANGLADESH_WITH_ADAPTATION_OPTIONS; http://www.climatehotmap.org/global-warming-effects/sea-level.html; http://www.soest.hawaii.edu/coasts/sealevel/; http://www.cakex.org/tools/sea-level-rise-and-coastal-flooding-impacts-viewer; http://static.weadapt.org/placemarks/files/225/golam_sarwar.pdf (the websites were accessed on 19 August 2013). RTBF (Belgian TV) produced evidence of such impacts in a video aired on 6 January 2010 and called '*De plein fouet, le climat vu du Sud*' of the series called '*Planète en question*'.

39 The Hyogo Framework defines its scope as 'encompassing disasters caused by hazards of natural origin and related environmental and technological hazards and risks. It thus reflects a holistic and multi-hazard approach to disaster risk management and the relationship between natural events or phenomena and disasters, which can have a significant impact on social, economic, cultural and environmental systems, as stressed in the Yokohama Strategy (United Nations, 1994; section I, part B, letter I, p. 8).' It further recommends that 'an

integrated, multi-hazard approach to disaster risk reduction should be factored into policies, planning, and programming related to sustainable development, relief, rehabilitation, and recovery activities in post-disaster and post-conflict situations in disaster-prone countries', and it refers to a similar recommendation in the Johannesburg Plan of Implementation, World Summit on Sustainable Development (Johannesburg, South Africa, 26 August–4 September 2002, paragraphs 37, 65; available at: http://www.un.org/jsummit/html/documents/summit_docs/131302_wssd_report_reissued.pdf; accessed on 19 Augsut 2013).

40 Joint submission by the Inter-agency Standing Committee (IASC) and ISDR to the UNFCCC Ad-hoc Working Group on Long Term Cooperative Action, http://www.unisdr.org/we/advocate/climate-change (accessed on 19 August 2013). Furthermore, the IPCC-SREX (2012) suggests that if disaster risk management is adopted in the short term, it will lead to adaptation to climate change in the longer term.

41 The full text of Bali Action Plan may be found at http://unfccc.int/resource/docs/2007/cop13/eng/06a01.pdf#page=3 (accessed on 19 August 2013).

42 http://unfccc.int/adaptation/items/4159.php (accessed on 19 August 2013).

43 http://unfccc.int/resource/docs/2012/cop18/eng/08a01.pdf (accessed on 19 August 2013), pp. 11 and 21.

44 'The late Brazilian adult education pioneer, Paulo Freire, thought of education as a collective study of reality and problem solving, and made policy recommendations using a Portuguese term roughly translatable as "consciousness-raising." In relation to disaster risk, "consciousness" is a useful term, going far deeper into the root causes of vulnerability than does the common expression "risk awareness". For example, in Turkey between 1995 and 2003, a series of deadly earthquakes was met with a crescendo of public outcry and a slowly deepening public understanding of what had to be demanded of the construction industry . . . This, too, is education' (Wisner, 2006, p. 8).

45 The health of homes, offices, schools, etc., could be compared to personal health demonstrating the relevance of prevention measures and building resilience, in the same way that people do for their bodies and minds. Rather than measuring progress on mortality reduction which would be more difficult, it is important to set targets on building resilience (including governance, institutional, educational, research, and technical activities), which is easier to measure.

References

IPCC-SREX (2012). *Special Report on Managing the Risks of Extreme Events and Disasters to Advance Climate Change Adaptation (SREX)*. Geneva: Intergovernmental Panel on Climate Change (IPCC). Available at: http://www.ipcc-wg2.gov/SREX (accessed on 19 August 2013).

National Research Council (2012). *Disaster Resilience: A National Imperative*. Washington, DC: The National Academies Press.

Sudmeier-Rieux, K. (2013). *Ecosystem Approach to DRR: Basic Concepts and Recommendations to Governments, with a Special Focus on Europe*. Strasbourg: Council of Europe (EUR-OPA Agreement).

United Nations (1994). *The Yokohama Strategy for a Safer World: Guidelines for Natural Disaster Prevention, Preparedness and Mitigation and Its Plan of Action ("Yokohama Strategy")*. Geneva: Secretariat of the International Decade for Natural Disaster Reduction (IDNDR, 1990–1999). Available at: http://www.preventionweb.net/english/professional/publications/v.php?id=8241 (accessed on 19 August 2013).

United Nations (2005). *Hyogo Framework for Action (2005–2015): Building the Resilience of Nations and Communities to Disasters*. Geneva: UNISDR. Available at: http://www.unisdr.org/we/coordinate/hfa (accessed on 19 August 2013).

United Nations (2009). *Global Assessment Report on Disaster Risk Reduction: Risk and Poverty in a Changing Climate, Invest Today for a Safer Tomorrow*. Geneva: UNISDR. Available at: http://www.preventionweb.net/english/hyogo/gar/report/index.php?id=9413 (accessed on 19 August 2013).

United Nations (2012). *The Future We Want, Outcome of the UNCSD (Rio+20)*, in UN General Assembly Resolution No 66/288 of 11 September 2012. New York: United Nations.

Wisner, B. (2006). *Let Our Children Teach Us!* Geneva: UNISDR and Bangalore: Books for Change. Available at: http://www.unisdr.org/2005/task-force/working%20groups/knowledge-education/docs/Let-our-Children-Teach-Us.pdf. Accessed on 10 August 2013.

Wisner, B. and Walker, P. (2005). *Beyond Kobe: A Proactive Look at the World Conference on Disaster Reduction (18–22 January 2005, Kobe, Japan)*. Boston: Tufts University. Available at: www.unisdr.org/wcdr/thematic-sessions/Beyond-Kobe-may-2005.pdf. Accessed on 10 August 2013.

Wisner, B., Gaillard, J. C. and Kelman, I. (2012). *The Routledge Handbook of Hazards and Disaster Risk Reduction*. London & New York: Taylor & Francis Group.

27

Disaster policy and climate change: how much more of the same?

Stephen Dovers and John Handmer

27.1 Introduction

Over recent decades, research and policy have steadily moved from viewing natural disasters as just that – 'natural', imposed on humans by outside forces – to understanding disasters as phenomena more often caused by the vulnerability of societies, settlements, and assets that are constructed or shaped by human decisions. Disasters are increasingly understood as something we have a strong degree of control over, through active policy choices or a failure to make choices. There has also been increasing understanding of the nature of disasters and of human vulnerability to them through strengthened research efforts and through better documentation of events and experiences: vulnerability to disasters, and thus disaster risk, grows with the size and spread of human settlements, and the value and exposure of built and other assets. More recently the prospect of climate change, and specifically an increased frequency and intensity of severe climatic events, has accelerated reconsideration of disasters and how society should or could deal with them.

This chapter examines ways in which disasters have been understood and responded to over time, from a policy and institutional perspective. The central question raised in the chapter is the degree to which current policy approaches and institutional capacities are sufficient for the future. Is a further incremental evolution of policy, such as has occurred over recent decades, enough? How much will the future be more of the same? The chapter extracts major themes from the disaster literature (e.g., Handmer and Dovers, 2013) and brings these into a closer consideration of the interplay between 'traditional' disaster thinking and a future where climate change is proposed to have a significant impact on disasters (e.g., IPCC, 2012).

We first summarise major trends in disaster policy and emergency management over recent decades, and then consider the implications of climate change for this already changed and changing agenda. We consider possible future implications, and advance the prospect of deeper integration of disaster policy across other policy sectors ('mainstreaming'), within the broader agenda of sustainable development.

We start with a note on the terminology used in this chapter, as the use of terms varies considerably across countries, disciplines, and agencies. *Disasters* here are considered to be larger scale events (whether of natural cause or not) which overwhelm capacities. *Disaster policy* (and related, broader institutional settings) comprises the higher level, strategic settings within which societies understand and respond to disaster events. *Emergency management* refers to the practical end of disaster response, expressed in specific agencies, practices, and processes to attend to more regular and tractable emergency situations (e.g. regular floods, industrial accidents, expected cyclones, etc.) as well as rarer, more difficult large scale events. The international terms of disaster risk and disaster risk reduction (DRR) are within this definitional scope; however, we cover both the 'routine' or traditional emergency management end of the spectrum as well as the greater magnitude 'non-routine' or 'complex unbounded' disaster events (see further below). Increasing vulnerability to routine events and to serious disasters increases what is often termed 'disaster risk'. The terminology focuses attention on the central question here, which is the adequacy of current policy and institutional arrangements to cope with both existing and unpredictable but likely more demanding future situations.

27.2 The evolution of disaster policy

An extensive literature documents the overriding importance of human factors – social, economic, and political – in

Extreme Natural Hazards, Disaster Risks and Societal Implications, eds. Alik Ismail-Zadeh, Jaime Urrutia Fucugauchi, Andrzej Kijko, Kuniyoshi Takeuchi and Ilya Zaliapin. Published by Cambridge University Press.

generating the conditions that can lead to disasters, and to more expectable emergency situations, and that can exacerbate the impacts of natural phenomena (White, 1946; Turner, 1978; Perrow, 1984; Quarantelli, 1998, Mileti, 1999; Mitchell, 1999; Pelling, 2003). The central role of vulnerability created by settlement, political, and economic circumstances has been increasingly understood. Also, it is increasingly recognised that human impacts on the Earth are changing global rather than local natural systems: the vulnerability of societies and communities is not influenced only by local or national decisions, but by global conditions and changes in social, technological, informational, and economic forces. Natural phenomena, in contrast, over which we have little or no direct control may make relatively modest contributions to disaster impacts: the risk of a disaster (in terms of impacts) occurring may be higher due to increased vulnerability even when no change has taken place in the frequency or intensity of natural phenomena (IPCC, 2012). This statement must be qualified for those whose livelihoods depend entirely upon enterprises closely influenced by climatic variability, and for exceptionally severe events such as the 2004 Asian tsunami, which may have very serious impacts at the local level, especially in poor countries and poor regions of otherwise wealthy countries. The European heatwaves of 2003, 2007, and 2010, the 2011 Japanese earthquake and tsunami, as well as Hurricane Katrina in New Orleans, highlight that rich countries also suffer extreme impacts.

In disaster and emergency management research and increasingly in policy, thinking has shifted from being dominated by a passive approach – accepting disasters as 'acts of God' – with the matching attitude that there is little that humans can do, to a more proactive approach that accepts the role of humans in creating vulnerability. This invites a greater focus on the purposeful development of institutions, policies, and emergency management practice aimed at reducing vulnerability and enhancing resilience, rather than simply responding to events. If the conditions for vulnerability to disasters are often in large part created by human choices (such as to settle or build in particular places, in a particular manner), then it is human choices that can address the situation. This is consistent with a broader theme in discourses of sustainable development: human culture, through its institutions, governance, and patterns of production and consumption, has created situations that are socially, environmentally, and/or economically unsustainable (Boyden, 1987). It follows that it is only through human culture (i.e. institutional etc. change) that greater sustainability can be pursued.

Disasters can result from an extreme event from nature triggering destruction and disruption to people and economies. This is our normal view of disaster, but extreme events do not necessarily result in disasters and disasters, in terms of impacts on humans, do not need extreme events, natural or otherwise. It is the impact on communities, economies, and ecosystems that defines a disaster, not the size of the natural event. Disasters frequently result from minor natural events, or some sequence of minor events, intersecting with susceptible human systems, resulting in major losses in the absence of major natural events. The capacity of a society or community to withstand and recover may be eroded over time through the cumulative impacts of a series of relatively minor or 'routine' events, such as a number of moderate floods occurring with greater than usual frequency. Furthermore, a gradual change in 'normal' conditions may also result in major impacts, for example thawing permafrost in the far northern hemisphere resulting in very severe damage to buildings and all types of infrastructure, or slow trends in atmospheric warming that gradually exceed the adaptive capacity of cropping systems. Human agency combines with natural variability to define each disaster event. Climate change, discussed below, may shift the frequency and intensity of some natural disturbances from 'routine' matters currently handled competently to a more difficult level. For example, there has been a shift in drought policy in a number of countries from treating drought as an exceptional occurrence requiring emergency assistance to policy settings to encourage landholder risk management (Cockfield and Dovers, 2013). Significant increases in drought occurrence would place such policy regimes under considerable pressure.

The shift in approach summarised above is a more positive perspective on disasters than the view that they were imposed on populations who were simply victims, where human intelligence and agency can be actively turned towards making people and their settlements safer and more sustainable. However, recognising the importance of human agency may also encourage attribution of blame, whether deserved or not, so the shift is not entirely positive. It is certainly the case that changes in political attitudes and the extension of media and communications capacity have seen the search for 'those responsible' begin almost as soon as a disaster event unfolds. A search for lessons and better ways of doing things after an event is of course worthwhile, but sensationalised, short term media excitement is usually less than productive. An issue arising from this situation, not dealt with further here, is the most appropriate form of post-disaster inquiry that can yield positive lessons for the future as well as, if necessary, reveal negligence or culpability.

Modern emergency management is a well-established system of regulations, capacities, and agencies involving many actors, all with particular backgrounds and reasons for involvement. Often, key organisations have their origins in the 'civil defence' or 'home guard' capacities developed during World War II, which later evolved towards domestic threats rather than external military ones. Working with the career uniformed services and established welfare groups, such as the Red Cross and the Salvation Army, they supported the home front. The Cold War and the possibility of nuclear attack legitimised the maintenance of civil defence. These war-related organisations found themselves increasingly busy with everyday events and crises arising from natural agents and from industrial, transport, and technological failures. These events affected and concerned many more people than actual or possible military or security risks. During the 1970s, most civil defence organisations in developed countries shifted focus to an emergency management, rather than war-related, emphasis and in doing so went from a focus solely on consequences to a more comprehensive approach that included prevention as well as response. Interestingly, there has been in recent years a rejoining of military and disaster/humanitarian concerns, with military resources being utilised more often to support post-disaster response and recovery (but very rarely prevention and preparedness).

It is easy to see emergency management only in terms of obvious, visible tasks of response and the short term thinking that informs this. However, we need to be aware that the key groups include many dedicated more to recovery and support than the immediate task of short term response. This is especially the case in the non-governmental and government welfare sector, who may act quickly but whose work extends much longer after an event, assisting communities and individuals. We are not critical of a strong response focus – that is what society, media, and politicians want when a crisis erupts, and not to respond when lives and communities are at risk would be morally and politically unacceptable. However, we argue that a much greater emphasis on strategic policy is needed to increase the focus on prevention through tackling exposure and vulnerability, while maintaining a high level of response capability. The imperative of adapting to climate change is in keeping with this position, as it stretches thinking forward in time.

A schema representing these trends in emergency management is presented in Table 27.1. We argue that these trends, which have been largely evolutionary and incremental, not only need to continue but to intensify. The progress of the trends and changes varies by jurisdiction, by sector, and by particular agencies: in some cases the changes have been swift, in others slow and progressive, in others less noticeable. Thus the rate of change and reform needed varies also, from consolidation and extension of recent reforms to considerable efforts in a short time. In general, the trends have been evident from the post-WWII period, intensifying in the last two decades.

Table 27.1. *Trends in emergency management (Handmer and Dovers, 2013)*

From	To
Broad framing of the basic issue:	
Disasters as 'other' or 'acts of God'	Disasters as generated by humans
Event driven	Situational and less visible creeping hazards included
Policy context:	
Lack of visibility and profile	Legal liability
	Rising expectations and critical scrutiny
	Impacts of counter-terrorism and security
Problem ownership and framing:	
Acceptance/individual decision-making	Community vulnerability and sustainability
Owned by government	Should be shared with those at risk
Local	Local–global
Choice	Institutional constraints
Style:	
Secret	Open
Paramilitary	Dominantly civilian
Uncertainty ignored or treated as quantifiable	Uncertainty is acknowledged
Policy emphasis:	
Accept or reduce loss	Manage vulnerability or increase resilience
Focus on the hazard and event	Focus on community safety and consequences
Solutions as separate	Solutions found in organization of society and the development process
Based on direct emergency management experience	Based on evidence, and whole of society priorities

To help shift the sector away from a sole focus on response, the concept of a comprehensive approach was formalised in the early 1980s in Australia and the USA via the framework of 'PPRR': prevention, preparation, response, and recovery. This was to emphasise the need to deal with all aspects of regular emergency management – not just response – and the importance of the overall state of vulnerability or resilience of societies and communities. Internationally, increased attention to communities' and nations' vulnerability to disaster is consistent with this shifting national focus, and led to the goal of DRR – a more proactive stance. In many countries and within international aid and development agencies, disasters and development were increasingly seen as closely linked (Cuny, 1983). The severe social and economic impacts of disasters can undermine the development process, setting countries back years, and poor human development is a major determinant of vulnerability to disasters. As a result, the concept and policy goal of DRR is often seen as a key part of development, a link recognised through a series of international meetings and agreements (e.g. at the World Summit on Sustainable Development in 2002, see further below).

A significant attempt to broaden the approach to emergency management has been through the introduction of a risk-based framework and approach, known in Australia and New Zealand as emergency risk management (ERM), based on the generic Australian–New Zealand Risk Management Standard (Standards Australia, 2004), now incorporated into an international standard (ISO31000: 2009). This approach offers a broader view of risk and the required response by spreading attention away from the event itself to include more explicit consideration of the socio-economic and environmental context of the risk, clear identification of what is at risk, what creates the risk, how that risk might be ameliorated, and monitoring and learning from the risk management experience.

This all highlights the desirability of emergency management or DRR serving a number of aims – with enough flexibility to cope with shifts in community and political priorities: to contribute to the aims of human development and economic stability. However, by themselves risk based approaches and PPRR support this aim, but do not achieve it. Achievement means viewing disasters as a policy and societal problem as much more than a matter for emergency managers only, but as a matter that is a whole-of-government and indeed whole-of-society concern. Addressing the preconditions of vulnerability to disasters requires attention by a wide range of policy sectors, agencies, and communities of interest. The event is the result of social and economic conditions and they may threaten our sustainability, and to that extent are issues of human and national security. A major challenge in disaster and emergency management is 'mainstreaming' disasters to ensure that the necessary actions are taken as a matter of course by policy agencies and actors across health, transport, communications, energy supply, water and sanitation, defence, land use planning, and more.

In summary, current trends in disaster research and management reflect a range of interests, some of them common to other public policy areas, such as climate change adaptation and sustainable development:

- seeking to put emergency management into the policy mainstream and away from a marginal or specialist activity by reframing problems;
- seeking to deal with causes rather than symptoms, emphasising the need for learning and greater efforts in strategic policy development;
- the need for appropriate institutional structures to deliver long-term solutions;
- sharing ownership of the problem, and responsibility for solutions, with those at risk and working to reduce vulnerability.

Among many in the social and policy sciences, learning to live with hazards and sustainable development have come to be seen as interlocking aims (Mileti, 1999). The World Summit on Sustainable Development (WSSD), the ten year follow-on to the first United Nations Conference on Environment and Development (UNCED) of 1992, was held in 2002 in Johannesburg. The summit made disaster reduction one of its central themes. To be very vulnerable is not sustainable – economically, environmentally, or socially.

27.3 Beyond 'natural' disasters

Disaster and emergency management is now expected to operate at the front line of climate change adaptation. The impacts of climate change are predicted to manifest in more frequent and severe extreme climate and weather events, and these are already a major contributor to disasters. However, it is not only extreme climate and weather that are of concern, and care is needed in drawing climate change and disaster research policy and practice together so as not to begin to ignore non-climate related events (technological and industrial failures, pandemics, etc.). Here, we are interested primarily in the impacts or consequences on human systems of unexpected events (climate driven, non-climate driven, and various combinations).

To the extent that climate change is expressed through extreme weather and climate, it falls to emergency management to plan and prepare for the possibilities of additional and different extremes, and to respond and deal with the consequences. This is, in principle, the usual business of emergency management. The expectation that emergency management is to play a major role in adaptation to climate change is new. This is evident in the attention the area is attracting internationally, particularly through the Intergovernmental Panel on Climate Change SREX (Special Report on Extremes) reports (IPCC, 2011; 2012), and nationally such as through Australia's National Climate Change Adaptation Research Facility (NCCARF) programme, an activity which in different forms is replicated in many countries.

A key policy question is to what extent government should be involved in leading climate adaptation, whether directly or through enabling and supporting non-government actors and the private sector. The Australian Productivity Commission's draft report on *Barriers to Effective Climate Change Adaptation* provides an illustration of arguments around this question (Productivity Commission, 2012). This report is written from a traditional economic perspective, which would be expected to argue against a role for government and an enhanced role for the market. The report does this and downplays the need for government involvement in climate change adaptation. However, the Commission made an important exception – emergency management – where it argued for a major role for government. One reason economists and governments emphasise emergency management is that many emergency management measures are no- or low-regrets approaches. That is, if the changes in climate and weather turn out to be less problematic than predicted, the efforts are not wasted as managing existing risks and vulnerabilities better is in most cases desirable.

Through attention to climate change, disasters are being recast as a security issue of national and international strategic importance, a position more prominent on policy agendas than ever before. Note, however, that the impacts of climate change include more than only climate-driven disasters, but also reductions in water availability or agricultural yield and so on.

The Stern Review (Stern *et al.*, 2006) argues that society has to act now or face 'devastating' economic consequences arising from climate change. On 17 April 2007 the UK put climate change on the agenda of the UN Security Council arguing that it threatened global security. This was not supported by all Security Council members, but follows a number of reports carrying similar warnings. In a similar vein, Australia's *National Strategy for Disaster Resilience* (Australia, 2011) asserts that 'climate change represents a most fundamental security challenge'. Such reports and commentary are strongest on the changes reflected in technical climate measurements, the causes of change, and future possibilities. However, for many people, such as those living in arctic, sub-arctic, and permafrost regions, changes to their environments have already been dramatic (CIEL 2007; IPCC 2012), destroying infrastructure and homes, threatening livelihoods and in some cases the viability and existence of communities.

Changes or increased variability in the climate are an important part of the future, However, other trends may be just as or more important: widespread environmental and water contamination is likely in future, partly caused by disruption through weather events, but also simply – and more predictably – from an expanding population and industrialisation that places more people closer to more hazardous industries and contaminants, and places people and built assets in locations made hazardous by natural phenomena, areas that may have previously been avoided.

It is trends in the exposure and vulnerability of people that are or should be firmly within the ambit of emergency management and climate change adaptation (IPCC, 2012). If they are not, emergency management will be confined to responding to emergencies of escalating scale and complexity without the necessary increased efforts to avoid exposure and risk. The work of emergency managers already largely involves addressing the results of either poor decisions or non-decisions by other policy sectors (e.g., land use planning, infrastructure development): the future may be defined by this role of cleaning up after others, or by a more active engagement with key trends so as to lessen the impacts of events in future.

Some of the key trends that will influence vulnerability to disasters in future include:

- Population increase, but more importantly the distribution of populations. Globally, there will be an increase in the number of people exposed to all types of hazards, with an increasing number living in hazardous locations with marginal livelihoods and therefore a scarcity of coping resources.
- Urbanisation, which creates intense concentrations of humans and economic activity, and an increased potential of impacts by most types of disaster events through greater impacts on more people. Conversely, though, there may at the same time be a greater capacity with urbanisation to plan and deal with disaster events precisely because of a concentration of resources.
- Armed and political conflict, which may be over resources, ethnic divisions, ideologies, or even without discernible reason. Conflict decreases resilience by devastating livelihoods, infrastructure, and sources of food and water, often forces mass displacement, and redirects resources from productive use to military and/or short term humanitarian crisis response. Conflict undermines emergency management, public health, communications, education, and agricultural

development, and deflects attention away from strategic issues such as responding to longer term environmental change.

- Often associated with conflict, the breakdown of governance and institutions decreases the capacity for organised emergency management and undermines the resilience of economies and communities, whether through reduced capacity or rising corruption.
- Uncertainty, and the difficulty of policy action in the face of uncertainty, can paralyse decision-making. It appears that there is increasing uncertainty on many fronts, including climatic, economic, social values and expectations, and political and legal conditions, with the impact of economic uncertainty illustrated dramatically by the recent and ongoing (at the time of writing) global financial crisis.
- Inequity and uneven distribution of economic power which, in addition to the trends listed above, create more vulnerable groups within society through changing employment conditions, declining access to health care, and displacement for major development projects.

We can ask how well disaster policy and emergency management meets expectations already, without considering future exacerbations, and then ask whether it has the operational, policy, and institutional capacity to cope with likely greater demands in future. Taking a broad definition of emergency management to include all involved (government and non-government), we argue that it does extraordinarily well in many countries, notwithstanding that things at times go wrong even in countries with high levels of resources and capacities. The picture is varied, however, with much emphasis on and thus evaluation of success focused only on very visible events with high profile responses. Much less attention is given to other parts of emergency management function: to less visible disasters in poor regions, and to prevention, preparedness, long-term recovery, and forward looking climate change adaptation. Subtle or systemic development policies aimed at increasing the general resilience of the many poor communities that are most affected by regular climatic or other events that are in themselves not dramatic and do not make for good headlines in the current media and political environment. When such actions are taken with an eye to an uncertain future comprising multiple, uncertain threats to communities, the story is harder still to communicate.

Addressing prevention, and improving resilience or reducing vulnerability, often requires fundamental change in policy, land use, and the distribution of economic and other resources. That may challenge powerful interests and may disrupt the status quo, and is likely to be resisted. Failure to address this basic challenge of incorporating DRR into human and economic development in many parts of the world will certainly result in large losses of life, livelihoods, and economic assets and activity.

27.4 Climate change and disaster policy

Here, we accept the strong scientific consensus that human-induced climate change is already occurring, that the rate of change will increase in the absence of at present unsupported levels of action to reduce greenhouse gas emissions, and that climate change will lead to increasing frequency and intensity of climate-driven natural disturbances and events such as wildfires, floods, cyclones, storms, coastal inundations, and heatwaves (IPCC, 2011, 2012). Given that a large proportion of disaster events and the work of emergency management relate to climate variability and extreme climatic events, there is clear relevance to disaster policy and emergency management. In this context, there are two important implications for disasters that are considered here:

1. the ability to cope with likely but inherently uncertain increases in the frequency and intensity of climate-related disaster and emergency events; and
2. a recasting of the policy agenda and discourse of disasters within the broader context of climate change adaptation, to avoid unconnected efforts that are maintained within separate agencies, portfolios, and disciplines.

(Note: climate change policy has two dimensions: *mitigation* (reducing greenhouse gas emissions), and *adaptation* (to future climate change and increased variability and to that already 'locked in' due to already increased greenhouse gases in the atmosphere). While disasters are relevant in discussions of mitigation, as a major forthcoming area of impact in the absence of policy action, there is not a direct policy link. This chapter deals with the far more directly relevant issue of adaptation.)

To frame thinking around the impact on disaster policy capacities in a changing climate, the following schema matches generalised prospects of global warming represented as a simple typology (Dovers, 2009), with three general categories of policy problems in disasters and emergency management (routine, non-routine, complex unbounded, from Handmer and Dovers (2013)). The degrees of warming are indicative only, and the schema illustrative and general, but note that the most common interpretation of the 1992 UN Framework Convention on Climate Change instruction to avoid 'dangerous' climate change is to keep global warming at less than 2 °C:

I. *More of the same, 'routine disaster problems', roughly up to 2 °C warming*. Not dissimilar from existing variability faced by societies in living memory and with which societies generally cope through existing emergency management capacities but could adapt better with known strategies. Some shift in the parameters of climate and climate-related events and impacts, perhaps increasing in frequency and magnitude, but within institutional and societal memory.

II. *'Non-routine disaster problems', roughly 2–4 °C warming.* Significantly exacerbated climate variability and impacts: droughts, floods, cyclones, heatwaves, vector-borne diseases, etc. Not events and impacts outside the historical experience of standard emergency management or disaster response capacities, but very challenging to capacities and institutions, where societies face a distinctly different climate and variability and events intensify, but within limits that allow adaptation even if this is very difficult.

III. *'Complex, unbounded disaster problems', roughly >4 °C.* Climate change and variability beyond individual human experience, institutional memory, and societal capacity, and well beyond the scope of existing disaster policy and emergency management capacities, threatening the production bases of economies, inundation of major coastal settlements, widespread health impacts, the integrity of ecosystems and biodiversity, and the stability of policy and institutional systems. A radically different operating environment for communities and societies.

This schema is illustrative, and the degrees of impact and ability to cope will vary from place to place and unevenly over time. Disasters may be non-routine in one place but routine in another, relative to previous experience and capacities. However, to encourage thinking about the relationship between disasters and climate change, such a simple schema is useful. Climate adaptation has received less and later attention than climate mitigation, and has particular attributes as a policy problem (Dovers and Hezri, 2010). Climate adaptation is a diffuse policy challenge, with few systemic or generalisable policy 'levers' that will consistently shape and guide responses over many, very different climatic, biophysical, and political contexts. Adaptation requires investment of resources in the near term, in the face of uncertainty regarding both the magnitude of risks and impacts and their precise location of occurrence, for benefits that will occur in future. A lack of clarity of division between public and private benefits, stretched scales of time and space, deep uncertainty, and scattered policy and property rights and responsibilities are attributes that create difficult policy problems (Dovers, 1997). Whether the issue is investing in housing stock rehabilitation against heatwaves, or foreshore protection against rising sea levels and storm surge, the political, personal, and business case for near term investment may be hard to make.

Justifying investment in climate adaptation measures is indeed difficult, and this difficulty is relevant to disasters. In climate science and projections, scientific standards of proof and evidence apply, and allow for a common language in debating uncertainty. In mitigation policy, a small range of systemic policy instruments – trading systems, carbon taxes, technology substitution, etc. – can be reasonably well modelled and the emissions reduction predicted. The situation with adaptation policy is very different in two ways. Predicting with any certainty the location or impact of a climatic event is almost impossible with the current state of down-scaled climate change models, so picking where to invest in adaptation measures is complex and will always be contested. The decision to invest in preventative measures in one place will, in a world of limited resources, generally imply that another place will not receive support: there are strong equity dimensions to adaptation actions.

Second, it is difficult to evidence or predict the efficacy of most adaptation measures. Will a levee be high and strong enough; will a warning system work; will a fire or cyclone awareness campaign change behaviours; will a strategic climate adaptation assessment process influence the final proposals emerging from an infrastructure agency; will a new building code to handle increased weather events be observed? And so on. The efficacy of many adaptation measures cannot really be predicted robustly.

It is evident in the adaptation literature, and in adaptation policy discussions, that too little attention has been paid to existing bodies of research and policy theory and practice. This includes relevant disciplines such as law, public policy, social psychology, and institutional design, and relevant areas of practice such as public health, environmental policy, and emergency management (Dovers and Hezri, 2010). Cross-disciplinary and cross-policy sector learning is a current and future task that can better inform both climate adaptation and disaster policy.

Under a future characterised by level II and especially level III in the schema above, it would be expected that institutional systems will struggle to cope. Our institutional systems are a product of the problems faced and knowledge held in the past: that is the nature of durable and influential institutions (Connor and Dovers, 2004). Climate variability and a frequency and intensity of climate disasters outside societal and institutional memory will inevitably challenge current understanding and capacity, and that challenge is intensified by the requirement to 'mainstream' both adaptation and disaster policy across sectors and portfolios that have given these little attention previously. Institutional change is a complex matter and the possibilities and imperatives vary greatly from country to country; however, coping with a significant increase in disaster risk, exacerbated by climate change, represents a sizable task of cultural and institutional change.

These complexities and difficulties affecting climate adaptation apply equally to disasters and emergency management, which is unsurprising as the prospect of more frequent and/or intense climate-driven disasters are central to climate adaptation. The two main implications for and avenues for progressing policy responses are similar and will be discussed now.

First is the argument for 'no-regrets' or at least low-regrets options. These are options that, even in the absence of projected human-induced climate change, offer social, environmental, or economic benefits for other reasons. No-regrets options exist for mitigation, such as energy efficiency for business cost reasons or national energy security, as well as a way to reduce greenhouse gas emissions. In adaptation, there are more abundant possibilities, some examples of which will suffice here. With the prospect of increased heatwaves, dwelling design and insulation also addresses economic efficiency (reduced electricity costs, thermal comfort, reduce heat-related mortality). Better connection of remnant fauna habitats already justified now to preserve biodiversity is made more pressing with the likelihood of shifting climates and the tolerance of wildlife species. Water efficiency improvements to cope with worse drought in future should yield agricultural production efficiencies, lessen infrastructure costs with reduced need for additional supply, and provide more water to river and wetland ecosystems. Better communications, housing, and livelihood opportunities for poor, remote communities are entirely justified in political and moral terms, and would reduce the vulnerability of these communities to the impacts of increased climate variability. This list is lengthy, but more broadly, if we do not cope well enough with disasters now – and there is considerable evidence that that is the case – then improving disaster policy and response may be now more justifiable than ever as we combine the existing arguments with those concerned with climate adaptation. Can we achieve level I on the schema above, and perhaps progress well into level II, on the basis of no- or low-regrets options, at least in many contexts?

If climate change eventuates at the lower end of the I–III scale above, then improvements in capacities to cope with disasters will mean that societies will be better placed to handle 'more of the same'. However, even without climate change, an increase in disaster impacts is almost certainly going to be experienced due to trends other than climate change noted above: increased capacity will be required climate change notwithstanding. Should climate change produce impacts in level III above, then that is well beyond more-of-the-same, and will require a vastly accelerated evolution of DRR capacities.

On no- or low-regrets options, while climate change adaptation benefits invite consideration of measures that would also achieve better preparedness for disasters that will doubtless occur climate change notwithstanding, there is another basis for no-regrets arguments for improving disaster preparedness capacities: human development. This argument is less often made explicitly, although as noted above disasters and development have been combined on international policy agendas at times. Particularly in poorer regions, basic development gains – in education, communication, health services, livelihoods – will, all other things being equal, result in households, communities, and economies more capable of coping with disasters. While there will always be disasters that overwhelm any coping capacity, to be somewhat less vulnerable to smaller, 'routine' disasters is almost always a product of higher levels of local scale, appropriate human and economic development. The development benefits of reduced vulnerability may be a less powerful argument in richer countries where poverty and insecure livelihoods are not (or are not seen as) major issues. In many places, however, the combination of the climate adaptation and human development no-regrets arguments linked to disaster preparedness and lowered vulnerability presents a far more powerful case than disasters on their own.

The second important implication of connecting disaster policy to discussions of climate change is that disasters become even more embedded in and connected to other policy sectors and processes, extending the broader trend discussed earlier of disaster policy and emergency management shifting from an isolated to a more open and connected policy sector, profession, and area of operational practice. Also, the slow, punctuated onset of climate change forces thinking about disasters over a longer period of time, where risk or vulnerability reduction strategies need to be considered over many decades, in line with projections of climate change and variability. Linking disasters to human development similarly demands a more purposeful, longer term approach where disaster policy must intersect with other concerns more closely. This inevitably makes disaster policy more complicated, difficult, and contested as it comes into closer contact with social and economic concerns, but potentially more effective in reducing vulnerability. The policy agenda has been reframed.

The literature and practice of 'mainstreaming' disasters is thin, and that dealing with what is known as 'climate policy integration' (CPI) is scant, a product largely of a relatively recent appearance on policy agendas (Kok *et al.*, 2008; Ahmad, 2009; Gupta and van der Grijp, 2010; Fankhauser and Schmidt-Traub, 2011). It can be expected that CPI will be a major area of scholarship and policy discussion in coming years, while it is less clear that disaster mainstreaming will be. It is likely that disaster mainstreaming will be largely attended to as a subset of CPI on the adaptation front, and we would argue that disaster researchers and practitioners need to engage energetically with CPI discussions to ensure, firstly, that disaster mainstreaming is given due attention and, second, that CPI discussions can benefit from the longer history of scholarship and professional and policy practice available in the disasters field.

Given the recent nature of both CPI and disaster mainstreaming thinking, there are clear benefits in seeking policy sectors where longer and larger bodies of knowledge may exist. A leading candidate is environmental policy integration (EPI), which as part of the sustainable development agenda has been the subject of research and policy attention since the 1990s and therefore may inform CPI and disasters (Ahmad, 2009). While mostly derived from Western and especially northern European contexts, EPI has a considerable literature and body of evaluated policy experiments (e.g., Lenschow, 2002; Ross and Dovers, 2008; Jordan and Lenschow, 2010). Environmental policy shares with disasters the attribute of existing mainly to address the negative consequences of actions (or inactions) in *other* policy sectors (for example, land use planning, industrial safety and control). More broadly, the challenge of 'horizontal' policy integration (whole-of-government, joined-up-government) is a long standing issue in public policy and institutional analysis (Peters, 1998) and there is considerable scope for CPI and disaster mainstreaming to explore lessons and institutional ideas across other policy sectors.

27.5 The future: disaster policy in broader context

The preceding discussion has noted a broadening agenda and widening of the scope of research and policy focus in disaster policy and emergency management, and suggests that this evolution needs to continue for better disaster preparedness, but also for climate change adaptation and basic human development. However, the mainstreaming of disaster policy and integration of climate adaptation into other policy sectors is a new and difficult task. This final section mounts an argument for how these too-often disparate issues can be connected. Both disasters and adaptation are embedded in a complex web of social, environmental, and economic concerns, and there is only one policy agenda that seeks to integrate social, environment, and economic concerns: sustainable development.

First promoted by the World Commission on Environment and Development (WCED, 1987), sustainable development was articulated as a global and national policy agenda at the United Nations Conference on Environment and Development in Rio de Janeiro in 1992, the 'Rio Earth Summit' (see UN, 1992). That meeting produced the wide-ranging *Rio Declaration*, and international agreements on forests, biodiversity, and, importantly, the UN Framework Convention on Climate Change. The overarching goal of sustainable development brings onto one policy and societal agenda a wide range of 'environmental' issues (forests, oceans, climate, biodiversity, water and rivers, etc.) with issues of poverty, human development, technology, and security. These issues are accepted as interrelated, and requiring integration of long and short term perspectives, and of social, environmental, and economic imperatives in decision-making.

The progress with the policy agenda of sustainable development has been mixed, and was reviewed – and strongly reaffirmed at an aspirational level – at the 2012 'Rio +20' UN Conference on Sustainable Development (UN, 2012). This complex story is not within the scope of the present discussion, but one point is important. Impatience with progress with sustainable development, and critiques of the size and complexity of the policy agenda, overlook two important points. First, sustainable development is a higher order social goal, akin to equity, the rule of law, democracy, or justice. Such goals are not quickly advanced or arguably ever finally resolved. Second, it is inarguable that the constituent issues of sustainable development, social, environmental, and economic, including disasters, are closely intertwined in both cause and effect and thus require policy attention to be alert to and act in accordance with this interdependency. As difficult, vague, or slow as understanding of and progress with sustainable development may be, there is no evident alternative framing of this range of inter-related issues available. As Harrison (1992) argued, and not unbelievably so, sustainable development is the 'universally agreed goal of human progress'. Thus, we would argue, as we seek to connect disaster policy with related issues of climate change, population change, poverty, vulnerability, land management, urbanisation, and more, sustainable development is the appropriate broader agenda that must be comprehended and utilised.

Relevant to disasters and to the themes of this chapter, between 1992 and 2012, climate change, as only one of the agenda items of sustainable development, has come to dominate arguments over environment and development, and indeed disasters (Ahmad, 2009; Dovers and Hezri, 2010). Somewhat belatedly, the issue of sustainable development entered the main forum of climate change discussions, via the IPCC in its more recent assessment reports (Ahmad, 2009). Moreover, mitigation dominated early debates, with adaptation a more recent issue as it became clear that some climate change – and thus its impacts – was locked into the system. This, we believe, was understandable but also unfortunate, as both disasters and sustainable development (and other constituent issues) became separated in research, public discussion, and policy thinking. Opportunities for no- or low-regrets measures, synergies in policy, and lesson drawing across issues and sectors have been less realised than they otherwise might have been.

Sustainable development is a higher order, long term social goal and, despite frustration and difficulties with progress, will not go away as a framework agenda and inevitable imperative. Indeed reviews of progress have confirmed the inevitability of having to focus on an integrated agenda. The formal outcomes of the UN Conference on Sustainable Development (Rio +20) in June 2012 (UN, 2012) made clear the links between climate change and disasters, within the broader context of sustainable development and land management in various paragraphs (111, 135, 178, 210, 280). At paragraphs 186–9, disasters are discussed more specifically:

> **186**. We reaffirm our commitment to the Hyogo Framework for Action 2005–2015: Building the Resilience of Nations and Communities to Disasters and call for States, the United Nations system, the international financial institutions, subregional, regional and international organizations and civil society to accelerate implementation of the Framework and the achievement of its goals . . .
>
> **187**. We recognize the importance of early warning systems as part of effective disaster risk reduction at all levels in order to reduce economic and social damages, including the loss of human life, and in this regard encourage States to integrate such systems into their national disaster risk reduction strategies and plans . . . We commit to undertake and strengthen in a timely manner risk assessment and disaster risk reduction instruments.
>
> **188**. We stress the importance of stronger interlinkages among disaster risk reduction, recovery and long-term development planning, and call for more coordinated and comprehensive strategies that integrate disaster risk reduction and climate change adaptation considerations into public and private investment, decision-making and the planning of humanitarian and development actions, in order to reduce risk, increase resilience and provide a smoother transition between relief, recovery and development. In this regard, we recognize the need to integrate a gender perspective into the design and implementation of all phases of disaster risk management.
>
> **189**. We call for all relevant stakeholders, including Governments, international, regional and subregional organizations, the private sector and civil society, to take appropriate and effective measures, taking into account the three dimensions of sustainable development, including through strengthening coordination and cooperation to reduce exposure to risk for the protection of people, and infrastructure and other national assets, from the impact of disasters, in line with the Hyogo Framework for Action and any post-2015 framework for disaster risk reduction.

The Rio +20 outcomes are significant in two ways. First, they restate the importance of DRR as a priority of the vast majority of nations and of the international community. This may be viewed as a rhetorical commitment only, but it is a significant statement nonetheless – few international policy statements enjoy such widespread support among countries and fewer again place disasters as central to sustainable development and human futures. Second, and in many ways more significant, is that disasters are rarely mentioned without being closely linked to sustainable development and climate change. These high level policy commitments imply a strong definitional shift that has been evolving for some time, as traced through this chapter: disasters cannot be treated without close reference to other policy sectors.

What prospects, then, for such mainstreaming? The insertion of one imperative into another policy domain is not easy, and it is doubly hard to insert a concern such as disasters across multiple sectors, agencies, and issues. Specialisation in government and in the agencies of government is a very useful thing: specialists are good at their speciality. But they may not be good at comprehending the implications of their decisions for other sectors – the classic example is land use planning which ignores natural hazards and creates vulnerability. Part of the evolution of emergency management we have noted above has been to break out of the 'expert specialist' mindset and to embrace the perspectives of others, including local communities. Many questions still exist. For example, what is the future of disaster policy and emergency management, where are the other key sectors that create vulnerability, and who can reduce the vulnerability?

There are two risks with pursuing a 'mainstreaming' approach via the sustainable development agenda. The first danger is simply the effort and time involved, which may frustrate those in the disaster policy community and divert effort from core tasks. That we believe is a risk worth taking, as the benefits of a mainstreamed disaster policy agenda are large, as demonstrated by other policy concerns which have sought incorporation into other multiple sectors, such as gender equity, occupational health and safety, and environment. The alternative is for those involved in disaster policy and practice to accept a position marginal to policy debates, and to continue to serve as those cleaning up after ill-considered decisions in other domains, such as land use planning, infrastructure development, health services provision, and communications.

The second risk is, as noted above, that only climate-driven hazards will gain attention, at the expense of other hazards. Disasters are not only climate driven, and research, policy, and operational practice need to comprehend, prepare for, and respond to industrial, social, technological, and other hazards as well. That is, however, one benefit of joining disasters more firmly to the agenda of sustainable development, alongside issues of technology, pollution, human health, and development and overall governance debates around human–natural system interactions. On one agenda rather than as separate and potentially competing imperatives, breadth and connectivity between issues is an essential consideration. Hence, we argue that the ability of human societies to better understand and respond to disasters requires that disaster policy and emergency management constantly and persistently pursue linkages with climate change adaptation, within the broader agenda of sustainable development. Disasters are important and always will be, but are too quickly forgotten between events and marginalised as political and policy concerns. Climate change is important, topical, and thoroughly linked to disasters. Sustainable development is the only political and policy agenda that forces such connected issues together, however difficult that may be. The combination of the three is potentially powerful, and may serve to provide a much stronger impetus for change – in cultural attitudes and thus policies and institutions terms – than has been the case in the past, or would be the case if the three are considered in isolation.

References

Ahmad, I. H. (2009). *Climate Policy Integration: Towards Operationalization. DESA Working Paper 73*. New York: UN Department of Economic and Social Affairs.

Australia (2011) *National Strategy for Disaster Resilience*. Canberra: Australian Government.

Boyden, S. V. (1987). *Western Civilization in Biological Perspective: Patterns in Biohistory*. Oxford: Clarendon Press.

CIEL (Centre for International Environmental Law) (2007). *Climate Change Program*. http://www.ciel.org/Climate/programclimate.html.

Cockfield, C. and Dovers, S. (2013). The science and policy of climate variability and climate change: intersections and possibilities. In: *Drought, Risk Management and Policy: Decision Making under Uncertainty*. eds., Botterill, L. and Cockfield, C. Boca Raton: CRC Press, pp. 29–44.

Connor, R. and Dovers, S. (2004). *Institutional Change for Sustainable Development*. Cheltenham: Edward Elgar.

Cuny, F. C. (1983). *Disasters and Development*. New York: Oxford University Press.

Dovers, S. (1997). Sustainability: demands on policy. *Journal of Public Policy*, **16**, 303–318.

Dovers, S. (2009). Normalizing adaptation. *Global Environmental Change*, **19**, 4–6.

Dovers, S. R. and Hezri, A. A. (2010). Institutions and policy processes: the means to the ends of adaptation. *Wiley Interdisciplinary Reviews: Climate Change*, **1**, 212–231.

Fankhauser, S. and Schmidt-Traub, G. (2011). From adaptation to climate resistant development: the costs of climate-proofing the Millennium Development Goals. *Climate and Development*, **3**, 94–113.

Gupta, J. and van der Grijp, N. (eds.) (2010). *Mainstreaming Climate Change in Development Cooperation*. Cambridge: Cambridge University Press.

Handmer, J. and Dovers, S. (2013). *Handbook of Disaster Policies and Institutions: Improving Emergency Management and Climate Adaptation*, second edn. London: Earthscan.

Harrison, P. (1992). *The Third Revolution: Population, Environment, and a Sustainable World*. Harmondsworth: Penguin.

IPCC (Intergovernmental Panel on Climate Change) (2011). *Special Report on Managing the Risks of Extreme Events and Disasters to Advance Climate Change Adaptation: Summary for Policy Makers*. Geneva: IPCC.

IPCC (Intergovernmental Panel on Climate Change) (2012). *Special Report on Managing the Risks of Extreme Events and Disasters to Advance Climate Change Adaptation*. Geneva: IPCC.

Jordan, A. and Lenschow, A. (2010). Environmental policy integration: a state of the art review. *Environmental Policy and Governance*, **20**, 147–158.

Kok, M., Metz, B., Verhagen, J. and van Rooijen, S. (2008). Integrating climate and development policies: national and international benefits. *Climate Policy*, **8**, 103–118.

Lenschow, A. (ed.) (2002). *Environmental Policy Integration: Greening Sectoral Policies in Europe*. London: Earthscan.

Mileti, D. (1999). *Disasters by Design: A Reassessment of Natural Hazards in the United States*. Washington, DC: Joseph Henry Press.

Mitchell, J. K. (1999). *Crucibles of Disaster: Megacities*. Tokyo: United Nations University.

Pelling, M. (2003). *The Vulnerability of Cities: Natural Disasters and Social Resilience*. London: Earthscan.

Perrow, C. (1984). *Normal Accidents: Living with High-Risk Technologies*. New York: Basic Books.

Peters, G. (1998). Managing horizontal government. *Public Administration*, **76**, 295–311.

Productivity Commission (2012). *Barriers to Effective Climate Change Adaptation (Draft Report) April 2012*. Canberra: Productivity Commission.

Quarantelli, E. L. (ed.) (1998). *What is a Disaster? Perspectives on the Question*. London: Routledge.

Ross, A. and Dovers, S. (2008). Making the harder yards: environmental policy integration in Australia. *Australian Journal of Public Administration*, **67**, 245–60.

Standards Australia (2004). *Australian Standard/New Zealand Standard 4360–2004: Risk Management*. Sydney: Standards

Stern N. (2006). *The Economics of Climate Change: the Stern Review*. Cambridge: Cambridge University Press.

Turner, B. (1978). *Man-Made Disasters*. London: Wykeham.

UN (United Nations) (1992). *Agenda 21: The Programme of Action from Rio*. New York: UN.

UN (United Nations) (2012). *The Future We Want: Rio +20 United Nations Conference on Sustainable Development, Outcomes of the Conference*. A/CONF.216/L.1. New York: UN.

WCED (World Commission on Environment and Development) (1987). *Our Common Future*. Oxford: Oxford University Press.

White, G. (1945). *Human Adjustment to Floods*. Chicago, IL: University of Chicago Press.

28

Vulnerability, impacts, and adaptation to sea level related hazards taking an ecosystem-based approach

Keith Alverson

28.1 Introduction

The least developed countries are the most vulnerable to climate change. This mantra is heard again and again – most notably perhaps at meetings of the United Nations Framework Convention on Climate Change (UNFCCC). Indeed partly in response to such thinking, much of the international funding available for climate change adaptation is only accessible to the least developed countries. One seemingly logical extrapolation from this stated correlation is that a perfectly reasonable national adaptation plan might be no more than a development plan. In other words, if a country can find a way to get off the 'least developed' list, for example by focusing its effort on gross domestic profit (GDP) growth, this would have the additional benefit, even without thinking at all about climate change per se, of the country simultaneously coming off the 'most vulnerable' list. The perceived correlation between vulnerability and development has also, ironically, lead international adaptation financing institutions to adopt the troubling concept of 'additionality', whereby in order to obtain funds least developed countries must show how their planned actions are in some way specifically addressing climate impacts, above and beyond business as usual development. In this chapter, I focus on sea level hazards to argue for an ecosystem-based approach to increased resilience in the face of climate change (Alverson, 2012). This is an approach that may at times complement, but may also be orthogonal to, traditional GDP focused development, and one that can benefit all countries, and local communities therein, irrespective of where they may lie on the development scale.

In 2011 I travelled to Bangkok for the second Asia Pacific Climate Change Adaptation Forum, which was then called off at the last minute due to the worst flooding in the past 50 years. Bangkok, an urban centre of great wealth in a relatively rich country, turns out to be quite vulnerable to climate impacts, sustaining 815 deaths, massive displacements of population, and US$45 billion in economic damage, including lasting damage to its automobile and electronics supply chains. In 2012, Hurricane Sandy slammed into New York City. As of this writing, the US House of Representatives had authorised a Federal aid package for US$50.7 billion for recovery efforts, and the Governor of New York had proposed to employ as much as US$400 million of these funds (about three quarters of 1%) to reduce vulnerability to future such events by purchasing homes wrecked in the storm, having them demolished, and preserving the flood-prone land permanently as undeveloped coastline.

As these examples clearly show, while development can ameliorate some aspects of vulnerability to sea level related hazards, we certainly should not be fooled into thinking that development pathways followed in the past, sustainable or not, comprise the entire solution, or even the most efficient solution, to this climate change adaptation challenge. At the very least, developing countries can find, and are finding, new development pathways that avoid some of the maladaptation that has already occurred in the developed world. Additionally, developing and developed countries alike can benefit from improving coastal zone planning and coastal ecosystem management as an integral part of policies to help reduce vulnerability and increase resilience to sea level hazards.

Human activities have pushed greenhouse gas levels in the atmosphere far outside the envelope of natural variability over the past million years. As one unambiguous consequence, global average temperatures have rapidly risen beyond the natural variability of about the past 4000 years (Marcott *et al.*, 2013). According to one popular argument (Rockström *et al.*, 2009) the past 10 000 years of Holocene

Extreme Natural Hazards, Disaster Risks and Societal Implications, eds. Alik Ismail-Zadeh, Jaime Urrutia Fucugauchi, Andrzej Kijko, Kuniyoshi Takeuchi and Ilya Zaliapin. Published by Cambridge University Press.

climatic stability allowed ecosystems and civilisations to develop and thrive. As we move beyond natural 'planetary boundaries' we risk pushing the planet beyond the safe operating space for humanity. Though not without substantial conceptual flaws, this global Earth system boundary approach has shed valuable light on the most dramatic human global scale impacts on the planet, highlighting where global mitigation efforts make sense. Unfortunately, for the climate change problem, such mitigation efforts have not succeeded. In 2011 atmospheric greenhouse gas levels rose to a record high, and did so at a record rate of increase, despite emissions reductions in Europe and the USA. Adaptation to climate change impacts in general, and sea level related hazards in particular, will require a very different paradigm. Instead of global efforts, we need to focus on the local scale to identify impacts and vulnerability, and to develop adaptation strategies.

Patterns of climate impacts and societal vulnerability are highly localised. At the local level the Holocene was rich with dramatic climatic variability and change. Reconstructions of local and regional natural climate variability over the past few millennia show that the kind of climate impacts that today pose the largest adaptation challenges – such as droughts, floods, sea level rise – have been far more dramatic in the past than during the instrumental period (Alverson *et al.*, 2003). Although there is no doubt that we live in both unusual and challenging times, at the local level, anthropogenic changes are not happening against a backdrop of 10 000 years of Holocene climatic stability. The dramatic and dynamic past history of regional climate variability has enabled ecosystems and societies to develop an inherent resilience that we can now benefit from in the face of dire global anthropogenic changes. Sea level rise provides one interesting case. We have clear observational evidence that global average sea level has been rising at a rate of just over 3 mm y^{-1} over the past couple of decades (Fig. 28.1). Less well publicised, though equally well established from observations, is that sea level rise is not globally uniform. In the western equatorial Pacific the average rise over this same period is larger than 10 mm y^{-1}, while the north-eastern Pacific sea level has actually been going down over this same time period (Fig. 28.2). Furthermore, local land subsidence is often several orders of magnitude larger than sea level rise. Land subsidence in Bangkok associated with groundwater extraction has been over a metre, dwarfing the few centimetres of climate change related signal.

If one looks in more detail at the full range of local changes impacting vulnerability to sea level during the Anthropocene period, the climate signal is often relatively small. It is instructive, for example, to consider what has

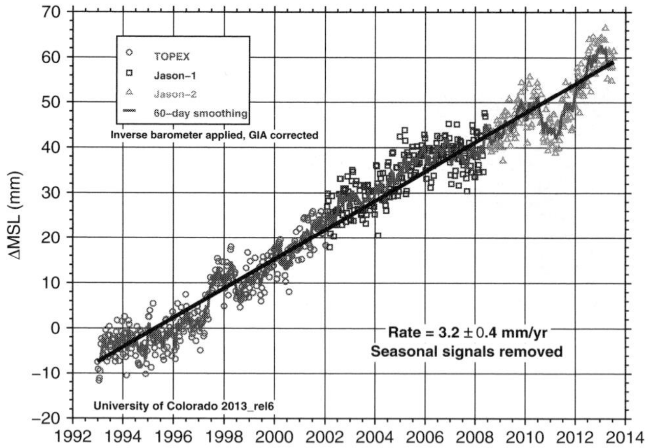

Figure 28.1. Global average sea level rise over the past 20 years as measured by satellite altimetry (Nerem *et al.*, 2010). Source: http://sealevel.colorado.edu.

changed in Bangkok over the 50 years since such massive flooding apparently last occurred. Climatic change has led to a mean sea level rise of a few centimetres. Over the same period, groundwater extraction has caused land subsidence of more than a metre and the sediment input from the Mekong that naturally builds up the delta has been dramatically impacted due to dams, canals, and other changes to the upstream river flow. The population has quadrupled, from just over 2 million in 1960, to over 8 million today. Over the same period there has been an incredible growth of wealth and industry, transforming the city into a substantial regional economic centre including for the banking, automotive, and semiconductor industries, linking Bangkok with the global economy. Clearly, these many non-climatic developments dominate the likely impacts of, and vulnerability and adaptive capacity to, sea level rise for the population in and around Bangkok.

If one were to listen uncritically to the political rhetoric in the UNFCCC negotiations one might be tempted to believe that vulnerability to sea level rise correlates with national boundaries, and is most dire in least developed countries and small island developing states. In this same context, sea level rise is often vastly oversimplified as a uniform, exogenous encroachment on coastal areas, accompanied by woefully simplistic blue colouring on maps of areas that would be inundated given some amount of sea level rise. However, as discussed in the example for Bangkok, in scientific terms, the actual patterns of impact and vulnerability are far more difficult to map out. They certainly do not correlate with national boundaries, national development indicators, or height above sea level contours. Vulnerability is usually not the

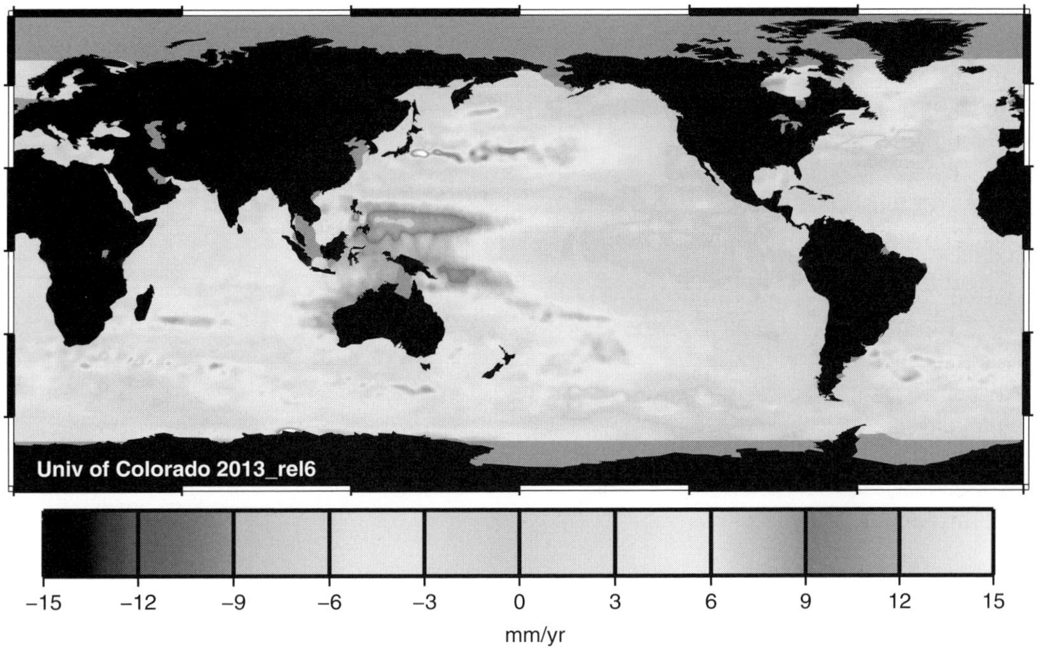

Figure 28.2. Global pattern of sea level rise trends over the past 20 years as measured by satellite altimetry (Nerem *et al.*, 2010). Source: http://sealevel.colorado.edu.

result of the slow encroachment of the sea directly, but is instead due to destruction and loss of life associated with extreme events and ensuing proximate effects such as disease, lack of access to drinking water, and economic losses. There is no reason to expect patterns of vulnerability, impacts, and adaptive capacity to map geographically at all, with large highly vulnerable and highly protected populations living side by side, defined not by how high above the flood line their dwelling is, but by their wealth, age, gender, or any number of other socio-economic variables.

As is so often the case, the developed world has numerous adaptation measures already in place, including, for example, the Thames barrier in London and the MOSE gates in Venice. Subsequent to Hurricane Sandy, New York City is now more seriously considering constructing similar movable flood barriers that can be deployed in advance of storm surges, despite the anticipated cost of 10–20 billion US dollars. However, it is not at all clear that such high technology expensive projects can be scaled up to less high profile coastal areas, or that they would be useful at all in the developing world. For the vast majority of the world's coasts, it is simply not economically or technically feasible to engineer our way out of vulnerability to sea level rise. For this reason, concrete adaptation actions can rarely literally be about pouring 'concrete'. Instead, a wide range of ecosystem-based approaches, wherein the natural resilience of a wide range of coastal ecosystems is harnessed to reduce human vulnerability, can be put in place.

28.2 Ecosystem based protection from sea level related hazards

Low lying coral atolls are not, of themselves, vulnerable to sea level rise. They are a perfect example of natural ecosystem resilience that has developed in response to a local climatic variability over many millennia. It was Charles Darwin, in one of the many, though less well-publicized insights derived whilst aboard the *HMS Beagle*, who first determined that coral atolls are not static, but grow upward to account for either land subsidence or rising seas. In fact, the many small island developing states now threatened by anthropogenic rises of a few millimetres per year withstood a rise of over 120 metres, at a rate on the order of 10 cm y^{-1}, at the end of the last glacial period, as coral growth kept pace with sea level rise (Woodroffe and Biribo, 2011). Of course, conditions today are very different. Relatively dense human populations now live on the islands, and the coral ecosystems that have protected them in the past are under serious threat from myriad local pressures and ocean acidification. Many of the events commonly attributed to sea level rise impinging on the islands can, in fact, be caused by local human activities, such as land reclamation and mining for building materials, which causes degradation of beaches and reefs, and the construction of causeways between islets (Donner, 2012). Thus sea level hazard adaptation measures for these islands today must be primarily based on understanding the vulnerability of local populations, including curtailing maladaptive practices to ensure the health and resilience of coral reef ecosystems.

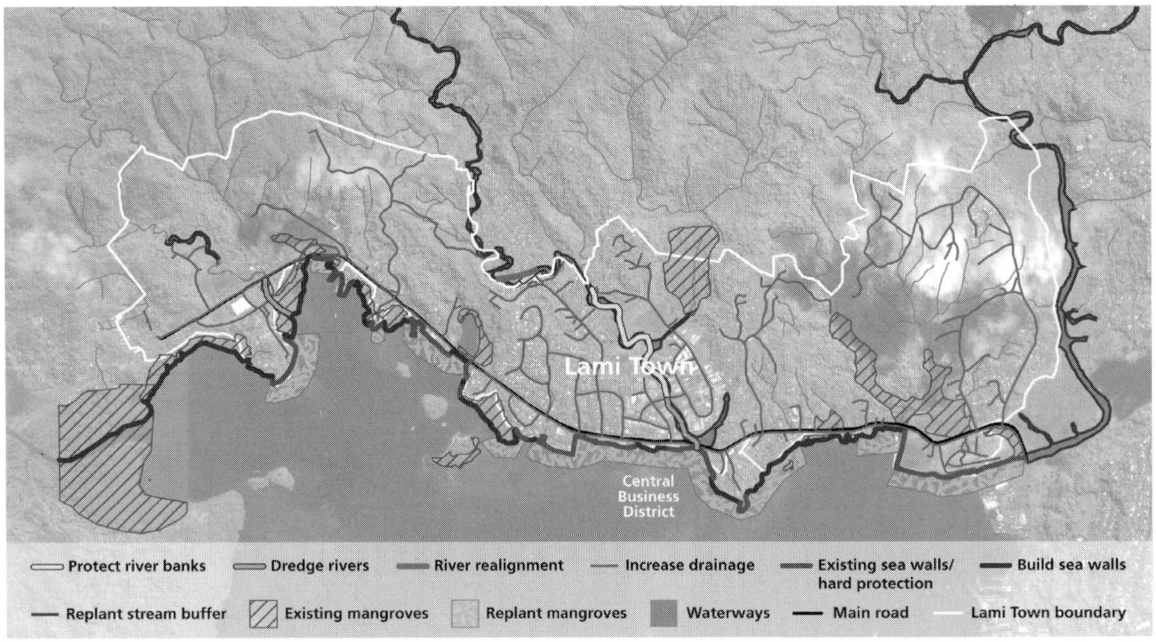

Figure 28.3. Proposed integrated ecosystem-based and engineering approach to reducing vulnerability to sea level hazards in the face of climate change induced sea level rise in Lami Town, Fiji. Source: UNEP and SPREP (Rao *et al.*, 2012). For colour version, see Plates section.

Rao *et al.* (2012) provided an integrative study of coastal vulnerability for Lami Town, Fiji, with comparative cost–benefit analysis of various ecosystem-based options, policy and social options, and engineering options for adaptation. In the case of Fiji, climate change is projected to increase flooding and erosion threat due to: (1) sea level, which has been rising at about double the global average, at 6 mm y^{-1} since 1993; (2) increased intensity and frequency of extreme rainfall events; and (3) fewer but more severe landfalls of tropical cyclones. Three types of protective natural resources were identified including 8.8×10^5 m^2 of mangrove forest, 1.387×10^7 m^2 of coral reef, and 3.3×10^6 m^2 of mudflats. At the same time a wide range of social and policy options, such as regulating land tenure, relocation, and flood warning systems, as well as engineering options, such as river bank reinforcement and seawalls, were considered. In this study, in part due to its very large spatial area, reduction of coral reef extraction provided by far the most cost effective protection to sea level hazards in terms of the cost of damage avoided per dollar spent. Indeed, arguably halting destruction of coral is no cost at all, even providing a benefit to the local economy, and thus a clear 'no regrets' option. On the converse side though, coral reefs were expected to avoid only a small fraction of the expected damage, indicating the need for an integrated, multi-option approach, including mangrove protection and restoration (Fig. 28.3).

One of the most well-known ecosystem-based adaptation options in the developing world over the past few years has been the protection, restoration, and/or sustainable management of mangrove forests. These forests protect the shoreline and communities from storm surges and sea level rise. Intact coastal habitats, including but not limited to mangroves, have been clearly shown to reduce damage from coastal inundations associated with sea level rise and storms (Arkema *et al.*, 2013). The efficacy of ecosystems, including mangroves, in protecting from tsunamis has been a matter of some debate with some authors arguing that bottom topography is by far the most dominant consideration in determining the location of tsunami run-up, and the fact that mangroves grow preferentially in sheltered areas hinders the study of their effectiveness in protecting from tsunamis (Chatenoux and Peduzzi, 2007). In addition to sea level protection, mangroves are an excellent example of a no regrets solution providing multiple benefits including securing human livelihoods by providing ecosystem goods such as food, timber, and medicine (Alongi, 2002; Gilmann *et al.*, 2008). Furthermore, mangroves increase fishery and aquaculture yields since many commercial fish species depend on mangrove habitat during their juvenile stages, and shrimp–mangrove integrated farming systems, for example in the Mekong Delta of Vietnam, have increased the productivity of aquaculture facilities.

Cost–benefit analyses of mangrove restoration projects have shown that rehabilitation can provide net economic benefit even when only considering the direct use of products such as timber, fish, and honey by local communities, with considerable additional value from shoreline protection (Badola and Hussain, 2005). Moreover, mangrove restoration can sometimes be more cost effective than maintaining hard structures, particularly over long relatively undeveloped coastline. In Vietnam, for example, planting 12 000 hectares of mangroves cost US$1.1 million but saved an estimated US$7.3 million per year in dyke maintenance (International Federation of the Red Cross, 2012). In Lami Town, Fiji, the estimated cost over 20 years of replanting mangroves was less than 5 Fijian dollars per square metre while building seawalls costs more than 2000 Fijian dollars per metre (Rao *et al.*, 2012). Although on a unit area or length basis, ecosystem-based options are generally orders of magnitude cheaper than engineering options, protection effectiveness also plays an important role, indicating the value of an integrated response with hard infrastructure being most useful in protecting high value, high traffic areas such as ports, and ecosystem-based approaches over extensive exposed coastlines. It is important to note here that the additional value of mangroves beyond protection from sea level hazards is not only relevant in developing countries. The annual economic value to the fisheries industry derived from mangroves in the Gulf of California, for example, has been estimated to be US$37 500 per hectare ($10^4$ m^2) of mangrove fringe (Arburto-Oropeza *et al.*, 2008).

Finally, and most importantly, mangrove forests are highly resilient, with the ability to recover after extreme events with a high degree of ecological stability. Like corals, mangrove forests can actively raise the forest floor in response to sea level rise, as their presence enables accumulation of sediment above the tidal range (Kumara *et al.*, 2010). Although their ability to do this effectively depends on the rate of sea level rise, studies suggest that mangroves today are keeping pace with sea level rise (Alongi, 2002) and that restoration and management can, in principle, enable mangrove ecosystems to continue to protect against sea level related hazards even with the additional sea level rise associated with climate change.

28.3 Coastal zone management

As coastal cities have grown in size and number, a wide range of engineering approaches to decreasing their vulnerability to sea level rise have been developed – from dykes and channels in Bangkok to reservoirs and levees in New Orleans, to seawalls in Japan, and movable barriers in London and Venice. All of these structures clearly reduce vulnerability to sea level related hazards. However, the unintended consequence of such high profile, and expensive, infrastructure projects can be a false sense of security, degradation of natural coastal zone protections, and ultimately a lack of resilience. New Orleans provides a good example. As long as the system of levees and reservoirs protected the city, people felt secure even as natural processes of sediment delivery failed to build up the delta and barrier islands as they had in the past. When the barriers were eventually breached by an extreme event – Hurricane Katrina – the city and its environment had lost the resilience that integrated coastal zone and river management might have provided had they been an integral component in a more holistic approach. After the 2011 Tohoku earthquake in Japan, people were so confident in the tsunami protection seawalls that had been put in place that on receiving tsunami warnings they retreated to the top of the seawall for protection, and to watch the event. The seawall's eventual collapse led to many unnecessary fatalities. Almost invariably, urban areas can benefit from natural resilience building provided by ecosystems in peri-urban areas. Such ecosystems approaches can include wetland restoration in riparian areas upstream, or dune, sediment, and barrier island restoration and management. Nicholls (2011) provides a nice review of land use oriented 'soft' defences.

Given the recognition that ecosystem degradation is a major driver of disaster risk, and that climate change is magnifying these same risks, there have been a number of studies looking at how to harness ecosystem services for disaster risk reduction. The United Nations Environmental Program's (UNEP) study 'Risk and Vulnerability Assessment Methodology Development Project' (RiVAMP) carried out pilot assessment in Negril, Jamaica (UNEP, 2010). In this study, one of the primary risk factors was shown to be beach erosion, which leads to enhanced vulnerability of coastal communities and infrastructure to storm surges associated with tropical storms that frequently make landfall on the island. The findings showed that healthy coral reef ecosystems effectively dissipate near shore wave energy and thus mitigate against beach erosion, while illegal sand mining, unsustainable resource use, removal of mangroves, sea grasses, and other coastal vegetation have all contributed to beach degradation and thus increased vulnerability to storm surge hazards. As shown in this study, coastal ecosystems serve as natural protective barriers, including through flood abatement, slope stabilisation, and coastal protection against storm surges. At the same time, ecosystems help build the resilience of local communities in the aftermath of hazard impacts, supporting livelihoods, for example by providing natural habitats for replenishing commercial fish stocks and sources of building materials and firewood.

28.4 False paradigms in adaptation

Having made the case, I hope, for incorporating ecosystem-based resilience as an integral component of plans to reduce human vulnerability to sea level rise related hazards, let me briefly turn to highlight what I think are a few of the more common fallacies that underpin many climate change adaptation and disaster risk reduction projects.

First, let me tackle the concept of 'additionality'. In order for countries to obtain funding for climate adaptation and disaster risk reduction work from major multilateral donors, including the Global Environment and Adaptation Funds, they must demonstrate that the planned interventions are distinct from business as usual development work. This concept has political support from both developed/donor and developing/recipient countries. Recipients hope that insisting on such a distinction will help ensure they receive 'new' funds, additional to traditional development assistance, rather than 'relabelled' funds that were in fact already being provided. Donors too are keen to demonstrate that they are indeed providing climate change adaptation support, or at the very least adding some kind of 'climate proofing' to their existing development investments. However, scientifically, this concept of 'additionality' is not justifiable. Primarily this is because the specific climate drivers that are being adapted to are often not large compared to social and economic drivers of vulnerability. Put another way, coastal communities are vulnerable to extreme events irrespective of background sea level rise, and the magnitude and extent of this vulnerability is often dominated by local, non-climatic drivers. Thus, very often the easiest and most efficient way to decrease vulnerability and/or increase resilience in the future is to look at the range of past variations and extreme events, and develop plans to better deal with them in the future. Bangkok and New York, for example, do not require future projections of regional hydrology or sea level rise in order to better prepare for potential future flooding of the extent and magnitude they experienced 2011 and 2012 respectively.

This leads to a second perplexingly ingrained paradigm in much climate change adaptation work – the perceived need to base planning on 'downscaled' and/or 'improved' model forecasts. The hopes that accurate regional climate forecasts can underpin specific adaptation measures are, for the most part, misplaced. In terms of global models, the range of projections for regional hydrological trends, including also projected numbers and magnitudes of cyclones, shows no sign of shrinking, and indeed there are good reasons to think the range may increase as more degrees of freedom, associated with adding additional forcing factors, are included in them (Maslin and Austin, 2012). Downscaling such models indeed provides some numbers with variability on horizontal scales that are more compatible with adaptation and disaster risk reduction work. However, there is little evidence that these high-resolution fields actually provide more accurate climate projections. The take home message is that instead of reducing disaster risk by implementing tailored additional actions targeting specific climate projections, it is usually easier and more effective to build resilience to a plausible range of variability, including extreme events.

28.5 Conclusions

Adaptation to climate change is one of the defining global environmental, social, and economic challenges of the twenty-first century, recognised by heads of state at the Rio +20 conference on sustainable development in 2012 as 'an immediate and urgent global priority'. One of the most robust aspects of this climate change is rising sea levels, and the associated increase in vulnerability to sea level related hazards. At the root of the dealing with this challenge is a troubling disparity between the global scale of the anthropogenic causes of global sea level rise and the local scale of extreme events and adaptation measures. Climate change is only one of many factors impacting on coastal vulnerability, thus disaster risk reduction measures will always require broader interdisciplinary perspectives. Some of the most vexing sea level hazard risk reduction challenges are not technical or scientific at their core, but will be dominated by social, economic, and legal considerations. However, one thing is quite clear. Ecosystems have evolved a great resilience to past sea level rise and extreme events. In our efforts to reduce our own vulnerability to these same forces, it behooves us to harness this powerful ecosystem-based resilience as an integral element of our climate change adaptation and disaster risk reduction efforts.

References

Alongi, D. M. (2002). Present state and future of the world's mangrove forests. *Environmental Conservation*, **29**(3), 331–349.

Alverson, K. (2012). Vulnerability, impacts and adaptation to sea level rise: taking an ecosystem-based approach. *Oceanography*, **25**(3), 156–160.

Alverson, K., Bradley, R. and Pedersen, T., eds. (2003). *Paleoclimate, Global Change and the Future*. International Geosphere Biosphere Programme Book Series. Berlin: Springer.

Arburto-Oropeza, O., Ezcurra, E., Danemann, G. *et al.* (2008). Mangroves in the Gulf of California increase fishery yields. *Proceedings of the National Academy of Sciences*, **105**, 10456–10459.

Arkema, K. K., Guannel, G., Verutes, G. *et al.* (2013). Coastal habitats shield people and property from sea-level rise and storms. *Nature Climate Change*, doi:10.1038/nclimate1944.

Badola, R., and Hussain, S. A. (2005). Valuing ecosystem functions: an empirical study on the storm protection function of Bhitarkanika mangrove ecosystem, India. *Environmental Conservation*, **32**, 85–92.

Chatenoux, B., and Peduzzi, P. (2007). Impacts from the 2004 Indian Ocean Tsunami: analyzing the potential protecting role of environmental features. *Natural Hazards*, **40**, 289–304.

Donner, S. (2012). Sea level rise and the ongoing Battle of Tarawa. *EOS Transactions American Geophysical Union*, **93**(17), 169.

Gilmann, E. L., Ellison, J., Duke, N. C. and Field, C. (2008). Threats to mangroves from climate change and adaptation options: a review. *Aquatic Botany*, **89**(2), 237–250.

International Federation of the Red Cross (2012). *Mangrove Plantation in Vietnam: Measuring Impact and Cost Benefit*. Technical Report. Available at: http://www.ifrc.org/Global/Publications/disasters/reducing_risks/Case-study-Vietnam.pdf. Accessed on 8 August 2013.

Kumara, M. P., Jayatissa, L. P., Krauss, K. W., Phillips, P. H. and Huxham, M. (2010). High mangrove density enhances surface accretion, surface elevation change, and tree survival in coastal areas susceptible to sea-level rise. *Oecologia*, **164**, 545–553.

Marcott, S. A., Shakun, J. D., Clark, P. U. and Mix, A. C. (2013). Reconstruction of regional and global temperature for the past 11,300 years. *Science*, **339**, 1198–1201.

Maslin, M. and Austin, P. (2012). Climate models at their limit? *Nature*, **486**, 183–184.

Nerem, R. S., Chambers, D., Choe, C. and Mitchum, G. T. (2010). Estimating mean sea level change from the TOPEX and Jason altimeter missions. *Marine Geodesy*, **33**(1): 435–446

Nicholls, R. (2011). Planning for the impacts of sea level rise. *Oceanography*, **24**(2), 144–157.

Rao, N. S., Carruthers, T. J. B., Anderson, P. *et al.* (2012). An economic analysis of ecosystem-based adaptation and engineering options for climate change adaptation in Lami Town Fiji. *Technical Report by the Secretariat of the Pacific Regional Environment Program*. UNEP Regional Seas Reports and Studies No. 190, 28 pp.

Rockström, J., Steffen, W., Noone, K. *et al.* (2009). A safe operating space for humanity. *Nature*, **461**, 472–475.

UNEP (2010). *Risk and Vulnerability Assessment Methodology Development Project (RiVAMP). Linking Ecosystems to Risk and Vulnerability Reduction. The Case of Jamaica*. Technical Report. Available at: http://postconflict.unep.ch/publications/RiVAMP.pdf. Accessed on 8 August 2013.

Woodroffe, C. D., and Biribo, N. (2011). Atolls. In *Encyclopedia of Modern Coral Reefs: Structure, Form and Process*, ed. Hopley, D. Berlin: Springer, pp. 51–71.

29

Extreme geohazards: risk management from a (re)insurance perspective

Anselm Smolka

29.1 Recent natural disasters: need for action

The loss data on great natural disasters since 1950, the period for which complete global coverage for such disasters can be assumed, show a dramatic increase in catastrophe losses over the last few decades. This is discussed in Chapter 30. In particular, the hurricane seasons of 2004 and 2005 with Hurricane Katrina illustrated the increasing loss potential of natural disasters. Katrina also demonstrated the worldwide implications of natural disasters in the form of the severe shortage in oil-producing and refining capacities and the ensuing sharp increase in the price of oil on the global market. The South Asian earthquake and tsunami in 2004 affected less developed regions, but claimed an excessive number of victims on tsunami-stricken coastlines around the Indian Ocean. Such lessons have been repeated by the series of earthquakes in 2010/11. It started with the Haiti earthquake in January 2010, and was followed by the Maule earthquake in Chile in February 2010, the Canterbury earthquake series in New Zealand in 2010/11, and the Tohoku earthquake in Japan in March 2011. Then, the Thai floods from October to December 2011 served as a strong wake-up call in respect of the vulnerability of modern industrial parks, where tens of billions of US dollars are concentrated in very small areas. More recently, the world's financial centre, New York, was hit by a severe storm surge accompanying an extremely large storm system that developed from former Hurricane Sandy and affected 15 northeastern states. Losses in the two-digit billion US dollar range have become a reality, even outside of the leading industrialised countries, and nowadays a much higher fraction of these losses is insured than in the past. Before the 2010 Maule earthquake, Santiago was last affected by the 1985 Valparaiso $M8$ earthquake. Whereas the total economic loss in 1985 was about 25 times as high in 2010, the insured loss increased by a factor of 100 in 2010. Comparing the 1995 Kobe and the 2011 Tohoku earthquakes, the statistics of the earthquake losses shows a factor 3 increase for the economic loss, and a factor 13 increase for the insured loss.

These disasters required, and have indeed provoked, a re-think on several aspects of natural disaster risk management, which has not yet produced final conclusions. The earthquakes of 2010 and 2011 not only shook the ground, they also shattered what may be called a false sense of security or complacency about how to assess and manage the risk from natural disasters.

This chapter starts with a portrayal of the salient features of disasters since 2010 in Section 29.2. Features of a holistic approach to risk management of natural disasters are described in Section 29.3. This approach comprises the steps of risk identification, risk evaluation, risk control, and risk financing. Specific emphasis is placed on the role of insurance and reinsurance. Section 29.4 focuses on how this framework, or parts of it, performed in the light of these disasters, and on the particular challenges that need to be addressed to improve risk reduction. Section 29.5 discusses extreme events and Section 29.6 presents conclusions.

29.2 Natural disasters 2010–2011

29.2.1 The Haiti earthquake

On 12 January 2010, a magnitude 7.1 earthquake occurred along the southern section of the plate boundary between the Caribbean and the North American plate in the vicinity of the capital of Haiti, Port-au-Prince, causing very heavy destruction in the capital region and its surroundings. Figures for the number of victims range widely from about 52 000 to 316 000. Irrespective of the actual figure, there is no doubt that this event was a terrible disaster that struck a

Extreme Natural Hazards, Disaster Risks and Societal Implications, eds. Alik Ismail-Zadeh, Jaime Urrutia Fucugauchi, Andrzej Kijko, Kuniyoshi Takeuchi and Ilya Zaliapin. Published by Cambridge University Press.

state whose administration had already been struggling for many years. The event had not been unforeseen: in a scientific publication from 2008, the high probability for such an earthquake had been forecast based on earthquake history and measurements of ongoing plate motion (Manaker *et al.*, 2008), but in view of the lamentable overall sociopolitical situation of the country, its public administration had other priorities. And the operational use of forecasts (International Commission on Earthquake Forecasting for Civil Protection, 2011) is still more a concept than a reality, even in well-developed countries, as proven by the case of the Aquila earthquake in Italy in 2009.

Of an estimated total direct loss of US$8 billion due to the Haiti earthquake, only a very minor fraction, in the range of US$200 million, was insured. The country was flooded with donor capital, but in the end any efforts to organise a concerted action for immediate disaster relief, for a well-organised reconstruction process, and to restore a functioning public administration failed. Notwithstanding partial successes and the admirable engagement of many individuals and aid organisations, disaster recovery after the Haiti earthquake must be seen as a missed opportunity. This is all the more embarrassing as the chance of further earthquakes happening even closer to the city is high. Based on historical experience, the 12 January earthquake marks the starting point in a new earthquake cycle (Manaker *et al.*, 2008).

29.2.2 The Maule earthquake

On 27 February 2010, a magnitude 8.8 earthquake located offshore along the plate boundary between the Nazca and the South American plate affected the central region of Chile from Concepción to Viña del Mar, including the country's capital Santiago. Due to advanced earthquake building regulations, the number of building collapses and therefore of victims was low. Direct economic losses reached US$30 billion, of which US$8 billion was insured. This marks a dramatic difference to the situation with the 1985 Valparaiso *M*8 earthquake, where the insured loss amounted to just US$80 million. The Maule earthquake was a success story in respect of saving lives and disaster recovery (Siembieda *et al.*, 2012). But the amount of non-structural losses, i.e. losses to non-load-bearing walls and other building elements like, for example, windows, facades, and suspended ceilings as well as to contents, machinery, and equipment was enormous. The fact that average loss percentages in Santiago were higher than in 1985 speaks for itself and raises the question of whether future efforts should be directed more towards controlling non-structural losses, especially in a situation where the primary goal of life safety has already been achieved to a satisfactory degree.

29.2.3 The Canterbury earthquake sequence

Nobody anticipated that the magnitude 7.1 Darfield earthquake of 4 September 2010, with its epicentre on the previously unidentified Greendale fault 40 km west of Christchurch, would mark the beginning of an earthquake sequence. This sequence culminated in the *M*6.2 Lyttelton earthquake of 22 February 2011, which destroyed the city centre of Christchurch, and the sequence is still ongoing as of the time of writing. The Canterbury earthquake sequence, and in particular the Lyttelton quake, raised many questions and subjected the world's oldest public earthquake insurance scheme, New Zealand's Earthquake Commission, as well as the entire insurance market to a stern test. The earthquake activated an old geological structure which had probably last been active at least 5000 years ago (Villamor *et al.*, 2011). Ground motions of the 22 February shock exceeded the design level prescribed by the building code by a factor of up to 3. The high ground motions led to widespread ground deformation, especially liquefaction to an extent never witnessed before. Accordingly, the degree of destruction in the Central Business District of Christchurch has been unprecedented for a country with such an advanced level of earthquake-resistive building practice. From a strictly engineering point of view, the goal of the building code was achieved, as demonstrated by the low number of victims (185, most of whom perished in two collapsed buildings). But the amount of total material losses was staggering for a city of this size, with a total so far of US$24 billion, most of which was insured.

The question of what society may consider an acceptable risk may well be reconsidered in the light of the Christchurch experience, especially in low and moderate seismicity regions. The USA had already moved in this direction before the earthquake by stipulating in the building code a lower probability level in low seismicity regions east of the Rocky Mountains. The often-heard argument that such a procedure would increase construction costs unacceptably remains to be proven. In fact, construction techniques are available that allow higher design levels, and at the same time allow the introduction of loss control features to reduce non-structural losses (Pampanin, 2011). Christchurch can be expected to become a natural laboratory for planning and building a safer city that will be more resilient to future natural events.

The reconstruction effort in Christchurch was hampered by the long and ongoing aftershock activity, peaking in the 22 February, 13 June, and 23 December shocks in 2012. Earthquake sequences pose particular challenges, but have not been in the focus of modern disaster management and recovery planning. There are good arguments for believing that an extended seismic crisis is more damaging for a region or city than a single, stronger main shock.

Surely this is true for the insurance industry. It had, and at the time of writing still has, extreme problems in settling the several hundred thousand claims caused by the various events, because of the difficulty in assessing so many losses and assigning them to specific events, and also because of the complicated interfacing with the underlying governmental coverage for residential buildings. The reputation of the industry has suffered severely from these problems.

29.2.4 The Tohoku earthquake

The news about the Christchurch earthquake of 22 February 2011 had hardly been digested when a magnitude 9 earthquake released tectonic stresses accumulated over centuries at the plate interface between the subducting Pacific and the overlying Eurasian/Okhotsk plate. The quake on 11 March was followed by a gigantic tsunami reaching heights of more than 30 metres on the northeast coast of Honshu in Japan. The tsunami in turn gave rise to one of the greatest technological disasters worldwide: the failure of three blocks of the Fukushima Dai-ichi nuclear power plant. As in the case of Haiti, the event was not entirely unforeseen. A previous very large tsunami had been documented for the Sendai region in a publication in 2001 (Minoura *et al.*, 2001), and large tsunamis had been observed in 1933 and 1896 in the northern part of the affected coast. But due to ambiguity and contradictions in the data, the official Japanese hazard map had not incorporated the possibility of large earthquakes happening along this section of the coast. Whereas the degree of destruction in the tsunami-stricken areas was close to 100%, there were no spectacular building failures due to earthquake shock thanks to the advanced earthquake building practice in Japan. However, as in Chile and New Zealand, the total amount of losses accumulating from mainly non-structural earthquake losses was again substantial. The total direct losses amount to between US$230 billion and US$314 billion, stemming from a region which contributes only 7% to the country's GDP. About 15% of the loss, US$35–40 billion, was insured (Munich Reinsurance Company, 2012). Thanks to favourable meteorological conditions, Tokyo escaped the worst in respect of the nuclear accident. But additional stress has been loaded upon faults in the Kanto area on- and off-shore, which increases the risk of an earthquake affecting the capital region in the foreseeable future (Toda *et al.*, 2011a).

Fukushima was not the first nuclear power plant affected by an earthquake in Japan. The world's largest nuclear power plant Kashiwazaki Kariwa was damaged by a moderate-sized earthquake in 2007 (International Atomic Energy Agency, 2008). In view of advanced seismological and engineering knowledge, plants constructed more than 20 years ago have to be considered as underdesigned, not only in Japan. Even if it was ultimately human error that caused the failure in Fukushima, it must be acknowledged that the chance of human error is much higher in an extreme situation, as exemplified by the Tohoku earthquake and tsunami, notwithstanding regular catastrophe drills. The ensuing disruption to the revival of the nuclear power industry worldwide was not the only global consequence of the Tohoku earthquake. It showed how exposed supply chains are nowadays. Globally operating companies all over the world, especially in the electronics and automobile sectors, suffered for months from a supply shortage for critical components, and were forced to look elsewhere for substitutes.

As in Chile and New Zealand, the need for loss-controlling construction techniques, i.e. techniques which also reduce material damage to non-structural building elements in addition to addressing structural integrity, became apparent. By far the majority of losses in the large area affected by earthquake shaking and, to a minor extent, liquefaction, were non-structural. But these non-structural losses added up to such high amounts that, contrary to intuition, and at least in respect of insured losses, the monetary earthquake losses were about twice as high as tsunami losses.

29.2.5 The Thai flood

Some of the Japanese firms affected by the Tohoku earthquake transferred production from their damaged facilities to areas abroad. Among these areas were industrial parks north of Bangkok in Thailand. In late summer, a flood situation developed there that eventually culminated in October and November 2011 in the largest flood since 1951. Hundreds of plants were submerged up to second floor level, so that average loss percentages reached 50–70%, a figure only seen before in isolated cases, and not for a population of risks. The long duration of the flood conditions, together with the humid tropical climate, contributed to the severity of the losses, as even machinery that had been moved to higher levels began to corrode. Possibly the greatest surprise was the fact that a producer of hard-disk drives in the affected area accounted for 25% of the world's hard disk drive production, which needless to say led to supply problems throughout the world. To an even greater degree than the Tohoku earthquake, the Thai flood was a global event in terms of the supply-chain problems it caused. Bangkok itself was more or less spared, but the total direct losses added up to nearly US$46 billion (World Bank, 2011), of which some US$16 billion was insured (Munich Reinsurance Company, 2012).

29.2.6 Hurricane Sandy

After a series of earthquake disasters and the 2011 Thai flood, the 2012 Hurricane Sandy served as a reminder of the loss potential of large windstorms. Sandy was a very exceptional event. It started as a tropical storm very late in the season in the Caribbean, south of Jamaica, and quickly reached hurricane status. On its way northwards, it struck Jamaica, eastern Cuba, Haiti, and the Bahamas. Haiti, which had not yet recovered from the 2010 earthquake, was hit severely once again. Whereas almost all the other tropical storms of the 2012 season later headed northeastwards to the open sea, Sandy's path was blocked by a high pressure system over Greenland, which redirected its track by 90 degrees right to the New Jersey coast. Its slow forward motion and very large wind field built up a heavy storm surge, which coincided with the high tide of the full moon as it reached the New Jersey coastline. This very unfortunate combination of circumstances led to the highest surge levels ever observed in New York. And even worse, the remnants of Sandy combined with a cold front arriving from the west, which re-intensified the storm system over land. The surge scenario had been anticipated in a study from 2011 (Rosenzweig *et al.*, 2011) and even before, but no preventive action had been taken, a very similar situation to that of New Orleans with Katrina. The world's financial centre, Wall Street, had to close down completely for two days, and overall 8.5 million people were without power for periods of between several days and two weeks. Access to Manhattan via tunnels was compromised for a full week, and full restoration of electric power took several weeks. The storm has demonstrated in exemplary fashion the vulnerability of the northeastern USA, and it must be kept in mind that the wind speeds of Sandy on land were rather mild, barely gaining hurricane strength. As with liquefaction in Christchurch and the tsunami in Tohoku, it was again a secondary peril, storm surge, which played a prominent role in the loss pattern.

29.3 Risk management in the context of natural disasters

From an insurance perspective, risk consists of three components: the hazard, the vulnerability of objects exposed to a hazard, and the value of the exposed objects. The hazard is usually defined as the exceedance probability of an event of a specified minimum size, e.g. the wind velocity in the case of windstorms or the ground shaking in the case of earthquakes. The vulnerability is expressed as the expected average loss, as a percentage of the replacement value, and depends on the hazard. For disaster prevention, planning, and response in general, this financial definition of vulnerability is supplemented by loss of life. It is well documented that loss of life decreases with the degree of economic development, whereas material losses exhibit the opposite behaviour, as exemplified by the 2010 Haiti and the 2004 Sumatra earthquakes on the one hand, and the earthquakes in Chile, New Zealand, and Japan in 2010–2011 on the other. Resilience has now been introduced as a fourth component. Resilience extends the concept of physical vulnerability to the overall capability of a region or society to withstand a disaster and to cope with its consequences. It is described in the form of various socio-economic indices and their combination (e.g. Cardona, 2005).

29.3.1 Risk identification

The growing importance of natural disasters for the insurance industry was already recognised in the early 1970s. This was a consequence of great natural disasters that occurred in less developed regions and that had an unexpected impact on the international reinsurance market. Examples include the 1972 earthquake in Managua, Nicaragua (US$100 million insured loss), and the 1972 tropical cyclone Tracy in Darwin, Australia (US$285 million insured loss). Therefore, for the purposes of risk identification, global natural disaster databases were developed (EM-DAT, 2011; Swiss Re, 2011; Wirtz *et al.*, 2012). Other successful products were the World Map for the Earthquake Hazard, developed by an international consortium of earthquake scientists (Giardini *et al.*, 1999) and the World Map of Natural Hazards (Munich Reinsurance Company, 1979, 2008). Such maps mark the transition from pure hazard assessment to risk evaluation, as they describe probabilities for one component of risk, which is the hazard (see Fig. 29.1).

29.3.2 Risk evaluation

The first fully probabilistic earthquake risk models that allowed the calculation of average annual losses (AALs or base rates) and probable maximum losses (PMLs) for distributed risk portfolios were developed by Sauter and Shah (1978) and for internal use in the reinsurance sector in 1987 (Rauch and Smolka, 1992). These models, as well as the suite of similar and increasingly refined models developed later by specialised companies such as Risk Management Solutions, Applied Insurance Research, and EQECAT, followed the principles of probabilistic single-site risk assessment (Cornell, 1968), and extended those principles to multiple locations. The use of such risk models has since become commonplace in the insurance industry. In such models, data on the hazard (e.g. earthquake intensities, wind speeds, or flood heights and their respective distribution in space and time), on the exposed

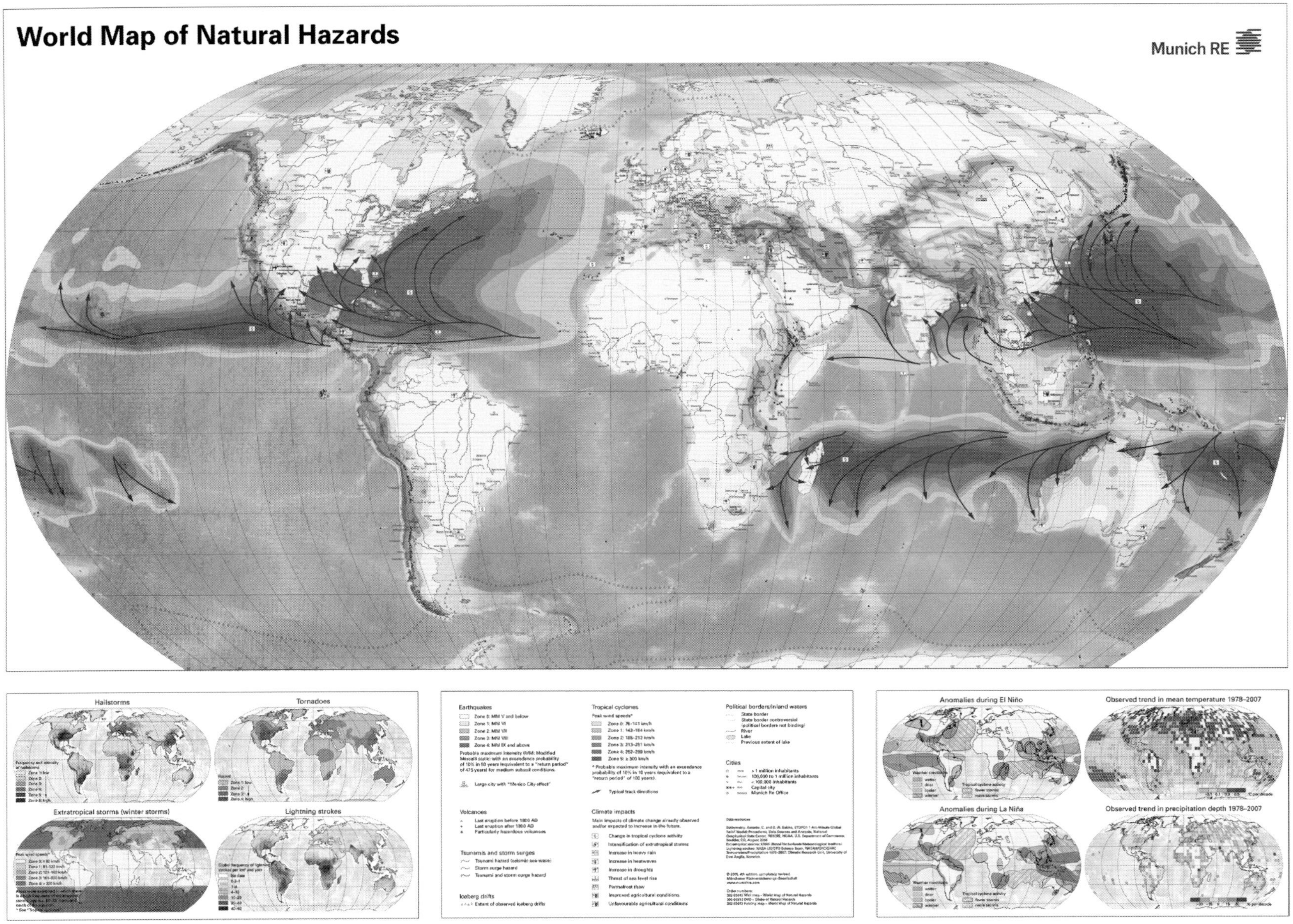

Figure 29.1. World Map of Natural Hazards. Earthquake hazard is shown in yellow–brownish colours and has been classified into five grades according to the minimum intensity (modified Mercalli scale) to be reached or exceeded once in 475 years. Darkest colour signifies the highest hazard. The hazard of tropical windstorms is presented in shades of green; again, the darkest colour corresponds to the highest hazard. Classification is according to the five-degree Saffir–Simpson scale. Green arrows represent the main cyclone tracks. Other hazards shown are extratropical storms (grey shading) and active volcanoes (small black symbols). For colour version see Plates section.

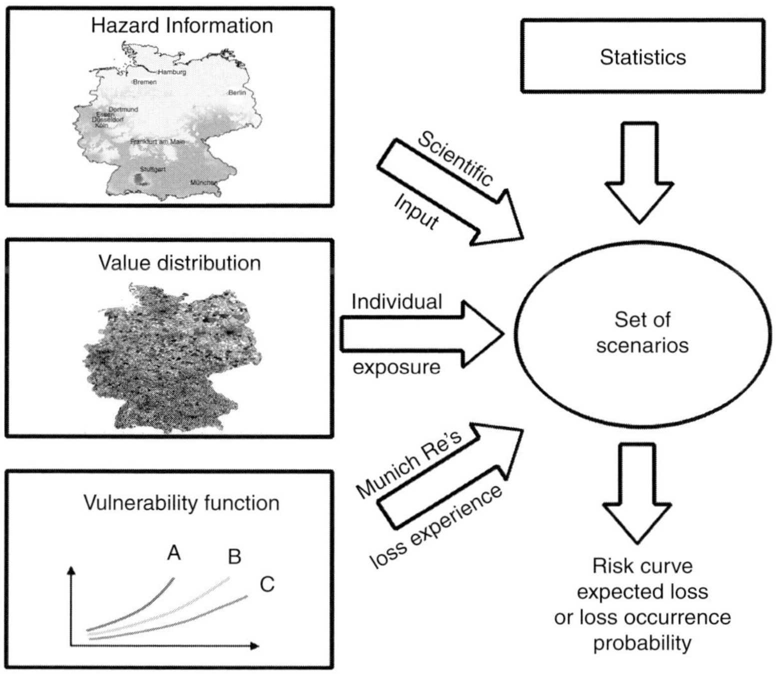

Figure 29.2. The Munich Reinsurance Company risk model MRHazard. Data on hazard, exposed values, and vulnerability are combined probabilistically. This results in an event set that serves as the basis for calculating insurance rates and probable maximum losses.

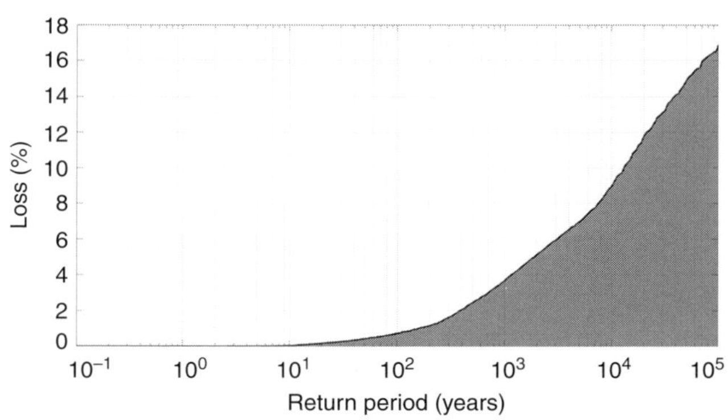

Figure 29.3. MRHazard example of a PML (or loss exceedance) curve. The PML curve is the graphical representation of the event set produced by MRHazard. It shows the losses to be expected as a function of annual occurrence probability or its inverse, that is, the recurrence period (in years). The curve in the graph is an example for a sample portfolio in Japan.

objects (their distribution per site and construction type and/or type of use), and on their vulnerabilities are combined and evaluated. The outcome of this procedure is a stochastic event set that describes the full probability distribution of potential event losses in the investigated region. Figure 29.2 shows schematically how such models work.

Figure 29.3 is the graphical presentation of the calculation result, called a loss exceedance curve. The example shows that a loss of 2.4% (or higher) of the total values in the considered region has to be expected on average once in 100 years, and a loss of 6.5% once in 1000 years. In insurance parlance, such a single value is called the PML. The PMLs form the basis for measures to guarantee the financial stability of a company in the form of reinsurance protection and limitations to primary coverage. The other important application of such risk models is the calculation of an adequate rate, i.e. the net insurance rate that would be necessary to compensate for all possible losses over time. This is the AAL. In mathematical diction, this is the integral over the full probability distribution.

In late 2005, the Organisation for Economic Development and Co-operation (OECD) initiated a global effort aimed at constructing an open-source Global Earthquake Model (GEM, Fig. 29.4) in accordance with uniform standards for data capturing, data processing, and calculation procedures. This model is now being developed under the guidance of the GEM Foundation, a public–private partnership consisting mainly of financial companies, national governments that are often represented by national research organisations, and international and donor organisations. The final goal of this collaborative initiative is to provide a global earthquake model, data, (software) tools, and other resources by the end of 2014, which can then be used, improved, and shared to develop risk reduction programmes on national, regional, and local scales.

Such models are also used for the ex-post estimation of event losses. But it must be kept in mind that simulations of up to hundreds of thousands of event scenarios for a

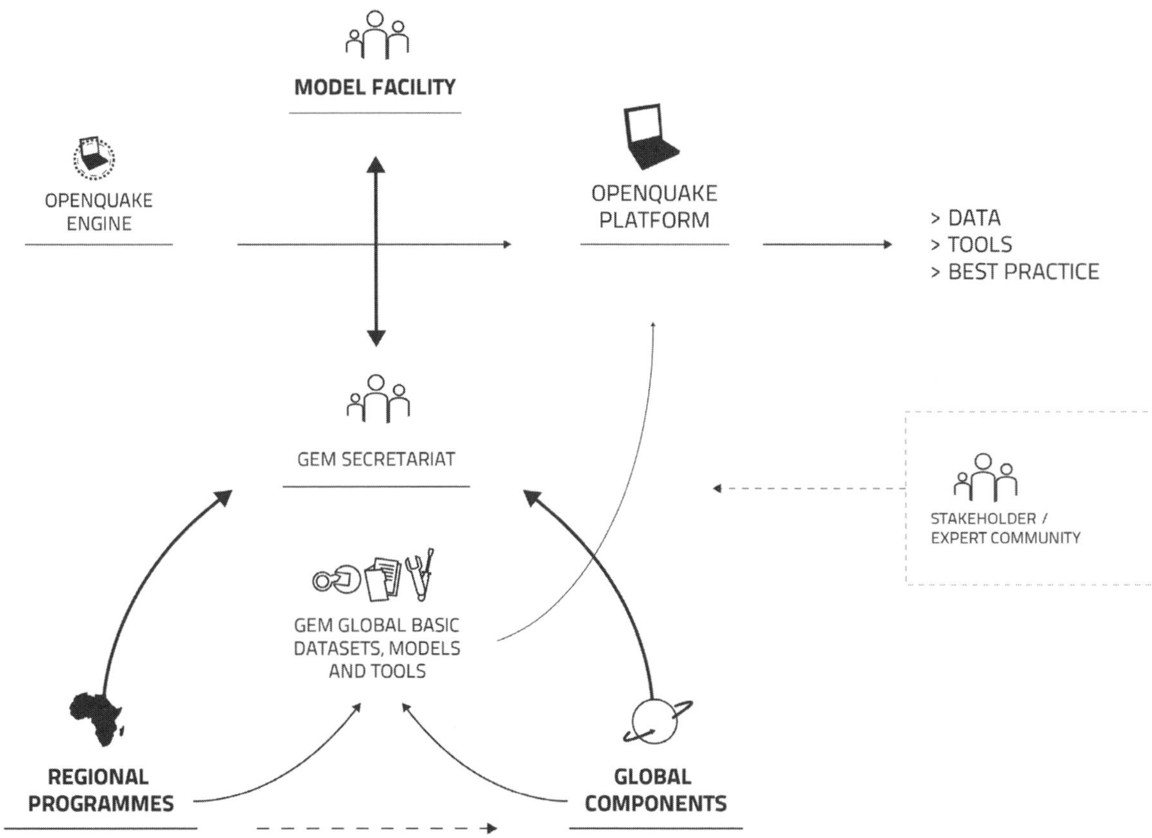

Figure 29.4. The GEM is an example of a public–private partnership which aims at assessing hazard and risk according to uniform standards to be collaboratively used as input to decisions that increase resilience to earthquakes worldwide.

probabilistic risk model have to focus on main phenomena and on average assumptions. Therefore, at least for the moment, they are unable to capture the wide range of peculiarities shown by specific events. Even when secondary effects such as liquefaction and tsunami for earthquakes, or inland flood for tropical cyclones, are included in a new generation of models, the uncertainties in modelling the loss arising from a specific scenario event will remain large. This calls for supplementing probabilistic methods with a deterministic approach at least for rare extreme events.

29.3.3 Risk control

The next step after identifying and evaluating the risks is risk control. In a broad sense, risk can be controlled by land use and building regulations, and by disaster preparedness measures that enhance resilience. As regards the financial dimension of losses and their insurance, the insurance industry has various measures at its disposal to keep the risk within affordable limits. They are briefly introduced below:

(i) Calculation and application of a risk-adequate price (i.e. the net insurance rate).
(ii) Accumulation control (i.e. monitoring the insured liabilities within a country or region).
(iii) Self-participation of the insured party in the loss, in the form of deductibles that reduce the number and amount of small insurance claims in the event of large disasters.
(iv) Liability limits that reduce the exposure from very large insurance claims from the top (i.e. losses are compensated for only up to the limit).
(v) Clear policy wordings in terms of event definitions and insured amounts (i.e. replacement value and its precise meaning vs. actual value).
(vi) Exclusion of particularly exposed areas or certain hazards. An example would be regions that are frequently flooded and therefore do not fulfil one of the principal preconditions of insurability, i.e. the lack of foreseeability of an event.
(vii) Improved claims settlement, which requires the formulation of contingency plans for regulating sometimes hundreds of thousands or even millions of claims in a single disaster, as observed in the

Maule, Tohoku, and Canterbury earthquake series in 2010/11.

(viii) Reinsurance and retrocession. Whereas the aforementioned measures work on the level of the original insurance policy, reinsurance and retrocession (i.e. the equivalent of reinsurance for the reinsurance sector itself) serve to limit the exposure from the large number of claims to be expected in great natural disasters by means of reinsurance contracts.

(ix) Catastrophe reserves. Premiums from natural disaster insurance cannot be considered as 'earned' premiums after one year. Great disasters happen rarely, as is well proven by all the recent earthquakes, which had occurrence probabilities of a few hundred years at least. Therefore, the relevant premiums have to be set aside to form financial reserves that are then tapped when the event occurs.

(x) Loss prevention.

The trend towards dramatically growing losses shows that it is essential to strengthen efforts in the areas of loss prevention and mitigation, i.e. taking proactive measures to reduce losses from future disasters by means of land use planning, construction techniques, and contingency plans. Losses in the ten-billion-dollar range, which used to be restricted to big, mature insurance markets, like the USA, Europe, and Japan, have now become possible in many parts of the world, as demonstrated by the Chile and New Zealand earthquakes and the Thai floods. How such measures are implemented, and by whom, is the subject of the following section.

29.3.4 Risk financing: the principle of risk partnership

Coping with loss burdens, such as those observed in the more recent disasters, and with still higher ones, such as may be anticipated when a megacity is visited by a large earthquake or hurricane at some point in the future, represents a formidable challenge that requires the cooperation of all parties involved, i.e. the potentially affected private persons and industries, the financial sector, and the state. Within the context of the role of insurance in disaster relief, we can distinguish between the insured persons or entities, primary insurers, reinsurers, capital markets, donor organisations, and governments/public authorities.

Each of these parties has its own tasks and responsibilities in managing the risk arising from natural hazards. Beyond the pure financing of future losses, much more effort needs to be invested in a pro-active strategy than has been the case up to now, i.e. in reducing and preventing future losses. Such a strategy is not simply a matter of financial resources. Good and advanced planning and coordination at all levels, from households and industrial companies to public institutions and authorities, is required. What precisely are the tasks of these parties? Householders and business owners (the insureds) can do a great deal to reduce the risk to their property by proper maintenance and by securing sensitive items like equipment, electronic installations, and machinery. In industrial businesses, emergency planning can help to prevent or minimise losses from future disasters. Finally, some of the financial risk has to be borne by the insureds in order to keep alive the interest in loss reduction. Typical forms of self-participation are deductibles, preferably expressed as a percentage of the sum insured, and coinsurance, i.e. a percentage participation in each and every loss. Primary insurers have to provide and secure capacity by

- charging technically adequate rates (i.e. rates that suffice to cover the expected future net losses);
- applying appropriate underwriting guidelines;
- accumulation control and portfolio management;
- establishing reserves for natural perils; and
- limiting their liability according to their financial strength and reinsurance protection.

Reinsurers are often the main risk carriers in natural disaster losses, making proper risk management by:

- balancing the risk over time and regions;
- providing technical support to the clients in rating considerations and assessments of PMLs;
- controlling and limiting liabilities (setting cession/occurrence limits, budgeting, retrocession; cession limits are based on sums insured, occurrence limits on losses). Depending on the maturity of a market, 50–90% or even more of insured losses may be reinsured.

Capital markets came on the scene in the 1990s. This type of alternative risk transfer (ART) can be seen as a supplement to reinsurance rather than competition. The principle behind ART is to transfer (financial) catastrophe risk to the capital markets, instead of taking out traditional reinsurance. The main function is to provide additional capacity for top-rank losses. Typical ART products are so-called catastrophe bonds (known as cat bonds), swaps, and derivatives. In popular terms, a cat bond means betting against the occurrence of a disaster within a specified timeframe. Swaps involve an exchange of risks from specified disasters between two companies, e.g. earthquake in Tokyo against windstorm in Europe. Derivatives are commonly used for weather-related risks, e.g. average temperature or precipitation falling below (or exceeding) a specific threshold level that would imply a financial loss to a business owner.

Donor organisations and development banks can foster and co-finance specific traditional (insurance) or non-traditional (capital markets) risk transfer schemes for governments or the private sector. The World Bank has been instrumental in setting up pool solutions for natural disaster insurance (e.g. the Turkish Pool TCIP and the Romanian Pool PRAC) and in providing schemes for disaster relief and reconstruction for governments (e.g. FONDEN/Mexico and the Caribbean Catastrophe Reinsurance Facility).

In the insurance context, the state has to act as a reinsurer of last resort for very rare, extraordinary losses and/or uninsurable risks that exceed the capacity of the private sector. However, the main task of the state lies in the field of risk management and risk reduction and involves:

- designing and enforcing land use and building regulations;
- securing the service ability of critical facilities and infrastructure;
- developing emergency plans that define precisely the responsibilities and the coordination of the authorities involved; and
- granting tax exemption for private insurers' catastrophe reserves.

Within this context, the role of the insurance sector is well established. The capital markets, on the other hand, still have to prove that they are willing to provide reliable and continuous capacity when investors have lost their money after truly large disasters, but first experiences have been encouraging. However, almost all the ART programmes have so far been placed for highly developed countries. The complexity of the programmes and the connected costs, as well as investor attitudes, usually require mature insurance markets. The state should create an environment where the greatest possible use is made of private resources for disaster recovery, combined with the availability of protection for as many people as possible. Linking the availability of such protection to the observance of building regulations can provide an efficient mechanism for code enforcement, especially where new constructions are concerned. This notwithstanding, mechanisms aimed at code compliance may serve to encourage rehabilitation measures as well. Another important role in the 'mitigation cycle' could be (and increasingly is) played by mortgage banks requiring natural disaster insurance as a precondition for a loan. Mortgage earthquake insurance worked efficiently in the case of the Maule earthquake 2010.

29.3.5 Disaster insurance: a tool for risk mitigation?

The use of disaster insurance as a motivating tool within the framework of loss mitigation programmes has become an increasingly prominent subject of discussion (e.g. Doherty *et al.*, 2008). So far, however, implementation of this concept lags far behind its potential. The reasons are manifold. The public at large is often unaware of insurance mechanisms or has an idealistic perception of the function of insurance. In the insurance sector, competition and a short-term financial perspective do not create a favourable environment for actively promoting prevention and mitigation measures, as the time-scale for any positive outcome tends to be too long.

Notwithstanding various attempts elsewhere, a project aimed at loss prevention, such as the community classification scheme of the insurance-sponsored Institute of Business and Home Safety (IBHS) in the USA, where communities are classified according to code compliance, remains the only one of its kind. Meanwhile, the IBHS has also started a testing facility demonstrating the usefulness of simple construction measures for loss prevention.

The classic example of a successful approach to loss prevention in property insurance is the inspection of insured objects by fire engineers employed by insurance companies in order to make recommendations on enhanced fire protection. The level of fire protection is a well-established criterion for rating and PML assessment. As far as earthquake risk is concerned, similar initiatives were taken by private firms as a consequence of the shrinkage of insurance capacity after the 1994 Northridge earthquake in California. On the basis of risk management surveys, physical earthquake protection was improved and the insurance coverage bought was adjusted to the demands of rare catastrophic losses.

Nevertheless, in natural hazards insurance, and especially in earthquake insurance, other features encourage loss reduction, namely risk-adjusted premiums and self-participation by the insured party. Risk-adjusted premiums involve tariff schemes that reflect the actual risk level commensurate with the location and the constructional characteristics of the insured object. Such tariff schemes are increasingly being used on a global scale, building on the support provided by advanced loss-modelling software. But the correct application of such schemes poses a problem, and in actual practice rates are usually still dictated simply by competition. Sometimes, for instance, rebates are given for alleged compliance with anti-seismic building regulations. Often, however, the alleged code compliance has not been checked by insurance engineers or agents and does not, in fact, exist. Therefore, awarding a rebate for code compliance can be counterproductive and even unjustified. This is particularly true for the majority of existing codes, whose principal goal is avoiding loss of life rather than reducing monetary loss. The Maule event as well as the Canterbury and the Tohoku earthquakes have vividly illustrated the importance of non-structural losses. All the three countries affected can be considered leaders in earthquake engineering, and, in fact, the codes accomplished their main goal of protecting lives. But the amount of material losses has been huge and run into tens (in Tohoku even hundreds) of billions of US dollars.

Coming back to self-participation as a means to motivate loss prevention, there are three different types of it. First, deductibles are expressed as a percentage of the sum insured or as a flat amount. Typical deductibles in earthquake-prone countries start at 2% and go up to 15% in highly exposed regions like California. But even in regions of moderate seismicity, deductibles of 10% are used if insurance penetration and, consequently, potential catastrophe losses are high (e.g. in Israel). Insurance payments only commence in excess of the deductible. Second, (proportional) coinsurance is again expressed as a percentage of the sum insured. Under this arrangement, the insured party carries a fixed proportion of each and every loss. Typical values range from 10 to 25%, and reach a level of 70–85% in Tokyo Bay. Third, first loss co-insurance or liability limits can be expressed as a percentage or a flat amount. Here, the insurer pays either from the first penny, or after a deductible up to a certain limit. The loss amount that exceeds the limit is retained by the insured for its own account.

All of these elements can be combined and are accompanied by corresponding premium rebates. The greatest incentive to take loss prevention and reduction measures is provided by proportional co-insurance of at least 10%, or by deductibles of 5% or more, as the insured party has to carry a substantial portion of any loss on its own. The effectiveness of the above-mentioned elements depends to a critical degree on the actual spread of insurance. In this sense, a distinction can be made between 'free' insurance markets and countries where earthquake coverage is obligatory or semi-obligatory.

In an unregulated market that is completely exposed to competition, it is a delicate task to find the right balance between tariff elements geared to loss prevention and acceptability for the consumer, with the result that a sufficient spread of insurance is achieved or maintained. A common reaction in such cases is the 'zero option', i.e. no insurance and no loss reduction. This option is neither in the interest of the general public, who ultimately have to pay the ensuing losses without having set aside reserves beforehand, nor in the interest of the insurance industry, which wishes to generate business. As a matter of fact, insurance conditions that are unattractive, or in extreme cases completely unaffordable, result in a situation where typically less than 10% of people have any earthquake insurance at all. This applies, with few exceptions, to both developed and less developed insurance markets. As a consequence of market penetration being so low, any attempts to foster loss prevention by means of insurance become futile.

A much better environment for using insurance as a direct incentive, or as an indirect contributor to loss reduction programmes, is provided by insurance markets where the coverage has been made mandatory by government legislation, or is at least widespread. In these markets, attempts to educate and raise the awareness of the consumer by means of brochures and videos reach many more people, and consequently have a greater chance of success than in free markets with low insurance penetration. Mortgage banks can play an efficient role in fostering high market penetration without the support of legal measures, by requiring disaster insurance as a precondition for a loan, as practised in Israel, Colombia, Chile, and an increasing number of other countries. In addition, and even more importantly, all of the above cited direct measures like deductibles can be brought to fruition without leaving room for the 'zero option', provided they are used of course. If, instead of this, full coverage without substantial deductibles is granted under mandatory schemes, the goal of loss prevention is missed again.

29.3.6 Public–private partnerships in natural disaster risk management

The foregoing discussion has identified several mechanisms for mitigating losses from natural disasters. The challenge is to integrate these mechanisms into a secure and tight network of risk reduction measures.

Insurance pools. Various solutions invoke the splitting of responsibilities between the parties. In New Zealand, for instance, the government-owned Earthquake Commission (EQC) provides basic insurance coverage for every household up to an annually adjusted, actual building value. Additional coverage for the replacement cost can be obtained in the private market, as well as insurance for commercial and industrial risks and for business interruption. In Japan, too, residential risks are covered to a large extent by the state-run Japan Reinsurance Company, whereas large businesses buy insurance in the private market, and the corresponding reinsurance is supplemented to a small extent by ART instruments like cat bonds. The concept of 'basic coverage' was also introduced by the California Earthquake Authority (CEA) in the aftermath of the Northridge earthquake in 1994, when earthquake insurance was difficult to obtain for homeowners. The CEA is an example of an often-used concept in natural disaster insurance schemes: the insurance pool. An insurance pool involves every company participating in disaster losses in proportion to its market share in premiums. This concept ensures that companies avoid being hit too severely, or even going bankrupt because of a disproportionately high loss burden from specific events. Examples of pool solutions are the Algerian Catastrophe Insurance Scheme, the Taiwan Residential Earthquake Insurance Facility (TREIF), and the Turkish Catastrophe Insurance Pool (TCIP), whereas schemes such as the US National Flood Insurance Program, the Florida Hurricane Insurance Pool, and pools in Switzerland, France, and Spain have been in existence for a longer time. Pools can be supported either by private or by state reinsurance. They are also being considered now in several other countries in Europe, Latin America, and Asia. Losses exceeding the capacity of the above-mentioned programmes usually become the responsibility of governments. The Turkish scheme (TCIP) can be considered as a model case and is therefore introduced here in further detail (Yunak, 2004). It was designed with the aid of the Contractual Savings and Insurance Practice Operations Section of the World Bank. The principal elements of the TCIP are:

- mandatory scheme: no post-disaster loans to affected parties without insurance;
- 2% deductible;
- a US$55 000 limit;
- rating scheme graded according to hazard zone and risk type; and
- initially complete risk transfer to global reinsurance.

The scheme covers dwellings and small commercial risks, whereas larger commercial and industrial risks and high value residential buildings are covered by the private market. What has been almost completely missing so far is efficient incentives for code enforcement. The TCIP represents an innovative concept for a less developed marketplace. It started operations in late 2000. Code enforcement is meant to be linked to the availability of insurance protection and/or governmental disaster assistance. However, establishing an efficient mechanism to guarantee code enforcement has so far failed (Gülkan, 2001). The TCIP has achieved a country-wide penetration rate of 25%, and even over 40% in the Istanbul region. Current efforts by the pool authorities focus on turning the decree into a law in order to further enhance penetration. Enjoying the status of a law is important, because only then can non-complying homeowners be sanctioned.

Despite all the practical problems and limitations, the Turkish scheme illustrates the potentially helpful role of international organisations like the World Bank in designing and backing new, proactive strategies for risk prevention and reduction. Current projects sometimes extend beyond the classical domain of property insurance by also addressing low-cost housing, public infrastructure, and crop insurance. Mexico's FONDEN is an example where the government has successfully secured coverage against natural disasters for its infrastructure. The Caribbean Catastrophe Reinsurance Facility (CCRIF) provides money to governments for the recovery phase after a disaster. Similar schemes are in the making for South Pacific Islands and for southeastern Europe.

Microinsurance: insurance for the poor. A further development towards making insurance accessible to more people is microinsurance. The subject is only briefly discussed here, as it is addressed in Chapter 30. Poor people typically suffer most from natural disasters. They tend to live in the most vulnerable buildings. At the same time, they usually do not have any basic understanding of what insurance means, and even if they had, they would lack access to insurance products. Microinsurance aims specifically at protecting the life and essential property of low-income people. In very general terms, microinsurance can be considered as a sort of cooperative or mutual insurance scheme on a small scale. The challenges of microinsurance are manifold. Addressing the local needs and conditions is indispensable. Therefore, microinsurance schemes usually take the form of a partnership between an internationally operating insurer, who contributes the insurance know-how, and a local institution that looks after the requirements of its local clients. Often the clients are illiterate and make their living in the informal economy, which requires innovative distribution channels and payment modes. A good summary can be found in Churchill *et al.* (2006).

29.4 Risk management: lessons from recent disasters

What are the lessons from recent disasters, in relation to the framework discussed above? Without claiming completeness, some of the pertinent issues are listed here:

- Worst case events and the question of acceptable risk (Canterbury and Tohoku earthquakes, Hurricane Sandy).
- Transient hazard changes (Haiti, Canterbury, and Tohoku earthquakes).
- Event clustering (all events, and Canterbury earthquakes in particular).
- Time-dependent hazard and earthquake forecasts (Haiti and Tohoku earthquakes).
- Role of secondary effects (liquefaction; tsunami; the Canterbury and Tohoku earthquakes; Hurricane Sandy).
- Concatenated events or event chains, NaTech events (combinations of natural and technological disasters, Tohoku earthquake).
- Life safety vs. material losses (Haiti vs. Maule, Canterbury, and Tohoku earthquakes).
- Control of material losses (Maule, Canterbury, and Tohoku earthquakes).
- Reconstruction in unsafe environments (Haiti and Canterbury earthquakes).

- The sociopolitical dimension of disaster risk (Haiti earthquake).
- Human error (Tohoku earthquake and Thai floods).
- The globalisation of disaster risk (Maule and Tohoku earthquakes, and Thai floods).
- Public–private partnerships for risk financing and reduction (Canterbury and Tohoku earthquakes).

Some of these issues will be discussed now, following the various steps in the risk management chain.

29.4.1 Risk identification

Maps like the World Map of Natural Hazards (Fig. 29.1) provide a good template for recording hazards. They tend to focus, however, on such main hazards as ground shaking for earthquakes, or wind speed for windstorms. However, it has been clearly illustrated how important concomitant effects such as ground failure, liquefaction, and tsunami for earthquakes (landslides and fire following earthquake may complete this list), or storm surge for windstorm can be. In order to capture risk, however, all three components of risk must be known. At the time they occurred, the 1972 Managua earthquake and the 1974 Cyclone Tracy were great surprises, as they produced disproportionally high losses in regions of low exposure. Capturing exposure may seem a trivial task in the era of satellite-based earth observation systems. But, in fact, the losses from the Thai flood, or the 2010–2011 Canterbury earthquake sequence, have been no less of a surprise than Managua and Darwin in the 1970s. Industrial parks as exposure hot spots have developed more or less unnoticed in many parts of the world, and only now have become the subject of intense investigations and data collection. Beyond exposure, even vulnerability is not a stable element. The Maule earthquake hit buildings in Santiago and at the coast, whose characteristics were completely different when compared to the building inventory affected by the previous *M*8 earthquake in 1985, and whose vulnerability turned out to be comparatively higher than for the 1985 inventory. This is a typical situation in all rapidly growing regions of the world.

29.4.2 Risk evaluation

Maximum magnitudes. After the Tohoku earthquake, there was much debate about the official Japanese hazard map and the fact that a magnitude 9 scenario had not been considered in the underlying hazard model. The possibility of such high magnitudes around the whole circum-Pacific region had already been discussed following the Sumatra earthquake in 2004 (McCaffrey 2008). *M*9s have now been introduced in risk models to a large extent. But it should be mentioned that introducing *M*9s is at the expense of the frequency of lower magnitude events. Therefore, their effect on country loss exceedance curves is not necessarily high, also because their epicentres are located offshore, with the possible exception of the Himalayan subduction zone. Tsunamis which invariably accompany *M*9 events may have a higher impact on the risk curve than the shaking due to top magnitude events.

Secondary effects. The significance of secondary effects has already been mentioned, and more effort definitely needs to be invested in addressing them in probabilistic risk models. The more effects are included, however, the higher the uncertainty of the results. It seems advisable to reduce such uncertainty by introducing more physics-based methods into the modelling, at least for events in the tail of the distribution. Forward modelling of earthquake wave propagation, for example, is becoming operational and can significantly reduce the uncertainty for ground motion modelling (Guidotti *et al.*, 2012). Furthermore, techniques for introducing extreme seismic events into modelling of probabilistic seismic hazard are under development (Sokolov and Ismail-Zadeh, 2012).

The principal challenge for risk modelling is to filter out the generally applicable features from the many peculiarities which a specific event exhibits. The degree of liquefaction observed in the Christchurch area was truly exceptional, as was the role played by the storm surge in Hurricane Sandy. The question modellers have to ask themselves is: 'To what extent is this applicable to a densely populated area elsewhere in the world?' The same goes for tsunamis. Not many countries have coastlines that are so heavily developed and, at the same time, extremely hazardous as Japan. Even there, the contribution of tsunami damage to the total (insured) loss was no higher than 25–30% in the Tohoku earthquake. But tourism has obviously increased the risk from tsunami, and in terms of human losses, tsunami is definitely an important hazard, as evidenced by the 2004 Sumatra earthquake, the 2011 Tohoku earthquake, and projections of loss of life for an *M*9 scenario at the Nankai trough (Cabinet Office's Central Disaster Prevention Council, 2012).

Temporal seismic hazard change. Here, a distinction needs to be made between two effects that play over different time-scales. The first is long-term time dependence due to earthquake cycles; the second is short-term transient hazard change due to stress transfer, ensuing triggered seismicity, and aftershocks. Whereas the consideration of time dependence has become usual practice in regions where sufficiently reliable data are available (California, Mexico, Turkey, Japan), transient hazard change had so far been neglected in hazard and risk modelling. In the classic approach, earthquake catalogues are declustered before they are processed statistically. This seemed to be justified by past observations, which did not hint at significantly increasing the ground motion probability, and in consequence losses from aftershock activity. The Haiti and Chile earthquakes, and particularly the Canterbury earthquake sequence and the Tohoku earthquake (and on a smaller scale the Emilia Romagna earthquakes in Italy 2012), have compromised such a procedure and have lent urgency to the question of how risk temporarily changes after significant events. Techniques to calculate temporal changes have become available (Toda *et al.*, 2011b). Of course, such temporal changes also have serious societal consequences for the affected population, and the need for making earthquake forecasting operational (International Commission on Earthquake Forecasting for Civil Protection, 2011) has been dramatically proven by the L'Aquila case where scientists have been declared guilty for giving misleading information to the public.

Clustering. Earthquake clustering is a related problem, and the Canterbury earthquake sequence heightened consciousness regarding the particular problems posed by such sequences, problems that already start with risk assessment. Situations like New Madrid 1811/12, and the Bocono fault in Venezuela 1812, the Sea of Marmara 1766, the Nankai trough 1854 and 1944/46, and in more recent times, Friuli 1976, come to mind. Clustering is more common for meteorological perils, see the extratropical cyclone seasons in Europe 1990 and 1999, and the 2005 hurricane season in the North Atlantic basin. It is therefore considered in windstorm models, but is tricky to treat in earthquake models, not in terms of occurrence (pure hazard), but in terms of losses (physical vulnerability regarding predamage, event definitions, and apportionment of losses for insurance purposes). The '*annus horribilis*' 2011 is an example of cross-peril clustering with several earthquakes, a highly active tornado season in the USA, and the Thai floods.

PLA. This has nothing to do with physical risk, but with resilience and the sociopolitical dimension of disasters. With PLA, all the factors are summarised that inflate a loss beyond its physically based dimension (Boissonade and Muirwood, 2008). It includes demand surge and business interruption, which can be due to denial of access (Christchurch Central Business District), physical inaccessibility (Thai flood), power failure (Hurricane Sandy), and supply-chain problems (Tohoku earthquake and Thai flood). In the insurance context, additional factors are lack of loss adjustment capacity and delays in loss adjustment (Christchurch), lump sum adjustments for mass claims (Christchurch, Chile), unclear policy wordings (Christchurch), and even political pressure aimed at *ex-gratia* payments or at waiving deductibles.

29.4.3 Risk control

Loss prevention. It is appropriate to start with a universal lesson before entering into a discussion of the insurance-specific aspects. This lesson is the role that non-structural losses have played in recent events. In times of dwindling financial resources, society can afford even less than before such high material losses coming from non-structural damage. Simply insuring them is not a sustainable solution, as insurance also depends on the economic environment. Chile, New Zealand, and Japan have shown that the primary goal of earthquake building codes – to avoid collapse and ensuing loss of life – has been achieved there. Together with California, these countries are pioneers in terms of earthquake-resistant construction. Much remains to be done elsewhere, as evidenced by extreme cases like the 2010 Haiti or the 2008 Wenchuan earthquakes. But overall, the time has come to introduce broadly based loss control features into building regulations. A well-known technique is base isolation, which, however, will probably remain restricted to highvalue structures (although there is one exception: permanent housing provided to the more than 10 000 people left homeless by the 2009 L'Aquila earthquake features base isolation). Precast Seismic Structural System (PRESSS) techniques (New Zealand Concrete Society, 2010; Pampanin, 2011) are less well known, but are available and cost-effective. An unresolved problem is posed, of course, by the existing building stock, and by heritage buildings in particular.

Risk-adequate price. It is a common experience after large disasters that prices charged *ex ante* turn out to have been insufficient, as important price drivers had been neglected. Unfortunately, such price drivers are often only revealed by the event (i.e. new building practices, underinsurance, PLA in general). Small markets are a particular problem. The Canterbury earthquake sequence will consume about 20 years of the 2011 property premium, including non-earthquake coverage, of course, for the whole of New Zealand.

Accumulation control. The case of industrial parks, which have sprung up in many parts of the world, including regions where one would not normally expect them, has already been mentioned. It is a common observation that disasters expose deficiencies in accumulation control. A particular case that deserves mention is that of multi-location policies, i.e. policies where several or many locations are covered under one policy, and the full sum insured is listed under the headquarters location. While it is feasible to split the values up for material damage, this is not possible for business interruption since, by the nature of the cover, only one value can be declared for all locations taken together.

Self-participation. The Canterbury earthquakes provided a textbook example of the efficiency of deductibles or the lack thereof. The basic coverage granted by the EQC for residential buildings and their contents in New Zealand stipulates a first loss limit of NZ$100 000 for buildings and NZ$20 000 for contents. The deductible is 0.5% of the limit, equivalent to NZ$500 and NZ$100 in each case. Such low deductibles fail to serve the two principal goals of self-participation. First, the number of claims to be administrated remained virtually unreduced, and EQC was already inundated with over 150 000 claims following the Darfield earthquake. Of these, only 20% had been visited by the time the Lyttelton earthquake struck 5.5 months later, producing a similar volume of claims. Second, such a low deductible offers no incentive to take loss prevention measures. 1–2% of the total replacement value should be a minimum deductible and had been in force in Chile when the Maule earthquake hit, serving well to keep the insurance payout within appropriate limits.

Policy wording. Three elements have to be considered in this context: event definition, sum insured, and underinsurance.

- *Event definition* Hurricane Sandy presented an interesting case. Simply because it was officially categorised as a 'post-tropical storm' shortly before landfall the question arose whether hurricane deductibles could be applied that would be substantially higher than regular policy deductibles. This has been rejected by several state supervisory authorities. In consequence, insurance payments may inflate considerably because of the very large area affected by Sandy and the ensuing high number of claims resulting from the storm. By contrast, in the Tohoku earthquake, the problem lay in spatially separate triggered events several hundred kilometres away from the main shock. Separating out these triggered events has a direct impact on reinsurance payments, as the priority under cat treaties counts per event. The same applies in primary insurance for per-location deductibles under multi-location policies. While careful wording helps to avoid unpleasant surprises, it cannot solve the problem of apportionment to specific events in event sequences (as observed in Christchurch), a process that is always arbitrary and usually impossible to verify.
- *Sum insured* Most commonly the sum insured is stipulated as the replacement value. But what does replacement value mean? Reinstatement to the original condition *ex ante*? Or are post-disaster code upgrades included? And what about the replacement value of heritage buildings? In New Zealand, policy wordings were very soft, or generous, at the time when the first earthquake of the Canterbury earthquake series, the Darfield earthquake, occurred. Therefore loss amounts which exceeded the replacement value were not unusual, and, in the end, indemnifiable. In respect of mortgage policies, it is important to know whether the declared sum insured refers to the full mortgage, or to the remaining amount. This was not at all clear for many mortgages affected by the Maule earthquake.
- *Claims settlement* The key to an efficient loss adjustment process is to have contingency plans that guarantee access to a sufficient number of loss adjusters of a professional standard. Under normal conditions, the EQC works with a very small staff of only 20 or so employees. The contingency plan consisted in increasing this number according to needs by the use of subcontractors. This plan worked, but not without difficulties, since the contracted partner did not have enough qualified adjusters either, and had to train people who had never seen earthquake damage before. Small markets in particular need the assistance of international, specialised loss adjusting firms. In Chile, one problem was that every claim required a sign-off from a local adjuster, which often delayed the payout unnecessarily.

29.4.4 Risk financing and risk partnership

Public–private partnerships (see also Chapter 30) have been tested by the Canterbury earthquake sequence, the Tohoku earthquake, and by Hurricane Sandy.

In New Zealand, risk is shared by the EQC, a government scheme that provides basic coverage for homeowners and is reinsured in the global market, and by private insurers who look after commercial (including municipal) and industrial risks, and provide excess coverage for homeowners. The overlap in coverage for residential buildings, together with the endless series of events, created numerous and complex problems in terms of loss adjustment. In the end, this led to considerable delays in settling the claims, and in turn to reputational problems for EQC, as well as for the private sector. These problems could have been reduced *ex ante* by introducing adequate policy deductibles (reducing the overall number of claims) and by adjusting the EQC first loss limits in such a way as to minimise overlaps with private excess coverage. Of course, increasing the limits would augment the liability of EQC, but this effect could be countered by increasing the deductibles accordingly. The prices for reinsurance and primary coverage have increased by multiples. The focus should be on ensuring higher self-participation on the part of the insured in order to foster loss prevention, on restricting overall insurance payouts, and in this way keeping prices at an affordable level.

In Japan, Japan Earthquake Reinsurance (JER) again provides coverage for homeowners and small businesses. The risk is kept entirely in the country with minor participation by the local market, and with no foreign reinsurance whatsoever. Additional coverage for residential buildings is provided by Mutuals, but there is no overlap in the sense that, unlike New Zealand, JER and the Mutuals can act independently, a fact that accelerated payouts considerably. Large commercial and industrial risks are served by private insurers and the policies often carry very low limits. In effect, only about 15% of the total material loss was insured in Japan, and there is a strong argument that the potential of the global private insurance market is not presently exploited to the full extent possible. In New Zealand, in contrast, less than 20% of the total bill remains with the government.

Sandy, and Katrina in 2005, also illustrated a problem in terms of overlapping coverage: in the USA, basic flood coverage for homeowners and small businesses is provided by the government-owned National Flood Insurance Program (NFIP) up to a limit of US$250 000. Private insurance carriers provide excess coverage and coverage for windstorm. As tropical cyclones are often accompanied by storm surge and heavy precipitation, the prevailing loss agent (wind or water) is unavoidably contentious. The NFIP is almost unsustainable in financial terms, and plans are therefore being drafted to privatise flood insurance entirely.

One can conclude that risk sharing in Japan worked well, although it has not been structured into the framework of a genuine partnership, and the capacity of the global market could be tapped to a greater extent. In New Zealand, the future of the EQC is uncertain. The fate of similar schemes, such as the Turkish TCIP or the Taiwanese TREIF after a large event, remains to be seen.

29.4.5 The question of acceptable risk

Beyond the realm of insurance, the high number of victims caused by the South Asia tsunami and the Haiti earthquake, the global consequences of the Tohoku earthquake and the Thai flood, and the extreme rarity of an event such as the Canterbury earthquake sequence and the Tohoku event chain, culminating in the Fukushima nuclear accident, have lent urgency to the question of what can be considered an acceptable risk for society on a regional, national, and global level. Is it wise to deny or simply overlook the possibility of natural events that release forces exceeding the probability levels of 10% in 50 years commonly used in building codes, or which produce losses beyond the once in 200 years threshold prescribed by many supervisory authorities? And how reliable are probabilistic assessments for accidents in industrial facilities? The Christchurch earthquake of 22 February 2011, which resulted in 'deconstructing' 70% of the city's Central Business District, and the Tohoku earthquake and tsunami, with the ensuing nuclear accident, may tell a different story. In a study prepared for the Cabinet Office of the Prime Minister of Japan, the projected number of victims for a similar earthquake and tsunami at the Nankai trough is 313 000, and only favourable meteorological conditions prevented a major radiation disaster after the tsunami-induced Fukushima accident in the Tokyo metropolitan region. The question of what constitutes acceptable risk must be posed with even greater urgency for truly great, extreme natural events as discussed in the following and final section.

29.5 Extreme natural events

What are 'extreme natural events'? A strict definition is difficult, as there is a smooth transition from 'normal' and 'big' events to exceptional ones. There are two criteria for an 'extreme event': the occurrence probability and the area affected. Regarding probability, it is common practice in natural disaster insurance to steer the business on the basis of loss scenarios whose regional frequency is once in 250–1000 years, at maximum. Examples include all the events discussed above, an earthquake in Tokyo, a double-landfall hurricane affecting Florida and the US Gulf Coast, or a great winter storm affecting northwest and central Europe. Only in recent years have scenarios materialised in regions beyond these highly mature insurance markets, viz. the Canterbury earthquakes in 2010/11 and the Thai flood in late 2011. Both of these produced losses in the two-digit billion US dollar range. As to the area affected, one may consider events that have serious regional or even global consequences as extreme. A good example would be a great volcanic eruption with an impact on the global climate. Finally, to turn an extreme event into a disaster it must affect a populated area. It does not make sense from a risk perspective to consider the regional frequency for events with a global impact. Both their global impact and their global probability have to be taken into account. In conclusion, and in an insurance context, we define extreme events as the occurrences whose probability is beyond the common planning horizon, i.e. below once in 1000 years, in both a global or a regional reference frame. Extreme natural events present a particular challenge. Nevertheless, the same principles and procedures can be applied as those already specified in the context of classic natural hazards concerning the risk management of extreme disasters. This means that the hazards have to be identified, and the related risks evaluated. Then the potential losses have to be controlled, prevented, and mitigated as far as possible. The particular challenge of very rare, extreme events lies in the fact that such disasters have not been much experienced in the documented human history. However, their impact could be huge, if they happened in our modern society. Therefore, their effects must be constructed on the basis of geological and geomorphological investigations and through theoretical modelling of the physical processes involved. Both these undertakings are fraught with considerable uncertainties. It is only partially possible to draw analogies from known events in the recent past. For instance, the South Asian tsunami of 26 December 2004 may well serve as a blueprint for damage patterns from similar future events in other regions. Another example is Hurricane Katrina of August 2005: the degree of destruction it inflicted on the US Gulf Coast would probably not be exceeded even by a strong tsunami. But, as the Boxing Day 2004 tsunami in South Asia demonstrated, the length of coastline affected could be greater. There now follows a brief summary of knowns and unknowns and a description of a rational approach to treating extreme events in accordance with the risk management principles.

29.5.1 Hazard identification

A wide range of extreme natural events pose a threat to the insurance industry: great volcanic eruptions are called 'super-eruptions': some historical examples are the Toba eruption on Sumatra 74 000 years ago, with an eruption volume of about 3000 km^3 (as compared to 18 km^3 in the famous Krakatoa eruption of 1883) and the Oruanui eruption in New Zealand 26 500 years ago with 500 km^3 eruption volume (e.g. Mason *et al.*, 2004; Self, 2006). Giant submarine landslides occur on continental slopes (Masson *et al.*, 2006). The best-documented cases are the Storegga slides west of Norway, the last of which occurred 7000 years ago. Its volume was about 2500 km^3 (e.g. Harbitz, 1992). Giant landslides also result from flank collapses of volcanic islands in the ocean or on coasts. In principle, every volcano in the ocean is an unstable edifice whose over-steepened slopes collapse from time to time in catastrophic slides or slide series. About 20 such catastrophic episodes have been documented for the Canary Islands within the last 2 million years (e.g. Masson *et al.*, 2006), and several have been documented from the Cape Verdes, the volcanic islands of the Caribbean (Le Friant, 2001), the Hawaiian Islands, and Reunion. For meteorite impact, objects with a diameter of 100 m, which is the smallest size necessary to reach the Earth's surface, hit the Earth on average once every several thousand years. Ten-kilometre objects, like the one which is blamed for the mass extinction marking the boundary between the Cretaceous and the Tertiary epochs 65 million years ago, happen once in about 100 million years. In the only documented meteorite fall, the Tunguska event of 1908, the object was too small to reach the Earth's surface and exploded in the atmosphere. The shock wave nevertheless flattened about 2000 km^2 of forest (e.g. Morrison, 2006).

A consequential effect common to all these hazards when they happen in oceans is tsunami generation. Other events, which are only briefly mentioned here, are the sudden release of methane hydrates buried under the ocean floor (such an event may have caused the Storegga slide mentioned above), abrupt climate change, cosmic storms, i.e. episodes of exceptionally high magnetic particle flux from the Sun, and reversals of the Earth's magnetic poles. Understanding atmospheric and hydrospheric hazards emanating from climate change, for example, perturbations in ocean currents like the Gulf Stream (e.g. Bryden *et al.*, 2005), is important and even directly relevant to modelling the global climate effects of great eruptions and impacts. Furthermore, cosmic storms may be highly threatening to a civilisation that relies so heavily on electronic devices as is the case nowadays (National Research Council, 2008).

29.5.2 Risk assessment

Risk assessment involves making a (probabilistic) assessment of the hazard and combining the hazard with data on the vulnerability of objects and communities. A quantitative, probabilistic assessment of the hazard suffers from the lack of reliable and cross-checked event catalogues. The situation is best for earthquake-induced tsunamis (National Geophysical Data Center (NGDC): tsunami database, http://www.ngdc.noaa.gov/seg/hazard/tsu_db.shtml), but there is room for improvement even here. This applies to an increasing degree to other events like volcanic flank collapses in ocean islands and submarine slides. Also, the inventory of great volcanic eruptions has grown (Mason *et al.*, 2004; Deligne *et al.*, 2010), but it can hardly be considered complete. The Global Volcano Model (GVM) project, has been set up in order to achieve a quantum leap in reducing the risk from volcanic eruptions of all sizes on a global scale by coordinating existing global and regional initiatives and addressing many components of risk assessment and management, including the forecasting of eruptions. Much more geological and geophysical fieldwork is needed in order to improve the existing inventories of great eruptions and submarine slides. The new technique of side-scan SONAR for mapping oceanic topography (Marine Geoscience Data System, 2013) has great potential for enhancing inventories of past submarine slides. Regarding meteorite impact, the Spaceguard survey has improved the data inventory for so-called near-earth asteroids (NEAs) substantially (Morrison, 2006), but unfortunately the programme has faced setbacks due to lack of funding.

Empirical evidence for vulnerability is usually lacking due to the rare occurrence of extreme events. But it is often possible to draw analogies from smaller-sized or similar events, such as 'normal' volcanic eruptions, explosions, or firestorms. Disasters like the South Asian and the Tohoku tsunamis in 2004 and 2011, respectively, offer a unique chance to collect empirical loss data. There is, however, a further complicating effect over and beyond the lack or scarcity of empirical loss data: losses from extreme natural events are often the end result of a combination of primary and secondary hazards in event chains, as already demonstrated by the 2011 Tohoku earthquake. Furthermore, losses can accumulate from many countries and from different lines of business. To some extent, this was demonstrated by the 2004 Sumatra earthquake and the ensuing tsunami, where losses occurred in all sorts of insurance classes. Identification of all the possible loss agents is of prime importance to avoid hidden loss accumulations. For example, the final loss from a meteorite impact on land stems from various components. It starts with the shock wave and then moves on to falling objects and wildfires, and finally extends to all the consequences of the ensuing climatic perturbations and their effects on the built environment, as well as on food production. Estimating the vulnerability of the built environment to these various hazards and assessing consequential losses under non-property-related insurance lines represents a particular and formidable challenge in view of the lack of actual loss experience.

29.5.3 Risk control

There is a common misconception about the disasters associated with extreme natural events, namely that control and prevention measures would be useless because of the sheer size of such events. Such an attitude denies the fact that there is a seamless transition from more common to extreme hazards. Furthermore, it has been demonstrated by Ward and Asphaug (2000) that the total risk of meteorite impact is dominated by objects with a diameter of some 200–300 m, i.e. just big enough to survive the plunge into the upper atmosphere. The frequency of such an impact is about once in 10 000 years. This principle can also claim validity in respect of other hazards, such as large volcanic eruptions, and is dictated by the exponential decrease in frequency with increasing event size. The decrease is even more pronounced in the extreme tail of the frequency distribution.

The list of measures for risk control presented in Section 29.3 is also suited to handling extreme events, but with somewhat differing weights regarding the various possible choices. As to calculating technical rates, extreme events are so rare that calculating an insurance rate ends up with values close to zero. They are, however, the ultimate expression of low-probability/high-consequence events, and it is therefore absolutely imperative to limit the liability arising from them.

This implies the need for exact knowledge of what is covered under existing insurance contracts. However trivial this may sound, it is very important because unclear and misleading terms are not unusual in insurance policy wordings. In addition, the impact of extreme disasters on the capital markets has to be considered and planned for, because it reduces the value of the assets and the capital strength of insurance companies. The combined effect of high material losses and reduced asset values was seen after the terrorist attack of 11 September 2001, although there was no causal connection between the two phenomena in this case.

Arguably, such considerations are philosophical in view of the extremely low likelihood of a meteorite impact in a densely populated and well-developed region. As risk assessment is also fraught with large uncertainties, as demonstrated above, a general exclusion from coverage may turn out to be the ultima ratio for most of these events, at least given the present state of knowledge. But drawing a line between what at first sight appears to be an insurable peril, e.g. a volcanic eruption, and what becomes an uninsurable one once the event exceeds a certain threshold, is a problem that has not yet been considered systematically, let alone solved, by the insurance industry. And what about prevention, which plays such an important role in connection with 'classic' natural perils? Indeed, many prevention measures that work in the case of earthquakes, for example, will also have a positive effect on other perils, at least to some extent. Chapman (2003) provides an excellent discussion of options for handling meteorite impact scenarios. He makes the interesting argument that this most threatening natural peril seems to be the only one where prevention of the event itself, not only of the ensuing losses, seems feasible, by diverting the celestial body away from its dangerous track through human intervention. But what appears technically feasible can have serious political and human implications, ranging from intentional misuse of the technology to unintentional failure of the mission (Morrison, 2006).

Whereas the theoretical possibility of completely preventing an extreme event is unique to the impact hazard, all the aforementioned extreme disasters have one thing in common: the (usually ample) warning time. These times range from a few minutes for coastal sites close to tsunami sources to several years for impacts. Great volcanic crises are somewhere in between with warning times of weeks to months. A practical problem, however, is that there are no criteria to assess the magnitude of a pending eruption from the precursory activity. And surprise eruptions are still possible due to gaps in the observational network. In an insurance context, the possibility of early warning is of limited value as regards reducing the risk of material losses. People can be evacuated, although the complexity of such an operation is unimaginable in the case of a truly large event that may affect an area the size of a country or continent. But moving buildings and infrastructure is impossible.

29.6 Conclusions

The series of natural disasters since 2010 has proven that we have entered into an era of new loss dimensions. Many questions regarding the various steps within the framework of risk management for natural disasters have been raised by those events, and the question of what constitutes an acceptable risk for society has been posed with renewed urgency. At the same time, valuable data and information have been provided to judge the adequacy of existing measures. Notwithstanding the enormous progress made in the last decade and before in respect of scientific and engineering knowledge, and in spite of the existence of tools to assess, and of mechanisms to manage the risk from natural disasters, these events have shown deficiencies in all these fields. In hazard assessment, secondary perils and temporal hazard changes have to be addressed. Regarding vulnerability, controlling physical losses is gaining more and more significance. As to risk assessment, the possibility of event chains including technological disasters and the globalisation of risk have to be considered. Finally, risk management has to take into account event clusters, and again globalisation and chain reactions on the one hand and resilience on the other hand. Local resilience depends on many, and not at least on socio-political, factors and circumstances. Public–private partnerships have proven to be an effective mechanism to cope with large disasters, but can be largely improved by tapping more into the financial capacity of the private sector, and by creating incentives for loss mitigation and prevention.

Turning to the challenge of very rare extreme events, it is, in principle, no different from the challenge posed by more common natural disasters like earthquakes, windstorms, and floods. Complete exclusions from coverage must probably play a greater role in the case of extreme hazards. An important first step is to identify and eliminate ambiguous policy wordings. They can produce unexpected losses, for which the appropriate premium may not have been charged. An example is the term 'tidal wave', which could lead to unintentional tsunami coverage under flood policies. As regards proactive risk prevention, nothing specific can be achieved over and beyond the measures used for the more common risks. The unique feature that some or even ample warning time will usually be available does not mean much in terms of reducing and preventing material losses. A systematic discussion on how the insurance sector should handle the risk of extreme disasters has not yet taken place, and is indeed overdue.

Beyond the field of insurance, an all-encompassing policy for extreme events would have to take into account the fact that mankind (and the financial sector) also deals with other urgent issues, and more frequent disasters. Consequently, a policy for handling extreme disasters must be weighed against the need to deal with more urgent phenomena, such as hurricanes on the scale of Katrina and Sandy or the Thai floods, or a future great earthquake affecting the Kanto plain or the Kansai region in Japan. Or, turning to completely different types of risk, the risks inherent in the capital markets, as proven by the post-2008 financial crises, or political and socio-economic crises like the civil wars, droughts, and famines in Africa, or the revolutions and their consequences in the Arabian world.

List of abbreviations

AAL	average annual loss
AIR	Applied Insurance Research
ART	alternative risk transfer
CBD	central business district
CCRIF	Caribbean Catastrophe Reinsurance Facility
CEA	California Earthquake Agency
EQC	Earthquake Commission
FONDEN	Fondo Desastres Naturales
GEM	Global earthquake model
GVM	Global volcano model
IBHS	Institute for Business and Home Safety
JER	Japan Earthquake Reinsurance
NEA	Near-Earth asteroid
NFIP	National Flood Insurance Program
NGDC	National Geophysical Data Center
OECD	Organisation for Economic Cooperation and Development
PLA	Post-loss amplification
PML	Probable maximum loss
PRAC	Romanian Catastrophe Insurance Scheme
PRESSS	Precast seismic structural system
RMS	Risk Management Solutions
TCIP	Turkish Catastrophe Insurance Pool
TREIF	Taiwan Residential Earthquake Insurance Facility

References

Boissonnade, A. C. and Muirwood, R. (2008). Modelling loss amplification after devastating disasters. In *Abstracts of the 2008 Joint Assembly of the American Geophysical Union*. Abstract #U53A-01. Washington DC: American Geophysical Union.

Bryden, H. L., Longworth, H. R. and Cunningham, S. A. (2005). Slowing of the Atlantic meridional overturning circulation at 25° N. *Nature*, **438**, 655–657.

Cabinet Office's Central Disaster Prevention Council (2012). *Estimated Damage and Death from a Nankai Temblor.* http://www.kantei.go.jp/foreign/policy/index/bousai/index_e.html

Cardona, O. D. (2005). A system of indicators for disaster risk management in the Americas. In *Proceedings of the 250th Anniversary of the 1755 Lisbon Earthquake*. Lisbon: Laboratorio Nacional de Engenharia Civil, pp. 80–89.

Chapman, C. R. (2003). How a near-earth object impact might affect society. In *Proceedings of the OECD Workshop on Near-Earth Objects: Risks, Policies and Actions, Frascati, Italy.* Paris: Organisation for Economic Development and Cooperation, Global Science Forum. Available at: http://www.oecd.org/science/sci-tech/2493218.pdf. Accessed on 16 August 2013.

Churchill, C., Reinhard, D. and Qureshi, Z. (2006). *Microinsurance – Making Insurance Work for the Poor.* Report Summary Microinsurance Conference. Munich: Schloss Hohenkammer.

Cornell, C. A. (1968). Engineering seismic risk analysis. *Bulletin of the Seismological Society of America*, **58**, 1583–1606.

Deligne, N. I., Coles, S. G. and Sparks, R. S. J. (2010). Recurrence rates of large explosive volcanic eruptions. *Journal of Geophysical Research*, **115**, B06203.

Doherty, N. A., Grace, M. F., Klein, R. W. *et al.* (2008). *Managing Large-scale Risks in a New Era of Catastrophes – Executive Summary.* Philadelphia: Wharton Risk Management and Decision Process Center. Available at: http://corpsriskanalysisgateway.us/data/docs/ref/WhartonRisk-Center_ ManagingLargeScaleRisks_2008.pdf. Accessed on 16 August 2013.

EM-DAT (2011). The International Disaster Database. Available at: http://emdat.be. Accessed on 16 August 2013.

Giardini, D., Grünthal, G., Shedlock, K. M. and Zhang, P. (1999). The GSHAP Global Seismic Hazard Map. *Annali di Geofisica*, **42**, 1225–1228.

Guidotti, R., Mazzieri, I., Stupazzini, M. and Dagna, P. (2012). 3D numerical simulation of the site-city interaction during the 22 February 2011 Mw 6.2 Christchurch earthquake. In *Electronic Proceedings of the 15th World Conference on Earthquake Engineering, 24–28 Lisbon, Portugal, September 2012.* Tokyo: International Association of Earthquake Engineering.

Gülkan, P. (2001). Rebuilding the Sea of Marmara region: recent structural revisions in Turkey to mitigate disasters. In *Proceedings of the Wharton-World Bank Conference on Challenges in Managing Catastrophic Risks: Lessons for the US and Emerging Economies*, 8–10 January, ed. H. Kunreuther. Washington, DC: World Bank.

Harbitz, C. B. (1992). Model simulations of tsunamis generated by the Storegga slides. *Marine Geology*, **105**, 1–21.

International Atomic Energy Agency (2008). IAEA fact-finding mission to review safety assessment at Kashiwazaki-Kariwa nuclear power plant. *Media Advisory*, 4 November 2008.

International Commission on Earthquake Forecasting for Civil Protection (2011). Operational earthquake forecasting. *Annals of Geophysics*, **54** (4), doi:10.4401/ag-5350.

Le Friant, A. (2001). Les destabilisations de flanc des volcans actifs de l'arc des Petits Antilles: origins et consequences. Doctoral thesis Université de Paris VII, 377 pp.

Manaker, D. M., Calais, E., Freed, A. M. *et al.* (2008). Interseismic plate coupling and strain partitioning in the northeastern Caribbean. *Geophysical Journal International*, **174**, 889–903.

Marine Geoscience Data System (2013). *Global Multi-Resolution Data Portal. Lamont Doherty Earth Observatory.* Available at: http://www.marine-geo.org/portals/gmrt. Accessed on 16 August 2013.

Mason, B. G., Pyle, D. M. and Oppenheimer, C. (2004). The size and frequency of the largest explosive eruptions on Earth. *Bulletin of Volcanology*, **66**, 735–748.

Masson, D. G., Harbitz, C. B., Wynn, R. B., Pedersen, G. and Løvholt, F. (2006). Submarine landslides: processes, triggers and hazard prediction. *Philosophical Transactions of the Royal Society A*, **364**, 2009–2039.

McCaffrey, R. (2008). Global frequency of magnitude 9 earthquakes. *Geology*, **36**, 263–266.

Minoura, K., Imamura, F., Sugawara, D., Kono, Y. and Iwashita, T. (2001). The 869 Jogan tsunami deposit and recurrence interval of large-scale tsunami on the Pacific coast of northeast Japan. *Journal of Natural Disaster Science*, **23**, 83–88.

Morrison, D. (2006). Asteroid and comet impacts: the ultimate environmental catastrophe. *Philosophical Transactions of the Royal Society A*, **364**, doi:10.1098/rsta.2006.181.

Munich Reinsurance Company (1979). *World Map of Natural Hazards*, 1st edn. Munich, Germany: Munich Re.

Munich Reinsurance Company (2008). *World Map of Natural Hazards*, 4th edn. Munich, Germany: Munich Re.

Munich Reinsurance Company (2012). *TopicsGeo – Annual Disaster Review 2011.* Munich, Germany: Munich Re.

National Research Council (2008). *Severe Space Weather Events – Understanding Societal and Economic Impacts.* Workshop Report. Washington, DC: The National Academies Press.

New Zealand Concrete Society (2010). *PRESSS Design Handbook*, ed., Pampanin, S. Auckland: New Zealand Concrete Society.

Pampanin, S. (2011). Alternative seismic design philosophies and solutions for precast concrete buildings: emerging solutions for damage-resisting structures (Part 2/3). *CPI Concrete Plant International*, 202–209.

Rauch, E. and Smolka, A. (1992). Earthquake insurance: seismic exposure and portfolio dispersal – their influence on the probable maximum loss. In *Proceedings of the 10th World Conference on Earthquake Engineering, Madrid, Spain.*

Tokyo: International Association of Earthquake Engineering, pp. 6101–6104.

Rosenzweig, C., Solecki, W., DeGaetano, A. *et al.*, eds. (2011). *Responding to Climate Change in New York State: The ClimAID Integrated Assessment for Effective Climate Change Adaptation*. Technical Report. Albany, NY: New York State Energy Research and Development Authority (NYSERDA).

Sauter, F. F. and Shah, H. C. (1978). *Estudio de Seguro Contra Terremoto*. San José, Costa Rica: Instituto Nacional de Costa Rica.

Self, S. (2006). The effects and consequences of very large explosive volcanic eruptions. *Philosophical Transactions of the Royal Society A*, **364**, doi:10.1098/rsta.2006.1814.

Siembieda, W., Johnson, L. and Franco, G. (2012). Rebuild fast but rebuild better: Chile's initial recovery following the 27 February 2010 earthquake and tsunami. *Earthquake Spectra*, **28**, 621–641.

Sokolov, V. and Ismail-Zadeh, A. (2012). Extreme seismic events and hazard assessment. In *Abstract Book of the IUGG GRC Conference on Extreme Natural Events and Their Impacts*, 8–11 December 2012, Orange, CA. Orange, CA: Chapman University.

Swiss Re (2011). Sigma No. 1/2011, p 35. http://swissre.com

Toda, S., Stein, R. S. and Lin, J. (2011a). Widespread seismicity excitation throughout central Japan following the 2011 $M = 9.0$ Tohoku earthquake, and its interpretation by Coulomb stress transfer. *Geophysical Research Letters*, **38**, L00G03.

Toda, S., Stein, R., Sevilgen, V. and Lin, J. (2011b). *Coulomb 3.3 Graphic-rich Deformation and Stress-change Software for Earthquake, Tectonic, and Volcano Research and Teaching-user Guide*. US Geological Survey Open-File Report 2011-1060, ix.

Villamor, P., Barrell, D., Litchfield, N. *et al.* (2011). Greendale Fault investigation of surface rupture characteristics for fault avoidance zonation. *GNS Science Consultancy Report* 2011–121.

Ward, S. N. and Asphaug, E. (2000). Asteroid impact tsunami: a probabilistic hazard assessment. *Icarus*, **145**, 64–78.

Wirtz, A. Kron, W., Löw, P. and Steuer, M. (2012). The need for data: natural disasters and the challenges of database management. *Natural Hazards*, accepted.

World Bank (2011). The World Bank supports Thailand's post-floods recovery effort. *World Bank News and Views*, 13 Dec. 2011.

Yunak, H. (2004). Turkish Catastrophe Insurance Pool. In *Proc. AIO/UNCTAD Conference on Insurance and Reinsurance of Catastrophic Risks for Long-Term Development in Africa. Casablanca, Morocco*. Douala, Cameroon: African Insurance Organisation.

30

Hitting the poor: public–private partnership as an option

Impact of natural catastrophes on economies at various stages of development

Angelika Wirtz, Petra Löw, Thomas Mahl, and Sibel Yildirim

30.1 Introduction

NatCatSERVICE, Munich Re's global loss database, confirms that the number of catastrophes due to natural extreme events has been increasing since the 1980s. The annual average number of disaster events since 2002 is greater than 800. Since 1980, more than 2.2 million people have lost their lives due to the natural catastrophes; the relevant financial losses are estimated to be about US$3500 billion (in 2011 values). The two deadliest disasters were caused by earthquakes/tsunamis: the 2004 Aceh-Sumatra earthquake followed by a tsunami, and the 2010 Haiti earthquake. Each claimed more than 220 000 lives. In terms of absolute overall losses, rich countries like the USA or Japan carry the largest burden. The two costliest disasters around the world were caused by an earthquake and a hurricane: the 2011 Great East Japan earthquake with a direct overall loss of US$210 billion and the 2005 Hurricane Katrina, which hit New Orleans, USA, with a direct overall loss of US$125 billion (both in original values). The relative load on the economies is a different matter. Measured in terms of gross domestic product (GDP) and purchasing power parity, the impact on country groups with a low per-capita income is disproportionately high.

For the people in developing and emerging countries, natural catastrophes often lead to a loss of their basis of existence and/or prospects for the future. Appropriate insurance protection can be of assistance. If we divide the world into an 'insured' world and a 'non-insured' world, we see that insurance penetration is low to non-existent in the poorer country groups.

Micro-insurance is designed for people with low and irregular incomes, giving them protection against risks and helping to minimise losses due to, for example, accident, death, and illness. The development and promotion of these insurances is still at an early stage, although the number of micro-insurance policies in force is constantly increasing. All those involved should increase their efforts to provide people in the less-economically developed regions of the world with access to affordable insurance solutions. Public–private partnership programmes and further risk transfer options for governments play an important role in this context: they are described and discussed in this chapter.

30.2 Professional management of loss data

Databases recording the impacts of natural catastrophes are important data sources for researchers, governments, and non-governmental organisations (NGOs) as well as for the insurance and finance industry. At present, there are three global loss databases: NatCatSERVICE (by Munich Re), Sigma (by Swiss Re), and EM-Dat (by CRED, the Centre for Research on the Epidemiology of Disasters). To some extent, the three databases are using different standards and definitions. For instance, the thresholds for an event to be entered in the databases are different. Sigma defines catastrophes on the basis of economic losses adjusted for inflation and a minimum of 20 dead or missing people (Swiss Re, 2010). EM-Dat assigns catastrophe status to all events that meet at least one of the following criteria: 10 fatalities, 100 persons affected, declaration of a state of emergency or request for international help (Below *et al.*, 2009). While NatCatSERVICE registers all events that have caused any injury or damage and, depending on financial or humanitarian impact, assigns them to one of six catastrophe classes ranging from 'small-scale loss' to 'great natural catastrophe' (Munich Re, 2009).

The Munich Re database currently includes around 30 000 data sets. The oldest relates to the eruption of

Extreme Natural Hazards, Disaster Risks and Societal Implications, eds. Alik Ismail-Zadeh, Jaime Urrutia Fucugauchi, Andrzej Kijko, Kuniyoshi Takeuchi and Ilya Zaliapin. Published by Cambridge University Press.

Mount Vesuvius on 24 August 79 AD and the consequent destruction of Herculaneum and Pompeii. The event was reported in a detailed account by Pliny the Younger. But historical records are rare: the database contains details of only 1500 catastrophes that occurred between the years 79 and 1899. These records, however, cannot be used as a basis for trend analysis because they are incomplete and available only for certain regions and types of event. Earthquakes in China, for example, are very well described in records dating back as far as 2000 years. In America, systematic record-keeping began only around the middle of the nineteenth century. Complete and reliable global data sets exist for events since 1980. They permit analyses at global, continental, or country levels for the last three decades.

Based on the United Nations definition, natural catastrophes are classified as great if a region's ability to help itself is distinctly overtaxed, making supraregional or international assistance necessary. As a rule, this is the case when there are thousands of fatalities, hundreds of thousands are left homeless, and/or overall losses are of exceptional proportions given the economic circumstances of the country concerned. The NatCatSERVICE database records 310 such disasters between 1950 and 2011. The only years in which no great natural catastrophes occurred were 1952, 1958, and 2009 (Munich Re, 2010).

30.2.1 Peril terminology and classification

Hurricanes, typhoons, tsunamis, and heat waves, just to name some of the natural hazards, are stored in the database appropriately, although the characterisation of some events is not easy. Consider a torrential inundating rain as a result of a typhoon. Should it be recorded as a flood or a storm? What about a tsunami that claims ten thousands of lives? Should it be viewed as an inundation or – in line with its triggering event – as an earthquake? These are questions that database operators address on a daily basis (Kron *et al.*, 2012).

Uniform standards, operability, and terminology are essential to enable natural hazard events to be assessed and compared with other databases. In 2007, Munich Re, CRED, and Swiss Re defined a common terminology in consultation with the United Nations Development Programme (UNDP), the Asian Disaster Reduction Centre, and the United Nations International Strategy for Disaster Reduction (UNISDR). Since then, natural hazard events have been divided into four 'hazard families': *geophysical*, *meteorological*, *hydrological*, and *climatological* events (Fig. 30.1). These families are subdivided into main events (e.g., storms), which are divided, in their turn, into perils (e.g., tropical cyclone, extra-tropical cyclone, convective storm) and sub-perils (e.g., tornado, hailstorm). Other 'hazard families' concern biological or extraterrestrial events (such as meteorite strikes).

Databases need to register disasters strictly according to what triggered them. A typhoon that impacts mainly through flooding, for instance, appears in the database as a 'meteorological event/storm/tropical cyclone/flood'. Analyses are thus possible at all levels. The number of meteorological events or the number of floods triggered by a tropical storm can be established. Such a system is vital for comparing entries from different databases.

30.2.2 Economic and insured losses

Financial loss is the most important parameter for NatCatSERVICE, and it is recorded in the database as an insured loss and an economic loss. The magnitude of such losses is registered for each event. In most cases, credible information about the number of persons injured or killed is available very soon in the aftermath of a catastrophe. Figures for insured losses are also quite reliable, because

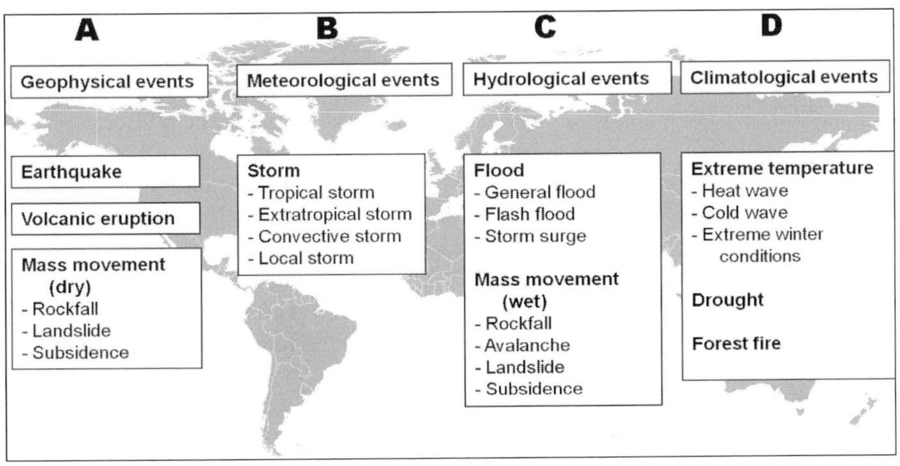

Figure 30.1. Peril hierarchy and terminology. (Source: NatCatSERVICE, Munich Re (2012a).)

they reflect claims actually paid by insurance companies. Economic losses, however, are more complex to assess.

Very often media reports publish loss estimates soon after the catastrophes, or even during the event. These early estimates can hardly be correct. Moreover, sometimes the damage is overestimated in the hope of mobilising more aid. As a general rule, government agencies, the United Nations, the European Union, universities, the World Bank and other development banks do detailed loss analyses only after particularly large or spectacular natural catastrophes. So in the case of many smaller events, the initially published figures are accepted as correct in the long run or even for ever.

The term 'economic loss' does not have a uniform definition. It is important to differentiate between 'direct losses', 'indirect losses', and 'secondary/consequential losses'. Direct losses are immediately visible and countable (consider the loss of homes, household property, schools, vehicles, machinery, livestock, etc.) (Munich Re, 2001). They are always calculated on the basis of replacement cost including repairs. However, it is quite challenging to estimate the value of historical buildings. Examples of indirect losses include higher transport costs due to damaged infrastructure, loss of jobs, and loss of rental income. Consequential losses (secondary costs) concern economic impact, for instance in terms of diminished tax revenues, lower economic output, reduced GDP, or a currency's lower exchange rate.

NatCatSERVICE has registered more than 20 000 loss events between 1980 and 2011. The exact volume of economic losses was verified officially by governments, statistic and finance authorities, etc. in a quarter of these cases. In the other events, the scale of loss has been estimated on the basis of insurance claims and available loss parameters. The loss information includes direct (tangible) losses, whereas secondary effects and consequential losses are not taken into account.

30.2.3 Sources and data quality

It has become much easier to investigate data on natural events largely because of the Internet. At the same time, it has become even more important to ensure that sources are robust and sound. NatCatSERVICE draws on around 200 sources that have been identified as the first rate for a particular region and/or type of event. The six groups of main sources are: (i) insurance industry information; (ii) investigations and analyses from Munich Re and its clients; (iii) meteorological and seismological services; (iv) reports and evaluations by aid organisations or NGOs, governments, the European Union, the United Nations, the World Bank and other development banks; (v) scientific analyses and studies; and (vi) reports by news agencies.

Despite first-class sources, the analysis process is sometimes fraught with problems. Typical challenges include false reports, incorrect conversion factors, and double counting of casualties. Such data are often copied and further disseminated. Therefore, database operators must strictly test the quality of the figures they obtain. The validation process checks the quality and number of the sources referred to as well as the plausibility of the loss figures and the description of the event. Data sets that fail to meet the quality standard are not used.

30.3 Natural disasters worldwide since 1980

Since the 1980s the frequency of natural catastrophes has been continuously increasing (Fig. 30.2). In the period of 1980–1989 an average number of 405 events per year was registered, it rose to 650 events in the 1990s, to 780 events for the period of 2000–2009, and to 800 between 2001 and 2011. The analysis of the annual numbers of the loss associated with natural hazard events reveals that climatological events such as wildfires, droughts, and extreme temperatures contribute to the losses at most since 1980 (Fig. 30.3). The percentage breakdown of events by the main types of natural hazards (geophysical/meteorological/hydrological/climatological events) shows that 72% of the total losses due to all natural catastrophes is due to storms and floods (Fig. 30.4). Meanwhile the number of fatalities shows nearly 40% of the entire death toll from 1980 until December 2011 (about 900 000 people) was caused by earthquakes and concatenated events (e.g., tsunami). The large portion of economic (39%) and insured losses (72%) is associated with storms, hurricanes and typhoons being on the top (Munich Re, 2012a, b).

The breakdown of all worldwide natural catastrophes from 1980 until 2011 by continents is shown in Fig. 30.5. More than half of all fatalities (1.2 million) as well as one third of the events (6500) occurred in Asia. As far as the economic losses are concerned, Asia is also leading with 42% while North America ranks second with 37%. The breakdown of the insured losses demonstrates the high insurance penetration in North America and especially in the USA. Although North America ranks second in economic losses, the continent leads the insured losses by far, with nearly two-thirds of all insured losses US$560 billion (in values of 2011), and the biggest part of it can be attributed to meteorological events in USA. Asia shows low insurance penetration: although Asia had to cope with economic losses of US$1500 billion (in values of 2011), the insured portion was only US$110 billion. The biggest part of the insured loss can be attributed to the richest Asian country; in Japan alone the insured loss was over US$80 billion.

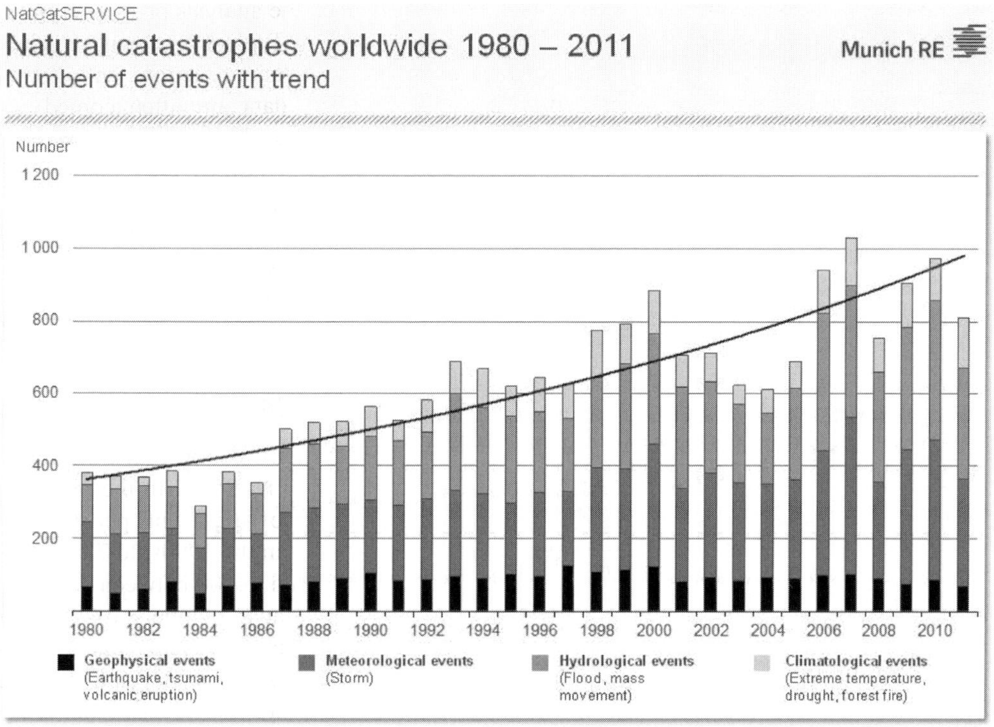

Figure 30.2. Annual number and trend line of natural catastrophes. (Source: NatCatSERVICE, Munich Re (2012b).)

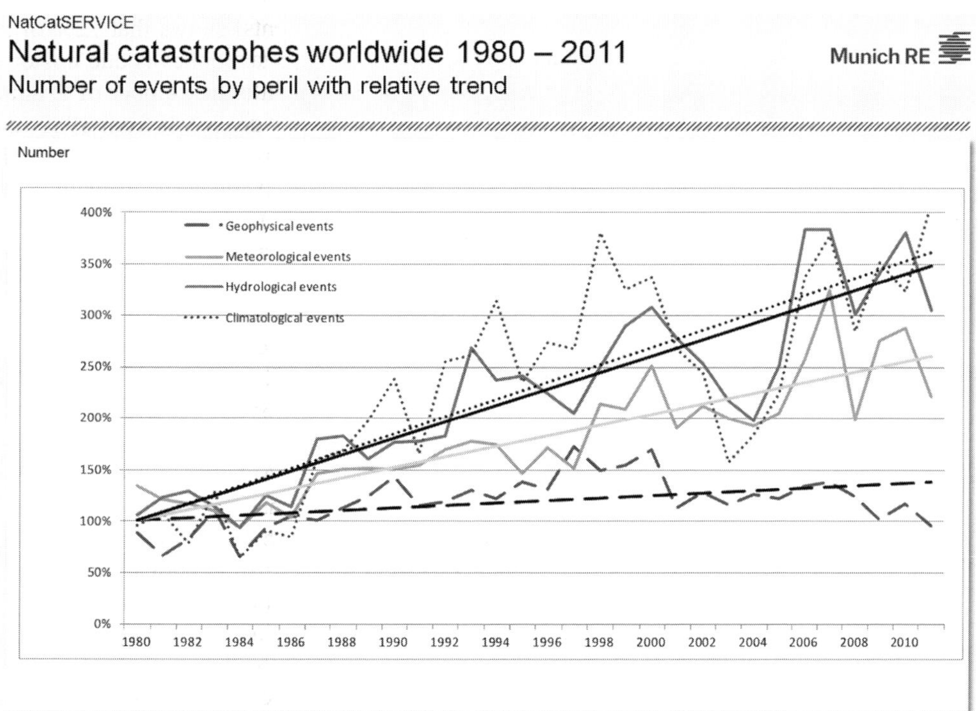

Figure 30.3. Annual numbers and relative trends of loss relevant natural events broken down to the different perils. (Source: NatCatSERVICE, Munich Re (2012a).)

For Africa there is a different aspect of natural catastrophes. Although only 9% of all events occur in Africa, nearly one-third of all fatalities are registered on this continent. The spread of loss events and fatalities per continent shows that continents with mainly poorer countries have more fatalities related to loss events. Continents with mainly richer countries (Europe, North America, and Australia) do not show this tendency. Concerning the

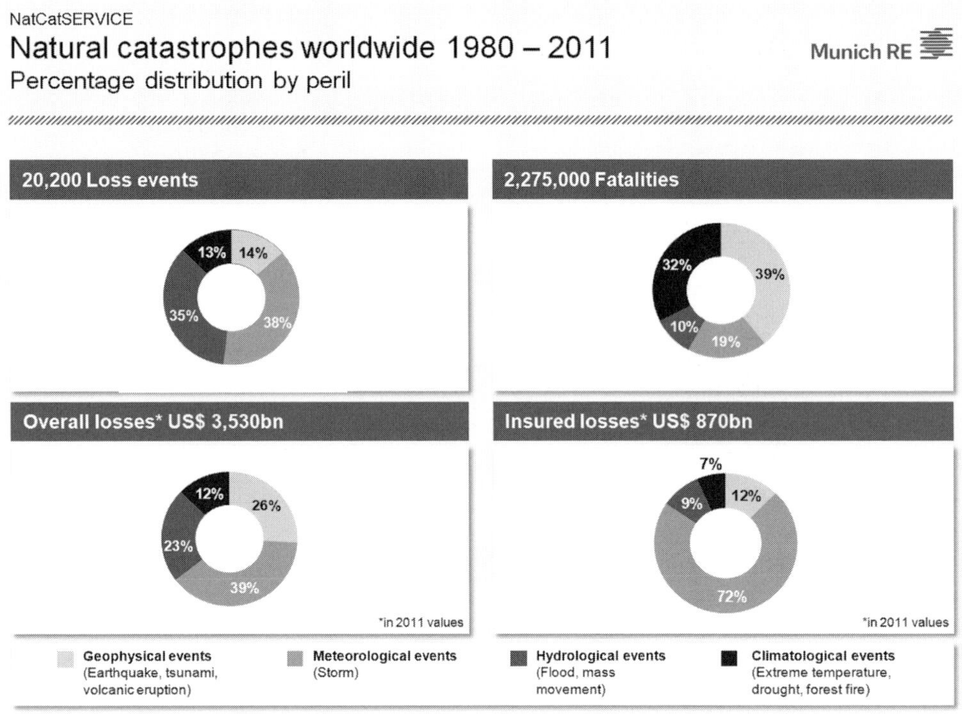

Figure 30.4. Percentage distribution of perils broken down to loss events, fatalities, overall and insured losses. (Source: NatCatSERVICE, Munich Re (2012b).)

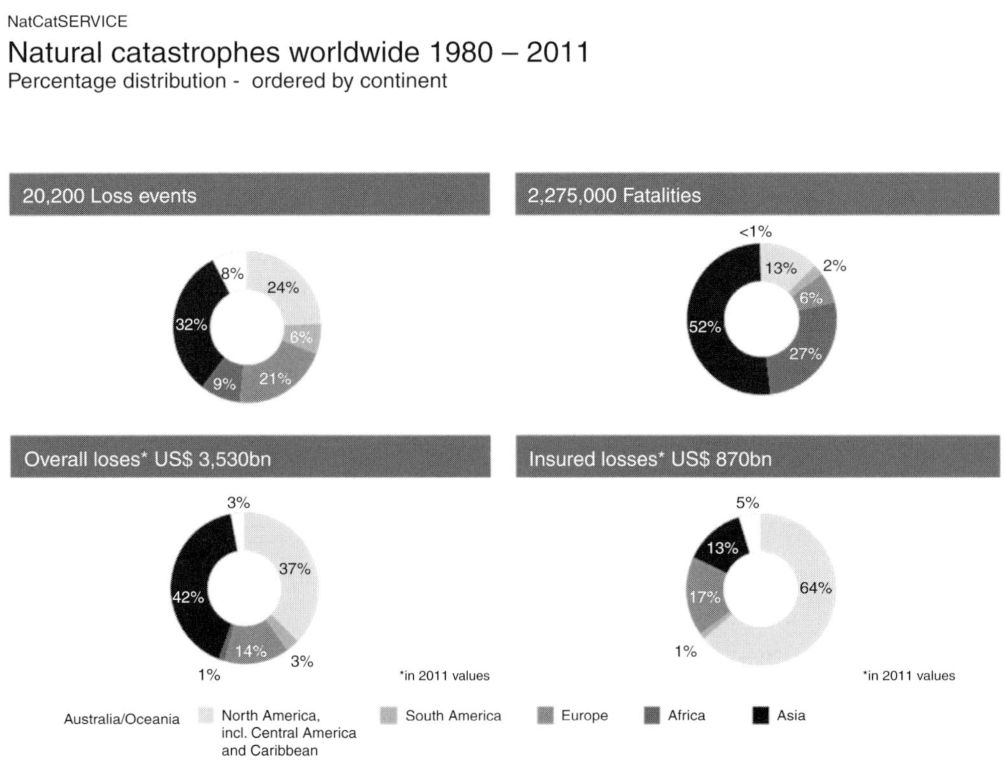

Figure 30.5. Percentage distribution of continents broken down to loss events, fatalities, overall and insured losses. (Source: NatCatSERVICE, Munich Re (2012b).)

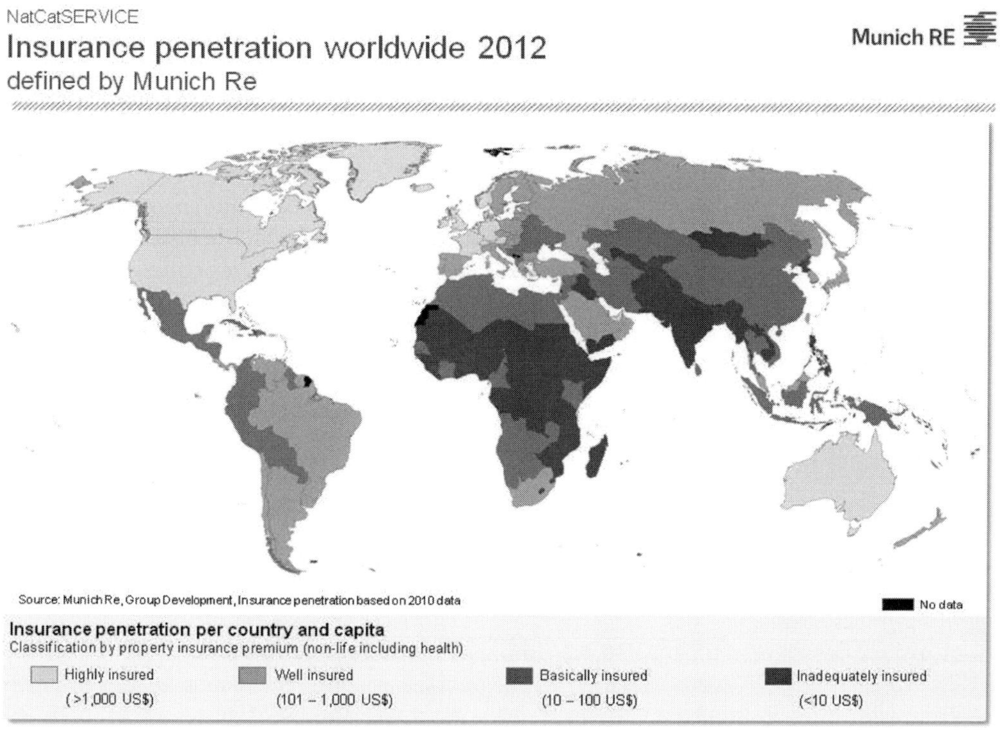

Figure 30.6. Insurance penetration worldwide from highly to inadequately insured (Source: NatCatSERVICE, Munich Re (2012b).)

losses the monitoring is different. The relative share of insured losses is higher than the relative share of the overall losses for continents with mainly industrialised countries. This can be related to the fact that developing countries have no or a low insurance penetration.

30.4 The insured and non-insured world

The analysis of worldwide natural catastrophes since 1980 and the breakdown by continents demonstrate that there is an uneven insurance penetration rate around the globe (Figs. 30.5 and 30.6). The most obvious reason for this inequality is that many people in developing countries cannot afford insurance cover. They need their money for basic needs. Also there are countries where – due to the political situation – insurance does not (or almost not) exist. Myanmar can be considered as an example: when Cyclone Nargis devastated a significant part of the country in May 2008, there was no insurance cover for the affected people.

There are various ways of calculating insurance premiums. One possibility would be to take the total insurance premium per country: this would lead to considerable distortions in particularly highly populated and large countries. A better approximation for the percentage of insured people or households is obtained using the insurance premiums per capita. However, even in rich countries with sufficient insurance premiums on average, there will be a large number of people, who cannot afford any cover at all and remain uninsured. Figure 30.6 shows the global distribution of property insurance premium per capita in US dollars, including the government pool solutions in some countries.

Four groups of countries can be distinguished:

- Inadequately insured (property insurance premiums US$ <10 per person per year): material assets not or inadequately insured.
- Basically insured (property insurance premiums from US$10 to US$100 per person per year): most vital basic insurance just about provided.
- Well insured (property insurance premiums from US$101 to US$1000 per person per year): material assets (households) adequately to very well insured.
- Highly insured (property insurance premiums exceed US$1001 per person per year): this group is to be considered to be extremely well insured; households take out several insurance policies. The USA and most countries in Europe are highly insured.

30.4.1 Analysis of impacts in the insured and non-insured world

While all natural catastrophes from 1980 to 2011 were more or less equally distributed, 76% of all fatalities (more than 1 billion people) were in countries belonging to the under-insured (inadequately and basically insured) group. The two deadliest events were the 2010 Haiti earthquake and the 2004 Indian Ocean earthquake/tsunami. Haiti and nearly all tsunami-affected countries belong to this group.

The well and highly insured countries are spared from catastrophic events as far as fatalities are concerned. There is one exception, the heat wave of 2003 in Europe with a death toll of 70 000 (Robine *et al.*, 2007). All countries affected by this extreme event are in the group of well and highly insured countries.

Roughly a third of all economic losses resulting from earthquakes, volcanic eruptions, storms, floods, and catastrophes related to extreme temperatures in the period 1980–2011 (US$1000 billion) occurred in the inadequately and basically insured groups. More than two-thirds of losses (US$2500 billion) occurred in wealthy industrialised countries, in part due to the high degree of industrialisation and comparatively high living standards.

The overwhelming majority of insured losses resulting from natural disasters from 1980 to 2011 (80%) were inevitably paid out in the countries with high insurance penetration; a sum of nearly US$700 billion. There are no major changes to be expected in this distribution of insured losses within the different country groups in the near future. But in the longer term the share of premiums from emerging countries will increase. In India, for example, it is forecast that the country's insurance market will increase to some US$150 billion (five times its current volume) over the next ten years (Munich Re, 2008). This growth will be driven above all by rising demand from India's middle class, currently accounting for some 300 million people. Between 2001 and 2006, the annual increase in premiums averaged roughly 24% in life insurance and 11% in property insurance. Based on total population, market penetration is still comparatively low in what is the second most heavily populated country in the world. The average premium, as a percentage of GDP, is about 0.6% for property insurance and 4% for life insurance. It is estimated that 90% of the Indian population has no property insurance protection. In terms of absolute premium volume, however, the country already ranks fifth in Asia after Japan, South Korea, China, and Taiwan and fifteenth in the world (2006) (Munich Re, 2008).

30.4.2 Impact on different economies

How are economies at different stages of development affected by natural catastrophes, in terms of both the number of victims and monetary losses? In order to compare the different economies, countries were categorised into four groups according to the World Bank's per-capita income groups (Fig. 30.7).

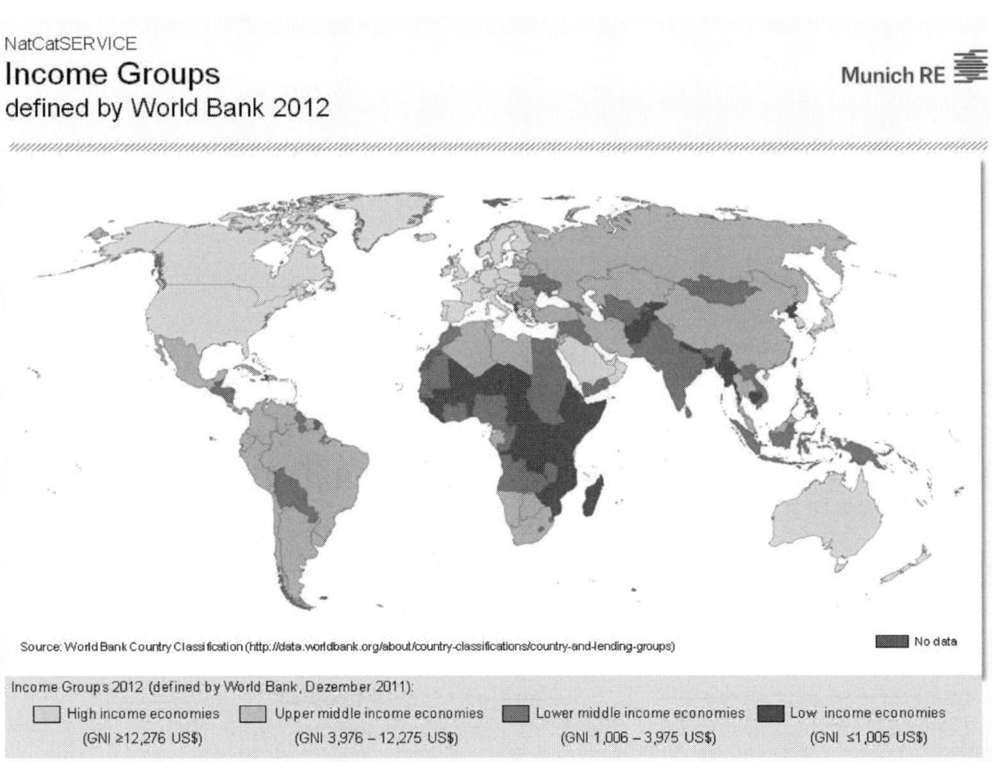

Figure 30.7. Map of income groups from high- to low-income economies. (Source: World Bank (2012).)

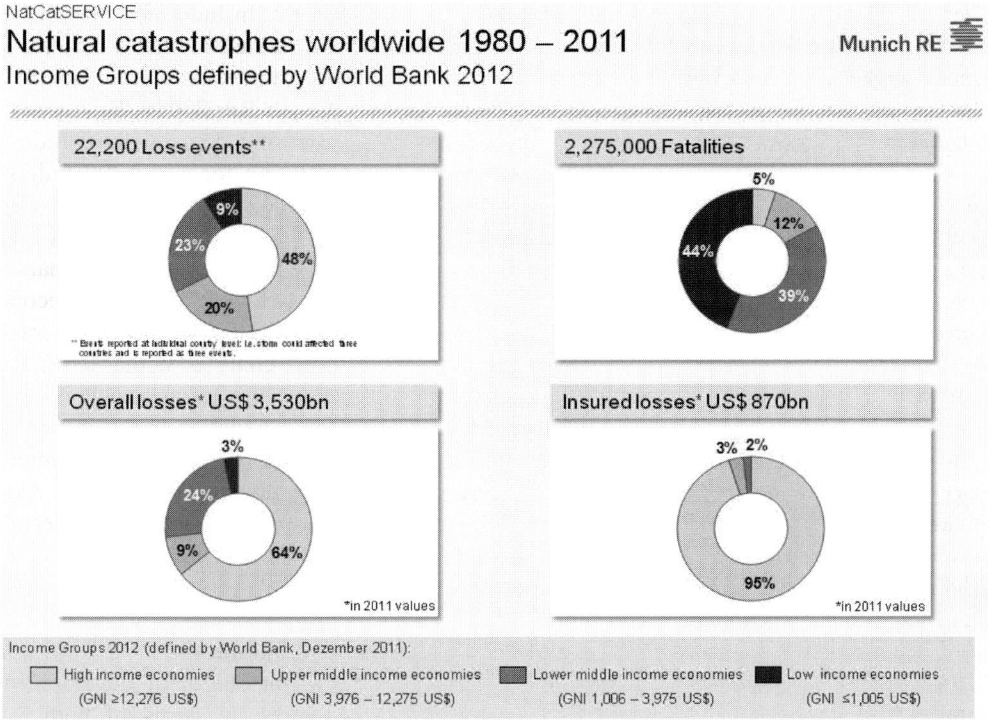

Figure 30.8. Percentage distribution of income groups broken down to loss events, fatalities, overall and insured losses. (Source: NatCatSERVICE, Munich Re (2012b).)

The wealthiest countries are most frequently hit by natural catastrophes (Fig. 30.8): 10 500 of the total of 22 200 events were recorded in this category. The most affected country is the USA with 3100 events. A total of 4500 events were recorded in the upper middle-income economies, with Russia, Mexico, and Turkey particularly to the fore. The group of lower middle-income countries accounted for 5100 events and the group of low income countries for 2000 events.

In terms of fatalities, the tendency is the opposite. The worst-hit countries are found to be the poorest countries with more than 1 million fatalities. The number of victims roughly reflects the stage of development reached in the regions affected. Many industrialised countries have relatively advanced building and safety codes, efficient disaster management systems, and early-warning facilities, which significantly mitigate disasters due to natural extreme events. However, in the high-income countries the lives of more than 100 000 people have been claimed, the 2003 heat wave in Europe alone being responsible for 70 000 fatalities (Fig. 30.8).

Insured losses are dominated by economies with a high per-capita income, as this is where insurance penetration and asset values are the highest. Examples show, however, that losses in markets considered to be in the uninsured category or in the low-income group of countries have also increased over the years. The floods in Mumbai, India, in July 2005 (due to the heaviest 24-hour-rainfall in the state of Maharashtra ever recorded in India) produced insured losses amounting US$700 million. The floods became the costliest natural catastrophe in the history of the Indian insurance market. The main reason was that, unlike the rest of the country, Mumbai has a high property insurance penetration. The average insurance premium per person in the state of Maharashtra in 2005 was US$12, compared to about US$4 in India as an average.

Figure 30.9 reflects the economic and insured losses as a percentage of the GDP by income groups, which are defined by the World Bank. To calculate these impacts and to take into account the changes between income groups of countries over the years, an income group for every country and every year was allocated. The income group's GDP per year was compared with the annual aggregated economic and insured losses, and the ratio of the GDP was calculated. Finally, the 32-year average was determined. The analysis shows (see Fig. 30.9) that the high-income countries carry the largest burden in absolute figures in terms of economic losses, accounting for about two-thirds of the global losses. The relative burden on countries regarding the economic losses in relation to GDP presents a quite different picture. The impact of absolute losses related to GDP is disproportionately high for country groups with low per-capita income. As far as the insured losses are concerned, the rich countries have in absolute as well as in relative figures the highest amount of insured losses. The analysis again reflects the low

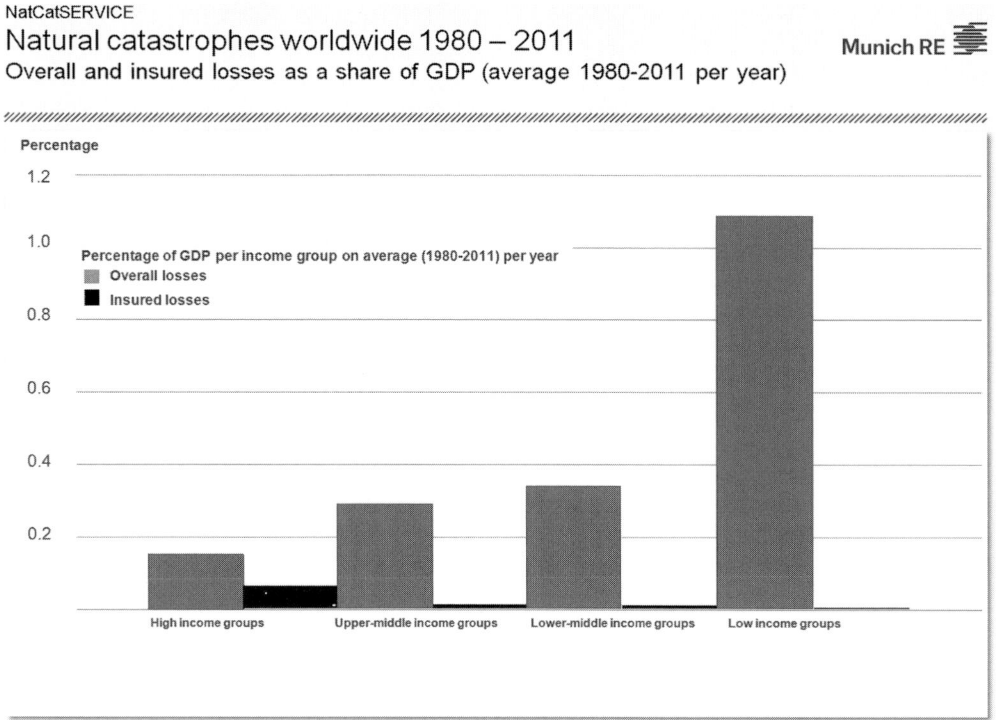

Figure 30.9. Percentage of GDP per income group in a long-term average (1980–2011) per year. (Source: World Bank, NatCatSERVICE, Munich Re (2012a).)

insurance penetration of poorer countries and shows that the poorer the countries, the lower are the insured losses, but the higher are the economic losses (both related to the country's GDP).

The above analyses showed that it is vital for the people in developing and emerging countries to have access to financial compensation after natural disasters. Affordable solutions should be implemented.

30.5 The role of reinsurance

As reinsurance is conducted worldwide and across all lines of insurance, it provides an optimal distribution of risk. It helps primary insurers to increase their underwriting capacities and protects them against extreme financial losses. It also enables them to insure risks they would otherwise lack the financial capacity to assume because it would be too risky and too costly for insurers to bear.

It is common practice for primary insurers (very often for local insurers) to assume a larger share of a risk than they can safely retain, and thus reinsurance is acquired as a means of transferring part of the risk to others, thereby increasing the pool of capital available to support the risk. In effect, the primary insurer carriers take a deductible equal to the difference between the customer's personal deductible and the level of loss beyond which the reinsurer(s) begins to pay. Additional reinsurance-type support can be obtained through retrocession, which is essentially reinsurance bought and sold among reinsurers. Reinsurance is typically bought in 'layers', each of which covers a certain band of costs above what the primary insurer can pay. The higher layers of cover are used only if the losses are sufficiently large. This practice helps distribute the risk among many reinsurers (i.e. the sellers of each layer of coverage).

The principle of reinsurance enables local insurers to assume only a small portion of a risk while the bulk of the risk and possible loss is transferred to the international market. The example of Hurricane Gilbert in 1988 illustrates this point. When Hurricane Gilbert hit the Caribbean islands in September 1988, Jamaica was seriously affected. Gusts of up to 320 km per hour and waves up to 8 metres high were reported. 50% of all houses suffered serious damages. The banana crop was totally destroyed. The overall insured loss at that time was US$720 million. However, the majority (US$690 million) was paid out from the global reinsurers, only US$30 million had to be borne by local insurers. Further examples are the earthquake of 2001 in El Salvador and the floods of 2002 in the Czech Republic. In both events, 95% of the insured losses were borne by the reinsurance industry. Thus, keeping only a small portion of the loss in the country concerned, while the remainder is shared among international groups and markets, permits the economies affected by catastrophes to recover much more quickly from major loss events.

The problem is that many people simply cannot afford the 'luxury' of insurance, and unfortunately these are frequently the very people that live in hazard-prone areas; as an example, the 2008 Wenchuan earthquake in China hit a rural and relatively poor region. The insurance companies received claims in the amount of US$300 million, whereas the direct loss was more than US$80 billion.

30.6 Integrated risk management

As learnt from the statistics, *developing countries with their fragile risk mitigating infrastructure, poor building fabric of houses, and absence of warning systems are more vulnerable to natural disasters than industrialised countries*. Although industrialised as well as developing countries can rely on different ad hoc financial resources to overcome the negative impact of a disaster in the short term, both country groups have to cope with the adverse effects on their own once the available funds have been consumed. In the long term, developing countries require a longer recovery period as the majority of their population face those risks without protection.

The financial responsibilities of governments in the aftermath of an extreme event depend heavily on the existing protection of the private sector. Governments act as an insurer of last resort. Particularly in countries with low insurance penetration expectations regarding support of the public sector are extremely high. Considering these interdependences a comprehensive integrated risk management needs to be developed in order to reduce and limit the governmental liabilities.

30.7 Increasing insurance penetration

A commonly cited reason for low demand in insurance, and particularly for low penetration of natural catastrophe insurance, is a lack of comprehension of its benefit. Considering the limited financial literacy awareness of households in developing countries, local people prefer to retain the risk instead of accepting an insurance cost, especially when the insurance cover is formulated by a complex wording they hardly understand. As a consequence, a lot of households rely on informal risk sharing schemes among several social groups to manage their risks. Obviously in the case of systemic risks, like natural catastrophes, these group arrangements may fail. With new regulations some governments have started to formalise risk sharing activities by encouraging and supporting the primary insurance market to develop simplified natural catastrophe insurance products for low-income households. In addition to the support of selected NGOs several developing countries have started insurance literacy awareness campaigns to raise demand. Based on past experiences, governmental support is meanwhile automatically expected by low-income people in the aftermath of natural catastrophes. Consequently, close cooperation between the public and private sectors to transfer responsibility is required in the future to limit the public sector's liability. Otherwise, the private insurance market will not be able to offer a sustainable product as it will be competing with compensation provided by the public sector for free.

30.8 Public–private partnership

Governments can reduce their contingent liability by 'transferring' part of the risk (except the general risk mitigation and financing) to the private insurance market. This can be done by creating an environment that promotes and motivates an insurer to offer attractive products to the designated target group. However, in the case of systemic risks, like natural catastrophes, public–private partnership solutions are the most appropriate instruments to coordinate the respective parties' liabilities and exposures. Governments play an important role in this respect by facilitating a platform to aggregate those risks. Subject to the geographical spread of risk and the size of the government-supported catastrophe pool, premium discounting diversification effects can be achieved within the pool. In addition, due to the greater scale and aggregation, the risk transfer to the reinsurance market becomes more efficient.

Most of the existing pools, for example Consorcio de Compensacion de Seguros (the Insurance Compensation Consortium) in Spain or the Taiwan Residential Earthquake Insurance Fund in Taiwan, focus on special hazards and predefined regions. The principle of insurance is on an indemnity basis by protecting mainly property risk. Along with reinsurance, protection capacity is provided by the domestic insurance market via a consortium.

30.9 Innovative developments

The generation of insurance pools was formerly based on traditional insurance techniques by providing coverage on the basis of the principle of indemnity. While this coverage is focused on the direct sustained losses, immediate payout and indirect losses are not covered.

Index insurance solutions, however, offer countries new opportunities of risk transfer because they provide immediate payout after a loss. Particularly for the micro-insurance market, where simple, transparent, moral hazard resistant products are required, an index solution appears to be promising. Reinsurers have researched and invested in the development of index insurance, often in collaboration with research institutes and primary insurers.

One of the main issues of index-based insurance is the 'basis risk'. The 'basis risk' solution is important, as there is the possibility that a particular policyholder or group of policyholders will not receive payment despite having suffered damage. Focusing on a bigger portfolio, where risk exposure to loss spreads over a much larger geographic region, could reduce the 'basis risk'. Index solutions on the national macro-level seem to be ideal to complement traditional indemnity-based insurance products. Similar to the traditional pool solutions they allow governments or domestic insurers to access international reinsurance capacity and service.

The Caribbean governments are some of the pioneers in the field of index risk insurance for governments: after the 2004 hurricane season they bought one of the first index insurances protecting against hurricane risks. With the assistance of donors and supranational organisations, a special-purpose vehicle called the Caribbean Catastrophe Risk Insurance Facility (CCRIF) was developed allowing the Caribbean governments to purchase their individual hurricane and earthquake protection on the basis of a calibrated index (Munich Re, 2007).

30.9.1 Example 1: The Caribbean Catastrophe Risk Insurance Facility Scheme (CCRIF)

CRIFF allows the governments of the Caribbean Community, an organisation of 15 Caribbean nations and dependencies, to purchase insurance coverage to finance immediate post-disaster recovery needs. The facility acts as a risk aggregator. CCRIF allows participating countries to pool their country-specific risks into a single, better diversified portfolio, which results in a substantial reduction in premium cost of about 45–50%. Claims payments depend on parametric triggers. Index-based (or parametric) insurance instruments pay claims based on the occurrence of a predefined event rather than an assessment of actual losses. This measurement, made remotely by an independent agency, allows for transparent, low settlement costs and quick-disbursing contracts.

The facility was created with initial funding from donors. This allowed the facility to cover start-up costs, retain some of the risk, and access the reinsurance markets where it is most efficient. The facility transfers the risks it cannot retain itself to the international financial markets. This is done through reinsurance and a catastrophic swap. The accumulation of reserves over time should lessen the facility's dependence on outside risk transfer and smooth the catastrophe reinsurance pricing cycle.

The facility maintains a financial protection to survive 1-in-1000-year events. Should the total insured loss exceed its claims-paying capacity, payouts will be prorated based on the total amount of expected claims compared to the remaining available funds.

CCRIF was established as an independent legal entity. It was created as an insurance captive managed by a specialised firm under the supervision of a board of directors composed of representatives from the donors and participating countries. This board is supported by the technical advice of a facility supervisor.

Participating countries pay an annual premium commensurate with their own specific risk exposure. Parametric insurance products are priced for each country, based on its individual risk profile. Annual premiums typically vary from US$0.2 million to US$4 million, for coverage ranging from US$10 to US$50 million.

30.9.2 Example 2: The Turkish Catastrophe Insurance Pool (TCIP)

Another example of an insurance instrument based on public–private partnership is the TCIP. Turkey is located in one of the most seismically active regions of the world with a majority of the population living in the earthquake-prone areas. Table 30.1 presents the six largest earthquakes in Turkey since 1980 and illustrates the enormous losses involved.

The persistent potential for large-scale natural disasters and the very low level of catastrophe insurance penetration among households have become a fiscal and social issue for the Turkish government. This issue led to the establishment of the TCIP in 1999. The scheme was established with the assistance of the World Bank before the Izmit earthquake of August 1999 (Gurenko *et al.*, 2007). The Management Board of the pool consists of seven members from the academic, public, and private sectors. To keep the administrative costs low and to create an efficient operational structure, most TCIP functions and operations are

Table 30.1. *Largest earthquakes in Turkey since 1980*

Year	Region	Fatalities	Overall losses in US$ million (original values)
2003	Bingöl	176	400
1999	Izmit	17000	12000
1999	Düzce	845	500
1998	Adana, Ceyhan	144	550
1992	Erzinan	650	750
1983	Horasan	1350	800

Source: NatCatSERVICE, Munich Re (2010, 2012a, b).

outsourced to the insurance and reinsurance industry, using their existing infrastructure. Currently, 32 insurance companies are entitled to distribute TCIP policies. The Undersecretary of the Treasury is responsible for supervising the programme and auditing all TCIP operations and accounts. TCIP and its revenues are exempt from all taxes, levies, and charges.

The programme has four principal objectives. Above all it provides earthquake insurance at affordable but actuarially sound rates for all registered urban dwellings. Its aim is to limit the government's financial exposure, to build long-term reserves to finance future earthquake losses, and to encourage risk reduction and mitigation practices in residential construction. Since the programme began in 1999, the insurance penetration for catastrophe cover has more than tripled. TCIP now provides coverage to more than 2 million Turkish homeowners. Following several small- and medium-scale earthquakes over the last few years, it has demonstrated its ability to pay claims quickly and fairly. Because of its low cost structure, it makes catastrophe insurance affordable for middle-income homeowners.

30.9.3 Example 3: Munich Climate Insurance Initiative (MCII)

The third example of public–private initiatives focuses on the risk of climate change. The MCII was founded in April 2005 by Munich Re together with other organisations. It consists of insurers, climate experts, economists, and independent organisations (Munich Climate Insurance Initiative, 2008).

MCII established cooperation between representatives of different organisations, such as the European Climate Forum (ECF), Germanwatch, the International Institute for Applied Systems Analysis (IIASA), Munich Re Company and Munich Re Foundation, the Potsdam Institute for Climate Impact Research, the United Nations Framework Convention on Climate Change (UNFCCC), the United Nations University Institute for Environment and Human Security (UNU-EHS), the World Bank, and other renowned institutions, companies, and experts.

The main objectives of the MCII are to develop insurance-related solutions to help manage the impacts of climate change, especially in developing countries. It will conduct and support pilot projects for the application of insurance-related solutions, in partnerships and through existing organisations. Furthermore, it will promote insurance approaches in cooperation with other organisations and initiatives within existing frameworks such as the United Nations system, international financial institutions, international donors, and the private sector. Finally, the aim is to identify and promote loss-reduction measures for climate-related events. One major objective of the MCII is to introduce climate insurance concepts in the post-Kyoto negotiations in line with the commitments of the Bali Roadmap.

These three examples of public–private partnership (CCRIF, TCIP, and MCII) show different approaches for financing losses from natural catastrophes, but provide governments and victims with immediate access to funds, which facilitates recovery and reconstruction measures. Such solutions greatly help governments to cope with the financial impact of natural catastrophes.

30.10 Conclusions

Data quality. Quality standards for disaster loss data should be high enough to provide data reliability for many users including researchers and decision-makers. Records and loss assessment methods should be consistent over time and transparent to users. Therefore, all databases should be reviewed continuously. Database operators need to develop more benchmarks for the sake of better comparability and higher transparency and credibility of data sets. An international hierarchy and terminology for natural hazards already exists. Current initiatives are working on geo-coding standards and loss estimation methods for natural catastrophes.

Data analyses. Analyses of natural catastrophes show that the number of catastrophes has been increasing continuously since the 1980s. Asia is extremely affected: one third of all global catastrophes and two-thirds of all fatalities have occurred in Asian countries. In terms of economic losses, the wealthy countries carry the largest burden. However, the relative burden on countries and the economic losses in relation to GDP present a quite different picture. The impact on the country groups with low per-capita income is disproportionately high.

Insurance, risk transfer, and micro-insurance solutions. Insurance cover remains a very effective way of protecting property and belongings after devastating catastrophes. People in the poorest countries of the world have the chance to acquire micro-insurance. Governments can reduce the financial impact of devastating natural catastrophes via insurance in cooperation with the World Bank, development banks, and the international insurance industry, which will facilitate a quick process of reconstruction. The task of insurers and reinsurers is not only to identify risks and to develop insurance solutions, but also to make their knowledge and potential prevention measures available to the public at large.

Endnote

Data Sources. For the analyses of the global impact of natural disasters the data are taken from the Munich Re NatCatSERVICE. This database on global natural catastrophes is one of the central elements of Munich Re's Geo Risks Research unit, which was founded in 1974 and has currently more than 30 000 entries. It contains data on the major catastrophes of the last 2000 years and all post-1980 loss events. Analyses and statistics on current and historic natural disasters can be downloaded from www.munichre.com/natcatservice/downloadcenter.

References

Below, R., Wirtz, A. and Guha-Sapir, D. (2009). *Disaster Category Classification and Peril Terminology for Operational Purposes*. Working paper No. 264. Leuven: Munich Re & CRED. Available online: http://cred.be/sites/default/files/DisCatClass_264.pdf

Gurenko, N. E., Lester, R. R., Mahul, O. and Gonulal, S. O. (2007). *Earthquake Insurance in Turkey: History of the Turkish Catastrophe Insurance Pool*. Washington, DC: The World Bank.

Kron, W., Löw, P., Steuer, M. and Wirtz, A. (2012). How to deal properly with a natural catastrophe database – analysis of flood losses. *Natural Hazards and Earth System Sciences*, **12**, 535–550.

Munich Climate Insurance Initiative (2008), http://www.climate-insurance.org. Accessed 1 October 2012.

Munich Re (2001). *Annual Review of Natural Catastrophes 2000*. Munich, Germany: Munich Re.

Munich Re (2007). Press Release on 1 June 2007 'Insurance cover for countries in the Caribbean in place for hurricane season 2007'. http://www.munichre.com/en/media_relations/press_releases/2007/2007_12_7_press_release.aspx.

Munich Re (2008). *Topics Geo – Natural Catastrophes 2007: Analyses, Assessments, Positions*. Munich, Germany: Munich Re.

Munich Re (2009). *Topics Geo – Annual Review of Natural Catastrophes 2008*. Munich, Germany: Munich Re.

Munich Re (2010). *Topics Geo – Annual Review of Natural Catastrophes 2009*. Munich, Germany: Munich Re.

Munich Re (2012a). *NatCatSERVICE Data*. Munich, Germany: Munich Re.

Munich Re (2012b). *NatCatSERVICE Download-Center*, www.munichre.com/natcatservice/downloadcenter

Robine, J. M., Cheung, S. L., Le Roy, S., Van Oyen, H. and Herrmann, F. R. (2007). Report on excess mortality in Europe during summer 2003. EU Community Action Programme for Public Health. http://ec.europa.eu/health/ph_projects/2005/action1_2005_a2_15_en.pdf

Swiss Re (2010). *Sigma – Natural and Man-Made Catastrophes 2009*. Zurich, Switzerland: Swiss Re.

World Bank (2012). *Data & Statistics, Country Classification*, www.worldbank.org.

Index